THE MULTILINGUAL ENERGY DICTIONARY

THE MULTILINGUAL ENERGY DICTIONARY

Dr. Alan Isaacs, editor

Facts On File, Inc.
460 Park Avenue South, New York, N.Y. 10016

THE MULTILINGUAL ENERGY DICTIONARY

First published in 1981 by Facts On File, Inc.,
460 Park Avenue South, New York, N.Y. 10016.

Library of Congress Cataloging in Publication Data

Isaacs, Alan, 1925–
The multilingual energy dictionary.

English, French, German, Italian, Portuguese, and Spanish.
1. Power resources—Dictionaries—Polyglot.
2. Power (Mechanics)—Dictionaries—Polyglot.
3. Dictionaries, Polyglot. I. Title.
TJ163.16.I78 333.79'03 80-26793
ISBN 0-87196-430-9

10 9 8 7 6 5 4 3 2 1

Printed in the United States of America

A

abbondanza naturale *(f)* It
De natürliche Isotopenhäufigkeit *(f)*
En natural abundance
Es abundancia natural *(f)*
Fr abondance naturelle *(f)*
Pt abundância natural *(f)*

Abdichtung *(f) n* De
En seal (boiler)
Es cierre *(m)*
Fr joint *(m)*
It dispositivo di tenuta *(m)*
Pt vedação *(f)*

Abfalltreibstoffe *(pl) n* De
En waste fuels *(pl)*
Es combustibles de desecho *(m pl)*
Fr combustibles de récupération *(m pl)*
It combustibili di scarico *(m pl)*
Pt combustíveis perdidos *(m pl)*

Abfallverbrennung *(f) n* De
En refuse incineration
Es incineración de basuras *(f)*
Fr incinération de déchets *(f)*
It incenerimento dei rifiuti *(m)*
Pt incineração de refugo *(f)*

Abgas *(n) n* De
En exhaust
Es descarga *(f)*
Fr échappement *(m)*
It scarico *(m)*
Pt escape *(m)*

Abgase *(pl) n* De
En waste gases *(pl)*
Es gases de desecho *(m pl)*
Fr gaz perdus *(m pl)*
It gas di scarico *(m pl)*
Pt gases peridos *(m pl)*

Abgasvorwärmer *(m) n* De
En economizer
Es economizador *(m)*
Fr économiseur *(m)*
It economizzatore *(m)*
Pt economizador *(m)*

abgegebene Energie *(f)* De
En output energy
Es energía de salida *(f)*
Fr énergie produite *(f)*
It energia erogata *(f)*
Pt energia de saída *(f)*

Ablauf *(m) n* De
En tailrace
Es canal de descarga *(m)*
Fr bief *(m)*
It canale di scarico *(m)*
Pt água de descarga *(f)*

abondance naturelle *(f)* Fr
De natürliche Isotopenhäufigkeit *(f)*
En natural abundance
Es abundancia natural *(f)*
It abbondanza naturale *(f)*
Pt abundância natural *(f)*

Abraum *(m) n* De
En spoil (mining)
Es estériles *(m pl)*
Fr déblai *(m)*
It materiale di sterro *(m)*
Pt entulho *(m)*

Absauggebläse *(n) n* De
En extraction fan
Es ventilador de extracción *(m)*
Fr ventilateur d'extraction *(m)*
It ventilatore ad estrazione *(m)*
Pt ventoínha de extracção *(f)*

Abschalten *(n) n* De
En shutdown
Es parada *(f)*
Fr arrêt *(m)*
It arresto *(m)*
Pt paralização *(f)*

Abschaltung *(f) n* De
En scram (nuclear reactor)
Es parada de emergencia *(f)*
Fr arrêt d'urgence *(m)*
It spegnimento immediato *(m)*
Pt paragem de emergência *(f)*

Abschirmung *(f) n* De
En shield
Es blindaje *(m)*
Fr bouclier *(m)*
It schermo *(m)*
Pt protector *(m)*

absolute Feuchtigkeit *(f)* De
En absolute humidity
Es humedad absoluta *(f)*
Fr humidité absolue *(f)*
It umidità assoluta *(f)*
Pt humidade absoluta *(f)*

absolute humidity En
De absolute Feuchtigkeit *(f)*
Es humedad absoluta *(f)*
Fr humidité absolue *(f)*
It umidità assoluta *(f)*
Pt humidade absoluta *(f)*

absoluter Nullpunkt *(m)* De
En absolute zero
Es cero absoluto *(m)*
Fr zéro absolu *(m)*
It zero assoluto *(m)*
Pt zero absoluto *(m)*

absolute Temperatur *(f)* De
En absolute temperature
Es temperatura absoluta *(f)*
Fr température absolue *(f)*
It temperatura assoluta *(f)*
Pt temperatura absoluta *(f)*

absolute temperature En
De absolute Temperatur *(f)*
Es temperatura absoluta *(f)*
Fr température absolue *(f)*
It temperatura assoluta *(f)*
Pt temperatura absoluta *(f)*

absolute zero En
De absoluter Nullpunkt *(m)*
Es cero absoluto *(m)*
Fr zéro absolu *(m)*
It zero assoluto *(m)*
Pt zero absoluto *(m)*

absorbedor directo *(m)* Es
De Direktabsorber *(m)*
En direct absorber
Fr absorbeur direct *(m)*
It assorbitore diretto *(m)*
Pt absorvedor directo *(m)*

absorbeur direct *(m)* Fr
De Direktabsorber *(m)*
En direct absorber
Es absorbedor directo *(m)*
It assorbitore diretto *(m)*
Pt absorvedor directo *(m)*

absorção *(f) n* Pt
De Absorption *(f)*
En absorption
Es absorción *(f)*
Fr absorption *(f)*
It assorbimento *(m)*

absorção atmosférica *(f)* Pt
De Luftabsorption *(f)*
En atmospheric absorption
Es absorción atmosférica *(f)*
Fr absorption atmosphérique *(f)*
It assorbimento atmosferico *(m)*

absorção molecular *(f)* Pt
De molekulare Absorption *(f)*
En molecular absorption
Es absorción molecular *(f)*
Fr absorption moléculaire *(f)*
It assorbimento molecolare *(m)*

absorción *(f)* *n* Es
De Absorption *(f)*
En absorption
Fr absorption *(f)*
It assorbimento *(m)*
Pt absorção *(f)*

absorción atmosférica *(f)* Es
De Luftabsorption *(f)*
En atmospheric absorption
Fr absorption atmosphérique *(f)*
It assorbimento atmosferico *(m)*
Pt absorção atmosférica *(f)*

absorción molecular *(f)* Es
De molekulare Absorption *(f)*
En molecular absorption
Fr absorption moléculaire *(f)*
It assorbimento molecolare *(m)*
Pt absorção molecular *(f)*

absorptance solaire *(f)* Fr
De Solarabsorptionsgrad *(m)*
En solar absorbtance
Es absorsancia solar *(f)*
It assorbimento solare *(m)*
Pt absorvância solar *(f)*

absorption *n* En, Fr *(f)*
De Absorption *(f)*
Es absorción *(f)*
It assorbimento *(m)*
Pt absorção *(f)*

Absorption *(f)* *n* De
En absorption
Es absorción *(f)*
Fr absorption *(f)*
It assorbimento *(m)*
Pt absorção *(f)*

absorption atmosphérique *(f)* Fr
De Luftabsorption *(f)*
En atmospheric absorption
Es absorción atmosférica *(f)*
It assorbimento atmosferico *(m)*
Pt absorção atmosférica *(f)*

absorption cycle En
De Absorptionsfolge *(f)*
Es ciclo de absorción *(f)*
Fr cycle d´absorption *(m)*
It ciclo di assorbimento *(m)*
Pt ciclo de absorção *(m)*

absorption dynamometer En
De Absorptions-dynamometer *(n)*
Es dinamómetro de absorción *(m)*
Fr dynamomètre à absorption *(m)*
It dinamometro ad assorbimento *(m)*
Pt dinamómetro de absorção *(m)*

absorption moléculaire *(f)* Fr
De molekulare Absorption *(f)*
En molecular absorption
Es absorción molecular *(f)*
It assorbimento molecolare *(m)*
Pt absorção molecular *(f)*

Absorptions-dynamometer *(n)* *n* De
En absorption dynamometer
Es dinamómetro de absorción *(m)*
Fr dynamomètre à absorption *(m)*
It dinamometro ad assorbimento *(m)*
Pt dinamómetro de absorção *(m)*

Absorptionsfenster *(n)* *n* De
En absorption window
Es ventanilla de absorción *(f)*
Fr fenêtre d´absorption *(f)*
It finestra di assorbimento *(f)*
Pt janela de absorção *(f)*

Absorptionsfolge *(f)* *n* De
En absorption cycle
Es ciclo de absorción *(f)*
Fr cycle d´absorption *(m)*
It ciclo di assorbimento *(m)*
Pt ciclo de absorção *(m)*

absorption window En
De Absorptionsfenster *(n)*
Es ventanilla de absorción *(f)*
Fr fenêtre d´absorption *(f)*
It finestra di assorbimento *(f)*
Pt janela de absorção *(f)*

absorsancia solar *(f)* Es
De Solarabsorptionsgrad *(m)*
En solar absorbtance
Fr absorptance solaire *(f)*
It assorbimento solare *(m)*
Pt absorvância solar *(f)*

absorvância solar *(f)* Pt
De Solarabsorptionsgrad *(m)*
En solar absorbtance
Es absorsancia solar *(f)*
Fr absorptance solaire *(f)*
It assorbimento solare *(m)*

absorvedor directo *(m)* Pt
De Direktabsorber *(m)*
En direct absorber
Es absorbedor directo *(m)*
Fr absorbeur direct *(m)*
It assorbitore diretto *(m)*

abundancia natural *(f)* Es
De natürliche Isotopenhäufigkeit *(f)*
En natural abundance
Fr abondance naturelle *(f)*
It abbondanza naturale *(f)*
Pt abundância natural *(f)*

abundância natural *(f)* Pt
De natürliche Isotopenhäufigkeit *(f)*
En natural abundance
Es abundancia natural *(f)*
Fr abondance naturelle *(f)*
It abbondanza naturale *(f)*

Abwärme *(f)* *n* De
En waste heat
Es calor residual *(m)*
Fr chaleur perdue *(f)*
It calore perduto *(m)*
Pt calor residual *(m)*

Abwärmekessel *(m)* *n* De
En waste-heat boiler
Es caldera de calor residual *(f)*
Fr chaudière de récupération des chaleurs perdues *(f)*
It caldaia a recupero di calore perduto *(f)*
Pt caldeira aquecida com calor residual *(f)*

Abwärmerück-gewinnung *(f)* *n* De
En waste-heat recovery
Es recuperación de calor residual *(f)*
Fr récupération des chaleurs perdues *(f)*
It recupero di calore perduto *(m)*
Pt recuperação de calor residual *(f)*

Abwasser *(n)* *n* De
En effluent
Es efluente *(m)*
Fr effluent *(m)*

It effluente *(m)*
Pt efluente *(m)*

Abweichung *(f)* *n* De
En deviation
Es desviación *(f)*
Fr déviation *(f)*
It deviazione *(f)*
Pt desvio *(m)*

Abzug *(m)* *n* De
En flue
Es humero *(m)*
Fr carneau *(m)*
It condotto *(m)*
Pt fumeiro *(m)*

accensione *(f)* *n* It
De Zündung *(f)*
En ignition
Es encendido *(m)*
Fr allumage *(m)*
Pt ignição *(f)*

accensione ad impulso elettronico *(f)* It
De elektronische Impulszündung *(f)*
En electronic pulse ignition
Es encendido por impulsos electrónicos *(m)*
Fr allumage électronique à impulsions *(m)*
Pt ignição por impulso electrónico *(f)*

accensione a scintilla *(f)* It
De Funkenzündung *(f)*
En spark ignition
Es encendido por chispa *(m)*
Fr allumage par étincelle *(m)*
Pt ignição por faísca *(f)*

accensione doppia *(f)* It
De Doppelzündung *(f)*
En dual ignition
Es doble encendido *(m)*
Fr double allumage *(m)*
Pt ignição dupla *(f)*

accidente por pérdida de refrigerante *(m)* Es
De Unfall durch Kühlmittelverlust *(m)*
En loss of coolant accident (LOCA)
Fr accident par perte de fluide réfrigérant *(m)*
It incidente di perdita del refrigerante *(m)*
Pt acidente por perda de refrigerante *(m)*

accident par perte de fluide réfrigérant *(m)* Fr
De Unfall durch Kühlmittelverlust *(m)*
En loss of coolant accident (LOCA)
Es accidente por pérdida de refrigerante *(m)*
It incidente di perdita del refrigerante *(m)*
Pt acidente por perda de refrigerante *(m)*

accumulateur *(m)* *n* Fr
De Akkumulator *(m)*
En accumulator
Es acumulador *(m)*
It accumulatore *(m)*
Pt acumulador *(m)*

accumulateur de vapeur *(m)* Fr
De Dampfspeicher *(m)*
En steam accumulator
Es acumulador de vapor *(m)*
It accumulatore di vapore *(m)*
Pt acumulador de vapor *(m)*

accumulation à transition de phase *(f)* Fr
De Phasenumwandlungs-speicherung *(f)*
En phase-transition storage
Es almacenamiento de transición de fase *(m)*
It immagazzinaggio a transizione di fase *(m)*
Pt armazenagem de transição de fase *(f)*

accumulation d'air chaud *(f)* Fr
De Heißluftspeicherung *(f)*
En hot-air storage
Es almacenamiento de aire caliente *(m)*
It immagazzinaggio di aria calda *(m)*
Pt armazenagem de ar quente *(f)*

accumulation d'eau chaude *(f)* Fr
De Warmwasser-speicherung *(f)*
En hot-water storage
Es almacenamiento de agua caliente *(m)*
It immagazzinaggio di acqua calda *(m)*
Pt armazenagem de água quente *(f)*

accumulation de chaleur *(f)* Fr
De Wärmespeicherung *(f)*
En heat storage
Es almacenamiento de calor *(m)*
It immagazzinaggio di calore *(m)*
Pt armazenagem de calor *(f)*

accumulation de chaleur diurne *(f)* Fr
De tägliche Wärmespeicherung *(f)*
En diurnal heat storage
Es almacenamiento diurno del calor *(m)*
It immagazzinaggio di calore diurno *(m)*
Pt armazenagem diurna de calor *(f)*

accumulation de dépôts marins *(f)* Fr
De Seeverschmutzung *(f)*
En marine fouling
Es suciedad por depósitos marinos *(f)*
It incrostazione marina *(f)*
Pt depósitos marítimos *(m)*

accumulation d'énergie thermique *(n)* Fr
De Wärmeenergie-speicherung *(f)*
En thermal-energy storage
Es almacenamiento de energía térmica *(m)*
It immagazzinaggio di energia termica *(m)*
Pt armazenagem de energia térmica *(f)*

accumulation électrochimique *(f)* Fr
De elektrochemische Speicherung *(f)*
En electrochemical storage
Es almacenamiento electroquímico *(m)*
It immagazzinaggio elettrochimico *(m)*
Pt armazenagem electroquímica *(f)*

accumulation magnétique *(f)* Fr
De Magnetspeicher *(m)*
En magnetic storage
Es almacenamiento magnético *(m)*
It immagazzinaggio magnetico *(m)*
Pt armazenagem magnética *(f)*

accumulation mécanique *(f)* Fr
De mechanische Speicherung *(f)*
En mechanical storage
Es almacenamiento mecánico *(m)*
It immagazzinaggio meccanico *(m)*
Pt armazenagem mecânica *(f)*

accumulation pompée *(f)* Fr
De gepumpte Speicherung *(f)*
En pumped storage
Es almacenamiento bombeado *(m)*
It immagazzinaggi a pompa *(m)*
Pt armazenagem realizada à bomba *(f)*

accumulator *n* En
De Akkumulator *(m)*
Es acumulador *(m)*
Fr accumulateur *(m)*
It accumulatore *(m)*
Pt acumulador *(m)*

accumulatore *(m) n* It
De Akkumulator *(m)*
En accumulator
Es acumulador *(m)*
Fr accumulateur *(m)*
Pt acumulador *(m)*

accumulatore di vapore *(m)* It
De Dampfspeicher *(m)*
En steam accumulator
Es acumulador de vapor *(m)*
Fr accumulateur de vapeur *(m)*
Pt acumulador de vapor *(m)*

aceite de lutita *(m)* Es
De Schieferöl *(n)*
En shale oil
Fr huile de schiste *(f)*
It olio di schisto *(m)*
Pt petróleo xistoso *(m)*

aceite diesel *(m)* Es
De Dieselöl *(n)*
En diesel oil
Fr combustible diesel *(m)*
It nafta *(f)*
Pt óleo diesel *(m)*

aceite lubricante *(m)* Es
De Schmieröl *(n)*
En lubricating oil
Fr huile de graissage *(f)*
It olio lubrificante *(m)*
Pt óleo de lubrificação *(m)*

acetilene *(m) n* It
De Azetylen *(n)*
En acetylene
Es acetileno *(m)*
Fr acétylène *(m)*
Pt acetileno *(m)*

acetileno *(m) n* Es, Pt
De Azetylen *(n)*
En acetylene
Fr acétylène *(m)*
It acetilene *(m)*

acetylene *n* En
De Azetylen *(n)*
Es acetileno *(m)*
Fr acétylène *(m)*
It acetilene *(m)*
Pt acetileno *(m)*

acétylène *(m) n* Fr
De Azetylen *(n)*
En acetylene
Es acetileno *(m)*
It acetilene *(m)*
Pt acetileno *(m)*

acidente por perda de refrigerante *(m)* Pt
De Unfall durch Kühlmittelverlust *(m)*
En loss of coolant accident (LOCA)
Es accidente por pérdida de refrigerante *(m)*
Fr accident par perte de fluide réfrigérant *(m)*
It incidente di perdita del refrigerante *(m)*

acidificação *(f) n* Pt
De Säuerung *(f)*
En acidification
Es acidulación *(f)*
Fr acidification *(f)*
It acidificazione *(f)*

acidification *n* En, Fr *(f)*
De Säuerung *(f)*
Es acidulación *(f)*
It acidificazione *(f)*
Pt acidificação *(f)*

acidificazione *(f) n* It
De Säuerung *(f)*
En acidification
Es acidulación *(f)*
Fr acidification *(f)*
Pt acidificação *(f)*

acidulación *(f) n* Es
De Säuerung *(f)*
En acidification
Fr acidification *(f)*
It acidificazione *(f)*
Pt acidificação *(f)*

acondicionamiento de aire *(m)* Es
De Klimaanlage *(f)*
En air conditioning
Fr climatisation *(f)*
It condizionamento dell'aria *(m)*
Pt ar condicionado *(m)*

acoustic energy En
De Schallenergie *(f)*
Es energía acústica *(f)*
Fr énergie acoustique *(f)*
It energia acustica *(f)*
Pt energia acústica *(f)*

acqua calda *(f)* It
De Warmwasser *(n)*
En hot water
Es agua caliente *(f)*
Fr eau chaude *(f)*
Pt água quente *(f)*

acqua demineralizzata *(f)* It
De entmineralisiertes Wasser *(n)*
En demineralized water
Es agua desmineralizada *(f)*
Fr eau déminéralisée *(f)*
Pt água desmineralizada *(f)*

acqua di alimentazione *(f)* It
De Speisewasser *(n)*
En feedwater
Es agua de alimentación *(f)*
Fr eau d'alimentation *(f)*
Pt água de alimentação *(f)*

acqua di alimentazione del condensatore *(f)* It
De Kondensator-Speisewasser *(n)*
En condenser feedwater
Es agua de alimentación del condensador *(f)*
Fr eau d'alimentation de condenseur *(f)*
Pt água de alimentação de condensador *(f)*

acqua di raffreddamento *(f)* It
De Kühlwasser *(n)*
En cooling water
Es agua de refrigeración *(f)*
Fr eau de refroidissement *(f)*
Pt água de refrigeração *(f)*

acqua pesante *(f)* It
De Schwerwasser *(n)*
En heavy water
Es agua pesada *(f)*
Fr eau lourde *(f)*
Pt água pesada *(f)*

acqua salmastra *(f)* It
De Sole *(f)*
En brine
Es salmuera *(f)*
Fr saumure *(f)*
Pt salmoura *(f)*

acqua salmastra geotermica *(f)* It
De geothermische Sole *(f)*
En geothermal brine
Es salmuera geotérmica *(f)*
Fr saumure géothermique *(f)*
Pt salmoura geotérmica *(f)*

active waste En
De radioaktive Abfallstoffe *(pl)*
Es residuo activo *(m)*
Fr déchets actifs *(m pl)*
It rifiuti attivi *(m pl)*
Pt desperdícios activos *(m pl)*

acumulador *(m) n* Es, Pt
De Akkumulator *(m)*
En accumulator
Fr accumulateur *(m)*
It accumulatore *(m)*

acumulador de vapor *(m)* Es, Pt
De Dampfspeicher *(m)*
En steam accumulator
Fr accumulateur de vapeur *(m)*
It accumulatore di vapore *(m)*

additif *(m) n* Fr
De Zusatzstoff *(m)*
En additive
Es aditivo *(m)*
It additivo *(m)*
Pt aditivo *(m)*

additifs détergents *(m pl)* Fr
De Waschmittelzusätze *(pl)*
En detergent additives *(pl)*
Es aditivos detergentes *(m pl)*
It additivi detergenti *(m pl)*
Pt aditivos detergentes *(m pl)*

additive *n* En
De Zusatzstoff *(m)*
Es aditivo *(m)*
Fr additif *(m)*
It additivo *(m)*
Pt aditivo *(m)*

additivi detergenti *(m pl)* It
De Waschmittelzusätze *(pl)*
En detergent additives *(pl)*
Es aditivos detergentes *(m pl)*
Fr additifs détergents *(m pl)*
Pt aditivos detergentes *(m pl)*

additivo *(m) n* It
De Zusatzstoff *(m)*
En additive
Es aditivo *(m)*
Fr additif *(m)*
Pt aditivo *(m)*

adiabatic *adj* En
De adiabatisch
Es adiabático
Fr adiabatique
It adiabatico
Pt adiabático

adiabatico *adj* It
De adiabatisch
En adiabatic
Es adiabático
Fr adiabatique
Pt adiabático

adiabático *adj* Es, Pt
De adiabatisch
En adiabatic
Fr adiabatique
It adiabatico

adiabatique *adj* Fr
De adiabatisch
En adiabatic
Es adiabático
It adiabatico
Pt adiabático

adiabatisch *adj* De
En adiabatic
Es adiabático
Fr adiabatique
It adiabatico
Pt adiabático

adiatermico *adj* It
De adiathermal
En adiathermal
Es adiatérmico
Fr adiathermique
Pt adiatérmico

adiatérmico *adj* Es, Pt
De adiathermal
En adiathermal
Fr adiathermique
It adiatermico

adiathermal *adj* De, En
Es adiatérmico
Fr adiathermique
It adiatermico
Pt adiatérmico

adiathermique *adj* Fr
De adiathermal
En adiathermal
Es adiatérmico
It adiatermico
Pt adiatérmico

aditivo *(m) n* Es, Pt
De Zusatzstoff *(m)*
En additive
Fr additif *(m)*
It additivo *(m)*

aditivos detergentes *(m pl)* Es, Pt
De Waschmittelzusätze *(pl)*
En detergent additives *(pl)*
Fr additifs détergents *(m pl)*
It additivi detergenti *(m pl)*

admissão de ar *(f)* Pt
De Lufteinlaβ *(m)*
En air inlet
Es entrada de aire *(f)*
Fr admission d'air *(f)*
It presa d'aria *(f)*

admission d'air *(f)* Fr
De Lufteinlaβ *(m)*
En air inlet
Es entrada de aire *(f)*
It presa d'aria *(f)*
Pt admissão de ar *(f)*

advanced gas-cooled reactor (AGR) En
De fortgeschrittener gasgekühlter Reaktor *(m)*
Es reactor avanzado refrigerado por gas *(m)*
Fr reacteur avancé à refroidissement au gaz *(m)*
It reattore avanzato raffreddato a gas *(m)*
Pt reactor avançado arrefecido a gás *(m)*

aerated flame En
De belüftete Flamme *(f)*
Es llama aireada *(f)*
Fr flamme aérée *(f)*
It fiamma aerata *(f)*
Pt chama arejada *(f)*

aeration *n* En
De Lüftung *(f)*
Es aireación *(f)*
Fr aération *(f)*
It aerazione *(f)*
Pt arejamento *(m)*

aération *(f) n* Fr
De Lüftung *(f)*
En aeration
Es aireación *(f)*
It aerazione *(f)*
Pt arejamento *(m)*

aération fixe *(f)* Fr
De feststehende Belüftung *(f)*
En fixed aeration
Es aireación fija *(f)*
It aerazione fissa *(f)*
Pt arejamento fixo *(m)*

aerazione *(f) n* It
De Lüftung *(f)*
En aeration
Es aireación *(f)*
Fr aération *(f)*
Pt arejamento *(m)*

aerazione fissa *(f)* It
De feststehende Belüftung *(f)*
En fixed aeration
Es aireación fija *(f)*
Fr aération fixe *(f)*
Pt arejamento fixo *(m)*

aerodinamica *(f) n* It
De Aerodynamik *(f)*
En aerodynamics
Es aerodinámica *(f)*
Fr aérodynamique *(f)*
Pt aerodinámica *(f)*

aerodinámica *(f) n* Es, Pt
De Aerodynamik *(f)*
En aerodynamics
Fr aérodynamique *(f)*
It aerodinamica *(f)*

aerodynamics *n* En
De Aerodynamik *(f)*
Es aerodinámica *(f)*
Fr aérodynamique *(f)*
It aerodinamica *(f)*
Pt aerodinámica *(f)*

Aerodynamik *(f) n* De
En aerodynamics
Es aerodinámica *(f)*
Fr aérodynamique *(f)*
It aerodinamica *(f)*
Pt aerodinámica *(f)*

aérodynamique *(f) n* Fr
De Aerodynamik *(f)*
En aerodynamics
Es aerodinámica *(f)*
It aerodinamica *(f)*
Pt aerodinámica *(f)*

aerogenerador *(m) n* Es
De Luftgenerator *(m)*
En aerogenerator
Fr aérogénérateur *(m)*
It aerogeneratore *(m)*
Pt aerogerador *(m)*

aérogénérateur *(m) n* Fr
De Luftgenerator *(m)*
En aerogenerator
Es aerogenerador *(m)*
It aerogeneratore *(m)*
Pt aerogerador *(m)*

aerogenerator *n* En
De Luftgenerator *(m)*
Es aerogenerador *(m)*
Fr aérogénérateur *(m)*
It aerogeneratore *(m)*
Pt aerogerador *(m)*

aerogeneratore *(m) n* It
De Luftgenerator *(m)*
En aerogenerator
Es aerogenerador *(m)*
Fr aérogénérateur *(m)*
Pt aerogerador *(m)*

aerogerador *(m)* *n* Pt
De Luftgenerator *(m)*
En aerogenerator
Es aerogenerador *(m)*
Fr aérogénérateur *(m)*
It aerogeneratore *(m)*

aerosol *n* En; Es, It, Pt *(m)*
De Aerosol *(n)*
Fr aérosol *(m)*

Aerosol *(n)* *n* De
En aerosol
Es aerosol *(m)*
Fr aérosol *(m)*
It aerosol *(m)*
Pt aerosol *(m)*

aérosol *(m)* *n* Fr
De Aerosol *(n)*
En aerosol
Es aerosol *(m)*
It aerosol *(m)*
Pt aerosol *(m)*

agent chélateur *(m)* Fr
De Cheliermittel *(n)*
En chelating agent
Es agente quelatante *(m)*
It agente di chelazione *(m)*
Pt agente de chelação *(m)*

agente de chelação *(m)* Pt
De Cheliermittel *(n)*
En chelating agent
Es agente quelatante *(m)*
Fr agent chélateur *(m)*
It agente di chelazione *(m)*

agente di chelazione *(m)* It
De Cheliermittel *(n)*
En chelating agent
Es agente quelatante *(m)*
Fr agent chélateur *(m)*
Pt agente de chelação *(m)*

agente quelatante *(m)* Es
De Cheliermittel *(n)*
En chelating agent
Fr agent chélateur *(m)*
It agente di chelazione *(m)*
Pt agente de chelação *(m)*

agente tensioattivo *(m)* It
De grenzflächenaktiver Stoff *(m)*
En surfactant
Es surfactante *(m)*
Fr agent tensio-actif *(m)*
Pt agente tenso-activo *(m)*

agente tenso-activo *(m)* Pt
De grenzflächenaktiver Stoff *(m)*
En surfactant
Es surfactante *(m)*
Fr agent tensio-actif *(m)*
It agente tensioattivo *(m)*

agent tensio-actif *(m)* Fr
De grenzflächenaktiver Stoff *(m)*
En surfactant
Es surfactante *(m)*
It agente tensioattivo *(m)*
Pt agente tenso-activo *(m)*

agrietamiento hidráulico *(m)* Es
De hydraulische Rißbildung *(f)*
En hydraulic fracturing
Fr fracturation hydraulique *(f)*
It fratturazione idraulica *(f)*
Pt fracturação hidráulica *(f)*

agua caliente *(f)* Es
De Warmwasser *(n)*
En hot water
Fr eau chaude *(f)*
It acqua calda *(f)*
Pt água quente *(f)*

água de alimentação *(f)* Pt
De Speisewasser *(n)*
En feedwater
Es agua de alimentación *(f)*
Fr eau d'alimentation *(f)*
It acqua di alimentazione *(f)*

água de alimentação de condensador *(f)* Pt
De Kondensator-Speisewasser *(n)*
En condenser feedwater
Es agua de alimentación del condensador *(f)*
Fr eau d'alimentation de condenseur *(f)*
It acqua di alimentazione del condensatore *(f)*

agua de alimentación *(f)* Es
De Speisewasser *(n)*
En feedwater
Fr eau d'alimentation *(f)*
It acqua di alimentazione *(f)*
Pt água de alimentação *(f)*

agua de alimentación del condensador *(f)* Es
De Kondensator-Speisewasser *(n)*
En condenser feedwater
Fr eau d'alimentation de condenseur *(f)*
It acqua di alimentazione del condensatore *(f)*
Pt água de alimentação de condensador *(f)*

água de descarga *(f)* Pt
De Ablauf *(m)*
En tailrace
Es canal de descarga *(m)*
Fr bief *(m)*
It canale di scarico *(m)*

água de refrigeração *(f)* Pt
De Kühlwasser *(n)*
En cooling water
Es agua de refrigeración *(f)*
Fr eau de refroidissement *(f)*
It acqua di raffreddamento *(f)*

agua de refrigeración *(f)* Es
De Kühlwasser *(n)*
En cooling water
Fr eau de refroidissement *(f)*
It acqua di raffreddamento *(f)*
Pt água de refrigeração *(f)*

agua desmineralizada *(f)* Es
De entmineralisiertes Wasser *(n)*
En demineralized water
Fr eau déminéralisée *(f)*
It acqua demineralizzata *(f)*
Pt água desmineralizada *(f)*

água desmineralizada *(f)* Pt
De entmineralisiertes Wasser *(n)*
En demineralized water
Es agua desmineralizada *(f)*
Fr eau déminéralisée *(f)*
It acqua demineralizzata *(f)*

agua pesada *(f)* Es
De Schwerwasser *(n)*
En heavy water
Fr eau lourde *(f)*
It acqua pesante *(f)*
Pt água pesada *(f)*

água pesada *(f)* Pt
De Schwerwasser *(n)*
En heavy water
Es agua pesada *(f)*
Fr eau lourde *(f)*
It acqua pesante *(f)*

água quente *(f)* Pt
De Warmwasser *(n)*
En hot water
Es agua caliente *(f)*
Fr eau chaude *(f)*
It acqua calda *(f)*

ahorro energético *(m)* Es
De Energiesparen *(n)*
En energy saving
Fr économie d'énergie *(f)*
It risparmio di energia *(m)*
Pt poupança de energia *(f)*

aimant *(m)* *n* Fr
De Magnet *(m)*
En magnet
Es imán *(m)*
It magmete *(m)*
Pt ímã *(m)*

aimant superconducteur *(m)* Fr
De supraleitfähiger Magnet *(m)*
En superconducting magnet
Es imán superconductor *(m)*
It magnete superconduttore *(m)*
Pt ímã supercondutor *(m)*

air *n* En, Fr *(m)*
De Luft *(f)*
Es aire *(m)*
It aria *(f)*
Pt ar *(m)*

air atomizer En
De Luftzerstäuber *(m)*
Es atomizador de aire *(m)*
Fr atomiseur d'air *(m)*
It nebulizzatore d'aria *(m)*
Pt atomizador de ar *(m)*

air-blast burner En
De Windbrenner *(m)*
Es quemador de chorro de aire *(m)*
Fr brûleur à air soufflé *(m)*
It bruciatore a ventilazione forzata *(m)*
Pt queimador de corrente de ar *(m)*

air compressor En
De Luftkompressor *(m)*
Es compresor de aire *(m)*
Fr compresseur d'air *(m)*
It compressore d'aria *(m)*
Pt compressor de ar *(m)*

air conditioning En
De Klimaanlage *(f)*
Es acondicionamiento de aire *(m)*
Fr climatisation *(f)*
It condizionamento dell'aria *(m)*
Pt ar condicionado *(m)*

air-cooled *adj* En
De luftgekühlt
Es refrigerado por aire
Fr refroidi par air
It raffreddato ad aria
Pt arrefecido a ar

air drilling En
De Druckluftbohren *(n)*
Es perforación por aire comprimido *(f)*
Fr forage à l'air comprimé *(m)*
It trapanatura ad aria compressa *(f)*
Pt perfuração pneumática *(f)*

air dryer En
De Lufttrockner *(m)*
Es secador de aire *(m)*
Fr sécheur d'air *(m)*
It essiccatore ad aria *(m)*
Pt secador por ar *(m)*

aire *(m)* *n* Es
De Luft *(f)*
En air
Fr air *(m)*
It aria *(f)*
Pt ar *(m)*

aireación *(f)* *n* Es
De Lüftung *(f)*
En aeration
Fr aération *(f)*
It aerazione *(f)*
Pt arejamento *(m)*

aireación fija *(f)* Es
De feststehende Belüftung *(f)*
En fixed aeration
Fr aération fixe *(f)*
It aerazione fissa *(f)*
Pt arejamento fixo *(m)*

aire primario *(m)* Es
De primäre Luft *(f)*
En primary air
Fr air primaire *(m)*
It aria primaria *(f)*
Pt ar primário *(m)*

aire saturado *(m)* Es
De gesättigte Luft *(f)*
En saturated air
Fr air saturé *(m)*
It aria satura *(f)*
Pt ar saturado *(m)*

aire secundario *(m)* Es
De Sekundärluft *(f)*
En secondary air
Fr air secondaire *(m)*
It aria secondaria *(f)*
Pt ar secundário *(m)*

aire sobrante *(m)* Es
De Luftüberschuß *(m)*
En excess air
Fr excès d'air *(m)*
It eccesso d'aria *(m)*
Pt ar em excesso *(m)*

air heater En
De Lufterhitzer *(m)*
Es calentador de aire *(m)*
Fr réchauffeur d'air *(m)*
It riscaldatore ad aria *(m)*
Pt aquecedor de ar *(m)*

air inlet En
De Lufteinlaß *(m)*
Es entrada de aire *(f)*
Fr admission d'air *(f)*
It presa d'aria *(f)*
Pt admissão de ar *(f)*

air leakage En
De Luftleck *(n)*
Es fuga de aire *(f)*
Fr fuite d'air *(f)*
It fuga d'aria *(f)*
Pt fuga de ar *(f)*

air pollution En
De Luftverschmutzung *(f)*
Es contaminación del aire *(f)*
Fr pollution de l'air *(f)*
It inquinamento atmosferico *(m)*
Pt poluição do ar *(f)*

air primaire *(m)* Fr
De primäre Luft *(f)*
En primary air
Es aire primario *(m)*
It aria primaria *(f)*
Pt ar primário *(m)*

air saturé *(m)* Fr
De gesättigte Luft *(f)*
En saturated air
Es aire saturado *(m)*
It aria satura *(f)*
Pt ar saturado *(m)*

air secondaire *(m)* Fr
De Sekundärluft *(f)*
En secondary air
Es aire secundario *(m)*
It aria secondaria *(f)*
Pt ar secundário *(m)*

air vent En
De Entlüftungsschlitz *(m)*
Es toma de aire *(f)*
Fr évent *(m)*
It sfiato aria *(m)*
Pt respirador de ar *(m)*

aislamiento *(m)* *n* Es
De Isolierung *(f)*
En insulation
Fr isolement *(m)*
It isolamento *(m)*
Pt isolamento *(m)*

aislamiento con fibra *(m)* Es
Am fiber insulation
De Faserisolierung *(f)*
En fibre insulation
Fr isolement par fibre *(m)*
It isolamento con fibra *(m)*
Pt isolamento com fibras *(m)*

aislamiento con fibra de lana mineral *(m)* Es
Am mineral-wool fiber insulation
De Steinwollefaser-isolierung *(f)*
En mineral-wool fibre insulation
Fr isolement par fibre de laine minérale *(m)*
It isolamento con fibra di lana minerale *(m)*
Pt isolamento com fibras de lã mineral *(m)*

aislamiento con fibra de vidrio *(m)* Es
Am glass-fiber insulation
De Glasfaserisolierung *(f)*
En glass-fibre insulation

Fr isolement par fibre de verre *(m)*
It isolamento con fibra di vetro *(m)*
Pt isolamento com fibra de vidro *(m)*

aislamiento de galerías *(m)* Es
De Dachbodenisolierung *(f)*
En loft insulation
Fr isolement des combles *(m)*
It isolamento della soffitta *(m)*
Pt isolamento do sotão *(m)*

aislamiento de pared de cavidad *(m)* Es
De Isolierung von Hohlziegelmauerwerk *(f)*
En cavity wall insulation
Fr isolement des murs doubles *(m)*
It isolamento di muro a cassavuota *(m)*
Pt isolamento de paredes de cavidade *(m)*

aislamiento de paredes *(m)* Es
De Wandisolierung *(f)*
En wall insulation
Fr isolement des murs *(m)*
It isolamento delle pareti *(m)*
Pt isolamento de paredes *(m)*

aislamiento de poliuretano *(m)* Es
De Polyurethanisolierung *(f)*
En polyurethane insulation
Fr isolement au polyuréthane *(m)*
It isolamento con poliuretano *(m)*
Pt isolamento de poliuretano *(m)*

aislamiento de suelos *(m)* Es
De Bodenisolierung *(f)*
En floor insulation
Fr isolement du plancher *(m)*
It isolamento del pavimento *(m)*
Pt isolamento de pavimento *(m)*

aislamiento de techos *(m)* Es
De Dachisolierung *(f)*
En roof insulation
Fr isolement des toits *(m)*
It isolamento del tetto *(m)*
Pt isolamento da cobertura *(m)*

aislamiento de ventanas *(m)* Es
De Fensterisolierung *(f)*
En window insulation
Fr isolement des fenêtres *(m)*
It isolamento delle finestre *(m)*
Pt isolamento de janela *(m)*

Akkumulator *(m) n* De
En accumulator
Es acumulador *(m)*
Fr accumulateur *(m)*
It accumulatore *(m)*
Pt acumulador *(m)*

albedo *n* En; Es, It, Pt *(m)*
De Albedo *(n)*
Fr albédo *(m)*

Albedo *(n) n* De
En albedo
Es albedo *(m)*
Fr albédo *(m)*
It albedo *(m)*
Pt albedo *(m)*

albédo *(m) n* Fr
De Albedo *(n)*
En albedo
Es albedo *(m)*
It albedo *(m)*
Pt albedo *(m)*

alcatrão de baixa temperatura *(m)* Pt
De Tieftemperaturteer *(m)*
En low-temperature tar
Es alquitrán de baja temperatura *(m)*
Fr goudron à basse température *(m)*
It catrame a bassa temperatura *(m)*

alcatrão de forno de coque *(m)* Pt
De Koksofenteer *(m)*
En coke-oven tar
Es alquitrán de horno de coquización *(m)*
Fr goudron *(m)*
It catrame di forno da coke *(m)*

alcatrão de hulha *(m)* Pt
De Steinkohlenteer *(m)*
En coal tar
Es alquitrán de hulla *(m)*
Fr goudron de houille *(m)*
It catrame di carbon fossile *(m)*

alchilazione *(f) n* It
De Alkylierung *(f)*
En alkylation
Es alquilación *(f)*
Fr alcoylation *(f)*
Pt alquilação *(f)*

alcohol *n* En, Es *(m)*
De Alkohol *(m)*
Fr alcool *(m)*
It alcool *(m)*
Pt álcool *(m)*

alcohol rectificado *(m)* Es
De rektifizierter Alkohol *(m)*
En rectified spirit
Fr alcool rectifié *(m)*
It spirito rettificato *(m)*
Pt álcool rectificado *(m)*

alcool *(m) n* Fr, It
De Alkohol *(m)*
En alcohol
Es alcohol *(m)*
It alcool *(m)*
Pt álcool *(m)*

álcool *(m) n* Pt
De Alkohol *(m)*
En alcohol
Es alcohol *(m)*
Fr alcool *(m)*
It alcool *(m)*

álcool rectificado *(m)* Pt
De rektifizierter Alkohol *(m)*
En rectified spirit
Es alcohol rectificado *(m)*
Fr alcool rectifié *(m)*
It spirito rettificato *(m)*

alcool rectifié *(m)* Fr
De rektifizierter Alkohol *(m)*
En rectified spirit
Es alcohol rectificado *(m)*
It spirito rettificato *(m)*
Pt álcool rectificado *(m)*

alcoylation *(f) n* Fr
De Alkylierung *(f)*
En alkylation
Es alquilación *(f)*
It alchilazione *(f)*
Pt alquilação *(f)*

aleación *(f) n* Es
De Legierung *(f)*
En alloy
Fr alliage *(m)*
It lega *(f)*
Pt liga *(f)*

aleación de zirconio *(f)* Es
De Zirkonlegierung *(f)*
En zircalloy
Fr alliage de zirconium *(m)*
It lega di zirconio *(f)*
Pt liga de zircónio *(f)*

algae *pl n* En
De Algen *(pl)*
Es algas *(f pl)*
Fr algues *(f pl)*
It alghe *(f pl)*
Pt algas *(f pl)*

algas *(f pl) n* Es, Pt
De Algen *(pl)*
En algae
Fr algues *(f pl)*
It alghe *(f pl)*

Algen *(pl) n* De
En algae
Es algas *(f pl)*
Fr algues *(f pl)*
It alghe *(f pl)*
Pt algas *(f pl)*

alghe *(f pl) n* It
De Algen *(pl)*
En algae
Es algas *(f pl)*
Fr algues *(f pl)*
Pt algas *(f pl)*

algues *(f pl) n* Fr
De Algen *(pl)*
En algae
Es algas *(f pl)*
It alghe *(f pl)*
Pt algas *(f pl)*

alimentação por gravidade *(f)* Pt
De Schwerkraftspeisung *(f)*
En gravity-feed
Es alimentación por gravedad *(f)*
Fr alimentation par gravité *(f)*
It alimentazione a gravità *(f)*

alimentación por gravedad *(f)* Es
De Schwerkraftspeisung *(f)*
En gravity-feed
Fr alimentation par gravité *(f)*
It alimentazione a gravità *(f)*
Pt alimentação por gravidade *(f)*

alimentación principal *(f)* Es
De Netzstrom *(m)*
En mains supply
Fr alimentation secteur *(f)*
It alimentazione di rete *(f)*
Pt distribuição pela linha principal *(f)*

alimentado a gás Pt
De gasgefeuert
En gas-fired
Es alimentado por gas
Fr chauffé au gaz
It alimentato a gas

alimentado a óleo Pt
De ölgefeuert
En oil-fired
Es alimentado con petróleo
Fr chauffé au mazout
It alimentato a nafta

alimentado con petróleo Es
De ölgefeuert
En oil-fired
Fr chauffé au mazout
It alimentato a nafta
Pt alimentado a óleo

alimentado por gas Es
De gasgefeuert
En gas-fired
Fr chauffé au gaz
It alimentato a gas
Pt alimentado a gás

alimentation par gravité *(f)* Fr
De Schwerkraftspeisung *(f)*
En gravity-feed
Es alimentación por gravedad *(f)*
It alimentazione a gravità *(f)*
Pt alimentação por gravidade

alimentation secteur *(f)* Fr
De Netzstrom *(m)*
En mains supply
Es alimentación principal *(f)*
It alimentazione di rete *(f)*
Pt distribuição pela linha principal *(f)*

alimentato a gas It
De gasgefeuert
En gas-fired
Es alimentado por gas
Fr chauffé au gaz
Pt alimentado a gás

alimentato a nafta It
De ölgefeuert
En oil-fired
Es alimentado con petróleo
Fr chauffé au mazout
Pt alimentado a óleo

alimentazione a gravità *(f)* It
De Schwerkraftspeisung *(f)*
En gravity-feed
Es alimentación por gravedad *(f)*
Fr alimentation par gravité *(f)*
Pt alimentação por gravidade *(f)*

alimentazione di rete *(f)* It
De Netzstrom *(m)*
En mains supply
Es alimentación principal *(f)*
Fr alimentation secteur *(f)*
Pt distribuição pela linha principal *(f)*

alkaline secondary cell En
De sekundäre Alkalizelle *(f)*
Es elemento alcalino de acumulador *(m)*
Fr élément secondaire alcalin *(m)*
It cellula secondaria alcalina *(f)*
Pt célula secundária alcalina *(f)*

Alkohol *(m) n* De
En alcohol
Es alcohol *(m)*
Fr alcool *(m)*
It alcool *(m)*
Pt álcool *(m)*

alkylation *n* En
De Alkylierung *(f)*
Es alquilación *(f)*
Fr alcoylation *(f)*
It alchilazione *(f)*
Pt alquilação *(f)*

Alkylierung *(f) n* De
En alkylation
Es alquilación *(f)*
Fr alcoylation *(f)*
It alchilazione *(f)*
Pt alquilação *(f)*

al largo It
De auf See
En offshore
Es marino
Fr au large
Pt fora da praia

alliage *(m) n* Fr
De Legierung *(f)*
En alloy
Es aleación *(f)*
It lega *(f)*
Pt liga *(f)*

alliage de zirconium *(m)* Fr
De Zirkonlegierung *(f)*
En zircalloy
Es aleación de zirconio *(f)*
It lega di zirconio *(f)*
Pt liga de zircónio *(f)*

alloy *n* En
De Legierung *(f)*
Es aleación *(f)*
Fr alliage *(m)*
It lega *(f)*
Pt liga *(f)*

allumage *(m) n* Fr
De Zündung *(f)*
En ignition
Es encendido *(m)*
It accensione *(f)*
Pt ignição *(f)*

allumage électronique à impulsions *(m)* Fr
De elektronische Impulszündung *(f)*
En electronic pulse ignition
Es encendido por impulsos electrónicos *(m)*
It accensione ad impulso elettronico *(f)*
Pt ignição por impulso electrónico *(f)*

allumage par étincelle *(m)* Fr
De Funkenzündung *(f)*
En spark ignition
Es encendido por chispa *(m)*
It accensione a scintilla *(f)*
Pt ignição por faísca *(f)*

allumage spontané *(m)* Fr
De spontane Zündung *(f)*
En spontaneous ignition
Es encendido espontáneo *(m)*
It autoaccensione *(f)*

Pt ignição espontânea *(f)*

alma (magnética) *(f)* Es
De Kern (magnetischer) *(m)*
En core (magnetic)
Fr noyau (magnétique) *(m)*
It nucleo (magnetico) *(m)*
Pt núcleo (magnético) *(m)*

almacenamiento bombeado *(m)* Es
De gepumpte Speicherung *(f)*
En pumped storage
Fr accumulation pompée *(f)*
It immagazzinaggi a pompa *(m)*
Pt armazenagem realizada à bomba *(f)*

almacenamiento de agua caliente *(m)* Es
De Warmwasser-speicherung *(f)*
En hot-water storage
Fr accumulation d'eau chaude *(f)*
It immagazzinaggio di acqua calda *(m)*
Pt armazenagem de água quente *(f)*

almacenamiento de aire caliente *(m)* Es
De Heiβluftspeicherung *(f)*
En hot-air storage
Fr accumulation d'air chaud *(f)*
It immagazzinaggio di aria calda *(m)*
Pt armazenagem de ar quente *(f)*

almacenamiento de aire comprimido *(m)* Es
De Druckluftspeicherung *(f)*
En compressed-air storage
Fr stockage de l'air comprimé *(m)*
It immagazzinaggio di aria compressa *(m)*
Pt armazenagem de ar comprimido *(f)*

almacenamiento de calor *(m)* Es
De Wärmespeicherung *(f)*
En heat storage
Fr accumulation de chaleur *(f)*
It immagazzinaggio di calore *(m)*
Pt armazenagem de calor *(f)*

almacenamiento de energía térmica *(m)* Es
De Wärmeenergie-speicherung *(f)*
En thermal-energy storage
Fr accumulation d'énergie thermique *(n)*
It immagazzinaggio di energia termica *(m)*
Pt armazenagem de energia térmica *(f)*

almacenamiento de transición de fase *(m)* Es
De Phasenum-wandlungs-speicherung *(f)*
En phase-transition storage
Fr accumulation à transition de phase *(f)*
It immagazzinaggio a transizione di fase *(m)*
Pt armazenagem de transição de fase *(f)*

almacenamiento diurno del calor *(m)* Es
De tägliche Wärmespeicherung *(f)*
En diurnal heat storage
Fr accumulation de chaleur diurne *(f)*
It immagazzinaggio di calore diurno *(m)*
Pt armazenagem diurna de calor *(f)*

almacenamiento electroquímico *(m)* Es
De elektrochemische Speicherung *(f)*
En electrochemical storage
Fr accumulation électrochimique *(f)*
It immagazzinaggio elettrochimico *(m)*
Pt armazenagem electroquímica *(f)*

almacenamiento magnético *(m)* Es
De Magnetspeicher *(m)*
En magnetic storage
Fr accumulation magnétique *(f)*
It immagazzinaggio magnetico *(m)*
Pt armazenagem magnética *(f)*

almacenamiento mecánico *(m)* Es
De mechanische Speicherung *(f)*
En mechanical storage
Fr accumulation mécanique *(f)*
It immagazzinaggio meccanico *(m)*
Pt armazenagem mecânica *(f)*

almendrilla *(f)* *n* Es
De Nuβkohle *(f)*
En nut coal
Fr noix de charbon *(f)*
It carbone di pezzatura noce *(m)*
Pt carvão em nozes *(m)*

alpha particle En
De Alpha-Teilchen *(n)*
Es partícula alfa *(f)*
Fr particule alpha *(f)*
It particella alfa *(f)*
Pt partícula alfa *(f)*

Alpha-Teilchen *(n)* *n* De
En alpha particle
Es partícula alfa *(f)*
Fr particule alpha *(f)*
It particella alfa *(f)*
Pt partícula alfa *(f)*

alquilação *(f)* *n* Pt
De Alkylierung *(f)*
En alkylation
Es alquilación *(f)*
Fr alcoylation *(f)*
It alchilazione *(f)*

alquilación *(f)* *n* Es
De Alkylierung *(f)*
En alkylation
Fr alcoylation *(f)*
It alchilazione *(f)*
Pt alquilação *(f)*

alquitrán de baja temperatura *(m)* Es
De Tieftemperaturteer *(m)*
En low-temperature tar
Fr goudron à basse température *(m)*
It catrame a bassa temperatura *(m)*
Pt alcatrão de baixa temperatura *(m)*

alquitrán de horno de coquización *(m)* Es
De Koksofenteer *(m)*
En coke-oven tar
Fr goudron *(m)*
It catrame di forno da coke *(m)*
Pt alcatrão de forno de coque *(m)*

alquitrán de hulla *(m)* Es
De Steinkohlenteer *(m)*
En coal tar
Fr goudron de houille *(m)*
It catrame di carbon fossile *(m)*
Pt alcatrão de hulha *(m)*

alta tensão *(f)* Pt
De Hochspannung *(f)*
En high tension (HT)
Es alta tensión *(f)*
Fr haute tension *(f)*
It alta tensione (AT) *(f)*

alta tensión *(f)* Es
De Hochspannung *(f)*
En high tension (HT)
Fr haute tension *(f)*
It alta tensione (AT) *(f)*
Pt alta tensão *(f)*

alta tensione (AT) *(f)* It
De Hochspannung *(f)*
En high tension (HT)
Es alta tensión *(f)*
Fr haute tension *(f)*
Pt alta tensão *(f)*

alternador *(m)* *n* Es, Pt
De Wechselstrom-generator *(m)*
En alternator
Fr alternateur *(m)*
It alternatore *(m)*

alternateur *(m)* *n* Fr
De Wechselstrom-generator *(m)*
En alternator
Es alternador *(m)*
It alternatore *(m)*
Pt alternador *(m)*

alternating current (a.c.) En
De Wechselstrom *(m)*
Es corriente alterna (c.a.) *(f)*
Fr courant alternatif (c.a.) *(m)*
It corrente alternata (c.a.) *(f)*
Pt corrente alterna (c.a.) *(f)*

alternator *n* En
De Wechselstrom-generator *(m)*
Es alternador *(m)*
Fr alternateur *(m)*
It alternatore *(m)*
Pt alternador *(m)*

alternatore *(m)* *n* It
De Wechselstrom-generator *(m)*
En alternator
Es alternador *(m)*
Fr alternateur *(m)*
Pt alternador *(m)*

altezza manometrica *(f)* It
De Druckhöhe *(f)*
En pressure head
Es altura manométrica *(f)*
Fr hauteur de refoulement *(f)*
Pt altura manométrica *(f)*

altitud *(f)* *n* Es
De Höhe *(f)*
En altitude
Fr altitude *(f)*
It altitudine *(f)*
Pt altitude *(f)*

altitude *n* En; Fr, Pt *(f)*
De Höhe *(f)*
Es altitud *(f)*
It altitudine *(f)*

altitudine *(f)* *n* It
De Höhe *(f)*
En altitude
Es altitud *(f)*
Fr altitude *(f)*
Pt altitude *(f)*

alto forno *(m)* It
De Hochofen *(m)*
En blast furnace
Es alto horno *(m)*
Fr haut fourneau *(m)*
Pt alto-forno *(m)*

alto-forno *(m)* *n* Pt
De Hochofen *(m)*
En blast furnace
Es alto horno *(m)*
Fr haut fourneau *(m)*
It alto forno *(m)*

alto horno *(m)* Es
De Hochofen *(m)*
En blast furnace
Fr haut fourneau *(m)*
It alto forno *(m)*
Pt alto-forno *(m)*

altura manométrica *(f)* Es, Pt
De Druckhöhe *(f)*
En pressure head
Fr hauteur de refoulement *(f)*
It altezza manometrica *(f)*

ambiental *adj* Es
De Umwelt-
En environmental
Fr de l'environnement
It ambientale
Pt ambiente

ambientale *adj* It
De Umwelt-
En environmental
Es ambiental
Fr de l'environnement
Pt ambiente

ambiente *adj* Pt
De Umwelt-
En environmental
Es ambiental
Fr de l'environnement
It ambientale

amélioration *(f)* *n* Fr
De Aufkonzentrierung *(f)*
En upgrading
Es mejoramiento *(m)*
It promovimento a grado superiore *(m)*
Pt melhoramento *(m)*

amiante *(m)* *n* Fr
De Asbest *(m)*
En asbestos
Es amianto *(m)*
It amianto *(m)*
Pt amianto *(m)*

amiante bleu *(m)* Fr
De blauer Asbest *(m)*
En blue asbestos
Es crocidolita *(f)*
It amianto azzurro *(m)*
Pt amianto azul *(m)*

amianto *(m)* *n* Es, It, Pt
De Asbest *(m)*
En asbestos
Fr amiante *(m)*

amianto azul *(m)* Pt
De blauer Asbest *(m)*
En blue asbestos
Es crocidolita *(f)*
Fr amiante bleu *(m)*
It amianto azzurro *(m)*

amianto azzurro *(m)* It
De blauer Asbest *(m)*
En blue asbestos
Es crocidolita *(f)*
Fr amiante bleu *(m)*
Pt amianto azul *(m)*

ammeter *n* En
De Ammeter *(n)*
Es amperímetro *(m)*
Fr ampèremètre *(m)*
It amperometro *(m)*
Pt amperímetro *(m)*

Ammeter *(n)* *n* De
En ammeter
Es amperímetro *(m)*
Fr ampèremètre *(m)*
It amperometro *(m)*
Pt amperímetro *(m)*

ammortamento *(m)* *n* It
De Amortisation *(f)*
En amortization
Es amortización *(f)*
Fr amortissement *(m)*
Pt amortização *(f)*

Amortisation *(f)* *n* De
En amortization
Es amortización *(f)*
Fr amortissement *(m)*
It ammortamento *(m)*
Pt amortização *(f)*

amortissement *(m)* *n* Fr
De Amortisation *(f)*
En amortization
Es amortización *(f)*
It ammortamento *(m)*
Pt amortização *(f)*

amortização *(f)* *n* Pt
De Amortisation *(f)*
En amortization
Es amortización *(f)*
Fr amortissement *(m)*
It ammortamento *(m)*

amortización *(f)* *n* Es
De Amortisation *(f)*
En amortization
Fr amortissement *(m)*
It ammortamento *(m)*
Pt amortização *(f)*

amortization *n* En
De Amortisation *(f)*
Es amortización *(f)*
Fr amortissement *(m)*
It ammortamento *(m)*
Pt amortização *(f)*

ampèremètre *(m)* *n* Fr
De Ammeter *(n)*
En ammeter
Es amperímetro *(m)*
It amperometro *(m)*
Pt amperímetro *(m)*

amperímetro *(m)* *n* Es, Pt
De Ammeter *(n)*
En ammeter
Fr ampèremètre *(m)*
It amperometro *(m)*

amperometro *(m)* *n* It
De Ammeter *(n)*
En ammeter
Es amperímetro *(m)*

Fr ampèremètre *(m)*
Pt amperímetro *(m)*

análise Auger *(f)* Pt
De Auger-Analyse *(f)*
En Auger analysis
Es análisis Auger *(m)*
Fr analyse Auger *(f)*
It analisi Auger *(f)*

análise de gás *(f)* Pt
De Gasanalyse *(f)*
En gas analysis
Es análisis del gas *(m)*
Fr analyse des gaz *(f)*
It analisi del gas *(f)*

analisi Auger *(f)* It
De Auger-Analyse *(f)*
En Auger analysis
Es análisis Auger *(m)*
Fr analyse Auger *(f)*
Pt análise Auger *(f)*

analisi del gas *(f)* It
De Gasanalyse *(f)*
En gas analysis
Es análisis del gas *(m)*
Fr analyse des gaz *(f)*
Pt análise de gás *(f)*

análisis Auger *(m)* Es
De Auger-Analyse *(f)*
En Auger analysis
Fr analyse Auger *(f)*
It analisi Auger *(f)*
Pt análise Auger *(f)*

análisis del gas *(m)* Es
De Gasanalyse *(f)*
En gas analysis
Fr analyse des gaz *(f)*
It analisi del gas *(f)*
Pt análise de gás *(f)*

analyse Auger *(f)* Fr
De Auger-Analyse *(f)*
En Auger analysis
Es análisis Auger *(m)*
It analisi Auger *(f)*
Pt análise Auger *(f)*

analyse des gaz *(f)* Fr
De Gasanalyse *(f)*
En gas analysis
Es análisis del gas *(m)*
It analisi del gas *(f)*
Pt análise de gás *(f)*

anemometer *n* En
De Windmesser *(m)*
Es anemómetro *(m)*
Fr anémomètre *(m)*
It anemometro *(m)*
Pt anemómetro *(m)*

anémomètre *(m) n* Fr
De Windmesser *(m)*
En anemometer
Es anemómetro *(m)*
It anemometro *(m)*
Pt anemómetro *(m)*

anémomètre à ailettes *(m)* Fr
De Flügelrad-Windmesser *(m)*
En vane anemometer
Es anemómetro de catavientos *(m)*
It anemometro a mulinello *(m)*
Pt anemómetro de aletas *(m)*

anemometro *(m) n* It
De Windmesser *(m)*
En anemometer
Es anemómetro *(m)*
Fr anémomètre *(m)*
Pt anemómetro *(m)*

anemómetro *(m) n* Es, Pt
De Windmesser *(m)*
En anemometer
Fr anémomètre *(m)*
It anemometro *(m)*

anemometro a mulinello *(m)* It
De Flügelrad-Windmesser *(m)*
En vane anemometer
Es anemómetro de catavientos *(m)*
Fr anémomètre à ailettes *(m)*
Pt anemómetro de aletas *(m)*

anemómetro de aletas *(m)* Pt
De Flügelrad-Windmesser *(m)*
En vane anemometer
Es anemómetro de catavientos *(m)*
Fr anémomètre à ailettes *(m)*
It anemometro a mulinello *(m)*

anemómetro de catavientos *(m)* Es
De Flügelrad-Windmesser *(m)*
En vane anemometer
Fr anémomètre à ailettes *(m)*
It anemometro a mulinello *(m)*
Pt anemómetro de aletas *(m)*

aneroid barometer En
De Aneroidbarometer *(n)*
Es barómetro aneroide *(m)*
Fr baromètre anéroïde *(m)*
It barometro aneroide *(m)*
Pt barómetro aneróide *(m)*

Aneroidbarometer *(n) n* De
En aneroid barometer
Es barómetro aneroide *(m)*
Fr baromètre anéroïde *(m)*
It barometro aneroide *(m)*
Pt barómetro aneróide *(m)*

angereicherter Brennstoff *(m)* De
En enriched fuel
Es combustible enriquecido *(m)*
Fr combustible enrichi *(m)*
It combustibile arricchito *(m)*
Pt combustível enriquecido *(m)*

angle de phase *(m)* Fr
De Phasenwinkel *(m)*
En phase-angle
Es ángulo de fase *(m)*
It angolo di fase *(m)*
Pt ângulo de fase *(m)*

angle zénithal *(m)* Fr
De Zenitwinkel *(m)*
En zenith angle
Es ángulo cenital *(m)*
It angolo dello zenit *(m)*
Pt ângulo zenital *(m)*

angolo dello zenit *(m)* It
De Zenitwinkel *(m)*
En zenith angle
Es ángulo cenital *(m)*
Fr angle zénithal *(m)*
Pt ângulo zenital *(m)*

angolo di fase *(m)* It
De Phasenwinkel *(m)*
En phase-angle
Es ángulo de fase *(m)*
Fr angle de phase *(m)*
Pt ângulo de fase *(m)*

ángulo cenital *(m)* Es
De Zenitwinkel *(m)*
En zenith angle
Fr angle zénithal *(m)*
It angolo dello zenit *(m)*
Pt ângulo zenital *(m)*

ángulo de fase *(m)* Es
De Phasenwinkel *(m)*
En phase-angle
Fr angle de phase *(m)*
It angolo di fase *(m)*
Pt ângulo de fase *(m)*

ângulo de fase *(m)* Pt
De Phasenwinkel *(m)*
En phase-angle
Es ángulo de fase *(m)*
Fr angle de phase *(m)*
It angolo di fase *(m)*

ângulo zenital *(m)* Pt
De Zenitwinkel *(m)*
En zenith angle
Es ángulo cenital *(m)*
Fr angle zénithal *(m)*
It angolo dello zenit *(m)*

anhídrido carbónico *(m)* Es
De Kohlendioxid *(n)*
En carbon dioxide
Fr anhydride carbonique *(m)*
It anidride carbonica *(f)*
Pt anidrido carbónico *(m)*

anhydride carbonique *(m)* Fr
De Kohlendioxid *(n)*
En carbon dioxide
Es anhídrido carbónico *(m)*
It anidride carbonica *(f)*
Pt anidrido carbónico *(m)*

anidride carbonica *(f)* It
De Kohlendioxid *(n)*
En carbon dioxide
Es anhídrido carbónico *(m)*
Fr anhydride carbonique *(m)*
Pt anidrido carbónico *(m)*

anidrido carbónico *(m)* Pt
De Kohlendioxid *(n)*
En carbon dioxide
Es anhídrido carbónico *(m)*
Fr anhydride carbonique *(m)*
It anidride carbonica *(f)*

Anilin *(n) n* De
En aniline
Es anilina *(f)*
Fr aniline *(f)*
It anilina *(f)*
Pt anilina *(f)*

anilina *(f) n* Es, It, Pt
De Anilin *(n)*
En aniline
Fr aniline *(f)*

aniline *n* En, Fr *(f)*
De Anilin *(n)*
Es anilina *(f)*
It anilina *(f)*
Pt anilina *(f)*

aniline point En
De Anilinpunkt *(m)*
Es punto de anilina *(m)*
Fr point d'aniline *(m)*
It punto di anilina *(m)*
Pt ponto de anilina *(m)*

Anilinpunkt *(m) n* De
En aniline point
Es punto de anilina *(m)*
Fr point d'aniline *(m)*
It punto di anilina *(m)*
Pt ponto de anilina *(m)*

Anker *(m) n* De
En armature
Es inducido *(m)*
Fr induit *(m)*
It armatura *(f)*
Pt armadura *(f)*

Anreicherung *(f) n* De
En enrichment
Es enriquecimiento *(m)*
Fr enrichissement *(m)*
It arricchimento *(m)*
Pt enriquecimento *(m)*

anthracene *n* En
De Anthrazen *(n)*
Es antraceno *(m)*
Fr anthracène *(m)*
It antracene *(m)*
Pt antraceno *(m)*

anthracène *(m) n* Fr
De Anthrazen *(n)*
En anthracene
Es antraceno *(m)*
It antracene *(m)*
Pt antraceno *(m)*

anthracite *n* En, Fr *(m)*
De Anthrazit *(m)*
Es antracita *(f)*
It antracite *(m)*
Pt antracite *(m)*

Anthrazen *(n) n* De
En anthracene
Es antraceno *(m)*
Fr anthracène *(m)*
It antracene *(m)*
Pt antraceno *(m)*

Anthrazit *(m) n* De
En anthracite
Es antracita *(f)*
Fr anthracite *(m)*
It antracite *(m)*
Pt antracite *(m)*

antiappannante *(m) n* It
De Belüftung *(f)*
En demister
Es desempañador *(m)*
Fr antibrouillard *(m)*
Pt desembaciador *(m)*

antibrouillard *(m) n* Fr
De Belüftung *(f)*
En demister
Es desempañador *(m)*
It antiappannante *(m)*
Pt desembaciador *(m)*

antidétonant *adj* Fr
De klopffest
En antiknock
Es antidetonante
It antidetonante
Pt antidetonante

antidetonante *adj* Es, It, Pt
De klopffest
En antiknock
Fr antidétonant

antiknock *adj* En
De klopffest
Es antidetonante
Fr antidétonant
It antidetonante
Pt antidetonante

antracene *(m) n* It
De Anthrazen *(n)*
En anthracene
Es antraceno *(m)*
Fr anthracène *(m)*
Pt antraceno *(m)*

antraceno *(m) n* Es, Pt
De Anthrazen *(n)*
En anthracene
Fr anthracène *(m)*
It antracene *(m)*

antracita *(f) n* Es
De Anthrazit *(m)*
En anthracite
Fr anthracite *(m)*
It antracite *(m)*
Pt antracite *(m)*

antracite *(m) n* It, Pt
De Anthrazit *(m)*
En anthracite
Es antracita *(f)*
Fr anthracite *(m)*

appareil de chauffage à arc *(m)* Fr
De Lichtbogen-Heizelement *(m)*
En arc heater
Es calentador de arco *(m)*
It riscaldatore ad arco *(m)*
Pt aquecedor a arco voltáico *(m)*

appareil de chauffage à conduit ouvert *(m)* Fr
De Heizelement mit offenem Abzug *(n)*
En open-flue heater
Es calentador de chimenea abierta *(m)*
It riscaldatore a condotto aperto *(m)*
Pt aquecedor de fumeiro aberto *(m)*

aquecedor *(m) n* Pt
De Heizelement *(n)*
En heater
Es calentador *(m)*
Fr réchauffeur *(m)*
It riscaldatore *(m)*

aquecedor a água de armazenagem *(m)* Pt
De Speicherwasser-heizung *(f)*
En storage water-heater
Es calentador de agua de acumulación *(m)*
Fr chauffe-eau à accumulation *(m)*
It riscaldatore di acqua a conservazione *(m)*

aquecedor a arco voltáico *(m)* Pt
De Lichtbogen-Heizelement *(m)*
En arc heater
Es calentador de arco *(m)*
Fr appareil de chauffage à arc *(m)*
It riscaldatore ad arco *(m)*

aquecedor auxiliado por ventoinha *(m)* Pt
De Heizelement mit Gebläse *(n)*
En fan-assisted heater
Es calentador auxiliado por ventilador *(m)*
Fr radiateur soufflant *(m)*
It riscaldatore a ventilatore *(m)*

aquecedor de água instantâneo *(m)* Pt
De Wasser-Schnellerhitzer *(m)*
En instantaneous water heater
Es calentador de agua instantáneo *(m)*
Fr chauffe-eau instantané *(m)*
It riscaldatore

istantaneo di acqua *(m)*

aquecedor de ar *(m)* Pt
De Lufterhitzer *(m)*
En air heater
Es calentador de aire *(m)*
Fr réchauffeur d'air *(m)*
It riscaldatore ad aria *(m)*

aquecedor de armazenagem *(m)* Pt
De Speicherheizung *(f)*
En storage heater
Es calentador para almacenamiento térmico *(m)*
Fr radiateur à accumulation *(m)*
It riscaldatore a conservazione *(m)*

aquecedor de ar quente *(m)* Pt
De Warmluftheizung *(f)*
En warm-air heater
Es calentador con aire caliente *(m)*
Fr chauffage à air chaud *(m)*
It riscaldatore ad aria tiepida *(m)*

aquecedor de banho *(m)* Pt
De Badheizelement *(n)*
En bath heater
Es calentador de baño *(m)*
Fr chauffe-bain *(m)*
It scaldabagno *(m)*

aquecedor de fumeiro aberto *(m)* Pt
De Heizelement mit offenem Abzug *(n)*
En open-flue heater
Es calentador de chimenea abierta *(m)*
Fr appareil de chauffage à conduit ouvert *(m)*
It riscaldatore a condotto aperto *(m)*

aquecedor de lingotes *(m)* Pt
De Blockerwärmer *(m)*
En billet heater
Es calentador de palanquillas *(m)*
Fr chauffe-billettes *(m)*
It riscaldatore di billette *(m)*

aquecedor hidráulico *(m)* Pt
De Wassererhitzer *(m)*
En water heater
Es calentador de agua *(m)*
Fr chauffe-eau *(m)*
It riscaldatore di acqua *(m)*

aquecedor por imersão *(m)* Pt
De Tauchsieder *(m)*
En immersion heater
Es calentador de inmersión *(m)*
Fr thermoplongeur *(m)*
It riscaldatore ad immersione *(m)*

aquecido a carvão Pt
De kohlenverbrennend
En coal-burning
Es caldeado con carbón
Fr chauffé au charbon
It scaldato a carbone

aquecimento *(m)* *n* Pt
De Heizung *(f)*
En heating
Es calefacción *(f)*
Fr chauffage *(m)*
It riscaldamento *(m)*

aquecimento central *(m)* Pt
De Zentralheizung *(f)*
En central heating
Es calefacción central *(f)*
Fr chauffage central *(m)*
It riscaldamento centrale *(m)*

aquecimento central a óleo *(m)* Pt
De Zentralheizung mit Ölfeuerung *(f)*
En oil-fired central heating
Es calefacción central con petróleo *(f)*
Fr chauffage central au mazout *(m)*
It riscaldamento centrale a nafta *(m)*

aquecimento de distrito *(m)* Pt
De Fernheizung *(f)*
En district heating
Es calefacción de distritos *(f)*
Fr chauffage urbain *(m)*
It riscaldamento di sezione *(m)*

aquecimento dieléctrico *(m)* Pt
De kapazitive Hochfrequenzer-wärmung *(f)*
En dielectric heating
Es caldeo dieléctrico *(m)*
Fr chauffage par pertes diélectriques *(m)*
It riscaldamento dielettrico *(m)*

aquecimento espacial *(m)* Pt
De Raumheizung *(f)*
En space heating
Es calefacción de espacios *(f)*
Fr chauffage de chambres *(m)*
It riscaldamento locale *(m)*

aquecimento indirecto *(m)* Pt
De indirekte Heizung *(f)*
En indirect heating
Es calefacción indirecto *(f)*
Fr chauffage indirect *(m)*
It riscaldamento indiretto *(m)*

aquecimento por frequência rádio *(m)* Pt
De Radiofrequenz-heizung *(f)*
En radiofrequency heating
Es calentamiento por radiofrecuencia *(m)*
Fr chauffage à radiofréquence *(m)*
It riscaldamento a radiofrequenza *(m)*

aquecimento por indução *(m)* Pt
De Induktionsheizung *(f)*
En induction heating
Es caldeo por inducción *(m)*
Fr chauffage par induction *(m)*
It riscaldamento ad induzione *(m)*

ar *(m)* *n* Pt
De Luft *(f)*
En air
Es aire *(m)*
Fr air *(m)*
It aria *(f)*

Arbeit *(f)* *n* De
En work
Es trabajo *(m)*
Fr travail *(m)*
It lavoro *(m)*
Pt trabalho *(m)*

Arbeitsfunktion *(f)* *n* De
En work function
Es función de trabajo *(f)*
Fr travail de sortie *(m)*
It funzione di lavoro *(f)*
Pt função de trabalho *(f)*

arc furnace En
De Lichtbogenofen *(m)*
Es horno de arco *(m)*
Fr four à arc *(m)*
It forno ad arco *(m)*
Pt forno a arco voltáico *(m)*

arc heater En
De Lichtbogen-Heizelement *(m)*
Es calentador de arco *(m)*
Fr appareil de chauffage à arc *(m)*
It riscaldatore ad arco *(m)*
Pt aquecedor a arco voltáico *(m)*

ar condicionado *(m)* Pt
De Klimaanlage *(f)*
En air conditioning
Es acondicionamiento de aire *(m)*
Fr climatisation *(f)*
It condizionamento dell'aria *(m)*

areia betuminosa *(f)* Pt
De bituminöser Sand *(m)*
En bituminous sand
Es arena bituminosa *(f)*

Fr sable bitumineux *(m)*
It sabbia bituminosa *(f)*

areias com alcatrão *(f pl)* Pt
De Ölsande *(pl)*
En tar sands *(pl)*
Es arenas impregnadas de brea *(f pl)*
Fr sables asphaltiques *(m pl)*
It sabbie impregnate di catrame *(f pl)*

arejamento *(m) n* Pt
De Lüftung *(f)*
En aeration
Es aireación *(f)*
Fr aération *(f)*
It aerazione *(f)*

arejamento fixo *(m)* Pt
De feststehende Belüftung *(f)*
En fixed aeration
Es aireación fija *(f)*
Fr aération fixe *(f)*
It aerazione fissa *(f)*

ar em excesso *(m)* Pt
De Luftüberschuβ *(m)*
En excess air
Es aire sobrante *(m)*
Fr excès d'air *(m)*
It eccesso d'aria *(m)*

arena bituminosa *(f)* Es
De bituminöser Sand *(m)*
En bituminous sand
Fr sable bitumineux *(m)*
It sabbia bituminosa *(f)*
Pt areia betuminosa *(f)*

arenas impregnadas de brea *(f pl)* Es
De Ölsande *(pl)*
En tar sands *(pl)*
Fr sables asphaltiques *(m pl)*
It sabbie impregnate di catrame *(f pl)*
Pt areias com alcatrão *(f pl)*

aria *(f) n* It
De Luft *(f)*
En air
Es aire *(m)*
Fr air *(m)*
Pt ar *(m)*

aria primaria *(f)* It
De primäre Luft *(f)*
En primary air
Es aire primario *(m)*
Fr air primaire *(m)*
Pt ar primário *(m)*

aria satura *(f)* It
De gesättigte Luft *(f)*
En saturated air
Es aire saturado *(m)*
Fr air saturé *(m)*
Pt ar saturado *(m)*

aria secondaria *(f)* It
De Sekundärluft *(f)*
En secondary air
Es aire secundario *(m)*
Fr air secondaire *(m)*
Pt ar secundário *(m)*

armadura *(f) n* Pt
De Anker *(m)*
En armature
Es inducido *(m)*
Fr induit *(m)*
It armatura *(f)*

armatura *(f) n* It
De Anker *(m)*
En armature
Es inducido *(m)*
Fr induit *(m)*
Pt armadura *(f)*

armature *n* En
De Anker *(m)*
Es inducido *(m)*
Fr induit *(m)*
It armatura *(f)*
Pt armadura *(f)*

armazenagem de água quente *(f)* Pt
De Warmwasser-speicherung *(f)*
En hot-water storage
Es almacenamiento de agua caliente *(m)*
Fr accumulation d'eau chaude *(f)*
It immagazzinaggio di acqua calda *(m)*

armazenagem de ar comprimido *(f)* Pt
De Druckluftspeicherung *(f)*
En compressed-air storage
Es almacenamiento de aire comprimido *(m)*
Fr stockage de l'air comprimé *(m)*
It immagazzinaggio di aria compressa *(m)*

armazenagem de ar quente *(f)* Pt
De Heiβluftspeicherung *(f)*
En hot-air storage
Es almacenamiento de aire caliente *(m)*
Fr accumulation d'air chaud *(f)*
It immagazzinaggio di aria calda *(m)*

armazenagem de calor *(f)* Pt
De Wärmespeicherung *(f)*
En heat storage
Es almacenamiento de calor *(m)*
Fr accumulation de chaleur *(f)*
It immagazzinaggio di calore *(m)*

armazenagem de energia térmica *(f)* Pt
De Wärmeenergie-speicherung *(f)*
En thermal-energy storage
Es almacenamiento de energía térmica *(m)*
Fr accumulation d'énergie thermique *(n)*
It immagazzinaggio di energia termica *(m)*

armazenagem de transição de fase *(f)* Pt
De Phasenum-wandlungs-speicherung *(f)*
En phase-transition storage
Es almacenamiento de transición de fase *(m)*
Fr accumulation à transition de phase *(f)*
It immagazzinaggio a transizione di fase *(m)*

armazenagem diurna de calor *(f)* Pt
De tägliche Wärmespeicherung *(f)*
En diurnal heat storage
Es almacenamiento diurno del calor *(m)*
Fr accumulation de chaleur diurne *(f)*
It immagazzinaggio di calore diurno *(m)*

armazenagem electroquímica *(f)* Pt
De elektrochemische Speicherung *(f)*
En electrochemical storage
Es almacenamiento electroquímico *(m)*
Fr accumulation électrochimique *(f)*
It immagazzinaggio elettrochimico *(m)*

armazenagem magnética *(f)* Pt
De Magnetspeicher *(m)*
En magnetic storage
Es almacenamiento magnético *(m)*
Fr accumulation magnétique *(f)*
It immagazzinaggio magnetico *(m)*

armazenagem mecânica *(f)* Pt
De mechanische Speicherung *(f)*
En mechanical storage
Es almacenamiento mecánico *(m)*
Fr accumulation mécanique *(f)*
It immagazzinaggio meccanico *(m)*

armazenagem realizada à bomba *(f)* Pt
De gepumpte Speicherung *(f)*
En pumped storage
Es almacenamiento bombeado *(m)*
Fr accumulation pompée *(f)*
It immagazzinaggi a pompa *(m)*

aromatic compound En
De aromatische Verbindung *(f)*
Es compuesto aromático *(m)*
Fr composé aromatique *(m)*
It composto aromatico *(m)*
Pt composto aromático *(m)*

aromatic crude oil En
De aromatisches Rohöl *(n)*
Es crudo aromático *(m)*
Fr pétrole brut aromatique *(m)*
It petrolio greggio aromatico *(m)*
Pt óleo aromático crú *(m)*

aromatisches Rohöl *(n)* De
En aromatic crude oil
Es crudo aromático *(m)*
Fr pétrole brut aromatique *(m)*
It petrolio greggio aromatico *(m)*
Pt óleo aromático crú *(m)*

aromatische Verbindung *(f)* De
En aromatic compound
Es compuesto aromático *(m)*
Fr composé aromatique *(m)*
It composto aromatico *(m)*
Pt composto aromático *(m)*

ar primário *(m)* Pt
De primäre Luft *(f)*
En primary air
Es aire primario *(m)*
Fr air primaire *(m)*
It aria primaria *(f)*

arranque *(m)* *n* Es, Pt
De Einschalten *(n)*
En start-up
Fr démarrage *(m)*
It avviamento *(m)*

arranque a frio *(m)* Pt
De Kaltstart *(m)*
En cold start
Es arranque en frío *(m)*
Fr démarrage à froid *(m)*
It avviamento a freddo *(m)*

arranque en frío *(m)* Es
De Kaltstart *(m)*
En cold start
Fr démarrage à froid *(m)*
It avviamento a freddo *(m)*
Pt arranque a frio *(m)*

arrefecido a ar Pt
De luftgekühlt
En air-cooled
Es refrigerado por aire
Fr refroidi par air
It raffreddato ad aria

arrefecido a hidrogénio Pt
De wasserstoffgekühlt
En hydrogen-cooled
Es enfriado con hidrógeno
Fr refroidi par l'hydrogène
It raffreddato ad idrogeno

arrefecimento ebuliente *(m)* Pt
De Heißwasserkühlung *(f)*
En ebullient cooling
Es enfriamiento desde la ebullición *(m)*
Fr refroidissement par ébullition *(m)*
It raffreddamento a ebollizione *(m)*

arrefecimento por radiação nocturna *(m)* Pt
De Nachtstrahlungs-kühlung *(f)*
En night-radiation cooling
Es refrigeración de radiación nocturna *(f)*
Fr refroidissement par rayonnement nocturne *(f)*
It raffreddamento a radiazioni notturne *(m)*

arresto *(m)* *n* It
De Abschalten *(n)*
En shutdown
Es parada *(f)*
Fr arrêt *(m)*
Pt paralização *(f)*

arrêt *(m)* *n* Fr
De Abschalten *(n)*
En shutdown
Es parada *(f)*
It arresto *(m)*
Pt paralização *(f)*

arrêt d'urgence *(m)* Fr
De Abschaltung *(f)*
En scram (nuclear reactor)
Es parada de emergencia *(f)*
It spegnimento immediato *(m)*
Pt paragem de emergência *(f)*

arricchimento *(m)* *n* It
De Anreicherung *(f)*
En enrichment
Es enriquecimiento *(m)*
Fr enrichissement *(m)*
Pt enriquecimento *(m)*

ar saturado *(m)* Pt
De gesättigte Luft *(f)*
En saturated air
Es aire saturado *(m)*
Fr air saturé *(m)*
It aria satura *(f)*

ar secundário *(m)* Pt
De Sekundärluft *(f)*
En secondary air
Es aire secundario *(m)*
Fr air secondaire *(m)*
It aria secondaria *(f)*

Asbest *(m)* *n* De
En asbestos
Es amianto *(m)*
Fr amiante *(m)*
It amianto *(m)*
Pt amianto *(m)*

asbestos *n* En
De Asbest *(m)*
Es amianto *(m)*
Fr amiante *(m)*
It amianto *(m)*
Pt amianto *(m)*

Asche *(f)* *n* De
En ash
Es ceniza *(f)*
Fr cendre *(f)*
It cenere *(f)*
Pt cinza *(f)*

asfalteni *(m pl)* *n* It
De Asphaltene *(pl)*
En asphaltenes
Es asfaltenos *(m pl)*
Fr asphaltènes *(m pl)*
Pt asfaltenos *(m pl)*

asfaltenos *(m pl)* *n* Es, Pt
De Asphaltene *(pl)*
En asphaltenes
Fr asphaltènes *(m pl)*
It asfalteni *(m pl)*

asfalto *(m)* *n* Es, It, Pt
De Asphalt *(m)*
En asphalt
Fr asphalte *(m)*

ash *n* En
De Asche *(f)*
Es ceniza *(f)*
Fr cendre *(f)*
It cenere *(f)*
Pt cinza *(f)*

asphalt *n* En
De Asphalt *(m)*
Es asfalto *(m)*
Fr asphalte *(m)*
It asfalto *(m)*
Pt asfalto *(m)*

Asphalt *(m)* *n* De
En asphalt
Es asfalto *(m)*
Fr asphalte *(m)*
It asfalto *(m)*
Pt asfalto *(m)*

asphalte *(m)* *n* Fr
De Asphalt *(m)*
En asphalt
Es asfalto *(m)*
It asfalto *(m)*
Pt asfalto *(m)*

Asphaltene *(pl)* *n* De
En asphaltenes
Es asfaltenos *(m pl)*
Fr asphaltènes *(m pl)*
It asfalteni *(m pl)*
Pt asfaltenos *(m pl)*

asphaltenes *pl n* En
De Asphaltene *(pl)*
Es asfaltenos *(m pl)*

Fr asphaltènes *(m pl)*
It asfalteni *(m pl)*
Pt asfaltenos *(m pl)*

asphaltènes *(m pl) n* Fr
De Asphaltene *(pl)*
En asphaltenes
Es asfaltenos *(m pl)*
It asfalteni *(m pl)*
Pt asfaltenos *(m pl)*

assorbimento *(m) n* It
De Absorption *(f)*
En absorption
Es absorción *(f)*
Fr absorption *(f)*
Pt absorção *(f)*

assorbimento atmosferico *(m)* It
De Luftabsorption *(f)*
En atmospheric absorption
Es absorción atmosférica *(f)*
Fr absorption atmosphérique *(f)*
Pt absorção atmosférica *(f)*

assorbimento molecolare *(m)* It
De molekulare Absorption *(f)*
En molecular absorption
Es absorción molecular *(f)*
Fr absorption moléculaire *(f)*
Pt absorção molecular *(f)*

assorbimento solare *(m)* It
De Solarabsorptionsgrad *(m)*
En solar absorbtance
Es absorsancia solar *(f)*
Fr absorptance solaire *(f)*
Pt absorvância solar *(f)*

assorbitore diretto *(m)* It
De Direktabsorber *(m)*
En direct absorber
Es absorbedor directo *(m)*
Fr absorbeur direct *(m)*
Pt absorvedor directo *(m)*

atenuação *(f) n* Pt
De Dämpfung *(f)*
En attenuation
Es atenuación *(f)*
Fr atténuation *(f)*
It attenuazione *(f)*

atenuación *(f) n* Es
De Dämpfung *(f)*
En attenuation
Fr atténuation *(f)*
It attenuazione *(f)*
Pt atenuação *(f)*

Äthan *(n) n* De
En ethane
Es etano *(m)*
Fr éthane *(m)*
It etano *(m)*
Pt etano *(m)*

Äthylalkohol *(m) n* De
En ethanol
Es etanol *(m)*
Fr éthanol *(m)*
It etanolo *(m)*
Pt etanol *(m)*

Äthylen *(n) n* De
En ethylene
Es etileno *(m)*
Fr éthylène *(m)*
It etilene *(m)*
Pt etileno *(m)*

Äthylenglykol *(n) n* De
En ethylene glycol
Es glicoletileno *(m)*
Fr éthylène glycol *(m)*
It glicole etilenico *(m)*
Pt glicol etilénico *(m)*

atmospheric absorption En
De Luftabsorption *(f)*
Es absorción atmosférica *(f)*
Fr absorption atmosphérique *(f)*
It assorbimento atmosferico *(m)*
Pt absorção atmosférica *(f)*

atmospheric pressure En
De Luftdruck *(m)*
Es presión atmosférica *(f)*
Fr pression atmosphérique *(f)*
It pressione atmosferica *(f)*
Pt pressão atmosférica *(f)*

Atomgewicht *(n) n* De
En atomic weight
Es peso atómico *(m)*
Fr poids atomique *(m)*
It peso atomico *(m)*
Pt peso atómico *(m)*

atomic mass En
De Atommasse *(f)*
Es masa atómica *(f)*
Fr masse atomique *(f)*
It massa dell'atomo *(f)*
Pt massa atómica *(f)*

atomic number En
De Atomnummer *(f)*
Es número atómico *(m)*
Fr nombre atomique *(m)*
It numero atomico *(m)*
Pt número atómico *(m)*

atomic weight En
De Atomgewicht *(n)*
Es peso atómico *(m)*
Fr poids atomique *(m)*
It peso atomico *(m)*
Pt peso atómico *(m)*

atomiseur *(m) n* Fr
De Zerstäuber *(m)*
En atomizer
Es atomizador *(m)*
It nebulizzatore *(m)*
Pt atomizador *(m*

atomiseur d'air *(m)* Fr
De Luftzerstäuber *(m)*
En air atomizer
Es atomizador de aire *(m)*
It nebulizzatore d'aria *(m)*
Pt atomizador de ar *(m)*

atomizador *(m) n* Es, Pt
De Zerstäuber *(m)*
En atomizer
Fr atomiseur *(m)*
It nebulizzatore *(m)*

atomizador de aire *(m)* Es
De Luftzerstäuber *(m)*
En air atomizer
Fr atomiseur d'air *(m)*
It nebulizzatore d'aria *(m)*
Pt atomizador de ar *(m)*

atomizador de ar *(m)* Pt
De Luftzerstäuber *(m)*
En air atomizer
Es atomizador de aire *(m)*
Fr atomiseur d'air *(m)*
It nebulizzatore d'aria *(m)*

atomizer *n* En
De Zerstäuber *(m)*
Es atomizador *(m)*
Fr atomiseur *(m)*
It nebulizzatore *(m)*
Pt atomizador *(m)*

Atommasse *(f) n* De
En atomic mass
Es masa atómica *(f)*
Fr masse atomique *(f)*
It massa dell'atomo *(f)*
Pt massa atómica *(f)*

Atomnummer *(f) n* De
En atomic number
Es número atómico *(m)*
Fr nombre atomique *(m)*
It numero atomico *(m)*
Pt número atómico *(m)*

attenuation *n* En
De Dämpfung *(f)*
Es atenuación *(f)*
Fr atténuation *(f)*
It attenuazione *(f)*
Pt atenuação *(f)*

atténuation *(f) n* Fr
De Dämpfung *(f)*
En attenuation
Es atenuación *(f)*
It attenuazione *(f)*
Pt atenuação *(f)*

attenuazione *(f) n* It
De Dämpfung *(f)*
En attenuation
Es atenuación *(f)*
Fr atténuation *(f)*
Pt atenuação *(f)*

Aufkonzentrierung *(f) n* De
En upgrading

Es mejoramiento *(m)*
Fr amélioration *(f)*
It promovimento a grado superiore *(m)*
Pt melhoramento *(m)*

auf See De
En offshore
Es marino
Fr au large
It al largo
Pt fora da praia

Aufzehrsystem *(n)* *n* De
En cleanup system
Es sistema de limpieza *(m)*
Fr système de nettoyage *(m)*
It sistema di depurazione *(m)*
Pt sistema de limpeza *(m)*

Auger-Analyse *(f)* *n* De
En Auger analysis
Es análisis Auger *(m)*
Fr analyse Auger *(f)*
It analisi Auger *(f)*
Pt análise Auger *(f)*

Auger analysis En
De Auger-Analyse *(f)*
Es análisis Auger *(m)*
Fr analyse Auger *(f)*
It analisi Auger *(f)*
Pt análise Auger *(f)*

au large Fr
De auf See
En offshore
Es marino
It al largo
Pt fora da praia

Ausbeutefaktor *(m)* *n* De
En ultimate recovery
Es producción final *(f)*
Fr récupération finale *(f)*
It recupero ultimo *(m)*
Pt recuperação final *(f)*

Ausbruch *(m)* *n* De
En blowout
Es reventón *(m)*
Fr éruption *(f)*
It eruzione *(f)*
Pt estouro *(m)*

Auskocher *(m)* *n* De
En reboiler
Es recaldera *(f)*
Fr rebouilleur *(m)*
It recaldaia *(f)*
Pt recaldeira *(f)*

Auspuffgas *(n)* *n* De
En exhaust
Es gases de escape *(m pl)*
Fr échappement *(m)*
It scarico *(m)*
Pt escape *(m)*

Ausschußbrennstoff *(m)* *n* De
En reject fuel
Es combustible recusable *(m)*
Fr combustible de rebut *(m)*
It combustibile di scarto *(m)*
Pt combustível rejeitado *(m)*

Außenarbeit *(f)* *n* De
En external work
Es trabajo externo *(m)*
Fr travail extérieur *(m)*
It lavoro esterno *(m)*
Pt trabalho externo *(m)*

Austauscher *(m)* *n* De
En exchanger
Es intercambiador *(m)*
Fr échangeur *(m)*
It scambiatore *(m)*
Pt permutador *(m)*

autoaccensione *(f)* *n* It
De spontane Zündung *(f)*
En spontaneous ignition
Es encendido espontáneo *(m)*
Fr allumage spontané *(m)*
Pt ignição espontânea *(f)*

automezzo con motore a nafta *(m)* It
De Straßenfahrzeug mit Dieselmotor *(n)*
En diesel engine road vehicle (DERV)
Es vehículo para carretera con motor diesel *(m)*
Fr véhicule diesel routier *(m)*
Pt veículo rodoviário de motor diesel *(m)*

autotransformador *(m)* *n* Es, Pt
De Autotransformator *(m)*
En autotransformer
Fr autotransformateur *(m)*
It autotrasformatore *(m)*

autotransformateur *(m)* *n* Fr
De Autotransformator *(m)*
En autotransformer
Es autotransformador *(m)*
It autotrasformatore *(m)*
Pt autotransformador *(m)*

Autotransformator *(m)* *n* De
En autotransformer
Es autotransformador *(m)*
Fr autotransformateur *(m)*
It autotrasformatore *(m)*
Pt autotransformador *(m)*

autotransformer *n* En
De Autotransformator *(m)*
Es autotransformador *(m)*
Fr autotransformateur *(m)*
It autotrasformatore *(m)*
Pt autotransformador *(m)*

autotrasformatore *(m)* *n* It
De Autotransformator *(m)*
En autotransformer
Es autotransformador *(m)*
Fr autotransformateur *(m)*
Pt autotransformador *(m)*

average solar flux En
De durchschnittlicher Sonnenfluß *(m)*
Es flujo solar medio *(m)*
Fr flux solaire moyen *(m)*
It flusso solare medio *(m)*
Pt fluxo solar médio *(m)*

aviation fuel En
De Flugkraftstoff *(m)*
Es gasolina de aviación *(f)*
Fr carburation aviation *(m)*
It combustibile per aviazione *(m)*
Pt combustível de aviação *(m)*

avviamento *(m)* *n* It
De Einschalten *(n)*
En start-up
Es arranque *(m)*
Fr démarrage *(m)*
Pt arranque *(m)*

avviamento a freddo *(m)* It
De Kaltstart *(m)*
En cold start
Es arranque en frío *(m)*
Fr démarrage à froid *(m)*
Pt arranque a frio *(m)*

avviatore *(m)* *n* It
De Starter *(m)*
En starter motor
Es motor de arranque *(m)*
Fr démarreur *(m)*
Pt motor de arranque *(m)*

avvolgimento *(m)* *n* It
De Wicklung *(f)*
En winding
Es devanado *(m)*
Fr enroulement *(m)*
Pt enrolamento *(m)*

axial flow turbine En
De Axialturbine *(f)*
Es turbina de flujo axial *(f)*
Fr turbine axiale *(f)*
It turbina a flusso assiale *(f)*
Pt turbina de fluxo axial *(f)*

Axialturbine *(f)* *n* De
En axial flow turbine
Es turbina de flujo axial *(f)*
Fr turbine axiale *(f)*
It turbina a flusso assiale *(f)*
Pt turbina de fluxo axial *(f)*

Azetylen *(n)* *n* De
En acetylene
Es acetileno *(m)*
Fr acétylène *(m)*
It acetilene *(m)*
Pt acetileno *(m)*

azimut *(m)* *n* Es, It
De Azimuth *(m)*
En azimuth
Fr azimuth *(m)*
Pt azimute *(m)*

azimute *(m)* *n* Pt
De Azimuth *(m)*
En azimuth
Es azimut *(m)*
Fr azimuth *(m)*
It azimut *(m)*

azimuth *n* En, Fr *(m)*
De Azimuth *(m)*
Es azimut *(m)*
It azimut *(m)*
Pt azimute *(m)*

Azimuth *(m)* *n* De
En azimuth
Es azimut *(m)*
Fr azimuth *(m)*
It azimut *(m)*
Pt azimute *(m)*

azote *(m)* *n* Fr
De Stickstoff *(m)*
En nitrogen
Es nitrógeno *(m)*
It azoto *(m)*
Pt azoto *(m)*

azoto *(m)* *n* It, Pt
De Stickstoff *(m)*
En nitrogen
Es nitrógeno *(m)*
Fr azote *(m)*

azufre *(m)* *n* Es
Am sulfur
De Schwefel *(m)*
En sulphur
Fr soufre *(m)*
It zolfo *(m)*
Pt enxofre *(m)*

B

bacino di raffreddamento *(m)* It
De Kühlteich *(m)*
En cooling pond
Es piscina de refrigeración *(f)*
Fr piscine de refroidissement *(f)*
Pt tanque refrigerante *(m)*

back electromotive force En
De Gegenelektro-motorische Kraft *(f)*
Es fuerza contra-electromotriz *(f)*
Fr force contreélectro-motrice *(f)*
It forza contro-elettromotrice *(f)*
Pt força contra-electromotriz *(f)*

background heat En
De Grundheizung *(f)*
Es calor de fondo *(m)*
Fr chaleur ambiante *(f)*
It calore di fondo *(m)*
Pt calor de fundo *(m)*

background radiation En
De Nulleffektstrahlung *(f)*
Es radiación ambiente *(f)*
Fr rayonnement provoquant le mouvement propre *(m)*
It effetto di fondo *(m)*
Pt radiação de fundo *(f)*

Backkohle *(f)* *n* De
En caking coal
Es hulla grasa *(f)*
Fr houille collante *(f)*
It carbone agglutinante *(m)*
Pt hulha gorda *(f)*

back-pressure turbine En
De Gegendruckturbine *(f)*
Es turbina de contrapresión *(f)*
Fr turbine à contre-pression *(f)*
It turbina a contropressione *(f)*
Pt turbina de contrapressão *(f)*

Badheizelement *(n)* *n* De
En bath heater
Es calentador de baño *(m)*
Fr chauffe-bain *(m)*
It scaldabagno *(m)*
Pt aquecedor de banho *(m)*

baffle *n* En
De Prallplatte *(f)*
Es deflector *(m)*
Fr déflecteur *(m)*
It parafiamma *(m)*
Pt deflector *(m)*

bagaço *(m)* *n* Pt
De Brennrohr *(n)*
En bagasse
Es bagazo *(m)*
Fr bagasse *(f)*
It bagasse esaurite *(f)*

bagasse *n* En, Fr *(f)*
De Brennrohr *(n)*
Es bagazo *(m)*
It bagasse esaurite *(f)*
Pt bagaço *(m)*

bagasse esaurite *(f)* It
De Brennrohr *(n)*
En bagasse
Es bagazo *(m)*
Fr bagasse *(f)*
Pt bagaço *(m)*

bagazo *(m)* *n* Es
De Brennrohr *(n)*
En bagasse
Fr bagasse *(f)*
It bagasse esaurite *(f)*
Pt bagaço *(m)*

balance de matériel et énergie *(f)* Fr
De Stoff- und Energiebilanz *(f)*
En material and energy balance
Es equilibrio materia-energía *(m)*
It bilancio materiale-energia *(m)*
Pt equilíbrio de material e energia *(m)*

balanced-flue convector En
De Konvektor mit entlastetem Abzug *(m)*
Es convector de chimenea equilibrada *(m)*
Fr convecteur à carneau équilibré *(m)*
It convettore a focolare equilibrato *(m)*
Pt convector de fluxo equilibrado *(m)*

balance energético *(m)* Es
De Energiebilanz *(f)*
En energy balance
Fr équilibre énergétique *(m)*
It equilibrio di energia *(m)*
Pt equilíbrio energético *(m)*

balastro puro *(m)* Pt
De sauberer Ballast *(m)*
En clean ballast
Es lastre limpio *(m)*
Fr ballast propre *(m)*
It zavorra pulita *(f)*

ballast propre *(m)* Fr
De sauberer Ballast *(m)*
En clean ballast
Es lastre limpio *(m)*
It zavorra pulita *(f)*
Pt balastro puro *(m)*

barco de gas *(m)* Es
De Gasschiff *(n)*
En gas ship
Fr navire méthanier *(m)*
It nave del gas *(f)*
Pt barco de gás *(m)*

barco de gás *(m)* Pt
De Gasschiff *(n)*
En gas ship
Es barco de gas *(m)*
Fr navire méthanier *(m)*
It nave del gas *(f)*

barco nuclear *(m)* Es, Pt
De Kernkraftschiff *(n)*
En nuclear ship
Fr navire nucléaire *(m)*
It nave nucleare *(f)*

barco de gas *(m)* Es
De Gasschiff *(n)*
En gas ship
Fr navire méthanier *(m)*
It nave del gas *(f)*
Pt barco de gás *(m)*

bare collector (solar) En
De Glattrohr-Kollektor *(m)*
Es colector desnudo *(m)*
Fr collecteur nu *(m)*
It collettore nudo *(m)*
Pt colector descoberto *(m)*

baril *(m) n* Fr
De Faβ *(n)*
En barrel
Es barril *(m)*
It barile *(m)*
Pt barril *(m)*

barile *(m) n* It
De Faβ *(n)*
En barrel
Es barril *(m)*
Fr baril *(m)*
Pt barril *(m)*

barografo *(m) n* It
De Barograph *(m)*
En barograph
Es barógrafo *(m)*
Fr barographe *(m)*
Pt barógrafo *(m)*

barógrafo *(m) n* Es, Pt
De Barograph *(m)*
En barograph
Fr barographe *(m)*
It barografo *(m)*

barograph *n* En
De Barograph *(m)*
Es barógrafo *(m)*
Fr barographe *(m)*
It barografo *(m)*
Pt barógrafo *(m)*

Barograph *(m) n* De
En barograph
Es barógrafo *(m)*
Fr barographe *(m)*
It barografo *(m)*
Pt barógrafo *(m)*

barographe *(m) n* Fr
De Barograph *(m)*
En barograph
Es barógrafo *(m)*
It barografo *(m)*
Pt barógrafo *(m)*

barometer *n* En
De Barometer *(n)*
Es barómetro *(m)*
Fr baromètre *(m)*
It barometro *(m)*
Pt barómetro *(m)*

Barometer *(n) n* De
En barometer
Es barómetro *(m)*
Fr baromètre *(m)*
It barometro *(m)*
Pt barómetro *(m)*

baromètre *(m) n* Fr
De Barometer *(n)*
En barometer
Es barómetro *(m)*
It barometro *(m)*
Pt barómetro *(m)*

baromètre anéroïde *(m)* Fr
De Aneroidbarometer *(n)*
En aneroid barometer
Es barómetro aneroide *(m)*
It barometro aneroide *(m)*
Pt barómetro aneróide *(m)*

barometro *(m) n* It
De Barometer *(n)*
En barometer
Es barómetro *(m)*
Fr baromètre *(m)*
Pt barómetro *(m)*

barómetro *(m) n* Es
De Barometer *(n)*
En barometer
Fr baromètre *(m)*
It barometro *(m)*
Pt barómetro *(m)*

barómetro *(m) n* Pt
De Barometer *(n)*
En barometer
Es barómetro *(m)*
Fr baromètre *(m)*
It barometro *(m)*

barometro aneroide *(m)* It
De Aneroidbarometer *(n)*
En aneroid barometer
Es barómetro aneroide *(m)*
Fr baromètre anéroïde *(m)*
Pt barómetro aneróide *(m)*

barómetro aneroide *(m)* Es
De Aneroidbarometer *(n)*
En aneroid barometer
Fr baromètre anéroïde *(m)*
It barometro aneroide *(m)*
Pt barómetro aneróide *(m)*

barómetro aneróide *(m)* Pt
De Aneroidbarometer *(n)*
En aneroid barometer
Es barómetro aneroide *(m)*
Fr baromètre anéroïde *(m)*
It barometro aneroide *(m)*

barra colectora *(f)* Es, Pt
De Leitungsschiene *(f)*
En busbar
Fr barre omnibus *(f)*
It sbarra *(f)*

barra combustibile *(f)* It
De Brennstoffstab *(m)*
En fuel rod
Es barra de combustible *(f)*
Fr barre de combustible *(f)*
Pt vara de combustível *(f)*

barra de combustible *(f)* Es
De Brennstoffstab *(m)*
En fuel rod
Fr barre de combustible *(f)*
It barra combustibile *(f)*
Pt vara de combustível *(f)*

barra di controllo *(f)* It
De Kontrollstab *(m)*
En control rod
Es varilla de control *(f)*
Fr barre de commande *(f)*
Pt haste de controle *(f)*

barrage *(m) n* Fr
De Damm *(m)*
En dam
Es presa *(f)*
It diga *(f)*
Pt barragem *(f)*

barrage marémoteur *(m)* Fr
De Gezeitendamm *(m)*
En tidal barrage
Es presa de marea *(f)*
It sbarramento di marea *(m)*
Pt barragem de marés *(f)*

barragem *(f) n* Pt
De Damm *(m)*
En dam
Es presa *(f)*
Fr barrage *(m)*
It diga *(f)*

barragem de marés *(f)* Pt
De Gezeitendamm *(m)*
En tidal barrage
Es presa de marea *(f)*
Fr barrage marémoteur *(m)*
It sbarramento di marea *(m)*

barre de combustible *(f)* Fr
De Brennstoffstab *(m)*
En fuel rod
Es barra de combustible *(f)*
It barra combustibile *(f)*
Pt vara de combustível *(f)*

barre de commande *(f)* Fr
De Kontrollstab *(m)*
En control rod
Es varilla de control *(f)*
It barra di controllo *(f)*
Pt haste de controle *(f)*

barreira de vapor *(f)* Pt
Am vapor barrier
De Dampfsperre *(f)*
En vapour barrier
Es barrera del vapor *(f)*
Fr écran d'étanchéité à la vapeur *(m)*
It schermo per il vapore *(m)*

barrel *n* En
De Faβ *(n)*
Es barril *(m)*
Fr baril *(m)*
It barile *(m)*
Pt barril *(m)*

barre omnibus *(f)* Fr
De Leitungsschiene *(f)*
En busbar
Es barra colectora *(f)*
It sbarra *(f)*
Pt barra colectora *(f)*

barrera del vapor *(f)* Es
Am vapor barrier
De Dampfsperre *(f)*
En vapour barrier
Fr écran d'étanchéité à la vapeur *(m)*
It schermo per il vapore *(m)*
Pt barreira de vapor *(f)*

barril *(m) n* Es, Pt
De Faβ *(n)*
En barrel
Fr baril *(m)*
It barile *(m)*

basal metabolic rate En
De Grundumsatz *(m)*
Es índice metabólico basal *(m)*
Fr métabolisme basal *(m)*
It tasso metabolico basale *(m)*
Pt índice de metabolismo basal *(m)*

base de carbón puro *(f)* Es
De Reinkohlengrundlage *(f)*
En pure-coal basis
Fr base de charbon pur *(f)*
It base di carbone puro *(f)*
Pt base de carvão puro *(f)*

base de carvão puro *(f)* Pt
De Reinkohlengrundlage *(f)*
En pure-coal basis
Es base de carbón puro *(f)*
Fr base de charbon pur *(f)*
It base di carbone puro *(f)*

base de charbon pur *(f)* Fr
De Reinkohlengrundlage *(f)*
En pure-coal basis
Es base de carbón puro *(f)*
It base di carbone puro *(f)*
Pt base de carvão puro *(f)*

base di carbone puro *(f)* It
De Reinkohlengrundlage *(f)*
En pure-coal basis
Es base de carbón puro *(f)*
Fr base de charbon pur *(f)*
Pt base de carvão puro *(f)*

base load En
De Grundkraft *(f)*
Es carga base *(f)*
Fr charge minimale *(f)*
It carico base *(m)*
Pt carga fundamental *(f)*

bassin solaire salant *(m)* Fr
De Solar-Salzteich *(m)*
En saline solar pond
Es piscina solar salina *(f)*
It laghetto solare salino *(m)*
Pt lago solar de sal *(m)*

bateria *(f) n* Pt
De Batterie *(f)*
En battery
Es batería *(f)*
Fr batterie *(f)*
It batteria *(f)*

batería *(f) n* Es
De Batterie *(f)*
En battery
Fr batterie *(f)*
It batteria *(f)*
Pt bateria *(f)*

batería de ácido-plomo *(f)* Es
De Blei-Säure-Batterie *(f)*
En lead-acid battery
Fr batterie au plomb *(f)*
It batteria acida al piombo *(f)*
Pt bateria de chumbo-ácido *(f)*

batería de cadmio *(f)* Es
De Kadmiumbatterie *(f)*
En cadmium battery
Fr pile au cadmium *(f)*
It batteria al cadmio *(f)*
Pt bateria de cádmio *(f)*

bateria de cádmio *(f)* Pt
De Kadmiumbatterie *(f)*
En cadmium battery
Es batería de cadmio *(f)*
Fr pile au cadmium *(f)*
It batteria al cadmio *(f)*

bateria de chumbo-ácido *(f)* Pt
De Blei-Säure-Batterie *(f)*
En lead-acid battery
Es batería de ácido-plomo *(f)*
Fr batterie au plomb *(f)*
It batteria acida al piombo *(f)*

bateria de cloreto de zinco *(f)* Pt
De Zinkchloridbatterie *(f)*
En zinc-chloride battery
Es batería de cloruro de zinc *(f)*
Fr batterie au chlorure de zinc *(f)*
It batteria al cloruro di zinco *(f)*

batería de cloruro de zinc *(f)* Es
De Zinkchloridbatterie *(f)*
En zinc-chloride battery
Fr batterie au chlorure de zinc *(f)*
It batteria al cloruro di zinco *(f)*
Pt bateria de cloreto de zinco *(f)*

bath heater En
De Badheizelement *(n)*
Es calentador de baño *(m)*
Fr chauffe-bain *(m)*
It scaldabagno *(m)*
Pt aquecedor de banho *(m)*

batteria *(f) n* It
De Batterie *(f)*
En battery
Es batería *(f)*
Fr batterie *(f)*
Pt bateria *(f)*

batteria acida al piombo *(f)* It
De Blei-Säure-Batterie *(f)*
En lead-acid battery
Es batería de ácido-plomo *(f)*
Fr batterie au plomb *(f)*
Pt bateria de chumbo-ácido *(f)*

batteria al cadmio *(f)* It
De Kadmiumbatterie *(f)*
En cadmium battery
Es batería de cadmio *(f)*
Fr pile au cadmium *(f)*
Pt bateria de cádmio *(f)*

batteria al cloruro di zinco *(f)* It
De Zinkchloridbatterie *(f)*
En zinc-chloride battery
Es batería de cloruro de zinc *(f)*
Fr batterie au chlorure de zinc *(f)*
Pt bateria de cloreto de zinco *(f)*

batterie *(f) n* Fr
De Batterie *(f)*
En battery

Es batería *(f)*
It batteria *(f)*
Pt bateria *(f)*

Batterie *(f) n* De
En battery
Es batería *(f)*
Fr batterie *(f)*
It batteria *(f)*
Pt bateria *(f)*

batterie au chlorure de zinc *(f)* Fr
De Zinkchloridbatterie *(f)*
En zinc-chloride battery
Es batería de cloruro de zinc *(f)*
It batteria al cloruro di zinco *(f)*
Pt bateria de cloreto de zinco *(f)*

batterie au plomb *(f)* Fr
De Blei-Säure-Batterie *(f)*
En lead-acid battery
Es batería de ácido-plomo *(f)*
It batteria acida al piombo *(f)*
Pt bateria de chumbo-ácido *(f)*

Batterieladegerät *(n) n* De
En battery charger
Es cargador de batería *(m)*
Fr chargeur de batterie *(m)*
It caricabatteria *(f)*
Pt carregador de bateria *(m)*

battery *n* En
De Batterie *(f)*
Es batería *(f)*
Fr batterie *(f)*
It batteria *(f)*
Pt bateria *(f)*

battery charger En
De Batterieladegerät *(n)*
Es cargador de batería *(m)*
Fr chargeur de batterie *(m)*
It caricabatteria *(f)*
Pt carregador de bateria *(m)*

Beleuchtung *(f) n* De
En illumination
Es iluminación *(f)*
Fr éclairage *(m)*
It illuminazione *(f)*
Pt iluminação *(f)*

belüftete Flamme *(f)* De
En aerated flame
Es llama aireada *(f)*
Fr flamme aérée *(f)*
It fiamma aerata *(f)*
Pt chama arejada *(f)*

Belüftung *(f) n* De
En demister
Es desempañador *(m)*
Fr antibrouillard *(m)*
It antiappannante *(m)*
Pt desembaciador *(m)*

Belüftung *(f) n* De
En ventilation
Es ventilación *(f)*
Fr ventilation *(f)*
It ventilazione *(f)*
Pt ventilação *(f)*

bemannte Ölbohrinsel *(f)* De
En manned platform (oil)
Es plataforma atendida por personal *(f)*
Fr plate-forme occupée *(f)*
It piattaforma abitata *(f)*
Pt plataforma com tripulação *(f)*

benceno *(m) n* Es
De Benzol *(n)*
En benzene
Fr benzène *(m)*
It benzolo *(m)*
Pt benzeno *(m)*

benzene *n* En
De Benzol *(n)*
Es benceno *(m)*
Fr benzène *(m)*
It benzolo *(m)*
Pt benzeno *(m)*

benzène *(m) n* Fr
De Benzol *(n)*
En benzene
Es benceno *(m)*
It benzolo *(m)*
Pt benzeno *(m)*

benzeno *(m) n* Pt
De Benzol *(n)*
En benzene
Es benceno *(m)*
Fr benzène *(m)*
It benzolo *(m)*

Benzin *(n) n* De
Am gasoline
En petrol
Es gasolina *(f)*
Fr essence *(f)*
It benzina *(f)*
Pt gasolina *(f)*

benzina *(f) n* It
Am gasoline
De Benzin *(n)*
En petrol
Es gasolina *(f)*
Fr essence *(f)*
Pt gasolina *(f)*

Benzinmotor *(m) n* De
En petrol engine
Es motor de gasolina *(m)*
Fr moteur à essence *(m)*
It motore a benzina *(m)*
Pt motor a gasolina *(m)*

Benzol *(n) n* De
En benzene
Es benceno *(m)*
Fr benzène *(m)*
It benzolo *(m)*
Pt benzeno *(m)*

benzolo *(m) n* It
De Benzol *(n)*
En benzene
Es benceno *(m)*
Fr benzène *(m)*
Pt benzeno *(m)*

Bergbau *(m) n* De
En mining
Es minería *(f)*
Fr exploitation minière *(f)*
It estrazione di minerali *(f)*
Pt mineração *(f)*

Bergwerk *(n) n* De
En mine
Es mina *(f)*
Fr mine *(f)*
It miniera *(f)*
Pt mina *(f)*

beschwerter Beton *(m)* De
En loaded concrete
Es hormigón cargado *(m)*
Fr béton chargé *(m)*
It cemento pesante *(m)*
Pt betão carregado *(m)*

bestrahlt *adj* De
En irradiated
Es irradiado
Fr irradié
It irradiato
Pt irradiado

Bestrahlung *(f) n* De
En irradiation
Es irradiación *(f)*
Fr irradiation *(f)*
It irradiazione *(f)*
Pt irradiação *(f)*

beta decay En
De Beta-Zerfall *(m)*
Es desintegración beta *(f)*
Fr désintégration bêta *(f)*
It disintegrazione beta *(f)*
Pt deterioração beta *(f)*

betão carregado *(m)* Pt
De beschwerter Beton *(m)*
En loaded concrete
Es hormigón cargado *(m)*
Fr béton chargé *(m)*
It cemento pesante *(m)*

beta particle En
De Beta-Partikel *(f)*
Es partícula beta *(f)*
Fr particule bêta *(f)*
It particella beta *(f)*
Pt partícula beta *(f)*

Beta-Partikel *(f) n* De
En beta particle
Es partícula beta *(f)*
Fr particule bêta *(f)*
It particella beta *(f)*
Pt partícula beta *(f)*

Beta-Zerfall *(m) n* De
En beta decay
Es desintegración beta *(f)*

Fr désintégration bêta *(f)*
It disintegrazione beta *(f)*
Pt deterioração beta *(f)*

béton chargé *(m)* Fr
De beschwerter Beton *(m)*
En loaded concrete
Es hormigón cargado *(m)*
It cemento pesante *(m)*
Pt betão carregado *(m)*

Betondruckbehälter *(m)* *n* De
En concrete pressure vessel
Es recipiente de presión de hormigón *(m)*
Fr récipient de pression en béton *(m)*
It recipiente a pressione di betone *(m)*
Pt recipiento de pressão de betão *(m)*

Betriebskosten *(pl)* *n* De
En running costs
Es costes de explotación *(m pl)*
Fr frais d'exploitation *(m pl)*
It costi di esercizio *(m pl)*
Pt custo de exploração *(m)*

betume *(m)* *n* Pt
De Bitumen *(n)*
En bitumen
Es bitún *(m)*
Fr bitume *(m)*
It bitume *(m)*

bief *(m)* *n* Fr
De Ablauf *(m)*
En tailrace
Es canal de descarga *(m)*
It canale di scarico *(m)*
Pt água de descarga *(f)*

biergol *(m)* *n* Fr
De Zweifachtreibstoff *(m)*
En bipropellant
Es bipropulsante *(m)*
It bipropellente *(m)*
Pt combustível bipropulsor *(m)*

bilame *(m)* *n* Fr
De Bimetallstreifen *(m)*
En bimetallic strip
Es tira bimetálica *(f)*
It nastro bimetallico *(m)*
Pt tira bimetálica *(f)*

bilan calorifique *(m)* Fr
De Wärmeausgleich *(m)*
En heat balance
Es equilibrio térmico *(m)*
It bilancio termico *(m)*
Pt equilíbrio térmico *(m)*

bilancio di massa *(m)* It
De Massenausgleich *(m)*
En mass balance
Es equilibrio de masa *(m)*
Fr équilibre de masse *(m)*
Pt equilíbrio de massa *(m)*

bilancio materiale-energia *(m)* It
De Stoff- und Energiebilanz *(f)*
En material and energy balance
Es equilibrio materia-energía *(m)*
Fr balance de matériel et énergie *(f)*
Pt equilíbrio de material e energia *(m)*

bilancio termico *(m)* It
De Wärmeausgleich *(m)*
En heat balance
Es equilibrio térmico *(m)*
Fr bilan calorifique *(m)*
Pt equilíbrio térmico *(m)*

billet furnace En
De Blockofen *(m)*
Es horno de palanquillas *(m)*
Fr four à billettes *(m)*
It forno per billette *(m)*
Pt forno de lingotes *(m)*

billet heater En
De Blockerwärmer *(m)*
Es calentador de palanquillas *(m)*
Fr chauffe-billettes *(m)*
It riscaldatore di billette *(m)*
Pt aquecedor de lingotes *(m)*

bimetallic strip En
De Bimetallstreifen *(m)*
Es tira bimetálica *(f)*
Fr bilame *(m)*
It nastro bimetallico *(m)*
Pt tira bimetálica *(f)*

Bimetallstreifen *(m)* *n* De
En bimetallic strip
Es tira bimetálica *(f)*
Fr bilame *(m)*
It nastro bimetallico *(m)*
Pt tira bimetálica *(f)*

binding energy En
De Bindungsenergie *(f)*
Es energía de enlace *(f)*
Fr énergie de liaison *(f)*
It energia di legatura *(f)*
Pt energia de fixação *(f)*

Bindungsenergie *(f)* *n* De
En binding energy
Es energía de enlace *(f)*
Fr énergie de liaison *(f)*
It energia di legatura *(f)*
Pt energia de fixação *(f)*

biochemical fuel cell En
De biochemische Brennstoffzelle *(f)*
Es célula de combustible bioquímico *(f)*
Fr pile à combustible biochimique *(f)*
It cellula combustibile biochimica *(f)*
Pt célula de combustível bioquímico *(f)*

biochemische Brennstoffzelle *(f)* De
En biochemical fuel cell
Es célula de combustible bioquímico *(f)*
Fr pile à combustible biochimique *(f)*
It cellula combustibile biochimica *(f)*
Pt célula de combustível bioquímico *(f)*

biodegradabile *adj* It
De biologisch abbaubar
En biodegradable
Es biodegradable
Fr biodégradable
Pt biodegradável

biodegradable *adj* En, Es
De biologisch abbaubar
Fr biodégradable
It biodegradabile
Pt biodegradável

biodégradable *adj* Fr
De biologisch abbaubar
En biodegradable
Es biodegradable
It biodegradabile
Pt biodegradável

biodegradável *adj* Pt
De biologisch abbaubar
En biodegradable
Es biodegradable
Fr biodégradable
It biodegradabile

biogas *n* En; Es, It *(m)*
De Biogas *(n)*
Fr biogaz *(m)*
Pt biogás *(m)*

Biogas *(n)* *n* De
En biogas
Es biogas *(m)*
Fr biogaz *(m)*
It biogas *(m)*
Pt biogás *(m)*

biogás *(m)* *n* Pt
De Biogas *(n)*
En biogas
Es biogas *(m)*
Fr biogaz *(m)*
It biogas *(m)*

biogaz *(m)* *n* Fr
De Biogas *(n)*
En biogas
Es biogas *(m)*
It biogas *(m)*
Pt biogás *(m)*

biological shield En
De biologischer Schild *(m)*
Es blindaje biológico *(m)*

Fr bouclier biologique (m)
It schermo biologico (m)
Pt protector biológico (m)

biologisch abbaubar De
En biodegradable
Es biodegradable
Fr biodégradable
It biodegradabile
Pt biodegradável

biologischer Schild *(m)* De
En biological shield
Es blindaje biológico *(m)*
Fr bouclier biologique *(m)*
It schermo biologico *(m)*
Pt protector biológico *(m)*

biomasa *(f) n* Es
De Lebendmasse *(f)*
En biomass
Fr biomasse *(f)*
It biomassa *(f)*
Pt biomassa *(f)*

biomass *n* En
De Lebendmasse *(f)*
Es biomasa *(f)*
Fr biomasse *(f)*
It biomassa *(f)*
Pt biomassa *(f)*

biomassa *(f) n* It, Pt
De Lebendmasse *(f)*
En biomass
Es biomasa *(f)*
Fr biomasse *(f)*

biomasse *(f) n* Fr
De Lebendmasse *(f)*
En biomass
Es biomasa *(f)*
It biomassa *(f)*
Pt biomassa *(f)*

bio-oxidação *(f) n* Pt
De Bio-Oxidation *(f)*
En bio-oxidation
Es bio-oxidación *(f)*
Fr bio-oxydation *(f)*
It biossidazione *(f)*

bio-oxidación *(f) n* Es
De Bio-Oxidation *(f)*
En bio-oxidation
Fr bio-oxydation *(f)*
It biossidazione *(f)*
Pt bio-oxidação *(f)*

bio-oxidation *n* En
De Bio-Oxidation *(f)*
Es bio-oxidación *(f)*
Fr bio-oxydation *(f)*
It biossidazione *(f)*
Pt bio-oxidação *(f)*

Bio-Oxidation *(f) n* De
En bio-oxidation
Es bio-oxidación *(f)*
Fr bio-oxydation *(f)*
It biossidazione *(f)*
Pt bio-oxidação *(f)*

bio-oxydation *(f) n* Fr
De Bio-Oxidation *(f)*
En bio-oxidation
Es bio-oxidación *(f)*
It biossidazione *(f)*
Pt bio-oxidação *(f)*

biosíntese *(f) n* Pt
De Biosynthese *(f)*
En biosynthesis
Es biosíntesis *(f)*
Fr biosynthèse *(f)*
It biosintesi *(f)*

biosintesi *(f) n* It
De Biosynthese *(f)*
En biosynthesis
Es biosíntesis *(f)*
Fr biosynthèse *(f)*
Pt biosíntese *(f)*

biosíntesis *(f) n* Es
De Biosynthese *(f)*
En biosynthesis
Fr biosynthèse *(f)*
It biosintesi *(f)*
Pt biosíntese *(f)*

biossidazione *(f) n* It
De Bio-Oxidation *(f)*
En bio-oxidation
Es bio-oxidación *(f)*
Fr bio-oxydation *(f)*
Pt bio-oxidação *(f)*

biossido di uranio *(m)* It
De Uraniumdioxid *(n)*
En uranium dioxide
Es dióxido de uranio *(m)*
Fr bioxyde d'uranium *(m)*
Pt dióxido de urânio

Biosynthese *(f) n* De
En biosynthesis
Es biosíntesis *(f)*
Fr biosynthèse *(f)*
It biosintesi *(f)*
Pt biosíntese *(f)*

biosynthèse *(f) n* Fr
De Biosynthese *(f)*
En biosynthesis
Es biosíntesis *(f)*
It biosintesi *(f)*
Pt biosíntese *(f)*

biosynthesis *n* En
De Biosynthese *(f)*
Es biosíntesis *(f)*
Fr biosynthèse *(f)*
It biosintesi *(f)*
Pt biosíntese *(f)*

bióxido de zirconio *(m)* Es
De Zirkonerde *(f)*
En zirconia
Fr zircone *(f)*
It zirconia *(f)*
Pt zircónia *(f)*

bioxyde d'uranium *(m)* Fr
De Uraniumdioxid *(n)*
En uranium dioxide
Es dióxido de uranio *(m)*
It biossido di uranio *(m)*
Pt dióxido de urânio

bipropellant *n* En
De Zweifachtreibstoff *(m)*
Es bipropulsante *(m)*
Fr biergol *(m)*
It bipropellente *(m)*
Pt combustível bipropulsor *(m)*

bipropellente *(m) n* It
De Zweifachtreibstoff *(m)*
En bipropellant
Es bipropulsante *(m)*
Fr biergol *(m)*
Pt combustível bipropulsor *(m)*

bipropulsante *(m) n* Es
De Zweifachtreibstoff *(m)*
En bipropellant
Fr biergol *(m)*
It bipropellente *(m)*
Pt combustível bipropulsor *(m)*

bit *n* En
De Bohrmeißel *(m)*
Es broca *(f)*
Fr trépan *(m)*
It punta *(f)*
Pt broca *(f)*

bitume *(m) n* Fr
De Bitumen *(n)*
En bitumen
Es bitún *(m)*
It bitume *(m)*
Pt betume *(m)*

bitume *(m) n* It
De Bitumen *(n)*
En bitumen
Es bitún *(m)*
Fr bitume *(m)*
Pt betume *(m)*

bitumen *n* En
De Bitumen *(n)*
Es bitún *(m)*
Fr bitume *(m)*
It bitume *(m)*
Pt betume *(m)*

Bitumen *(n) n* De
En bitumen
Es bitún *(m)*
Fr bitume *(m)*
It bitume *(m)*
Pt betume *(m)*

bituminöse Kohle *(f)* De
En bituminous coal
Es carbón bituminoso *(m)*
Fr charbon bitumineux *(m)*
It carbone bituminoso *(m)*
Pt carvão betuminoso *(m)*

bituminöser Sand *(m)* De
En bituminous sand
Es arena bituminosa *(f)*
Fr sable bitumineux *(m)*

It sabbia bituminosa *(f)*
Pt areia betuminosa *(f)*

bituminous coal En
De bituminöse Kohle *(f)*
Es carbón bituminoso *(m)*
Fr charbon bitumineux *(m)*
It carbone bituminoso *(m)*
Pt carvão betuminoso *(m)*

bituminous sand En
De bituminöser Sand *(m)*
Es arena bituminosa *(f)*
Fr sable bitumineux *(m)*
It sabbia bituminosa *(f)*
Pt areia betuminosa *(f)*

bitún *(m) n* Es
De Bitumen *(n)*
En bitumen
Fr bitume *(m)*
It bitume *(m)*
Pt betume *(m)*

black body En
De Planckscher Strahler *(m)*
Es cuerpo negro *(m)*
Fr corps noir *(m)*
It corpo nero *(m)*
Pt corpo negro *(m)*

black-body radiation En
De Plancksche Strahlung *(f)*
Es radiación de cuerpo negro *(f)*
Fr rayonnement du corps noir *(m)*
It radiazione corpo nero *(f)*
Pt radiação de corpo negro *(f)*

black coal En
De Schwarzkohle *(f)*
Es carbón negro *(m)*
Fr houille *(f)*
It carbone nero *(m)*
Pt carvão negro *(m)*

blanket (nuclear reactor) *n* En
De Brutzone *(f)*
Es zona fértil *(f)*
Fr couche fertile *(f)*
It copertura *(f)*
Pt cortina *(f)*

blast furnace En
De Hochofen *(m)*
Es alto horno *(m)*
Fr haut fourneau *(m)*
It alto forno *(m)*
Pt alto-forno *(m)*

blauer Asbest *(m)* De
En blue asbestos
Es crocidolita *(f)*
Fr amiante bleu *(m)*
It amianto azzurro *(m)*
Pt amianto azul *(m)*

Bleigehalt *(m) n* De
En lead content
Es contenido de plomo *(m)*
Fr teneur en plomb *(f)*
It contenuto di piombo *(m)*
Pt teor de chumbo *(m)*

Blei-Säure-Batterie *(f) n* De
En lead-acid battery
Es batería de ácido-plomo *(f)*
Fr batterie au plomb *(f)*
It batteria acida al piombo *(f)*
Pt bateria de chumbo-ácido *(f)*

blindaje *(m) n* Es
De Abschirmung *(f)*
En shield
Fr bouclier *(m)*
It schermo *(m)*
Pt protector *(m)*

blindaje biológico *(m)* Es
De biologischer Schild *(m)*
En biological shield
Fr bouclier biologique *(m)*
It schermo biologico *(m)*
Pt protector biológico *(m)*

Blockerwärmer *(m) n* De
En billet heater
Es calentador de palanquillas *(m)*
Fr chauffe-billettes *(m)*
It riscaldatore di billette *(m)*
Pt aquecedor de lingotes *(m)*

Blockofen *(m) n* De
En billet furnace
Es horno de palanquillas *(m)*
Fr four à billettes *(m)*
It forno per billette *(m)*
Pt forno de lingotes *(m)*

blowout *n* En
De Ausbruch *(m)*
Es reventón *(m)*
Fr éruption *(f)*
It eruzione *(f)*
Pt estouro *(m)*

blue asbestos En
De blauer Asbest *(m)*
Es crocidolita *(f)*
Fr amiante bleu *(m)*
It amianto azzurro *(m)*
Pt amianto azul *(m)*

bobina de campo *(f)* Es
De Feldspule *(f)*
En field coil
Fr bobinage de champ *(m)*
It bobina di campo *(f)*
Pt bobina de campo *(f)*

bobina de campo *(f)* Pt
De Feldspule *(f)*
En field coil
Es bobina de campo *(f)*
Fr bobinage de champ *(m)*
It bobina di campo *(f)*

bobina de indução *(f)* Pt
De Induktionsspule *(f)*
En induction coil
Es bobina de inducción *(f)*
Fr bobine d'induction *(f)*
It rocchetto d'induzione *(m)*

bobina de inducción *(f)* Es
De Induktionsspule *(f)*
En induction coil
Fr bobine d'induction *(f)*
It rocchetto d'induzione *(m)*
Pt bobina de indução *(f)*

bobina di campo *(f)* It
De Feldspule *(f)*
En field coil
Es bobina de campo *(f)*
Fr bobinage de champ *(m)*
Pt bobina de campo *(f)*

bobinage de champ *(m)* Fr
De Feldspule *(f)*
En field coil
Es bobina de campo *(f)*
It bobina di campo *(f)*
Pt bobina de campo *(f)*

bobina mobile It
De Drehspule-
En moving-coil
Es bobina móvil
Fr bobine mobile
Pt bobina móvel

bobina móvel Pt
De Drehspule-
En moving-coil
Es bobina móvil
Fr bobine mobile
It bobina mobile

bobina móvil Es
De Drehspule-
En moving-coil
Fr bobine mobile
It bobina mobile
Pt bobina móvel

bobine d'induction *(f)* Fr
De Induktionsspule *(f)*
En induction coil
Es bobina de inducción *(f)*
It rocchetto d'induzione *(m)*
Pt bobina de indução *(f)*

bobine mobile Fr
De Drehspule-
En moving-coil
Es bobina móvil
It bobina mobile
Pt bobina móvel

bodengestreut *adj* De
En ground-scattered
Es dispersado en tierra
Fr diffusé au sol
It diffuso sul suolo
Pt disperso pelo solo

Bodenisolierung *(f) n* De
En floor insulation
Es aislamiento de suelos *(m)*
Fr isolement du plancher *(m)*
It isolamento del pavimento *(m)*
Pt isolamento de pavimento *(m)*

Bodenschätze *(pl) n* De
En resources
Es recursos *(m pl)*
Fr ressources *(f pl)*
It risorse *(f pl)*
Pt recursos *(m pl)*

Bohren *(n) n* De
En drilling
Es perforación *(f)*
Fr forage *(m)*
It trivellazione *(f)*
Pt perfuração *(f)*

Bohrkonstruktionen *(pl) n* De
En drilling structures
Es estructuras de perforación *(f pl)*
Fr structures de forage *(f pl)*
It strutture di trivellazione *(f pl)*
Pt estructuras de perfuração *(f pl)*

Bohrmeiβel *(m) n* De
En bit
Es broca *(f)*
Fr trépan *(m)*
It punta *(f)*
Pt broca *(f)*

Bohrschlamm *(m) n* De
En drilling mud
Es lodo de perforación *(m)*
Fr boue de forage *(f)*
It fango di trivellazione *(m)*
Pt lodo de perfuração *(m)*

Bohrung *(f) n* De
En well
Es pozo *(m)*
Fr puits *(m)*
It pozzo *(m)*
Pt poço *(m)*

boiler *n* En
De Kessel *(m)*
Es caldera *(f)*
Fr chaudière *(f)*
It caldaia *(f)*
Pt caldeira *(f)*

boiler auxiliary plant En
De Kesselhilfsanlage *(f)*
Es planta auxiliar de calderas *(f)*
Fr chaufferie auxiliaire *(f)*
It impianto ausiliario caldaia *(m)*
Pt instalação auxiliar de caldeira *(f)*

boiling point En
De Siedepunkt *(m)*
Es punto de ebullición *(f)*
Fr point d'ébullition *(f)*
It punto di ebollizione *(m)*
Pt ponto de ebulição *(m)*

boiling-water reactor (BWR) En
De Siedereaktor *(m)*
Es reactor de agua en ebullición *(m)*
Fr réacteur à eau bouillante *(m)*
It reattore ad acqua bollente *(m)*
Pt reactor de água em ebulição *(m)*

bolometer *n* En
De Bolometer *(n)*
Es bolómetro *(m)*
Fr bolomètre *(m)*
It bolometro *(m)*
Pt bolómetro *(m)*

Bolometer *(n) n* De
En bolometer
Es bolómetro *(m)*
Fr bolomètre *(m)*
It bolometro *(m)*
Pt bolómetro *(m)*

bolomètre *(m) n* Fr
De Bolometer *(n)*
En bolometer
Es bolómetro *(m)*
It bolometro *(m)*
Pt bolómetro *(m)*

bolometro *(m) n* It
De Bolometer *(n)*
En bolometer
Es bolómetro *(m)*
Fr bolomètre *(m)*
Pt bolómetro *(m)*

bolómetro *(m) n* Es, Pt
De Bolometer *(n)*
En bolometer
Fr bolomètre *(m)*
It bolometro *(m)*

bomba *(f) n* Es, Pt
De Pumpe *(f)*
En pump
Fr pompe *(f)*
It pompa *(f)*

bomba all'idrogeno *(f)* It
De Wasserstoffbombe *(f)*
En hydrogen bomb
Es bomba de hidrógeno *(f)*
Fr bombe à hydrogène *(f)*
Pt bomba de hidrogénio *(f)*

bomba de alimentação *(f)* Pt
De Speisepumpe *(f)*
En feed pump
Es bomba de alimentación *(f)*
Fr pompe d'alimentation *(f)*
It pompa di alimentazione *(f)*

bomba de alimentación *(f)* Es
De Speisepumpe *(f)*
En feed pump
Fr pompe d'alimentation *(f)*
It pompa di alimentazione *(f)*
Pt bomba de alimentação *(f)*

bomba de calor *(f)* Es
De Wärmepumpe *(f)*
En heat pump
Fr thermopompe *(f)*
It pompa di calore *(f)*
Pt bomba térmica *(f)*

bomba de hidrogénio *(f)* Pt
De Wasserstoffbombe *(f)*
En hydrogen bomb
Es bomba de hidrógeno *(f)*
Fr bombe à hydrogène *(f)*
It bomba all'idrogeno *(f)*

bomba de hidrógeno *(f)* Es
De Wasserstoffbombe *(f)*
En hydrogen bomb
Fr bombe à hydrogène *(f)*
It bomba all'idrogeno *(f)*
Pt bomba de hidrogénio *(f)*

bomba de vacío *(f)* Es
De Vakuumpumpe *(f)*
En vacuum pump
Fr pompe à vide *(f)*
It depressore *(m)*
Pt bomba de vácuo *(f)*

bomba de vácuo *(f)* Pt
De Vakuumpumpe *(f)*
En vacuum pump
Es bomba de vacío *(f)*
Fr pompe à vide *(f)*
It depressore *(m)*

bomba térmica *(f)* Pt
De Wärmepumpe *(f)*
En heat pump
Es bomba de calor *(f)*
Fr thermopompe *(f)*
It pompa di calore *(f)*

bomb calorimeter En
De Bombenkalorimeter *(n)*
Es calorímetro de bomba *(m)*
Fr bombe calorimétrique *(f)*
It calorimetro a bomba *(m)*
Pt calorímetro de bomba *(m)*

bombe à hydrogène *(f)* Fr
De Wasserstoffbombe *(f)*
En hydrogen bomb
Es bomba de hidrógeno *(f)*
It bomba all'idrogeno *(f)*
Pt bomba de hidrogénio *(f)*

bombe calorimétrique *(f)* Fr
De Bombenkalorimeter *(n)*
En bomb calorimeter
Es calorímetro de bomba *(m)*
It calorimetro a bomba *(m)*
Pt calorímetro de bomba *(m)*

Bombenkalorimeter *(n)* *n* De
En bomb calorimeter
Es calorímetro de bomba *(m)*
Fr bombe calorimétrique *(f)*
It calorimetro a bomba *(m)*
Pt calorímetro de bomba *(m)*

borne (eléctrico) *(m)* *n* Pt
De Klemme (elektrische) *(f)*
En terminal (electric)
Es terminal (eléctrico) *(m)*
Fr borne (éléctrique) *(f)*
It terminale (elettrico) *(m)*

borne (éléctrique) *(f)* *n* Fr
De Klemme (elektrische) *(f)*
En terminal (electric)
Es terminal (eléctrico) *(m)*
It terminale (elettrico) *(m)*
Pt borne (eléctrico) *(m)*

botella magnética *(f)* Es
De magnetische Flasche *(f)*
En magnetic bottle
Fr bouteille magnétique *(f)*
It bottiglia magnetica *(f)*
Pt garrafa magnética *(f)*

bottiglia magnetica *(f)* It
De magnetische Flasche *(f)*
En magnetic bottle
Es botella magnética *(f)*
Fr bouteille magnétique *(f)*
Pt garrafa magnética *(f)*

boucle de retour *(f)* Fr
De geschlossener Regelkreis *(m)*
En closed loop
Es bucle cerrado *(m)*
It circuito chiuso *(m)*
Pt circuito fechado *(m)*

boucle ouverte *(f)* Fr
De offener Kreislauf *(m)*
En open loop
Es bucle abierto *(m)*
It circuito aperto *(m)*
Pt circuito aberto *(m)*

bouclier *(m)* *n* Fr
De Abschirmung *(f)*
En shield
Es blindaje *(m)*
It schermo *(m)*
Pt protector *(m)*

bouclier biologique *(m)* Fr
De biologischer Schild *(m)*
En biological shield
Es blindaje biológico *(m)*
It schermo biologico *(m)*
Pt protector biológico *(m)*

bouclier thermique *(m)* Fr
De Wärmeschutz *(m)*
En heat shield; thermal shield
Es pantalla térmica *(f)*
It schermo termico *(m)*
Pt protector térmico *(m)*

boue de forage *(f)* *n* Fr
De Bohrschlamm *(m)*
En drilling mud
Es lodo de perforación *(m)*
It fango di trivellazione *(m)*
Pt lodo de perfuração *(m)*

Bourdon gauge En
De Bourdonmanometer *(n)*
Es manómetro de Bourdon *(m)*
Fr manomètre de Bourdon *(m)*
It manometro Bourdon *(m)*
Pt manómetro de Bourdon *(m)*

Bourdonmanometer *(n)* *n* De
En Bourdon gauge
Es manómetro de Bourdon *(m)*
Fr manomètre de Bourdon *(m)*
It manometro Bourdon *(m)*
Pt manómetro de Bourdon *(m)*

bouteille magnétique *(f)* Fr
De magnetische Flasche *(f)*
En magnetic bottle
Es botella magnética *(f)*
It bottiglia magnetica *(f)*
Pt garrafa magnética *(f)*

brai *(m)* *n* Fr
De Pech *(n)*
En pitch
Es brea *(f)*
It pece *(f)*
Pt pez *(f)*

Braunkohle *(f)* De
En brown coal; lignite
Es lignito *(m)*
Fr lignite *(m)*
It lignite *(m)*
Pt carvão castanho *(m)*; lenhite *(f)*

brea *(f)* *n* Es
De Pech *(n)*
En pitch
Fr brai *(m)*
It pece *(f)*
Pt pez *(f)*

break-even point En
De Rentabilitätsgrenze *(f)*
Es punto comparativo *(m)*
Fr point mort *(m)*
It punto di pareggio *(m)*
Pt ponto de equilíbrio *(m)*

Brechung *(f)* *n* De
En refraction
Es refracción *(f)*
Fr réfraction *(f)*
It rifrazione *(f)*
Pt refracção *(f)*

Brechungsindex *(m)* *n* De
En refractive index
Es índice de refracción *(m)*
Fr indice de réfraction *(m)*
It indice di rifrazione *(m)*
Pt índice de refracção *(m)*

breeder reactor En
De Brutreaktor *(m)*
Es reactor reproductor *(m)*
Fr réacteur surrégénérateur *(m)*
It reattore autofissilizzante *(m)*
Pt reactor reproductor *(m)*

Breiteneffekt *(m)* *n* De
En latitude effect
Es efecto de latitud *(m)*
Fr effet de la latitude *(m)*
It effetto latitudine *(m)*
Pt efeito da latitude *(m)*

Brennelement *(n)* *n* De
En fuel element
Es elemento combustible *(m)*
Fr élément combustible *(m)*
It elemento combustibile *(m)*
Pt elemento combustível *(m)*

Brenner *(m)* *n* De
En burner
Es quemador *(m)*
Fr brûleur *(m)*
It bruciatore *(m)*
Pt queimador *(m)*

Brenner mit verzögerter Luftmischung *(m)* De
En delayed-mixing burner
Es quemador de mezcla retardada *(m)*
Fr brûleur a mélange différé *(m)*
It bruciatore di

mescolata ritardato *(m)*
Pt queimador misturador retardado *(m)*

Brenner-Treiböl *(n) n* De
En burner fuel-oil
Es fuel-oil para quemador *(m)*
Fr pétrole lampant *(m)*
It nafta per bruciatori *(f)*
Pt fuel-oil para queimadores *(m)*

Brennkammer *(f) n* De
En combustion chamber
Es cámara de combustión *(f)*
Fr chambre de combustion *(f)*
It camera di combustione *(f)*
Pt câmara de combustão *(f)*

Brennofen *(m) n* De
En kiln
Es estufa *(f)*
Fr four *(m)*
It forno *(m)*
Pt forno de requeima *(m)*

Brennrohr *(n) n* De
En bagasse
Es bagazo *(m)*
Fr bagasse *(f)*
It bagasse esaurite *(f)*
Pt bagaço *(m)*

Brennstoff *(m) n* De
En fuel
Es combustible *(m)*
Fr combustible *(m)*
It combustibile *(m)*
Pt combustível *(m)*

Brennstoffkanal *(m) n* De
En fuel channel
Es canal de combustible *(m)*
Fr canal de combustible *(m)*
It canale del combustibile *(m)*
Pt canal de combustível *(m)*

Brennstoffstab *(m) n* De
En fuel rod
Es barra de combustible *(f)*
Fr barre de combustible *(f)*
It barra combustibile *(f)*
Pt vara de combustível *(f)*

Brennstoffverbrauch *(m) n* De
En fuel consumption
Es consumo de combustible *(m)*
Fr consommation de combustible *(f)*
It consumo di combustibile *(m)*
Pt consumo de combustível *(m)*

Brennstoffzelle *(f) n* De
En fuel cell
Es célula de combustible *(f)*
Fr pile à combustible *(f)*
It pila a combustibile *(f)*
Pt célula de combustível *(f)*

Brennstoffzyklus *(m) n* De
En fuel cycle
Es ciclo de los combustibles *(m)*
Fr cycle du combustible *(m)*
It ciclo del combustibile *(m)*
Pt ciclo de combustível *(m)*

brightness *n* En
De Helligkeit *(f)*
Es brillo *(m)*
Fr brillance *(f)*
It luminosità *(f)*
Pt brilho *(m)*

brilho *(m) n* Pt
De Helligkeit *(f)*
En brightness
Es brillo *(m)*
Fr brillance *(f)*
It luminosità *(f)*

brilho efectivo *(m)* Pt
De effektive Helligkeit *(f)*
En effective brightness
Es brillo efectivo *(m)*
Fr brillance efficace *(f)*
It luminosità effettiva *(f)*

brillance *(f) n* Fr
De Helligkeit *(f)*
En brightness
Es brillo *(m)*
It luminosità *(f)*
Pt brilho *(m)*

brillance efficace *(f)* Fr
De effektive Helligkeit *(f)*
En effective brightness
Es brillo efectivo *(m)*
It luminosità effettiva *(f)*
Pt brilho efectivo *(m)*

brillo *(m) n* Es
De Helligkeit *(f)*
En brightness
Fr luminance *(f)*
It luminosità *(f)*
Pt brilho *(m)*

brillo efectivo *(m)* Es
De effektive Helligkeit *(f)*
En effective brightness
Fr luminance efficace *(f)*
It luminosità effettiva *(f)*
Pt brilho efectivo *(m)*

brine *n* En
De Sole *(f)*
Es salmuera *(f)*
Fr saumure *(f)*
It acqua salmastra *(f)*
Pt salmoura *(f)*

brique réfractaire *(f)* Fr
De feuerfester Stein *(m)*
En firebrick
Es ladrillo refractario *(m)*
It mattone refrattario *(m)*
Pt tijolo refractário *(m)*

brise-bise *(m) n* Fr
Am draft excluder
De Luftzugschutz *(m)*
En draught excluder
Es eliminador de corrientes *(m)*
It esclusore d'aria *(m)*
Pt eliminador de corrente de ar *(m)*

broca *(f) n* Es, Pt
De Bohrmeißel *(m)*
En bit
Fr trépan *(m)*
It punta *(f)*

brown coal En
De Braunkohle *(f)*; Lignit *(n)*
Es lignito *(m)*
Fr lignite *(m)*
It lignite *(m)*
Pt carvão castanho *(m)*; lenhite *(f)*

bruciatore *(m) n* It
De Brenner *(m)*
En burner
Es quemador *(m)*
Fr brûleur *(m)*
Pt queimador *(m)*

bruciatore a corrente d'aria naturale *(m)* It
Am natural-draft burner
De natürlicher Luftstrombrenner *(m)*
En natural-draught burner
Es quemador de tiro natural *(m)*
Fr brûleur à tirage naturel *(m)*
Pt queimador de corrente de ar natural *(m)*

bruciatore a getto *(m)* It
De Düsenbrenner *(m)*
En jet burner
Es quemador de chorro *(m)*
Fr brûleur-jet *(m)*
Pt queimador de jacto *(m)*

bruciatore a ventilazione forzata *(m)* It
De Windbrenner *(m)*
En air-blast burner
Es quemador de chorro de aire *(m)*
Fr brûleur à air soufflé *(m)*
Pt queimador de corrente de ar *(m)*

bruciatore circolare *(m)* It
De Kreisbrenner *(m)*
En circular burner
Es quemador circular *(m)*

Fr brûleur circulaire *(m)*
Pt queimador circular *(m)*

bruciatore di fiamma direzionale *(m)* It
De flammengerichteter Brenner *(m)*
En directional flame burner
Es quemador de llama direccional *(m)*
Fr brûleur à flamme dirigée *(m)*
Pt queimador de chama direccional *(m)*

bruciatore di mescolata ritardato *(m)* It
De Brenner mit verzögerter Luftmischung *(m)*
En delayed-mixing burner
Es quemador de mezcla retardada *(m)*
Fr brûleur a mélange différé *(m)*
Pt queimador misturador retardado *(m)*

bruciatore post-aerato *(m)* It
De nachbelüfteter Brenner *(m)*
En post-aerated burner
Es quemador posaireado *(m)*
Fr brûleur à post-aération *(m)*
Pt queimador pós-arejado *(m)*

bruciatore pre-aerato *(m)* It
De vorbelüfteter Brenner *(m)*
En pre-aerated burner
Es quemador preaireado *(m)*
Fr brûleur à pré-aération *(m)*
Pt queimador pré-arejado *(m)*

brûleur *(m) n* Fr
De Brenner *(m)*
En burner
Es quemador *(m)*
It bruciatore *(m)*
Pt queimador *(m)*

brûleur à air soufflé *(m)* Fr
De Windbrenner *(m)*
En air-blast burner
Es quemador de chorro de aire *(m)*
It bruciatore a ventilazione forzata *(m)*
Pt queimador de corrente de ar *(m)*

brûleur à flamme dirigée *(m)* Fr
De flammengerichteter Brenner *(m)*
En directional flame burner
Es quemador de llama direccional *(m)*
It bruciatore di fiamma direzionale *(m)*
Pt queimador de chama direccional *(m)*

brûleur a mélange différé *(m)* Fr
De Brenner mit verzögerter Luftmischung *(m)*
En delayed-mixing burner
Es quemador de mezcla retardada *(m)*
It bruciatore di mescolata ritardato *(m)*
Pt queimador misturador retardado *(m)*

brûleur à post-aération *(m)* Fr
De nachbelüfteter Brenner *(m)*
En post-aerated burner
Es quemador posaireado *(m)*
It bruciatore post-aerato *(m)*
Pt queimador pós-arejado *(m)*

brûleur à pré-aération *(m)* Fr
De vorbelüfteter Brenner *(m)*
En pre-aerated burner
Es quemador preaireado *(m)*
It bruciatore pre-aerato *(m)*
Pt queimador pré-arejado *(m)*

brûleur à tirage naturel *(m)* Fr
Am natural-draft burner
De natürlicher Luftstrombrenner *(m)*
En natural-draught burner
Es quemador de tiro natural *(m)*
It bruciatore a corrente d'aria naturale *(m)*
Pt queimador de corrente de ar natural *(m)*

brûleur circulaire *(m)* Fr
De Kreisbrenner *(m)*
En circular burner
Es quemador circular *(m)*
It bruciatore circolare *(m)*
Pt queimador circular *(m)*

brûleur-jet *(m) n* Fr
De Düsenbrenner *(m)*
En jet burner
Es quemador de chorro *(m)*
It bruciatore a getto *(m)*
Pt queimador de jacto *(m)*

brut *adj* Fr
De brutto
En gross
Es bruto
It lordo
Pt bruto

brut à base naphténique *(m)* Fr
De naphthenisches Rohöl *(n)*
En naphthenic crude oil
Es crudo nafténico *(m)*
It petrolio greggio naftenico *(m)*
Pt petróleo crú nafténico *(m)*

brut de synthèse *(m)* Fr
De synthetisches Rohöl *(n)*
En synthetic crude oil (SCO)
Es crudo sintético *(m)*
It petrolio greggio sintetico *(m)*
Pt petróleo crú sintético *(m)*

brut léger *(m)* Fr
De leichtes Rohöl *(n)*
En light crude oil
Es crudo ligero *(m)*
It petrolio greggio leggero *(m)*
Pt petróleo crú leve *(m)*

bruto *adj* Es
De brutto
En gross
Fr brut
It lordo
Pt bruto

bruto *adj* Pt
De brutto
En gross
Es bruto
Fr brut
It lordo

Brutreaktor *(m) n* De
En breeder reactor
Es reactor reproductor *(m)*
Fr réacteur surrégénérateur *(m)*
It reattore autofissilizzante *(m)*
Pt reactor reproductor *(m)*

brutto *adj* De
En gross
Es bruto
Fr brut
It lordo
Pt bruto

Brutzone *(f) n* De
En blanket
Es zona fértil *(f)*
Fr couche fertile *(f)*
It copertura *(f)*
Pt cortina *(f)*

bucket calorimeter En
De Schalenkalorimeter *(n)*
Es calorímetro de cubeta *(m)*
Fr calorimètre à godet *(m)*
It calorimetro di secchio *(m)*

Pt calorímetro de balde *(m)*

bucle abierto *(m)* Es
De offener Kreislauf *(m)*
En open loop
Fr boucle ouverte *(f)*
It circuito aperto *(m)*
Pt circuito aberto *(m)*

bucle cerrado *(m)* Es
De geschlossener Regelkreis *(m)*
En closed loop
Fr boucle de retour *(f)*
It circuito chiuso *(m)*
Pt circuito fechado *(m)*

bulk density En
De Schüttdichte *(f)*
Es densidad en masa *(f)*
Fr densité apparente *(f)*
It densità apparente *(m)*
Pt densidade em massa *(f)*

bulk oil-carrier En
De Ölsammelschiff *(n)*
Es petrolero a granel *(m)*
Fr pétrolier vraquier *(m)*
It petroliera all'ingrosso *(f)*
Pt petroleiro a granel *(m)*

burner *n* En
De Brenner *(m)*
Es quemador *(m)*
Fr brûleur *(m)*
It bruciatore *(m)*
Pt queimador *(m)*

burner fuel-oil En
De Brenner-Treiböl *(n)*
Es fuel-oil para quemador *(m)*
Fr pétrole lampant *(m)*
It nafta per bruciatori *(f)*
Pt fuel-oil para queimadores *(m)*

busbar *n* En
De Leitungsschiene *(f)*
Es barra colectora *(f)*
Fr barre omnibus *(f)*
It sbarra *(f)*
Pt barra colectora *(f)*

Butadien *(n)* *n* De
En butadiene
Es butadieno *(m)*
Fr butadiène *(m)*
It butadiene *(m)*
Pt butadieno *(m)*

butadiene *n* En; It *(m)*
De Butadien *(n)*
Es butadieno *(m)*
Fr butadiène *(m)*
Pt butadieno *(m)*

butadiène *(m)* *n* Fr
De Butadien *(n)*
En butadiene
Es butadieno *(m)*
It butadiene *(m)*
Pt butadieno *(m)*

butadieno *(m)* *n* Es, Pt
De Butadien *(n)*
En butadiene
Fr butadiène *(m)*
It butadiene *(m)*

Butan *(n)* *n* De
En butane
Es butano *(m)*
Fr butane *(m)*
It butano *(m)*
Pt butano *(m)*

butane *n* En; Fr *(m)*
De Butan *(n)*
Es butano *(m)*
It butano *(m)*
Pt butano *(m)*

butano *(m)* *n* Es, It, Pt
De Butan *(n)*
En butane
Fr butane *(m)*

Buten *(n)* *n* De
En butene
Es buteno *(m)*
Fr butène *(m)*
It butene *(m)*
Pt buteno *(m)*

butene *n* En; It *(m)*
De Buten *(n)*
Es buteno *(m)*
Fr butène *(m)*
Pt buteno *(m)*

butène *(m)* *n* Fr
De Buten *(n)*
En butene
Es buteno *(m)*
It butene *(m)*
Pt buteno *(m)*

buteno *(m)* *n* Es, Pt
De Buten *(n)*
En butene
Fr butène *(m)*
It butene *(m)*

C

cable-tool drilling En
De Seilbohren *(n)*
Es perforación a cable *(f)*
Fr forage au câble *(m)*
It trivellazione con utensile a fune *(f)*
Pt perfuração a cabo *(f)*

cadmium battery En
De Kadmiumbatterie *(f)*
Es batería de cadmio *(f)*
Fr pile au cadmium *(f)*
It batteria al cadmio *(f)*
Pt bateria de cádmio *(f)*

caduta di pressione *(f)* It
De Druckabfall *(m)*
En pressure drop
Es caída de presión *(f)*
Fr chute de pression *(f)*
Pt queda de pressão *(f)*

caduta di tensione *(f)* It
De Spannungsunterschied *(m)*
En potential difference
Es diferencia de potencial *(f)*
Fr différence de potentiel *(f)*
Pt diferença de potencial *(f)*

caída de presión *(f)* Es
De Druckabfall *(m)*
En pressure drop
Fr chute de pression *(f)*
It caduta di pressione *(f)*
Pt queda de pressão *(f)*

caking coal En
De Backkohle *(f)*
Es hulla grasa *(f)*
Fr houille collante *(f)*
It carbone agglutinante *(m)*
Pt hulha gorda *(f)*

Calcit *(n)* *n* De
En calcite
Es calcita *(f)*
Fr calcite *(f)*
It calcite *(f)*
Pt calcite *(f)*

calcita *(f)* *n* Es
De Calcit *(n)*
En calcite
Fr calcite *(f)*
It calcite *(f)*
Pt calcite *(f)*

calcite *n* En; Fr, It, Pt *(f)*
De Calcit *(n)*
Es calcita *(f)*

caldaia *(f)* *n* It
De Kessel *(m)*
En boiler
Es caldera *(f)*
Fr chaudière *(f)*
Pt caldeira *(f)*

caldaia a processo diretto *(f)* It
De Zwangsdurchlaufkessel *(m)*
En once-through boiler
Es caldera de proceso directo *(f)*
Fr chaudière sans recyclage *(f)*
Pt caldeira de uma só passagem *(f)*

caldaia a recupero di calore perduto *(f)* It
De Abwärmekessel *(m)*
En waste-heat boiler
Es caldera de calor residual *(f)*
Fr chaudière de récupération des chaleurs perdues *(f)*
Pt caldeira aquecida com calor residual *(f)*

caldaia a sezioni *(f)* It
De Teilkammerkessel *(m)*
En sectional boiler

Es caldera compartimentada *(f)*
Fr chaudière sectionnelle *(f)*
Pt caldeira de secções *(f)*

caldaia a tubo d'acqua *(f)* It
De Wasserrohrkessel *(m)*
En water-tube boiler
Es caldera de tubos de agua *(f)*
Fr chaudière tubulaire *(f)*
Pt caldeira de tubos de água *(f)*

caldaia a tubo di fumo *(f)* It
De Heizrohrkessel *(m)*
En fire-tube boiler
Es caldera pirotubular *(f)*
Fr chaudière à tube de fumée *(f)*
Pt caldeira de tubo de fumaça *(f)*

caldaia cilindrica *(f)* It
De Schalenkessel *(m)*
En shell boiler
Es caldera de coraza *(f)*
Fr chaudière à paroi *(f)*
Pt caldeira de camisa *(f)*

caldaia radiante *(f)* It
De Strahlungsheizkessel *(m)*
En radiant boiler
Es caldera radiante *(f)*
Fr chaudière radiante *(f)*
Pt convector de radiação *(m)*

caldeado con carbón Es
De kohlenverbrennend
En coal-burning
Fr chauffé au charbon
It scaldato a carbone
Pt aquecido a carvão

caldeira *(f)* *n* Pt
De Kessel *(m)*
En boiler
Es caldera *(f)*
Fr chaudière *(f)*
It caldaia *(f)*

caldeira aquecida com calor residual *(f)* Pt
De Abwärmekessel *(m)*
En waste-heat boiler
Es caldera de calor residual *(f)*
Fr chaudière de récupération des chaleurs perdues *(f)*
It caldaia a recupero di calore perduto *(f)*

caldeira de camisa *(f)* Pt
De Schalenkessel *(m)*
En shell boiler
Es caldera de coraza *(f)*
Fr chaudière à paroi *(f)*
It caldaia cilindrica *(f)*

caldeira de secções *(f)* Pt
De Teilkammerkessel *(m)*
En sectional boiler
Es caldera compartimentada *(f)*
Fr chaudière sectionnelle *(f)*
It caldaia a sezioni *(f)*

caldeira de tubo de fumaça *(f)* Pt
De Heizrohrkessel *(m)*
En fire-tube boiler
Es caldera pirotubulár *(f)*
Fr chaudière à tube de fumée *(f)*
It caldaia a tubo di fumo *(f)*

caldeira de tubos de água *(f)* Pt
De Wasserrohrkessel *(m)*
En water-tube boiler
Es caldera de tubos de agua *(f)*
Fr chaudière tubulaire *(f)*
It caldaia a tubo d'acqua *(f)*

caldeira de uma só passagem *(f)* Pt
De Zwangsdurchlauf-kessel *(m)*
En once-through boiler
Es caldera de proceso directo *(f)*
Fr chaudière sans recyclage *(f)*
It caldaia a processo diretto *(f)*

caldeo dieléctrico *(m)* Es
De kapazitive Hochfrequenzer-wärmung *(f)*
En dielectric heating
Fr chauffage par pertes diélectriques *(m)*
It riscaldamento dielettrico *(m)*
Pt aquecimento dieléctrico *(m)*

caldeo por inducción *(m)* Es
De Induktionsheizung *(f)*
En induction heating
Fr chauffage par induction *(m)*
It riscaldamento ad induzione *(m)*
Pt aquecimento por indução *(m)*

caldera *(f)* *n* Es
De Kessel *(m)*
En boiler
Fr chaudière *(f)*
It caldaia *(f)*
Pt caldeira *(f)*

caldera compartimentada *(f)* Es
De Teilkammerkessel *(m)*
En sectional boiler
Fr chaudière sectionnelle *(f)*
It caldaia a sezioni *(f)*
Pt caldeira de secções *(f)*

caldera de calor residual *(f)* Es
De Abwärmekessel *(m)*
En waste-heat boiler
Fr chaudière de récupération des chaleurs perdues *(f)*
It caldaia a recupero di calore perduto *(f)*
Pt caldeira aquecida com calor residual *(f)*

caldera de coraza *(f)* Es
De Schalenkessel *(m)*
En shell boiler
Fr chaudière à paroi *(f)*
It caldaia cilindrica *(f)*
Pt caldeira de camisa *(f)*

caldera de proceso directo *(f)* Es
De Zwangsdurchlauf-kessel *(m)*
En once-through boiler
Fr chaudière sans recyclage *(f)*
It caldaia a processo diretto *(f)*
Pt caldeira de uma só passagem *(f)*

caldera de tubos de agua *(f)* Es
De Wasserrohrkessel *(m)*
En water-tube boiler
Fr chaudière tubulaire *(f)*
It caldaia a tubo d'acqua *(f)*
Pt caldeira de tubos de água *(f)*

caldera pirotubular *(f)* Es
De Heizrohrkessel *(m)*
En fire-tube boiler
Fr chaudière à tube de fumée *(f)*
It caldaia a tubo di fumo *(f)*
Pt caldeira de tubo de fumaça *(f)*

caldera radiante *(f)* Es
De Strahlungsheizkessel *(m)*
En radiant boiler
Fr chaudière radiante *(f)*
It caldaia radiante *(f)*
Pt convector de radiação *(m)*

calefacción *(f)* *n* Es
De Heizung *(f)*
En heating
Fr chauffage *(m)*
It riscaldamento *(m)*
Pt aquecimento *(m)*

calefacción central *(f)* Es
De Zentralheizung *(f)*
En central heating
Fr chauffage central *(m)*
It riscaldamento centrale *(m)*
Pt aquecimento central *(m)*

calefacción central con petróleo *(f)* Es
De Zentralheizung mit Ölfeuerung *(f)*
En oil-fired central heating
Fr chauffage central au mazout *(m)*

It riscaldamento centrale a nafta *(m)*
Pt aquecimento central a óleo *(m)*

calefacción de distritos *(f)* Es
De Fernheizung *(f)*
En district heating
Fr chauffage urbain *(m)*
It riscaldamento di sezione *(m)*
Pt aquecimento de distrito *(m)*

calefacción de espacios *(f)* Es
De Raumheizung *(f)*
En space heating
Fr chauffage de chambres *(m)*
It riscaldamento locale *(m)*
Pt aquecimento espacial *(m)*

calefacción indirecto *(f)* Es
De indirekte Heizung *(f)*
En indirect heating
Fr chauffage indirect *(m)*
It riscaldamento indiretto *(m)*
Pt aquecimento indirecto *(m)*

calentador *(m) n* Es
De Heizelement *(n)*
En heater
Fr réchauffeur *(m)*
It riscaldatore *(m)*
Pt aquecedor *(m)*

calentador auxiliado por ventilador *(m)* Es
De Heizelement mit Gebläse *(n)*
En fan-assisted heater
Fr radiateur soufflant *(m)*
It riscaldatore a ventilatore *(m)*
Pt aquecedor auxiliado por ventoinha *(m)*

calentador con aire caliente *(m)* Es
De Warmluftheizung *(f)*
En warm-air heater
Fr chauffage à air chaud *(m)*
It riscaldatore ad aria tiepida *(m)*
Pt aquecedor de ar quente *(m)*

calentador de agua *(m)* Es
De Wassererhitzer *(m)*
En water heater
Fr chauffe-eau *(m)*
It riscaldatore di acqua *(m)*
Pt aquecedor hidráulico *(m)*

calentador de agua de acumulación *(m)* Es
De Speicherwasser-heizung *(f)*
En storage water-heater
Fr chauffe-eau à accumulation *(m)*
It riscaldatore di acqua a conservazione *(m)*
Pt aquecedor a água de armazenagem *(m)*

calentador de agua instantáneo *(m)* Es
De Wasser-Schnellerhitzer *(m)*
En instantaneous water heater
Fr chauffe-eau instantané *(m)*
It riscaldatore istantaneo di acqua *(m)*
Pt aquecedor de água instantâneo *(m)*

calentador de aire *(m)* Es
De Lufterhitzer *(m)*
En air heater
Fr réchauffeur d'air *(m)*
It riscaldatore ad aria *(m)*
Pt aquecedor de ar *(m)*

calentador de arco *(m)* Es
De Lichtbogen-Heizelement *(m)*
En arc heater
Fr appareil de chauffage à arc *(m)*
It riscaldatore ad arco *(m)*
Pt aquecedor a arco voltáico *(m)*

calentador de baño *(m)* Es
De Badheizelement *(n)*
En bath heater
Fr chauffe-bain *(m)*
It scaldabagno *(m)*
Pt aquecedor de banho *(m)*

calentador de chimenea abierta *(m)* Es
De Heizelement mit offenem Abzug *(n)*
En open-flue heater
Fr appareil de chauffage à conduit ouvert *(m)*
It riscaldatore a condotto aperto *(m)*
Pt aquecedor de fumeiro aberto *(m)*

calentador de inmersión *(m)* Es
De Tauchsieder *(m)*
En immersion heater
Fr thermoplongeur *(m)*
It riscaldatore ad immersione *(m)*
Pt aquecedor por imersão *(m)*

calentador de palanquillas *(m)* Es
De Blockerwärmer *(m)*
En billet heater
Fr chauffe-billettes *(m)*
It riscaldatore di billette *(m)*
Pt aquecedor de lingotes *(m)*

calentador para almacenamiento térmico *(m)* Es
De Speicherheizung *(f)*
En storage heater
Fr radiateur à accumulation *(m)*
It riscaldatore a conservazione *(m)*
Pt aquecedor de armazenagem *(m)*

calentamiento por radiofrecuencia *(m)* Es
De Radiofrequenz-heizung *(f)*
En radiofrequency heating
Fr chauffage à radiofréquence *(m)*
It riscaldamento a radiofrequenza *(m)*
Pt aquecimento por frequência rádio *(m)*

calidad *(f) n* Es
De Qualität *(f)*
En quality
Fr qualité *(f)*
It qualità *(f)*
Pt qualidade *(f)*

calidad del vapor *(f)* Es
De Dampfqualität *(f)*
En steam quality
Fr qualité de vapeur *(f)*
It qualità del vapore *(f)*
Pt qualidade do vapor *(f)*

caloduc *(m) n* Fr
De Wärmeübertragungs-rohr *(n)*
En heat pipe
Es tubo calefactor *(m)*
It tubo di calore *(m)*
Pt tubo de calor *(m)*

calor *(m) n* Es, Pt
De Wärme *(f)*
En heat
Fr chaleur *(f)*
It calore *(m)*

calor de combustão *(m)* Pt
De Verbrennungswärme *(f)*
En heat of combustion
Es calor de combustión *(m)*
Fr chaleur de combustion *(f)*
It calore di combustione *(m)*

calor de combustión *(m)* Es
De Verbrennungswärme *(f)*
En heat of combustion
Fr chaleur de combustion *(f)*
It calore di combustione *(m)*
Pt calor de combustão *(m)*

calor de decomposição *(m)* Pt
De Zersetzungswärme *(f)*
En decay heat

Es calor de desintegración *(m)*
Fr chaleur de désintégration *(f)*
It calore di disintegrazione *(m)*

calor de desintegración *(m)* Es
De Zersetzungswärme *(f)*
En decay heat
Fr chaleur de désintégration *(f)*
It calore di disintegrazione *(m)*
Pt calor de descomposição *(m)*

calor de fondo *(m)* Es
De Grundheizung *(f)*
En background heat
Fr chaleur ambiante *(f)*
It calore di fondo *(m)*
Pt calor de fundo *(m)*

calor de fundo *(m)* Pt
De Grundheizung *(f)*
En background heat
Es calor de fondo *(m)*
Fr chaleur ambiante *(f)*
It calore di fondo *(m)*

calor de reacção *(m)* Pt
De Reaktionswärme *(f)*
En heat of reaction
Es calor de reacción *(m)*
Fr chaleur de réaction *(f)*
It calore di reazione *(m)*

calor de reacción *(m)* Es
De Reaktionswärme *(f)*
En heat of reaction
Fr chaleur de réaction *(f)*
It calore di reazione *(m)*
Pt calor de reacção *(m)*

calore *(m) n* It
De Wärme *(f)*
En heat
Es calor *(m)*
Fr chaleur *(f)*
Pt calor *(m)*

calore di combustione *(m)* It
De Verbrennungswärme *(f)*
En heat of combustion
Es calor de combustión *(m)*
Fr chaleur de combustion *(f)*
Pt calor de combustão *(m)*

calore di disintegrazione *(m)* It
De Zersetzungswärme *(f)*
En decay heat
Es calor de desintegración *(m)*
Fr chaleur de désintégration *(f)*
Pt calor de descomposição *(m)*

calore di fondo *(m)* It
De Grundheizung *(f)*
En background heat
Es calor de fondo *(m)*
Fr chaleur ambiante *(f)*
Pt calor de fundo *(m)*

calore di reazione *(m)* It
De Reaktionswärme *(f)*
En heat of reaction
Es calor de reacción *(m)*
Fr chaleur de réaction *(f)*
Pt calor de reacção *(m)*

calore latente *(m)* It
De gebundene Wärme *(f)*
En latent heat
Es calor latente *(m)*
Fr chaleur latente *(f)*
Pt calor latente *(m)*

calore latente specifico *(m)* It
De spezifische gebundene Wärme *(f)*
En specific latent heat
Es calor latente específico *(m)*
Fr chaleur latente spécifique *(f)*
Pt calor latente específico *(m)*

calore non disponibile *(m)* It
De nicht verfügbare Wärme *(f)*
En unavailable heat
Es calor no disponible *(m)*
Fr chaleur non disponible *(f)*
Pt calor não disponível *(m)*

calore perduto *(m)* It
De Abwärme *(f)*
En waste heat
Es calor residual *(m)*
Fr chaleur perdue *(f)*
Pt calor residual *(m)*

calor e potência combinados *(m)* Pt
De kombinierte Wärme und Energie *(f)*
En combined heat and power
Es calor y potencia combinados *(m)*
Fr chaleur et puissance combinées *(f)*
It potenza e calore combinati *(f)*

calore sensibile *(m)* It
De spübare Wärme *(f)*
En sensible heat
Es calor sensible *(m)*
Fr chaleur sensible *(f)*
Pt calor sensível *(m)*

calore utile *(m)* It
De Wärmeleistung *(f)*
En useful heat
Es calor útil *(m)*
Fr chaleur utile *(f)*
Pt calor útil *(m)*

calorifero *(m) n* It
De Heizschlange *(f)*
En calorifier
Es calorífero *(m)*
Fr calorifiant *(m)*
Pt calorificador *(m)*

calorífero *(m) n* Es
De Heizschlange *(f)*
En calorifier
Fr calorifiant *(m)*
It calorifero *(m)*
Pt calorificador *(m)*

calorifiant *(m) n* Fr
De Heizschlange *(f)*
En calorifier
Es calorífero *(m)*
It calorifero *(m)*
Pt calorificador *(m)*

calorificador *(m) n* Pt
De Heizschlange *(f)*
En calorifier
Es calorífero *(m)*
Fr calorifiant *(m)*
It calorifero *(m)*

calorific value En
De Heizwert *(m)*
Es potencia calorífica *(f)*
Fr valeur calorifique *(f)*
It potere calorifico *(m)*
Pt valor calorífico *(m)*

calorifier *n* En
De Heizschlange *(f)*
Es calorífero *(m)*
Fr calorifiant *(m)*
It calorifero *(m)*
Pt calorificador *(m)*

calorifugeage de tuyaux *(m) n* Fr
De Rohrummantellung *(f)*
En pipe lagging
Es revestimiento de tuberías *(m)*
It rivestimento isolante di tubazioni *(m)*
Pt revestimento de canos *(m)*

calorimeter *n* En
De Kalorimeter *(n)*
Es calorímetro *(m)*
Fr calorimètre *(m)*
It calorimetro *(m)*
Pt calorímetro *(m)*

calorimètre *(m) n* Fr
De Kalorimeter *(n)*
En calorimeter
Es calorímetro *(m)*
It calorimetro *(m)*
Pt calorímetro *(m)*

calorimètre à godet *(m)* Fr
De Schalenkalorimeter *(n)*
En bucket calorimeter
Es calorímetro de cubeta *(m)*
It calorimetro di secchio *(m)*
Pt calorímetro de balde *(m)*

calorimetria *(f) n* It, Pt
De Wärmemengenmessung *(f)*
En calorimetry
Es calorimetría *(f)*
Fr calorimétrie *(f)*

calorimetría *(f) n* Es
De Wärmemengenmessung *(f)*
En calorimetry
Fr calorimétrie *(f)*
It calorimetria *(f)*
Pt calorimetria *(f)*

calorimétrie *(f) n* Fr
De Wärmemengenmessung *(f)*
En calorimetry
Es calorimetría *(f)*
It calorimetria *(f)*
Pt calorimetria *(f)*

calorimetro *(m) n* It
De Kalorimeter *(n)*
En calorimeter
Es calorímetro *(m)*
Fr calorimètre *(m)*
Pt calorímetro *(m)*

calorímetro *(m) n* Es, Pt
De Kalorimeter *(n)*
En calorimeter
Fr calorimètre *(m)*
It calorimetro *(m)*

calorimetro a bomba *(m)* It
De Bombenkalorimeter *(n)*
En bomb calorimeter
Es calorímetro de bomba *(m)*
Fr bombe calorimétrique *(f)*
Pt calorímetro de bomba *(m)*

calorímetro de balde *(m)* Pt
De Schalenkalorimeter *(n)*
En bucket calorimeter
Es calorímetro de cubeta *(m)*
Fr calorimètre à godet *(m)*
It calorimetro di secchio *(m)*

calorímetro de bomba *(m)* Es, Pt
De Bombenkalorimeter *(n)*
En bomb calorimeter
Fr bombe calorimétrique *(f)*
It calorimetro a bomba *(m)*

calorímetro de cubeta *(m)* Es
De Schalenkalorimeter *(n)*
En bucket calorimeter
Fr calorimètre à godet *(m)*
It calorimetro di secchio *(m)*
Pt calorímetro de balde *(m)*

calorimetro di secchio *(m)* It
De Schalenkalorimeter *(n)*
En bucket calorimeter
Es calorímetro de cubeta *(m)*
Fr calorimètre à godet *(m)*
Pt calorímetro de balde *(m)*

calorimetry *n* En
De Wärmemengenmessung *(f)*
Es calorimetría *(f)*
Fr calorimétrie *(f)*
It calorimetria *(f)*
Pt calorimetria *(f)*

calor latente *(m)* Es, Pt
De gebundene Wärme *(f)*
En latent heat
Fr chaleur latente *(f)*
It calore latente *(m)*

calor latente específico *(m)* Es, Pt
De spezifische gebundene Wärme *(f)*
En specific latent heat
Fr chaleur latente spécifique *(f)*
It calore latente specifico *(m)*

calor latente molar *(m)* Es, Pt
De molare gebundene Wärme *(f)*
En molar latent heat
Fr chaleur latente molaire *(f)*
It calore latente molare *(m)*

calore latente molare *(m)* It
De molare gebundene Wärme *(f)*
En molar latent heat
Es calor latente molar *(m)*
Fr chaleur latente molaire *(f)*
Pt calor latente molar *(m)*

calor não disponível *(m)* Pt
De nicht verfügbare Wärme *(f)*
En unavailable heat
Es calor no disponible *(m)*
Fr chaleur non disponible *(f)*
It calore non disponibile *(m)*

calor no disponible *(m)* Es
De nicht verfügbare Wärme *(f)*
En unavailable heat
Fr chaleur non disponible *(f)*
It calore non disponibile *(m)*
Pt calor não disponível *(m)*

calor residual *(m)* Es, Pt
De Abwärme *(f)*
En waste heat
Fr chaleur perdue *(f)*
It calore perduto *(m)*

calor sensible *(m)* Es
De spürbare Wärme *(f)*
En sensible heat
Fr chaleur sensible *(f)*
It calore sensibile *(m)*
Pt calor sensível *(m)*

calor sensível *(m)* Pt
De spürbare Wärme *(f)*
En sensible heat
Es calor sensible *(m)*
Fr chaleur sensible *(f)*
It calore sensibile *(m)*

calor útil *(m)* Es, Pt
De Wärmeleistung *(f)*
En useful heat
Fr chaleur utile *(f)*
It calore utile *(m)*

calor y potencia combinados *(m)* Es
De kombinierte Wärme und Energie *(f)*
En combined heat and power
Fr chaleur et puissance combinées *(f)*
It potenza e calore combinati *(f)*
Pt calor e potência combinados *(m)*

camada de carvão *(f)* Pt
De Kohlenflöz *(n)*
En coal seam
Es capa de carbón *(f)*
Fr filon houiller *(m)*
It filone di carbone *(m)*

câmara de combustão *(f)* Pt
De Brennkammer *(f)*
En combustion chamber
Es cámara de combustión *(f)*
Fr chambre de combustion *(f)*
It camera di combustione *(f)*

cámara de combustión *(f)* Es
De Brennkammer *(f)*
En combustion chamber
Fr chambre de combustion *(f)*
It camera di combustione *(f)*
Pt câmara de combustão *(f)*

camera di combustione *(f)* It
De Brennkammer *(f)*
En combustion chamber
Es cámara de combustión *(f)*
Fr chambre de combustion *(f)*
Pt câmara de combustão *(f)*

camera di manovra *(f)* It
De Schaltraum *(m)*
En control room
Es sala de control *(f)*
Fr salle de commande *(f)*
Pt sala de comando *(f)*

camino *(m) n* It
De Schornstein *(m)*
En chimney
Es chimenea *(f)*
Fr cheminée *(f)*
Pt chaminé *(f)*

campo *(m) n* Es, It, Pt
De Feld *(n)*
En field
Fr champ *(m)*

campo de gravidade *(m)* Pt
De Gravitationsfeld *(n)*
En gravitational field
Es campo gravitacional *(m)*
Fr champ de gravitation *(m)*
It campo gravitazionale *(m)*

campo eléctrico *(m)* Es, Pt
De elektrisches Feld *(n)*
En electric field
Fr champ électrique *(m)*
It campo elettrico *(m)*

campo electromagnético *(m)* Es, Pt
De elektromagnetisches Feld *(n)*
En electromagnetic field
Fr champ électromagnétique *(m)*
It campo elettromagnetico *(m)*

campo elettrico *(m)* It
De elektrisches Feld *(n)*
En electric field
Es campo eléctrico *(m)*
Fr champ électrique *(m)*
Pt campo eléctrico *(m)*

campo elettromagnetico *(m)* It
De elektromagnetisches Feld *(n)*
En electromagnetic field
Es campo electromagnético *(m)*
Fr champ électromagnétique *(m)*
Pt campo electromagnético *(m)*

campo gravitacional *(m)* Es
De Gravitationsfeld *(n)*
En gravitational field
Fr champ de gravitation *(m)*
It campo gravitazionale *(m)*
Pt campo de gravidade *(m)*

campo gravitazionale *(m)* It
De Gravitationsfeld *(n)*
En gravitational field
Es campo gravitacional *(m)*
Fr champ de gravitation *(m)*
Pt campo de gravidade *(m)*

campo magnetico *(m)* It
De Magnetfeld *(n)*
En magnetic field
Es campo magnético *(m)*
Fr champ magnétique *(m)*
Pt campo magnético *(m)*

campo magnético *(m)* Es, Pt
De Magnetfeld *(n)*
En magnetic field
Fr champ magnétique *(m)*
It campo magnetico *(m)*

canal de combustible *(m)* Es, Fr
De Brennstoffkanal *(m)*
En fuel channel
It canale del combustibile *(m)*
Pt canal de combustível *(m)*

canal de combustível *(m)* Pt
De Brennstoffkanal *(m)*
En fuel channel
Es canal de combustible *(m)*
Fr canal de combustible *(m)*
It canale del combustibile *(m)*

canal de descarga *(m)* Es
De Ablauf *(m)*
En tailrace
Fr bief *(m)*
It canale di scarico *(m)*
Pt água de descarga *(f)*

canale del combustibile *(m)* It
De Brennstoffkanal *(m)*
En fuel channel
Es canal de combustible *(m)*
Fr canal de combustible *(m)*
Pt canal de combustível *(m)*

canale di scarico *(m)* It
De Ablauf *(m)*
En tailrace
Es canal de descarga *(m)*
Fr bief *(m)*
Pt água de descarga *(f)*

canalización circular *(f)* Es
De Ringleitung *(f)*
En ring main
Fr réseau secteur *(m)*
It conduttori ad anello *(m)*
Pt condutor em anel fechado *(m)*

canel *(m) n* Es
De Mattkohle *(f)*
En cannel coal
Fr houille grasse *(f)*
It carbone a lunga fiamma *(m)*
Pt carvão cannel *(m)*

cannel coal En
De Mattkohle *(f)*
Es canel *(m)*
Fr houille grasse *(f)*
It carbone a lunga fiamma *(m)*
Pt carvão cannel *(m)*

cantidad escalar *(f)* Es
De Skalarquantität *(f)*
En scalar quantity
Fr quantité scalaire *(f)*
It grandezza scalare *(f)*
Pt quantidade escalar *(f)*

cantidad vectorial *(f)* Es
De Vektorquantität *(f)*
En vector quantity
Fr quantité vectorielle *(f)*
It quantità di vettore *(f)*
Pt quantidade de vector *(f)*

capacidad de calor específico *(f)* Es
De spezifische Wärmekapazität *(f)*
En specific heat capacity
Fr capacité calorifique spécifique *(f)*
It capacità termica specifica *(f)*
Pt capacidade de calor específico *(f)*

capacidad de calor molar *(f)* Es
De molare Wärmekapazität *(f)*
En molar heat capacity
Fr capacité de chaleur molaire *(f)*
It capacità termica molare *(f)*
Pt termo-capacidade molar *(f)*

capacidade de calor específico *(f)* Pt
De spezifische Wärmekapazität *(f)*
En specific heat capacity
Es capacidad de calor específico *(f)*
Fr capacité calorifique spécifique *(f)*
It capacità termica specifica *(f)*

capacidade geradora *(f)* Pt
De Erzeugungskapazität *(f)*
En generating capacity
Es capacidad generadora *(f)*
Fr capacité génératrice *(f)*
It capacità di generazione *(f)*

capacidade geradora global *(f)* Pt
De Gesamterzeugungs-kapazität *(f)*
En overall generating capacity
Es capacidad generadora total *(f)*
Fr capacité génératrice globale *(f)*
It capacità totale di generazione *(f)*

capacidade térmica *(f)* Pt
De Wärmekapazität *(f)*
En heat capacity
Es capacidad térmica *(f)*
Fr capacité calorifique *(f)*
It capacità termica *(f)*

capacidad generadora *(f)* Es
De Erzeugungskapazität *(f)*
En generating capacity
Fr capacité génératrice *(f)*
It capacità di generazione *(f)*
Pt capacidade geradora *(f)*

capacidad generadora total *(f)* Es
De Gesamterzeugungs-kapazität *(f)*
En overall generating capacity
Fr capacité génératrice globale *(f)*
It capacità totale di generazione *(f)*
Pt capacidade geradora global *(f)*

capacidad térmica *(f)* Es
De Wärmekapazität *(f)*
En heat capacity
Fr capacité calorifique *(f)*
It capacità termica *(f)*
Pt capacidade térmica *(f)*

capacità di generazione *(f)* It
De Erzeugungskapazität *(f)*
En generating capacity
Es capacidad generadora *(f)*
Fr capacité génératrice *(f)*
Pt capacidade geradora *(f)*

capacitance *n* En; Fr *(f)*
De Kapazitanz *(f)*
Es capacitancia *(f)*
It capacitanza *(f)*
Pt capacitância *(f)*

capacitancia *(f) n* Es
De Kapazitanz *(f)*
En capacitance
Fr capacitance *(f)*
It capacitanza *(f)*
Pt capacitância *(f)*

capacitância *(f) n* Pt
De Kapazitanz *(f)*
En capacitance
Es capacitancia *(f)*
Fr capacitance *(f)*
It capacitanza *(f)*

capacitanza *(f) n* It
De Kapazitanz *(f)*
En capacitance
Es capacitancia *(f)*
Fr capacitance *(f)*
Pt capacitância *(f)*

capacità termica *(f)* It
De Wärmekapazität *(f)*
En heat capacity
Es capacidad térmica *(f)*
Fr capacité calorifique *(f)*
Pt capacidade térmica *(f)*

capacità termica molare *(f)* It
De molare Wärmekapazität *(f)*
En molar heat capacity
Es capacidad de calor molar *(f)*
Fr capacité de chaleur molaire *(f)*
Pt termo-capacidade molar *(f)*

capacità termica specifica *(f)* It
De spezifische Wärmekapazität *(f)*
En specific heat capacity
Es capacidad de calor específico *(f)*
Fr capacité calorifique spécifique *(f)*
Pt capacidade de calor específico *(f)*

capacità totale di generazione *(f)* It
De Gesamterzeugungs-kapazität *(f)*
En overall generating capacity
Es capacidad generadora total *(f)*
Fr capacité génératrice globale *(f)*
Pt capacidade geradora global *(f)*

capacité calorifique *(f)* Fr
De Wärmekapazität *(f)*
En heat capacity
Es capacidad térmica *(f)*
It capacità termica *(f)*
Pt capacidade térmica *(f)*

capacité calorifique spécifique *(f)* Fr
De spezifische Wärmekapazität *(f)*
En specific heat capacity
Es capacidad de calor específico *(f)*
It capacità termica specifica *(f)*
Pt capacidade de calor específico *(f)*

capacité de chaleur molaire *(f)* Fr
De molare Wärmekapazität *(f)*
En molar heat capacity
Es capacidad de calor molar *(f)*
It capacità termica molare *(f)*
Pt termo-capacidade molar *(f)*

capacité génératrice *(f)* Fr
De Erzeugungskapazität *(f)*
En generating capacity
Es capacidad generadora *(f)*
It capacità di generazione *(f)*
Pt capacidade geradora *(f)*

capacité génératrice globale *(f)* Fr
De Gesamterzeugungs-kapazität *(f)*
En overall generating capacity
Es capacidad generadora total *(f)*
It capacità totale di generazione *(f)*
Pt capacidade geradora global *(f)*

capacitor *n* En; Pt *(m)*
De Kondensator (elektrischer) *(m)*
Es condensador (eléctrico) *(m)*
Fr condensateur *(m)*
It condensatore (elettrico) *(m)*

capa de carbón *(f)* Es
De Kohlenflöz *(n)*
En coal seam
Fr filon houiller *(m)*
It filone di carbone *(m)*
Pt camada de carvão *(f)*

capitalisation *(f) n* Fr
De Kapitalausstattung *(f)*
En capitalization
Es capitalización *(f)*
It capitalizzazione *(f)*
Pt capitalização *(f)*

capitalização *(f) n* Pt
De Kapitalausstattung *(f)*
En capitalization
Es capitalización *(f)*
Fr capitalisation *(f)*
It capitalizzazione *(f)*

capitalización *(f) n* Es
De Kapitalausstattung *(f)*
En capitalization
Fr capitalisation *(f)*
It capitalizzazione *(f)*
Pt capitalização *(f)*

capitalization *n* En
De Kapitalausstattung *(f)*
Es capitalización *(f)*
Fr capitalisation *(f)*
It capitalizzazione *(f)*
Pt capitalização *(f)*

capitalizzazione *(f) n* It
De Kapitalausstattung *(f)*
En capitalization
Es capitalización *(f)*
Fr capitalisation *(f)*
Pt capitalização *(f)*

cap rock En
De Deckgestein *(n)*
Es estrato impermeable de cobertura *(m)*
Fr roche couverture *(f)*
It strato impermeabile di copertura *(m)*
Pt rocha encaixante *(f)*

características de descarga *(f pl)* Es, Pt
De Entladungsmerkmale *(pl)*
En discharge characteristics
Fr caractéristiques de décharge *(f pl)*
It caratteristiche di scarica *(f pl)*

caractéristiques de décharge *(f pl)* Fr
De Entladungsmerkmale *(pl)*
En discharge characteristics
Es características de descarga *(f pl)*
It caratteristiche di scarica *(f pl)*
Pt características de descarga *(f pl)*

caratteristiche di scarica *(f pl)* It
De Entladungsmerkmale *(pl)*
En discharge characteristics
Es características de descarga *(f pl)*
Fr caractéristiques de décharge *(f pl)*
Pt características de descarga *(f pl)*

carbón *(m) n* Es
De Kohle *(f)*
En coal
Fr charbon *(m)*
It carbone *(m)*
Pt carvão *(m)*

carbonaceous fuel En
De kohlenstoffhaltiger Brennstoff *(m)*
Es combustible carbonoso *(m)*
Fr combustible charbonneux *(m)*
It combustibile carbonioso *(m)*
Pt combustível carbonoso *(m)*

carbón bituminoso *(m)* Es
De bituminöse Kohle *(f)*
En bituminous coal
Fr charbon bitumineux *(m)*
It carbone bituminoso *(m)*
Pt carvão betuminoso *(m)*

carbon black En
De Rußschwarz *(n)*
Es negro de carbón *(m)*
Fr noir de carbone *(m)*
It nerofumo di gas *(m)*
Pt negro de carvão *(m)*

carbon content En
De Kohlenstoffgehalt *(m)*
Es contenido de carbono *(m)*
Fr teneur en carbone *(f)*
It contenuto di carbonio *(m)*
Pt teor de carbono *(m)*

carbón de coquización *(m)* Es
De Kokskohle *(f)*
En coking coal
Fr charbon cokéfiant *(m)*
It carbone cokificante *(m)*
Pt carvão de coqueificação *(m)*

carbon dioxide En
De Kohlendioxid *(n)*
Es anhídrido carbónico *(m)*
Fr anhydride carbonique *(m)*
It anidride carbonica *(f)*
Pt anidrido carbónico *(m)*

carbón duro *(m)* Es
De Steinkohle *(f)*
En hard coal
Fr charbon dure *(m)*
It carbone duro *(m)*
Pt carvão duro *(m)*

carbone *(m) n* It
De Kohle *(f)*
En coal
Es carbón *(m)*
Fr charbon *(m)*
Pt carvão *(m)*

carbone agglutinante *(m)* It
De Backkohle *(f)*
En caking coal
Es hulla grasa *(f)*
Fr houille collante *(f)*
Pt hulha gorda *(f)*

carbone a lunga fiamma *(m)* It
De Mattkohle *(f)*
En cannel coal
Es canel *(m)*
Fr houille grasse *(f)*
Pt carvão cannel *(m)*

carbone bituminoso *(m)* It
De bituminöse Kohle *(f)*
En bituminous coal
Es carbón bituminoso *(m)*
Fr charbon bitumineux *(m)*
Pt carvão betuminoso *(m)*

carbone cokificante *(m)* It
De Kokskohle *(f)*
En coking coal
Es carbón de coquización *(m)*
Fr charbon cokéfiant *(m)*
Pt carvão de coqueificação *(m)*

carbone di pezzatura noce *(m)* It
De Nußkohle *(f)*
En nut coal
Es almendrilla *(f)*
Fr noix de charbon *(f)*
Pt carvão em nozes *(m)*

carbone dolce *(m)* It
De Holzkohle *(f)*
En charcoal
Es carbón vegetal *(m)*
Fr charbon de bois *(m)*
Pt carvão vegital *(m)*

carbone duro *(m)* It
De Steinkohle *(f)*
En hard coal
Es carbón duro *(m)*
Fr charbon dure *(m)*
Pt carvão duro *(m)*

carbone fisso *(m)* It
De nicht flüchtiger Kohlenstoff *(m)*
En fixed carbon
Es carbón fijo *(m)*
Fr carbone fixe *(m)*
Pt carvão fixo *(m)*

carbone fixe *(m)* Fr
De nicht flüchtiger Kohlenstoff *(m)*
En fixed carbon
Es carbón fijo *(m)*
It carbone fisso *(m)*
Pt carvão fixo *(m)*

carbone metabituminoso *(m)* It
De metabituminöse Kohle *(f)*
En metabituminous coal
Es carbón metabituminoso *(m)*
Fr charbon métabitumineux *(m)*
Pt carvão metabetuminoso *(m)*

carbone nero *(m)* It
De Schwarzkohle *(f)*
En black coal
Es carbón negro *(m)*
Fr houille *(f)*
Pt carvão negro *(m)*

carbone non agglutinante *(m)* It
De nicht anbackende Kohle *(f)*
En noncaking coal
Es carbón inaglutinable *(m)*
Fr houille maigre *(f)*
Pt carvaõ não aglutinante *(m)*

carbone polverizzato *(m)* It
De pulverisierte Kohle *(f)*
En pulverized coal
Es carbón pulverizado *(m)*
Fr charbon pulvérisé *(m)*
Pt carvão pulverizado *(m)*

carbón equivalente *(m)* Es
De Kohlenersatz *(m)*
En coal equivalent
Fr équivalent charbon *(m)*
It equivalente del carbone *(m)*
Pt equivalente carbónico *(m)*

carbone sapropelico *(m)* It
De Sapropelkohle *(f)*
En sapropelic coal
Es carbón sapropélico *(m)*
Fr charbon sapropélique *(m)*
Pt carvão sapropélico *(m)*

carbone semibituminoso *(m)* It
De halbbituminöse Kohle *(f)*
En semibituminous coal
Es carbón semibituminoso *(m)*
Fr houille demi-grasse *(f)*
Pt carvão semibetuminoso *(m)*

carbone senza fumo *(m)* It
De rauchlose Kohle *(f)*
En smokeless coal
Es carbón fumífugo *(m)*
Fr charbon sans fumée *(m)*
Pt carvão ardendo sem produzir fumo *(m)*

carbones no aglomerantes *(m pl)* Es
De nicht klumpende Kohlen *(pl)*
En nonagglomerating coals
Fr charbons non-agglomérants *(m pl)*
It carboni non agglomeranti *(m pl)*
Pt carvões não-aglomerantes *(m pl)*

carbone tout-venant *(m)* It
De Normalkohle *(f)*
En run-of-mine coal
Es carbón todouno *(m)*
Fr charbon tout-venant *(m)*
Pt carvão tal como sai da mina *(m)*

carbone umico *(m)* It
De Humuskohle *(f)*
En humic coal
Es carbón húmico *(m)*
Fr charbon humique *(m)*
Pt carvão humico *(m)*

carbón fijo *(m)* Es
De nicht flüchtiger Kohlenstoff *(m)*
En fixed carbon
Fr carbone fixe *(m)*
It carbone fisso *(m)*
Pt carvão fixo *(m)*

carbón fumífugo *(m)* Es
De rauchlose Kohle *(f)*
En smokeless coal
Fr charbon sans fumée *(m)*
It carbone senza fumo *(m)*
Pt carvão ardendo sem produzir fumo *(m)*

carbón húmico *(m)* Es
De Humuskohle *(f)*
En humic coal
Fr charbon humique *(m)*
It carbone umico *(m)*
Pt carvão humico *(m)*

carbonifère *adj* Fr
De kohlenhaltig
En carboniferous
Es carbonífero
It carbonifero
Pt carbonífero

carbonifero *adj* It
De kohlenhaltig
En carboniferous
Es carbonífero
Fr carbonifère
Pt carbonífero

carbonífero *adj* Es, Pt
De kohlenhaltig
En carboniferous
Fr carbonifère
It carbonifero

carboniferous *adj* En
De kohlenhaltig
Es carbonífero
Fr carbonifère
It carbonifero
Pt carbonífero

carbón inaglutinable *(m)* Es
De nicht anbackende Kohle *(f)*
En noncaking coal
Fr houille maigre *(f)*
It carbone non agglutinante *(m)*
Pt carvaõ não aglutinante *(m)*

carboni non agglomeranti *(m pl)* It
De nicht klumpende Kohlen *(pl)*
En nonagglomerating coals
Es carbones no aglomerantes *(m pl)*
Fr charbons non-agglomérants *(m pl)*
Pt carvões não-aglomerantes *(m pl)*

carbonisation *(f) n* Fr
De Karbonisation *(f)*
En carbonization
Es carbonización *(f)*
It carbonizzazione *(f)*
Pt carbonização *(f)*

carbonização *(f) n* Pt
De Karbonisation *(f)*
En carbonization
Es carbonización *(f)*
Fr carbonisation *(f)*
It carbonizzazione *(f)*

carbonización *(f) n* Es
De Karbonisation *(f)*
En carbonization
Fr carbonisation *(f)*
It carbonizzazione *(f)*
Pt carbonização *(f)*

carbonization *n* En
De Karbonisation *(f)*
Es carbonización *(f)*
Fr carbonisation *(f)*
It carbonizzazione *(f)*
Pt carbonização *(f)*

carbonizzazione *(f) n* It
De Karbonisation *(f)*
En carbonization
Es carbonización *(f)*
Fr carbonisation *(f)*
Pt carbonização *(f)*

carbón metabituminoso *(m)* Es
De metabituminöse kohle *(f)*
En metabituminous coal
Fr charbon métabitumineux *(m)*
It carbone metabituminoso *(m)*
Pt carvão metabetuminoso *(m)*

carbon monoxide En
De Kohlenmonoxid *(n)*
Es óxido de carbono *(m)*
Fr oxyde de carbone *(m)*
It ossido di carbonio *(m)*
Pt óxido de carbono *(m)*

carbón negro *(m)* Es
De Schwarzkohle *(f)*
En black coal
Fr houille *(f)*
It carbone nero *(m)*
Pt carvão negro *(m)*

carbón pulverizado *(m)* Es
De pulverisierte Kohle *(f)*
En pulverized coal
Fr charbon pulvérisé *(m)*
It carbone polverizzato *(m)*
Pt carvão pulverizado *(m)*

carbon refractory En
De kohlenstoffhaltiger feuerfester Stoff *(m)*
Es refractario de carbón *(m)*

Fr réfractaire au carbone *(m)*
It refrattario al carbonio *(m)*
Pt refractário ao carbono *(m)*

carbon residue En
De Koksrückstand *(m)*
Es residuo de carbón *(m)*
Fr résidu de carbone *(m)*
It residuo di carbonio *(m)*
Pt resíduo de carbono *(m)*

carbón sapropélico *(m)* Es
De Sapropelkohle *(f)*
En sapropelic coal
Fr charbon sapropélique *(m)*
It carbone sapropelico *(m)*
Pt carvão sapropélico *(m)*

carbón semibituminoso *(m)* Es
De halbbituminöse Kohle *(f)*
En semibituminous coal
Fr houille demi-grasse *(f)*
It carbone semibituminoso *(m)*
Pt carvão semibetuminoso *(m)*

carbón todouno *(m)* Es
De Normalkohle *(f)*
En run-of-mine coal
Fr charbon tout-venant *(m)*
It carbone tout-venant *(m)*
Pt carvão tal como sai da mina *(m)*

carbón vegetal *(m)* Es
De Holzkohle *(f)*
En charcoal
Fr charbon de bois *(m)*
It carbone dolce *(m)*
Pt carvão vegital *(m)*

carburador *(m) n* Es, Pt
De Vergaser *(n)*
En carburettor
Fr carburateur *(m)*
It carburatore *(m)*

carburante per missili *(m)* It
De Raketentreibstoff *(m)*
En rocket fuel
Es combustible para cohetes *(m)*
Fr combustible pour fusées *(m)*
Pt combustível para foguetões *(m)*

carburateur *(m) n* Fr
De Vergaser *(n)*
En carburettor
Es carburador *(m)*
It carburatore *(m)*
Pt carburador *(m)*

carburation aviation *(m)* Fr
De Flugkraftstoff *(m)*
En aviation fuel
Es gasolina de aviación *(f)*
It combustibile per aviazione *(m)*
Pt combustível de aviação *(m)*

carburatore *(m) n* It
De Vergaser *(n)*
En carburettor
Es carburador *(m)*
Fr carburateur *(m)*
Pt carburador *(m)*

carburéacteur *(m) n* Fr
De Düsentreibstoff *(m)*
En jet fuel
Es combustible de propulsión a chorro *(m)*
It combustibile per aviogetti *(m)*
Pt combustível de jactos *(m)*

carburettor *n* En
De Vergaser *(n)*
Es carburador *(m)*
Fr carburateur *(m)*
It carburatore *(m)*
Pt carburador *(m)*

carga (eléctrica) *(f) n* Es, Pt
De Ladung (elektrische) *(f)*
En charge (electrical)
Fr charge (électrique) *(f)*
It carica (elettrica) *(f)*

carga (mecánica) *(f) n* Es, Pt
De Last (mechanische) *(f)*
En load (mechanical)
Fr charge (méchanique) *(f)*
It carico (meccanico) *(m)*

carga base *(f)* Es
De Grundkraft *(f)*
En base load
Fr charge minimale *(f)*
It carico base *(m)*
Pt carga fundamental *(f)*

carga de ponta *(f)* Pt
De Spitzenbelastung *(f)*
En peak load
Es carga de punta *(f)*
Fr charge de pointe *(f)*
It carico di punta *(m)*

carga de punta *(f)* Es
De Spitzenbelastung *(f)*
En peak load
Fr charge de pointe *(f)*
It carico di punta *(m)*
Pt carga de ponta *(f)*

cargador de batería *(m)* Es
De Batterieladegerät *(n)*
En battery charger
Fr chargeur de batterie *(m)*
It caricabatteria *(f)*
Pt carregador de bateria *(m)*

carga fundamental *(f)* Pt
De Grundkraft *(f)*
En base load
Es carga base *(f)*
Fr charge minimale *(f)*
It carico base *(m)*

carica (elettrica) *(f) n* It
De Charge (elektrische) *(f)*
En charge (electrical)
Es carga (eléctrica) *(f)*
Fr charge (électrique) *(f)*
Pt carga (eléctrica) *(f)*

caricabatteria *(f) n* It
De Batterieladegerät *(n)*
En battery charger
Es cargador de batería *(m)*
Fr chargeur de batterie *(m)*
Pt carregador de bateria *(m)*

caricamento ritardato *(m)* It
De verzögerte Verkokung *(f)*
En delayed coking
Es coquización retardada *(f)*
Fr cokéification différée *(f)*
Pt coqueação retardada *(f)*

carico (meccanico) *(m) n* It
De Last (mechanische) *(f)*
En load (mechanical)
Es carga (mecánica) *(f)*
Fr charge (méchanique) *(f)*
Pt carga (mecânica) *(f)*

carico base *(m)* It
De Grundkraft *(f)*
En base load
Es carga base *(f)*
Fr charge minimale *(f)*
Pt carga fundamental *(f)*

carico di punta *(m)* It
De Spitzenbelastung *(f)*
En peak load
Es carga de punta *(f)*
Fr charge de pointe *(f)*
Pt carga de ponta *(f)*

carneau *(m) n* Fr
De Abzug *(m)*
En flue
Es humero *(m)*
It condotto *(m)*
Pt fumeiro *(m)*

carneaux multiples *(m pl)* Fr
De Sammel-Abgasleitungen *(pl)*
En multiple flues *(pl)*
Es conductos de humo múltiples *(m pl)*
It condotti multipli *(m pl)*

Pt fumeiros múltiplos *(m pl)*

Carnot cycle En
De Carnotscher Kreisprozeβ *(m)*
Es ciclo de Carnot *(m)*
Fr cycle de Carnot *(m)*
It ciclo di Carnot *(m)*
Pt ciclo de Carnot *(m)*

Carnotscher Kreisprozeβ *(m)* De
En Carnot cycle
Es ciclo de Carnot *(m)*
Fr cycle de Carnot *(m)*
It ciclo di Carnot *(m)*
Pt ciclo de Carnot *(m)*

carregador de bateria *(m)* Pt
De Batterieladegerät *(n)*
En battery chargor
Es cargador de batería *(m)*
Fr chargeur de batterie *(m)*
It caricabatteria *(f)*

cartel de petróleo *(m)* Es, Pt
De Ölkartell *(n)*
En oil cartel
Fr cartel pétrolier *(m)*
It cartello del petrolio *(m)*

cartello del petrolio *(m)* It
De Ölkartell *(n)*
En oil cartel
Es cartel de petróleo *(m)*
Fr cartel pétrolier *(m)*
Pt cartel de petróleo *(m)*

cartel pétrolier *(m)* Fr
De Ölkartell *(n)*
En oil cartel
Es cartel de petróleo *(m)*
It cartello del petrolio *(m)*
Pt cartel de petróleo *(m)*

carvão *(m) n* Pt
De Kohle *(f)*
En coal
Es carbón *(m)*
Fr charbon *(m)*
It carbone *(m)*

carvão ardendo sem produzir fumo *(m)* Pt
De rauchlose Kohle *(f)*
En smokeless coal
Es carbón fumífugo *(m)*
Fr charbon sans fumée *(m)*
It carbone senza fumo *(m)*

carvão betuminoso *(m)* Pt
De bituminöse Kohle *(f)*
En bituminous coal
Es carbón bituminoso *(m)*
Fr charbon bitumineux *(m)*
It carbone bituminoso *(m)*

carvão cannel *(m)* Pt
De Mattkohle *(f)*
En cannel coal
Es canel *(m)*
Fr houille grasse *(f)*
It carbone a lunga fiamma *(m)*

carvão castanho *(m)* Pt
De Braunkohle *(f)*; Lignit *(n)*
En brown coal; lignite
Es lignito *(m)*
Fr lignite *(m)*
It lignite *(m)*

carvão de coqueificação *(m)* Pt
De Kokskohle *(f)*
En coking coal
Es carbón de coquización *(m)*
Fr charbon cokéfiant *(m)*
It carbone cokificante *(m)*

carvão duro *(m)* Pt
De Steinkohle *(f)*
En hard coal
Es carbón duro *(m)*
Fr charbon dure *(m)*
It carbone duro *(m)*

carvão em nozes *(m)* Pt
De Nuβkohle *(f)*
En nut coal
Es almendrilla *(f)*
Fr noix de charbon *(f)*
It carbone di pezzatura noce *(m)*

carvão fixo *(m)* Pt
De nicht flüchtiger Kohlenstoff *(m)*
En fixed carbon
Es carbón fijo *(m)*
Fr carbone fixe *(m)*
It carbone fisso *(m)*

carvão humico *(m)* Pt
De Humuskohle *(f)*
En humic coal
Es carbón húmico *(m)*
Fr charbon humique *(m)*
It carbone umico *(m)*

carvão metabetuminoso *(m)* Pt
De metabituminöse Kohle *(f)*
En metabituminous coal
Es carbón metabituminoso *(m)*
Fr charbon métabitumineux *(m)*
It carbone metabituminoso *(m)*

carvaõ não aglutinante *(m)* Pt
De nicht anbackende Kohle *(f)*
En noncaking coal
Es carbón inaglutinable *(m)*
Fr houille maigre *(f)*
It carbone non agglutinante *(m)*

carvão negro *(m)* Pt
De Schwarzkohle *(f)*
En black coal
Es carbón negro *(m)*
Fr houille *(f)*
It carbone nero *(m)*

carvão pulverizado *(m)* Pt
De pulverisierte Kohle *(f)*
En pulverized coal
Es carbón pulverizado *(m)*
Fr charbon pulvérisé *(m)*
It carbone polverizzato *(m)*

carvão sapropélico *(m)* Pt
De Sapropelkohle *(f)*
En sapropelic coal
Es carbón sapropélico *(m)*
Fr charbon sapropélique *(m)*
It carbone sapropelico *(m)*

carvão semibetuminoso *(m)* Pt
De halbbituminöse Kohle *(f)*
En semibituminous coal
Es carbón semibituminoso *(m)*
Fr houille demi-grasse *(f)*
It carbone semibituminoso *(m)*

carvão tal como sai da mina *(m)* Pt
De Normalkohle *(f)*
En run-of-mine coal
Es carbón todouno *(m)*
Fr charbon tout-venant *(m)*
It carbone tout-venant *(m)*

carvão vegital *(m)* Pt
De Holzkohle *(f)*
En charcoal
Es carbón vegetal *(m)*
Fr charbon de bois *(m)*
It carbone dolce *(m)*

carvões não-aglomerantes *(m pl)* Pt
De nicht klumpende Kohlen *(pl)*
En nonagglomerating coals
Es carbones no aglomerantes *(m pl)*
Fr charbons non-agglomérants *(m pl)*
It carboni non agglomeranti *(m pl)*

cascade cycle En
De Kaskadenprozeβ *(m)*
Es ciclo en cascada *(m)*
Fr cycle en cascade *(m)*
It ciclo di cascata *(m)*
Pt ciclo de cascada *(m)*

catalisador *(m) n* Pt
De Katalysator *(m)*
En catalyst

Es catalizador *(m)*
Fr catalyseur *(m)*
It catalizzatore *(m)*

catálise *(f) n* Pt
De Katalyse *(f)*
En catalysis
Es catálisis *(f)*
Fr catalyse *(f)*
It catalisi *(f)*

catalisi *(f) n* It
De Katalyse *(f)*
En catalysis
Es catálisis *(f)*
Fr catalyse *(f)*
Pt catálise *(f)*

catálisis *(f) n* Es
De Katalyse *(f)*
En catalysis
Fr catalyse *(f)*
It catalisi *(f)*
Pt catálise *(f)*

catalizador *(m) n* Es
De Katalysator *(m)*
En catalyst
Fr catalyseur *(m)*
It catalizzatore *(m)*
Pt catalisador *(m)*

catalizzatore *(m) n* It
De Katalysator *(m)*
En catalyst
Es catalizador *(m)*
Fr catalyseur *(m)*
Pt catalisador *(m)*

catalyse *(f) n* Fr
De Katalyse *(f)*
En catalysis
Es catálisis *(f)*
It catalisi *(f)*
Pt catálise *(f)*

catalyseur *(m) n* Fr
De Katalysator *(m)*
En catalyst
Es catalizador *(m)*
It catalizzatore *(m)*
Pt catalisador *(m)*

catalysis *n* En
De Katalyse *(f)*
Es catálisis *(f)*
Fr catalyse *(f)*
It catalisi *(f)*
Pt catálise *(f)*

catalyst *n* En
De Katalysator *(m)*
Es catalizador *(m)*
Fr catalyseur *(m)*
It catalizzatore *(m)*
Pt catalisador *(m)*

catalytic combustion En
De katalytische Verbrennung *(f)*
Es combustión catalítica *(f)*
Fr combustion catalytique *(f)*
It combustione per catalisi *(f)*
Pt combustão catalítica *(f)*

catalytic cracking En
De katalytisches Kracken *(n)*
Es crácking catalítico *(m)*
Fr craquage catalytique *(m)*
It piroscissione per catalisi *(f)*
Pt cracking catalítico *(m)*

catalytic incineration En
De katalytische Veraschung *(f)*
Es incineración catalítica *(f)*
Fr incinération catalytique *(f)*
It incenerimento catalitico *(m)*
Pt incineração catalítica *(f)*

catalytic reforming En
De katalytische Reformierung *(f)*
Es reformación catalítica *(f)*
Fr reformage catalytique *(m)*
It riforma catalitica *(f)*
Pt reforma catalítica *(f)*

cathode *n* En
De Kathode *(f)*
Es cátodo *(m)*
Fr cathode *(f)*
It catodo *(m)*
Pt cátodo *(m)*

cathode *(f) n* Fr
De Kathode *(f)*
En cathode
Es cátodo *(m)*
It catodo *(m)*
Pt cátodo *(m)*

cathode-ray tube En
De Kathodenstrahlröhre *(f)*
Es tubo de rayos catódicos *(m)*
Fr tube à rayons cathodiques *(m)*
It tubo a raggi catodici *(m)*
Pt válvula de raios catódicos *(f)*

catodo *(m) n* It
De Kathode *(f)*
En cathode
Es cátodo *(m)*
Fr cathode *(f)*
Pt cátodo *(m)*

cátodo *(m) n* Es, Pt
De Kathode *(f)*
En cathode
Fr cathode *(f)*
It catodo *(m)*

catrame a bassa temperatura *(m)* It
De Tieftemperaturteer *(m)*
En low-temperature tar
Es alquitrán de baja temperatura *(m)*
Fr goudron à basse température *(m)*
Pt alcatrão de baixa temperatura *(m)*

catrame di carbon fossile *(m)* It
De Steinkohlenteer *(m)*
En coal tar
Es alquitrán de hulla *(m)*
Fr goudron de houille *(m)*
Pt alcatrão de hulha *(m)*

catrame di forno da coke *(m)* It
De Koksofenteer *(m)*
En coke-oven tar
Es alquitrán de horno de coquización *(m)*
Fr goudron *(m)*
Pt alcatrão de forno de coque *(m)*

caustic washing En
De Neutralisieren *(n)*
Es lavado cáustico *(m)*
Fr lavage à la soude caustique *(m)*
It lavaggio caustico *(m)*
Pt lavagem caústica *(f)*

cavity wall insulation En
De Isolierung von Hohlziegelmauerwerk *(f)*
Es aislamiento de pared de cavidad *(m)*
Fr isolement des murs doubles *(m)*
It isolamento di muro a cassavuota *(m)*
Pt isolamento de paredes de cavidade *(m)*

cell (battery) *n* En
De Zelle (Batterie) *(f)*
Es elemento (batería) *(m)*
Fr élément (batterie) *(m)*
It elemento (batteria) *(m)*
Pt célula (bateria) *(f)*

cellula al nichel-cadmio *(f)* It
De Nickel-Kadmiumzelle *(f)*
En nickel-cadmium cell
Es pila de níquel-cadmio *(f)*
Fr pile au nickel-cadmium *(f)*
Pt célula de níquel-cádmio *(f)*

cellula al nichel-ferro *(f)* It
De Nickel-Eisenzelle *(f)*
En nickel-iron cell
Es pila de níquel-hierro *(f)*
Fr pile au nickel-fer *(f)*
Pt célula de níquel-ferro *(f)*

cellula combustibile biochimica *(f)* It
De biochemische Brennstoffzelle *(f)*
En biochemical fuel cell
Es célula de combustible bioquímico *(f)*

Fr pile à combustible biochimique *(f)*
Pt célula de combustível bioquímico *(f)*

cellula combustibile idrogeno-aria *(f)* It
De Wasserstoff-Luft-brennstoffzelle *(f)*
En hydrogen-air fuel cell
Es célula de combustible de hidrógeno-aire *(f)*
Fr pile à combustible hydrogène-air *(f)*
Pt célula de combustível hidrogénio-ar *(f)*

cellula di argento-cadmio *(f)* It
De Silber-Kadmiumzelle *(f)*
En silver-cadmium cell
Es pila de plata-cadmio *(f)*
Fr pile à l'argent-cadmium *(f)*
Pt célula de prata-cádmio *(f)*

cellula di argento-zinco *(f)* It
De Silber-Zinkzelle *(f)*
En silver-zinc cell
Es pila de plata-zinc *(f)*
Fr pile à l'argent-zinc *(f)*
Pt célula de prata-zinco *(f)*

cellula di silicone *(f)* It
De Silikonzelle *(f)*
En silicon cell
Es célula de silicio *(f)*
Fr pile au silicium *(f)*
Pt célula de silíco *(f)*

cellula fotoelettrica *(f)* It
De Photozelle *(f)*
En photocell
Es fotocélula *(f)*
Fr cellule photoélectrique *(f)*
Pt fotocélula *(f)*

cellula galvanica *(f)* It
De galvanische Zelle *(f)*
En galvanic cell
Es elemento galvánico *(m)*
Fr pile galvanique *(f)*
Pt célula galvânica *(f)*

cellula metallo-aria *(f)* It
De Metall-Luftzelle *(f)*
En metal-air cell
Es célula de metal-aire *(f)*
Fr pile métal-air *(f)*
Pt célula de metal-ar *(f)*

cellula secondaria alcalina *(f)* It
De sekundäre Alkalizelle *(f)*
En alkaline secondary cell
Es elemento alcalino de acumulador *(m)*
Fr élément secondaire alcalin *(m)*
Pt célula secundária alcalina *(f)*

cellula solare *(f)* It
De Solarzelle *(f)*
En solar cell
Es célula solar *(f)*
Fr cellule solaire *(f)*
Pt célula solar *(f)*

cellula solare all'arseniuro di gallio *(f)* It
De Galliumarsenid-Solarzelle *(f)*
En gallium arsenide solar cell
Es célula solar de arseniuro de galio *(f)*
Fr cellule solaire à l'arséniure de gallium *(f)*
Pt célula solar de arseneto de gálio *(f)*

cellule photoélectrique *(f)* Fr
De Photozelle *(f)*
En photocell
Es fotocélula *(f)*
It cellula fotoelettrica *(f)*
Pt fotocélula *(f)*

cellule solaire *(f)* Fr
De Solarzelle *(f)*
En solar cell
Es célula solar *(f)*
It cellula solare *(f)*
Pt célula solar *(f)*

cellule solaire à l'arséniure de gallium *(f)* Fr
De Galliumarsenid-Solarzelle *(f)*
En gallium arsenide solar cell
Es célula solar de arseniuro de galio *(f)*
It cellula solare all'arseniuro di gallio *(f)*
Pt célula solar de arseneto de gálio *(f)*

cellulosa *(f) n* It
De Zellulose *(f)*
En cellulose
Es celulosa *(f)*
Fr cellulose *(f)*
Pt celulose *(f)*

cellulose *n* En; Fr *(f)*
De Zellulose *(f)*
Es celulosa *(f)*
It cellulosa *(f)*
Pt celulose *(f)*

celosía *(f) n* Es
De Spaltstoffgitter *(n)*
En lattice
Fr treillis *(m)*
It traliccio *(m)*
Pt treliça *(f)*

célula (bateria) *(f) n* Pt
De Zelle (Batterie) *(f)*
En cell (battery)
Es elemento (batería) *(m)*
Fr élément (batterie) *(m)*
It elemento (batteria) *(m)*

célula de combustible *(f)* Es
De Brennstoffzelle *(f)*
En fuel cell
Fr pile à combustible *(f)*
It pila a combustibile *(f)*
Pt célula de combustível *(f)*

célula de combustible bioquímico *(f)* Es
De biochemische Brennstoffzelle *(f)*
En biochemical fuel cell
Fr pile à combustible biochimique *(f)*
It cellula combustibile biochimica *(f)*
Pt célula de combustível bioquímico *(f)*

célula de combustible de hidrógeno-aire *(f)* Es
De Wasserstoff-Luft-brennstoffzelle *(f)*
En hydrogen-air fuel cell
Fr pile à combustible hydrogène-air *(f)*
It cellula combustibile idrogeno-aria *(f)*
Pt célula de combustível hidrogénio-ar *(f)*

célula de combustível *(f)* Pt
De Brennstoffzelle *(f)*
En fuel cell
Es célula de combustible *(f)*
Fr pile à combustible *(f)*
It pila a combustibile *(f)*

célula de combustível bioquímico *(f)* Pt
De biochemische Brennstoffzelle *(f)*
En biochemical fuel cell
Es célula de combustible bioquímico *(f)*
Fr pile à combustible biochimique *(f)*
It cellula combustibile biochimica *(f)*

célula de combustível hidrogénio-ar *(f)* Pt
De Wasserstoff-Luft-brennstoffzelle *(f)*
En hydrogen-air fuel cell
Es célula de combustible de hidrógeno-aire *(f)*
Fr pile à combustible hydrogène-air *(f)*
It cellula combustibile idrogeno-aria *(f)*

célula de metal-aire *(f)* Es
De Metall-Luftzelle *(f)*
En metal-air cell
Fr pile métal-air *(f)*
It cellula metallo-aria *(f)*
Pt célula de metal-ar *(f)*

célula de metal-ar *(f)* Pt
De Metall-Luftzelle *(f)*
En metal-air cell
Es célula de metal-aire *(f)*

Fr pile métal-air *(f)*
It cellula metallo-aria *(f)*

célula de níquel-cádmio *(f)* Pt
De Nickel-Kadmiumzelle *(f)*
En nickel-cadmium cell
Es pila de níquel-cadmio *(f)*
Fr pile au nickel-cadmium *(f)*
It cellula al nichel-cadmio *(f)*

célula de níquel-ferro *(f)* Pt
De Nickel-Eisenzelle *(f)*
En nickel-iron cell
Es pila de níquel-hierro *(f)*
Fr pile au nickel-fer *(f)*
It cellula al nichel-ferro *(f)*

célula de prata-cádmio *(f)* Pt
De Silber-Kadmiumzelle *(f)*
En silver-cadmium cell
Es pila de plata-cadmio *(f)*
Fr pile à l'argent-cadmium *(f)*
It cellula di argento-cadmio *(f)*

célula de prata-zinco *(f)* Pt
De Silber-Zinkzelle *(f)*
En silver-zinc cell
Es pila de plata-zinc *(f)*
Fr pile à l'argent-zinc *(f)*
It cellula di argento-zinco *(f)*

célula de silicio *(f)* Es
De Silikonzelle *(f)*
En silicon cell
Fr pile au silicium *(f)*
It cellula di silicone *(f)*
Pt célula de silíco *(f)*

célula de silíco *(f)* Pt
De Silikonzelle *(f)*
En silicon cell
Es célula de silicio *(f)*
Fr pile au silicium *(f)*
It cellula di silicone *(f)*

célula galvânica *(f)* Pt
De galvanische Zelle *(f)*
En galvanic cell
Es elemento galvánico *(m)*
Fr pile galvanique *(f)*
It cellula galvanica *(f)*

célula secundária alcalina *(f)* Pt
De sekundäre Alkalizelle *(f)*
En alkaline secondary cell
Es elemento alcalino de acumulador *(m)*
Fr élément secondaire alcalin *(m)*
It cellula secondaria alcalina *(f)*

célula solar *(f)* Es, Pt
De Solarzelle *(f)*
En solar cell
Fr cellule solaire *(f)*
It cellula solare *(f)*

célula solar de arseneto de gálio *(f)* Pt
De Galliumarsenid-Solarzelle *(f)*
En gallium arsenide solar cell
Es célula solar de arseniuro de galio *(f)*
Fr cellule solaire à l'arséniure de gallium *(f)*
It cellula solare all'arseniuro di gallio *(f)*

célula solar de arseniuro de galio *(f)* Es
De Galliumarsenid-Solarzelle *(f)*
En gallium arsenide solar cell
Fr cellule solaire à l'arséniure de gallium *(f)*
It cellula solare all'arseniuro di gallio *(f)*
Pt célula solar de arseneto de gálio *(f)*

celulosa *(f) n* Es
De Zellulose *(f)*
En cellulose
Fr cellulose *(f)*
It cellulosa *(f)*
Pt celulose *(f)*

celulose *(f) n* Pt
De Zellulose *(f)*
En cellulose
Es celulosa *(f)*
Fr cellulose *(f)*
It cellulosa *(f)*

cemento pesante *(m)* It
De beschwerter Beton *(m)*
En loaded concrete
Es hormigón cargado *(m)*
Fr béton chargé *(m)*
Pt betão carregado *(m)*

cendre *(f) n* Fr
De Asche *(f)*
En ash
Es ceniza *(f)*
It cenere *(f)*
Pt cinza *(f)*

cendre de charbon *(f)* Fr
De Steinkohlenasche *(f)*
En coal ash
Es ceniza de carbón *(f)*
It cenere di carbone *(f)*
Pt cinza de carvão *(f)*

cendre de pétrole *(f)* Fr
De Ölasche *(f)*
En oil ash
Es escoria vanadosa de petróleo *(f)*
It cenere di petrolio *(f)*
Pt cinza de óleo *(f)*

cendres volantes *(f)* Fr
De Flugasche *(f)*
En fly ash
Es ceniza en suspensión *(f)*
It cenere ventilata *(f)*
Pt cinza muito fina *(f)*

cenere *(f) n* It
De Asche *(f)*
En ash
Es ceniza *(f)*
Fr cendre *(f)*
Pt cinza *(f)*

cenere di carbone *(f)* It
De Steinkohlenasche *(f)*
En coal ash
Es ceniza de carbón *(f)*
Fr cendre de charbon *(f)*
Pt cinza de carvão *(f)*

cenere di petrolio *(f)* It
De Ölasche *(f)*
En oil ash
Es escoria vanadosa de petróleo *(f)*
Fr cendre de pétrole *(f)*
Pt cinza de óleo *(f)*

cenere ventilata *(f)* It
De Flugasche *(f)*
En fly ash
Es ceniza en suspensión *(f)*
Fr cendres volantes *(f)*
Pt cinza muito fina *(f)*

cénit *(m) n* Es
De Zenit *(n)*
En zenith
Fr zénith *(m)*
It zenit *(m)*
Pt zénite *(m)*

ceniza *(f) n* Es
De Asche *(f)*
En ash
Fr cendre *(f)*
It cenere *(f)*
Pt cinza *(f)*

ceniza de carbón *(f)* Es
De Steinkohlenasche *(f)*
En coal ash
Fr cendre de charbon *(f)*
It cenere di carbone *(f)*
Pt cinza de carvão *(f)*

ceniza en suspensión *(f)* Es
De Flugasche *(f)*
En fly ash
Fr cendres volantes *(f)*
It cenere ventilata *(f)*
Pt cinza muito fina *(f)*

central de energia a carvão *(f)* Pt
De kohlegefeuertes Kraftwerk *(n)*
En coal-fired power station
Es central eléctrica de carbón *(f)*
Fr centrale électrique au charbon *(f)*
It centrale elettrica a carbone *(f)*

central de energia nuclear *(f)* Pt
De Kernkraftwerk *(n)*
En nuclear power station
Es central nuclear *(f)*
Fr centrale nucléaire *(f)*
It centrale nucleare *(f)*

central de energia solar híbrida *(f)* Pt
De Hybrid-Solaranlage *(f)*
En hybrid solar plant
Es planta solar híbrida *(f)*
Fr centrale solaire hybride *(f)*
It impianto solare ibrido *(m)*

central de energia térmica *(f)* Pt
De Wärmekraftwerk *(n)*
En thermal power station
Es central térmica *(f)*
Fr centrale thermique *(f)*
It centrale di energia termica *(f)*

centrale di energia termica *(f)* It
De Wärmekraftwerk *(n)*
En thermal power station
Es central térmica *(f)*
Fr centrale thermique *(f)*
Pt central de energia térmica *(f)*

centrale électrique *(f)* Fr
De Kraftwerk *(n)*
En power station
Es central eléctrica *(f)*
It centrale elettrica *(f)*
Pt central energética *(f)*

centrale électrique au charbon *(f)* Fr
De kohlgefeuertes Kraftwerk *(n)*
En coal-fired power station
Es central eléctrica de carbón *(f)*
It centrale elettrica a carbone *(f)*
Pt central de energia a carvão *(f)*

centrale elettrica *(f)* It
De Kraftwerk *(n)*
En power station
Es central eléctrica *(f)*
Fr centrale électrique *(f)*
Pt central energética *(f)*

centrale elettrica a carbone *(f)* It
De kohlegefeuertes Kraftwerk *(n)*
En coal-fired power station
Es central eléctrica de carbón *(f)*
Fr centrale électrique au charbon *(f)*
Pt central de energia a carvão *(f)*

centrale hydroélectrique *(f)* Fr
De hydroelektrisches Kraftwerk *(n)*
En hydroelectric power station
Es central hidroeléctrica *(f)*
It centrale idroelettrica *(f)*
Pt central hidroeléctrica *(f)*

centrale idroelettrica *(f)* It
De hydroelektrisches Kraftwerk *(n)*
En hydroelectric power station
Es central hidroeléctrica *(f)*
Fr centrale hydroélectrique *(f)*
Pt central hidroeléctrica *(f)*

central eléctrica *(f)* Es
De Kraftwerk *(n)*
En power station
Fr centrale électrique *(f)*
It centrale elettrica *(f)*
Pt central energética *(f)*

central eléctrica de carbón *(f)* Es
De kohlegefeuertes Kraftwerk *(n)*
En coal-fired power station
Fr centrale électrique au charbon *(f)*
It centrale elettrica a carbone *(f)*
Pt central de energia a carvão *(f)*

central energética *(f)* Pt
De Kraftwerk *(n)*
En power station
Es central eléctrica *(f)*
Fr centrale électrique *(f)*
It centrale elettrica *(f)*

centrale nucléaire *(f)* Fr
De Kernkraftwerk *(n)*
En nuclear power station
Es central nuclear *(f)*
It centrale nucleare *(f)*
Pt central de energia nuclear *(f)*

centrale nucleare *(f)* It
De Kernkraftwerk *(n)*
En nuclear power station
Es central nuclear *(f)*
Fr centrale nucléaire *(f)*
Pt central de energia nuclear *(f)*

centrale solaire hybride *(f)* Fr
De Hybrid-Solaranlage *(f)*
En hybrid solar plant
Es planta solar híbrida *(f)*
It impianto solare ibrido *(m)*
Pt central de energia solar híbrida *(f)*

centrale thermique *(f)* Fr
De Wärmekraftwerk *(n)*
En thermal power station
Es central térmica *(f)*
It centrale di energia termica *(f)*
Pt central de energia térmica *(f)*

central heating En
De Zentralheizung *(f)*
Es calefacción central *(f)*
Fr chauffage central *(m)*
It riscaldamento centrale *(m)*
Pt aquecimento central *(m)*

central hidroeléctrica *(f)* Es, Pt
De hydroelektrisches Kraftwerk *(n)*
En hydroelectric power station
Fr centrale hydroélectrique *(f)*
It centrale idroelettrica *(f)*

central nuclear *(f)* Es
De Kernkraftwerk *(n)*
En nuclear power station
Fr centrale nucléaire *(f)*
It centrale nucleare *(f)*
Pt central de energia nuclear *(f)*

central térmica *(f)* Es
De Wärmekraftwerk *(n)*
En thermal power station
Fr centrale thermique *(f)*
It centrale di energia termica *(f)*
Pt central de energia térmica *(f)*

centrifuga *(f) n* It
De Zentrifuge *(f)*
En centrifuge
Es centrífuga *(f)*
Fr centrifugeuse *(f)*
Pt centrifugadora *(f)*

centrífuga *(f) n* Es
De Zentrifuge *(f)*
En centrifuge
Fr centrifugeuse *(f)*
It centrifuga *(f)*
Pt centrifugadora *(f)*

centrifugadora *(f) n* Pt
De Zentrifuge *(f)*
En centrifuge
Es centrífuga *(f)*
Fr centrifugeuse *(f)*
It centrifuga *(f)*

centrifugal compressor En
De radialer Turboverdichter *(m)*
Es compresor centrífugo *(m)*
Fr compresseur centrifuge *(m)*
It compressore centrifugo *(m)*
Pt compressor centrífugo *(m)*

centrifugal force En
De Zentrifugalkraft *(f)*
Es fuerza contrífuga *(f)*
Fr force centrifuge *(f)*
It forza centrifuga *(f)*
Pt força centrífuga *(f)*

centrifuge *n* En
De Zentrifuge *(f)*
Es centrífuga *(f)*
Fr centrifugeuse *(f)*
It centrifuga *(f)*
Pt centrifugadora *(f)*

centrifugeuse *(f) n* Fr
De Zentrifuge *(f)*
En centrifuge
Es centrífuga *(f)*
It centrifuga *(f)*
Pt centrifugadora *(f)*

centripetal force En
De Zentripetalkraft *(f)*
Es fuerza centrípeta *(f)*
Fr force centripète *(f)*
It forza centripeta *(f)*
Pt força centrípeda *(f)*

centro de produção solar *(f)* Pt
De Solarfarm *(f)*
En solar farm
Es granja solar *(f)*
Fr ferme solaire *(f)*
It podere solare *(m)*

cera mineral *(f)* Es
De Ozokerit *(m)*
En mineral wax
Fr cire minérale *(f)*
It cera minerale *(f)*
Pt cera mineral *(f)*

cera mineral *(f)* Pt
De Ozokerit *(m)*
En mineral wax
Es cera mineral *(f)*
Fr cire minérale *(f)*
It cera minerale *(f)*

cera minerale *(f)* It
De Ozokerit *(m)*
En mineral wax
Es cera mineral *(f)*
Fr cire minérale *(f)*
Pt cera mineral *(f)*

cermet refractory En
De feuerfestes Cermet *(n)*
Es refractario de cerametal *(m)*
Fr réfractaire aux cermets *(m)*
It refrattario al cermete *(m)*
Pt refractário metalocerâmico *(m)*

cernita *(f) n* It
De Klassierung *(f)*
En grading (coal)
Es clasificación *(f)*
Fr criblage *(m)*
Pt granulometria *(f)*

cero absoluto *(m)* Es
De absoluter Nullpunkt *(m)*
En absolute zero
Fr zéro absolu *(m)*
It zero assoluto *(m)*
Pt zero absoluto *(m)*

cetane number En
De Cetanzahl *(f)*
Es número cetano *(m)*
Fr indice de cétane *(m)*
It numero di cetano *(m)*
Pt número cetânico *(m)*

Cetanzahl *(f) n* De
En cetane number
Es número cetano *(m)*
Fr indice de cétane *(m)*
It numero di cetano *(m)*
Pt número cetânico *(m)*

chain reaction En
De Kettenreaktion *(f)*
Es reacción en cadena *(f)*
Fr réaction en chaîne *(f)*
It reazione a catena *(f)*
Pt reacção em cadeia *(f)*

chaleur *(f) n* Fr
De Wärme *(f)*
En heat
Es calor *(m)*
It calore *(m)*
Pt calor *(m)*

chaleur ambiante *(f)* Fr
De Grundheizung *(f)*
En background heat
Es calor de fondo *(m)*
It calore di fondo *(m)*
Pt calor de fundo *(m)*

chaleur de combustion *(f)* Fr
De Verbrennungswärme *(f)*
En heat of combustion
Es calor de combustión *(m)*
It calore di combustione *(m)*
Pt calor de combustão *(m)*

chaleur de désintégration *(f)* Fr
De Zersetzungswärme *(f)*
En decay heat
Es calor de desintegración *(m)*
It calore di disintegrazione *(m)*
Pt calor de descomposição *(m)*

chaleur de réaction *(f)* Fr
De Reaktionswärme *(f)*
En heat of reaction
Es calor de reacción *(m)*
It calore di reazione *(m)*
Pt calor de reacção *(m)*

chaleur et puissance combinées *(f)* Fr
De kombinierte Wärme und Energie *(f)*
En combined heat and power
Es calor y potencia combinados *(m)*
It potenza e calore combinati *(f)*
Pt calor e potência combinados *(m)*

chaleur latente *(f)* Fr
De gebundene Wärme *(f)*
En latent heat
Es calor latente *(m)*
It calore latente *(m)*
Pt calor latente *(m)*

chaleur latente molaire *(f)* Fr
De molare gebundene Wärme *(f)*
En molar latent heat
Es calor latente molar *(m)*
It calore latente molare *(m)*
Pt calor latente molar *(m)*

chaleur latente spécifique *(f)* Fr
De spezifische gebundene Wärme *(f)*
En specific latent heat
Es calor latente específico *(m)*
It calore latente specifico *(m)*
Pt calor latente específico *(m)*

chaleur non disponible *(f)* Fr
De nicht verfügbare Wärme *(f)*
En unavailable heat
Es calor no disponible *(m)*
It calore non disponibile *(m)*
Pt calor não disponível *(m)*

chaleur perdue *(f)* Fr
De Abwärme *(f)*
En waste heat
Es calor residual *(m)*
It calore perduto *(m)*
Pt calor residual *(m)*

chaleur sensible *(f)* Fr
De spürbare Wärme *(f)*
En sensible heat
Es calor sensible *(m)*
It calore sensibile *(m)*
Pt calor sensível *(m)*

chaleur utile *(f)* Fr
De Wärmeleistung *(f)*
En useful heat
Es calor útil *(m)*
It calore utile *(m)*
Pt calor útil *(m)*

chalumeau à plasma *(m)* Fr
De Plasmabrenner *(m)*
En plasma torch
Es soplete para plasma *(m)*
It torcia a plasma *(f)*
Pt maçarico de plasma *(m)*

chama *(f) n* Pt
De Flamme *(f)*
En flame
Es llama *(f)*
Fr flamme *(f)*
It fiamma *(f)*

chama arejada *(f)* Pt
De belüftete Flamme *(f)*
En aerated flame
Es llama aireada *(f)*

Fr flamme aérée *(f)*
It fiamma aerata *(f)*

chama não arejada *(f)* Pt
De unbelüftete Flamme *(f)*
En nonaerated flame
Es llama no aireada *(f)*
Fr flamme non aérée *(f)*
It fiamma non aerata *(f)*

chama piloto *(f)* Pt
De Zündflamme *(f)*
En pilot flame
Es llama piloto *(f)*
Fr veilleuse *(f)*
It semprevivo *(m)*

chambre de combustion *(f)* Fr
De Brennkammer *(f)*
En combustion chamber
Es cámara de combustión *(f)*
It camera di combustione *(f)*
Pt câmara de combustão *(f)*

chaminé *(f)* *n* Pt
De Schornstein *(m)*
En chimney
Es chimenea *(f)*
Fr cheminée *(f)*
It camino *(m)*

champ *(m)* *n* Fr
De Feld *(n)*
En field
Es campo *(m)*
It campo *(m)*
Pt campo *(m)*

champ de gravitation *(m)* Fr
De Gravitationsfeld *(n)*
En gravitational field
Es campo gravitacional *(m)*
It campo gravitazionale *(m)*
Pt campo de gravidade *(m)*

champ électrique *(m)* Fr
De elektrisches Feld *(n)*
En electric field
Es campo eléctrico *(m)*
It campo elettrico *(m)*
Pt campo eléctrico *(m)*

champ électromagnétique *(m)* Fr
De elektromagnetisches Feld *(n)*
En electromagnetic field
Es campo electromagnético *(m)*
It campo elettromagnetico *(m)*
Pt campo electromagnético *(m)*

champ magnétique *(m)* Fr
De Magnetfeld *(n)*
En magnetic field
Es campo magnético *(m)*
It campo magnetico *(m)*
Pt campo magnético *(m)*

channel induction furnace En
De Kanalinduktionsofen *(m)*
Es horno de inducción de canal *(m)*
Fr four à induction à chenal *(m)*
It forno ad induzione a canale *(m)*
Pt forno de indução em canal *(m)*

charbon *(m)* *n* Fr
De Kohle *(f)*
En coal
Es carbón *(m)*
It carbone *(m)*
Pt carvão *(m)*

charbon bitumineux *(m)* Fr
De bituminöse Kohle *(f)*
En bituminous coal
Es carbón bituminoso *(m)*
It carbone bituminoso *(m)*
Pt carvão betuminoso *(m)*

charbon cokéfiant *(m)* Fr
De Kokskohle *(f)*
En coking coal
Es carbón de coquización *(m)*
It carbone cokificante *(m)*
Pt carvão de coqueificação *(m)*

charbon de bois *(m)* Fr
De Holzkohle *(f)*
En charcoal
Es carbón vegetal *(m)*
It carbone dolce *(m)*
Pt carvão vegital *(m)*

charbon dure *(m)* Fr
De Steinkohle *(f)*
En hard coal
Es carbón duro *(m)*
It carbone duro *(m)*
Pt carvão duro *(m)*

charbon humique *(m)* Fr
De Humuskohle *(f)*
En humic coal
Es carbón húmico *(m)*
It carbone umico *(m)*
Pt carvão humico *(m)*

charbon métabitumineux *(m)* Fr
De metabituminöse Kohle *(f)*
En metabituminous coal
Es carbón metabituminoso *(m)*
It carbone metabituminoso *(m)*
Pt carvão metabetuminoso *(m)*

charbon pulvérisé *(m)* Fr
De pulverisierte Kohle *(f)*
En pulverized coal
Es carbón pulverizado *(m)*
It carbone polverizzato *(m)*
Pt carvão pulverizado *(m)*

charbon sans fumée *(m)* Fr
De rauchlose Kohle *(f)*
En smokeless coal
Es carbón fumífugo *(m)*
It carbone senza fumo *(m)*
Pt carvão ardendo sem produzir fumo *(m)*

charbon sapropélique *(m)* Fr
De Sapropelkohle *(f)*
En sapropelic coal
Es carbón sapropélico *(m)*
It carbone sapropelico *(m)*
Pt carvão sapropélico *(m)*

charbons non-agglomérants *(m pl)* Fr
De nicht klumpende Kohlen *(pl)*
En nonagglomerating coals
Es carbones no aglomerantes *(m pl)*
It carboni non agglomeranti *(m pl)*
Pt carvões não-aglomerantes *(m pl)*

charbon tout-venant *(m)* Fr
De Normalkohle *(f)*
En run-of-mine coal
Es carbón todouno *(m)*
It carbone tout-venant *(m)*
Pt carvão tal como sai da mina *(m)*

charcoal *n* En
De Holzkohle *(f)*
Es carbón vegetal *(m)*
Fr charbon de bois *(m)*
It carbone dolce *(m)*
Pt carvão vegital *(m)*

charge (electrical) *n* En; Fr *(f)*
De Ladung (elektrische) *(f)*
Es carga (eléctrica) *(f)*
It carica (elettrica) *(f)*
Pt carga (eléctrica) *(f)*

charge (mécanique) *(f)* *n* Fr
De Last (mechanische) *(f)*
En load (mechanical)
Es carga (mecánica) *(f)*
It carico (meccanico) *(m)*
Pt carga (mecânica) *(f)*

charge de pointe *(f)* Fr
De Spitzenbelastung *(f)*
En peak load
Es carga de punta *(f)*

It carico di punta *(m)*
Pt carga de ponta *(f)*

charge minimale *(f)* Fr
De Grundkraft *(f)*
En base load
Es carga base *(f)*
It carico base *(m)*
Pt carga fundamental *(f)*

chargeur de batterie *(m)* Fr
De Batterieladegerät *(n)*
En battery charger
Es cargador de batería *(m)*
It caricabatteria *(f)*
Pt carregador de bateria *(m)*

chaudière *(f) n* Fr
De Kessel *(m)*
En boiler
Es caldera *(f)*
It caldaia *(f)*
Pt caldeira *(f)*

chaudière à paroi *(f)* Fr
De Schalenkessel *(m)*
En shell boiler
Es caldera de coraza *(f)*
It caldaia cilindrica *(f)*
Pt caldeira de camisa *(f)*

chaudière à tube de fumée *(f)* Fr
De Heizrohrkessel *(m)*
En fire-tube boiler
Es caldera pirotubular *(f)*
It caldaia a tubo di fumo *(f)*
Pt caldeira de tubo de fumaça *(f)*

chaudière de récupération des chaleurs perdues *(f)* Fr
De Abwärmekessel *(m)*
En waste-heat boiler
Es caldera de calor residual *(f)*
It caldaia a recupero di calore perduto *(f)*
Pt caldeira aquecida com calor residual *(f)*

chaudière radiante *(f)* Fr
De Strahlungsheizkessel *(m)*
En radiant boiler
Es caldera radiante *(f)*
It caldaia radiante *(f)*
Pt convector de radiação *(m)*

chaudière sans recyclage *(f)* Fr
De Zwangsdurchlaufkessel *(m)*
En once-through boiler
Es caldera de proceso directo *(f)*
It caldaia a processo diretto *(f)*
Pt caldeira de uma só passagem *(f)*

chaudière sectionnelle *(f)* Fr
De Teilkammerkessel *(m)*
En sectional boiler
Es caldera compartimentada *(f)*
It caldaia a sezioni *(f)*
Pt caldeira de secções *(f)*

chaudière tubulaire *(f)* Fr
De Wasserrohrkessel *(m)*
En water-tube boiler
Es caldera de tubos de agua *(f)*
It caldaia a tubo d'acqua *(f)*
Pt caldeira de tubos de água *(f)*

chauffage *(m) n* Fr
De Heizung *(f)*
En heating
Es calefacción *(f)*
It riscaldamento *(m)*
Pt aquecimento *(m)*

chauffage à air chaud *(m)* Fr
De Warmluftheizung *(f)*
En warm-air heater
Es calentador con aire caliente *(m)*
It riscaldatore ad aria tiepida *(m)*
Pt aquecedor de ar quente *(m)*

chauffage à radiofréquence *(m)* Fr
De Radiofrequenzheizung *(f)*
En radiofrequency heating
Es calentamiento por radiofrecuencia *(m)*
It riscaldamento a radiofrequenza *(m)*
Pt aquecimento por frequência rádio *(m)*

chauffage central *(m)* Fr
De Zentralheizung *(f)*
En central heating
Es calefacción central *(f)*
It riscaldamento centrale *(m)*
Pt aquecimento central *(m)*

chauffage central au mazout *(m)* Fr
De Zentralheizung mit Ölfeuerung *(f)*
En oil-fired central heating
Es calefacción central con petróleo *(f)*
It riscaldamento centrale a nafta *(m)*
Pt aquecimento central a óleo *(m)*

chauffage de chambres *(m)* Fr
De Raumheizung *(f)*
En space heating
Es calefacción de espacios *(f)*
It riscaldamento locale *(m)*
Pt aquecimento espacial *(m)*

chauffage indirect *(m)* Fr
De indirekte Heizung *(f)*
En indirect heating
Es calefacción indirecto *(f)*
It riscaldamento indiretto *(m)*
Pt aquecimento indirecto *(m)*

chauffage par induction *(m)* Fr
De Induktionsheizung *(f)*
En induction heating
Es caldeo por inducción *(m)*
It riscaldamento ad induzione *(m)*
Pt aquecimento por indução *(m)*

chauffage par pertes diélectriques *(m)* Fr
De kapazitive Hochfrequenzerwärmung *(f)*
En dielectric heating
Es caldeo dieléctrico *(m)*
It riscaldamento dielettrico *(m)*
Pt aquecimento dieléctrico *(m)*

chauffage urbain *(m)* Fr
De Fernheizung *(f)*
En district heating
Es calefacción de distritos *(f)*
It riscaldamento di sezione *(m)*
Pt aquecimento de distrito *(m)*

chauffé au charbon Fr
De kohlenverbrennend
En coal-burning
Es caldeado con carbón
It scaldato a carbone
Pt aquecido a carvão

chauffé au gaz Fr
De gasgefeuert
En gas-fired
Es alimentado por gas
It alimentato a gas
Pt alimentado a gás

chauffé au mazout Fr
De ölgefeuert
En oil-fired
Es alimentado con petróleo
It alimentato a nafta
Pt alimentado a óleo

chauffe-bain *(m) n* Fr
De Badheizelement *(n)*
En bath heater
Es calentador de baño *(m)*
It scaldabagno *(m)*
Pt aquecedor de banho *(m)*

chauffe-billettes *(m) (n)* Fr
De Blockerwärmer *(m)*
En billet heater

Es calentador de palanquillas *(m)*
It riscaldatore di billette *(m)*
Pt aquecedor de lingotes *(m)*

chauffe-eau *(m) n* Fr
De Wassererhitzer *(m)*
En water heater
Es calentador de agua *(m)*
It riscaldatore di acqua *(m)*
Pt aquecedor hidráulico *(m)*

chauffe-eau à accumulation *(m)* Fr
De Speicherwasser-heizung *(f)*
En storage water-heater
Es calentador de agua de acumulación *(m)*
It riscaldatore di acqua a conservazione *(m)*
Pt aquecedor a água de armazenagem *(m)*

chauffe-eau instantané *(m)* Fr
De Wasser-Schneller-hitzer *(m)*
En instantaneous water heater
Es calentador de agua instantáneo *(m)*
It riscaldatore istantaneo di acqua *(m)*
Pt aquecedor de água instantâneo *(m)*

chaufferie auxiliaire *(f)* Fr
De Kesselhilfsanlage *(f)*
En boiler auxiliary plant
Es planta auxiliar de calderas *(f)*
It impianto ausiliario caldaia *(m)*
Pt instalação auxiliar de caldeira *(f)*

chelating agent En
De Cheliermittel *(n)*
Es agente quelatante *(m)*
Fr agent chélateur *(m)*
It agente di chelazione *(m)*
Pt agente de chelação *(m)*

Cheliermittel *(n) n* De
En chelating agent
Es agente quelatante *(m)*
Fr agent chélateur *(m)*
It agente di chelazione *(m)*
Pt agente de chelação *(m)*

chemical energy En
De chemische Energie *(f)*
Es energía química *(f)*
Fr énergie chimique *(f)*
It energia chimica *(f)*
Pt energia química *(f)*

chemical feedstock En
De chemisches Stangenmaterial *(n)*
Es material químico de carga *(m)*
Fr stock d'alimentation chimique *(m)*
It materiale di alimentazione chimico *(m)*
Pt stock de abastecimento químico *(m)*

chemical fuel En
De chemischer Brennstoff *(m)*
Es combustible químico *(m)*
Fr combustible chimique *(m)*
It combustibile chimico *(m)*
Pt combustível químico *(m)*

cheminée *(f) n* Fr
De Schornstein *(m)*
En chimney
Es chimenea *(f)*
It camino *(m)*
Pt chaminé *(f)*

chemische Energie *(f)* De
En chemical energy
Es energía química *(f)*
Fr énergie chimique *(f)*
It energia chimica *(f)*
Pt energia química *(f)*

chemischer Brennstoff *(m)* De
En chemical fuel
Es combustible químico *(m)*
Fr combustible chimique *(m)*
It combustibile chimico *(m)*
Pt combustível químico *(m)*

chemisches Stangenmaterial *(n)* De
En chemical feedstock
Es material químico de carga *(m)*
Fr stock d'alimentation chimique *(m)*
It materiale di alimentazione chimico *(m)*
Pt stock de abastecimento químico *(m)*

chimenea *(f) n* Es
De Schornstein *(m)*
En chimney
Fr cheminée *(f)*
It camino *(m)*
Pt chaminé *(f)*

chimica delle radiazioni *(f)* It
De Strahlungschemie *(f)*
En radiation chemistry
Es química de la radiación *(f)*
Fr radiochimie *(f)*
Pt química das radiações *(f)*

chimney *n* En
De Schornstein *(m)*
Es chimenea *(f)*
Fr cheminée *(f)*
It camino *(m)*
Pt chaminé *(f)*

chorro *(m) n* Es
De Düse *(f)*
En jet
Fr jet *(m)*
It getto *(m)*
Pt jacto *(m)*

chromatographie gazeuse *(f)* Fr
De Gaschromatographie *(f)*
En gas chromatography
Es cromatografía de gases *(f)*
It cromatografia in fase gassosa *(f)*
Pt cromatografia gasosa *(f)*

chute de pression *(f)* Fr
De Druckabfall *(m)*
En pressure drop
Es caída de presión *(f)*
It caduta di pressione *(f)*
Pt queda de pressão *(f)*

ciclo aberto *(m)* Pt
De offener Zyklus *(m)*
En open cycle
Es ciclo abierto *(m)*
Fr circuit ouvert *(m)*
It ciclo aperto *(m)*

ciclo abierto *(m)* Es
De offener Zyklus *(m)*
En open cycle
Fr circuit ouvert *(m)*
It ciclo aperto *(m)*
Pt ciclo aberto *(m)*

ciclo aperto *(m)* It
De offener Zyklus *(m)*
En open cycle
Es ciclo abierto *(m)*
Fr circuit ouvert *(m)*
Pt ciclo aberto *(m)*

ciclo calore-lavoro *(m)* It
De Wärmearbeitszyklus *(m)*
En heat-work cycle
Es ciclo de calor-trabajo *(m)*
Fr cycle de chaleur-travail *(m)*
Pt ciclo de calor-trabalho *(m)*

ciclo cerrado *(m)* Es
De geschlossener Kreislauf *(m)*
En closed cycle
Fr circuit fermé *(m)*
It ciclo chiuso *(m)*
Pt ciclo fechado *(m)*

ciclo chiuso *(m)* It
De geschlossener Kreislauf *(m)*
En closed cycle
Es ciclo cerrado *(m)*

Fr circuit fermé *(m)*
Pt ciclo fechado *(m)*

ciclo combinado *(m)* Es
De Kombinationsbetrieb *(m)*
En combined cycle
Fr cycle mixte *(m)*
It ciclo combinato *(m)*
Pt ciclo combinado *(m)*

ciclo combinado *(m)* Pt
De Kombinationsbetrieb *(m)*
En combined cycle
Es ciclo combinado *(m)*
Fr cycle mixte *(m)*
It ciclo combinato *(m)*

ciclo combinato *(m)* It
De Kombinationsbetrieb *(m)*
En combined cycle
Es ciclo combinado *(m)*
Fr cycle mixte *(m)*
Pt ciclo combinado *(m)*

ciclo de absorção *(m)* Pt
De Absorptionsfolge *(f)*
En absorption cycle
Es ciclo de absorción *(f)*
Fr cycle d'absorption *(m)*
It ciclo di assorbimento *(m)*

ciclo de absorción *(f)* Es
De Absorptionsfolge *(f)*
En absorption cycle
Fr cycle d'absorption *(m)*
It ciclo di assorbimento *(m)*
Pt ciclo de absorção *(m)*

ciclo de calor-trabajo *(m)* Es
De Wärmearbeitszyklus *(m)*
En heat-work cycle
Fr cycle de chaleur-travail *(m)*
It ciclo calore-lavoro *(m)*
Pt ciclo de calor-trabalho *(m)*

ciclo de calor-trabalho *(m)* Pt
De Wärmearbeitszyklus *(m)*
En heat-work cycle
Es ciclo de calor-trabajo *(m)*
Fr cycle de chaleur-travail *(m)*
It ciclo calore-lavoro *(m)*

ciclo de Carnot *(m)* Es
De Carnotscher Kreisprozeß *(m)*
En Carnot cycle
Fr cycle de Carnot *(m)*
It ciclo di Carnot *(m)*

ciclo de cascada *(m)* Pt
De Kaskadenprozeß *(m)*
En cascade cycle
Es ciclo en cascada *(m)*
Fr cycle en cascade *(m)*
It ciclo di cascata *(m)*

ciclo de combustível *(m)* Pt
De Brennstoffzyklus *(m)*
En fuel cycle
Es ciclo de los combustibles *(m)*
Fr cycle du combustible *(m)*
It ciclo del combustibile *(m)*

ciclo de compresión del vapor *(m)* Es
Am vapor compression cycle
De umgekehrter Carnot-Prozeß *(m)*
En vapour compression cycle
Fr cycle de compression de la vapeur *(m)*
It ciclo di compressione del vapore *(m)*
Pt ciclo de compressão de vapor *(m)*

ciclo de compressão de vapor *(m)* Pt
Am vapor compression cycle
De umgekehrter Carnot-Prozeß *(m)*
En vapour compression cycle
Es ciclo de compresión del vapor *(m)*
Fr cycle de compression de la vapeur *(m)*
It ciclo di compressione del vapore *(m)*

ciclo del combustibile *(m)* It
De Brennstoffzyklus *(m)*
En fuel cycle
Es ciclo de los combustibles *(m)*
Fr cycle du combustible *(m)*
Pt ciclo de combustível *(m)*

ciclo de los combustibles *(m)* Es
De Brennstoffzyklus *(m)*
En fuel cycle
Fr cycle du combustible *(m)*
It ciclo del combustibile *(m)*
Pt ciclo de combustível *(m)*

ciclo de Otto *(m)* Es
De Otto-Zyklus *(m)*
En Otto cycle
Fr cycle d'Otto *(m)*
It ciclo di Otto *(m)*
Pt ciclo de Otto *(m)*

ciclo de Otto *(m)* Pt
De Otto-Zyklus *(m)*
En Otto cycle
Es ciclo de Otto *(m)*
Fr cycle d'Otto *(m)*
It ciclo di Otto *(m)*

ciclo de Rankine *(m)* Es
De Clausius-Rankine-Prozeß *(m)*
En Rankine cycle
Fr cycle de Rankine *(m)*
It ciclo di Rankine *(m)*
Pt ciclo ranquinizado *(m)*

ciclo de Stirling *(m)* Es, Pt
De Stirling-Zyklus *(m)*
En Stirling cycle
Fr cycle de Stirling *(m)*
It ciclo di Stirling *(m)*

ciclo de vapor *(m)* Es, Pt
De Dampfzyklus *(m)*
En steam cycle
Fr cycle de la vapeur *(m)*
It ciclo di vapore *(m)*

ciclo di assorbimento *(m)* It
De Absorptionsfolge *(f)*
En absorption cycle
Es ciclo de absorción *(f)*
Fr cycle d'absorption *(m)*
Pt ciclo de absorção *(m)*

ciclo di Carnot *(m)* It
De Carnotscher Kreisprozeß *(m)*
En Carnot cycle
Es ciclo de Carnot *(m)*
Fr cycle de Carnot *(m)*
Pt ciclo de Carnot *(m)*

ciclo di cascata *(m)* It
De Kaskadenprozeß *(m)*
En cascade cycle
Es ciclo en cascada *(m)*
Fr cycle en cascade *(m)*
Pt ciclo de cascada *(m)*

ciclo di compressione del vapore *(m)* It
Am vapor compression cycle
De umgekehrter Carnot-Proze *(m)*
En vapour compression cycle
Es ciclo de compresión del vapor *(m)*
Fr cycle de compression de la vapeur *(m)*
Pt ciclo de compressão de vapor *(m)*

ciclo di Otto *(m)* It
De Otto-Zyklus *(m)*
En Otto cycle
Es ciclo de Otto *(m)*
Fr cycle d'Otto *(m)*
Pt ciclo de Otto *(m)*

ciclo di Rankine *(m)* It
De Clausius-Rankine-Prozeß *(m)*
En Rankine cycle
Es ciclo de Rankine *(m)*
Fr cycle de Rankine *(m)*
Pt ciclo ranquinizado *(m)*

ciclo di Stirling *(m)* It
De Stirling-Zyklus *(m)*
En Stirling cycle
Es ciclo de Stirling *(m)*
Fr cycle de Stirling *(m)*
Pt ciclo de Stirling *(m)*

ciclo di vapore *(m)* It
De Dampfzyklus *(m)*
En steam cycle
Es ciclo de vapor *(m)*
Fr cycle de la vapeur *(m)*
Pt ciclo de vapor *(m)*

ciclo en cascada *(m)* Es
De Kaskadenprozeβ *(m)*
En cascade cycle
Fr cycle en cascade *(m)*
It ciclo di cascata *(m)*
Pt ciclo de cascada *(m)*

cicloesano *(m) n* It
De Zyklohexan *(n)*
En cyclohexane
Es ciclohexano *(m)*
Fr cyclohexane *(m)*
Pt ciclohexano *(m)*

ciclo fechado *(m)* Pt
De geschlossener Kreislauf *(m)*
En closed cycle
Es ciclo cerrado *(m)*
Fr circuit fermé *(m)*
It ciclo chiuso *(m)*

ciclohexano *(m) n* Es, Pt
De Zyklohexan *(n)*
En cyclohexane
Fr cyclohexane *(m)*
It cicloesano *(m)*

ciclón *(m) n* Es
De Zyklon *(m)*
En cyclone
Fr cyclone *(m)*
It ciclone *(m)*
Pt ciclone *(m)*

ciclone *(m) n* It, Pt
De Zyklon *(m)*
En cyclone
Es ciclón *(m)*
Fr cyclone *(m)*

ciclo ranquinizado *(m)* Pt
De Clausius-Rankine-Prozeβ *(m)*
En Rankine cycle
Es ciclo de Rankine *(m)*
Fr cycle de Rankine *(m)*
It ciclo di Rankine *(m)*

ciclo reversibile *(m)* It
De umkehrbarer Zyklus *(m)*
En reversible cycle
Es ciclo reversible *(m)*
Fr cycle réversible *(m)*
Pt ciclo reversível *(m)*

ciclo reversible *(m)* Es
De umkehrbarer Zyklus *(m)*
En reversible cycle
Fr cycle réversible *(m)*
It ciclo reversibile *(m)*
Pt ciclo reversível *(m)*

ciclo reversível *(m)* Pt
De umkehrbarer Zyklus *(m)*
En reversible cycle
Es ciclo reversible *(m)*
Fr cycle réversible *(m)*
It ciclo reversibile *(m)*

ciclo sencillo *(m)* Es
De Einzelgang *(m)*
En single cycle
Fr simple cycle *(m)*
It ciclo unico *(m)*
Pt ciclo único *(m)*

ciclo unico *(m)* It
De Einzelgang *(m)*
En single cycle
Es ciclo sencillo *(m)*
Fr simple cycle *(m)*
Pt ciclo único *(m)*

ciclo único *(m)* Pt
De Einzelgang *(m)*
En single cycle
Es ciclo sencillo *(m)*
Fr simple cycle *(m)*
It ciclo unico *(m)*

cierre *(m) n* Es
De Abdichtung *(f)*
En seal
Fr joint *(m)*
It dispositivo di tenuta *(m)*
Pt vedação *(f)*

cinematica *(f) n* It
De Kinematik *(f)*
En kinematics
Es cinemática *(f)*
Fr cinématique *(f)*
Pt cinemática *(f)*

cinemática *(f) n* Es, Pt
De Kinematik *(f)*
En kinematics
Fr cinématique *(f)*
It cinematica *(f)*

cinématique *(f) n* Fr
De Kinematik *(f)*
En kinematics
Es cinemática *(f)*
It cinematica *(f)*
Pt cinemática *(f)*

cinza *(f) n* Pt
De Asche *(f)*
En ash
Es ceniza *(f)*
Fr cendre *(f)*
It cenere *(f)*

cinza de carvão *(f)* Pt
De Steinkohlenasche *(f)*
En coal ash
Es ceniza de carbón *(f)*
Fr cendre de charbon *(f)*
It cenere di carbone *(f)*

cinza de óleo *(f)* Pt
De Ölasche *(f)*
En oil ash
Es escoria vanadosa de petróleo *(f)*
Fr cendre de pétrole *(f)*
It cenere di petrolio *(f)*

cinza muito fina *(f)* Pt
De Flugasche *(f)*
En fly ash
Es ceniza en suspensión *(f)*
Fr cendres volantes *(f)*
It cenere ventilata *(f)*

circuit *n* En; Fr *(m)*
De Schaltung *(f)*
Es circuito *(m)*
It circuito *(m)*
Pt circuito *(m)*

circuit breaker En
De Leitungsschalter *(m)*
Es disyuntor *(m)*
Fr disjoncteur *(m)*
It interruttore automatico *(m)*
Pt corta-circuito *(m)*

circuit de refroidissement *(m)* Fr
De Kühlmittelkreislauf *(m)*
En cooling circuit
Es circuito de refrigeración *(m)*
It circuito di raffreddamento *(m)*
Pt circuito refrigerante *(m)*

circuit équivalent *(m)* Fr
De Ersatzschaltung *(f)*
En equivalent circuit
Es circuito equivalente *(m)*
It circuito equivalente *(m)*
Pt circuito equivalente *(m)*

circuit fermé *(m)* Fr
De geschlossener Kreislauf *(m)*
En closed cycle
Es ciclo cerrado *(m)*
It ciclo chiuso *(m)*
Pt ciclo fechado *(m)*

circuito *(m) n* Es, It, Pt
De Schaltung *(f)*
En circuit
Fr circuit *(m)*

circuito aberto *(m)* Pt
De offener Kreislauf *(m)*
En open loop
Es bucle abierto *(m)*
Fr boucle ouverte *(f)*
It circuito aperto *(m)*

circuito aperto *(m)* It
De offener Kreislauf *(m)*
En open loop
Es bucle abierto *(m)*
Fr boucle ouverte *(f)*
Pt circuito aberto *(m)*

circuito chiuso *(m)* It
De geschlossener Regelkreis *(m)*
En closed loop
Es bucle cerrado *(m)*
Fr boucle de retour *(f)*
Pt circuito fechado *(m)*

circuito de refrigeración *(m)* Es
De Kühlmittelkreislauf *(m)*
En cooling circuit
Fr circuit de refroidissement *(m)*
It circuito di raffreddamento *(m)*

Pt circuito refrigerante *(m)*

circuito di raffreddamento *(m)* It
De Kühlmittelkreislauf *(m)*
En cooling circuit
Es circuito de refrigeración *(m)*
Fr circuit de refroidissement *(m)*
Pt circuito refrigerante *(m)*

circuito equivalente *(m)* Es, It, Pt
De Ersatzschaltung *(f)*
En equivalent circuit
Fr circuit équivalent *(m)*

circuito fechado *(m)* Pt
De geschlossener Regelkreis *(m)*
En closed loop
Es bucle cerrado *(m)*
Fr boucle de retour *(f)*
It circuito chiuso *(m)*

circuito refrigerante *(m)* Pt
De Kühlmittelkreislauf *(m)*
En cooling circuit
Es circuito de refrigeración *(m)*
Fr circuit de refroidissement *(m)*
It circuito di raffreddamento *(m)*

circuit ouvert *(m)* Fr
De offener Zyklus *(m)*
En open cycle
Es ciclo abierto *(m)*
It ciclo aperto *(m)*
Pt ciclo aberto *(m)*

circular burner En
De Kreisbrenner *(m)*
Es quemador circular *(m)*
Fr brûleur circulaire *(m)*
It bruciatore circolare *(m)*
Pt queimador circular *(m)*

cire minérale *(f)* Fr
De Ozokerit *(m)*
En mineral wax
Es cera mineral *(f)*
It cera minerale *(f)*
Pt cera mineral *(f)*

cladding *n* En
De Umhüllung *(f)*
Es encamisado *(m)*
Fr gainage *(m)*
It incamiciatura *(f)*
Pt revestimento *(m)*

clasificación *(f) n* Es
De Klassierung *(f)*
En grading
Fr criblage *(m)*
It cernita *(f)*
Pt granulometria *(f)*

Clausius-Rankine-Prozeß *(m) n* De
En Rankine cycle
Es ciclo de Rankine *(m)*
Fr cycle de Rankine *(m)*
It ciclo di Rankine *(m)*
Pt ciclo ranquinizado *(m)*

clay treatment En
De Lehmbehandlung *(f)*
Es tratamiento con arcilla *(m)*
Fr traitement à l'argile *(m)*
It trattamento all'argilla *(m)*
Pt tratamento de argila *(m)*

clean-air zone En
De Sauberluftzone *(f)*
Es zona de aire limpio *(f)*
Fr zone d'air pur *(f)*
It zona aria pulita *(f)*
Pt zona de ar puro *(f)*

clean ballast En
De sauberer Ballast *(m)*
Es lastre limpio *(m)*
Fr ballast propre *(m)*
It zavorra pulita *(f)*
Pt balastro puro *(m)*

cleanup system En
De Aufzehrsystem *(n)*
Es sistema de limpieza *(m)*
Fr système de nettoyage *(m)*
It sistema di depurazione *(m)*
Pt sistema de limpeza *(m)*

clima *(m) n* Es, It, Pt
De Klima *(n)*
En climate
Fr climat *(m)*

climat *(m) n* Fr
De Klima *(n)*
En climate
Es clima *(m)*
It clima *(m)*
Pt clima *(m)*

climate *n* En
De Klima *(n)*
Es clima *(m)*
Fr climat *(m)*
It clima *(m)*
Pt clima *(m)*

climatisation *(f) n* Fr
De Klimaanlage *(f)*
En air conditioning
Es acondicionamiento de aire *(m)*
It condizionamento dell'aria *(m)*
Pt ar condicionado *(m)*

closed cycle En
De geschlossener Kreislauf *(m)*
Es ciclo cerrado *(m)*
Fr circuit fermé *(m)*
It ciclo chiuso *(m)*
Pt ciclo fechado *(m)*

closed loop En
De geschlossener Regelkreis *(m)*
Es bucle cerrado *(m)*
Fr boucle de retour *(f)*
It circuito chiuso *(m)*
Pt circuito fechado *(m)*

cloud point En
De Paraffinausscheidungspunkt *(m)*
Es punto de opacidad *(m)*
Fr point de trouble *(m)*
It punto di intorbidimento *(m)*
Pt ponto de névoa *(m)*

coal *n* En
De Kohle *(f)*
Es carbón *(m)*
Fr charbon *(m)*
It carbone *(m)*
Pt carvão *(m)*

coal ash En
De Steinkohlenasche *(f)*
Es ceniza de carbón *(f)*
Fr cendre de charbon *(f)*
It cenere di carbone *(f)*
Pt cinza de carvão *(f)*

coal-burning *adj* En
De kohlenverbrennend
Es caldeado con carbón
Fr chauffé au charbon
It scaldato a carbone
Pt aquecido a carvão

coal equivalent En
De Kohlenersatz *(m)*
Es carbón equivalente *(m)*
Fr équivalent charbon *(m)*
It equivalente del carbone *(m)*
Pt equivalente carbónico *(m)*

coal-fired power station En
De kohlegefeuertes Kraftwerk *(n)*
Es central eléctrica de carbón *(f)*
Fr centrale électrique au charbon *(f)*
It centrale elettrica a carbone *(f)*
Pt central de energia a carvão *(f)*

coal gas En
De Steinkohlengas *(n)*
Es gas de carbón *(m)*
Fr gaz de houille *(m)*
It gas illuminante *(m)*
Pt gás de carvão *(m)*

coal measure En
De Kohlenlager *(n)*
Es formación hullera *(f)*
Fr couche de houille *(f)*
It giacimento carbonifero *(m)*
Pt medida de carvão *(f)*

coal reserves *pl n* En
De Kohlereserven *(pl)*
Es reservas carboníferas *(f pl)*
Fr réserves de houille *(f pl)*
It riserves di carbone *(f pl)*
Pt reservas de carvão *(f pl)*

coal seam En
De Kohlenflöz *(n)*
Es capa de carbón *(f)*
Fr filon houiller *(m)*
It filone di carbone *(m)*
Pt camada de carvão *(f)*

coal tar En
De Steinkohlenteer *(m)*
Es alquitrán de hulla *(m)*
Fr goudron de houille *(m)*
It catrame di carbon fossile *(m)*
Pt alcatrão de hulha *(m)*

coal-tar fuels *pl n* En
De Steinkohlenteer-Treibstoffe *(pl)*
Es combustibles de carbón-alquitrán *(m pl)*
Fr combustibles de goudron de houille *(m pl)*
It combustibili a base di catrame *(m pl)*
Pt combustíveis de alcatrão de hulha *(m pl)*

cocina solar *(f)* Es
De Solarherd *(m)*
En solar cooker
Fr fourneau solaire *(m)*
It fornello solare *(m)*
Pt fervedor solar *(m)*

cock *n* En
De Hahn *(m)*
Es espita *(f)*
Fr robinet *(m)*
It rubinetto *(m)*
Pt torneira *(f)*

coefficient de transfert de chaleur *(m)* Fr
De Wärmeübertragungszahl *(f)*
En heat-transfer coefficient
Es coeficiente de transferencia térmica *(m)*
It coefficiente di scambio di calore *(m)*
Pt coeficiente de transferência térmica *(m)*

coefficiente di scambio di calore *(m)* It
De Wärmeübertragungszahl *(f)*
En heat-transfer coefficient
Es coeficiente de transferencia térmica *(m)*
Fr coefficient de transfert de chaleur *(m)*
Pt coeficiente de transferência térmica *(m)*

coeficiente de transferencia térmica *(m)* Es
De Wärmeübertragungszahl *(f)*
En heat-transfer coefficient
Fr coefficient de transfert de chaleur *(m)*
It coefficiente di scambio di calore *(m)*
Pt coeficiente de transferência térmica *(m)*

coeficiente de transferência térmica *(m)* Pt
De Wärmeübertragungszahl *(f)*
En heat-transfer coefficient
Es coeficiente de transferencia térmica *(m)*
Fr coefficient de transfert de chaleur *(m)*
It coefficiente di scambio di calore *(m)*

coeur (réacteur nucléaire) *(m) n* Fr
De Spaltzone (Kernkraftreaktor) *(f)*
En core (nuclear reactor)
Es núcleo (reactor nuclear) *(m)*
It cuore (reattore nucléaire) *(m)*
Pt núcleo (reactor nuclear) *(m)*

cohete nuclear *(m)* Es
De Kernkraftrakete *(f)*
En nuclear rocket
Fr fusée nucléaire *(f)*
It razzo nucleare *(m)*
Pt foguetão nuclear *(m)*

coil-in-bath heat exchanger En
De Wärmeaustauscher bestehend aus Rohrschlangen im Bad *(m)*
Es intercambiador de calor con serpentín en baño *(m)*
Fr échangeur de chaleur à serpentin immergé *(m)*
It scambiatore di calore a serpentino in bagno *(m)*
Pt termo-permutador de bobina em banho *(m)*

coke *n* En; Fr, It *(m)*
De Koks *(m)*
Es coque *(m)*
Pt coque *(m)*

coke de brai *(m)* Fr
De Pechkoks *(m)*
En pitch coke
Es coque con ligante *(m)*
It coke di pece *(m)*
Pt coque de pez *(m)*

coke di pece *(m)* It
De Pechkoks *(m)*
En pitch coke
Es coque con ligante *(m)*
Fr coke de brai *(m)*
Pt coque de pez *(m)*

coke formato *(m)* It
De geformter Koks *(m)*
En formed coke
Es coque conformado *(m)*
Fr coke formé *(m)*
Pt coque formado *(m)*

coke formé *(m)* Fr
De geformter Koks *(m)*
En formed coke
Es coque conformado *(m)*
It coke formato *(m)*
Pt coque formado *(m)*

cokéification différée *(f)* Fr
De verzögerte Verkokung *(f)*
En delayed coking
Es coquización retardada *(f)*
It caricamento ritardato *(m)*
Pt coqueação retardada *(f)*

coke metallurgico *(m)* It
De metallurgischer Koks *(m)*
En metallurgical coke
Es coque metalúrgico *(m)*
Fr coke métallurgique *(m)*
Pt coque metalúrgico *(m)*

coke métallurgique *(m)* Fr
De metallurgischer Koks *(m)*
En metallurgical coke
Es coque metalúrgico *(m)*
It coke metallurgico *(m)*
Pt coque metalúrgico *(m)*

coke-oven tar En
De Koksofenteer *(m)*
Es alquitrán de horno de coquización *(m)*
Fr goudron *(m)*
It catrame di forno da coke *(m)*
Pt alcatrão de forno de coque *(m)*

coking coal En
De Kokskohle *(f)*
Es carbón de coquización *(m)*
Fr charbon cokéfiant *(m)*
It carbone cokificante *(m)*
Pt carvão de coqueificação *(m)*

coking oil En
De Koksöl *(n)*
Es petróleo de coquización *(m)*
Fr huile cokéfiante *(f)*
It olio cokificante *(m)*
Pt petróleo de coqueificação *(m)*

colas *(f pl) n* Pt
De Haldenabfall *(m)*
En tailings *(pl)*
Es desechos *(m pl)*
Fr produits de queue *(m pl)*
It residui di scarto *(m pl)*

cold start En
De Kaltstart *(m)*
Es arranque en frío *(m)*
Fr démarrage à froid *(m)*
It avviamento a freddo *(m)*
Pt arranque a frio *(m)*

colector *(m) n* Es
De Kommutator *(m)*
En commutator
Fr collecteur *(m)*
It commutatore *(m)*
Pt comutador *(m)*

colector de janela única *(m)* Pt
De Einzelfenster-Kollektor *(m)*
En single-window collector
Es colector de ventana única *(m)*
Fr collecteur à simple fenêtre *(m)*
It raccogliatore a finestra unica *(m)*

colector descoberto *(m)* Pt
De Glattrohr- Kollektor *(m)*
En bare collector
Es colector desnudo *(m)*
Fr collecteur nu *(m)*
It collettore nudo *(m)*

colector desnudo *(m)* Es
De Glattrohr-Kollektor *(m)*
En bare collector
Fr collecteur nu *(m)*
It collettore nudo *(m)*
Pt colector descoberto *(m)*

colector de ventana única *(m)* Es
De Einzelfenster-Kollektor *(m)*
En single-window collector
Fr collecteur à simple fenêtre *(m)*
It raccogliatore a finestra unica *(m)*
Pt colector de janela única *(m)*

collecteur *(m) n* Fr
De Kommutator *(m)*
En commutator
Es colector *(m)*
It commutatore *(m)*
Pt comutador *(m)*

collecteur à simple fenêtre *(m)* Fr
De Einzelfenster-Kollektor *(m)*
En single-window collector
Es colector de ventana única *(m)*
It raccogliatore a finestra unica *(m)*
Pt colector de janela única *(m)*

collecteur nu *(m)* Fr
De Glattrohr-Kollektor *(m)*
En bare collector
Es colector desnudo *(m)*
It collettore nudo *(m)*
Pt colector descoberto *(m)*

collettore nudo *(m)* It
De Glattrohr-Kollektor *(m)*
En bare collector
Es colector desnudo *(m)*
Fr collecteur nu *(m)*
Pt colector descoberto *(m)*

colonna di distillazione *(f)* It
De Destillationskolonne *(f)*
En distillation column
Es torre de destilación *(f)*
Fr colonne de distillation *(f)*
Pt coluna de destilação *(f)*

colonne de distillation *(f)* Fr
De Destillationskolonne *(f)*
En distillation column
Es torre de destilación *(f)*
It colonna di distillazione *(f)*
Pt coluna de destilação *(f)*

coluna de destilação *(f)* Pt
De Destillationskolonne *(f)*
En distillation column
Es torre de destilación *(f)*
Fr colonne de distillation *(f)*
It colonna di distillazione *(f)*

comando manual *(m)* Pt
De manuelle Steuerung *(f)*
En manual control
Es control manual *(m)*
Fr commande manuelle *(f)*
It comando manuale *(m)*

comando manuale *(m)* It
De manuelle Steuerung *(f)*
En manual control
Es control manual *(m)*
Fr commande manuelle *(f)*
Pt comando manual *(m)*

combined cycle *n* En
De Kombinationsbetrieb *(m)*
Es ciclo combinado *(m)*
Fr cycle mixte *(m)*
It ciclo combinato *(m)*
Pt ciclo combinado *(m)*

combined heat and power En
De kombinierte Wärme und Energie *(f)*
Es calor y potencia combinados *(m)*
Fr chaleur et puissance combinées *(f)*
It potenza e calore combinati *(f)*
Pt calor e potência combinados *(m)*

combustão *(f) n* Pt
De Verbrennung *(f)*
En combustion
Es combustión *(f)*
Fr combustion *(f)*
It combustione *(f)*

combustão catalítica *(f)* Pt
De katalytische Verbrennung *(f)*
En catalytic combustion
Es combustión catalítica *(f)*
Fr combustion catalytique *(f)*
It combustione per catalisi *(f)*

combustão de ciclo duplo *(f)* Pt
De Zweikreisverbrennung *(f)*
En dual-cycle combustion
Es combustión de ciclo doble *(f)*
Fr combustion à deux temps *(f)*
It combustione a ciclo doppio *(f)*

combustão de duas fases *(f)* Pt
De Zweistufenverbrennung *(f)*
En two-stage combustion
Es combustión en dos etapas *(f)*
Fr combustion à deux étages *(f)*
It combustione a due stadi *(f)*

combustão de leito fluidificado *(f)* Pt
De Fließbettverbrennung *(f)*
En fluidized-bed combustion
Es combustión de lecho fluidizado *(f)*
Fr combustion à lit fluidisé *(f)*

It combustione a letto fluidizzato *(f)*

combustibile *(m) n* It
De Brennstoff *(m)*
En fuel
Es combustible *(m)*
Fr combustible *(m)*
Pt combustível *(m)*

combustibile arricchito *(m)* It
De angereicherter Brennstoff *(m)*
En enriched fuel
Es combustible enriquecido *(m)*
Fr combustible enrichi *(m)*
Pt combustível enriquecido *(m)*

combustibile carbonioso *(m)* It
De kohlenstoffhaltiger Brennstoff *(m)*
En carbonaceous fuel
Es combustible carbonoso *(m)*
Fr combustible charbonneux *(m)*
Pt combustível carbonoso *(m)*

combustibili a base di catrame *(m pl)* It
De Steinkohlenteer-Treibstoffe *(pl)*
En coal-tar fuels *(pl)*
Es combustibles de carbón-alquitrán *(m pl)*
Fr combustibles de goudron de houille *(m pl)*
Pt combustíveis de alcatrão de hulha *(m pl)*

combustibile chimico *(m)* It
De chemischer Brennstoff *(m)*
En chemical fuel
Es combustible químico *(m)*
Fr combustible chimique *(m)*
Pt combustível químico *(m)*

combustibile diesel per motori marini *(m)* It
De Schiffsdieseltreibstoff *(m)*
En marine diesel-fuel
Es combustible diesel para marina *(m)*
Fr combustible diesel marin *(m)*
Pt diesel-fuel marítimo *(m)*

combustibile di scarto *(m)* It
De Ausschußbrennstoff *(m)*
En reject fuel
Es combustible recusable *(m)*
Fr combustible de rebut *(m)*
Pt combustível rejeitado *(m)*

combustibile fossile *(m)* It
De fossiler Brennstoff *(m)*
En fossil fuel
Es combustible fósil *(m)*
Fr combustible fossile *(m)*
Pt combustível fóssil *(m)*

combustibile per aviazione *(m)* It
De Flugkraftstoff *(m)*
En aviation fuel
Es gasolina de aviación *(f)*
Fr carburation aviation *(m)*
Pt combustível de aviação *(m)*

combustibile per aviogetti *(m)* It
De Düsentreibstoff *(m)*
En jet fuel
Es combustible de propulsión a chorro *(m)*
Fr carburéacteur *(m)*
Pt combustível de jactos *(m)*

combustibile solido *(m)* It
De Festkraftstoff *(m)*
En solid fuel
Es combustible sólido *(m)*
Fr combustible solide *(m)*
Pt combustível sólido *(m)*

combustibili di scarico *(m pl)* It
De Abfalltreibstoffe *(pl)*
En waste fuels *(pl)*
Es combustibles de desecho *(m pl)*
Fr combustibles de récupération *(m pl)*
Pt combustíveis perdidos *(m pl)*

combustibili residui *(m pl)* It
De Rückstandsöl *(n)*
En residual fuels *(pl)*
Es combustibles residuales *(m pl)*
Fr fuel résiduel *(m)*
Pt combustíveis residuais *(m pl)*

combustible *(m) n* Es, Fr
De Brennstoff *(m)*
En fuel
It combustibile *(m)*
Pt combustível *(m)*

combustible carbonoso *(m)* Es
De kohlenstoffhaltiger Brennstoff *(m)*
En carbonaceous fuel
Fr combustible charbonneux *(m)*
It combustibile carbonioso *(m)*
Pt combustível carbonoso *(m)*

combustible charbonneux *(m)* Fr
De kohlenstoffhaltiger Brennstoff *(m)*
En carbonaceous fuel
Es combustible carbonoso *(m)*
It combustibile carbonioso *(m)*
Pt combustível carbonoso *(m)*

combustible chimique *(m)* Fr
De chemischer Brennstoff *(m)*
En chemical fuel
Es combustible químico *(m)*
It combustibile chimico *(m)*
Pt combustível químico *(m)*

combustible de propulsión a chorro *(m)* Es
De Düsentreibstoff *(m)*
En jet fuel
Fr carburéacteur *(m)*
It combustibile per aviogetti *(m)*
Pt combustível de jactos *(m)*

combustible de rebut *(m)* Fr
De Ausschußbrennstoff *(m)*
En reject fuel
Es combustible recusable *(m)*
It combustibile di scarto *(m)*
Pt combustível rejeitado *(m)*

combustible diesel *(m)* Fr
De Dieselöl *(n)*
En diesel oil
Es aceite diesel *(m)*
It nafta *(f)*
Pt óleo diesel *(m)*

combustible diesel marin *(m)* Fr
De Schiffsdieseltreibstoff *(m)*
En marine diesel-fuel
Es combustible diesel para marina *(m)*
It combustibile diesel per motori marini *(m)*
Pt diesel-fuel marítimo *(m)*

combustible diesel para marina *(m)* Es
De Schiffsdieseltreibstoff *(m)*
En marine diesel-fuel
Fr combustible diesel marin *(m)*
It combustibile diesel per motori marini *(m)*
Pt diesel-fuel marítimo *(m)*

combustible enrichi *(m)* Fr
De angereicherter Brennstoff *(m)*
En enriched fuel
Es combustible enriquecido *(m)*
It combustibile arricchito *(m)*
Pt combustível enriquecido *(m)*

combustible enriquecido *(m)* Es
De angereicherter Brennstoff *(m)*
En enriched fuel
Fr combustible enrichi *(m)*
It combustibile arricchito *(m)*
Pt combustível enriquecido *(m)*

combustible fósil *(m)* Es
De fossiler Brennstoff *(m)*
En fossil fuel
Fr combustible fossile *(m)*
It combustibile fossile *(m)*
Pt combustível fóssil *(m)*

combustible fossile *(m)* Fr
De fossiler Brennstoff *(m)*
En fossil fuel
Es combustible fósil *(m)*
It combustibile fossile *(m)*
Pt combustível fóssil *(m)*

combustible para cohetes *(m)* Es
De Raketentreibstoff *(m)*
En rocket fuel
Fr combustible pour fusées *(m)*
It carburante per missili *(m)*
Pt combustível para foguetões *(m)*

combustible pour fusées *(m)* Fr
De Raketentreibstoff *(m)*
En rocket fuel
Es combustible para cohetes *(m)*
It carburante per missili *(m)*
Pt combustível para foguetões *(m)*

combustible químico *(m)* Es
De chemischer Brennstoff *(m)*
En chemical fuel
Fr combustible chimique *(m)*
It combustibile chimico *(m)*
Pt combustível químico *(m)*

combustible recusable *(m)* Es
De Ausschußbrennstoff *(m)*
En reject fuel
Fr combustible de rebut *(m)*
It combustibile di scarto *(m)*
Pt combustível rejeitado *(m)*

combustibles de carbón-alquitrán *(m pl)* Es
De Steinkohlenteer-Treibstoffe *(pl)*
En coal-tar fuels *(pl)*
Fr combustibles de goudron de houille *(m pl)*
It combustibili a base di catrame *(m pl)*
Pt combustíveis de alcatrão de hulha *(m pl)*

combustibles de desecho *(m pl)* Es
De Abfalltreibstoffe *(pl)*
En waste fuels *(pl)*
Fr combustibles de récupération *(m pl)*
It combustibili di scarico *(m pl)*
Pt combustíveis perdidos *(m pl)*

combustibles de goudron de houille *(m pl)* Fr
De Steinkohlenteer-Treibstoffe *(pl)*
En coal-tar fuels *(pl)*
Es combustibles de carbón-alquitrán *(m pl)*
It combustibili a base di catrame *(m pl)*
Pt combustíveis de alcatrão de hulha *(m pl)*

combustibles de récupération *(m pl)* Fr
De Abfalltreibstoffe *(pl)*
En waste fuels *(pl)*
Es combustibles de desecho *(m pl)*
It combustibili di scarico *(m pl)*
Pt combustíveis perdidos *(m pl)*

combustible solide *(m)* Fr
De Festkraftstoff *(m)*
En solid fuel
Es combustible sólido *(m)*
It combustibile solido *(m)*
Pt combustível sólido *(m)*

combustible sólido *(m)* Es
De Festkraftstoff *(m)*
En solid fuel
Fr combustible solide *(m)*
It combustibile solido *(m)*
Pt combustível sólido *(m)*

combustibles residuales *(m pl)* Es
De Rückstandsöl *(n)*
En residual fuels *(pl)*
Fr fuel résiduel *(m)*
It combustibili residui *(m pl)*
Pt combustíveis residuais *(m pl)*

combustion *n* En; Fr *(f)*
De Verbrennung *(f)*
Es combustión *(f)*
It combustione *(f)*
Pt combustão *(f)*

combustión *(f) n* Es
De Verbrennung *(f)*
En combustion
Fr combustion *(f)*
It combustione *(f)*
Pt combustão *(f)*

combustion à deux étages *(f)* Fr
De Zweistufenverbrennung *(f)*
En two-stage combustion
Es combustión en dos etapas *(f)*
It combustione a due stadi *(f)*
Pt combustão de duas fases *(f)*

combustion à deux temps *(f)* Fr
De Zweikreisverbrennung *(f)*
En dual-cycle combustion
Es combustión de ciclo doble *(f)*
It combustione a ciclo doppio *(f)*
Pt combustão de ciclo duplo *(f)*

combustion à lit fluidisé *(f)* Fr
De Fließbettverbrennung *(f)*
En fluidized-bed combustion
Es combustión de lecho fluidizado *(f)*
It combustione a letto fluidizzato *(f)*
Pt combustão de leito fluidificado *(f)*

combustión catalítica *(f)* Es
De katalytische Verbrennung *(f)*
En catalytic combustion
Fr combustion catalytique *(f)*
It combustione per catalisi *(f)*
Pt combustão catalítica *(f)*

combustion catalytique *(f)* Fr
De katalytische Verbrennung *(f)*

En catalytic combustion
Es combustión catalítica *(f)*
It combustione per catalisi *(f)*
Pt combustão catalítica *(f)*

combustion chamber En
De Brennkammer *(f)*
Es cámara de combustión *(f)*
Fr chambre de combustion *(f)*
It camera di combustione *(f)*
Pt câmara de combustão *(f)*

combustión de ciclo doble *(f)* Es
De Zweikreisverbrennung *(f)*
En dual-cycle combustion
Fr combustion à deux temps *(f)*
It combustione a ciclo doppio *(f)*
Pt combustão de ciclo duplo *(f)*

combustión de lecho fluidizado *(f)* Es
De Fließbettverbrennung *(f)*
En fluidized-bed combustion
Fr combustion à lit fluidisé *(f)*
It combustione a letto fluidizzato *(f)*
Pt combustão de leito fluidificado *(f)*

combustione *(f) n* It
De Verbrennung *(f)*
En combustion
Es combustión *(f)*
Fr combustion *(f)*
Pt combustão *(f)*

combustione a ciclo doppio *(f)* It
De Zweikreisverbrennung *(f)*
En dual-cycle combustion
Es combustión de ciclo doble *(f)*
Fr combustion à deux temps *(f)*
Pt combustão de ciclo duplo *(f)*

combustione a due stadi *(f)* It
De Zweistufenverbrennung *(f)*
En two-stage combustion
Es combustión en dos etapas *(f)*
Fr combustion à deux étages *(f)*
Pt combustão de duas fases *(f)*

combustione a letto fluidizzato *(f)* It
De Fließbettverbrennung *(f)*
En fluidized-bed combustion
Es combustión de lecho fluidizado *(f)*
Fr combustion à lit fluidisé *(f)*
Pt combustão de leito fluidificado *(f)*

combustión en dos etapas *(f)* Es
De Zweistufenverbrennung *(f)*
En two-stage combustion
Fr combustion à deux étages *(f)*
It combustione a due stadi *(f)*
Pt combustão de duas fases *(f)*

combustione per catalisi *(f)* It
De katalytische Verbrennung *(f)*
En catalytic combustion
Es combustión catalítica *(f)*
Fr combustion catalytique *(f)*
Pt combustão catalítica *(f)*

combustíveis de alcatrão de hulha *(m pl)* Pt
De Steinkohlenteer-Treibstoffe *(pl)*
En coal-tar fuels *(pl)*
Es combustibles de carbón-alquitrán *(m pl)*
Fr combustibles de goudron de houille *(m pl)*
It combustibili a base di catrame *(m pl)*

combustíveis perdidos *(m pl)* Pt
De Abfalltreibstoffe *(pl)*
En waste fuels *(pl)*
Es combustibles de desecho *(m pl)*
Fr combustibles de récupération *(m pl)*
It combustibili di scarico *(m pl)*

combustíveis propulsores líquidos *(m pl)* Pt
De flüssige Treibstoffe *(pl)*
En liquid propellants *(pl)*
Es propulsantes líquidos *(m pl)*
Fr propergols liquides *(m pl)*
It propellenti liquidi *(m pl)*

combustíveis residuais *(m pl)* Pt
De Rückstandsöl *(n)*
En residual fuels *(pl)*
Es combustibles residuales *(m pl)*
Fr fuel résiduel *(m)*
It combustibili residui *(m pl)*

combustível *(m) n* Pt
De Brennstoff *(m)*
En fuel
Es combustible *(m)*
Fr combustible *(m)*
It combustibile *(m)*

combustível bipropulsor *(m)* Pt
De Zweifachtreibstoff *(m)*
En bipropellant
Es bipropulsante *(m)*
Fr biergol *(m)*
It bipropellente *(m)*

combustível carbonoso *(m)* Pt
De kohlenstoffhaltiger Brennstoff *(m)*
En carbonaceous fuel
Es combustible carbonoso *(m)*
Fr combustible charbonneux *(m)*
It combustibile carbonioso *(m)*

combustível de aviação *(m)* Pt
De Flugkraftstoff *(m)*
En aviation fuel
Es gasolina de aviación *(f)*
Fr carburation aviation *(m)*
It combustibile per aviazione *(m)*

combustível de jactos *(m)* Pt
De Düsentreibstoff *(m)*
En jet fuel
Es combustible de propulsión a chorro *(m)*
Fr carburéacteur *(m)*
It combustibile per aviogetti *(m)*

combustível enriquecido *(m)* Pt
De angereicherter Brennstoff *(m)*
En enriched fuel
Es combustible enriquecido *(m)*
Fr combustible enrichi *(m)*
It combustibile arricchito *(m)*

combustível fóssil *(m)* Pt
De fossiler Brennstoff *(m)*
En fossil fuel
Es combustible fósil *(m)*
Fr combustible fossile *(m)*
It combustibile fossile *(m)*

combustível para foguetões *(m)* Pt
De Raketentreibstoff *(m)*
En rocket fuel
Es combustible para cohetes *(m)*
Fr combustible pour fusées *(m)*

It carburante per missili *(m)*

combustível propulsor *(m)* Pt
De Treibstoff *(m)*
En propellant
Es propulsante *(m)*
Fr propergol *(m)*
It propellente *(m)*

combustível propulsor de foguetão híbrido *(m)* Pt
De Hybrid-Raketen-treibstoff *(m)*
En hybrid rocket propellant
Es propulsante de cohete híbrido *(m)*
Fr propergol hybride *(m)*
It propellente ibrido per missili *(m)*

combustível químico *(m)* Pt
De chemischer Brennstoff *(m)*
En chemical fuel
Es combustible químico *(m)*
Fr combustible chimique *(m)*
It combustibile chimico *(m)*

combustível rejeitado *(m)* Pt
De Ausschußbrennstoff *(m)*
En reject fuel
Es combustible recusable *(m)*
Fr combustible de rebut *(m)*
It combustibile di scarto *(m)*

combustível sólido *(m)* Pt
De Festkraftstoff *(m)*
En solid fuel
Es combustible sólido *(m)*
Fr combustible solide *(m)*
It combustibile solido *(m)*

commande de niveau *(f)* Fr
De Füllstandsregler *(m)*
En level control
Es control de nivel *(m)*
It regolatore di livello *(m)*
Pt controle de nível *(m)*

commande manuelle *(f)* Fr
De manuelle Steuerung *(f)*
En manual control
Es control manual *(m)*
It comando manuale *(m)*
Pt comando manual *(m)*

commercial fast reactor (CFR) En
De kommerzieller Schnellreaktor *(m)*
Es reactor rápido comercial *(m)*
Fr réacteur rapide industriel *(m)*
It reattore rapido commerciale *(m)*
Pt reactor rápido comercial *(m)*

commutator *n* En
De Kommutator *(m)*
Es colector *(m)*
Fr collecteur *(m)*
It commutatore *(m)*
Pt comutador *(m)*

commutatore *(m) n* It
De Kommutator *(m)*
En commutator
Es colector *(m)*
Fr collecteur *(m)*
Pt comutador *(m)*

composé aromatique *(m)* Fr
De aromatische Verbindung *(f)*
En aromatic compound
Es compuesto aromático *(m)*
It composto aromatico *(m)*
Pt composto aromático *(m)*

composto aromatico *(m)* It
De aromatische Verbindung *(f)*
En aromatic compound
Es compuesto aromático *(m)*
Fr composé aromatique *(m)*
Pt composto aromático *(m)*

composto aromático *(m)* Pt
De aromatische Verbindung *(f)*
En aromatic compound
Es compuesto aromático *(m)*
Fr composé aromatique *(m)*
It composto aromatico *(m)*

compresor centrífugo *(m)* Es
De radialer Turboverdichter *(m)*
En centrifugal compressor
Fr compresseur centrifuge *(m)*
It compressore centrifugo *(m)*
Pt compressor centrífugo *(m)*

compresor de aire *(m)* Es
De Luftkompressor *(m)*
En air compressor
Fr compresseur d'air *(m)*
It compressore d'aria *(m)*
Pt compressor de ar *(m)*

compressed-air storage En
De Druckluftspeicherung *(f)*
Es almacenamiento de aire comprimido *(m)*
Fr stockage de l'air comprimé *(m)*
It immagazzinaggio di aria compressa *(m)*
Pt armazenagem de ar comprimido *(f)*

compresseur centrifuge *(m)* Fr
De radialer Turboverdichter *(m)*
En centrifugal compressor
Es compresor centrífugo *(m)*
It compressore centrifugo *(m)*
Pt compressor centrífugo *(m)*

compresseur d'air *(m)* Fr
De Luftkompressor *(m)*
En air compressor
Es compresor de aire *(m)*
It compressore d'aria *(m)*
Pt compressor de ar *(m)*

compression ratio En
De Kompressions-verhältnis *(n)*
Es relación de compresión *(f)*
Fr taux de compression *(m)*
It rapporto di compressione *(m)*
Pt taxa de compressão *(f)*

compressor centrífugo *(m)* Pt
De radialer Turboverdichter *(m)*
En centrifugal compressor
Es compresor centrífugo *(m)*
Fr compresseur centrifuge *(m)*
It compressore centrifugo *(m)*

compressor de ar *(m)* Pt
De Luftkompressor *(m)*
En air compressor
Es compresor de aire *(m)*
Fr compresseur d'air *(m)*
It compressore d'aria *(m)*

compressore centrifugo *(m)* It
De radialer Turboverdichter *(m)*
En centrifugal compressor
Es compresor centrífugo *(m)*
Fr compresseur centrifuge *(m)*
Pt compressor centrífugo *(m)*

compressore d'aria *(m)* It
De Luftkompressor *(m)*
En air compressor
Es compresor de aire *(m)*
Fr compresseur d'air *(m)*
Pt compressor de ar *(m)*

comprimento de onda *(m)* Pt
De Wellenlänge *(f)*
En wavelength
Es longitud de onda *(f)*
Fr longueur d'onde *(f)*
It lunghezza d'onda *(f)*

compteur *(m) n* Fr
De Meßgerät *(n)*
En meter
Es medidor *(m)*
It metro *(m)*
Pt contador *(m)*

compteur à piston rotatif *(m)* Fr
De Drehverschiebungsmesser *(m)*
En rotary-displacement meter
Es contador de desplazamiento rotativo *(m)*
It metro di spostamento rotante *(m)*
Pt contador de deslocação rotativa *(m)*

compteur Geiger *(m)* Fr
De Geiger-Zähler *(m)*
En Geiger counter
Es contador de Geiger *(m)*
It contatore Geiger *(m)*
Pt contador Geiger *(m)*

compteur humide *(m)* Fr
De Naßmesser *(m)*
En wet meter
Es medidor húmedo *(m)*
It contatore a liquido *(m)*
Pt medidor de humidade *(m)*

compteur Venturi *(m)* Fr
De Venturimesser *(m)*
En Venturi meter
Es medidor Venturi *(m)*
It venturimetro *(m)*
Pt contador de Venturi *(m)*

compuesto aromático *(m)* Es
De aromatische Verbindung *(f)*
En aromatic compound
Fr composé aromatique *(m)*
It composto aromatico *(m)*
Pt composto aromático *(m)*

comutador *(m) n* Pt
De Kommutator *(m)*
En commutator
Es colector *(m)*
Fr collecteur *(m)*
It commutatore *(m)*

comutador cronométrico *(m)* Pt
De Zeitschalter *(m)*
En time switch
Es temporizador *(m)*
Fr minuterie *(f)*
It interruttore a tempo *(m)*

concentric-tube heat exchanger En
De Wärmeaustauscher mit konzentrischen Röhren *(m)*
Es intercambiador de calor de tubos concéntricos *(m)*
Fr échangeur de chaleur à tubes concentriques (m)
It scambiatore di calore a tubo concentrico *(m)*
Pt termo-permutador de tubo concêntrico *(m)*

concreção *(f) n* Pt
De Sintern *(n)*
En sintering
Es sinterización *(f)*
Fr frittage *(m)*
It sinterizzazione *(f)*

concrete pressure vessel En
De Betondruckbehälter *(m)*
Es recipiente de presión de hormigón *(m)*
Fr récipient de pression en béton *(m)*
It recipiente a pressione di betone *(m)*
Pt recipiento de pressão de betão *(m)*

condensação *(f) n* Pt
De Kondensation *(f)*
En condensation
Es condensación *(f)*
Fr condensation *(f)*
It condensazione *(f)*

condensación *(f) n* Es
De Kondensation *(f)*
En condensation
Fr condensation *(f)*
It condensazione *(f)*
Pt condensação *(f)*

condensador (eléctrico) *(m) n* Es
De Kondensator (elektrischer) *(m)*
En capacitor
Fr condensateur *(m)*
It condensatore (elettrico) *(m)*
Pt capacitor *(m)*

condensador (químico) *(m) n* Es, Pt
De Kondensator (chemischer) *(m)*
En condenser (chemical)
Fr condenseur *(m)*
It condensatore (chimico) *(m)*

condensateur *(m) n* Fr
De Kondensator (elektrischer) *(m)*
En capacitor
Es condensador (eléctrico) *(m)*
It condensatore (elettrico) *(m)*
Pt capacitor *(m)*

condensation *n* En; Fr *(f)*
De Kondensation *(f)*
Es condensación *(f)*
It condensazione *(f)*
Pt condensação *(f)*

condensatore (elettrico) *(m) n* It
De Kondensator (elektrischer) *(m)*
En capacitor
Es condensador (eléctrico) *(m)*
Fr condensateur *(m)*
Pt capacitor *(m)*

condensatore (chimico) *(m) n* It
De Kondensator (chemischer) *(m)*
En condenser (chemical)
Es condensador (químico) *(m)*
Fr condenseur *(m)*
Pt condensador *(m)*

condensazione *(f) n* It
De Kondensation *(f)*
En condensation
Es condensación *(f)*
Fr condensation *(f)*
Pt condensação *(f)*

condenser (chemical) *n* En
De Kondensator (chemischer) *(m)*
Es condensador (químico) *(m)*
Fr condenseur *(m)*
It condensatore (chimico) *(m)*
Pt condensador *(m)*

condenser feedwater En
De Kondensator-Speisewasser *(n)*
Es agua de alimentación del condensador *(f)*
Fr eau d'alimentation de condenseur *(f)*
It acqua di alimentazione del condensatore *(f)*
Pt água de alimentação de condensador *(f)*

condenseur *(m) n* Fr
De Kondensator (chemischer) *(m)*
En condenser (chemical)
Es condensador (químico) *(m)*
It condensatore (chimico) *(m)*
Pt condensador *(m)*

condizionamento dell'aria *(m)* It
De Klimaanlage *(f)*
En air conditioning
Es acondicionamiento de aire *(m)*
Fr climatisation *(f)*
Pt ar condicionado *(m)*

condotti multipli *(m pl)* It
De Sammel-Abgas-leitungen *(pl)*
En multiple flues *(pl)*
Es conductos de humo múltiples *(m pl)*
Fr carneaux multiples *(m pl)*
Pt fumeiros múltiplos *(m pl)*

condotto *(m) n* It
De Abzug *(m)*
En flue
Es humero *(m)*
Fr carneau *(m)*
Pt fumeiro *(m)*

condotto marino *(m)* It
De Seerohrleitung *(f)*
En marine pipeline
Es conducción submarina *(f)*
Fr pipeline marin *(m)*
Pt oleoduto marítimo *(m)*

condução *(f) n* Pt
De Leitung *(f)*
En conduction
Es conducción *(f)*
Fr conduction *(f)*
It conduzione *(f)*

condução por furos *(f)* Pt
De Defektleitung *(f)*
En hole conduction
Es conducción por lagunas *(f)*
Fr conduction par les trous *(f)*
It conduzione dei buchi *(f)*

conducción *(f) n* Es
De Leitung *(f)*
En conduction
Fr conduction *(f)*
It conduzione *(f)*
Pt condução *(f)*

conducción por lagunas *(f)* Es
De Defektleitung *(f)*
En hole conduction
Fr conduction par les trous *(f)*
It conduzione dei buchi *(f)*
Pt condução por furos *(f)*

conducción submarina *(f)* Es
De Seerohrleitung *(f)*
En marine pipeline
Fr pipeline marin *(m)*
It condotto marino *(m)*
Pt oleoduto marítimo *(m)*

conduction *n* En; Fr *(f)*
De Leitung *(f)*
Es conducción *(f)*
It conduzione *(f)*
Pt condução *(f)*

conduction par les trous *(f)* Fr
De Defektleitung *(f)*
En hole conduction
Es conducción por lagunas *(f)*
It conduzione dei buchi *(f)*
Pt condução por furos *(f)*

conductividad térmica *(f)* Es
De Wärmeleitfähigkeit *(f)*
En thermal conductivity
Fr conductivité thermique *(f)*
It conduttività termica *(f)*
Pt condutividade térmica *(f)*

conductivité thermique *(f)* Fr
De Wärmeleitfähigkeit *(f)*
En thermal conductivity
Es conductividad térmica *(f)*
It conduttività termica *(f)*
Pt condutividade térmica *(f)*

conductos de humo múltiples *(m pl)* Es
De Sammel-Abgas-leitungen *(pl)*
En multiple flues *(pl)*
Fr carneaux multiples *(m pl)*
It condotti multipli *(m pl)*
Pt fumeiros múltiplos *(m pl)*

condutividade térmica *(f)* Pt
De Wärmeleitfähigkeit *(f)*
En thermal conductivity
Es conductividad térmica *(f)*
Fr conductivité thermique *(f)*
It conduttività termica *(f)*

condutor em anel fechado *(m)* Pt
De Ringleitung *(f)*
En ring main
Es canalización circular *(f)*
Fr réseau secteur *(m)*
It conduttori ad anello *(m)*

conduttività termica *(f)* It
De Wärmeleitfähigkeit *(f)*
En thermal conductivity
Es conductividad térmica *(f)*
Fr conductivité thermique *(f)*
Pt condutividade térmica *(f)*

conduttori ad anello *(m)* It
De Ringleitung *(f)*
En ring main
Es canalización circular *(f)*
Fr réseau secteur *(m)*
Pt condutor em anel fechado *(m)*

conduzione *(f) n* It
De Leitung *(f)*
En conduction
Es conducción *(f)*
Fr conduction *(f)*
Pt condução *(f)*

conduzione dei buchi *(f)* It
De Defektleitung *(f)*
En hole conduction
Es conducción por lagunas *(f)*
Fr conduction par les trous *(f)*
Pt condução por furos *(f)*

confinamiento *(m) n* Es
De Containment *(m)*
En containment
Fr confinement *(m)*
It contenimento *(m)*
Pt contenção *(f)*

confinement *(m) n* Fr
De Sicherheitshülle *(f)*
En containment
Es confinamiento *(m)*
It contenimento *(m)*
Pt contenção *(f)*

conglomerate rocks *(pl)* En
De klastische Gesteine *(pl)*
Es rocas conglomeradas *(f pl)*
Fr roches conglomérés *(f pl)*
It rocce conglomerate *(f pl)*
Pt rochas conglomeradas *(f pl)*

congruent melting En
De kongruentes Schmelzen *(n)*
Es fusión congruente *(f)*
Fr fusion congruente *(f)*
It fusione congruente *(f)*
Pt fusão congruente *(f)*

conjuncteur *(m) n* Fr
De Schütz *(m)*
En contactor
Es contactor *(m)*
It contattore *(m)*
Pt contactor *(m)*

conservação *(f) n* Pt
De Erhaltung *(f)*
En conservation
Es conservación *(f)*
Fr conservation *(f)*
It conservazione *(f)*

conservação da energia *(f)* Pt
De Energieerhaltung *(f)*

En conservation of energy
Es conservación de energía *(f)*
Fr conservation de l'énergie *(f)*
It conservazione dell'energia *(f)*

conservação de energia-massa *(f)* Pt
De Massenenergie-Erhaltung *(f)*
En mass-energy conservation
Es conservación masa-energía *(f)*
Fr conservation d'energie-masse *(f)*
It conservazione dell'energia-massa *(f)*

conservación *(f) n* Es
De Erhaltung *(f)*
En conservation
Fr conservation *(f)*
It conservazione *(f)*
Pt conservação *(f)*

conservación de energía *(f)* Es
De Energieerhaltung *(f)*
En conservation of energy
Fr conservation de l'énergie *(f)*
It conservazione dell'energia *(f)*
Pt conservação da energia *(f)*

conservación masa-energía *(f)* Es
De Massenenergie-Erhaltung *(f)*
En mass-energy conservation
Fr conservation d'énergie-masse *(f)*
It conservazione dell'energia-massa *(f)*
Pt conservação de energia-massa *(f)*

conservation *n* En; Fr *(f)*
De Erhaltung *(f)*
Es conservación *(f)*
It conservazione *(f)*
Pt conservação *(f)*

conservation de l'énergie *(f)* Fr
De Energieerhaltung *(f)*
En conservation of energy
Es conservación de energía *(f)*
It conservazione dell'energia *(f)*
Pt conservação da energia *(f)*

conservation d'énergie-masse *(f)* Fr
De Massenenergie-Erhaltung *(f)*
En mass-energy conservation
Es conservación masa-energía *(f)*
It conservazione dell'energia-massa *(f)*
Pt conservação de energia-massa *(f)*

conservation of energy En
De Energieerhaltung *(f)*
Es conservación de energía *(f)*
Fr conservation de l'énergie *(f)*
It conservazione dell'energia *(f)*
Pt conservação da energia *(f)*

conservazione *(f) n* It
De Erhaltung *(f)*
En conservation
Es conservación *(f)*
Fr conservation *(f)*
Pt conservação *(f)*

conservazione dell'energia *(f)* It
De Energieerhaltung *(f)*
En conservation of energy
Es conservación de energía *(f)*
Fr conservation de l'énergie *(f)*
Pt conservação da energia *(f)*

conservazione dell'energia-massa *(f)* It
De Massenenergie-Erhaltung *(f)*
En mass-energy conservation
Es conservación masa-energía *(f)*
Fr conservation d'énergie-masse *(f)*
Pt conservação de energia-massa *(f)*

consol process En
De Konsolverfahren *(n)*
Es proceso consol *(m)*
Fr procédé consol *(m)*
It processo consol *(m)*
Pt processo consol *(m)*

consommation de combustible *(f)* Fr
De Brennstoffverbrauch *(m)*
En fuel consumption
Es consumo de combustible *(m)*
It consumo di combustibile *(m)*
Pt consumo de combustível *(m)*

consommation d'énergie *(f)* Fr
De Energieverbrauch *(m)*
En energy consumption
Es consumo de energía *(m)*
It consumo di energia *(m)*
Pt consumo de energia *(m)*

consommation d'énergie par personne *(f)* Fr
De Stromverbrauch pro Kopf *(m)*
En per capita energy consumption
Es consumo de energía per capita *(m)*
It consumo di energia pro capite *(m)*
Pt consumo de energia per capita *(m)*

consommation de pétrole *(f)* Fr
De Olverbrauch *(m)*
En oil consumption
Es consumo de petróleo *(m)*
It consumo di petrolio *(m)*
Pt consumo de petróleo *(m)*

consommation de pointe *(f)* Fr
De Spitzenbedarf *(m)*
En peak demand
Es demanda punta *(f)*
It domanda di punta *(f)*
Pt procura de ponta *(f)*

consommation domestique d'énergie *(f)* Fr
De Haushalt-Energieverbrauch *(m)*
En domestic energy consumption
Es consumo doméstico de energia *(m)*
It consumo di energia per usi domestici *(m)*
Pt consumo doméstico de energia *(m)*

consommation industrielle d'énergie *(f)* Fr
De Energieverbrauch der Industrie *(m)*
En industrial energy consumption
Es consumo industrial de energía *(m)*
It consumo industriale di energia *(m)*
Pt consumo industrial de energia *(m)*

constante de descomposição *(f)* Pt
De Zersetzungskonstante *(f)*
En decay constant
Es constante de desintegración *(f)*
Fr constante de désintégration *(f)*
It constante di disintegrazione *(f)*

constante de desintegración *(f)* Es
De Zersetzungskonstante *(f)*
En decay constant
Fr constante de désintégration *(f)*
It constante di disintegrazione *(f)*

Pt constante de descomposição *(f)*

constante de désintégration *(f)* Fr
De Zersetzungskonstante *(f)*
En decay constant
Es constante de desintegración *(f)*
It constante di disintegrazione *(f)*
Pt constante de descomposição *(f)*

constante de gravitación *(f)* Es
De Gravitationskonstante *(f)*
En gravitational constant
Fr constante de gravitation *(f)*
It costante gravitazionale *(f)*
Pt constante gravitacional *(f)*

constante de gravitation *(f)* Fr
De Gravitationskonstante *(f)*
En gravitational constant
Es constante de gravitación *(f)*
It costante gravitazionale *(f)*
Pt constante gravitacional *(f)*

constante di disintegrazione *(f)* It
De Zersetzungskonstante *(f)*
En decay constant
Es constante de desintegración *(f)*
Fr constante de désintégration *(f)*
Pt constante de descomposição *(f)*

constante dieléctrica *(f)* Es
De Dielektrizitätskonstante *(f)*
En dielectric constant
Fr constante diélectrique *(f)*
It costante dielettrica *(f)*
Pt constante dieléctrica *(f)*

constante dieléctrica *(f)* Pt
De Dielektrizitätskonstante *(f)*
En dielectric constant
Es constante dieléctrica *(f)*
Fr constante diélectrique *(f)*
It costante dielettrica *(f)*

constante diélectrique *(f)* Fr
De Dielektrizitätskonstante *(f)*
En dielectric constant
Es constante dieléctrica *(f)*
It costante dielettrica *(f)*
Pt constante dieléctrica *(f)*

constante eléctrica *(f)* Es, Pt
De Elektrizitätskonstante *(f)*
En electric constant
Fr constante électrique *(f)*
It constante elettrica *(f)*

constante électrique *(f)* Fr
De Elektrizitätskonstante *(f)*
En electric constant
Es constante eléctrica *(f)*
It constante elettrica *(f)*
Pt constante eléctrica *(f)*

constante elettrica *(f)* It
De Elektrizitätskonstante *(f)*
En electric constant
Es constante eléctrica *(f)*
Fr constante électrique *(f)*
Pt constante eléctrica *(f)*

constante gravitacional *(f)* Pt
De Gravitationskonstante *(f)*
En gravitational constant
Es constante de gravitación *(f)*
Fr constante de gravitation *(f)*
It costante gravitazionale *(f)*

constante magnética *(f)* Es, Pt
De Magnetkonstante *(f)*
En magnetic constant
Fr constante magnétique *(f)*
It costante magnetica *(f)*

constante magnétique *(f)* Fr
De Magnetkonstante *(f)*
En magnetic constant
Es constante magnética *(f)*
It costante magnetica *(f)*
Pt constante magnética *(f)*

constante solaire *(f)* Fr
De Solarkonstante *(f)*
En solar constant
Es constante solar *(f)*
It costante solare *(f)*
Pt constante solar *(f)*

constante solar *(f)* Es, Pt
De Solarkonstante *(f)*
En solar constant
Fr constante solaire *(f)*
It costante solare *(f)*

constante universal de gas *(f)* Es
De universelle Gaskonstante *(f)*
En universal gas constant
Fr constante universelle des gaz *(f)*
It costante universale dei gas *(f)*
Pt constante universal de gás *(f)*

constante universal de gás *(f)* Pt
De universelle Gaskonstante *(f)*
En universal gas constant
Es constante universal de gas *(f)*
Fr constante universelle des gaz *(f)*
It costante universale dei gas *(f)*

constante universelle des gaz *(f)* Fr
De universelle Gaskonstante *(f)*
En universal gas constant
Es constante universal de gas *(f)*
It costante universale dei gas *(f)*
Pt constante universal de gás *(f)*

consumo de combustible *(m)* Es
De Brennstoffverbrauch *(m)*
En fuel consumption
Fr consommation de combustible *(f)*
It consumo di combustibile *(m)*
Pt consumo de combustível *(m)*

consumo de combustível *(m)* Pt
De Brennstoffverbrauch *(m)*
En fuel consumption
Es consumo de combustible *(m)*
Fr consommation de combustible *(f)*
It consumo di combustibile *(m)*

consumo de energia *(m)* Pt
De Energieverbrauch *(m)*
En energy consumption
Es consumo de energía *(m)*
Fr consommation d'énergie *(f)*
It consumo di energia *(m)*

consumo de energía *(m)* Es
De Energieverbrauch *(m)*
En energy consumption
Fr consommation d'énergie *(f)*
It consumo di energia *(m)*

Pt consumo de energia *(m)*

consumo de energia per capita *(m)* Pt
De Stromverbrauch pro Kopf *(m)*
En per capita energy consumption
Es consumo de energía per capita *(m)*
Fr consommation d'énergie par personne *(f)*
It consumo di energia pro capite *(m)*

consumo de energía per capita *(m)* Es
De Stromverbrauch pro Kopf *(m)*
En per capita energy consumption
Fr consommation d'énergie par personne *(f)*
It consumo di energia pro capite *(m)*
Pt consumo de energia per capita *(m)*

consumo de petróleo *(m)* Es, Pt
De Ölverbrauch *(m)*
En oil consumption
Fr consommation de pétrole *(f)*
It consumo di petrolio *(m)*

consumo di combustibile *(m)* It
De Brennstoffverbrauch *(m)*
En fuel consumption
Es consumo de combustible *(m)*
Fr consommation de combustible *(f)*
Pt consumo de combustível *(m)*

consumo di energia *(m)* It
De Energieverbrauch *(m)*
En energy consumption
Es consumo de energía *(m)*
Fr consommation d'énergie *(f)*
Pt consumo de energia *(m)*

consumo di energia per usi domestici *(m)* It
De Haushalt-Energieverbrauch *(m)*
En domestic energy consumption
Es consumo doméstico de energía *(m)*
Fr consommation domestique d'énergie *(f)*
Pt consumo doméstico de energia *(m)*

consumo di energia pro capite *(m)* It
De Stromverbrauch pro Kopf *(m)*
En per capita energy consumption
Es consumo de energía per capita *(m)*
Fr consommation d'énergie par personne *(f)*
Pt consumo de energia per capita *(m)*

consumo di petrolio *(m)* It
De Ölverbrauch *(m)*
En oil consumption
Es consumo de petróleo *(m)*
Fr consommation de pétrole *(f)*
Pt consumo de petróleo *(m)*

consumo doméstico de energia *(m)* Pt
De Haushalt-Energieverbrauch *(m)*
En domestic energy consumption
Es consumo doméstico de energía *(m)*
Fr consommation doméstique d'énergie *(f)*
It consumo di energia per usi domestici *(m)*

consumo doméstico de energía *(m)* Es
De Haushalt-Energieverbrauch *(m)*
En domestic energy consumption
Fr consommation doméstique d'énergie *(f)*
It consumo di energia per usi domestici *(m)*
Pt consumo doméstico de energia *(m)*

consumo industrial de energia *(m)* Pt
De Energieverbrauch der Industrie *(m)*
En industrial energy consumption
Es consumo industrial de energía *(m)*
Fr consommation industrielle d'énergie *(f)*
It consumo industriale di energia *(m)*

consumo industrial de energía *(m)* Es
De Energieverbrauch der Industrie *(m)*
En industrial energy consumption
Fr consommation industrielle d'énergie *(f)*
It consumo industriale di energia *(m)*
Pt consumo industrial de energia *(m)*

consumo industriale di energia *(m)* It
De Energieverbrauch der Industrie *(m)*
En industrial energy consumption
Es consumo industrial de energía *(m)*
Fr consommation industrielle d'énergie *(f)*
Pt consumo industrial de energia *(m)*

contactor *n* En; Es, Pt *(m)*
De Schütz *(m)*
Fr conjuncteur *(m)*
It contattore *(m)*

contact potential En
De Kontaktspannung *(f)*
Es potencial de contacto *(m)*
Fr potentiel de contact *(m)*
It potenziale di contatto *(m)*
Pt potencial de contacto *(m)*

contador *(m) n* Pt
De Meßgerät *(n)*
En meter
Es medidor *(m)*
Fr compteur *(m)*
It metro *(m)*

contador de deslocação rotativa *(m)* Pt
De Drehverschiebungsmesser *(m)*
En rotary-displacement meter
Es contador de desplazamiento rotativo *(m)*
Fr compteur à piston rotatif *(m)*
It metro di spostamento rotante *(m)*

contador de desplazamiento rotativo *(m)* Es
De Drehverschiebungsmesser *(m)*
En rotary-displacement meter
Fr compteur à piston rotatif *(m)*
It metro di spostamento rotante *(m)*
Pt contador de deslocação rotativa *(m)*

contador de Geiger *(m)* Es
De Geiger-Zähler *(m)*
En Geiger counter
Fr compteur Geiger *(m)*
It contatore Geiger *(m)*
Pt contador Geiger *(m)*

contador de Venturi *(m)* Pt
De Venturimesser *(m)*
En Venturi meter
Es medidor Venturi *(m)*
Fr compteur Venturi *(m)*
It venturimetro *(m)*

contador Geiger *(m)* Pt
De Geiger-Zähler *(m)*
En Geiger counter
Es contador de Geiger *(m)*
Fr compteur Geiger *(m)*
It contatore Geiger *(m)*

containment *n* En
De Containment *(m)*
Es confinamiento *(m)*
Fr confinement *(m)*
It contenimento *(m)*
Pt contenção *(f)*

Containment *(m) n* De
En containment
Es confinamiento *(m)*
Fr confinement *(m)*
It contenimento *(m)*
Pt contenção *(f)*

contaminación *(f) n* Es
De Verschmutzung *(f)*
En pollution
Fr pollution *(f)*
It inquinamento *(m)*
Pt poluição *(f)*

contaminación del aire *(f)* Es
De Luftverschmutzung *(f)*
En air pollution
Fr pollution de l'air *(f)*
It inquinamento atmosferico *(m)*
Pt poluição do ar *(f)*

contaminated waste En
De kontaminierter Abfall *(m)*
Es residuo contaminado *(m)*
Fr déchets contaminés *(m pl)*
It rifiuti contaminati *(m pl)*
Pt desperdício contaminado *(m)*

contatore a liquido *(m)* It
De Naßmesser *(m)*
En wet meter
Es medidor húmedo *(m)*
Fr compteur humide *(m)*
Pt medidor de humidade *(m)*

contatore a secco *(m)* It
De Trockenmeßgerät *(n)*
En dry meter
Es medidor seco *(m)*
Fr dessicomètre *(m)*
Pt medidor de seco *(m)*

contatore Geiger *(m)* It
De Geiger-Zähler *(m)*
En Geiger counter
Es contador de Geiger *(m)*
Fr compteur Geiger *(m)*
Pt contador Geiger *(m)*

contattore *(m) n* It
De Schütz *(m)*
En contactor
Es contactor *(m)*
Fr conjuncteur *(m)*
Pt contactor *(m)*

contenção *(f) n* Pt
De Containment *(m)*
En containment
Es confinamiento *(m)*
Fr confinement *(m)*
It contenimento *(m)*

contenido de carbono *(m)* Es
De Kohlenstoffgehalt *(m)*
En carbon content
Fr teneur en carbone *(f)*
It contenuto di carbonio *(m)*
Pt teor de carbono *(m)*

contenido de plomo *(m)* Es
De Bleigehalt *(m)*
En lead content
Fr teneur en plomb *(f)*
It contenuto di piombo *(m)*
Pt teor de chumbo *(m)*

contenimento *(m) n* It
De Containment *(m)*
En containment
Es confinamiento *(m)*
Fr confinement *(m)*
Pt contenção *(f)*

contenuto di carbonio *(m)* It
De Kohlenstoffgehalt *(m)*
En carbon content
Es contenido de carbono *(m)*
Fr teneur en carbone *(f)*
Pt teor de carbono *(m)*

contenuto di piombo *(m)* It
De Bleigehalt *(m)*
En lead content
Es contenido de plomo *(m)*
Fr teneur en plomb *(f)*
Pt teor de chumbo *(m)*

contour mining En
De Konturenabbau *(m)*
Es minería siguiendo el perfil del terreno *(f)*
Fr exploitation minière des contours *(f)*
It scavo a contorno *(m)*
Pt mineração de curvas de nível *(f)*

contre-réaction *(f) n* Fr
De negative Rückkoppelung *(f)*
En negative feedback
Es realimentación negativa *(f)*
It controreazione *(f)*
Pt retôrno negativo *(m)*

control de nivel *(m)* Es
De Füllstandsregler *(m)*
En level control
Fr commande de niveau *(f)*
It regolatore di livello *(m)*
Pt controle de nível *(m)*

controle de nível *(m)* Pt
De Füllstandsregler *(m)*
En level control
Es control de nivel *(m)*
Fr commande de niveau *(f)*
It regolatore di livello *(m)*

control manual *(m)* Es
De manuelle Steuerung *(f)*
En manual control
Fr commande manuelle *(f)*
It comando manuale *(m)*
Pt comando manual *(m)*

control rod En
De Kontrollstab *(m)*
Es varilla de control *(f)*
Fr barre de commande *(f)*
It barra di controllo *(f)*
Pt haste de controle *(f)*

control room En
De Schaltraum *(m)*
Es sala de control *(f)*
Fr salle de commande *(f)*
It camera di manovra *(f)*
Pt sala de comando *(f)*

control system En
De Steuersystem *(n)*
Es sistema de control *(m)*
Fr système de commande *(m)*
It sistema di controllo *(m)*
Pt sistema de controle *(m)*

controreazione *(f) n* It
De negative Rückkoppelung *(f)*
En negative feedback
Es realimentación negativa *(f)*
Fr contre-réaction *(f)*
Pt retôrno negativo *(m)*

convecção *(f) n* Pt
De Konvektion *(f)*
En convection
Es convección *(f)*
Fr convection *(f)*
It convezione *(f)*

convecção forçada *(f)* Pt
De Gebläsekonvektion *(f)*
En forced convection
Es convección forzada *(f)*
Fr convection forcée *(f)*
It convezione forzata *(f)*

convecção natural *(f)* Pt
De natürliche Konvektion *(f)*
En natural convection
Es convección natural *(f)*
Fr convection naturelle *(f)*
It convezione naturale *(f)*

convección *(f) n* Es
De Konvektion *(f)*
En convection
Fr convection *(f)*
It convezione *(f)*
Pt convecção *(f)*

convección forzada *(f)* Es
De Gebläsekonvektion *(f)*
En forced convection
Fr convection forcée *(f)*
It convezione forzata *(f)*
Pt convecção forçada *(f)*

convección natural *(f)* Es
De natürliche Konvektion *(f)*
En natural convection
Fr convection naturelle *(f)*
It convezione naturale *(f)*
Pt convecção natural *(f)*

convecteur à carneau équilibré *(m)* Fr
De Konvektor mit entlastetem Abzug *(m)*
En balanced-flue convector
Es convector de chimenea equilibrada *(m)*
It convettore a focolare equilibrato *(m)*
Pt convector de fluxo equilibrado *(m)*

convecteur radiant *(m)* Fr
De Strahlungskonvektor *(m)*
En radiant convector
Es convector radiante *(m)*
It convettore radiante *(m)*
Pt convector de radiação *(m)*

convecteur sans carneau *(m)* Fr
De Konvektor ohne Abzug *(m)*
En flueless convector
Es convector sin tubo *(m)*
It convettore senza condotto del fumo *(m)*
Pt convector sem fumeiro *(m)*

convection *n* En; Fr *(f)*
De Konvektion *(f)*
Es convección *(f)*
It convezione *(f)*
Pt convecção *(f)*

convection current En
De Konvektionsstrom *(m)*
Es corriente de convección *(f)*
Fr courant de convection *(m)*
It corrente di convezione *(f)*
Pt corrente de convecção *(f)*

convection forcée *(f)* Fr
De Gebläsekonvektion *(f)*
En forced convection
Es convección forzada *(f)*
It convezione forzata *(f)*
Pt convecção forçada *(f)*

convection naturelle *(f)* Fr
De natürliche Konvektion *(f)*
En natural convection
Es convección natural *(f)*
It convezione naturale *(f)*
Pt convecção natural *(f)*

convection recuperator En
De Konvektions-rekuperator *(m)*
Es recuperador de convección *(m)*
Fr récupérateur à convection *(m)*
It recuperatore di convezione *(m)*
Pt recuperador por convecção *(m)*

convector de chimenea equilibrada *(m)* Es
De Konvektor mit entlastetem Abzug *(m)*
En balanced-flue convector
Fr convecteur à carneau équilibré *(m)*
It convettore a focolare equilibrato *(m)*
Pt convector de fluxo equilibrado *(m)*

convector de fluxo equilibrado *(m)* Pt
De Konvektor mit entlastetem Abzug *(m)*
En balanced-flue convector
Es convector de chimenea equilibrada *(m)*
Fr convecteur à carneau équilibré *(m)*
It convettore a focolare equilibrato *(m)*

convector de radiação *(m)* Pt
De Strahlungsheizkessel *(m)*
En radiant boiler
Es caldera radiante *(f)*
Fr chaudière radiante *(f)*
It caldaia radiante *(f)*

convector de radiação *(m)* Pt
De Strahlungskonvektor *(m)*
En radiant convector
Es convector radiante *(m)*
Fr convecteur radiant *(m)*
It convettore radiante *(m)*

convector radiante *(m)* Es
De Strahlungskonvektor *(m)*
En radiant convector
Fr convecteur radiant *(m)*
It convettore radiante *(m)*
Pt convector de radiação *(m)*

convector sem fumeiro *(m)* Pt
De Konvektor ohne Abzug *(m)*
En flueless convector
Es convector sin tubo *(m)*
Fr convecteur sans carneau *(m)*
It convettore senza condotto del fumo *(m)*

convector sin tubo *(m)* Es
De Konvektor ohne Abzug *(m)*
En flueless convector
Fr convecteur sans carneau *(m)*
It convettore senza condotto del fumo *(m)*
Pt convector sem fumeiro *(m)*

conversão de energia a bordo *(m)* Pt
De Energieumwandlung an Bord *(f)*
En on-board energy conversion
Es conversión de energía a bordo *(f)*
Fr transformation d'énergie à bord *(f)*
It conversione di energia a bordo *(f)*

conversão directa *(f)* Pt
De Direktumwandlung *(f)*
En direct conversion
Es conversión directa *(f)*
Fr conversion directe *(f)*
It conversione diretta *(f)*

conversión de energía a bordo *(f)* Es
De Energieumwandlung an Bord *(f)*
En on-board energy conversion
Fr transformation d'énergie à bord *(f)*
It conversione di energia a bordo *(f)*
Pt conversão de energia a bordo *(m)*

conversión directa *(f)* Es
De Direktumwandlung *(f)*
En direct conversion (solar)
Fr conversion directe *(f)*
It conversione diretta *(f)*
Pt conversão directa *(f)*

conversion directe *(f)* Fr
De Direktumwandlung *(f)*
En direct conversion
Es conversión directa *(f)*

It conversione diretta *(f)*
Pt conversão directa *(f)*

conversione di energia a bordo *(f)* It
De Energieumwandlung an Bord *(f)*
En on-board energy conversion
Es conversión de energia a bordo *(f)*
Fr transformation d'énergie à bord *(f)*
Pt conversão de energia a bordo *(m)*

conversione diretta *(f)* It
De Direktumwandlung *(f)*
En direct conversion
Es conversión directa *(f)*
Fr conversion directe *(f)*
Pt conversão directa *(f)*

conversor de potencial *(m)* Pt
De Spannungswandler *(m)*
En potential converter
Es convertidor de potencial *(m)*
Fr convertisseur de tension *(m)*
It convertitore di tensione *(m)*

converter reactor En
De Konverter *(m)*
Es reactor convertidor *(m)*
Fr réacteur convertisseur *(m)*
It reattore convertitore *(m)*
Pt reactor de conversão *(m)*

convertidor de potencial *(m)* Es
De Spannungswandler *(m)*
En potential converter
Fr convertisseur de tension *(m)*
It convertitore di tensione *(m)*
Pt conversor de potencial *(m)*

convertisseur *(m)* *n* Fr
De Motor-Generator *(m)*
En motor-generator
Es motor-generador *(m)*
It gruppo motore-dinamo *(m)*
Pt moto-gerador *(m)*

convertisseur de tension *(m)* Fr
De Spannungswandler *(m)*
En potential converter
Es convertidor de potencial *(m)*
It convertitore di tensione *(m)*
Pt conversor de potencial *(m)*

convertitore di tensione *(m)* It
De Spannungswandler *(m)*
En potential converter
Es convertidor de potencial *(m)*
Fr convertisseur de tension *(m)*
Pt conversor de potencial *(m)*

convettore a focolare equilibrato *(m)* It
De Konvektor mit entlastetem Abzug *(m)*
En balanced-flue convector
Es convector de chimenea equilibrada *(m)*
Fr convecteur à carneau équilibré *(m)*
Pt convector de fluxo equilibrado *(m)*

convettore radiante *(m)* It
De Strahlungskonvektor *(m)*
En radiant convector
Es convector radiante *(m)*
Fr convecteur radiant *(m)*
Pt convector de radiação *(m)*

convettore senza condotto del fumo *(m)* It
De Konvektor ohne Abzug *(m)*
En flueless convector
Es convector sin tubo *(m)*
Fr convecteur sans carneau *(m)*
Pt convector sem fumeiro *(m)*

convezione *(f)* *n* It
De Konvektion *(f)*
En convection
Es convección *(f)*
Fr convection *(f)*
Pt convecção *(f)*

convezione forzata *(f)* It
De Gebläsekonvektion *(f)*
En forced convection
Es convección forzada *(f)*
Fr convection forcée *(f)*
Pt convecção forçada *(f)*

convezione naturale *(f)* It
De natürliche Konvektion *(f)*
En natural convection
Es convección natural *(f)*
Fr convection naturelle *(f)*
Pt convecção natural *(f)*

coolant *n* En
De Kühlmittel *(n)*
Es refrigerante *(m)*
Fr réfrigérant *(m)*
It refrigerante *(m)*
Pt refrigerante *(m)*

cooling circuit En
De Kühlmittelkreislauf *(m)*
Es circuito de refrigeración *(m)*
Fr circuit de refroidissement *(m)*
It circuito di raffreddamento *(m)*
Pt circuito refrigerante *(m)*

cooling pond En
De Kühlteich *(m)*
Es piscina de refrigeración *(f)*
Fr piscine de refroidissement *(f)*
It bacino di raffreddamento *(m)*
Pt tanque refrigerante *(m)*

cooling water En
De Kühlwasser *(n)*
Es agua de refrigeración *(f)*
Fr eau de refroidissement *(f)*
It acqua di raffreddamento *(f)*
Pt água de refrigeração *(f)*

copertura *(f)* *n* It
De Brutzone *(f)*
En blanket
Es zona fértil *(f)*
Fr couche fertile *(f)*
Pt cortina *(f)*

copper loss En
De Kupferverlust *(m)*
Es pérdida en el cobre *(f)*
Fr perte dans le cuivre *(f)*
It perdita nel rame *(f)*
Pt perda no cobre *(f)*

coque *(m)* *n* Es
De Koks *(m)*
En coke
Fr coke *(m)*
It coke *(m)*
Pt coque *(m)*

coque *(m)* *n* Pt
De Koks *(m)*
En coke
Es coque *(m)*
Fr coke *(m)*
It coke *(m)*

coqueação retardada *(f)* Pt
De verzögerte Verkokung *(f)*
En delayed coking
Es coquización retardada *(f)*
Fr cokéification différée *(f)*
It caricamento ritardato *(m)*

coque conformado *(m)* Es
De geformter Koks *(m)*
En formed coke
Fr coke formé *(m)*
It coke formato *(m)*
Pt coque formado *(m)*

coque con ligante *(m)* Es
De Pechkoks *(m)*
En pitch coke
Fr coke de brai *(m)*
It coke di pece *(m)*
Pt coque de pez *(m)*

coque de pez *(m)* Pt
De Pechkoks *(m)*
En pitch coke
Es coque con ligante *(m)*
Fr coke de brai *(m)*
It coke di pece *(m)*

coque formado *(m)* Pt
De geformter Koks *(m)*
En formed coke
Es coque conformado *(m)*
Fr coke formé *(m)*
It coke formato *(m)*

coque metalúrgico *(m)* Es, Pt
De metallurgischer Koks *(m)*
En metallurgical coke
Fr coke métallurgique *(m)*
It coke metallurgico *(m)*

coquización retardada *(f)* Es
De verzögerte Verkokung *(f)*
En delayed coking
Fr cokéification différée *(f)*
It caricamento ritardato *(m)*
Pt coqueação retardada *(f)*

core (nuclear reactor) *n* En
De Spaltzone (Kernkraftreaktor) *(f)*
Es núcleo (reactor nuclear) *(m)*
Fr coeur (réacteur nucléaire) *(m)*
It cuore (reattore nucléaire) *(m)*
Pt núcleo (reactor nuclear) *(m)*

core (magnetic) *n* En
De Kern (magnetischer) *(m)*
Es alma (magnética) *(f)*
Fr noyau (magnétique) *(m)*
It nucleo (magnetico) *(m)*
Pt núcleo (magnético) *(m)*

corpo negro *(m)* Pt
De Planckscher Strahler *(m)*
En black body
Es cuerpo negro *(m)*
Fr corps noir *(m)*
It corpo nero *(m)*

corpo nero *(m)* It
De Planckscher Strahler *(m)*
En black body
Es cuerpo negro *(m)*
Fr corps noir *(m)*
Pt corpo negro *(m)*

corps noir *(m)* Fr
De Planckscher Strahler *(m)*
En black body
Es cuerpo negro *(m)*
It corpo nero *(m)*
Pt corpo negro *(m)*

corrente *(f)* *n* It, Pt
De Strom *(m)*
En current
Es corriente *(f)*
Fr courant *(m)*

corrente alterna (c.a.) *(f)* Pt
De Wechselstrom *(m)*
En alternating current (a.c.)
Es corriente alterna (c.a.) *(f)*
Fr courant alternatif (c.a.) *(m)*
It corrente alternata (c.a.) *(f)*

corrente alternata (c.a.) *(f)* It
De Wechselstrom *(m)*
En alternating current (a.c.)
Es corriente alterna (c.a.) *(f)*
Fr courant alternatif (c.a.) *(m)*
Pt corrente alterna (c.a.) *(f)*

corrente continua (c.c.) *(f)* It
De Gleichstrom *(m)*
En direct current (d.c.)
Es corriente continua (c.c.) *(f)*
Fr courant continu (c.c.) *(m)*
Pt corrente directa (c.d.) *(f)*

corrente de convecção *(f)* Pt
De Konvektionsstrom *(m)*
En convection current
Es corriente de convección *(f)*
Fr courant de convection *(m)*
It corrente di convezione *(f)*

corrente di convezione *(f)* It
De Konvektionsstrom *(m)*
En convection current
Es corriente de convección *(f)*
Fr courant de convection *(m)*
Pt corrente de convecção *(f)*

corrente directa (c.d.) *(f)* Pt
De Gleichstrom *(m)*
En direct current (d.c.)
Es corriente continua (c.c.) *(f)*
Fr courant continu (c.c.) *(m)*
It corrente continua (c.c.) *(f)*

corrente indotta *(f)* It
De induzierter Strom *(m)*
En induced current
Es corriente inducida *(f)*
Fr courant induit *(m)*
Pt corrente induzida *(f)*

corrente induzida *(f)* Pt
De induzierter Strom *(m)*
En induced current
Es corriente inducida *(f)*
Fr courant induit *(m)*
It corrente indotta *(f)*

corrente parasítico *(m)* Pt
De Wirbelstrom *(m)*
En eddy current
Es corriente de Foucault *(f)*
Fr courant de Foucault *(m)*
It corrente parassita *(f)*

corrente parassita *(f)* It
De Wirbelstrom *(m)*
En eddy current
Es corriente de Foucault *(f)*
Fr courant de Foucault *(m)*
Pt corrente parasítico *(m)*

corriente *(f)* *n* Es
De Strom *(m)*
En current
Fr courant *(m)*
It corrente *(f)*
Pt corrente *(f)*

corriente alterna (c.a.) *(f)* Es
De Wechselstrom *(m)*
En alternating current (a.c.)
Fr courant alternatif (c.a.) *(m)*
It corrente alternata (c.a.) *(f)*
Pt corrente alterna (c.a.) *(f)*

corriente continua (c.c.) *(f)* Es
De Gleichstrom *(m)*
En direct current (d.c.)
Fr courant continu (c.c.) *(m)*
It corrente continua (c.c.) *(f)*
Pt corrente directa (c.d.) *(f)*

corriente de convección *(f)* Es
De Konvektionsstrom *(m)*
En convection current
Fr courant de convection *(m)*
It corrente di convezione *(f)*
Pt corrente de convecção *(f)*

corriente de Foucault *(f)* Es
De Wirbelstrom *(m)*
En eddy current
Fr courant de Foucault *(m)*
It corrente parassita *(f)*
Pt corrente parasítico *(m)*

corriente inducida *(f)* Es
De induzierter Strom *(m)*
En induced current
Fr courant induit *(m)*
It corrente indotta *(f)*
Pt corrente induzida *(f)*

corta-circuito *(m) n* Pt
De Leitungsschalter *(m)*
En circuit breaker
Es disyuntor *(m)*
Fr disjoncteur *(m)*
It interruttore automatico *(m)*

cortina *(f) n* Pt
De Brutzone *(f)*
En blanket
Es zona fértil *(f)*
Fr couche fertile *(f)*
It copertura *(f)*

costante dielettrica *(f)* It
De Dielektrizitäts-konstante *(f)*
En dielectric constant
Es constante dieléctrica *(f)*
Fr constante diélectrique *(f)*
Pt constante dieléctrica *(f)*

costante gravitazionale *(f)* It
De Gravitations-konstante *(f)*
En gravitational constant
Es constante de gravitación *(f)*
Fr constante de gravitation *(f)*
Pt constante gravitacional *(f)*

costante magnetica *(f)* It
De Magnetkonstante *(f)*
En magnetic constant
Es constante magnética *(f)*
Fr constante magnétique *(f)*
Pt constante magnética *(f)*

costante solare *(f)* It
De Solarkonstante *(f)*
En solar constant
Es constante solar *(f)*
Fr constante solaire *(f)*
Pt constante solar *(f)*

costante universale dei gas *(f)* It
De universelle Gaskonstante *(f)*
En universal gas constant
Es constante universal de gas *(f)*
Fr constante universelle des gaz *(f)*
Pt constante universal de gás *(f)*

coste de distribución *(m)* Es
De Vertriebskosten *(pl)*
En distribution cost
Fr frais de distribution *(m)*
It costo di distribuzione *(m)*
Pt custo de distribuição *(m)*

costes de explotación *(m pl)* Es
De Betriebskosten *(pl)*
En running costs
Fr frais d'exploitation *(m pl)*
It costi di esercizio *(m pl)*
Pt custo de exploração *(m)*

costi di esercizio *(m pl)* It
De Betriebskosten *(pl)*
En running costs
Es costes de explotación *(m pl)*
Fr frais d'exploitation *(m pl)*
Pt custo de exploração *(m)*

costo di distribuzione *(m)* It
De Vertriebskosten *(pl)*
En distribution cost
Es coste de distribución *(m)*
Fr frais de distribution *(m)*
Pt custo de distribuição *(m)*

couche de houille *(f)* Fr
De Kohlenlager *(n)*
En coal measure
Es formación hullera *(f)*
It giacimento carbonifero *(m)*
Pt medida de carvão *(f)*

couche fertile *(f)* Fr
De Brutzone *(f)*
En blanket
Es zona fértil *(f)*
It copertura *(f)*
Pt cortina *(f)*

couche hydrofuge *(f)* Fr
De Feuchtigkeits-sperrschicht *(f)*
En damp-proof course
Es hilada hidrófuga *(f)*
It trattamento impermeabilizzante *(m)*
Pt curso a prova de imfietrações *(m)*

courant *(m) n* Fr
De Strom *(m)*
En current
Es corriente *(f)*
It corrente *(f)*
Pt corrente *(f)*

courant alternatif (c.a.) *(m)* Fr
De Wechselstrom *(m)*
En alternating current (a.c.)
Es corriente alterna (c.a.) *(f)*
It corrente alternata (c.a.) *(f)*
Pt corrente alterna (c.a.) *(f)*

courant continu (c.c.) *(m)* Fr
De Gleichstrom *(m)*
En direct current (d.c.)
Es corriente continua (c.c.) *(f)*
It corrente continua (c.c.) *(f)*
Pt corrente directa (c.d.) *(f)*

courant de chaleur tellurique *(m)* Fr
De terrestrischer Wärmeflus *(m)*
En terrestrial heat flow
Es flujo térmico terrestre *(m)*
It flusso termico terrestre *(m)*
Pt fluxo de calor de terra *(m)*

courant de convection *(m)* Fr
De Konvektionsstrom *(m)*
En convection current
Es corriente de convección *(f)*
It corrente di convezione *(f)*
Pt corrente de convecção *(f)*

courant de Foucault *(m)* Fr
De Wirbelstrom *(m)*
En eddy current
Es corriente de Foucault *(f)*
It corrente parassita *(f)*
Pt corrente parasítico *(m)*

courant induit *(m)* Fr
De induzierter Strom *(m)*
En induced current
Es corriente inducida *(f)*
It corrente indotta *(f)*
Pt corrente induzida *(f)*

cracking *n* En; Pt *(m)*
De Kracken *(n)*
Es cráckíng *(m)*
Fr craquage *(m)*
It piroscissione *(f)*

cráckíng *(m) n* Es
De Kracken *(n)*
En cracking
Fr craquage *(m)*
It piroscissione *(f)*
Pt cracking *(m)*

cráckíng catalítico *(m)* Es
De katalytisches Kracken *(n)*
En catalytic cracking
Fr craquage catalytique *(m)*

It piroscissione per catalisi *(f)*
Pt cracking catalítico *(m)*

cracking catalítico *(m)* Pt
De katalytisches Kracken *(n)*
En catalytic cracking
Es cráking catalítico *(m)*
Fr craquage catalytique *(m)*
It piroscissione per catalisi *(f)*

cráking térmico *(m)* Es
De thermisches Kracken *(n)*
En thermal cracking
Fr craquage thermique *(m)*
It piroscissione termica *(f)*
Pt cracking térmico *(m)*

cracking térmico *(m)* Pt
De thermisches Kracken *(n)*
En thermal cracking
Es cráking térmico *(m)*
Fr craquage thermique *(m)*
It piroscissione termica *(f)*

craquage *(m) n* Fr
De Kracken *(n)*
En cracking
Es cráking *(m)*
It piroscissione *(f)*
Pt cracking *(m)*

craquage catalytique *(m)* Fr
De katalytisches Kracken *(n)*
En catalytic cracking
Es cráking catalítico *(m)*
It piroscissione per catalisi *(f)*
Pt cracking catalítico *(m)*

craquage thermique *(m)* Fr
De thermisches Kracken *(n)*
En thermal cracking
Es cráking térmico *(m)*
It piroscissione termica *(f)*
Pt cracking térmico *(m)*

cresol *n* En; Es, Pt *(m)*
De Kresol *(n)*
Fr crésol *(m)*
It cresolo *(m)*

crésol *(m) n* Fr
De Kresol *(n)*
En cresol
Es cresol *(m)*
It cresolo *(m)*
Pt cresol *(m)*

cresolo *(m) n* It
De Kresol *(n)*
En cresol
Es cresol *(m)*
Fr crésol *(m)*
Pt cresol *(m)*

criblage *(m) n* Fr
De Klassierung *(f)*
En grading (coal)
Es clasificación *(f)*
It cernita *(f)*
Pt granulometria *(f)*

criogenia *(f) n* It, Pt
De Tieftemperatur-technik *(f)*
En cryogenics
Es criogénica *(f)*
Fr cryogénie *(f)*

criogénica *(f) n* Es
De Tieftemperatur-technik *(f)*
En cryogenics
Fr cryogénie *(f)*
It criogenia *(f)*
Pt criogenia *(f)*

criostato *(m) n* It, Pt
De Kryostat *(m)*
En cryostat
Es crióstato *(m)*
Fr cryostat *(m)*

crióstato *(m) n* Es
De Kryostat *(m)*
En cryostat
Fr cryostat *(m)*
It criostato *(m)*
Pt criostato *(m)*

crise de l'énergie *(f)* Fr
De Energiekrise *(f)*
En energy crisis
Es crisis energética *(f)*
It crisi energetica *(f)*
Pt crise energética *(f)*

crise energética *(f)* Pt
De Energiekrise *(f)*
En energy crisis
Es crisis energética *(f)*
Fr crise de l'énergie *(f)*
It crisi energetica *(f)*

crisi energetica *(f)* It
De Energiekrise *(f)*
En energy crisis
Es crisis energética *(f)*
Fr crise de l'énergie *(f)*
Pt crise energética *(f)*

crisis energética *(f)* Es
De Energiekrise *(f)*
En energy crisis
Fr crise de l'énergie *(f)*
It crisi energetica *(f)*
Pt crise energética *(f)*

critical *adj* En
De kritisch
Es crítico
Fr critique
It critico
Pt crítico

critico *adj* It
De kritisch
En critical
Es crítico
Fr critique
Pt crítico

crítico *adj* Es, Pt
De kritisch
En critical
Fr critique
It critico

critique *adj* Fr
De kritisch
En critical
Es crítico
It critico
Pt crítico

crocidolita *(f)* Es
De blauer Asbest *(m)*
En blue asbestos
Fr amiante bleu *(m)*
It amianto azzurro *(m)*
Pt amianto azul *(m)*

cromatografía de gases *(f)* Es
De Gaschromatographie *(f)*
En gas chromatography
Fr chromatographie gazeuse *(f)*
It cromatografia in fase gassosa *(f)*
Pt cromatografia gasosa *(f)*

cromatografia gasosa *(f)* Pt
De Gaschromatographie *(f)*
En gas chromatography
Es cromatografía de gases *(f)*
Fr chromatographie gazeuse *(f)*
It cromatografia in fase gassosa *(f)*

cromatografia in fase gassosa *(f)* It
De Gaschromatographie *(f)*
En gas chromatography
Es cromatografía de gases *(f)*
Fr chromatographie gazeuse *(f)*
Pt cromatografia gasosa *(f)*

crude oil En
De Rohöl *(n)*
Es petróleo crudo *(m)*
Fr pétrole brut *(m)*
It petrolio grezzo *(m)*
Pt óleo crú *(m)*

crudo aromático *(m)* Es
De aromatisches Rohöl *(n)*
En aromatic crude oil
Fr pétrole brut aromatique *(m)*
It petrolio greggio aromatico *(m)*
Pt óleo aromático crú *(m)*

crudo ligero *(m)* Es
De leichtes Rohöl *(n)*
En light crude oil
Fr brut léger *(m)*

It petrolio greggio leggero *(m)*
Pt petróleo crú leve *(m)*

crudo nafténico *(m)* Es
De naphthenisches Rohöl *(n)*
En naphthenic crude oil
Fr brut à base naphténique *(m)*
It petrolio greggio naftenico *(m)*
Pt petróleo crú nafténico *(m)*

crudo parafínico *(m)* Es
De paraffinisches Rohöl *(n)*
En paraffinic crude oil
Fr pétrole brut paraffinique *(m)*
It petrolio grezzo paraffinico *(m)*
Pt óleo crú parafínico *(m)*

crudo sintético *(m)* Es
De synthetisches Rohöl *(n)*
En synthetic crude oil (SCO)
Fr brut de synthèse *(m)*
It petrolio greggio sintetico *(m)*
Pt petróleo crú sintético *(m)*

cryogenics *n* En
De Tieftemperaturtechnik *(f)*
Es criogénica *(f)*
Fr cryogénie *(f)*
It criogenia *(f)*
Pt criogenia *(f)*

cryogénie *(f) n* Fr
De Tieftemperaturtechnik *(f)*
En cryogenics
Es criogénica *(f)*
It criogenia *(f)*
Pt criogenia *(f)*

cryostat *n* En; Fr *(m)*
De Kryostat *(m)*
Es crióstato *(m)*
It criostato *(m)*
Pt criostato *(m)*

cuadratura *(f) n* Es
De Flächeninhaltsbestimmung *(f)*
En quadrature
Fr quadrature *(f)*
It quadratura *(f)*
Pt quadratura *(f)*

cuando *(m) n* Es
De Quantum *(n)*
En quantum
Fr quantum *(m)*
It quantum *(m)*
Pt quantum *(m)*

cubilot *(m) n* Fr
De Kuppel *(f)*
En cupola
Es cúpula *(f)*
It cupola *(f)*
Pt cúpula *(f)*

cuerpo negro *(m)* Es
De Planckscher Strahler *(m)*
En black body
Fr corps noir *(m)*
It corpo nero *(m)*
Pt corpo negro *(m)*

cuore (reattore nucleare) *(m)* It
De Spaltzone (Kernkraftreaktor) *(f)*
En core (nuclear reactor)
Es núcleo (reactor nuclear) *(m)*
Fr coeur (réacteur nucléaire) *(m)*
Pt núcleo (reactor nuclear) *(m)*

cupola *n* En; It *(f)*
De Kuppel *(f)*
Es cúpula *(f)*
Fr cubilot *(m)*
Pt cúpula *(f)*

cúpula *(f) n* Es, Pt
De Kuppel *(f)*
En cupola
Fr cubilot *(m)*
It cupola *(f)*

current *n* En
De Strom *(m)*
Es corriente *(f)*
Fr courant *(m)*
It corrente *(f)*
Pt corrente *(f)*

curso a prova de imfietrações *(m)* Pt
De Feuchtigkeitssperrschicht *(f)*
En damp-proof course
Es hilada hidrófuga *(f)*
Fr couche hydrofuge *(f)*
It trattamento impermeabilizzante *(m)*

custo de distribuição *(m)* Pt
De Vertriebskosten *(pl)*
En distribution cost
Es coste de distribución *(m)*
Fr frais de distribution *(m)*
It costo di distribuzione *(m)*

custo de exploração *(m)* Pt
De Betriebskosten *(pl)*
En running costs
Es costes de explotación *(m pl)*
Fr frais d'exploitation *(m pl)*
It costi di esercizio *(m pl)*

cycle d'absorption *(m)* Fr
De Absorptionsfolge *(f)*
En absorption cycle
Es ciclo de absorción *(f)*
It ciclo di assorbimento *(m)*
Pt ciclo de absorção *(m)*

cycle de Carnot *(m)* Fr
De Carnotscher Kreisprozeβ *(m)*
En Carnot cycle
Es ciclo de Carnot *(m)*
It ciclo di Carnot *(m)*
Pt ciclo de Carnot *(m)*

cycle de chaleur-travail *(m)* Fr
De Wärmearbeitszyklus *(m)*
En heat-work cycle
Es ciclo de calor-trabajo *(m)*
It ciclo calore-lavoro *(m)*
Pt ciclo de calor-trabalho *(m)*

cycle de compression de la vapeur *(m)* Fr
Am vapor compression cycle
De umgekehrter Carnot-Prozeβ *(m)*
En vapour compression cycle
Es ciclo de compresión del vapor *(m)*
It ciclo di compressione del vapore *(m)*
Pt ciclo de compressão de vapor *(m)*

cycle de la vapeur *(m)* Fr
De Dampfzyklus *(m)*
En steam cycle
Es ciclo de vapor *(m)*
It ciclo di vapore *(m)*
Pt ciclo de vapor *(m)*

cycle de Rankine *(m)* Fr
De Clausius-Rankine-Prozeβ *(m)*
En Rankine cycle
Es ciclo de Rankine *(m)*
It ciclo di Rankine *(m)*
Pt ciclo ranquinizado *(m)*

cycle de Stirling *(m)* Fr
De Stirling-Zyklus *(m)*
En Stirling cycle
Es ciclo de Stirling *(m)*
It ciclo di Stirling *(m)*
Pt ciclo de Stirling *(m)*

cycle d'Otto *(m)* Fr
De Otto-Zyklus *(m)*
En Otto cycle
Es ciclo de Otto *(m)*
It ciclo di Otto *(m)*
Pt ciclo de Otto *(m)*

cycle du combustible *(m)* Fr
De Brennstoffzyklus *(m)*
En fuel cycle
Es ciclo de los combustibles *(m)*
It ciclo del combustibile *(m)*
Pt ciclo de combustível *(m)*

cycle en cascade *(m)* Fr
De Kaskadenprozeβ *(m)*
En cascade cycle
Es ciclo en cascada *(m)*

It ciclo di cascata *(m)*
Pt ciclo de cascada *(m)*

cycle mixte *(m)* Fr
De Kombinationsbetrieb *(m)*
En combined cycle
Es ciclo combinado *(m)*
It ciclo combinato *(m)*
Pt ciclo combinado *(m)*

cycle réversible *(m)* Fr
De umkehrbarer Zyklus *(m)*
En reversible cycle
Es ciclo reversible *(m)*
It ciclo reversibile *(m)*
Pt ciclo reversível *(m)*

cyclohexane *n* En; Fr *(m)*
De Zyklohexan *(n)*
Es ciclohexano *(m)*
It cicloesano *(m)*
Pt ciclohexano *(m)*

cyclone *n* En; Fr *(m)*
De Zyklon *(m)*
Es ciclón *(m)*
It ciclone *(m)*
Pt ciclone *(m)*

cyclone furnace En
De Wirbelofen *(m)*
Es hogar de turbulencia *(m)*
Fr four à cyclone *(m)*
It forno a ciclone *(m)*
Pt forno de ciclone *(m)*

D

Dachbodenisolierung *(f)* *n* De
En loft insulation
Es aislamiento de galerías *(m)*
Fr isolement des combles *(m)*
It isolamento della soffitta *(m)*
Pt isolamento do sotão *(m)*

Dachisolierung *(f)* *n* De
En roof insulation
Es aislamiento de techos *(m)*
Fr isolement des toits *(m)*
It isolamento del tetto *(m)*
Pt isolamento da cobertura *(m)*

dam *n* En
De Damm *(m)*
Es presa *(f)*
Fr barrage *(m)*
It diga *(f)*
Pt barragem *(f)*

Damm *(m)* *n* De
En dam
Es presa *(f)*
Fr barrage *(m)*
It diga *(f)*
Pt barragem *(f)*

Dampf *(m)* *n* De
Am steam; vapor
En steam; vapour
Es vapor *(m)*
Fr vapeur *(f)*
It vapore *(m)*
Pt vapor *(m)*

Dampfdichte *(f)* *n* De
Am vapor density
En vapour density
Es densidad del vapor *(f)*
Fr densité de vapeur *(f)*
It densità del vapore *(f)*
Pt densidade de vapor *(f)*

Dampfdruck *(m)* *n* De
Am vapor pressure
En vapour pressure
Es presión del vapor *(f)*
Fr tension de vapeur *(f)*
It tensione di vapore *(f)*
Pt pressão de vapor *(f)*

Dampfmaschine *(f)* *n* De
En steam engine
Es motor de vapor *(m)*
Fr machine à vapeur *(f)*
It macchina a vapore *(f)*
Pt motor a vapor *(m)*

Dampfqualität *(f)* *n* De
En steam quality
Es calidad del vapor *(f)*
Fr qualité de vapeur *(f)*
It qualità del vapore *(f)*
Pt qualidade do vapor *(f)*

Dampfspeicher *(m)* *n* De
En steam accumulator
Es acumulador de vapor *(m)*
Fr accumulateur de vapeur *(m)*
It accumulatore di vapore *(m)*
Pt acumulador de vapor *(m)*

Dampfsperre *(f)* *n* De
Am vapor barrier
En vapour barrier
Es barrera del vapor *(f)*
Fr écran d'étanchéité à la vapeur *(m)*
It schermo per il vapore *(m)*
Pt barreira de vapor *(f)*

Dampftafel *(f)* *n* De
En steam table
Es tabla de vapor *(f)*
Fr table à vapeur *(f)*
It tavola di vapore *(f)*
Pt tabela de vapor *(m)*

Dampfturbine *(f)* *n* De
En steam turbine
Es turbina de vapor *(f)*
Fr turbine à vapeur *(f)*
It turbina a vapore *(f)*
Pt turbina a vapor *(f)*

Dämpfung *(f)* *n* De
En attenuation
Es atenuación *(f)*
Fr atténuation *(f)*
It attenuazione *(f)*
Pt atenuação *(f)*

Dampfzyklus *(m)* *n* De
En steam cycle
Es ciclo de vapor *(m)*
Fr cycle de la vapeur *(m)*
It ciclo di vapore *(m)*
Pt ciclo de vapor *(m)*

damp-proof course En
De Feuchtigkeits-sperrschicht *(f)*
Es hilada hidrófuga *(f)*
Fr couche hydrofuge *(f)*
It trattamento impermeabilizzante *(m)*
Pt curso a prova de imfietrações *(m)*

danger de radiation *(m)* Fr
De Strahlungsgefahr *(f)*
En radiation hazard
Es riesgo de radiación *(m)*
It pericolo di radiazioni *(m)*
Pt perigo de radiação *(m)*

danger d'incendie *(m)* Fr
De Feuergefahr *(f)*
En fire hazard
Es riesgo de incendio *(m)*
It pericolo di incendio *(m)*
Pt perigo de incêndio *(m)*

dealkylation *n* En
De Dealkylierung *(f)*
Es desalquilación *(f)*
Fr désalcoylation *(f)*
It disalcalizzazione *(f)*
Pt desalquilação *(f)*

Dealkylierung *(f)* *n* De
En dealkylation
Es desalquilación *(f)*
Fr désalcoylation *(f)*
It disalcalizzazione *(f)*
Pt desalquilação *(f)*

deasfaltazione *(f)* *n* It
De Deasphaltieren *(n)*
En deasphalting
Es desasfaltación *(f)*
Fr désasphaltage *(m)*
Pt desasfaltação *(f)*

Deasphaltieren *(n)* *n* De
En deasphalting
Es desasfaltación *(f)*
Fr désasphaltage *(m)*
It deasfaltazione *(f)*
Pt desasfaltação *(f)*

deasphalting *n* En
De Deasphaltieren *(n)*
Es desasfaltación *(f)*
Fr désasphaltage *(m)*
It deasfaltazione *(f)*
Pt desasfaltação *(f)*

débitmètre *(m) n* Fr
De Durchflußmesser *(m)*
En flowmeter
Es medidor de flujo *(m)*
It flussometro *(m)*
Pt fluxómetro *(m)*

déblai *(m) n* Fr
De Abraum *(m)*
En spoil (mining)
Es estériles *(m pl)*
It materiale di sterro *(m)*
Pt entulho *(m)*

déblais de roche *(m pl)* Fr
De Felsabraum *(m)*
En rock spoils
Es desperdicios de roca *(m pl)*
It materiale di sterro di roccia *(m)*
Pt detritos de rocha *(m pl)*

débutanisation *(f) n* Fr
De Debutanisierung *(f)*
En debutanization
Es desbutanización *(f)*
It debutanizzazione *(f)*
Pt desbutanização *(f)*

Debutanisierung *(f) n* De
En debutanization
Es desbutanización *(f)*
Fr débutanisation *(f)*
It debutanizzazione *(f)*
Pt desbutanização *(f)*

debutanization *n* En
De Debutanisierung *(f)*
Es desbutanización *(f)*
Fr débutanisation *(f)*
It debutanizzazione *(f)*
Pt desbutanização *(f)*

debutanizzazione *(f) n* It
De Debutanisierung *(f)*
En debutanization
Es desbutanización *(f)*
Fr débutanisation *(f)*
Pt desbutanização *(f)*

decay constant En
De Zersetzungs-konstante *(f)*
Es constante de desintegración *(f)*
Fr constante de désintégration *(f)*
It constante di disintegrazione *(f)*
Pt constante de descomposição *(f)*

decay heat En
De Zersetzungswärme *(f)*
Es calor de desintegración *(m)*
Fr chaleur de désintégration *(f)*
It calore di disintegrazione *(m)*
Pt calor de descomposição *(m)*

décentralisation *(f) n* Fr
De Dezentralisierung *(f)*
En decentralization
Es descentralización *(f)*
It decentralizzazione *(f)*
Pt descentralização *(f)*

decentralization *n* En
De Dezentralisierung *(f)*
Es descentralización *(f)*
Fr décentralisation *(f)*
It decentralizzazione *(f)*
Pt descentralização *(f)*

decentralizzazione *(f) n* It
De Dezentralisierung *(f)*
En decentralization
Es descentralización *(f)*
Fr décentralisation *(f)*
Pt descentralização *(f)*

décharge *(f) n* Fr
De Entladung *(f)*
En discharge
Es descarga (bateria) *(f)*
It scarica *(f)*
Pt descarga (batería) *(f)*

déchets actifs *(m pl)* Fr
De radioaktive Abfallstoffe *(pl)*
En active waste
Es residuo activo *(m)*
It rifiuti attivi *(m pl)*
Pt desperdícios activos *(m pl)*

déchets contaminés *(m pl)* Fr
De kontaminierter Abfall *(m)*
En contaminated waste
Es residuo contaminado *(m)*
It rifiuti contaminati *(m pl)*
Pt desperdício contaminado *(m)*

Deckgestein *(n) n* De
En cap rock
Es estrato impermeable de cobertura *(m)*
Fr roche couverture *(f)*
It strato impermeabile di copertura *(m)*
Pt rocha encaixante *(f)*

deep mining En
De Tiefbau *(m)*
Es explotación en profundidad *(f)*
Fr exploitation des niveaux inférieurs *(f)*
It scavo profondo *(m)*
Pt mineração profunda *(f)*

défaut de masse *(m)* Fr
De Massendefekt *(m)*
En mass defect
Es defecto de masa *(m)*
It difetto di massa *(m)*
Pt defeito de massa *(m)*

defecto de masa *(m)* Es
De Massendefekt *(m)*
En mass defect
Fr défaut de masse *(m)*
It difetto di massa *(m)*
Pt defeito de massa *(m)*

defeito de massa *(m)* Pt
De Massendefekt *(m)*
En mass defect
Es defecto de masa *(m)*
Fr défaut de masse *(m)*
It difetto di massa *(m)*

Defektleitung *(f) n* De
En hole conduction
Es conducción por lagunas *(f)*
Fr conduction par les trous *(f)*
It conduzione dei buchi *(f)*
Pt condução por furos *(f)*

déflecteur *(m) n* Fr
De Prallplatte *(f)*
En baffle
Es deflector *(m)*
It parafiamma *(m)*
Pt deflector *(m)*

déflecteur de courant d'air *(m)* Fr
Am draft diverter
De Luftzugablenkung *(f)*
En draught diverter
Es desviador de corrientes *(m)*
It deviatore d'aria *(m)*
Pt deflector de corrente de ar *(m)*

deflector *(m) n* Es, Pt
De Prallplatte *(f)*
En baffle
Fr déflecteur *(m)*
It parafiamma *(m)*

deflector de corrente de ar *(m)* Pt
Am draft diverter
De Luftzugablenkung *(f)*
En draught diverter
Es desviador de corrientes *(m)*
Fr déflecteur de courant d'air *(m)*
It deviatore d'aria *(m)*

dehumidifier *n* En
De Entfeuchter *(m)*
Es deshumidificador *(m)*
Fr déshumidificateur *(m)*
It deumidificatore *(m)*
Pt deshumidificador *(m)*

deidratazione *(f) n* It
De Entwässerung *(f)*
En dewatering
Es desecación *(f)*
Fr dénoyage *(m)*
Pt deshidratação *(f)*

delayed coking En
De verzögerte Verkokung *(f)*
Es coquización retardada *(f)*
Fr cokéification différée *(f)*
It caricamento ritardato *(m)*
Pt coqueação retardada *(f)*

delayed-mixing burner En
De Brenner mit

verzögerter Luftmischung *(m)*
Es quemador de mezcla retardada *(m)*
Fr brûleur a mélange différé *(m)*
It bruciatore di mescolata ritardato *(m)*
Pt queimador misturador retardado *(m)*

delayed neutron En
De verzögertes Neutron *(n)*
Es neutrón diferido *(m)*
Fr neutron retardé *(m)*
It neutrone ritardato *(m)*
Pt neutrão retardado *(m)*

de l'environnement Fr
De Umwelt-
En environmental
Es ambiental
It ambientale
Pt ambiente

délestage *(m) n* Fr
De Lastabgabe *(f)*
En load-shedding
Es restricción de la carga *(f)*
It spargimento di carico *(m)*
Pt restrição de carga *(f)*

délestage à sous-fréquence *(m)* Fr
De Unterfrequenz-Lastabschaltung *(f)*
En underfrequency load-shedding
Es restricción de la carga a baja frecuencia *(f)*
It spargimento di carico a sottofrequenza *(m)*
Pt restrição de carga a sobfrequência *(f)*

delle macroparticelle It
De Partikel-
En particulate
Es de macropartículas
Fr particulaire
Pt em partículas

de macropartículas Es
De Partikel-
En particulate
Fr particulaire
It delle macroparticelle
Pt em partículas

démagnétisation *(f) n* Fr
De Entmagnetisierung *(f)*
En demagnetization
Es desmagnetización *(f)*
It smagnetizzazione *(f)*
Pt desmagnetização *(f)*

demagnetization *n* En
De Entmagnetisierung *(f)*
Es desmagnetización *(f)*
Fr démagnétisation *(f)*
It smagnetizzazione *(f)*
Pt desmagnetização *(f)*

demanda punta *(f)* Es
De Spitzenbedarf *(m)*
En peak demand
Fr consommation de pointe *(f)*
It domanda di punta *(f)*
Pt procura de ponta *(f)*

démarrage *(m) n* Fr
De Einschalten *(n)*
En start-up
Es arranque *(m)*
It avviamento *(m)*
Pt arranque *(m)*

démarrage à froid *(m)* Fr
De Kaltstart *(m)*
En cold start
Es arranque en frío *(m)*
It avviamento a freddo *(m)*
Pt arranque a frio *(m)*

démarreur *(m) n* Fr
De Starter *(m)*
En starter motor
Es motor de arranque *(m)*
It avviatore *(m)*
Pt motor de arranque *(m)*

demetanizzazione *(f) n* It
De Demethanisierung *(f)*
En demethanization
Es desmetanización *(f)*
Fr déméthanisation *(f)*
Pt desmetanização *(f)*

déméthanisation *(f) n* Fr
De Demethanisierung *(f)*
En demethanization
Es desmetanización *(f)*
It demetanizzazione *(f)*
Pt desmetanização *(f)*

Demethanisierung *(f) n* De
En demethanization
Es desmetanización *(f)*
Fr déméthanisation *(f)*
It demetanizzazione *(f)*
Pt desmetanização *(f)*

demethanization *n* En
De Demethanisierung *(f)*
Es desmetanización *(f)*
Fr déméthanisation *(f)*
It demetanizzazione *(f)*
Pt desmetanização *(f)*

demineralized water En
De entmineralisiertes Wasser *(n)*
Es agua desmineralizada *(f)*
Fr eau déminéralisée *(f)*
It acqua demineralizzata *(f)*
Pt água desmineralizada *(f)*

demister *n* En
De Belüftung *(f)*
Es desempañador *(m)*
Fr antibrouillard *(m)*
It antiappannante *(m)*
Pt desembaciador *(m)*

demi-vie *(f) n* Fr
De Halbwertszeit *(f)*
En half-life
Es media vida *(f)*
It periodo di dimezzamento *(m)*
Pt meia vida *(f)*

dénoyage *(m) n* Fr
De Entwässerung *(f)*
En dewatering
Es desecación *(f)*
It deidratazione *(f)*
Pt deshidratação *(f)*

densidad *(f) n* Es
De Dichte *(f)*
En density
Fr densité *(f)*
It densità *(f)*
Pt densidade *(f)*

densidad del vapor *(f)* Es
Am vapor density
De Dampfdichte *(f)*
En vapour density
Fr densité de vapeur *(f)*
It densità del vapore *(f)*
Pt densidade de vapor *(f)*

densidad de potencia *(f)* Es
De Leistungsdichte *(f)*
En power density
Fr puissance volumique *(f)*
It densità di potenza *(f)*
Pt densidade de potência *(f)*

densidade *(f) n* Pt
De Dichte *(f)*
En density
Es densidad *(f)*
Fr densité *(f)*
It densità *(f)*

densidade de potência *(f)* Pt
De Leistungsdichte *(f)*
En power density
Es densidad de potencia *(f)*
Fr puissance volumique *(f)*
It densità di potenza *(f)*

densidade de vapor *(f)* Pt
Am vapor density
De Dampfdichte *(f)*
En vapour density
Es densidad del vapor *(f)*
Fr densité de vapeur *(f)*
It densità del vapore *(f)*

densidade em massa *(f)* Pt
De Schüttdichte *(f)*
En bulk density
Es densidad en masa *(f)*
Fr densité apparente *(f)*
It densità apparente *(m)*

densidad en masa *(f)* Es
De Schüttdichte *(f)*
En bulk density
Fr densité apparente *(f)*
It densità apparente *(m)*
Pt densidade em massa *(f)*

densità *(f) n* It
De Dichte *(f)*
En density
Es densidad *(f)*
Fr densité *(f)*
Pt densidade *(f)*

densità apparente *(m)* It
De Schüttdichte *(f)*
En bulk density
Es densidad en masa *(f)*
Fr densité apparente *(f)*
Pt densidade em massa *(f)*

densità del vapore *(f)* It
Am vapor density
De Dampfdichte *(f)*
En vapour density
Es densidad del vapor *(f)*
Fr densité de vapeur *(f)*
Pt densidade de vapor *(f)*

densità di potenza *(f)* It
De Leistungsdichte *(f)*
En power density
Es densidad de potencia *(f)*
Fr puissance volumique *(f)*
Pt densidade de potência *(f)*

densité *(f) n* Fr
De Dichte *(f)*
En density
Es densidad *(f)*
It densità *(f)*
Pt densidade *(f)*

densité apparente *(f)* Fr
De Schüttdichte *(f)*
En bulk density
Es densidad en masa *(f)*
It densità apparente *(m)*
Pt densidade em massa *(f)*

densité de vapeur *(f)* Fr
Am vapor density
De Dampfdichte *(f)*
En vapour density
Es densidad del vapor *(f)*
It densità del vapore *(f)*
Pt densidade de vapor *(f)*

density *n* En
De Dichte *(f)*
Es densidad *(f)*
Fr densité *(f)*
It densità *(f)*
Pt densidade *(f)*

depósito *(m) n* Es
De Sammelbecken *(n)*
En reservoir
Fr réservoir *(m)*
It serbatoio *(m)*
Pt reservatório *(m)*

depósito de calor *(m)* Es
De Wärmespeicher *(m)*
En heat reservoir
Fr réservoir de chaleur *(m)*
It serbatoio di calore *(m)*
Pt reservatório de calor *(m)*

depósito de carbón lenticular *(m)* Es
De linsenförmiges Kohlenlager *(n)*
En lenticular coal deposit
Fr dépôt de charbon lenticulaire *(m)*
It deposito di carbone lenticolare *(m)*
Pt depósito de carvão lenticular *(m)*

depósito de carvão lenticular *(m)* Pt
De linsenförmiges Kohlenlager *(n)*
En lenticular coal deposit
Es depósito de carbón lenticular *(m)*
Fr dépôt de charbon lenticulaire *(m)*
It deposito di carbone lenticolare *(m)*

deposito di carbone lenticolare *(m)* It
De linsenförmiges Kohlenlager *(n)*
En lenticular coal deposit
Es depósito de carbón lenticular *(m)*
Fr dépôt de charbon lenticulaire *(m)*
Pt depósito de carvão lenticular *(m)*

depósitos marítimos *(m)* Pt
De Seeverschmutzung *(f)*
En marine fouling
Es suciedad por depósitos marinos *(f)*
Fr accumulation de dépôts marins *(f)*
It incrostazione marina *(f)*

dépôt de charbon lenticulaire *(m)* Fr
De linsenförmiges Kohlenlager *(n)*
En lenticular coal deposit
Es depósito de carbón lenticular *(m)*
It deposito di carbone lenticolare *(m)*
Pt depósito de carvão lenticular *(m)*

depressore *(m) n* It
De Vakuumpumpe *(f)*
En vacuum pump
Es bomba de vacío *(f)*
Fr pompe à vide *(f)*
Pt bomba de vácuo *(f)*

dépressurisation *(f) n* Fr
De Umstellung auf normalen Luftdruck *(f)*
En depressurization
Es depresurización *(f)*
It depressurizzazione *(f)*
Pt despressurização *(f)*

depressurization *n* En
De Umstellung auf normalen Luftdruck *(f)*
Es depresurización *(f)*
Fr dépressurisation *(f)*
It depressurizzazione *(f)*
Pt despressurização *(f)*

depressurizzazione *(f) n* It
De Umstellung auf normalen Luftdruck *(f)*
En depressurization
Es depresurización *(f)*
Fr dépressurisation *(f)*
Pt despressurização *(f)*

depresurización *(f) n* Es
De Umstellung auf normalen Luftdruck *(f)*
En depressurization
Fr dépressurisation *(f)*
It depressurizzazione *(f)*
Pt despressurização *(f)*

dépropanisation *(f) n* Fr
De Entpropanisierung *(f)*
En depropanization
Es despropanización *(f)*
It depropanizzazione *(f)*
Pt despropanização *(f)*

depropanization *n* En
De Entpropanisierung *(f)*
Es despropanización *(f)*
Fr dépropanisation *(f)*
It depropanizzazione *(f)*
Pt despropanização *(f)*

depropanizzazione *(f) n* It
De Entpropanisierung *(f)*
En depropanization
Es despropanización *(f)*
Fr dépropanisation *(f)*
Pt despropanização *(f)*

de reserva Es, Pt
De Reserve
En standby
Fr en attente
It emergenza

desactivadores de metales *(m pl)* Es
De Metall-Deaktiviermungsmittel *(pl)*
En metal deactivators
Fr désactiveurs de métaux *(m pl)*
It disattivatori di metallo *(m pl)*
Pt desactivadores metálicos *(m pl)*

desactivadores metálicos *(m pl)* Pt
De Metall-Deaktivierungsmittel *(pl)*
En metal deactivators
Es desactivadores de metales *(m pl)*
Fr désactiveurs de métaux *(m pl)*
It disattivatori di metallo *(m pl)*

désactiveurs de métaux *(m pl)* Fr
De Metall-Deaktivierungsmittel *(pl)*
En metal deactivators
Es desactivadores de metales *(m pl)*
It disattivatori di metallo *(m pl)*
Pt desactivadores metálicos *(m pl)*

desalación *(f)* n Es
De Meerwasserentsalzung *(f)*
En desalination
Fr dessalage *(m)*
It desalinizzazione *(f)*
Pt dessalinização *(f)*

désalcoylation *(f)* n Fr
De Dealkylierung *(f)*
En dealkylation
Es desalquilación *(f)*
It disalcalizzazione *(f)*
Pt desalquilação *(f)*

desalination n En
De Meerwasserentsalzung *(f)*
Es desalación *(f)*
Fr dessalage *(m)*
It desalinizzazione *(f)*
Pt dessalinização *(f)*

desalinizzazione *(f)* n It
De Meerwasserentsalzung *(f)*
En desalination
Es desalación *(f)*
Fr dessalage *(m)*
Pt dessalinização *(f)*

desalquilação *(f)* n Pt
De Dealkylierung *(f)*
En dealkylation
Es desalquilación *(f)*
Fr désalcoylation *(f)*
It disalcalizzazione *(f)*

desalquilación *(f)* n Es
De Dealkylierung *(f)*
En dealkylation
Fr désalcoylation *(f)*
It disalcalizzazione *(f)*
Pt desalquilação *(f)*

desasfaltação *(f)* n Pt
De Deasphaltieren *(n)*
En deasphalting
Es desasfaltación *(f)*
Fr désasphaltage *(m)*
It deasfaltazione *(f)*

desasfaltación *(f)* n Es
De Deasphaltieren *(n)*
En deasphalting
Fr désasphaltage *(m)*
It deasfaltazione *(f)*
Pt desasfaltação *(f)*

désasphaltage *(m)* n Fr
De Deasphaltieren *(n)*
En deasphalting
Es desasfaltación *(f)*
It deasfaltazione *(f)*
Pt desasfaltação *(f)*

desbutanização *(f)* n Pt
De Debutanisierung *(f)*
En debutanization
Es desbutanización *(f)*
Fr débutanisation *(f)*
It debutanizzazione *(f)*

desbutanización *(f)* n Es
De Debutanisierung *(f)*
En debutanization
Fr débutanisation *(f)*
It debutanizzazione *(f)*
Pt desbutanização *(f)*

descarga (bateria) *(f)* n Es
De Entladung *(f)*
En discharge (battery)
Fr décharge *(f)*
It scarica *(f)*
Pt descarga *(f)*

descarga (batería) *(f)* n Pt
De Entladung *(f)*
En discharge (battery)
Es descarga *(f)*
Fr décharge *(f)*
It scarica *(f)*

descarga *(f)* n Pt
De Lastabgabe *(f)*
En load-shedding
Es restricción de la carga *(f)*
Fr délestage *(m)*
It spargimento di carico *(m)*

descarga de sobfrequência *(f)* Pt
De Unterfrequenz-Lastabschaltung *(f)*
En underfrequency load-shedding
Es restricción de la carga a baja frecuencia *(f)*
Fr délestage à sous-fréquence *(m)*
It spargimento di carico a sottofrequenza *(m)*

descentralização *(f)* n Pt
De Dezentralisierung *(f)*
En decentralization
Es descentralización *(f)*
Fr décentralisation *(f)*
It decentralizzazione *(f)*

descentralización *(f)* n Es
De Dezentralisierung *(f)*
En decentralization
Fr décentralisation *(f)*
It decentralizzazione *(f)*
Pt descentralização *(f)*

desecación *(f)* n Es
De Entwässerung *(f)*
En dewatering
Fr dénoyage *(m)*
It deidratazione *(f)*
Pt deshidratação *(f)*

desechos *(m pl)* n Es
De Haldenabfall *(m)*
En tailings *(pl)*
Fr produits de queue *(m pl)*
It residui di scarto *(m pl)*
Pt colas *(f pl)*

desembaciador *(m)* n Pt
De Belüftung *(f)*
En demister
Es desempañador *(m)*
Fr antibrouillard *(m)*
It antiappannante *(m)*

desempañador *(m)* n Es
De Belüftung *(f)*
En demister
Fr antibrouillard *(m)*
It antiappannante *(m)*
Pt desembaciador *(m)*

deshidratação *(f)* n Pt
De Entwässerung *(f)*
En dewatering
Es desecación *(f)*
Fr dénoyage *(m)*
It deidratazione *(f)*

deshumidificador *(m)* n Es, Pt
De Entfeuchter *(m)*
En dehumidifier
Fr déshumidificateur *(m)*
It deumidificatore *(m)*

déshumidificateur *(m)* n Fr
De Entfeuchter *(m)*
En dehumidifier
Es deshumidificador *(m)*
It deumidificatore *(m)*
Pt deshumidificador *(m)*

design ambientale integrato *(m)* It
De integrierte Umweltplanung *(f)*
En integrated environmental design (IED)
Es diseño ambiental integrado *(m)*
Fr étude de l'environment integreé *(f)*
Pt projecto de ambiente integrado *(m)*

desintegração *(f)* n Pt
De Spaltung *(f)*
En fission
Es fisión *(f)*
Fr fission *(f)*
It fissione *(f)*

desintegração nuclear *(f)* Pt
De Nuklearspaltung *(f)*
En nuclear fission
Es fisión nuclear *(f)*
Fr fission nucléaire *(f)*
It fissione nucleare *(f)*

desintegración beta *(f)* Es
De Beta-Zerfall *(m)*
En beta decay
Fr désintégration bêta *(f)*
It disintegrazione beta *(f)*
Pt deterioração beta *(f)*

désintégration bêta *(f)* Fr
De Beta-Zerfall *(m)*
En beta decay
Es desintegración beta *(f)*

It disintegrazione beta *(f)*
Pt deterioração beta *(f)*

desintegrável *adj* Pt
De spaltbar
En fissionable
Es fisionable
Fr fissionable
It fissionabile

desmagnetização *(f) n* Pt
De Entmagnetisierung *(f)*
En demagnetization
Es desmagnetización *(f)*
Fr démagnétisation *(f)*
It smagnetizzazione *(f)*

desmagnetización *(f) n* Es
De Entmagnetisierung *(f)*
En demagnetization
Fr démagnétisation *(f)*
It smagnetizzazione *(f)*
Pt desmagnetização *(f)*

desmetanização *(f) n* Pt
De Demethanisierung *(f)*
En demethanization
Es desmetanización *(f)*
Fr déméthanisation *(f)*
It demetanizzazione *(f)*

desmetanización *(f) n* Es
De Demethanisierung *(f)*
En demethanization
Fr déméthanisation *(f)*
It demetanizzazione *(f)*
Pt desmetanização *(f)*

desolforazione *(f) n* It
Am desulfurization
De Entschwefelung *(f)*
En desulphurization
Es desulfurización *(f)*
Fr désulfuration *(f)*
Pt dessulfurização *(f)*

desperdício contaminado *(m)* Pt
De kontaminierter Abfall *(m)*
En contaminated waste
Es residuo contaminado *(m)*
Fr déchets contaminés *(m pl)*
It rifiuti contaminati *(m pl)*

desperdícios activos *(m pl)* Pt
De radioaktive Abfallstoffe *(pl)*
En active waste
Es residuo activo *(m)*
Fr déchets actifs *(m pl)*
It rifiuti attivi *(m pl)*

desperdicios de roca *(m pl)* Es
De Felsabraum *(m)*
En rock spoils
Fr déblais de roche *(m pl)*
It materiale di sterro di roccia *(m)*
Pt detritos de rocha *(m pl)*

despressurização *(f) n* Pt
De Umstellung auf normalen Luftdruck *(f)*
En depressurization
Es depresurización *(f)*
Fr dépressurisation *(f)*
It depressurizzazione *(f)*

despropanização *(f) n* Pt
De Entpropanisierung *(f)*
En depropanization
Es despropanización *(f)*
Fr dépropanisation *(f)*
It depropanizzazione *(f)*

despropanización *(f) n* Es
De Entpropanisierung *(f)*
En depropanization
Fr dépropanisation *(f)*
It depropanizzazione *(f)*
Pt despropanização *(f)*

dessalage *(m) n* Fr
De Meerwasserentsalzung *(f)*
En desalination
Es desalación *(f)*
It desalinizzazione *(f)*
Pt dessalinização *(f)*

dessalinização *(f) n* Pt
De Meerwasserentsalzung *(f)*
En desalination
Es desalación *(f)*
Fr dessalage *(m)*
It desalinizzazione *(f)*

dessicomètre *(m) n* Fr
De Trockenmeßgerät *(n)*
En dry meter
Es medidor seco *(m)*
It contatore a secco *(m)*
Pt medidor de seco *(m)*

dessulfurização *(f) n* Pt
Am desulfurization
De Entschwefelung *(f)*
En desulphurization
Es desulfurización *(f)*
Fr désulfuration *(f)*
It desolforazione *(f)*

destilação *(f) n* Pt
De Destillation *(f)*
En distillation
Es destilación *(f)*
Fr distillation *(f)*
It distillazione *(f)*

destilação de efeito múltiplo *(f)* Pt
De Mehrfacheffekt-Destillation *(f)*
En multiple-effect distillation
Es destilación de efecto múltiple *(f)*
Fr distillation à effet multiple *(f)*
It distillazione ad effetto multiplo *(f)*

destilação de efeito simples *(f)* Pt
De Einzeleffekt-Destillation *(f)*
En single-effect distillation
Es destilación de simple efecto *(f)*
Fr distillation simple effet *(f)*
It distillazione ad effetto unico *(f)*

destilação fraccional *(f)* Pt
De fraktionierte Destillation *(f)*
En fractional distillation
Es destilación fraccionada *(f)*
Fr distillation fractionnée *(f)*
It distillazione a frazione *(f)*

destilación *(f) n* Es
De Destillation *(f)*
En distillation
Fr distillation *(f)*
It distillazione *(f)*
Pt destilação *(f)*

destilación de efecto múltiple *(f)* Es
De Mehrfacheffekt-Destillation *(f)*
En multiple-effect distillation
Fr distillation à effet multiple *(f)*
It distillazione ad effetto multiplo *(f)*
Pt destilação de efeito múltiplo *(f)*

destilación de simple efecto *(f)* Es
De Einzeleffekt-Destillation *(f)*
En single-effect distillation
Fr distillation simple effet *(f)*
It distillazione ad effetto unico *(f)*
Pt destilação de efeito simples *(f)*

destilación fraccionada *(f)* Es
De fraktionierte Destillation *(f)*
En fractional distillation
Fr distillation fractionnée *(f)*
It distillazione a frazione *(f)*
Pt destilação fraccional *(f)*

destilado *(m) n* Es, Pt
De Destillat *(n)*
En distillate
Fr distillat *(m)*
It distillato *(m)*

Destillat *(n) n* De
En distillate
Es destilado *(m)*
Fr distillat *(m)*
It distillato *(m)*
Pt destilado *(m)*

Destillation *(f) n* De
En distillation
Es destilación *(f)*
Fr distillation *(f)*

It distillazione *(f)*
Pt destilação *(f)*

Destillationskolonne *(f)* *n* De
En distillation column
Es torre de destilación *(f)*
Fr colonne de distillation *(f)*
It colonna di distillazione *(f)*
Pt coluna de destilação *(f)*

Destillatöl *(n)* *n* De
En distillate fuel oil
Es fuel-oil destilado *(m)*
Fr fuel-oil distillé *(m)*
It nafta distillata *(f)*
Pt fuel-oil de destilados *(m)*

désulfuration *(f)* *n* Fr
Am desulfurization
De Entschwefelung *(f)*
En desulphurization
Es desulfurización *(f)*
It desolforazione *(f)*
Pt dessulfurização *(f)*

desulfurización *(f)* *n* Es
Am desulfurization
De Entschwefelung *(f)*
En desulphurization
Fr désulfuration *(f)*
It desolforazione *(f)*
Pt dessulfurização *(f)*

desulfurization *n* Am
De Entschwefelung *(f)*
En desulphurization
Es desulfurización *(f)*
Fr désulfuration *(f)*
It desolforazione *(f)*
Pt dessulfurização *(f)*

desulphurization *n* En
Am desulfurization
De Entschwefelung *(f)*
Es desulfurización *(f)*
Fr désulfuration *(f)*
It desolforazione *(f)*
Pt dessulfurização *(f)*

desviación *(f)* *n* Es
De Abweichung *(f)*
En deviation
Fr déviation *(f)*
It deviazione *(f)*
Pt desvio *(m)*

desviación media *(f)* Es
De durchschnittliche Abweichung *(f)*
En mean deviation
Fr déviation moyenne *(f)*
It deviazione media *(f)*
Pt desvio médio *(m)*

desviador de corrientes *(m)* Es
Am draft diverter
De Luftzugablenkung *(f)*
En draught diverter
Fr déflecteur de courant d'air *(m)*
It deviatore d'aria *(m)*
Pt deflector de corrente de ar *(m)*

desvio *(m)* *n* Pt
De Abweichung *(f)*
En deviation
Es desviación *(f)*
Fr déviation *(f)*
It deviazione *(f)*

desvio médio *(m)* Pt
De durchschnittliche Abweichung *(f)*
En mean deviation
Es desviación media *(f)*
Fr déviation moyenne *(f)*
It deviazione media *(f)*

desvolatilizador *(m)* *n* Es, Pt
De Devolatisierer *(m)*
En devolatilizer
Fr dévolatiliseur *(m)*
It devolatilizzante *(f)*

detergent additives *(pl)* En
De Waschmittelzusätze *(pl)*
Es aditivos detergentes *(m pl)*
Fr additifs détergents *(m pl)*
It additivi detergenti *(m pl)*
Pt aditivos detergentes *(m pl)*

deterioração beta *(f)* Pt
De Beta-Zerfall *(m)*
En beta decay
Es desintegración beta *(f)*
Fr désintégration bêta *(f)*
It disintegrazione beta *(f)*

detonação *(f)* *n* Pt
De Detonation *(f)*
En detonation
Es detonación *(f)*
Fr détonation *(f)*
It detonazione *(f)*

detonación *(f)* *n* Es
De Detonation *(f)*
En detonation
Fr détonation *(f)*
It detonazione *(f)*
Pt detonação *(f)*

detonation *n* En
De Detonation *(f)*
Es detonación *(f)*
Fr détonation *(f)*
It detonazione *(f)*
Pt detonação *(f)*

Detonation *(f)* *n* De
En detonation
Es detonación *(f)*
Fr détonation *(f)*
It detonazione *(f)*
Pt detonação *(f)*

détonation *(f)* *n* Fr
De Detonation *(f)*
En detonation
Es detonación *(f)*
It detonazione *(f)*
Pt detonação *(f)*

detonazione *(f)* *n* It
De Detonation *(f)*
En detonation
Es detonación *(f)*
Fr détonation *(f)*
Pt detonação *(f)*

detritos de rocha *(m pl)* Pt
De Felsabraum *(m)*
En rock spoils
Es desperdicios de roca *(m pl)*
Fr déblais de roche *(m pl)*
It materiale di sterro di roccia *(m)*

deumidificatore *(m)* *n* It
De Entfeuchter *(m)*
En dehumidifier
Es deshumidificador *(m)*
Fr déshumidificateur *(m)*
Pt deshumidificador *(m)*

deuterio *(m)* *n* Es, Pt
De Deuterium *(n)*
En deuterium
Fr deuterium *(m)*
It deuterio *(m)*

deutério *(m)* *n* Pt
De Deuterium *(n)*
En deuterium
Es deuterio *(m)*
Fr deuterium *(m)*
It deuterio *(m)*

deuterium *n* En; Fr *(m)*
De Deuterium *(n)*
Es deuterio *(m)*
It deuterio *(m)*
Pt deutério *(m)*

Deuterium *(n)* *n* De
En deuterium
Es deuterio *(m)*
Fr deuterium *(m)*
It deuterio *(m)*
Pt deutério *(m)*

devanado *(m)* *n* Es
De Wicklung *(f)*
En winding
Fr enroulement *(m)*
It avvolgimento *(m)*
Pt enrolamento *(m)*

déversement *(m)* *n* Fr
De Kippen *(n)*
En dumping
Es vaciamiento *(m)*
It pressatura *(f)*
Pt esvaziamento *(m)*

deviation *n* En
De Abweichung *(f)*
Es desviación *(f)*
Fr déviation *(f)*
It deviazione *(f)*
Pt desvio *(m)*

déviation *(f)* *n* Fr
De Abweichung *(f)*
En deviation
Es desviación *(f)*
It deviazione *(f)*
Pt desvio *(m)*

déviation moyenne *(f)* Fr
De durchschnittliche Abweichung *(f)*
En mean deviation
Es desviación media *(f)*
It deviazione media *(f)*
Pt desvio médio *(m)*

deviatore d'aria *(m)* It
Am draft diverter
De Luftzugablenkung *(f)*
En draught diverter
Es desviador de corrientes *(m)*
Fr déflecteur de courant d'air *(m)*
Pt deflector de corrente de ar *(m)*

deviazione *(f) n* It
De Abweichung *(f)*
En deviation
Es desviación *(f)*
Fr déviation *(f)*
Pt desvio *(m)*

deviazione media *(f)* It
De durchschnittliche Abweichung *(f)*
En mean deviation
Es desviación media *(f)*
Fr déviation moyenne *(f)*
Pt desvio médio *(m)*

dévolatiliseur *(m) n* Fr
De Devolatisierer *(m)*
En devolatilizer
Es desvolatilizador *(m)*
It devolatilizzante *(f)*
Pt desvolatilizador *(m)*

devolatilizer *n* En
De Devolatisierer *(m)*
Es desvolatilizador *(m)*
Fr dévolatiliseur *(m)*
It devolatilizzante *(f)*
Pt desvolatilizador *(m)*

devolatilizzante *(f) n* It
De Devolatisierer *(m)*
En devolatilizer
Es desvolatilizador *(m)*
Fr dévolatiliseur *(m)*
Pt desvolatilizador *(m)*

Devolatisierer *(m) n* De
En devolatilizer
Es desvolatilizador *(m)*
Fr dévolatiliseur *(m)*
It devolatilizzante *(f)*
Pt desvolatilizador *(m)*

dewatering *n* En
De Entwässerung *(f)*
Es desecación *(f)*
Fr dénoyage *(m)*
It deidratazione *(f)*
Pt deshidratação *(f)*

dew point En
De Taupunkt *(m)*
Es punto de condensación *(m)*
Fr point de rosée *(m)*
It punto di rugiada *(m)*
Pt ponto de orvalho *(m)*

Dezentralisierung *(f) n* De
En decentralization
Es descentralización *(f)*
Fr décentralisation *(f)*
It decentralizzazione *(f)*
Pt descentralização *(f)*

diagrama de flujo energético *(m)* Es
De Energieflußdiagramm *(n)*
En energy-flow diagram
Fr schéma de passage de l'énergie *(m)*
It diagramma del flusso energetico *(m)*
Pt diagrama de fluxo de energia *(m)*

diagrama de fluxo de energia *(m)* Pt
De Energieflußdiagramm *(n)*
En energy-flow diagram
Es diagrama de flujo energético *(m)*
Fr schéma de passage de l'énergie *(m)*
It diagramma del flusso energetico *(m)*

diagrama de Sankey *(m)* Es, Pt
De Sankeydiagramm *(n)*
En Sankey diagram
Fr diagramme de Sankey *(m)*
It diagramma di Sankey *(m)*

diagrama de vapor de Mollier *(m)* Es, Pt
De H-S-Diagramm *(n)*
En Mollier steam diagram
Fr diagramme de Mollier *(m)*
It diagramma di Mollier *(m)*

diagrama indicador *(m)* Es, Pt
De Indikatordiagramm *(n)*
En indicator diagram
Fr schéma indicateur *(m)*
It diagramma del ciclo indicato *(m)*

Diagramm *(n) n* De
En graph
Es gráfica *(f)*
Fr graphique *(m)*
It grafico *(m)*
Pt gráfico *(m)*

diagramma del ciclo indicato *(m)* It
De Indikatordiagramm *(n)*
En indicator diagram
Es diagrama indicador *(m)*
Fr schéma indicateur *(m)*
Pt diagrama indicador *(m)*

diagramma del flusso energetico *(m)* It
De Energieflußdiagramm *(n)*
En energy-flow diagram
Es diagrama de flujo energético *(m)*
Fr schéma de passage de l'énergie *(m)*
Pt diagrama de fluxo de energia *(m)*

diagramma di Mollier *(m)* It
De H-S-Diagramm *(n)*
En Mollier steam diagram
Es diagrama de vapor de Mollier *(m)*
Fr diagramme de Mollier *(m)*
Pt diagrama de vapor de Mollier *(m)*

diagramma di Sankey *(m)* It
De Sankeydiagramm *(n)*
En Sankey diagram
Es diagrama de Sankey *(m)*
Fr diagramme de Sankey *(m)*
Pt diagrama de Sankey *(m)*

diagramme de Mollier *(m)* Fr
De H-S-Diagramm *(n)*
En Mollier steam diagram
Es diagrama de vapor de Mollier *(m)*
It diagramma di Mollier *(m)*
Pt diagrama de vapor de Mollier *(m)*

diagramme de Sankey *(m)* Fr
De Sankeydiagramm *(n)*
En Sankey diagram
Es diagrama de Sankey *(m)*
It diagramma di Sankey *(m)*
Pt diagrama de Sankey *(m)*

Dichte *(f) n* De
En density
Es densidad *(f)*
Fr densité *(f)*
It densità *(f)*
Pt densidade *(f)*

dielectric *n* En
De Dielektrikum *(n)*
Es dieléctrico *(m)*
Fr diélectrique *(m)*
It dielettrico *(m)*
Pt dieléctrico *(m)*

dielectric constant En
De Dielektrizitäts-konstante *(f)*
Es constante dieléctrica *(f)*
Fr constante diélectrique *(f)*
It costante dielettrica *(f)*
Pt constante dieléctrica *(f)*

dielectric heating En
De kapazitive Hochfrequenzer-wärmung *(f)*
Es caldeo dieléctrico *(m)*

Fr chauffage par pertes diélectriques *(m)*
It riscaldamento dielettrico *(m)*
Pt aquecimento dieléctrico *(m)*

dielectric loss En
De dielektrischer Verlust *(m)*
Es pérdida dieléctrica *(f)*
Fr perte diélectrique *(f)*
It perdita dielettrica *(f)*
Pt perda dieléctrica *(f)*

dieléctrico *(m) n* Es, Pt
De Dielektrikum *(n)*
En dielectric
Fr diélectrique *(m)*
It dielettrico *(m)*

dieléctrico de gas *(m)* Es
De Gas-Dielektrikum *(n)*
En gas dielectric
Fr diélectrique à gaz *(m)*
It dielettrico gas *(m)*
Pt dieléctrico gasoso *(m)*

dieléctrico gasoso *(m)* Pt
De Gas-Dielektrikum *(n)*
En gas dielectric
Es dieléctrico de gas *(m)*
Fr diélectrique à gaz *(m)*
It dielettrico gas *(m)*

diélectrique *(m) n* Fr
De Dielektrikum *(n)*
En dielectric
Es dieléctrico *(m)*
It dielettrico *(m)*
Pt dieléctrico *(m)*

diélectrique à gaz *(m)* Fr
De Gas-Dielektrikum *(n)*
En gas dielectric
Es dieléctrico de gas *(m)*
It dielettrico gas *(m)*
Pt dieléctrico gasoso *(m)*

Dielektrikum *(n) n* De
En dielectric
Es dieléctrico *(m)*
Fr diélectrique *(m)*
It dielettrico *(m)*
Pt dieléctrico *(m)*

dielektrischer Verlust *(m)* De
En dielectric loss
Es pérdida dieléctrica *(f)*
Fr perte diélectrique *(f)*
It perdita dielettrica *(f)*
Pt perda dieléctrica *(f)*

Dielektrizitäts-konstante *(f) n* De
En dielectric constant
Es constante dieléctrica *(f)*
Fr constante diélectrique *(f)*
It costante dielettrica *(f)*
Pt constante dieléctrica *(f)*

dielettrico *(m) n* It
De Dielektrikum *(n)*
En dielectric
Es dieléctrico *(m)*
Fr diélectrique *(m)*
Pt dieléctrico *(m)*

dielettrico gas *(m)* It
De Gas-Dielektrikum *(n)*
En gas dielectric
Es dieléctrico de gas *(m)*
Fr diélectrique à gaz *(m)*
Pt dieléctrico gasoso *(m)*

diesel engine En
De Dieselmotor *(m)*
Es motor diesel *(m)*
Fr moteur diesel *(m)*
It motore diesel *(m)*
Pt motor diesel *(m)*

diesel engine road vehicle (DERV) En
De Straβenfahrzeng mit Dieselmotor *(n)*
Es vehículo para carretera con motor diesel *(m)*
Fr véhicule diesel routier *(m)*
It automezzo con motore a nafta *(m)*
Pt veículo rodoviário de motor diesel *(m)*

diesel-fuel marítimo *(m)* Pt
De Schiffsdieseltreibstoff *(m)*
En marine diesel-fuel
Es combustible diesel para marina *(m)*
Fr combustible diesel marin *(m)*
It combustibile diesel per motori marini *(m)*

Dieselmotor *(m) n* De
En diesel engine
Es motor diesel *(m)*
Fr moteur diesel *(m)*
It motore diesel *(m)*
Pt motor diesel *(m)*

diesel oil En
De Dieselöl *(n)*
Es aceite diesel *(m)*
Fr combustible diesel *(m)*
It nafta *(f)*
Pt óleo diesel *(m)*

Dieselöl *(n) n* De
En diesel oil
Es aceite diesel *(m)*
Fr combustible diesel *(m)*
It nafta *(f)*
Pt óleo diesel *(m)*

diferença de potencial *(f)* Pt
De Spannungsunter-schied *(m)*
En potential difference
Es diferencia de potencial *(f)*
Fr différence de potentiel *(f)*
It caduta di tensione *(f)*

diferencia de potencial *(f)* Es
De Spannungsunter-schied *(m)*
En potential difference
Fr différence de potentiel *(f)*
It caduta di tensione *(f)*
Pt diferença de potencial *(f)*

difetto di massa *(m)* It
De Massendefekt *(m)*
En mass defect
Es defecto de masa *(m)*
Fr défaut de masse *(m)*
Pt defeito de massa *(m)*

différence de potentiel *(f)* Fr
De Spannungsunter-schied *(m)*
En potential difference
Es diferencia de potencial *(f)*
It caduta di tensione *(f)*
Pt diferença de potencial *(f)*

diffusé au sol Fr
De bodengestreut
En ground-scattered
Es dispersado en tierra
It diffuso sul suolo
Pt disperso pelo solo

diffuse radiation En
De Streustrahlung *(f)*
Es radiación difusa *(f)*
Fr rayonnement diffus *(m)*
It radiazione diffusa *(f)*
Pt radiação difusa *(f)*

diffusion *n* En; Fr *(f)*
De Diffusion *(f)*
Es difusión *(f)*
It diffusione *(f)*
Pt difusão *(f)*

Diffusion *(f) n* De
En diffusion
Es difusión *(f)*
Fr diffusion *(f)*
It diffusione *(f)*
Pt difusão *(f)*

diffusione *(f) n* It
De Diffusion *(f)*
En diffusion
Es difusión *(f)*
Fr diffusion *(f)*
Pt difusão *(f)*

diffusione gassosa *(f)* It
De Gasdiffusion *(f)*
En gaseous diffusion
Es difusión gaseosa *(f)*
Fr diffusion gazeuse *(f)*
Pt difusão gasosa *(f)*

diffusione termica *(f)* It
De Thermodiffusion *(f)*
En thermal diffusion
Es difusión térmica *(f)*
Fr diffusion thermique *(f)*
Pt difusão térmica *(f)*

diffusion gazeuse *(f)* Fr
De Gasdiffusion *(f)*
En gaseous diffusion
Es difusión gaseosa *(f)*
It diffusione gassosa *(f)*
Pt difusão gasosa *(f)*

diffusion thermique *(f)* Fr
De Thermodiffusion *(f)*
En thermal diffusion
Es difusión térmica *(f)*
It diffusione termica *(f)*
Pt difusão térmica *(f)*

diffuso sul suolo It
De bodengestreut
En ground-scattered
Es dispersado en tierra
Fr diffusé au sol
Pt disperso pelo solo

difusão *(f) n* Pt
De Diffusion *(f)*
En diffusion
Es difusión *(f)*
Fr diffusion *(f)*
It diffusione *(f)*

difusão gasosa *(f)* Pt
De Gasdiffusion *(f)*
En gaseous diffusion
Es difusión gaseosa *(f)*
Fr diffusion gazeuse *(f)*
It diffusione gassosa *(f)*

difusão térmica *(f)* Pt
De Thermodiffusion *(f)*
En thermal diffusion
Es difusión térmica *(f)*
Fr diffusion thermique *(f)*
It diffusione termica *(f)*

difusión *(f) n* Es
De Diffusion *(f)*
En diffusion
Fr diffusion *(f)*
It diffusione *(f)*
Pt difusão *(f)*

difusión gaseosa *(f)* Es
De Gasdiffusion *(f)*
En gaseous diffusion
Fr diffusion gazeuse *(f)*
It diffusione gassosa *(f)*
Pt difusão gasosa *(f)*

difusión térmica *(f)* Es
De Thermodiffusion *(f)*
En thermal diffusion
Fr diffusion thermique *(f)*
It diffusione termica *(f)*
Pt difusão térmica *(f)*

diga *(f) n* It
De Damm *(m)*
En dam
Es presa *(f)*
Fr barrage *(m)*
Pt barragem *(f)*

digester *n* En
De Digestor *(m)*
Es digestor *(m)*
Fr digesteur *(m)*
It digestore *(m)*
Pt digestor *(m)*

digesteur *(m) n* Fr
De Digestor *(m)*
En digester
Es digestor *(m)*
It digestore *(m)*
Pt digestor *(m)*

digestor *(m) n* Es, Pt
De Digestor *(m)*
En digester
Fr digesteur *(m)*
It digestore *(m)*

Digestor *(m) n* De
En digester
Es digestor *(m)*
Fr digesteur *(m)*
It digestore *(m)*
Pt digestor *(m)*

digestore *(m) n* It
De Digestor *(m)*
En digester
Es digestor *(m)*
Fr digesteur *(m)*
Pt digestor *(m)*

dilatação *(f) n* Pt
De Expansion *(f)*
En expansion
Es dilatación *(f)*
Fr dilatation *(f)*
It espansione *(f)*

dilatación *(f) n* Es
De Expansion *(f)*
En expansion
Fr dilatation *(f)*
It espansione *(f)*
Pt dilatação *(f)*

dilatation *(f) n* Fr
De Expansion *(f)*
En expansion
Es dilatación *(f)*
It espansione *(f)*
Pt dilatação *(f)*

dinamica *(f) n* It
De Dynamik *(f)*
En dynamics
Es dinámica *(f)*
Fr dynamique *(f)*
Pt dinâmica *(f)*

dinámica *(f) n* Es
De Dynamik *(f)*
En dynamics
Fr dynamique *(f)*
It dinamica *(f)*
Pt dinâmica *(f)*

dinâmica *(f) n* Pt
De Dynamik *(f)*
En dynamics
Es dinámica *(f)*
Fr dynamique *(f)*
It dinamica *(f)*

dinamo *(f) n* It
De Dynamo *(m)*
En dynamo
Es dínamo *(f)*
Fr dynamo *(f)*
Pt dínamo *(m)*

dínamo *(f) n* Es, Pt
De Dynamo *(m)*
En dynamo
Fr dynamo *(f)*
It dinamo *(f)*

dinamometro *(m) n* It
De Dynamometer *(n)*
En dynamometer
Es dinamómetro *(m)*
Fr dynamomètre *(m)*
Pt dinamómetro *(m)*

dinamómetro *(m) n* Es, Pt
De Dynamometer *(n)*
En dynamometer
Fr dynamomètre *(m)*
It dinamometro *(m)*

dinamometro ad assorbimento *(m)* It
De Absorptions-dynamometer *(n)*
En absorption dynamometer
Es dinamómetro de absorción *(m)*
Fr dynamomètre à absorption *(m)*
Pt dinamómetro de absorção *(m)*

dinamómetro de absorção *(m)* Pt
De Absorptions-dynamometer *(n)*
En absorption dynamometer
Es dinamómetro de absorción *(m)*
Fr dynamomètre à absorption *(m)*
It dinamometro ad assorbimento *(m)*

dinamómetro de absorción *(m)* Es
De Absorptions-dynamometer *(n)*
En absorption dynamometer
Fr dynamomètre à absorption *(m)*
It dinamometro ad assorbimento *(m)*
Pt dinamómetro de absorção *(m)*

dióxido de uranio *(m)* Es
De Uraniumdioxid *(n)*
En uranium dioxide
Fr bioxyde d'uranium *(m)*
It biossido di uranio *(m)*
Pt dióxido de urânio

dióxido de urânio Pt
De Uraniumdioxid *(n)*
En uranium dioxide
Es dióxido de uranio *(m)*
Fr bioxyde d'uranium *(m)*
It biossido di uranio *(m)*

direct absorber En
De Direktabsorber *(m)*
Es absorbedor directo *(m)*
Fr absorbeur direct *(m)*
It assorbitore diretto *(m)*

Pt absorvedor directo *(m)*

direct conversion En
De Direktumwandlung *(f)*
Es conversión directa *(f)*
Fr conversion directe *(f)*
It conversione diretta *(f)*
Pt conversão directa *(f)*

direct current (d.c.) En
De Gleichstrom *(m)*
Es corriente continua (c.c.) *(f)*
Fr courant continu (c.c.) *(m)*
It corrente continua (c.c.) *(f)*
Pt corrente directa (c.d.) *(f)*

directional drilling En
De gerichtetes Bohren *(n)*
Es perforación direccional *(f)*
Fr forage dirigé *(m)*
It trivellazione direzionale *(f)*
Pt perfuração direccional *(f)*

directional flame burner En
De flammengerichteter Brenner *(m)*
Es quemador de llama direccional *(m)*
Fr brûleur à flamme dirigée *(m)*
It bruciatore di fiamma direzionale *(m)*
Pt queimador de chama direccional *(m)*

Direktabsorber *(m) n* De
En direct absorber
Es absorbedor directo *(m)*
Fr absorbeur direct *(m)*
It assorbitore diretto *(m)*
Pt absorvedor directo *(m)*

Direktumwandlung *(f) n* De
En direct conversion
Es conversión directa *(f)*
Fr conversion directe *(f)*
It conversione diretta *(f)*
Pt conversão directa *(f)*

disalcalizzazione *(f) n* It
De Dealkylierung *(f)*
En dealkylation
Es desalquilación *(f)*
Fr désalcoylation *(f)*
Pt desalquilação *(f)*

disattivatori di metallo *(m pl)* It
De Metall-Deaktiviermungsmittel *(pl)*
En metal deactivators
Es desactivadores de metales *(m pl)*
Fr désactiveurs de métaux *(m pl)*
Pt desactivadores metálicos *(m pl)*

discharge *n* En
De Entladung *(f)*
Es descarga (bateria) *(f)*
Fr décharge *(f)*
It scarica *(f)*
Pt descarga (batería) *(f)*

discharge characteristics *pl n* En
De Entladungsmerkmale *(pl)*
Es características de descarga *(f pl)*
Fr caractéristiques de décharge *(f pl)*
It caratteristiche di scarica *(f pl)*
Pt características de descarga *(f pl)*

diseño ambiental integrado *(m)* Es
De integrierte Umweltplanung *(f)*
En integrated environmental design (IED)
Fr étude de l'environment integreé *(f)*
It design ambientale integrato *(m)*
Pt projecto de ambiente integrado *(m)*

disintegrazione beta *(f)* It
De Beta-Zerfall *(m)*
En beta decay
Es desintegración beta *(f)*
Fr désintégration bêta *(f)*
Pt deterioração beta *(f)*

disjoncteur *(m) n* Fr
De Leitungsschalter *(m)*
En circuit breaker
Es disyuntor *(m)*
It interruttore automatico *(m)*
Pt corta-circuito *(m)*

disolvente *(m) n* Es
De Lösungsmittel *(n)*
En solvent
Fr solvant *(m)*
It solvente *(m)*
Pt dissolvente *(m)*

dispersado en tierra Es
De bodengestreut
En ground-scattered
Fr diffusé au sol
It diffuso sul suolo
Pt disperso pelo solo

dispersion *(f) n* Fr
De Streuung *(f)*
En scattering
Es dispersión *(f)*
It dispersione *(f)*
Pt espalhamento *(m)*

dispersión *(f) n* Es
De Streuung *(f)*
En scattering
Fr dispersion *(f)*
It dispersione *(f)*
Pt espalhamento *(m)*

dispersione *(f) n* It
De Streuung *(f)*
En scattering
Es dispersión *(f)*
Fr dispersion *(f)*
Pt espalhamento *(m)*

disperso pelo solo Pt
De bodengestreut
En ground-scattered
Es dispersado en tierra
Fr diffusé au sol
It diffuso sul suolo

disponibilidade solar *(f)* Pt
De Solarverfügbarkeit *(f)*
En solar availability
Es disponibilidad solar *(f)*
Fr disponibilité solaire *(f)*
It disponibilità solare *(f)*

disponibilidad solar *(f)* Es
De Solarverfügbarkeit *(f)*
En solar availability
Fr disponibilité solaire *(f)*
It disponibilità solare *(f)*
Pt disponibilidade solar *(f)*

disponibilità solare *(f)* It
De Solarverfügbarkeit *(f)*
En solar availability
Es disponibilidad solar *(f)*
Fr disponibilité solaire *(f)*
Pt disponibilidade solar *(f)*

disponibilité solaire *(f)* Fr
De Solarverfügbarkeit *(f)*
En solar availability
Es disponibilidad solar *(f)*
It disponibilità solare *(f)*
Pt disponibilidade solar *(f)*

dispositivo di tenuta *(m)* It
De Abdichtung *(f)*
En seal
Es cierre *(m)*
Fr joint *(m)*
Pt vedação *(f)*

dissolvente *(m) n* Pt
De Lösungsmittel *(n)*
En solvent
Es disolvente *(m)*
Fr solvant *(m)*
It solvente *(m)*

distillat *(m) n* Fr
De Destillat *(n)*
En distillate
Es destilado *(m)*
It distillato *(m)*
Pt destilado *(m)*

distillate *n* En
De Destillat *(n)*
Es destilado *(m)*
Fr distillat *(m)*
It distillato *(m)*
Pt destilado *(m)*

distillate fuel oil En
De Destillatöl *(n)*
Es fuel-oil destilado *(m)*
Fr fuel-oil distillé *(m)*
It nafta distillata *(f)*
Pt fuel-oil de destilados *(m)*

distillation *n* En; Fr *(f)*
De Destillation *(f)*
Es destilación *(f)*
It distillazione *(f)*
Pt destilação *(f)*

distillation à effet multiple *(f)* Fr
De Mehrfacheffekt-Destillation *(f)*
En multiple-effect distillation
Es destilación de efecto múltiple *(f)*
It distillazione ad effetto multiplo *(f)*
Pt destilação de efeito múltiplo *(f)*

distillation column En
De Destillationskolonne *(f)*
Es torre de destilación *(f)*
Fr colonne de distillation *(f)*
It colonna di distillazione *(f)*
Pt coluna de destilação *(f)*

distillation fractionnée *(f)* Fr
De fraktionierte Destillation *(f)*
En fractional distillation
Es destilación fraccionada *(f)*
It distillazione a frazione *(f)*
Pt destilação fraccional *(f)*

distillation simple effet *(f)* Fr
De Einzeleffekt-Destillation *(f)*
En single-effect distillation
Es destilación de simple efecto *(f)*
It distillazione ad effetto unico *(f)*
Pt destilação de efeito simples *(f)*

distillato *(m)* *n* It
De Destillat *(n)*
En distillate
Es destilado *(m)*
Fr distillat *(m)*
Pt destilado *(m)*

distillazione *(f)* *n* It
De Destillation *(f)*
En distillation
Es destilación *(f)*
Fr distillation *(f)*
Pt destilação *(f)*

distillazione ad effetto multiplo *(f)* It
De Mehrfacheffekt-Destillation *(f)*
En multiple-effect distillation
Es destilación de efecto múltiple *(f)*
Fr distillation à effet multiple *(f)*
Pt destilação de efeito múltiplo *(f)*

distillazione ad effetto unico *(f)* It
De Einzeleffekt-Destillation *(f)*
En single-effect distillation
Es destilación de simple efecto *(f)*
Fr distillation simple effet *(f)*
Pt destilação de efeito simples *(f)*

distillazione a frazione *(f)* It
De fraktionierte Destillation *(f)*
En fractional distillation
Es destilación fraccionada *(f)*
Fr distillation fractionnée *(f)*
Pt destilação fraccional *(f)*

distribuição pela linha principal *(f)* Pt
De Netzstrom *(m)*
En mains supply
Es alimentación principal *(f)*
Fr alimentation secteur *(f)*
It alimentazione di rete *(f)*

distribution cost En
De Vertriebskosten *(pl)*
Es coste de distribución *(m)*
Fr frais de distribution *(m)*
It costo di distribuzione *(m)*
Pt custo de distribuição *(m)*

district heating En
De Fernheizung *(f)*
Es calefacción de distritos *(f)*
Fr chauffage urbain *(m)*
It riscaldamento di sezione *(m)*
Pt aquecimento de distrito *(m)*

disyuntor *(m)* *n* Es
De Leitungsschalter *(m)*
En circuit breaker
Fr disjoncteur *(m)*
It interruttore automatico *(m)*
Pt corta-circuito *(m)*

diurnal heat storage En
De tägliche Wärmespeicherung *(f)*
Es almacenamiento diurno del calor *(m)*
Fr accumulation de chaleur diurne *(f)*
It immagazzinaggio di calore diurno *(m)*
Pt armazenagem diurna de calor *(f)*

diviseur de tension *(m)* Fr
De Spannungsteiler *(m)*
En potential divider; voltage divider
Es divisor de potencial; divisor de tensión *(m)*
It partitore di tensione *(m)*
Pt divisor de potencial; divisor de tensão *(m)*

divisor de potencial *(m)* Es, Pt
De Spannungsteiler *(m)*
En potential divider; voltage divider
Fr diviseur de tension *(m)*
It partitore di tensione *(m)*

divisor de tensão *(m)* Pt
De Spannungsteiler *(m)*
En potential divider; voltage divider
Es divisor de potencial; divisor de tensión *(m)*
Fr diviseur de tension *(m)*
It partitore di tensione *(m)*

divisor de tensión *(m)* Es
De Spannungsteiler *(m)*
En potential divider; voltage divider
Fr diviseur de tension *(m)*
It partitore di tensione *(m)*
Pt divisor de potencial; divisor de tensão *(m)*

doble encendido *(m)* Es
De Doppelzündung *(f)*
En dual ignition
Fr double allumage *(m)*
It accensione doppia *(f)*
Pt ignição dupla *(f)*

doble encristalado de unidad sellada *(m)* Es
De luftdichte Doppelverglasungs-elemente *(pl)*
En sealed-unit double glazing
Fr double vitrage à éléments scellés *(m)*
It doppia vetrata sigillata *(f)*
Pt vitrificação dupla de unidade vedada *(f)*

doble encristalado de ventana secundaria *(m)* Es
De Doppelverglasung für sekundäre Fenster *(f)*

En secondary-window double glazing
Fr double vitrage secondaire *(f)*
It doppia vetrata di finestre secondarie *(f)*
Pt vitrificação dupla de janela secundária *(f)*

domanda di punta *(f)* It
De Spitzenbedarf *(m)*
En peak demand
Es demanda punta *(f)*
Fr consommation de pointe *(f)*
Pt procura de ponta *(f)*

domestic energy consumption En
De Haushalt-Energieverbrauch *(m)*
Es consumo doméstico de energía *(m)*
Fr consommation domestique d'énergie *(f)*
It consumo di energia per usi domestici *(m)*
Pt consumo doméstico de energia *(m)*

Doppeltreibstoffmotor *(m) n* De
En dual-fuel engine
Es motor de dos combustibles *(m)*
Fr moteur polycarburant *(m)*
It motore a due combustibili *(m)*
Pt motor de dois combustíveis *(m)*

Doppelverglasung *(f) n* De
En double glazing
Es encristalado doble *(m)*
Fr double vitrage *(m)*
It doppia vetratura *(f)*
Pt vitrificação dupla *(f)*

Doppelverglasung für sekundäre Fenster *(f)* De
En secondary-window double glazing
Es doble encristalado de ventana secundaria *(m)*
Fr double vitrage secondaire *(f)*
It doppia vetrata di finestre secondarie *(f)*
Pt vitrificação dupla de janela secundária *(f)*

Doppelzündung *(f) n* De
En dual ignition
Es doble encendido *(m)*
Fr double allumage *(m)*
It accensione doppia *(f)*
Pt ignição dupla *(f)*

doppia vetrata di finestre secondarie *(f)* It
De Doppelverglasung für sekundäre Fenster *(f)*
En secondary-window double glazing
Es doble encristalado de ventana secundaria *(m)*
Fr double vitrage secondaire *(f)*
Pt vitrificação dupla de janela secundária *(f)*

doppia vetrata sigillata *(f)* It
De luftdichte Doppelverglasungs-elemente *(pl)*
En sealed-unit double glazing
Es doble encristalado de unidad sellada *(m)*
Fr double vitrage à éléments scellés *(m)*
Pt vitrificação dupla de unidade vedada *(f)*

doppia vetratura *(f)* It
De Doppelverglasung *(f)*
En double glazing
Es encristalado doble *(m)*
Fr double vitrage *(m)*
Pt vitrificação dupla *(f)*

Doppler effect En
De Doppler-Effekt *(m)*
Es efecto Doppler *(m)*
Fr effet Doppler-Fizeau *(m)*
It effetto Doppler *(m)*
Pt efeito Doppler *(m)*

Doppler-Effekt *(m)* De
En Doppler effect
Es efecto Doppler *(m)*
Fr effet Doppler-Fizeau *(m)*
It effetto Doppler *(m)*
Pt efeito Doppler *(m)*

dose *n* En; Fr, It, Pt *(f)*
De Dosis *(f)*
Es dosis *(f)*

dose letal *(f)* Pt
De tödliche Dosis *(f)*
En lethal dose
Es dosis letal *(f)*
Fr dose mortelle *(f)*
It dose letale *(f)*

dose letale *(f)* It
De tödliche Dosis *(f)*
En lethal dose
Es dosis letal *(f)*
Fr dose mortelle *(f)*
Pt dose letal *(f)*

dose mortelle *(f)* Fr
De tödliche Dosis *(f)*
En lethal dose
Es dosis letal *(f)*
It dose letale *(f)*
Pt dose letal *(f)*

dosimeter *n* En
De Dosimeter *(n)*
Es dosímetro *(m)*
Fr dosimètre *(m)*
It dosimetro *(m)*
Pt dosímetro *(m)*

Dosimeter *(n) n* De
En dosimeter
Es dosímetro *(m)*
Fr dosimètre *(m)*
It dosimetro *(m)*
Pt dosímetro *(m)*

dosimètre *(m) n* Fr
De Dosimeter *(n)*
En dosimeter
Es dosímetro *(m)*
It dosimetro *(m)*
Pt dosímetro *(m)*

dosimetria *(f) n* It, Pt
De Dosimetrie *(f)*
En dosimetry
Es dosimetría *(f)*
Fr dosimétrie *(f)*

dosimetría *(f) n* Es
De Dosimetrie *(f)*
En dosimetry
Fr dosimétrie *(f)*
It dosimetria *(f)*
Pt dosimetria *(f)*

Dosimetrie *(f) n* De
En dosimetry
Es dosimetría *(f)*
Fr dosimétrie *(f)*
It dosimetria *(f)*
Pt dosimetria *(f)*

dosimétrie *(f) n* Fr
De Dosimetrie *(f)*
En dosimetry
Es dosimetría *(f)*
It dosimetria *(f)*
Pt dosimetria *(f)*

dosimetro *(m) n* It
De Dosimeter *(n)*
En dosimeter
Es dosímetro *(m)*
Fr dosimètre *(m)*
Pt dosímetro *(m)*

dosímetro *(m) n* Es, Pt
De Dosimeter *(n)*
En dosimeter
Fr dosimètre *(m)*
It dosimetro *(m)*

dosimetry *n* En
De Dosimetrie *(f)*
Es dosimetría *(f)*
Fr dosimétrie *(f)*
It dosimetria *(f)*
Pt dosimetria *(f)*

dosis *(f) n* Es
De Dosis *(f)*
En dose
Fr dose *(f)*
It dose *(f)*
Pt dose *(f)*

Dosis *(f) n* De
En dose
Es dosis *(f)*
Fr dose *(f)*
It dose *(f)*
Pt dose *(f)*

dosis letal *(f)* Es
De tödliche Dosis *(f)*
En lethal dose
Fr dose mortelle *(f)*
It dose letale *(f)*
Pt dose letal *(f)*

double allumage *(m)* Fr
De Doppelzündung *(f)*
En dual ignition
Es doble encendido *(m)*
It accensione doppia *(f)*
Pt ignição dupla *(f)*

double glazing En
De Doppelverglasung *(f)*
Es encristalado doble *(m)*
Fr double vitrage *(m)*
It doppia vetratura *(f)*
Pt vitrificação dupla *(f)*

double vitrage *(m)* Fr
De Doppelverglasung *(f)*
En double glazing
Es encristalado doble *(m)*
It doppia vetratura *(f)*
Pt vitrificação dupla *(f)*

double vitrage à éléments scellés *(m)* Fr
De luftdichte Doppelverglasungselemente *(pl)*
En sealed-unit double glazing
Es doble encristalado de unidad sellada *(m)*
It doppia vetrata sigillata *(f)*
Pt vitrificação dupla de unidade vedada *(f)*

double vitrage secondaire *(m)* Fr
De Doppelverglasung für sekundäre Fenster *(f)*
En secondary-window double glazing
Es doble encristalado de ventana secundaria *(m)*
It doppia vetrata di finestre secondarie *(f)*
Pt vitrificação dupla de janela secundária *(f)*

draft diverter Am
De Luftzugablenkung *(f)*
En draught diverter
Es desviador de corrientes *(m)*
Fr déflecteur de courant d'air *(m)*
It deviatore d'aria *(m)*
Pt deflector de corrente de ar *(m)*

draft excluder Am
De Luftzugschutz *(m)*
En draught excluder
Es eliminador de corrientes *(m)*
Fr brise-bise *(m)*
It esclusore d'aria *(m)*
Pt eliminador de corrente de ar *(m)*

draught diverter En
Am draft diverter
De Luftzugablenkung *(f)*
Es desviador de corrientes *(m)*
Fr déflecteur de courant d'air *(m)*
It deviatore d'aria *(m)*
Pt deflector de corrente de ar *(m)*

draught excluder En
Am draft excluder
De Luftzugschutz *(m)*
Es eliminador de corrientes *(m)*
Fr brise-bise *(m)*
It esclusore d'aria *(m)*
Pt eliminador de corrente de ar *(m)*

Drehbohren *(n) n* De
En rotary drilling
Es perforación rotativa *(f)*
Fr forage rotary *(m)*
It sondaggio a rotazione *(m)*
Pt perfuração rotativa *(f)*

Dreheisen- De
En moving-iron
Es núcleo giratorio
Fr fer mobile
It ferro rotante
Pt ímã móvel

Drehspule- De
En moving-coil
Es bobina móvil
Fr bobine mobile
It bobina mobile
Pt bobina móvel

Drehverschiebungsmesser *(m) n* De
En rotary-displacement meter
Es contador de desplazamiento rotativo *(m)*
Fr compteur à piston rotatif *(m)*
It metro di spostamento rotante *(m)*
Pt contador de deslocação rotativa *(m)*

Dreifachverglasung *(f) n* De
En triple glazing n
Es encristalado triple *(m)*
Fr triple vitrage *(m)*
It tripla vetrata *(f)*
Pt vitrificação tripla *(f)*

dreiphasig *adj* De
En three-phase
Es trifásico
Fr triphasé
It trifase
Pt trifásico

drilling *n* En
De Bohren *(n)*
Es perforación *(f)*
Fr forage *(m)*
It trivellazione *(f)*
Pt perfuração *(f)*

drilling mud En
De Bohrschlamm *(m)*
Es lodo de perforación *(m)*
Fr boue de forage *(f)*
It fango di trivellazione *(m)*
Pt lodo de perfuração *(m)*

drilling structures *(pl)* En
De Bohrkonstruktionen *(pl)*
Es estructuras de perforación *(f pl)*
Fr structures de forage *(f pl)*
It strutture di trivellazione *(f pl)*
Pt estructuras de perfuração *(f pl)*

Druck *(m) n* De
En pressure
Es presión *(f)*
Fr pression *(f)*
It pressione *(f)*
Pt pressão *(f)*

Druckabfall *(m) n* De
En pressure drop
Es caída de presión *(f)*
Fr chute de pression *(f)*
It caduta di pressione *(f)*
Pt queda de pressão *(f)*

Druckhöhe *(f) n* De
En pressure head
Es altura manométrica *(f)*
Fr hauteur de refoulement *(f)*
It altezza manometrica *(f)*
Pt altura manométrica *(f)*

Druckluftbohren *(n) n* De
En air drilling
Es perforación por aire comprimido *(f)*
Fr forage à l'air comprimé *(m)*
It trapanatura ad aria compressa *(f)*
Pt perfuração pneumática *(f)*

Druckluftspeicherung *(f) n* De
En compressed-air storage
Es almacenamiento de aire comprimido *(m)*
Fr stockage de l'air comprimé *(m)*
It immagazzinaggio di aria compressa *(m)*
Pt armazenagem de ar comprimido *(f)*

Druckröhrenreaktor *(m) n* De
En pressure-tube reactor
Es reactor de tubos de presión *(m)*
Fr réacteur à tube de force *(m)*
It reattore con tubo di pressione *(m)*
Pt reactor de tubo de pressão *(m)*

Druckwasserreaktor *(m) n* De
En pressurized-water reactor (PWR)
Es reactor de agua a presión *(m)*

Fr réacteur à eau sous pression *(m)*
It reattore ad acqua pressurizzata *(m)*
Pt reactor de água pressurizada *(m)*

dry-ash furnace En
De Trockenasche-Ofen *(m)*
Es hogar de cenizas pulverulentas *(m)*
Fr four à cendres sèches *(m)*
It forno di cenere secca *(m)*
Pt forno de cinze seca *(m)*

dryer *n* En
De Trockner *(m)*
Es secador *(m)*
Fr sécheur *(m)*
It essiccatore *(m)*
Pt secador *(m)*

dry meter En
De Trockenmeßgerät *(n)*
Es medidor seco *(m)*
Fr dessicomètre *(m)*
It contatore a secco *(m)*
Pt medidor de seco *(m)*

dual-cycle combustion En
De Zweikreisverbrennung *(f)*
Es combustión de ciclo doble *(f)*
Fr combustion à deux temps *(f)*
It combustione a ciclo doppio *(f)*
Pt combustão de ciclo duplo *(f)*

dual-fluid heat transfer En
De Wärmeübertragung mit zwei Flüssigkeiten *(f)*
Es transferencia térmica de dos flúidos *(f)*
Fr transfert de chaleur à double fluide *(m)*
It trasferimento di calore a doppio fluido *(m)*
Pt termotransferência de dois fluidos *(f)*

dual-fuel engine En
De Doppeltreibstoffmotor *(m)*
Es motor de dos combustibles *(m)*
Fr moteur polycarburant *(m)*
It motore a due combustibili *(m)*
Pt motor de dois combustíveis *(m)*

dual ignition En
De Doppelzündung *(f)*
Es doble encendido *(m)*
Fr double allumage *(m)*
It accensione doppia *(f)*
Pt ignição dupla *(f)*

dumping *n* En
De Kippen *(n)*
Es vaciamiento *(m)*
Fr déversement *(m)*
It pressatura *(f)*
Pt esvaziamento *(m)*

Durchflußmesser *(m) n* De
En flow meter
Es medidor de flujo *(m)*
Fr débitmètre *(m)*
It flussometro *(m)*
Pt fluxómetro *(m)*

Durchgangsdämpfung *(f)* De
En transmission loss
Es pérdida por transmisión *(f)*
Fr perte par transmission *(f)*
It perdita per trasmissione *(f)*
Pt perda por transmissão *(f)*

Durchschnitt *(m) n* De
En mean
Es media *(f)*
Fr moyen *(m)*
It medio *(f)*
Pt médio *(m)*

durchschnittliche Abweichung *(f)* De
En mean deviation
Es desviación media *(f)*
Fr déviation moyenne *(f)*
It deviazione media *(f)*
Pt desvio médio *(m)*

durchschnittlicher Sonnenfluß *(m)* De
En average solar flux
Es flujo solar medio *(m)*
Fr flux solaire moyen *(m)*
It flusso solare medio *(m)*
Pt fluxo solar médio *(m)*

durée de remboursement *(f)* Fr
De Rückzahldauer *(f)*
En payback period
Es período de reembolso *(m)*
It periodo di remborso *(m)*
Pt período de reembolso *(m)*

durée de vie des neutrons *(f)* Fr
De Neutronenlebensdauer *(f)*
En neutron lifetime
Es vida de neutrón *(f)*
It vita dei neutroni *(f)*
Pt vida de neutrão *(f)*

Düse *(f) n* De
En jet
Es chorro *(m)*
Fr jet *(m)*
It getto *(m)*
Pt jacto *(m)*

Düsenbrenner *(m) n* De
En jet burner
Es quemador de chorro *(m)*
Fr brûleur-jet *(m)*
It bruciatore a getto *(m)*
Pt queimador de jacto *(m)*

Düsentreibstoff *(m) n* De
En jet fuel
Es combustible de propulsión a chorro *(m)*
Fr carburéacteur *(m)*
It combustibile per aviogetti *(m)*
Pt combustível de jactos *(m)*

dust *n* En
De Staub *(m)*
Es polvo *(m)*
Fr poussière *(f)*
It polvere *(f)*
Pt pó *(m)*

dynamics *n* En
De Dynamik *(f)*
Es dinámica *(f)*
Fr dynamique *(f)*
It dinamica *(f)*
Pt dinâmica *(f)*

Dynamik *(f) n* De
En dynamics
Es dinámica *(f)*
Fr dynamique *(f)*
It dinamica *(f)*
Pt dinâmica *(f)*

dynamique *(f) n* Fr
De Dynamik *(f)*
En dynamics
Es dinámica *(f)*
It dinamica *(f)*
Pt dinâmica *(f)*

dynamo *n* En; Fr *(f)*
De Dynamo *(m)*
Es dínamo *(f)*
It dinamo *(f)*
Pt dínamo *(m)*

Dynamo *(m) n* De
En dynamo
Es dínamo *(f)*
Fr dynamo *(f)*
It dinamo *(f)*
Pt dínamo *(m)*

dynamometer *n* En
De Dynamometer *(n)*
Es dinamómetro *(m)*
Fr dynamomètre *(m)*
It dinamometro *(m)*
Pt dinamómetro *(m)*

Dynamometer *(n) n* De
En dynamometer
Es dinamómetro *(m)*
Fr dynamomètre *(m)*
It dinamometro *(m)*
Pt dinamómetro *(m)*

dynamomètre *(m) n* Fr
De Dynamometer *(n)*
En dynamometer
Es dinamómetro *(m)*

It dinamometro *(m)*
Pt dinamómetro *(m)*

dynamomètre à absorption *(m)* Fr
De Absorptions-dynamometer *(n)*
En absorption dynamometer
Es dinamómetro de absorción *(m)*
It dinamometro ad assorbimento *(m)*
Pt dinamómetro de absorção *(m)*

E

earth *n* En
Am ground
De Erde *(f)*
Es tierra *(f)*
Fr terre *(f)*
It terra *(f)*
Pt massa *(f)*

eau chaude *(f)* Fr
De Warmwasser *(n)*
En hot water
Es agua caliente *(f)*
It acqua calda *(f)*
Pt água quente *(f)*

eau d'alimentation *(f)* Fr
De Speisewasser *(n)*
En feedwater
Es agua de alimentación *(f)*
It acqua di alimentazione *(f)*
Pt água de alimentação *(f)*

eau d'alimentation de condenseur *(f)* Fr
De Kondensator-Speisewasser *(n)*
En condenser feedwater
Es agua de alimentación del condensador *(f)*
It acqua di alimentazione del condensatore *(f)*
Pt água de alimentação de condensador *(f)*

eau déminéralisée *(f)* Fr
De entmineralisiertes Wasser *(n)*
En demineralized water
Es agua desmineralizada *(f)*
It acqua demineralizzata *(f)*
Pt água desmineralizada *(f)*

eau de refroidissement *(f)* Fr
De Kühlwasser *(n)*
En cooling water
Es agua de refrigeración *(f)*
It acqua di raffreddamento *(f)*
Pt água de refrigeração *(f)*

eau lourde *(f)* Fr
De Schwerwasser *(n)*
En heavy water
Es agua pesada *(f)*
It acqua pesante *(f)*
Pt água pesada *(f)*

ebullient cooling En
De Heißwasserkühlung *(f)*
Es enfriamiento desde la ebullición *(m)*
Fr refroidissement par ébullition *(m)*
It raffreddamento a ebollizione *(m)*
Pt arrefecimento ebuliente *(m)*

eccesso d'aria *(m)* It
De Luftüberschuß *(m)*
En excess air
Es aire sobrante *(m)*
Fr excès d'air *(m)*
Pt ar em excesso *(m)*

eccitazione *(f) n* It
De Erregung *(f)*
En excitation
Es excitación *(f)*
Fr excitation *(f)*
Pt excitação *(f)*

échange d'ions *(m)* Fr
De Ionenaustausch *(m)*
En ion exchange
Es intercambio de iones *(m)*
It scambio ionico *(m)*
Pt permuta de ioēs *(f)*

échangeur *(m) n* Fr
De Austauscher *(m)*
En exchanger
Es intercambiador *(m)*
It scambiatore *(m)*
Pt permutador *(m)*

échangeur de chaleur *(m)* Fr
De Wärmeaustauscher *(m)*
En heat exchanger
Es intercambiador de calor *(m)*
It scambiatore di calore *(m)*
Pt termo-permutador *(m)*

échangeur de chaleur à calandre *(m)* Fr
De Röhrenwärme-austauscher *(m)*
En shell-and-tube heat exchanger
Es intercambiador de calor de coraza y tubos *(m)*
It scambiatore di calore a cilindro e tubi *(m)*
Pt termo-permutador de camisa e tubo *(m)*

échangeur de chaleur à plaque *(m)* Fr
De Plattenwärme-austauscher *(m)*
En plate heat-exchanger
Es intercambiador de calor de placas *(m)*
It scambiatore di calore a piastra *(m)*
Pt termo-permutador de placa *(m)*

échangeur de chaleur à serpentin immergé *(m)* Fr
De Wärmeaustauscher bestehend aus Rohrschlangen im Bad *(m)*
En coil-in-bath heat exchanger
Es intercambiador de calor con serpentín en baño *(m)*
It scambiatore di calore a serpentino in bagno *(m)*
Pt termo-permutador de bobina em banho *(m)*

échangeur de chaleur à serpentin *(m)* Fr
De Spiralröhren-Wärmeaustauscher *(m)*
En spiral-tube heat exchanger
Es intercambiador de calor de tubos en espiral *(m)*
It scambiatore di calore con tubo a spirale *(m)*
Pt termo-permutador de tubo em espiral *(m)*

échangeur de chaleur à tubes concentriques (m) Fr
De Wärmeaustauscher mit konzentrischen Röhren *(m)*
En concentric-tube heat exchanger
Es intercambiador de calor de tubos concéntricos *(m)*
It scambiatore di calore a tubo concentrico *(m)*
Pt termo-permutador de tubo concêntrico *(m)*

échangeur de chaleur en verre *(m)* Fr
De Glaswärme-austauscher *(m)*
En glass heat exchanger
Es intercambiador de calor de vidrio *(m)*
It scambiatore di calore di vetro *(m)*
Pt termo-permutador de vidro *(m)*

échappement *(m) n* Fr
De Auspuffgas *(n)*
En exhaust
Es gases de escape *(m pl)*
It scarico *(m)*
Pt escape *(m)*

échappement de gaz *(m)* Fr
De Gasaustritt *(m)*

En gas escape
Es fuga de gas *(f)*
It fuga di gas *(f)*
Pt escape de gás *(f)*

éclairage *(m) n* Fr
De Beleuchtung *(f)*
En illumination
Es iluminación *(f)*
It illuminazione *(f)*
Pt iluminação *(f)*

economia de hidrogénio *(f)* Pt
De Wasserstoffwirtschaft *(f)*
En hydrogen economy
Es economía de hidrógeno *(f)*
Fr économie d'hydrogène *(f)*
It economia dell'idrogeno *(f)*

economía de hidrógeno *(f)* Es
De Wasserstoffwirtschaft *(f)*
En hydrogen economy
Fr économie d'hydrogène *(f)*
It economia dell'idrogeno *(f)*
Pt economia de hidrogénio *(f)*

economia dell'idrogeno *(f)* It
De Wasserstoffwirtschaft *(f)*
En hydrogen economy
Es economía de hidrógeno *(f)*
Fr économie d'hydrogène *(f)*
Pt economia de hidrogénio *(f)*

economia energetica *(f)* It
De Energiewirtschaft *(f)*
En energy economics
Es economía energética *(f)*
Fr économie de l'énergie *(f)*
Pt economia energética *(f)*

economía energética *(f)* Es
De Energiewirtschaft *(f)*
En energy economics
Fr économie de l'énergie *(f)*
It economia energetica *(f)*
Pt economia energética *(f)*

economia energética *(f)* Pt
De Energiewirtschaft *(f)*
En energy economics
Es economía energética *(f)*
Fr économie de l'énergie *(f)*
It economia energetica *(f)*

économie de l'énergie *(f)* Fr
De Energiewirtschaft *(f)*
En energy economics
Es economía energética *(f)*
It economia energetica *(f)*
Pt economia energética *(f)*

économie d'énergie *(f)* Fr
De Energiesparen *(n)*
En energy saving
Es ahorro energético *(m)*
It risparmio di energia *(m)*
Pt poupança de energia *(f)*

économie d'hydrogène *(f)* Fr
De Wasserstoffwirtschaft *(f)*
En hydrogen economy
Es economía de hidrógeno *(f)*
It economia dell'idrogeno *(f)*
Pt economia de hidrogénio *(f)*

économiseur *(m) n* Fr
De Abgasvorwärmer *(m)*
En economizer
Es economizador *(m)*
It economizzatore *(m)*
Pt economizador *(m)*

economizador *(m) n* Es, Pt
De Abgasvorwärmer *(m)*
En economizer
Fr économiseur *(m)*
It economizzatore *(m)*

economizer *n* En
De Abgasvorwärmer *(m)*
Es economizador *(m)*
Fr économiseur *(m)*
It economizzatore *(m)*
Pt economizador *(m)*

economizzatore *(m) n* It
De Abgasvorwärmer *(m)*
En economizer
Es economizador *(m)*
Fr économiseur *(m)*
Pt economizador *(m)*

écran d'étanchéité à la vapeur *(m)* Fr
Am vapor barrier
De Dampfsperre *(f)*
En vapour barrier
Es barrera del vapor *(f)*
It schermo per il vapore *(m)*
Pt barreira de vapor *(f)*

efecto de estricción *(m)* Es
De Schnüreffekt *(m)*
En pinch effect
Fr effet de pincement *(m)*
It effetto di strizione *(m)*
Pt efeito de aperto *(m)*

efecto de latitud *(m)* Es
De Breiteneffekt *(m)*
En latitude effect
Fr effet de la latitude *(m)*
It effetto latitudine *(m)*
Pt efeito da latitude *(m)*

efecto Doppler *(m)* Es
De Doppler-Effekt *(m)*
En Doppler effect
Fr effet Doppler-Fizeau *(m)*
It effetto Doppler *(m)*
Pt efeito Doppler *(m)*

efecto termoeléctrico *(m)* Es
De thermoelektrischer Effekt *(m)*
En thermoelectric effect
Fr effet thermoélectrique *(m)*
It effetto termoelettrico *(m)*
Pt efeito termo-eléctrico *(m)*

efeito da latitude *(m)* Pt
De Breiteneffekt *(m)*
En latitude effect
Es efecto de latitud *(m)*
Fr effet de la latitude *(m)*
It effetto latitudine *(m)*

efeito de aperto *(m)* Pt
De Schnüreffekt *(m)*
En pinch effect
Es efecto de estricción *(m)*
Fr effet de pincement *(m)*
It effetto di strizione *(m)*

efeito Doppler *(m)* Pt
De Doppler-Effekt *(m)*
En Doppler effect
Es efecto Doppler *(m)*
Fr effet Doppler-Fizeau *(m)*
It effetto Doppler *(m)*

efeito termo-eléctrico *(m)* Pt
De thermoelektrischer Effekt *(m)*
En thermoelectric effect
Es efecto termoeléctrico *(m)*
Fr effet thermoélectrique *(m)*
It effetto termoelettrico *(m)*

effective brightness En
De effektive Helligkeit *(f)*
Es brillo efectivo *(m)*
Fr brillance efficace *(f)*
It luminosità effettiva *(f)*
Pt brilho efectivo *(m)*

effective value En
De Effektivwert *(m)*
Es valor eficaz *(m)*
Fr valeur efficace *(f)*
It valore efficace *(m)*
Pt valor eficaz *(m)*

effektive Helligkeit *(f)* De
En effective brightness
Es brillo efectivo *(m)*
Fr brillance efficace *(f)*

It luminosità effettiva *(f)*
Pt brilho efectivo *(m)*

Effektivwert *(m)* *n* De
En effective value
Es valor eficaz *(m)*
Fr valeur efficace *(f)*
It valore efficace *(m)*
Pt valor eficaz *(m)*

effet de la latitude *(m)* Fr
De Breiteneffekt *(m)*
En latitude effect
Es efecto de latitud *(m)*
It effetto latitudine *(m)*
Pt efeito da latitude *(m)*

effet de pincement *(m)* Fr
De Schnüreffekt *(m)*
En pinch effect
Es efecto de estricción *(m)*
It effetto di strizione *(m)*
Pt efeito de aperto *(m)*

effet Doppler-Fizeau *(m)* Fr
De Doppler-Effekt *(m)*
En Doppler effect
Es efecto Doppler *(m)*
It effetto Doppler *(m)*
Pt efeito Doppler *(m)*

effet thermoélectrique *(m)* Fr
De thermoelektrischer Effekt *(m)*
En thermoelectric effect
Es efecto termoeléctrico *(m)*
It effetto termoelettrico *(m)*
Pt efeito termo-eléctrico *(m)*

effetto di fondo *(m)* It
De Nulleffektstrahlung *(f)*
En background radiation
Es radiación ambiente *(f)*
Fr rayonnement provoquant le mouvement propre *(m)*
Pt radiação de fundo *(f)*

effetto di strizione *(m)* It
De Schnüreffekt *(m)*
En pinch effect
Es efecto de estricción *(m)*
Fr effet de pincement *(m)*
Pt efeito de aperto *(m)*

effetto Doppler *(m)* It
De Doppler-Effekt *(m)*
En Doppler effect
Es efecto Doppler *(m)*
Fr effet Doppler-Fizeau *(m)*
Pt efeito Doppler *(m)*

effetto latitudine *(m)* It
De Breiteneffekt *(m)*
En latitude effect
Es efecto de latitud *(m)*
Fr effet de la latitude *(m)*
Pt efeito da latitude *(m)*

effetto termoelettrico *(m)* It
De thermoelektrischer Effekt *(m)*
En thermoelectric effect
Es efecto termoeléctrico *(m)*
Fr effet thermoélectrique *(m)*
Pt efeito termo-eléctrico *(m)*

efficiency *n* En
De Leistungsfähigkeit *(f)*
Es eficiencia *(f)*
Fr rendement *(m)*
It efficienza *(f)*
Pt eficiência *(f)*

efficienza *(f)* *n* It
De Leistungsfähigkeit *(f)*
En efficiency
Es eficiencia *(f)*
Fr rendement *(m)*
Pt eficiência *(f)*

efficienza di generazione *(f)* It
De Erzeugungsleistung *(f)*
En generating efficiency
Es rendimiento de generación *(m)*
Fr rendement générateur *(m)*
Pt eficiência geradora *(f)*

effluent *n* En; Fr *(m)*
De Abwasser *(n)*
Es efluente *(m)*
It effluente *(m)*
Pt efluente *(m)*

effluente *(m)* *n* It
De Abwasser *(n)*
En effluent
Es efluente *(m)*
Fr effluent *(m)*
Pt efluente *(m)*

eficiencia *(f)* *n* Es
De Leistungsfähigkeit *(f)*
En efficiency
Fr rendement *(m)*
It efficienza *(f)*
Pt eficiência *(f)*

eficiência *(f)* *n* Pt
De Leistungsfähigkeit *(f)*
En efficiency
Es eficiencia *(f)*
Fr rendement *(m)*
It efficienza *(f)*

eficiência geradora *(f)* Pt
De Erzeugungsleistung *(f)*
En generating efficiency
Es rendimiento de generación *(m)*
Fr rendement générateur *(m)*
It efficienza di generazione *(f)*

eficiência global *(f)* Pt
De Gesamtleistungs-fähigkeit *(f)*
En overall efficiency
Es rendimiento global *(m)*
Fr rendement global *(m)*
It rendimento totale *(m)*

eficiência mecânica *(f)* Pt
De mechanischer Wirkungsgrad *(m)*
En mechanical efficiency
Es rendimiento mecánico *(m)*
Fr rendement mécanique *(m)*
It resa meccanica *(f)*

eficiência térmica *(f)* Pt
De thermischer Wirkungsgrad *(m)*
En thermal efficiency
Es rendimiento térmico *(m)*
Fr rendement thermique *(m)*
It rendimento termico *(m)*

efluente *(m)* *n* Es, Pt
De Abwasser *(n)*
En effluent
Fr effluent *(m)*
It effluente *(m)*

Eingabeenergie *(f)* *n* De
En input energy
Es energía de entrada *(f)*
Fr énergie d'entrée *(f)*
It energia immessa *(f)*
Pt energia de entrada *(f)*

Einpreßbohrung *(f)* *n* De
En injection well
Es pozo de inyección *(m)*
Fr puits d'injection *(m)*
It pozzo petrolifero sotto pressione *(m)*
Pt poço de injecção *(m)*

Einschalten *(n)* *n* De
En start-up
Es arranque *(m)*
Fr démarrage *(m)*
It avviamento *(m)*
Pt arranque *(m)*

Einzeleffekt-Destillation *(f)* *n* De
En single-effect distillation
Es destilación de simple efecto *(f)*
Fr distillation simple effet *(f)*
It distillazione ad effetto unico *(f)*
Pt destilação de efeito simples *(f)*

Einzelfenster-Kollektor *(m)* *n* De
En single-window collector
Es colector de ventana única *(m)*
Fr collecteur à simple fenêtre *(m)*
It raccogliatore a finestra unica *(m)*

Pt colector de janela única *(m)*

Einzelgang *(m) n* De
En single cycle
Es ciclo sencillo *(m)*
Fr simple cycle *(m)*
It ciclo unico *(m)*
Pt ciclo único *(m)*

electrão *(m) n* Pt
De Elektron *(n)*
En electron
Es electrón *(m)*
Fr électron *(m)*
It elettrone *(m)*

electrical energy En
De Elektroenergie *(f)*
Es energía eléctrica *(f)*
Fr énergie électrique *(f)*
It energia elettrica *(f)*
Pt energia eléctrica *(f)*

electric constant En
De Elektrizitätskonstante *(f)*
Es constante eléctrica *(f)*
Fr constante électrique *(f)*
It constante elettrica *(f)*
Pt constante eléctrica *(f)*

electric field En
De elektrisches Feld *(n)*
Es campo eléctrico *(m)*
Fr champ électrique *(m)*
It campo elettrico *(m)*
Pt campo eléctrico *(m)*

electric furnace En
De Elektroofen *(m)*
Es horno eléctrico *(m)*
Fr four électrique *(m)*
It forno elettrico *(m)*
Pt forno eléctrico *(m)*

electricidad *(f) n* Es
De Elektrizität *(f)*
En electricity
Fr électricité *(f)*
It elettricità *(f)*
Pt electricidade *(f)*

electricidad de generación nocturna *(f)* Es
De Nachtstrom *(m)*
En night-generated electricity
Fr électricité produite de nuit *(f)*
It elettricità generata di notte *(f)*
Pt electricidade de geração nocturna *(f)*

electricidade *(f) n* Pt
De Elektrizität *(f)*
En electricity
Es electricidad *(f)*
Fr électricité *(f)*
It elettricità *(f)*

electricidade de geração nocturna *(f)* Pt
De Nachtstrom *(m)*
En night-generated electricity
Es electricidad de generación nocturna *(f)*
Fr électricité produite de nuit *(f)*
It elettricità generata di notte *(f)*

electricidade fora de ponta *(f)* Pt
De Strom aus der Schwachlastzeit *(m)*
En off-peak electricity
Es electricidad fuera de horas punta *(f)*
Fr électricité hors-pointe *(f)*
It elettricità fuori di ore di punta *(f)*

electricidad fuera de horas punta *(f)* Es
De Strom aus der Schwachlastzeit *(m)*
En off-peak electricity
Fr électricité hors-pointe *(f)*
It elettricità fuori di ore di punta *(f)*
Pt electricidade fora de ponta *(f)*

électricité *(f) n* Fr
De Elektrizität *(f)*
En electricity
Es electricidad *(f)*
It elettricità *(f)*
Pt electricidade *(f)*

électricité hors-pointe *(f)* Fr
De Strom aus der Schwachlastzeit *(m)*
En off-peak electricity
Es electricidad fuera de horas punta *(f)*
It elettricità fuori di ore di punta *(f)*
Pt electricidade fora de ponta *(f)*

électricité produite de nuit *(f)* Fr
De Nachtstrom *(m)*
En night-generated electricity
Es electricidad de generación nocturna *(f)*
It elettricità generata di notte *(f)*
Pt electricidade de geração nocturna *(f)*

electricity *n* En
De Elektrizität *(f)*
Es electricidad *(f)*
Fr électricité *(f)*
It elettricità *(f)*
Pt electricidade *(f)*

electricity generation En
De Stromerzeugung *(f)*
Es generación de electricidad *(f)*
Fr production d'électricité *(f)*
It generazione elettrica *(f)*
Pt geração de electricidade *(f)*

electric motor En
De Elektromotor *(m)*
Es motor eléctrico *(m)*
Fr moteur électrique *(m)*
It motore elettrico *(m)*
Pt motor eléctrico *(m)*

electric vehicle En
De Elektrofahrzeug *(n)*
Es vehículo eléctrico *(m)*
Fr véhicule électrique *(m)*
It veicolo elettrico *(m)*
Pt veículo eléctrico *(m)*

electrificação *(f) n* Pt
De Elektrifizierung *(f)*
En electrification
Es electrificación *(f)*
Fr électrification *(f)*
It elettrificazione *(f)*

electrificación *(f) n* Es
De Elektrifizierung *(f)*
En electrification
Fr électrification *(f)*
It elettrificazione *(f)*
Pt electrificação *(f)*

electrification *n* En
De Elektrifizierung *(f)*
Es electrificación *(f)*
Fr électrification *(f)*
It elettrificazione *(f)*
Pt electrificação *(f)*

électrification *(f) n* Fr
De Elektrifizierung *(f)*
En electrification
Es electrificación *(f)*
It elettrificazione *(f)*
Pt electrificação *(f)*

électro-aimant *(m) n* Fr
De Elektromagnet *(m)*
En electromagnet
Es electroimán *(m)*
It elettromagnete *(m)*
Pt electroimã *(m)*

electrochemical *adj* En
De elektrochemisch
Es electroquímico
Fr électrochimique
It elettrochimico
Pt electroquímico

electrochemical storage En
De elektrochemische Speicherung *(f)*
Es almacenamiento electroquímico *(m)*
Fr accumulation électrochimique *(f)*
It immagazzinaggio elettrochimico *(m)*
Pt armazenagem electroquímica *(f)*

électrochimique *adj* Fr
De elektrochemisch
En electrochemical
Es electroquímico
It elettrochimico
Pt electroquímico

electrodinámica cuántica *(f)* Es
De Quantenelektrodynamik *(f)*
En quantum electrodynamics
Fr électrodynamique quantique *(f)*
It elettrodinamica dei quanti *(f)*
Pt electrodinâmica quântica *(f)*

electrodinâmica quântica *(f)* Pt
De Quantenelektrodynamik *(f)*
En quantum electrodynamics
Es electrodinámica cuántica *(f)*
Fr électrodynamique quantique *(f)*
It elettrodinamica dei quanti *(f)*

électrodynamique quantique *(f)* Fr
De Quantenelektrodynamik *(f)*
En quantum electrodynamics
Es electrodinámica cuántica *(f)*
It elettrodinamica dei quanti *(f)*
Pt electrodinâmica quântica *(f)*

electrogasdinámico *adj* Es
De elektrogasdynamisch
En electrogasdynamic (EGD)
Fr électrogazdynamique
It elettrogasdinamico
Pt electrogasodinâmico

electrogasdynamic (EGD) En
De elektrogasdynamisch
Es electrogasdinámico
Fr électrogazdynamique
It elettrogasdinamico
Pt electrogasodinâmico

electrogasodinâmico *adj* Pt
De elektrogasdynamisch
En electrogasdynamic (EGD)
Es electrogasdinámico
Fr électrogazdynamique
It elettrogasdinamico

électrogazdynamique *adj* Fr
De elektrogasdynamisch
En electrogasdynamic (EGD)
Es electrogasdinámico
It elettrogasdinamico
Pt electrogasodinâmico

electroimã *(m) n* Pt
De Elektromagnet *(m)*
En electromagnet
Es electroimán *(m)*
Fr électro-aimant *(m)*
It elettromagnete *(m)*

electroimán *(m) n* Es
De Elektromagnet *(m)*
En electromagnet
Fr électro-aimant *(m)*
It elettromagnete *(m)*
Pt electroimã *(m)*

electrólise *(f) n* Pt
De Elektrolyse *(f)*
En electrolysis
Es electrolisis *(f)*
Fr électrolyse *(f)*
It elettrolisi *(f)*

electrolisis *(f) n* Es
De Elektrolyse *(f)*
En electrolysis
Fr électrolyse *(f)*
It elettrolisi *(f)*
Pt electrólise *(f)*

electrolito *(m) n* Es
De Elektrolyt *(n)*
En electrolyte
Fr électrolyte *(m)*
It elettrolito *(m)*
Pt electrólito *(m)*

electrólito *(m) n* Pt
De Elektrolyt *(n)*
En electrolyte
Es electrolito *(m)*
Fr électrolyte *(m)*
It elettrolito *(m)*

électrolyse *(f) n* Fr
De Elektrolyse *(f)*
En electrolysis
Es electrolisis *(f)*
It elettrolisi *(f)*
Pt electrólise *(f)*

electrolysis *n* En
De Elektrolyse *(f)*
Es electrolisis *(f)*
Fr électrolyse *(f)*
It elettrolisi *(f)*
Pt electrólise *(f)*

electrolyte *n* En
De Elektrolyt *(n)*
Es electrolito *(m)*
Fr électrolyte *(m)*
It elettrolito *(m)*
Pt electrólito *(m)*

électrolyte *(m) n* Fr
De Elektrolyt *(n)*
En electrolyte
Es electrolito *(m)*
It elettrolito *(m)*
Pt electrólito *(m)*

electromagnet *n* En
De Elektromagnet *(m)*
Es electroimán *(m)*
Fr électro-aimant *(m)*
It elettromagnete *(m)*
Pt electroimã *(m)*

electromagnetic *adj* En
De elektromagnetisch
Es electromagnético
Fr électromagnétique
It elettromagnetico
Pt electromagnético

electromagnetic field En
De elektromagnetisches Feld *(n)*
Es campo electromagnético *(m)*
Fr champ électromagnétique *(m)*
It campo elettromagnetico *(m)*
Pt campo electromagnético *(m)*

electromagnético *adj* Es, Pt
De elektromagnetisch
En electromagnetic
Fr électromagnétique
It elettromagnetico

electromagnetic radiation En
De elektromagnetische Strahlung *(f)*
Es radiación electromagnética *(f)*
Fr rayonnement électromagnétique *(m)*
It radiazione elettromagnetica *(f)*
Pt radiação electromagnética *(f)*

electromagnetic spectrum En
De elektromagnetisches Spektrum *(n)*
Es espectro electromagnético *(m)*
Fr spectre électromagnétique *(m)*
It spettro elettromagnetico *(m)*
Pt espectro electromagnético *(m)*

électromagnétique *adj* Fr
De elektromagnetisch
En electromagnetic
Es electromagnético
It elettromagnetico
Pt electromagnético

electromotive force (e.m.f.) En
De elektromotorische Kraft (EMK) *(f)*
Es fuerza electromotriz (f.e.m.) *(f)*
Fr force électromotrice (f.é.m.) *(f)*
It forza elettromotrice (f.e.m.) *(f)*
Pt força electromotora (f.e.m.) *(f)*

electron *n* En
De Elektron *(n)*
Es electrón *(m)*
Fr électron *(m)*
It elettrone *(m)*
Pt electrão *(m)*

électron *(m) n* Fr
De Elektron *(n)*
En electron
Es electrón *(m)*
It elettrone *(m)*
Pt electrão *(m)*

electrón *(m) n* Es
De Elektron *(n)*
En electron

Fr électron *(m)*
It elettrone *(m)*
Pt electrão *(m)*

electronic pulse ignition En
De elektronische Impulszündung *(f)*
Es encendido por impulsos electrónicos *(m)*
Fr allumage électronique à impulsions *(m)*
It accensione ad impulso elettronico *(f)*
Pt ignição por impulso electrónico *(f)*

electronvoltio (eV) *(m)* Es
De Elektronenvolt (eV) *(n)*
En electronvolt (eV)
Fr électron-volt (eV) *(m)*
It volt-elettrone *(m)*
Pt volt-electrónico *(m)*

electronvolt (eV) En
De Elektronenvolt (eV) *(n)*
Es electronvoltio (eV) *(m)*
Fr électron-volt (eV) *(m)*
It volt-elettrone *(m)*
Pt volt-electrónico *(m)*

électron-volt (eV) *(m)* Fr
De Elektronenvolt (eV) *(n)*
En electronvolt (eV)
Es electronvoltio (eV) *(m)*
It volt-elettrone *(m)*
Pt volt-electrónico *(m)*

electroquímico *adj* Es, Pt
De elektrochemisch
En electrochemical
Fr électrochimique
It elettrochimico

Elektrifizierung *(f) n* De
En electrification
Es electrificación *(f)*
Fr électrification *(f)*
It elettrificazione *(f)*
Pt electrificação *(f)*

elektrisches Feld *(n)* De
En electric field
Es campo eléctrico *(m)*
Fr champ électrique *(m)*
It campo elettrico *(m)*
Pt campo eléctrico *(m)*

Elektrizität *(f) n* De
Es electricidad *(f)*
En electricity
Fr électricité *(f)*
It elettricità *(f)*
Pt electricidade *(f)*

Elektrizitätskonstante *(f) n* De
En electric constant
Es constante eléctrica *(f)*
Fr constante électrique *(f)*
It constante elettrica *(f)*
Pt constante eléctrica *(f)*

elektrochemisch *adj* De
En electrochemical
Es electroquímico
Fr électrochimique
It elettrochimico
Pt electroquímico

elektrochemische Speicherung *(f)* De
En electrochemical storage
Es almacenamiento electroquímico *(m)*
Fr accumulation électrochimique *(f)*
It immagazzinaggio elettrochimico *(m)*
Pt armazenagem electroquímica *(f)*

Elektroenergie *(f) n* De
En electrical energy
Es energía eléctrica *(f)*
Fr énergie électrique *(f)*
It energia elettrica *(f)*
Pt energia eléctrica *(f)*

Elektrofahrzeug *(n) n* De
En electric vehicle
Es vehículo eléctrico *(m)*
Fr véhicule électrique *(m)*
It veicolo elettrico *(m)*
Pt veículo eléctrico *(m)*

elektrogasdynamisch De
En electrogasdynamic (EGD)
Es electrogasdinámico
Fr électrogazdynamique
It elettrogasdinamico
Pt electrogasodinâmico

Elektrolyse *(f) n* De
En electrolysis
Es electrolisis *(f)*
Fr électrolyse *(f)*
It elettrolisi *(f)*
Pt electrólise *(f)*

Elektrolyt *(n) n* De
En electrolyte
Es electrolito *(m)*
Fr électrolyte *(m)*
It elettrolito *(m)*
Pt electrólito *(m)*

Elektromagnet *(m) n* De
En electromagnet
Es electroimán *(m)*
Fr électro-aimant *(m)*
It elettromagnete *(m)*
Pt electroimã *(m)*

elektromagnetisch *adj* De
En electromagnetic
Es electromagnético
Fr électromagnétique
It elettromagnetico
Pt electromagnético

elektromagnetisches Feld *(n)* De
En electromagnetic field
Es campo electromagnético *(m)*
Fr champ électromagnétique *(m)*
It campo elettromagnetico *(m)*
Pt campo electromagnético *(m)*

elektromagnetisches Spektrum *(n)* De
En electromagnetic spectrum
Es espectro electromagnético *(m)*
Fr spectre électromagnétique *(m)*
It spettro elettromagnetico *(m)*
Pt espectro electromagnético *(m)*

elektromagnetische Strahlung *(f)* De
En electromagnetic radiation
Es radiación electromagnética *(f)*
Fr rayonnement électromagnétique *(m)*
It radiazione elettromagnetica *(f)*
Pt radiação electromagnética *(f)*

Elektromotor *(m) n* De
En electric motor
Es motor eléctrico *(m)*
Fr moteur électrique *(m)*
It motore elettrico *(m)*
Pt motor eléctrico *(m)*

elektromotorische Kraft (EMK) *(f)* De
En electromotive force (e.m.f.)
Es fuerza electromotriz (f.e.m.) *(f)*
Fr force électromotrice (f.é.m.) *(f)*
It forza elettromotrice (f.e.m.) *(f)*
Pt força electromotora (f.e.m.) *(f)*

Elektron *(n) n* De
En electron
Es electrón *(m)*
Fr électron *(m)*
It elettrone *(m)*
Pt electrão *(m)*

Elektronenvolt (eV) *(n) n* De
En electronvolt (eV)
Es electronvoltio (eV) *(m)*
Fr électron-volt (eV) *(m)*
It volt-elettrone *(m)*
Pt volt-electrónico *(m)*

elektronische Impulszündung *(f)* De
En electronic pulse ignition
Es encendido por impulsos electrónicos *(m)*
Fr allumage électronique à impulsions *(m)*
It accensione ad

impulso elettronico *(f)*
Pt ignição por impulso electrónico *(f)*

Elektroofen *(m)* De
En electric furnace
Es horno eléctrico *(m)*
Fr four électrique *(m)*
It forno elettrico *(m)*
Pt forno eléctrico *(m)*

element (chemical) *n* En
De Element (chemisches) *(n)*
Es elemento (químico) *(m)*
Fr élément (chimique) *(m)*
It elemento (chimico) *(m)*
Pt elemento (químico) *(m)*

Element (chemisches) *(n)* *n* De
En element (chemical)
Es elemento (químico) *(m)*
Fr élément (chimique) *(m)*
It elemento (chimico) *(m)*
Pt elemento (químico) *(m)*

élément (batterie) *(m)* *n* Fr
De Zelle (Batterie) *(f)*
En cell (battery)
Es elemento (batería) *(m)*
It elemento (batteria) *(m)*
Pt célula (bateria) *(f)*

élément (chimique) *(m)* *n* Fr
De Element (chemisches) *(n)*
En element (chemical)
Es elemento (químico) *(m)*
It elemento (chimico) *(m)*
Pt elemento (químico) *(m)*

élément combustible *(m)* Fr
De Brennelement *(n)*
En fuel element
Es elemento combustible *(m)*
It elemento combustibile *(m)*
Pt elemento combustível *(m)*

elemento (batería; batteria) *(m)* *n* Es, It
De Zelle (Batterie) *(f)*
En cell (battery)
Fr élément (batterie) *(m)*
Pt célula (bateria) *(f)*

elemento (químico; chimico) *(m)* *n* Es, It, Pt
De Element (chemisches) *(n)*
En element (chemical)
Fr élément (chimique) *(m)*

elemento alcalino de acumulador *(m)* Es
De sekundäre Alkalizelle *(f)*
En alkaline secondary cell
Fr élément secondaire alcalin *(m)*
It cellula secondaria alcalina *(f)*
Pt célula secundária alcalina *(f)*

elemento combustibile *(m)* It
De Brennelement *(n)*
En fuel element
Es elemento combustible *(m)*
Fr élément combustible *(m)*
Pt elemento combustível *(m)*

elemento combustible *(m)* Es
De Brennelement *(n)*
En fuel element
Fr élément combustible *(m)*
It elemento combustibile *(m)*
Pt elemento combustível *(m)*

elemento combustível *(m)* Pt
De Brennelement *(n)*
En fuel element
Es elemento combustible *(m)*
Fr élément combustible *(m)*
It elemento combustibile *(m)*

elemento galvánico *(m)* Es
De galvanische Zelle *(f)*
En galvanic cell
Fr pile galvanique *(f)*
It cellula galvanica *(f)*
Pt célula galvânica *(f)*

élément secondaire alcalin *(m)* Fr
De sekundäre Alkalizelle *(f)*
En alkaline secondary cell
Es elemento alcalino de acumulador *(m)*
It cellula secondaria alcalina *(f)*
Pt célula secundária alcalina *(f)*

elettricità *(f)* *n* It
De Elektrizität *(f)*
En electricity
Es electricidad *(f)*
Fr électricité *(f)*
Pt electricidade *(f)*

elettricità fuori di ore di punta *(f)* It
De Strom aus der Schwachlastzeit *(m)*
En off-peak electricity
Es electricidad fuera de horas punta *(f)*
Fr électricité hors-pointe *(f)*
Pt electricidade fora de ponta *(f)*

elettricità generata di notte *(f)* It
De Nachtstrom *(m)*
En night-generated electricity
Es electricidad de generación nocturna *(f)*
Fr électricité produite de nuit *(f)*
Pt electricidade de geração nocturna *(f)*

elettrificazione *(f)* *n* It
De Elektrifizierung *(f)*
En electrification
Es electrificación *(f)*
Fr électrification *(f)*
Pt electrificação *(f)*

elettrochimico *adj* It
De elektrochemisch
En electrochemical
Es electroquímico
Fr électrochimique
Pt electroquímico

elettrodinamica dei quanti *(f)* It
De Quantenelektrodynamik *(f)*
En quantum electrodynamics
Es electrodinámica cuántica *(f)*
Fr électrodynamique quantique *(f)*
Pt electrodinâmica quântica *(f)*

elettrogasdinamico *adj* It
De elektrogasdynamisch
En electrogasdynamic (EGD)
Es electrogasdinámico
Fr électrogazdynamique
Pt electrogasodinâmico

elettrolisi *(f)* *n* It
De Elektrolyse *(f)*
En electrolysis
Es electrolisis *(f)*
Fr électrolyse *(f)*
Pt electrólise *(f)*

elettrolito *(m)* *n* It
De Elektrolyt *(n)*
En electrolyte
Es electrolito *(m)*
Fr électrolyte *(m)*
Pt electrólito *(m)*

elettromagnete *(m)* *n* It
De Elektromagnet *(m)*
En electromagnet
Es electroimán *(m)*
Fr électro-aimant *(m)*
Pt electroimã *(m)*

elettromagnetico *adj* It
De elektromagnetisch
En electromagnetic
Es electromagnético

Fr électromagnétique
Pt electromagnético

elettrone *(m) n* It
De Elektron *(n)*
En electron
Es electrón *(m)*
Fr électron *(m)*
Pt electrão *(m)*

eliminador de corrente de ar *(m)* Pt
Am draft excluder
De Luftzugschutz *(m)*
En draught excluder
Es eliminador de corrientes *(m)*
Fr brise-bise *(m)*
It esclusore d'aria *(m)*

eliminador de corrientes *(m)* Es
Am draft excluder
De Luftzugschutz *(m)*
En draught excluder
Fr brise-bise *(m)*
It esclusore d'aria *(m)*
Pt eliminador de corrente de ar *(m)*

elio *(m) n* It
De Helium *(n)*
En helium
Es helio *(m)*
Fr hélium *(m)*
Pt hélio *(m)*

eliostato *(m) n* It
De Heliostat *(m)*
En heliostat
Es heliostato *(m)*
Fr héliostat *(m)*
Pt heliostato *(m)*

emergenza *adj* It
De Reserve-
En standby
Es de reserva
Fr en attente
Pt de reserva

emettenza *(f) n* It
De Emittanz *(f)*
En emittance
Es emitancia *(f)*
Fr émittance *(f)*
Pt emitância *(f)*

emisividad *(f) n* Es
De Emissionsvermögen *(n)*
En emissivity
Fr pouvoir émissif *(m)*
It emissività *(f)*
Pt emissividade *(f)*

emission spectrum En
De Emissionsspektrum *(n)*
Es espectro de emisión *(m)*
Fr spectre d'émission *(m)*
It spettro delle emissioni *(m)*
Pt espectro de emissão *(m)*

Emissionsspektrum *(n) n* De
En emission spectrum
Es espectro de emisión *(m)*
Fr spectre d'émission *(m)*
It spettro delle emissioni *(m)*
Pt espectro de emissão *(m)*

Emissionsvermögen *(n) n* De
En emissivity
Es emisividad *(f)*
Fr pouvoir émissif *(m)*
It emissività *(f)*
Pt emissividade *(f)*

emissividade *(f) n* Pt
De Emissionsvermögen *(n)*
En emissivity
Es emisividad *(f)*
Fr pouvoir émissif *(m)*
It emissività *(f)*

emissività *(f) n* It
De Emissionsvermögen *(n)*
En emissivity
Es emisividad *(f)*
Fr pouvoir émissif *(m)*
Pt emissividade *(f)*

emissivity *n* En
De Emissionsvermögen *(n)*
Es emisividad *(f)*
Fr pouvoir émissif *(m)*
It emissività *(f)*
Pt emissividade *(f)*

emitancia *(f) n* Es
De Emittanz *(f)*
En emittance
Fr émittance *(f)*
It emettenza *(f)*
Pt emitância *(f)*

emitância *(f) n* Pt
De Emittanz *(f)*
En emittance
Es emitancia *(f)*
Fr émittance *(f)*
It emettenza *(f)*

emittance *n* En
De Emittanz *(f)*
Es emitancia *(f)*
Fr émittance *(f)*
It emettenza *(f)*
Pt emitância *(f)*

émittance *(f) n* Fr
De Emittanz *(f)*
En emittance
Es emitancia *(f)*
It emettenza *(f)*
Pt emitância *(f)*

Emittanz *(f) n* De
En emittance
Es emitancia *(f)*
Fr émittance *(f)*
It emettenza *(f)*
Pt emitância *(f)*

em partículas Pt
De Partikel-
En particulate
Es de macropartículas
Fr particulaire
It delle macroparticelle

en attente Fr
De Reserve-
En standby
Es de reserva
It emergenza
Pt de reserva

encamisado *(m) n* Es
De Umhüllung *(f)*
En cladding
Fr gainage *(m)*
It incamiciatura *(f)*
Pt revestimento *(m)*

encendido *(m) n* Es
De Zündung *(f)*
En ignition
Fr allumage *(m)*
It accensione *(f)*
Pt ignição *(f)*

encendido espontáneo *(m)* Es
De spontane Zündung *(f)*
En spontaneous ignition
Fr allumage spontané *(m)*
It autoaccensione *(f)*
Pt ignição espontânea *(f)*

encendido por chispa *(m)* Es
De Funkenzündung *(f)*
En spark ignition
Fr allumage par étincelle *(m)*
It accensione a scintilla *(f)*
Pt ignição por faísca *(f)*

encendido por impulsos electrónicos *(m)* Es
De elektronische Impulszündung *(f)*
En electronic pulse ignition
Fr allumage électronique à impulsions *(m)*
It accensione ad impulso elettronico *(f)*
Pt ignição por impulso electrónico *(f)*

encrassement *(m) n* Fr
De Verstopfung *(f)*
En fouling (boiler)
Es incrustación *(f)*
It incrostazione *(f)*
Pt incrustação *(m)*

encristalado doble *(m)* Es
De Doppelverglasung *(f)*
En double glazing
Fr double vitrage *(m)*
It doppia vetratura *(f)*
Pt vitrificação dupla *(f)*

encristalado triple *(m)* Es
De Dreifachverglasung *(f)*
En triple glazing n

Fr triple vitrage *(m)*
It tripla vetrata *(f)*
Pt vitrificação tripla *(f)*

endoénergétique *adj* Fr
De endoenergetisch
En endoergic
Es endoérgico
It endoergico
Pt endoérgico

endoenergetisch *adj* De
En endoergic
Es endoérgico
Fr endoénergétique
It endoergico
Pt endoérgico

endoergic *adj* En
De endoenergetisch
Es endoérgico
Fr endoénergétique
It endoergico
Pt endoérgico

endoergico *adj* It
De endoenergetisch
En endoergic
Es endoérgico
Fr endoénergétique
Pt endoérgico

endoérgico *adj* Es, Pt
De endoenergetisch
En endoergic
Fr endoénergétique
It endoergico

endotermico *adj* It
De endotherm
En endothermic
Es endotérmico
Fr endothermique
Pt endotérmico

endotérmico *adj* Es, Pt
De endotherm
En endothermic
Fr endothermique
It endotermico

endotherm *adj* De
En endothermic
Es endotérmico
Fr endothermique
It endotermico
Pt endotérmico

endothermic *adj* En
De endotherm
Es endotérmico
Fr endothermique
It endotermico
Pt endotérmico

endothermique *adj* Fr
De endotherm
En endothermic
Es endotérmico
It endotermico
Pt endotérmico

energia *(f) n* It, Pt
De Energie *(f)*
En energy
Es energía *(f)*
Fr énergie *(f)*

energía *(f) n* Es
De Energie *(f)*
En energy
Fr énergie *(f)*
It energia *(f)*
Pt energia *(f)*

energia acustica *(f)* It
De Schallenergie *(f)*
En acoustic energy
Es energía acústica *(f)*
Fr énergie acoustique *(f)*
Pt energia acústica *(f)*

energía acústica *(f)* Es
De Schallenergie *(f)*
En acoustic energy
Fr énergie acoustique *(f)*
It energia acustica *(f)*
Pt energia acústica *(f)*

energia acústica *(f)* Pt
De Schallenergie *(f)*
En acoustic energy
Es energía acústica *(f)*
Fr énergie acoustique *(f)*
It energia acustica *(f)*

energia chimica *(f)* It
De chemische Energie *(f)*
En chemical energy
Es energía química *(f)*
Fr énergie chimique *(f)*
Pt energia química *(f)*

energia cinetica *(f)* It
De kinetische Energie *(f)*
En kinetic energy
Es energía cinética *(f)*
Fr énergie cinétique *(f)*
Pt energia cinética *(f)*

energía cinética *(f)* Es
De kinetische Energie *(f)*
En kinetic energy
Fr énergie cinétique *(f)*
It energia cinetica *(f)*
Pt energia cinética *(f)*

energia cinética *(f)* Pt
De kinetische Energie *(f)*
En kinetic energy
Es energía cinética *(f)*
Fr énergie cinétique *(f)*
It energia cinetica *(f)*

energía de enlace *(f)* Es
De Bindungsenergie *(f)*
En binding energy
Fr énergie de liaison *(f)*
It energia di legatura *(f)*
Pt energia de fixação *(f)*

energia de entrada *(f)* Pt
De Eingabeenergie *(f)*
En input energy
Es energía de entrada *(f)*
Fr énergie d'entrée *(f)*
It energia immessa *(f)*

energía de entrada *(f)* Es
De Eingabeenergie *(f)*
En input energy
Fr énergie d'entrée *(f)*
It energia immessa *(f)*
Pt energia de entrada *(f)*

energia de fixação *(f)* Pt
De Bindungsenergie *(f)*
En binding energy
Es energía de enlace *(f)*
Fr énergie de liaison *(f)*
It energia di legatura *(f)*

energía de la onda *(f)* Es
De Wellenenergie *(f)*
En wave energy
Fr énergie ondulatoire *(f)*
It energia d'onda *(f)*
Pt energia ondulatória *(f)*

energia de rotação *(f)* Pt
De Rotationsenergie *(f)*
En rotational energy
Es energía rotacional *(f)*
Fr énergie de rotation *(f)*
It energia rotazionale *(f)*

energia de saída *(f)* Pt
De abgegebene Energie *(f)*
En output energy
Es energía de salida *(f)*
Fr énergie produite *(f)*
It energia erogata *(f)*

energía de salida *(f)* Es
De abgegebene Energie *(f)*
En output energy
Fr énergie produite *(f)*
It energia erogata *(f)*
Pt energia de saída *(f)*

energia di legatura *(f)* It
De Bindungsenergie *(f)*
En binding energy
Es energía de enlace *(f)*
Fr énergie de liaison *(f)*
Pt energia de fixação *(f)*

energia d'onda *(f)* It
De Wellenenergie *(f)*
En wave energy
Es energía de la onda *(f)*
Fr énergie ondulatoire *(f)*
Pt energia ondulatória *(f)*

energía eléctrica *(f)* Es
De Elektroenergie *(f)*
En electrical energy
Fr énergie électrique *(f)*
It energia elettrica *(f)*
Pt energia eléctrica *(f)*

energia eléctrica *(f)* Pt
De Elektroenergie *(f)*
En electrical energy
Es energía eléctrica *(f)*
Fr énergie électrique *(f)*
It energia elettrica *(f)*

energia elettrica *(f)* It
De Elektroenergie *(f)*
En electrical energy
Es energía eléctrica *(f)*
Fr énergie électrique *(f)*
Pt energia eléctrica *(f)*

energia erogata *(f)* It
De abgegebene Energie *(f)*
En output energy
Es energía de salida *(f)*
Fr énergie produite *(f)*
Pt energia de saída *(f)*

energia geotermica *(f)* It
De geothermische Energie *(f)*
En geothermal energy
Es energía geotérmica *(f)*
Fr énergie géothermique *(f)*
Pt energia geotérmica *(f)*

energía geotérmica *(f)* Es
De geothermische Energie *(f)*
En geothermal energy
Fr énergie géothermique *(f)*
It energia geotermica *(f)*
Pt energia geotérmica *(f)*

energia geotérmica *(f)* Pt
De geothermische Energie *(f)*
En geothermal energy
Es energía geotérmica *(f)*
Fr énergie géothermique *(f)*
It energia geotermica *(f)*

energía hidroeléctrica *(f)* Es
De hydroelektrische Energie *(f)*
En hydroelectric energy
Fr énergie hydroélectrique *(f)*
It energia idroelettrica *(f)*
Pt energia hidroeléctrica *(f)*

energia hidroeléctrica *(f)* Pt
De hydroelektrische Energie *(f)*
En hydroelectric energy
Es energía hidroeléctrica *(f)*
Fr énergie hydroélectrique *(f)*
It energia idroelettrica *(f)*

energia hidrogravitacional *(f)* Pt
De Wasserschwerkraft-Energie *(f)*
En hydrogravitational energy
Es energía hidrogravitacional *(f)*
Fr énergie d'hydrogravitation *(f)*
It energia idrogravitazionale *(f)*

energía hidrogravitacional *(f)* Es
De Wasserschwerkraft-Energie *(f)*
En hydrogravitational energy
Fr énergie d'hydrogravitation *(f)*
It energia idrogravitazionale *(f)*
Pt energia hidrogravitacional *(f)*

energia idroelettrica *(f)* It
De hydroelektrische Energie *(f)*
En hydroelectric energy
Es energía hidroeléctrica *(f)*
Fr énergie hydroélectrique *(f)*
Pt energia hidroeléctrica *(f)*

energia idrogravitazionale *(f)* It
De Wasserschwerkraft-Energie *(f)*
En hydrogravitational energy
Es energía hidrogravitacional *(f)*
Fr énergie d'hydrogravitation *(f)*
Pt energia hidrogravitacional *(f)*

energia immessa *(f)* It
De Eingabeenergie *(f)*
En input energy
Es energía de entrada *(f)*
Fr énergie d'entrée *(f)*
Pt energia de entrada *(f)*

energia interna *(f)* It, Pt
De innere Energie *(f)*
En internal energy
Es energía interna *(f)*
Fr énergie interne *(f)*

energía interna *(f)* Es
De innere Energie *(f)*
En internal energy
Fr énergie interne *(f)*
It energia interna *(f)*
Pt energia interna *(f)*

energia libera *(f)* It
De freigesetzte Energie *(f)*
En free energy
Es energía libre *(f)*
Fr énergie libre *(f)*
Pt energia livre *(f)*

energía libre *(f)* Es
De freigesetzte Energie *(f)*
En free energy
Fr énergie libre *(f)*
It energia libera *(f)*
Pt energia livre *(f)*

energia livre *(f)* Pt
De freigesetzte Energie *(f)*
En free energy
Es energía libre *(f)*
Fr énergie libre *(f)*
It energia libera *(f)*

energia luminosa *(f)* It, Pt
De Leuchtenergie *(f)*
En luminous energy
Es energía luminosa *(f)*
Fr énergie lumineuse *(f)*

energía luminosa *(f)* Es
De Leuchtenergie *(f)*
En luminous energy
Fr énergie lumineuse *(f)*
It energia luminosa *(f)*
Pt energia luminosa *(f)*

energía mecánica *(f)* Es
De mechanische Energie *(f)*
En mechanical energy
Fr énergie mécanique *(f)*
It energia meccanica *(f)*
Pt energia mecânica *(f)*

energia mecânica *(f)* Pt
De mechanische Energie *(f)*
En mechanical energy
Es energía mecánica *(f)*
Fr énergie mécanique *(f)*
It energia meccanica *(f)*

energia meccanica *(f)* It
De mechanische Energie *(f)*
En mechanical energy
Es energía mecánica *(f)*
Fr énergie mécanique *(f)*
Pt energia mecânica *(f)*

energia não renovável *(f)* Pt
De nicht erneuerbare Energie *(f)*
En nonrenewable energy
Es energía no recuperable *(f)*
Fr énergie non renouvelable *(f)*
It energia non rinnovabile *(f)*

energia non rinnovabile *(f)* It
De nicht erneuerbare Energie *(f)*
En nonrenewable energy
Es energía no recuperable *(f)*
Fr énergie non renouvelable *(f)*
Pt energia não renovável *(f)*

energía no recuperable *(f)* Es
De nicht erneuerbare Energie *(f)*
En nonrenewable energy
Fr énergie non renouvelable *(f)*
It energia non rinnovabile *(f)*
Pt energia não renovável *(f)*

energia nuclear *(f)* Pt
De Nuklearenergie *(f)*
En nuclear energy
Es energía nuclear *(f)*
Fr énergie nucléaire *(f)*
It energia nucleare *(f)*

energía nuclear *(f)* Es
De Nuklearenergie *(f)*
En nuclear energy
Fr énergie nucléaire *(f)*

It energia nucleare *(f)*
Pt energia nuclear *(f)*

energia nucleare *(f)* It
De Nuklearenergie *(f)*
En nuclear energy
Es energía nuclear *(f)*
Fr énergie nucléaire *(f)*
Pt energia nuclear *(f)*

energia ondulatória *(f)* Pt
De Wellenenergie *(f)*
En wave energy
Es energía de la onda *(f)*
Fr énergie ondulatoire *(f)*
It energia d'onda *(f)*

energia potencial *(f)* Pt
De potentielle Energie *(f)*
En potential energy
Es energía potencial *(f)*
Fr énergie potentielle *(f)*
It energia potenziale *(f)*

energía potencial *(f)* Es
De potentielle Energie *(f)*
En potential energy
Fr énergie potentielle *(f)*
It energia potenziale *(f)*
Pt energia potencial *(f)*

energia potenziale *(f)* It
De potentielle Energie *(f)*
En potential energy
Es energía potencial *(f)*
Fr énergie potentielle *(f)*
Pt energia potencial *(f)*

energia primaria *(f)* It
De primäre Energie *(f)*
En primary energy
Es energía primaria *(f)*
Fr énergie primaire *(f)*
Pt energia primária *(f)*

energía primaria *(f)* Es
De primäre Energie *(f)*
En primary energy
Fr énergie primaire *(f)*
It energia primaria *(f)*
Pt energia primária *(f)*

energia primária *(f)* Pt
De primäre Energie *(f)*
En primary energy
Es energía primaria *(f)*
Fr énergie primaire *(f)*
It energia primaria *(f)*

energía química *(f)* Es
De chemische Energie *(f)*
En chemical energy
Fr énergie chimique *(f)*
It energia chimica *(f)*
Pt energia química *(f)*

energia química *(f)* Pt
De chemische Energie *(f)*
En chemical energy
Es energía química *(f)*
Fr énergie chimique *(f)*
It energia chimica *(f)*

energía renovable *(f)* Es
De erneuerbare Energie *(f)*
En renewable energy
Fr énergie renouvelable *(f)*
It energia rinnovabile *(f)*
Pt energia renovável *(f)*

energia renovável *(f)* Pt
De erneuerbare Energie *(f)*
En renewable energy
Es energía renovable *(f)*
Fr énergie renouvelable *(f)*
It energia rinnovabile *(f)*

energia rinnovabile *(f)* It
De erneuerbare Energie *(f)*
En renewable energy
Es energía renovable *(f)*
Fr énergie renouvelable *(f)*
Pt energia renovável *(f)*

energía rotacional *(f)* Es
De Rotationsenergie *(f)*
En rotational energy
Fr énergie de rotation *(f)*
It energia rotazionale *(f)*
Pt energia de rotação *(f)*

energia rotazionale *(f)* It
De Rotationsenergie *(f)*
En rotational energy
Es energía rotacional *(f)*
Fr énergie de rotation *(f)*
Pt energia de rotação *(f)*

energia secondaria *(f)* It
De Sekundärenergie *(f)*
En secondary energy
Es energía secundaria *(f)*
Fr énergie secondaire *(f)*
Pt energia secundária *(f)*

energía secundaria *(f)* Es
De Sekundärenergie *(f)*
En secondary energy
Fr énergie secondaire *(f)*
It energia secondaria *(f)*
Pt energia secundária *(f)*

energia secundária *(f)* Pt
De Sekundärenergie *(f)*
En secondary energy
Es energía secundaria *(f)*
Fr énergie secondaire *(f)*
It energia secondaria *(f)*

energia solar *(f)* Pt
De Sonnenenergie *(f)*
En solar energy
Es energía solar *(f)*
Fr énergie solaire *(f)*
It energia solare *(f)*

energía solar *(f)* Es
De Sonnenenergie *(f)*
En solar energy
Fr énergie solaire *(f)*
It energia solare *(f)*
Pt energia solar *(f)*

energia solare *(f)* It
De Sonnenenergie *(f)*
En solar energy
Es energía solar *(f)*
Fr énergie solaire *(f)*
Pt energia solar *(f)*

energia termica *(f)* It
De Wärmeenergie *(f)*
En heat energy; thermal energy
Es energía térmica *(f)*
Fr énergie calorifique; énergie thermique *(f)*
Pt energia térmica *(f)*

energia térmica *(f)* Pt
De Wärmeenergie *(f)*
En heat energy; thermal energy
Es energía térmica *(f)*
Fr énergie calorifique; énergie thermique *(f)*
It energia termica *(f)*

energía térmica *(f)* Es
De Wärmeenergie *(f)*
En heat energy; thermal energy
Fr énergie calorifique; énergie thermique *(f)*
It energia termica *(f)*
Pt energia térmica *(f)*

energia termica oceanica *(f)* It
De Meereswärme-energie *(f)*
En ocean thermal energy
Es energía térmica oceánica *(f)*
Fr énergie thermique des océans *(f)*
Pt energia térmica oceânica *(f)*

energía térmica oceánica *(f)* Es
De Meereswärme-energie *(f)*
En ocean thermal energy
Fr énergie thermique des océans *(f)*
It energia termica oceanica *(f)*
Pt energia térmica oceânica *(f)*

energia térmica oceânica *(f)* Pt
De Meereswärme-energie *(f)*
En ocean thermal energy
Es energía térmica oceánica *(f)*
Fr énergie thermique des océans *(f)*
It energia termica oceanica *(f)*

energia total *(f)* Pt
De Gesamtenergie *(f)*
En total energy
Es energía total *(f)*
Fr énergie totale *(f)*
It energia totale *(f)*

energía total *(f)* Es
De Gesamtenergie *(f)*
En total energy
Fr énergie totale *(f)*
It energia totale *(f)*
Pt energia total *(f)*

energia totale *(f)* It
De Gesamtenergie *(f)*
En total energy
Es energía total *(f)*
Fr énergie totale *(f)*
Pt energia total *(f)*

Energie *(f) n* De
En energy
Es energía *(f)*
Fr énergie *(f)*

It energia *(f)*
Pt energia *(f)*

énergie *(f) n* Fr
De Energie *(f)*
En energy
Es energía *(f)*
It energia *(f)*
Pt energia *(f)*

énergie acoustique *(f)* Fr
De Schallenergie *(f)*
En acoustic energy
Es energía acústica *(f)*
It energia acustica *(f)*
Pt energia acústica *(f)*

Energiebilanz *(f) n* De
En energy balance
Es balance energético *(m)*
Fr équilibre énergétique *(m)*
It equilibrio di energia *(m)*
Pt equilíbrio energético *(m)*

énergie calorifique *(f)* Fr
De Wärmeenergie *(f)*
En heat energy; thermal energy
Es energía térmica *(f)*
It energia termica *(f)*
Pt energia térmica *(f)*

énergie chimique *(f)* Fr
De chemische Energie *(f)*
En chemical energy
Es energía química *(f)*
It energia chimica *(f)*
Pt energia química *(f)*

énergie cinétique *(f)* Fr
De kinetische Energie *(f)*
En kinetic energy
Es energía cinética *(f)*
It energia cinetica *(f)*
Pt energia cinética *(f)*

énergie de liaison *(f)* Fr
De Bindungsenergie *(f)*
En binding energy
Es energía de enlace *(f)*
It energia di legatura *(f)*
Pt energia de fixação *(f)*

énergie d'entrée *(f)* Fr
De Eingabeenergie *(f)*
En input energy
Es energía de entrada *(f)*
It energia immessa *(f)*
Pt energia de entrada *(f)*

énergie de rotation *(f)* Fr
De Rotationsenergie *(f)*
En rotational energy
Es energía rotacional *(f)*
It energia rotazionale *(f)*
Pt energia de rotação *(f)*

énergie d'hydrogravitation *(f)* Fr
De Wasserschwerkraft-Energie *(f)*
En hydrogravitational energy
Es energía hidrogravitacional *(f)*
It energia idrogravitazionale *(f)*
Pt energia hidrogravitacional *(f)*

énergie électrique *(f)* Fr
De Elektroenergie *(f)*
En electrical energy
Es energía eléctrica *(f)*
It energia elettrica *(f)*
Pt energia eléctrica *(f)*

Energieerhaltung *(f) n* De
En conservation of energy
Es conservación de energía *(f)*
Fr conservation de l'énergie *(f)*
It conservazione dell'energia *(f)*
Pt conservação da energia *(f)*

Energieflußdiagramm *(n) n* De
En energy-flow diagram
Es diagrama de flujo energético *(m)*
Fr schéma de passage de l'énergie *(m)*
It diagramma del flusso energetico *(m)*
Pt diagrama de fluxo de energia *(m)*

énergie géothermique *(f)* Fr
De geothermische Energie *(f)*
En geothermal energy
Es energía geotérmica *(f)*
It energia geotermica *(f)*
Pt energia geotérmica *(f)*

énergie hydroélectrique *(f)* Fr
De hydroelektrische Energie *(f)*
En hydroelectric energy
Es energía hidroeléctrica *(f)*
It energia idroelettrica *(f)*
Pt energia hidroeléctrica *(f)*

énergie interne *(f)* Fr
De innere Energie *(f)*
En internal energy
Es energía interna *(f)*
It energia interna *(f)*
Pt energia interna *(f)*

Energiekrise *(f) n* De
En energy crisis
Es crisis energética *(f)*
Fr crise de l'énergie *(f)*
It crisi energetica *(f)*
Pt crise energética *(f)*

énergie libre *(f)* Fr
De freigesetzte Energie *(f)*
En free energy
Es energía libre *(f)*
It energia libera *(f)*
Pt energia livre *(f)*

énergie lumineuse *(f)* Fr
De Leuchtenergie *(f)*
En luminous energy
Es energía luminosa *(f)*
It energia luminosa *(f)*
Pt energia luminosa *(f)*

énergie mécanique *(f)* Fr
De mechanische Energie *(f)*
En mechanical energy
Es energía mecánica *(f)*
It energia meccanica *(f)*
Pt energia mecânica *(f)*

énergie non renouvelable *(f)* Fr
De nicht erneuerbare Energie *(f)*
En nonrenewable energy
Es energía no recuperable *(f)*
It energia non rinnovabile *(f)*
Pt energia não renovável *(f)*

énergie nucléaire *(f)* Fr
De Nuklearenergie *(f)*
En nuclear energy
Es energía nuclear *(f)*
It energia nucleare *(f)*
Pt energia nuclear *(f)*

énergie ondulatoire *(f)* Fr
De Wellenenergie *(f)*
En wave energy
Es energía de la onda *(f)*
It energia d'onda *(f)*
Pt energia ondulatória *(f)*

Energiepolitik *(f) n* De
En energy policy
Es política energética *(f)*
Fr politique de l'énergie *(f)*
It politica energetica *(f)*
Pt política energética *(f)*

énergie potentielle *(f)* Fr
De potentielle Energie *(f)*
En potential energy
Es energía potencial *(f)*
It energia potenziale *(f)*
Pt energia potencial *(f)*

énergie primaire *(f)* Fr
De primäre Energie *(f)*
En primary energy
Es energía primaria *(f)*
It energia primaria *(f)*
Pt energia primária *(f)*

énergie produite *(f)* Fr
De abgegebene Energie *(f)*
En output energy
Es energía de salida *(f)*
It energia erogata *(f)*
Pt energia de saída *(f)*

Energiequellen *(pl) n* De
En energy resources
Es recursos energéticos *(m pl)*
Fr ressources d'énergie *(f pl)*
It risorse energetiche *(f pl)*

Pt recursos energéticos *(m pl)*

énergie renouvelable *(f)* Fr
De erneuerbare Energie *(f)*
En renewable energy
Es energía renovable *(f)*
It energia rinnovabile *(f)*
Pt energia renovável *(f)*

Energiereserven *(pl) n* De
En energy reserves
Es reservas energéticas *(f pl)*
Fr réserves d'énergie *(f pl)*
It riserve energetiche *(f pl)*
Pt reservas energéticas *(f pl)*

énergie secondaire *(f)* Fr
De Sekundärenergie *(f)*
En secondary energy
Es energía secundaria *(f)*
It energia secondaria *(f)*
Pt energia secundária *(f)*

énergie solaire *(f)* Fr
De Sonnenenergie *(f)*
En solar energy
Es energía solar *(f)*
It energia solare *(f)*
Pt energia solar *(f)*

Energiesparen *(n) n* De
En energy saving
Es ahorro energético *(m)*
Fr économie d'énergie *(f)*
It risparmio di energia *(m)*
Pt poupança de energia *(f)*

énergie thermique *(f)* Fr
De Wärmeenergie *(f)*
En heat energy; thermal energy
Es energía térmica *(f)*
It energia termica *(f)*
Pt energia térmica *(f)*

énergie thermique des océans *(f)* Fr
De Meereswärme-energie *(f)*
En ocean thermal energy
Es energía térmica oceánica *(f)*
It energia termica oceanica *(f)*
Pt energia térmica oceânica *(f)*

énergie totale *(f)* Fr
De Gesamtenergie *(f)*
En total energy
Es energía total *(f)*
It energia totale *(f)*
Pt energia total *(f)*

Energieumwandlung an Bord *(f)* De
En on-board energy conversion
Es conversión de energía a bordo *(f)*
Fr transformation d'énergie à bord *(f)*
It conversione di energia a bordo *(f)*
Pt conversão de energia a bordo *(m)*

Energieverbrauch *(m) n* De
En energy consumption
Es consumo de energía *(m)*
Fr consommation d'énergie *(f)*
It consumo di energia *(m)*
Pt consumo de energia *(m)*

Energieverbrauch der Industrie *(m)* De
En industrial energy consumption
Es consumo industrial de energía *(m)*
Fr consommation industrielle d'énergie *(f)*
It consumo industriale di energia *(m)*
Pt consumo industrial de energia *(m)*

Energiewirtschaft *(f) n* De
En energy economics
Es economía energética *(f)*
Fr économie de l'énergie *(f)*
It economia energetica *(f)*
Pt economia energética *(f)*

energy *n* En
De Energie *(f)*
Es energía *(f)*
Fr énergie *(f)*
It energia *(f)*
Pt energia *(f)*

energy balance En
De Energiebilanz *(f)*
Es balance energético *(m)*
Fr équilibre énergétique *(m)*
It equilibrio di energia *(m)*
Pt equilíbrio energético *(m)*

energy consumption En
De Energieverbrauch *(m)*
Es consumo de energía *(m)*
Fr consommation d'énergie *(f)*
It consumo di energia *(m)*
Pt consumo de energia *(m)*

energy crisis En
De Energiekrise *(f)*
Es crisis energética *(f)*
Fr crise de l'énergie *(f)*
It crisi energetica *(f)*
Pt crise energética *(f)*

energy economics En
De Energiewirtschaft *(f)*
Es economía energética *(f)*
Fr économie de l'énergie *(f)*
It economia energetica *(f)*
Pt economia energética *(f)*

energy-flow diagram En
De Energieflußdiagramm *(n)*
Es diagrama de flujo energético *(m)*
Fr schéma de passage de l'énergie *(m)*
It diagramma del flusso energetico *(m)*
Pt diagrama de fluxo de energia *(m)*

energy policy En
De Energiepolitik *(f)*
Es política energética *(f)*
Fr politique de l'énergie *(f)*
It politica energetica *(f)*
Pt política energética *(f)*

energy reserves *pl n* En
De Energiereserven *(pl)*
Es reservas energéticas *(f pl)*
Fr réserves d'énergie *(f pl)*
It riserve energetiche *(f pl)*
Pt reservas energéticas *(f pl)*

energy resources *pl n* En
De Energiequellen *(pl)*
Es recursos energéticos *(m pl)*
Fr ressources d'énergie *(f pl)*
It risorse energetiche *(f pl)*
Pt recursos energéticos *(m pl)*

energy saving En
De Energiesparen *(n)*
Es ahorro energético *(m)*
Fr économie d'énergie *(f)*
It risparmio di energia *(m)*
Pt poupança de energia *(f)*

enfriada con hidrógeno Es
De wasserstoffgekühlt
En hydrogen-cooled
Fr refroidi par l'hydrogène
It raffreddata ad idrogeno
Pt arrefecido a hidrogénio

enfriador intermedio *(m)* Es
De Zwischenkühler *(m)*
En intercooler
Fr refroidisseur intermédiaire *(m)*
It refrigeratore intermedio *(m)*
Pt inter-arrefecedor *(m)*

enfriamiento desde la ebullición *(m)* Es
De Heiβwasserkühlung *(f)*
En ebullient cooling
Fr refroidissement par ébullition *(m)*
It raffreddamento a ebollizione *(m)*
Pt arrefecimento ebuliente *(m)*

engine *n* En
De Motor *(m)*
Es motor *(m)*
Fr moteur *(m)*
It motore *(m)*
Pt motor *(m)*

enriched fuel En
De angereicherter Brennstoff *(m)*
Es combustible enriquecido *(m)*
Fr combustible enrichi *(m)*
It combustibile arricchito *(m)*
Pt combustível enriquecido *(m)*

enrichissement *(m) n* Fr
De Anreicherung *(f)*
En enrichment
Es enriquecimiento *(m)*
It arricchimento *(m)*
Pt enriquecimento *(m)*

enrichment *n* En
De Anreicherung *(f)*
Es enriquecimiento *(m)*
Fr enrichissement *(m)*
It arricchimento *(m)*
Pt enriquecimento *(m)*

enriquecimento *(m) n* Pt
De Anreicherung *(f)*
En enrichment
Es enriquecimiento *(m)*
Fr enrichissement *(m)*
It arricchimento *(m)*

enriquecimiento *(m) n* Es
De Anreicherung *(f)*
En enrichment
Fr enrichissement *(m)*
It arricchimento *(m)*
Pt enriquecimento *(m)*

enrolamento *(m) n* Pt
De Wicklung *(f)*
En winding
Es devanado *(m)*
Fr enroulement *(m)*
It avvolgimento *(m)*

enroulement *(m) n* Fr
De Wicklung *(f)*
En winding
Es devanado *(m)*
It avvolgimento *(m)*
Pt enrolamento *(m)*

entalpia *(f) n* It, Pt
De Enthalpie *(f)*
En enthalpy
Es entalpía *(f)*
Fr enthalpie *(f)*

entalpía *(f) n* Es
De Enthalpie *(f)*
En enthalpy
Fr enthalpie *(f)*
It entalpia *(f)*
Pt entalpia *(f)*

Entfeuchter *(m) n* De
En dehumidifier
Es deshumidificador *(m)*
Fr déshumidificateur *(m)*
It deumidificatore *(m)*
Pt deshumidificador *(m)*

enthalpie *(f) n* Fr
De Enthalpie *(f)*
En enthalpy
Es entalpía *(f)*
It entalpia *(f)*
Pt entalpia *(f)*

Enthalpie *(f) n* De
En enthalpy
Es entalpía *(f)*
Fr enthalpie *(f)*
It entalpia *(f)*
Pt entalpia *(f)*

enthalpy *n* En
De Enthalpie *(f)*
Es entalpía *(f)*
Fr enthalpie *(f)*
It entalpia *(f)*
Pt entalpia *(f)*

Entladung *(f) n* De
En discharge
Es descarga (bateria) *(f)*
Fr décharge *(f)*
It scarica *(f)*
Pt descarga (batería) *(f)*

Entladungsmerkmale *(pl) n* De
En discharge characteristics
Es características de descarga *(f pl)*
Fr caractéristiques de décharge *(f pl)*
It caratteristiche di scarica *(f pl)*
Pt características de descarga *(f pl)*

Entlüftungsschlitz *(m) n* De
En air vent
Es toma de aire *(f)*
Fr évent *(m)*
It sfiato aria *(m)*
Pt respirador de ar *(m)*

Entmagnetisierung *(f) n* De
En demagnetization
Es desmagnetización *(f)*
Fr démagnétisation *(f)*
It smagnetizzazione *(f)*
Pt desmagnetização *(f)*

entmineralisiertes Wasser *(n)* De
En demineralized water
Es agua desmineralizada *(f)*
Fr eau déminéralisée *(f)*
It acqua demineralizzata *(f)*
Pt água desmineralizada *(f)*

Entpropanisierung *(f) n* De
En depropanization
Es despropanización *(f)*
Fr dépropanisation *(f)*
It depropanizzazione *(f)*
Pt despropanização *(f)*

entrada de aire *(f)* Es
De Lufteinlaβ *(m)*
En air inlet
Fr admission d'air *(f)*
It presa d'aria *(f)*
Pt admissão de ar *(f)*

entropia *(f) n* It, Pt
De Entropie *(f)*
En entropy
Es entropía *(f)*
Fr entropie *(f)*

entropía *(f) n* Es
De Entropie *(f)*
En entropy
Fr entropie *(f)*
It entropia *(f)*
Pt entropia *(f)*

entropie *(f) n* Fr
De Entropie *(f)*
En entropy
Es entropía *(f)*
It entropia *(f)*
Pt entropia *(f)*

Entropie *(f) n* De
En entropy
Es entropía *(f)*
Fr entropie *(f)*
It entropia *(f)*
Pt entropia *(f)*

entropy *n* En
De Entropie *(f)*
Es entropía *(f)*
Fr entropie *(f)*
It entropia *(f)*
Pt entropia *(f)*

Entschwefelung *(f) n* De
Am desulfurization
En desulphurization
Es desulfurización *(f)*
Fr désulfuration *(f)*
It desolforazione *(f)*
Pt dessulfurização *(f)*

entulho *(m) n* Pt
De Abraum *(m)*
En spoil (mining)
Es estériles *(m pl)*
Fr déblai *(m)*
It materiale di sterro *(m)*

Entwässerung *(f) n* De
En dewatering
Es desecación *(f)*
Fr dénoyage *(m)*
It deidratazione *(f)*
Pt deshidratação *(f)*

envejecimiento térmico *(m)* Es
De thermische Alterung *(f)*
En thermal ageing

Fr vieillissement thermique *(m)*
It invecchiamento termico *(m)*
Pt envelhecimento térmico *(m)*

envelhecimento térmico *(m)* Pt
De thermische Alterung *(f)*
En thermal ageing
Es envejecimiento térmico *(m)*
Fr vieillissement thermique *(m)*
It invecchiamento termico *(m)*

environmental *adj* En
De Umwelt-
Es ambiental
Fr de l'environnement
It ambientale
Pt ambiente

enxofre *(m) n* Pt
Am sulfur
De Schwefel *(m)*
En sulphur
Es azufre *(m)*
Fr soufre *(m)*
It zolfo *(m)*

épuration humide *(f)* Fr
De Naßreinigen *(n)*
En wet scrubbing
Es lavado húmedo *(m)*
It lavaggio *(m)*
Pt limpeza com lavagem *(f)*

équilibre *(m) n* Fr
De Gleichgewicht *(n)*
En equilibrium
Es equilibrio *(m)*
It equilibrio *(m)*
Pt equilíbrio *(m)*

équilibre de masse *(m)* Fr
De Massenausgleich *(m)*
En mass balance
Es equilibrio de masa *(m)*
It bilancio di massa *(m)*
Pt equilíbrio de massa *(m)*

équilibre énergétique *(m)* Fr
De Energiebilanz *(f)*
En energy balance
Es balance energético *(m)*
It equilibrio di energia *(m)*
Pt equilíbrio energético *(m)*

equilibrio *(m) n* Es, It
De Gleichgewicht *(n)*
En equilibrium
Fr équilibre *(m)*
Pt equilíbrio *(m)*

equilíbrio *(m) n* Pt
De Gleichgewicht *(n)*
En equilibrium
Es equilibrio *(m)*
Fr équilibre *(m)*
It equilibrio *(m)*

equilibrio de masa *(m)* Es
De Massenausgleich *(m)*
En mass balance
Fr équilibre de masse *(m)*
It bilancio di massa *(m)*
Pt equilíbrio de massa *(m)*

equilíbrio de massa *(m)* Pt
De Massenausgleich *(m)*
En mass balance
Es equilibrio de masa *(m)*
Fr équilibre de masse *(m)*
It bilancio di massa *(m)*

equilíbrio de material e energia *(m)* Pt
De Stoff- und Energiebilanz *(f)*
En material and energy balance
Es equilibrio materia-energía *(m)*
Fr balance de matériel et énergie *(f)*
It bilancio materiale-energia *(m)*

equilibrio di energia *(m)* It
De Energiebilanz *(f)*
En energy balance
Es balance energético *(m)*
Fr équilibre énergétique *(m)*
Pt equilíbrio energético *(m)*

equilíbrio energético *(m)* Pt
De Energiebilanz *(f)*
En energy balance
Es balance energético *(m)*
Fr équilibre énergétique *(m)*
It equilibrio di energia *(m)*

equilibrio materia-energía *(m)* Es
De Stoff- und Energiebilanz *(f)*
En material and energy balance
Fr balance de matériel et énergie *(f)*
It bilancio materiale-energia *(m)*
Pt equilíbrio de material e energia *(m)*

equilibrio térmico *(m)* Es
De Wärmeausgleich *(m)*
En heat balance
Fr bilan calorifique *(m)*
It bilancio termico *(m)*
Pt equilíbrio térmico *(m)*

equilíbrio térmico *(m)* Pt
De Wärmeausgleich *(m)*
En heat balance
Es equilibrio térmico *(m)*
Fr bilan calorifique *(m)*
It bilancio termico *(m)*

equilibrium *n* En
De Gleichgewicht *(n)*
Es equilibrio *(m)*
Fr équilibre *(m)*
It equilibrio *(m)*
Pt equilíbrio *(m)*

équivalent charbon *(m)* Fr
De Kohlenersatz *(m)*
En coal equivalent
Es carbón equivalente *(m)*
It equivalente del carbone *(m)*
Pt equivalente carbónico *(m)*

equivalent circuit En
De Ersatzschaltung *(f)*
Es circuito equivalente *(m)*
Fr circuit équivalent *(m)*
It circuito equivalente *(m)*
Pt circuito equivalente *(m)*

equivalente carbónico *(m)* Pt
De Kohlenersatz *(m)*
En coal equivalent
Es carbón equivalente *(m)*
Fr équivalent charbon *(m)*
It equivalente del carbone *(m)*

equivalente del carbone *(m)* It
De Kohlenersatz *(m)*
En coal equivalent
Es carbón equivalente *(m)*
Fr équivalent charbon *(m)*
Pt equivalente carbónico *(m)*

equivalente del petrolio *(m)* It
De Ölersatz *(m)*
En oil equivalent
Es petróleo equivalente *(m)*
Fr équivalent pétrole *(m)*
Pt equivalente de petróleo *(m)*

equivalente de petróleo *(m)* Pt
De Ölersatz *(m)*
En oil equivalent
Es petróleo equivalente *(m)*
Fr équivalent pétrole *(m)*
It equivalente del petrolio *(m)*

équivalent pétrole *(m)* Fr
De Ölersatz *(m)*
En oil equivalent
Es petróleo equivalente *(m)*

It equivalente del petrolio *(m)*
Pt equivalente de petróleo *(m)*

Erdanziehungskraft *(f) n* De
En gravitational force
Es fuerza gravitatoria *(f)*
Fr force de gravitation *(f)*
It forza di gravità *(f)*
Pt força de gravidade *(f)*

Erde *(f) n* De
Am ground
En earth
Es tierra *(f)*
Fr terre *(f)*
It terra *(f)*
Pt massa *(f)*

Erdgas *(n) n* De
En natural gas
Es gas natural *(m)*
Fr gaz naturel *(m)*
It gas naturale *(m)*
Pt gás natural *(m)*

Erdgasersatz *(m) n* De
En substitute natural gas
Es gas natural de sustitución *(m)*
Fr gaz naturel de remplacement *(m)*
It sostituto di gas naturale *(m)*
Pt gás natural de substituição *(m)*

Erdöl *(n) n* De
En petroleum; rock oil
Es petróleo *(m)*
Fr pétrole *(m)*
It petrolio *(m)*
Pt petróleo *(m)*

Erhaltung *(f) n* De
En conservation
Es conservación *(f)*
Fr conservation *(f)*
It conservazione *(f)*
Pt conservação *(f)*

erneuerbare Energie *(f)* De
En renewable energy
Es energía renovable *(f)*
Fr énergie renouvelable *(f)*
It energia rinnovabile *(f)*
Pt energia renovável *(f)*

erosão *(f) n* Pt
De Erosion *(f)*
En erosion
Es erosión *(f)*
Fr érosion *(f)*
It erosione *(f)*

erosion *n* En
De Erosion *(f)*
Es erosión *(f)*
Fr érosion *(f)*
It erosione *(f)*
Pt erosão *(f)*

Erosion *(f) n* De
En erosion
Es erosión *(f)*
Fr érosion *(f)*
It erosione *(f)*
Pt erosão *(f)*

érosion *(f) n* Fr
De Erosion *(f)*
En erosion
Es erosión *(f)*
It erosione *(f)*
Pt erosão *(f)*

erosión *(f) n* Es
De Erosion *(f)*
En erosion
Fr érosion *(f)*
It erosione *(f)*
Pt erosão *(f)*

erosione *(f) n* It
De Erosion *(f)*
En erosion
Es erosión *(f)*
Fr érosion *(f)*
Pt erosão *(f)*

Erregung *(f) n* De
En excitation
Es excitación *(f)*
Fr excitation *(f)*
It eccitazione *(f)*
Pt excitação *(f)*

Ersatzschaltung *(f) n* De
En equivalent circuit
Es circuito equivalente *(m)*
Fr circuit équivalent *(m)*
It circuito equivalente *(m)*
Pt circuito equivalente *(m)*

Ertrag *(m) n* De
En yield
Es rendimiento *(m)*
Fr rendement *(m)*
It rendimento *(m)*
Pt rendimento *(m)*

éruption *(f) n* Fr
De Ausbruch *(m)*
En blowout
Es reventón *(m)*
It eruzione *(f)*
Pt estouro *(m)*

eruzione *(f) n* It
De Ausbruch *(m)*
En blowout
Es reventón *(m)*
Fr éruption *(f)*
Pt estouro *(m)*

erwiesene Reserven *(pl)* De
En proven reserves
Es reservas probadas *(f pl)*
Fr réserves prouvées *(f pl)*
It riserve dimostrate *(f pl)*
Pt reservas comprovadas *(f pl)*

Erzeugungskapazität *(f) n* De
En generating capacity
Es capacidad generadora *(f)*
Fr capacité génératrice *(f)*
It capacità di generazione *(f)*
Pt capacidade geradora *(f)*

Erzeugungsleistung *(f) n* De
En generating efficiency
Es rendimiento de generación *(m)*
Fr rendement générateur *(m)*
It efficienza di generazione *(f)*
Pt eficiência geradora *(f)*

esano *(m) n* It
De Hexan *(n)*
En hexane
Es hexano *(m)*
Fr hexane *(m)*
Pt hexano *(m)*

escape *(m) n* Pt
De Auspuffgas *(n)*
En exhaust
Es gases de escape *(m pl)*
Fr échappement *(m)*
It scarico *(m)*

escape de gás *(f)* Pt
De Gasaustritt *(m)*
En gas escape
Es fuga de gas *(f)*
Fr échappement de gaz *(m)*
It fuga di gas *(f)*

esclusore d'aria *(m)* It
Am draft excluder
De Luftzugschutz *(m)*
En draught excluder
Es eliminador de corrientes *(m)*
Fr brise-bise *(m)*
Pt eliminador de corrente de ar *(m)*

escoria *(f) n* Es
De Schlacke *(f)*
En slag
Fr laitier *(m)*
It scoria *(f)*
Pt escória *(f)*

escória *(f) n* Pt
De Schlacke *(f)*
En slag
Es escoria *(f)*
Fr laitier *(m)*
It scoria *(f)*

escoria vanadosa de petróleo *(f)* Es
De Ölasche *(f)*
En oil ash
Fr cendre de pétrole *(f)*
It cenere di petrolio *(f)*
Pt cinza de óleo *(f)*

esoergico *adj* It
De exoenergetisch
En exoergic
Es exoérgico
Fr exoénergétique
Pt exoérgico

esotermico *adj* It
De exotherm
En exothermic
Es exotérmico
Fr exothermique
Pt exotérmico

espalhamento *(m) n* Pt
De Streuung *(f)*
En scattering
Es dispersión *(f)*
Fr dispersion *(f)*
It dispersione *(f)*

espansione *(f) n* It
De Expansion *(f)*
En expansion
Es dilatación *(f)*
Fr dilatation *(f)*
Pt dilatação *(f)*

espanso di urea-formaldeide *(m)* It
De Harnstoff-Formaldehyd-schaumstoff *(m)*
En urea-formaldehyde foam
Es espuma de urea-formaldehido *(f)*
Fr mousse urée et formaldéhyde *(f)*
Pt espuma de ureia-formaldeído *(f)*

espectro *(m) n* Es, Pt
De Spektrum *(n)*
En spectrum
Fr spectre *(m)*
It spettro *(m)*

espectro de emisión *(m)* Es
De Emissionsspektrum *(n)*
En emission spectrum
Fr spectre d'émission *(m)*
It spettro delle emissioni *(m)*
Pt espectro de emissão *(m)*

espectro de emissão *(m)* Pt
De Emissionsspektrum *(n)*
En emission spectrum
Es espectro de emisión *(m)*
Fr spectre d'émission *(m)*
It spettro delle emissioni *(m)*

espectro electromagnético *(m)* Es
De elektromagnetisches Spektrum *(n)*
En electromagnetic spectrum
Fr spectre électromagnétique *(m)*
It spettro elettromagnetico *(m)*
Pt espectro electromagnético *(m)*

espectro electromagnético *(m)* Pt
De elektromagnetisches Spektrum *(n)*
En electromagnetic spectrum
Es espectro electromagnético *(m)*
Fr spectre électromagnétique *(m)*
It spettro elettromagnetico *(m)*

espectrómetro *(m) n* Es, Pt
De Spektrometer *(n)*
En spectrometer
Fr spectromètre *(m)*
It spettrometro *(m)*

espectroscopia *(f) n* Es, Pt
De Spektroskopie *(f)*
En spectroscopy
Fr spectroscopie *(f)*
It spettroscopia *(f)*

espectroscopia de masa *(f)* Es
De Massenspektro-skopie *(f)*
En mass spectroscopy
Fr spectroscopie de masse *(f)*
It spettroscopia di massa *(f)*
Pt espectroscopia de massa *(f)*

espectroscopia de massa *(f)* Pt
De Massenspektro-skopie *(f)*
En mass spectroscopy
Es espectroscopia de masa *(f)*
Fr spectroscopie de masse *(f)*
It spettroscopia di massa *(f)*

espectroscopio *(m) n* Es
De Spektroskop *(m)*
En spectroscope
Fr spectroscope *(m)*
It spettroscopio *(m)*
Pt espectroscópio *(m)*

espectroscópio *(m) n* Pt
De Spektroskop *(m)*
En spectroscope
Es espectroscopio *(m)*
Fr spectroscope *(m)*
It spettroscopio *(m)*

espectro visible *(m)* Es
De sichtbares Spektrum *(n)*
En visible spectrum
Fr spectre visible *(m)*
It spettro visibile *(m)*
Pt espectro visível *(m)*

espectro visível *(m)* Pt
De sichtbares Spektrum *(n)*
En visible spectrum
Es espectro visible *(m)*
Fr spectre visible *(m)*
It spettro visibile *(m)*

espita *(f) n* Es
De Hahn *(m)*
En cock
Fr robinet *(m)*
It rubinetto *(m)*
Pt torneira *(f)*

esplorazione geochimica *(f)* It
De geochemische Untersuchung *(f)*
En geochemical exploration
Es exploración geoquímica *(f)*
Fr exploration géochimique *(f)*
Pt exploração geoquímica *(f)*

esplorazione geofisica *(f)* It
De geophysische Untersuchung *(f)*
En geophysical exploration
Es exploración geofísica *(f)*
Fr exploration géophysique *(f)*
Pt exploração geofísica *(f)*

esplosione *(f) n* It
De Explosion *(f)*
En explosion
Es explosión *(f)*
Fr explosion *(f)*
Pt explosão *(f)*

esplosivi *(m pl) n* It
De Sprengstoffe *(pl)*
En explosives *(pl)*
Es explosivos *(m pl)*
Fr explosifs *(m pl)*
Pt explosivos *(m pl)*

esponenziale *adj* It
De exponential
En exponential
Es exponencial
Fr exponentiel
Pt exponencial

espuma de urea-formaldehido *(f)* Es
De Harnstoff-Formaldehyd-schaumstoff *(m)*
En urea-formaldehyde foam
Fr mousse urée et formaldéhyde *(f)*
It espanso di urea-formaldeide *(m)*
Pt espuma de ureia-formaldeído *(f)*

espuma de ureia-formaldeído *(f)* Pt
De Harnstoff-Formaldehyd-schaumstoff *(m)*
En urea-formaldehyde foam
Es espuma de urea-formaldehido *(f)*
Fr mousse urée et formaldéhyde *(f)*

It espanso di urea-formaldeide *(m)*

essai de gonflement *(m)* Fr
De Quellversuch *(m)*
En swelling test
Es prueba de esponjamiento *(f)*
It prova di ringonfiamento *(f)*
Pt teste de inchamento *(m)*

essai de résistance au choc *(m)* Fr
De Sturzfestigkeits-prüfung *(f)*
En shatter test
Es prueba de cohesión *(f)*
It prova di frangibilità *(f)*
Pt teste de fragmentação *(m)*

essai de sédimentation *(m)* Fr
De Rückstandsprüfung *(f)*
En sediment test
Es prueba de sedimentos *(f)*
It prova di sedimentazione *(f)*
Pt teste de sedimentos *(m)*

essence *(f) n* Fr
Am gasoline
De Benzin *(n)*
En petrol
Es gasolina *(f)*
It benzina *(f)*
Pt gasolina *(f)*

essiccatore *(m) n* It
De Trockner *(m)*
En dryer
Es secador *(m)*
Fr sécheur *(m)*
Pt secador *(m)*

essiccatore ad alta frequenza *(m)* It
De Hochfrequenz-Trockner *(m)*
En high-frequency dryer
Es secador de alta frecuencia *(m)*
Fr sécheur haute fréquence *(m)*
Pt secador de alta frequência *(m)*

essiccatore ad aria *(m)* It
De Lufttrockner *(m)*
En air dryer
Es secador de aire *(m)*
Fr sécheur d'air *(m)*
Pt secador por ar *(m)*

essiccatore di pellicola *(m)* It
De Filmtrockner *(m)*
En film dryer
Es secador de película *(m)*
Fr sécheur de pellicule *(m)*
Pt secador de película *(m)*

estadística *(f) n* Es
De Statistik *(f)*
En statistics
Fr statistiques *(f)*
It statistica *(f)*
Pt estatística *(f)*

estatística *(f) n* Pt
De Statistik *(f)*
En statistics
Es estadística *(f)*
Fr statistiques *(f)*
It statistica *(f)*

estequiométrico *adj* Pt
De stöchiometrisch
En stoichiometric
Es estoquiometrico
Fr stoechiométrique
It stechiometrico

estériles *(m pl) n* Es
De Abraum *(m)*
En spoil (mining)
Fr déblai *(m)*
It materiale di sterro *(m)*
Pt entulho *(m)*

estoquiometrico *adj* Es
De stöchiometrisch
En stoichiometric
Fr stoechiométrique
It stechiometrico
Pt estequiométrico

estouro *(m) n* Pt
De Ausbruch *(m)*
En blowout
Es reventón *(m)*
Fr éruption *(f)*
It eruzione *(f)*

estrato impermeable de cobertura *(m)* Es
De Deckgestein *(n)*
En cap rock
Fr roche couverture *(f)*
It strato impermeabile di copertura *(m)*
Pt rocha encaixante *(f)*

estrazione di minerali *(f)* It
De Bergbau *(m)*
En mining
Es minería *(f)*
Fr exploitation minière *(f)*
Pt mineração *(f)*

estructura alveolar *(f)* Es
De Wabenstruktur *(f)*
En honeycomb structure
Fr structure en nids d'abeilles *(f)*
It structura a nido d'ape *(f)*
Pt estrutura alveolar *(f)*

estructuras de perforación *(f pl)* Es
De Bohrkonstruktionen *(pl)*
En drilling structures
Fr structures de forage *(f pl)*
It strutture di trivellazione *(f pl)*
Pt estructuras de perfuração *(f pl)*

estructuras de perfuração *(f pl)* Pt
De Bohrkonstruktionen *(pl)*
En drilling structures
Es estructuras de perforación *(f pl)*
Fr structures de forage *(f pl)*
It strutture di trivellazione *(f pl)*

estrutura alveolar *(f)* Pt
De Wabenstruktur *(f)*
En honeycomb structure
Es estructura alveolar *(f)*
Fr structure en nids d'abeilles *(f)*
It structura a nido d'ape *(f)*

estuary tidal power En
De Gezeitenkraft in Flußmündungen *(f)*
Es potencia mareal de estuario *(f)*
Fr puissance marémotrice d'estuaire *(f)*
It potenza della marea di estuario *(f)*
Pt potência de marés de estuário *(f)*

estufa *(f) n* Es
De Brennofen; Ofen *(m)*
En kiln; oven
Fr four *(m)*
It forno *(m)*
Pt forno; forno de requeima *(m)*

esvaziamento *(m) n* Pt
De Kippen *(n)*
En dumping
Es vaciamiento *(m)*
Fr déversement *(m)*
It pressatura *(f)*

etano *(m) n* Es, It, Pt
De Äthan *(n)*
En ethane
Fr éthane *(m)*

etanol *(m) n* Es, Pt
De Äthylalkohol *(m)*
En ethanol
Fr éthanol *(m)*
It etanolo *(m)*

etanolo *(m) n* It
De Äthylalkohol *(m)*
En ethanol
Es etanol *(m)*
Fr éthanol *(m)*
Pt etanol *(m)*

ethane *n* En
De Äthan *(n)*
Es etano *(m)*
Fr éthane *(m)*
It etano *(m)*
Pt etano *(m)*

éthane *(m) n* Fr
De Äthan *(n)*
En ethane
Es etano *(m)*

It etano *(m)*
Pt etano *(m)*

ethanol *n* En
De Äthylalkohol *(m)*
Es etanol *(m)*
Fr éthanol *(m)*
It etanolo *(m)*
Pt etanol *(m)*

éthanol *(m) n* Fr
De Äthylalkohol *(m)*
En ethanol
Es etanol *(m)*
It etanolo *(m)*
Pt etanol *(m)*

ethylene *n* En
De Äthylen *(n)*
Es etileno *(m)*
Fr éthylène *(m)*
It etilene *(m)*
Pt etileno *(m)*

éthylène *(m) n* Fr
De Äthylen *(n)*
En ethylene
Es etileno *(m)*
It etilene *(m)*
Pt etileno *(m)*

ethylene glycol En
De Äthylenglykol *(n)*
Es glicoletileno *(m)*
Fr éthylène glycol *(m)*
It glicole etilenico *(m)*
Pt glicol etilénico *(m)*

éthylène glycol *(m)* Fr
De Äthylenglykol *(n)*
En ethylene glycol
Es glicoletileno *(m)*
It glicole etilenico *(m)*
Pt glicol etilénico *(m)*

etilene *(m) n* It
De Äthylen *(n)*
En ethylene
Es etileno *(m)*
Fr éthylène *(m)*
Pt etileno *(m)*

etileno *(m) n* Es, Pt
De Äthylen *(n)*
En ethylene
Fr éthylène *(m)*
It etilene *(m)*

étude de l'environment integreé *(f)* Fr
De integrierte Umweltplanung *(f)*
En integrated environmental design (IED)
Es diseño ambiental integrado *(m)*
It design ambientale integrato *(m)*
Pt projecto de ambiente integrado *(m)*

evaporação *(f) n* Pt
De Verdunstung *(f)*
En evaporation
Es evaporación *(f)*
Fr évaporation *(f)*
It evaporazione *(f)*

evaporación *(f) n* Es
De Verdunstung *(f)*
En evaporation
Fr évaporation *(f)*
It evaporazione *(f)*
Pt evaporação *(f)*

evaporation *n* En
De Verdunstung *(f)*
Es evaporación *(f)*
Fr évaporation *(f)*
It evaporazione *(f)*
Pt evaporação *(f)*

évaporation *(f) n* Fr
De Verdunstung *(f)*
En evaporation
Es evaporación *(f)*
It evaporazione *(f)*
Pt evaporação *(f)*

evaporazione *(f) n* It
De Verdunstung *(f)*
En evaporation
Es evaporación *(f)*
Fr évaporation *(f)*
Pt evaporação *(f)*

évent *(m) n* Fr
De Entlüftungsschlitz *(m)*
En air vent
Es toma de aire *(f)*
It sfiato aria *(m)*
Pt respirador de ar *(m)*

excès d'air *(m)* Fr
De Luftüberschuβ *(m)*
En excess air
Es aire sobrante *(m)*
It eccesso d'aria *(m)*
Pt ar em excesso *(m)*

excess air En
De Luftüberschuβ *(m)*
Es aire sobrante *(m)*
Fr excès d'air *(m)*
It eccesso d'aria *(m)*
Pt ar em excesso *(m)*

exchanger *n* En
De Austauscher *(m)*
Es intercambiador *(m)*
Fr échangeur *(m)*
It scambiatore *(m)*
Pt permutador *(m)*

excitação *(f) n* Pt
De Erregung *(f)*
En excitation
Es excitación *(f)*
Fr excitation *(f)*
It eccitazione *(f)*

excitación *(f) n* Es
De Erregung *(f)*
En excitation
Fr excitation *(f)*
It eccitazione *(f)*
Pt excitação *(f)*

excitation *n* En; Fr *(f)*
De Erregung *(f)*
Es excitación *(f)*
It eccitazione *(f)*
Pt excitação *(f)*

exhaust *n* En
De Auspuffgas *(n)*
Es gases de escape *(m pl)*
Fr échappement *(m)*
It scarico *(m)*
Pt escape *(m)*

existencia para emergencia *(f)* Es
De Vorrat *(m)*
En stockpile
Fr stock de réserves *(m)*
It riserva *(f)*
Pt pilha de armazenagem *(f)*

exoénergétique *adj* Fr
De exoenergetisch
En exoergic
Es exoérgico
It esoergico
Pt exoérgico

exoenergetisch *adj* De
En exoergic
Es exoérgico
Fr exoénergétique
It esoergico
Pt exoérgico

exoergic *adj* En
De exoenergetisch
Es exoérgico
Fr exoénergétique
It esoergico
Pt exoérgico

exoérgico *adj* Es, Pt
De exoenergetisch
En exoergic
Fr exoénergétique
It esoergico

exotérmico *adj* Es, Pt
De exotherm
En exothermic
Fr exothermique
It esotermico

exotherm *adj* De
En exothermic
Es exotérmico
Fr exothermique
It esotermico
Pt exotérmico

exothermic *adj* En
De exotherm
Es exotérmico
Fr exothermique
It esotermico
Pt exotérmico

exothermique *adj* Fr
De exotherm
En exothermic
Es exotérmico
It esotermico
Pt exotérmico

expanded polystyrene pellet En
De geschäumte Polystyrolkugel *(f)*
Es gránulo de poliestireno expandido *(m)*
Fr granule de polystyrène expansé *(m)*
It palline di polistirolo espanso *(f)*
Pt grânulo de

polistireno expandido *(m)*

expansion *n* En
De Expansion *(f)*
Es dilatación *(f)*
Fr dilatation *(f)*
It espansione *(f)*
Pt dilatação *(f)*

Expansion *(f) n* De
En expansion
Es dilatación *(f)*
Fr dilatation *(f)*
It espansione *(f)*
Pt dilatação *(f)*

exploding wires *(pl)* En
De Sprengdrähte *(pl)*
Es hilos explosivos *(m pl)*
Fr fils explosifs *(m pl)*
It fili esplosivi *(m pl)*
Pt fios de explosão *(m pl)*

exploitation des niveaux inférieurs *(f)* Fr
De Tiefbau *(m)*
En deep mining
Es explotación en profundidad *(f)*
It scavo profondo *(m)*
Pt mineração profunda *(f)*

exploitation minière *(f)* Fr
De Bergbau *(m)*
En mining
Es minería *(f)*
It estrazione di minerali *(f)*
Pt mineração *(f)*

exploitation minière des contours *(f)* Fr
De Konturenabbau *(m)*
En contour mining
Es minería siguiendo el perfil del terreno *(f)*
It scavo a contorno *(m)*
Pt mineração de curvas de nível *(f)*

exploitation minière souterraine *(f)* Fr
De Grubenbetrieb *(m)*
En underground mining
Es minería subterránea *(f)*
It scavo sotteraneo *(m)*
Pt mineração subterrânea *(f)*

exploração geofísica *(f)* Pt
De geophysische Untersuchung *(f)*
En geophysical exploration
Es exploración geofísica *(f)*
Fr exploration géophysique *(f)*
It esplorazione geofisica *(f)*

exploração geoquímica *(f)* Pt
De geochemische Untersuchung *(f)*
En geochemical exploration
Es exploración geoquímica *(f)*
Fr exploration géochimique *(f)*
It esplorazione geochimica *(f)*

exploración geofísica *(f)* Es
De geophysische Untersuchung *(f)*
En geophysical exploration
Fr exploration géophysique *(f)*
It esplorazione geofisica *(f)*
Pt exploração geofísica *(f)*

exploración geoquímica *(f)* Es
De geochemische Untersuchung *(f)*
En geochemical exploration
Fr exploration géochimique *(f)*
It esplorazione geochimica *(f)*
Pt exploração geoquímica *(f)*

exploration géochimique *(f)* Fr
De geochemische Untersuchung *(f)*
En geochemical exploration
Es exploración geoquímica *(f)*
It esplorazione geochimica *(f)*
Pt exploração geoquímica *(f)*

exploration géophysique *(f)* Fr
De geophysische Untersuchung *(f)*
En geophysical exploration
Es exploración geofísica *(f)*
It esplorazione geofisica *(f)*
Pt exploração geofísica *(f)*

explosão *(f) n* Pt
De Explosion *(f)*
En explosion
Es explosión *(f)*
Fr explosion *(f)*
It esplosione *(f)*

explosifs *(m pl) n* Fr
De Sprengstoffe *(pl)*
En explosives *(pl)*
Es explosivos *(m pl)*
It esplosivi *(m pl)*
Pt explosivos *(m pl)*

explosion *n* En; Fr *(f)*
De Explosion *(f)*
Es explosión *(f)*
It esplosione *(f)*
Pt explosão *(f)*

Explosion *(f) n* De
En explosion
Es explosión *(f)*
Fr explosion *(f)*
It esplosione *(f)*
Pt explosão *(f)*

explosión *(f) n* Es
De Explosion *(f)*
En explosion
Fr explosion *(f)*
It esplosione *(f)*
Pt explosão *(f)*

explosives *(pl) n* En
De Sprengstoffe *(pl)*
Es explosivos *(m pl)*
Fr explosifs *(m pl)*
It esplosivi *(m pl)*
Pt explosivos *(m pl)*

explosivos *(m pl) n* Es
De Sprengstoffe *(pl)*
En explosives *(pl)*
Fr explosifs *(m pl)*
It esplosivi *(m pl)*
Pt explosivos *(m pl)*

explosivos *(m pl) n* Pt
De Sprengstoffe *(pl)*
En explosives *(pl)*
Es explosivos *(m pl)*
Fr explosifs *(m pl)*
It esplosivi *(m pl)*

explotación en profundidad *(f)* Es
De Tiefbau *(m)*
En deep mining
Fr exploitation des niveaux inférieurs *(f)*
It scavo profondo *(m)*
Pt mineração profunda *(f)*

exponencial *adj* Es, Pt
De exponential
En exponential
Fr exponentiel
It esponenziale

exponential *adj* De, En
Es exponencial
Fr exponentiel
It esponenziale
Pt exponencial

exponentiel *adj* Fr
De exponential
En exponential
Es exponencial
It esponenziale
Pt exponencial

external work En
De Auβenarbeit *(f)*
Es trabajo externo *(m)*
Fr travail extérieur *(m)*
It lavoro esterno *(m)*
Pt trabalho externo *(m)*

extraction fan En
De Absauggebläse *(n)*
Es ventilador de extracción *(m)*
Fr ventilateur d'extraction *(m)*

It ventilatore ad estrazione *(m)*
Pt ventoínha de extracção *(f)*

extra-high voltage transmission En
De Höchstspannungs-transmission *(f)*
Es transmisión con tensión muy alta *(f)*
Fr transmission des ultra-hautes tensions *(f)*
It trasmissione a tensione extra elevata *(f)*
Pt transmissão de voltagem extra-alta *(f)*

F

fábrica piloto *(f)* Pt
De Versuchsanlage *(f)*
En pilot plant
Es planta piloto *(f)*
Fr installation pilote *(f)*
It impianto pilota *(m)*

facteur de charge *(m)* Fr
De Lastfaktor *(m)*
En load factor
Es factor de carga *(m)*
It fattore di carico *(m)*
Pt factor de carga *(m)*

facteur de puissance *(m)* Fr
De Leistungsfaktor *(m)*
En power factor
Es factor de potencia *(m)*
It fattore di potenza *(m)*
Pt factor de potência *(m)*

factor de carga *(m)* Es, Pt
De Lastfaktor *(m)*
En load factor
Fr facteur de charge *(m)*
It fattore di carico *(m)*

factor de potencia *(m)* Es
De Leistungsfaktor *(m)*
En power factor
Fr facteur de puissance *(m)*
It fattore di potenza *(m)*
Pt factor de potência *(m)*

factor de potência *(m)* Pt
De Leistungsfaktor *(m)*
En power factor
Es factor de potencia *(m)*
Fr facteur de puissance *(m)*
It fattore di potenza *(m)*

fall-out *n* En
De radioaktiver Niederschlag *(m)*
Es precipitación radiactiva *(f)*
Fr retombée radioactive *(f)*
It precipitazione radioattiva *(f)*
Pt precipitação radioactiva *(f)*

fan *n* En
De Gebläse *(n)*
Es ventilador *(m)*
Fr ventilateur *(m)*
It ventilatore *(m)*
Pt ventoinha *(f)*

fan-assisted heater En
De Heizelement mit Gebläse *(n)*
Es calentador auxiliado por ventilador *(m)*
Fr radiateur soufflant *(m)*
It riscaldatore a ventilatore *(m)*
Pt aquecedor auxiliado por ventoinha *(m)*

fango di trivellazione *(m)* It
De Bohrschlamm *(m)*
En drilling mud
Es lodo de perforación *(m)*
Fr boue de forage *(f)*
Pt lodo de perfuração *(m)*

fase *(f)* *n* Es, It, Pt
De Phase *(f)*
En phase
Fr phase *(f)*

Faserisolierung *(f)* *n* De
Am fiber insulation
En fibre insulation
Es aislamiento con fibra *(m)*
Fr isolement par fibre *(m)*
It isolamento con fibra *(m)*
Pt isolamento com fibras *(m)*

Faseroptik *(f)* *n* De
Am fiber optics
En fibre optics
Es óptica de las fibras *(f)*
Fr optique des fibres *(f)*
It fibra ottica *(f)*
Pt óptica de fibras *(f)*

Faß *(n)* *n* De
En barrel
Es barril *(m)*
Fr baril *(m)*
It barile *(m)*
Pt barril *(m)*

fast breeder reactor En
De Schnellbrüter *(m)*
Es reactor reproductor rápido *(m)*
Fr réacteur surrégénérateur rapide *(m)*
It reattore veloce autofissilizzante *(m)*
Pt reactor reproductor rápido *(m)*

fast neutron En
De schnelles Neutron *(n)*
Es neutrón rápido *(m)*
Fr neutron rapide *(m)*
It neutrone veloce *(m)*
Pt neutrão rápido *(m)*

fast reactor En
De Schnellreaktor *(m)*
Es reactor rápido *(m)*
Fr réacteur rapide *(m)*
It reattore veloce *(m)*
Pt reactor rápido *(m)*

fattore di carico *(m)* It
De Lastfaktor *(m)*
En load factor
Es factor de carga *(m)*
Fr facteur de charge *(m)*
Pt factor de carga *(m)*

fattore di potenza *(m)* It
De Leistungsfaktor *(m)*
En power factor
Es factor de potencia *(m)*
Fr facteur de puissance *(m)*
Pt factor de potência *(m)*

Faulschlammanlage *(f)* *n* De
En digester
Es digestor *(m)*
Fr digesteur *(m)*
It digestore *(m)*
Pt digestor *(m)*

feedback *n* En
De Rückkoppelung *(f)*
Es realimentación *(f)*
Fr rétroaction *(f)*
It retroazione *(f)*
Pt retôrno *(m)*

feed pump En
De Speisepumpe *(f)*
Es bomba de alimentación *(f)*
Fr pompe d'alimentation *(f)*
It pompa di alimentazione *(f)*
Pt bomba de alimentação *(f)*

feedstock *n* En
De Stangenmaterial *(n)*
Es material de carga *(m)*
Fr stock d'alimentation *(m)*
It materiale di alimentazione *(m)*
Pt stock de abastecimento *(m)*

feedwater *n* En
De Speisewasser *(n)*
Es agua de alimentación *(f)*
Fr eau d'alimentation *(f)*
It acqua di alimentazione *(f)*
Pt água de alimentação *(f)*

Feld *(n)* *n* De
En field
Es campo *(m)*
Fr champ *(m)*
It campo *(m)*
Pt campo *(m)*

Feldspule *(f)* *n* De
En field coil
Es bobina de campo *(f)*
Fr bobinage de champ *(m)*
It bobina di campo *(f)*
Pt bobina de campo *(f)*

Feldstärke *(f)* *n* De
En field strength
Es intensidad de campo *(f)*
Fr intensité de champ *(f)*
It intensità di campo *(f)*
Pt intensidade de campo *(f)*

Felsabraum *(m)* *n* De
En rock spoils
Es desperdicios de roca *(m pl)*
Fr déblais de roche *(m pl)*
It materiale di sterro di roccia *(m)*
Pt detritos de rocha *(m pl)*

fenêtre d'absorption *(f)* Fr
De Absorptionsfenster *(n)*
En absorption window
Es ventanilla de absorción *(f)*
It finestra di assorbimento *(f)*
Pt janela de absorção *(f)*

Fensterisolierung *(f)* *n* De
En window insulation
Es aislamiento de ventanas *(m)*
Fr isolement des fenêtres *(m)*
It isolamento delle finestre *(m)*
Pt isolamento de janela *(m)*

fermentação *(f)* *n* Pt
De Gärung *(f)*
En fermentation
Es fermentación *(f)*
Fr fermentation *(f)*
It fermentazione *(f)*

fermentación *(f)* *n* Es
De Gärung *(f)*
En fermentation
Fr fermentation *(f)*
It fermentazione *(f)*
Pt fermentação *(f)*

fermentation *n* En; Fr *(f)*
De Gärung *(f)*
Es fermentación *(f)*
It fermentazione *(f)*
Pt fermentação *(f)*

fermentazione *(f)* *n* It
De Gärung *(f)*
En fermentation
Es fermentación *(f)*
Fr fermentation *(f)*
Pt fermentação *(f)*

ferme solaire *(f)* Fr
De Solarfarm *(f)*
En solar farm
Es granja solar *(f)*
It podere solare *(m)*
Pt centro de produção solar *(f)*

fer mobile Fr
De Dreheisen-
En moving-iron
Es núcleo giratorio
It ferro rotante
Pt ímã móvel

Fernheizung *(f)* *n* De
En district heating
Es calefacción de distritos *(f)*
Fr chauffage urbain *(m)*
It riscaldamento di sezione *(m)*
Pt aquecimento de distrito *(m)*

ferrimagnetic *adj* En
De ferrimagnetisch
Es ferrimagnético
Fr ferrimagnétique
It ferrimagnetico
Pt ferrimagnético

ferrimagnetico *adj* It
De ferrimagnetisch
En ferrimagnetic
Es ferrimagnético
Fr ferrimagnétique
Pt ferrimagnético

ferrimagnético *adj* Es, Pt
De ferrimagnetisch
En ferrimagnetic
Fr ferrimagnétique
It ferrimagnetico

ferrimagnétique *adj* Fr
De ferrimagnetisch
En ferrimagnetic
Es ferrimagnético
It ferrimagnetico
Pt ferrimagnético

ferrimagnetisch *adj* De
En ferrimagnetic
Es ferrimagnético
Fr ferrimagnétique
It ferrimagnetico
Pt ferrimagnético

Ferrit *(n)* *n* De
En ferrite
Es ferrita *(f)*
Fr ferrite *(m)*
It ferrite *(f)*
Pt ferrite *(f)*

ferrita *(f)* *n* Es
De Ferrit *(n)*
En ferrite
Fr ferrite *(m)*
It ferrite *(f)*
Pt ferrite *(f)*

ferrite *n* En; Fr *(m)*; It, Pt *(f)*
De Ferrit *(n)*
Es ferrita *(f)*

ferromagnetic *adj* En
De ferromagnetisch
Es ferromagnético
Fr ferromagnétique
It ferromagnetico
Pt ferromagnético

ferromagnetico *adj* It
De ferromagnetisch
En ferromagnetic
Es ferromagnético
Fr ferromagnétique
Pt ferromagnético

ferromagnético *adj* Es, Pt
De ferromagnetisch
En ferromagnetic
Fr ferromagnétique
It ferromagnetico

ferromagnétique *adj* Fr
De ferromagnetisch
En ferromagnetic
Es ferromagnético
It ferromagnetico
Pt ferromagnético

ferromagnetisch *adj* De
En ferromagnetic
Es ferromagnético
Fr ferromagnétique
It ferromagnetico
Pt ferromagnético

ferro rotante It
De Dreheisen-
En moving-iron
Es núcleo giratorio
Fr fer mobile
Pt ímã móvel

fervedor solar *(m)* Pt
De Solarherd *(m)*
En solar cooker
Es cocina solar *(f)*
Fr fourneau solaire *(m)*
It fornello solare *(m)*

Festkraftstoff *(m)* *n* De
En solid fuel
Es combustible sólido *(m)*
Fr combustible solide *(m)*
It combustibile solido *(m)*
Pt combustível sólido *(m)*

feststehende Belüftung *(f)* De
En fixed aeration
Es aireación fija *(f)*
Fr aération fixe *(f)*
It aerazione fissa *(f)*
Pt arejamento fixo *(m)*

Feuchtigkeit *(f)* *n* De
En humidity
Es humedad *(f)*
Fr humidité *(f)*
It umidità *(f)*
Pt humidade *(f)*

Feuchtigkeits-sperrschicht *(f)* *n* De

En damp-proof course
Es hilada hidrófuga *(f)*
Fr couche hydrofuge *(f)*
It trattamento impermeabilizzante *(m)*
Pt curso a prova de imfietrações *(m)*

feuerfest *adj* De
En refractory
Es refractario
Fr réfractaire
It refrattario
Pt refractário

feuerfester Stein *(m)* De
En firebrick
Es ladrillo refractario *(m)*
Fr brique réfractaire *(f)*
It mattone refrattario *(m)*
Pt tijolo refractário *(m)*

feuerfestes Cermet *(n)* De
En cermet refractory
Es refractario de cerametal *(m)*
Fr réfractaire aux cermets *(m)*
It refrattario al cermete *(m)*
Pt refractário metaloceràmico *(m)*

Feuergefahr *(f) n* De
En fire hazard
Es riesgo de incendio *(m)*
Fr danger d'incendie *(m)*
It pericolo di incendio *(m)*
Pt perigo de incêndio *(m)*

feuergefährlich *adj* De
En flammable
Es inflamable
Fr inflammable
It infiammabile
Pt inflamável

Feuerkiste *(f) n* De
En firebox
Es hogar *(m)*
Fr foyer *(m)*
It focolaio *(m)*
Pt fornalha *(f)*

fiamma *(f) n* It
De Flamme *(f)*
En flame
Es llama *(f)*
Fr flamme *(f)*
Pt chama *(f)*

fiamma aerata *(f)* It
De belüftete Flamme *(f)*
En aerated flame
Es llama aireada *(f)*
Fr flamme aérée *(f)*
Pt chama arejada *(f)*

fiamma non aerata *(f)* It
De unbelüftete Flamme *(f)*
En nonaerated flame
Es llama no aireada *(f)*
Fr flamme non aérée *(f)*
Pt chama não arejada *(f)*

fiber insulation Am
De Faserisolierung *(f)*
En fibre insulation
Es aislamiento con fibra *(m)*
Fr isolement par fibre *(m)*
It isolamento con fibra *(m)*
Pt isolamento com fibras *(m)*

fiber optics Am
De Faseroptik *(f)*
En fibre optics
Es óptica de las fibras *(f)*
Fr optique des fibres *(f)*
It fibra ottica *(f)*
Pt óptica de fibras *(f)*

fibra ottica *(f)* It
Am fiber optics
De Faseroptik *(f)*
En fibre optics
Es óptica de las fibras *(f)*
Fr optique des fibres *(f)*
Pt óptica de fibras *(f)*

fibre insulation En
Am fiber insulation
De Faserisolierung *(f)*
Es aislamiento con fibra *(m)*
Fr isolement par fibre *(m)*
It isolamento con fibra *(m)*
Pt isolamento com fibras *(m)*

fibre optics En
Am fiber optics
De Faseroptik *(f)*
Es óptica de las fibras *(f)*
Fr optique des fibres *(f)*
It fibra ottica *(f)*
Pt óptica de fibras *(f)*

field *n* En
De Feld *(n)*
Es campo *(m)*
Fr champ *(m)*
It campo *(m)*
Pt campo *(m)*

field coil En
De Feldspule *(f)*
Es bobina de campo *(f)*
Fr bobinage de champ *(m)*
It bobina di campo *(f)*
Pt bobina de campo *(f)*

field strength En
De Feldstärke *(f)*
Es intensidad de campo *(f)*
Fr intensité de champ *(f)*
It intensità di campo *(f)*
Pt intensidade de campo *(f)*

filament *n* En; Fr *(m)*
De Glühfaden *(m)*
Es filamento *(m)*
It filamento *(m)*
Pt filamento *(m)*

filamento *(m) n* Es, It, Pt
De Glühfaden *(m)*
En filament
Fr filament *(m)*

filamento a filo caldo *(m)* It
De Heizfaden *(m)*
En hot-wire filament
Es filamento de hilo caliente *(m)*
Fr filament thermique *(m)*
Pt filamento de fio quente *(m)*

filamento de fio quente *(m)* Pt
De Heizfaden *(m)*
En hot-wire filament
Es filamento de hilo caliente *(m)*
Fr filament thermique *(m)*
It filamento a filo caldo *(m)*

filamento de hilo caliente *(m)* Es
De Heizfaden *(m)*
En hot-wire filament
Fr filament thermique *(m)*
It filamento a filo caldo *(m)*
Pt filamento de fio quente *(m)*

filament thermique *(m)* Fr
De Heizfaden *(m)*
En hot-wire filament
Es filamento de hilo caliente *(m)*
It filamento a filo caldo *(m)*
Pt filamento de fio quente *(m)*

fili esplosivi *(m pl)* It
De Sprengdrähte *(pl)*
En exploding wires
Es hilos explosivos *(m pl)*
Fr fils explosifs *(m pl)*
Pt fios de explosão *(m pl)*

film dryer En
De Filmtrockner *(m)*
Es secador de película *(m)*
Fr sécheur de pellicule *(m)*
It essiccatore di pellicola *(m)*
Pt secador de película *(m)*

Filmtrockner *(m) n* De
En film dryer
Es secador de película *(m)*
Fr sécheur de pellicule *(m)*
It essiccatore di pellicola *(m)*
Pt secador de película *(m)*

filone di carbone *(m)* It
De Kohlenflöz *(n)*
En coal seam
Es capa de carbón *(f)*

Fr filon houiller *(m)*
Pt camada de carvão *(f)*

filon houiller *(m)* Fr
De Kohlenflöz *(n)*
En coal seam
Es capa de carbón *(f)*
It filone di carbone *(m)*
Pt camada de carvão *(f)*

fils explosifs *(m pl)* Fr
De Sprengdrähte *(pl)*
En exploding wires
Es hilos explosivos *(m pl)*
It fili esplosivi *(m pl)*
Pt fios de explosão *(m pl)*

filtre particulaire *(m)* Fr
De Partikelfilter *(m)*
En particulate filter
Es filtro de macropartículas *(m)*
It filtro delle macroparticelle *(m)*
Pt filtro de partículas *(m)*

filtro delle macroparticelle *(m)* It
De Partikelfilter *(m)*
En particulate filter
Es filtro de macropartículas *(m)*
Fr filtre particulaire *(m)*
Pt filtro de partículas *(m)*

filtro de macropartículas *(m)* Es
De Partikelfilter *(m)*
En particulate filter
Fr filtre particulaire *(m)*
It filtro delle macroparticelle *(m)*
Pt filtro de partículas *(m)*

filtro de partículas *(m)* Pt
De Partikelfilter *(m)*
En particulate filter
Es filtro de macropartículas *(m)*
Fr filtre particulaire *(m)*
It filtro delle macroparticelle *(m)*

finestra di assorbimento *(f)* It
De Absorptionsfenster *(n)*
En absorption window
Es ventanilla de absorción *(f)*
Fr fenêtre d'absorption *(f)*
Pt janela de absorção *(f)*

fios de explosão *(m pl)* Pt
De Sprengdrähte *(pl)*
En exploding wires
Es hilos explosivos *(m pl)*
Fr fils explosifs *(m pl)*
It fili esplosivi *(m pl)*

firebox *n* En
De Feuerkiste *(f)*
Es hogar *(m)*
Fr foyer *(m)*
It focolaio *(m)*
Pt fornalha *(f)*

firebrick En
De feuerfester Stein *(m)*
Es ladrillo refractario *(m)*
Fr brique réfractaire *(f)*
It mattone refrattario *(m)*
Pt tijolo refractário *(m)*

fire hazard En
De Feuergefahr *(f)*
Es riesgo de incendio *(m)*
Fr danger d'incendie *(m)*
It pericolo di incendio *(m)*
Pt perigo de incêndio *(m)*

fire-tube boiler En
De Heizrohrkessel *(m)*
Es caldera pirotubular *(f)*
Fr chaudière à tube de fumée *(f)*
It caldaia a tubo di fumo *(f)*
Pt caldeira de tubo de fumaça *(f)*

fisión *(f) n* Es
De Spaltung *(f)*
En fission
Fr fission *(f)*
It fissione *(f)*
Pt desintegração *(f)*

fisionable *adj* Es
De spaltbar
En fissionable
Fr fissionable
It fissionabile
Pt desintegrável

fisión nuclear *(f)* Es
De Nuklearspaltung *(f)*
En nuclear fission
Fr fission nucléaire *(f)*
It fissione nucleare *(f)*
Pt desintegração nuclear *(f)*

físsil *adj* Pt
De thermisch spaltbar
En fissile
Es fissile
Fr fissile
It fissile

fissile *adj* En, Es, Fr, It
De thermisch spaltbar
Pt físsil

fission *n* En; Fr *(f)*
De Spaltung *(f)*
Es fisión *(f)*
It fissione *(f)*
Pt desintegração *(f)*

fissionabile *adj* It
De spaltbar
En fissionable
Es fisionable
Fr fissionable
Pt desintegrável

fissionable *adj* En, Fr
De spaltbar
Es fisionable
It fissionabile
Pt desintegrável

fissione *(f) n* It
De Spaltung *(f)*
En fission
Es fisión *(f)*
Fr fission *(f)*
Pt desintegração *(f)*

fissione nucleare *(f)* It
De Nuklearspaltung *(f)*
En nuclear fission
Es fisión nuclear *(f)*
Fr fission nucléaire *(f)*
Pt desintegração nuclear *(f)*

fission nucléaire *(f)* Fr
De Nuklearspaltung *(f)*
En nuclear fission
Es fisión nuclear *(f)*
It fissione nucleare *(f)*
Pt desintegração nuclear *(f)*

fission product En
De Spaltprodukt *(n)*
Es producto de fisión *(m)*
Fr produit de la fission *(m)*
It prodotto di fissione *(m)*
Pt produto de desintegração *(m)*

fixed aeration En
De feststehende Belüftung *(f)*
Es aireación fija *(f)*
Fr aération fixe *(f)*
It aerazione fissa *(f)*
Pt arejamento fixo *(m)*

fixed carbon En
De nicht flüchtiger Kohlenstoff *(m)*
Es carbón fijo *(m)*
Fr carbone fixe *(m)*
It carbone fisso *(m)*
Pt carvão fixo *(m)*

Flächeninhaltsbestimmung *(f) n* De
En quadrature
Es cuadratura *(f)*
Fr quadrature *(f)*
It quadratura *(f)*
Pt quadratura *(f)*

flame *n* En
De Flamme *(f)*
Es llama *(f)*
Fr flamme *(f)*
It fiamma *(f)*
Pt chama *(f)*

flame speed En
De Flammengeschwindigkeit *(f)*
Es velocidad de la llama *(f)*
Fr vitesse de flamme *(f)*
It velocità della fiamma *(f)*
Pt velocidade de chama *(f)*

flammable *adj* En
De feuergefährlich
Es inflamable
Fr inflammable
It infiammabile
Pt inflamável

flamme *(f) n* Fr
De Flamme *(f)*
En flame
Es llama *(f)*
It fiamma *(f)*
Pt chama *(f)*

Flamme *(f) n* De
En flame
Es llama *(f)*
Fr flamme *(f)*
It fiamma *(f)*
Pt chama *(f)*

flamme aérée *(f)* Fr
De belüftete Flamme *(f)*
En aerated flame
Es llama aireada *(f)*
It fiamma aerata *(f)*
Pt chama arejada *(f)*

flammengerichteter Brenner *(m)* De
En directional flame burner
Es quemador de llama direccional *(m)*
Fr brûleur à flamme dirigée *(m)*
It bruciatore di fiamma direzionale *(m)*
Pt queimador de chama direccional *(m)*

Flammengeschwindigkeit *(f) n* De
En flame speed
Es velocidad de la llama *(f)*
Fr vitesse de flamme *(f)*
It velocità della fiamma *(f)*
Pt velocidade de chama *(f)*

flamme non aérée *(f)* Fr
De unbelüftete Flamme *(f)*
En nonaerated flame
Es llama no aireada *(f)*
It fiamma non aerata *(f)*
Pt chama não arejada *(f)*

Flammpunkt *(m) n* De
En flash point
Es punto de deflagración *(m)*
Fr point d'éclair *(m)*
It punto di infiammabilità *(m)*
Pt ponto de fulgor *(m)*

flash point En
De Flammpunkt *(m)*
Es punto de deflagración *(m)*
Fr point d'éclair *(m)*
It punto di infiammabilità *(m)*
Pt ponto de fulgor *(m)*

Fließbettverbrennung *(f) n* De
En fluidized-bed combustion
Es combustión de lecho fluidizado *(f)*
Fr combustion à lit fluidisé *(f)*
It combustione a letto fluidizzato *(f)*
Pt combustão de leito fluidificado *(f)*

floor insulation En
De Bodenisolierung *(f)*
Es aislamiento de suelos *(m)*
Fr isolement du plancher *(m)*
It isolamento del pavimento *(m)*
Pt isolamento de pavimento *(m)*

flowmeter *n* En
De Durchflußmesser *(m)*
Es medidor de flujo *(m)*
Fr débitmètre *(m)*
It flussometro *(m)*
Pt fluxómetro *(m)*

flüchtiger Bestandteil *(m)* De
En volatile matter
Es materia volátil *(f)*
Fr matière volatile *(f)*
It sostanza volatile *(f)*
Pt matéria volátil *(f)*

flue *n* En
De Abzug *(m)*
Es humero *(m)*
Fr carneau *(m)*
It condotto *(m)*
Pt fumeiro *(m)*

flue gas En
De Abgas *(n)*
Es gas de chimenea *(m)*
Fr gaz de carneau *(m)*
It gas della combustione *(m)*
Pt gas de fumeiro *(m)*

flueless convector En
De Konvektor ohne Abzug *(m)*
Es convector sin tubo *(m)*
Fr convecteur sans carneau *(m)*
It convettore senza condotto del fumo *(m)*
Pt convector sem fumeiro *(m)*

Flugasche *(f) n* De
En fly ash
Es ceniza en suspensión *(f)*
Fr cendres volantes *(f)*
It cenere ventilata *(f)*
Pt cinza muito fina *(f)*

Flügelrad-Windmesser *(m) n* De
En vane anemometer
Es anemómetro de catavientos *(m)*
Fr anémomètre à ailettes *(m)*
It anemometro a mulinello *(m)*
Pt anemómetro de aletas *(m)*

Flugkraftstoff *(m) n* De
En aviation fuel
Es gasolina de aviación *(f)*
Fr carburation aviation *(m)*
It combustibile per aviazione *(m)*
Pt combustível de aviação *(m)*

fluide de transfert de chaleur *(m)* Fr
De Wärmeübertragungsflüssigkeit *(f)*
En heat-transfer fluid
Es flúido de transferencia térmica *(m)*
It fluido di scambio del calore *(m)*
Pt fluido de transferência térmica *(m)*

fluide moteur *(m)* Fr
De Treibflüssigkeit *(f)*
En working fluid
Es flúido motor *(m)*
It fluido operante *(m)*
Pt fluido operativo *(m)*

fluidized-bed combustion En
De Fließbettverbrennung *(f)*
Es combustión de lecho fluidizado *(f)*
Fr combustion à lit fluidisé *(f)*
It combustione a letto fluidizzato *(f)*
Pt combustão de leito fluidificado *(f)*

fluidized solid En
De verflüssigter Feststoff *(m)*
Es sólido fluidizado *(m)*
Fr solide fluidifié *(m)*
It solido fluidizzato *(m)*
Pt sólido fluidificado *(m)*

flúido de transferencia térmica *(m)* Es
De Wärmeübertragungsflüssigkeit *(f)*
En heat-transfer fluid
Fr fluide de transfert de chaleur *(m)*
It fluido di scambio del calore *(m)*
Pt fluido de transferência térmica *(m)*

fluido de transferência térmica *(m)* Pt
De Wärmeübertragungsflüssigkeit *(f)*
En heat-transfer fluid
Es flúido de transferencia térmica *(m)*
Fr fluide de transfert de chaleur *(m)*
It fluido di scambio del calore *(m)*

fluido di scambio del calore *(m)* It
De Wärmeübertragungs-flüssigkeit *(f)*
En heat-transfer fluid
Es flúido de transferencia térmica *(m)*
Fr fluide de transfert de chaleur *(m)*
Pt fluido de transferência térmica *(m)*

flúido motor *(m)* Es
De Treibflüssigkeit *(f)*
En working fluid
Fr fluide moteur *(m)*
It fluido operante *(m)*
Pt fluido operativo *(m)*

fluido operante *(m)* It
De Treibflüssigkeit *(f)*
En working fluid
Es flúido motor *(m)*
Fr fluide moteur *(m)*
Pt fluido operativo *(m)*

fluido operativo *(m)* Pt
De Treibflüssigkeit *(f)*
En working fluid
Es flúido motor *(m)*
Fr fluide moteur *(m)*
It fluido operante *(m)*

flujo de neutrones *(m)* Es
De Neutronenfluβ *(m)*
En neutron flux
Fr flux de neutrons *(m)*
It flusso di neutroni *(m)*
Pt fluxo de neutrões *(m)*

flujo solar *(m)* Es
De Sonnenfluβ *(m)*
En solar flux
Fr flux solaire *(m)*
It flusso solare *(m)*
Pt fluxo solar *(m)*

flujo solar cenital *(m)* Es
De Zenitsonnenfluβ *(m)*
En zenith solar flux
Fr flux solaire zénithal *(m)*
It flusso solare allo zenit *(m)*
Pt fluxo solar zenital *(m)*

flujo solar horizontal *(m)* Es
De horizontaler Sonnenfluβ *(m)*
En horizontal solar flux
Fr flux solaire horizontal *(m)*
It flusso solare orizzontale *(m)*
Pt fluxo solar horizontal *(m)*

flujo solar medio *(m)* Es
De durchschnittlicher Sonnenfluβ *(m)*
En average solar flux
Fr flux solaire moyen *(m)*
It flusso solare medio *(m)*
Pt fluxo solar médio *(m)*

flujo solar urbano *(m)* Es
De Stadtsonnenfluβ *(m)*
En urban solar flux
Fr flux solaire urbain *(m)*
It flusso solare urbano *(m)*
Pt fluxo solar urbano *(m)*

flujo térmico *(m)* Es
De Wärmefluβ *(m)*
En thermal flux
Fr flux thermique *(m)*
It flusso termico *(m)*
Pt fluxo térmico *(m)*

flujo térmico terrestre *(m)* Es
De terrestrischer Wärmefluβ *(m)*
En terrestial heat flow
Fr courant de chaleur tellurique *(m)*
It flusso termico terrestre *(m)*
Pt fluxo de calor de terra *(m)*

flüssige Kohlenwasser-stoffe *(pl)* De
En liquid hydrocarbons
Es hidrocarburos líquidos *(m pl)*
Fr hydrocarbures liquides *(m pl)*
It idrocarburi liquidi *(m pl)*
Pt hidrocarbonetos líquidos *(m pl)*

flüssige Treibstoffe *(pl)* *n* De
En liquid propellants *(pl)*
Es propulsantes líquidos *(m pl)*
Fr propergols liquides *(m pl)*
It propellenti liquidi *(m pl)*
Pt combustíveis propulsores líquidos *(m pl)*

Flüssigmetall-Schnellbrüter *(m)* De
En liquid-metal fast breeder reactor (LMFBR)
Es reactor reproductor rápido de metal líquido *(m)*
Fr réacteur surrégénérateur rapide à métal liquide *(m)*
It reattore veloce autofissilizzante a metallo liquido *(m)*
Pt reactor reproductor rápido de metal líquido *(m)*

flusso di neutroni *(m)* It
De Neutronenfluβ *(m)*
En neutron flux
Es flujo de neutrones *(m)*
Fr flux de neutrons *(m)*
Pt fluxo de neutrões *(m)*

flussometro *(m)* *n* It
De Durchfluβmesser *(m)*
En flowmeter
Es medidor de flujo *(m)*
Fr débitmètre *(m)*
Pt fluxómetro *(m)*

flusso solare *(m)* It
De Sonnenfluβ *(m)*
En solar flux
Es flujo solar *(m)*
Fr flux solaire *(m)*
Pt fluxo solar *(m)*

flusso solare allo zenit *(m)* It
De Zenitsonnenfluβ *(m)*
En zenith solar flux
Es flujo solar cenital *(m)*
Fr flux solaire zénithal *(m)*
Pt fluxo solar zenital *(m)*

flusso solare medio *(m)* It
De durchschnittlicher Sonnenfluβ *(m)*
En average solar flux
Es flujo solar medio *(m)*
Fr flux solaire moyen *(m)*
Pt fluxo solar médio *(m)*

flusso solare orizzontale *(m)* It
De horizontaler Sonnenfluβ *(m)*
En horizontal solar flux
Es flujo solar horizontal *(m)*
Fr flux solaire horizontal *(m)*
Pt fluxo solar horizontal *(m)*

flusso solare urbano *(m)* It
De Stadtsonnenfluβ *(m)*
En urban solar flux
Es flujo solar urbano *(m)*
Fr flux solaire urbain *(m)*
Pt fluxo solar urbano *(m)*

flusso termico *(m)* It
De Wärmefluβ *(m)*
En thermal flux
Es flujo térmico *(m)*
Fr flux thermique *(m)*
Pt fluxo térmico *(m)*

flusso termico terrestre *(m)* It
De terrestrischer Wärmefluβ *(m)*
En terrestial heat flow
Es flujo térmico terrestre *(m)*
Fr courant de chaleur tellurique *(m)*
Pt fluxo de calor de terra *(m)*

flux de neutrons *(m)* Fr
De Neutronenfluβ *(m)*
En neutron flux
Es flujo de neutrones *(m)*
It flusso di neutroni *(m)*
Pt fluxo de neutrões *(m)*

fluxo de calor de terra *(m)* Pt
De terrestrischer Wärmefluβ *(m)*
En terrestial heat flow
Es flujo térmico terrestre *(m)*
Fr courant de chaleur tellurique *(m)*
It flusso termico terrestre *(m)*

fluxo de neutrões *(m)* Pt
De Neutronenfluβ *(m)*
En neutron flux
Es flujo de neutrones *(m)*
Fr flux de neutrons *(m)*
It flusso di neutroni *(m)*

fluxómetro *(m) n* Pt
De Durchfluβmesser *(m)*
En flowmeter
Es medidor de flujo *(m)*
Fr débitmètre *(m)*
It flussometro *(m)*

fluxo solar *(m)* Pt
De Sonnenfluβ *(m)*
En solar flux
Es flujo solar *(m)*
Fr flux solaire *(m)*
It flusso solare *(m)*

fluxo solar horizontal *(m)* Pt
De horizontaler Sonnenfluβ *(m)*
En horizontal solar flux
Es flujo solar horizontal *(m)*
Fr flux solaire horizontal *(m)*
It flusso solare orizzontale *(m)*

fluxo solar médio *(m)* Pt
De durchschnittlicher Sonnenfluβ *(m)*
En average solar flux
Es flujo solar medio *(m)*
Fr flux solaire moyen *(m)*
It flusso solare medio *(m)*

fluxo solar urbano *(m)* Pt
De Stadtsonnenfluβ *(m)*
En urban solar flux
Es flujo solar urbano *(m)*
Fr flux solaire urbain *(m)*
It flusso solare urbano *(m)*

fluxo solar zenital *(m)* Pt
De Zenitsonnenfluβ *(m)*
En zenith solar flux
Es flujo solar cenital *(m)*
Fr flux solaire zénithal *(m)*
It flusso solare allo zenit *(m)*

fluxo térmico *(m)* Pt
De Wärmefluβ *(m)*
En thermal flux
Es flujo térmico *(m)*
Fr flux thermique *(m)*
It flusso termico *(m)*

flux solaire *(m)* Fr
De Sonnenfluβ *(m)*
En solar flux
Es flujo solar *(m)*
It flusso solare *(m)*
Pt fluxo solar *(m)*

flux solaire horizontal *(m)* Fr
De horizontaler Sonnenfluβ *(m)*
En horizontal solar flux
Es flujo solar horizontal *(m)*
It flusso solare orizzontale *(m)*
Pt fluxo solar horizontal *(m)*

flux solaire moyen *(m)* Fr
De durchschnittlicher Sonnenfluβ *(m)*
En average solar flux
Es flujo solar medio *(m)*
It flusso solare medio *(m)*
Pt fluxo solar médio *(m)*

flux solaire urbain *(m)* Fr
De Stadtsonnenfluβ *(m)*
En urban solar flux
Es flujo solar urbano *(m)*
It flusso solare urbano *(m)*
Pt fluxo solar urbano *(m)*

flux solaire zénithal *(m)* Fr
De Zenitsonnenfluβ *(m)*
En zenith solar flux
Es flujo solar cenital *(m)*
It flusso solare allo zenit *(m)*
Pt fluxo solar zenital *(m)*

flux thermique *(m)* Fr
De Wärmefluβ *(m)*
En thermal flux
Es flujo térmico *(m)*
It flusso termico *(m)*
Pt fluxo térmico *(m)*

fly ash En
De Flugasche *(f)*
Es ceniza en suspensión *(f)*
Fr cendres volantes *(f)*
It cenere ventilata *(f)*
Pt cinza muito fina *(f)*

flywheel *n* En
De Schwungrad *(n)*
Es volante *(m)*
Fr volant *(m)*
It volano *(m)*
Pt volante de inércia *(m)*

focolaio *(m) n* It
De Feuerkiste *(f)*
En firebox
Es hogar *(m)*
Fr foyer *(m)*
Pt fornalha *(f)*

foguetão nuclear *(m)* Pt
De Kernkraftrakete *(f)*
En nuclear rocket
Es cohete nuclear *(m)*
Fr fusée nucléaire *(f)*
It razzo nucleare *(m)*

fonctionnement sous charge partielle *(m)* Fr
De Teillastbetrieb *(m)*
En part-load operation
Es funcionamiento con carga reducida *(m)*
It funzionamento a carico parziale *(m)*
Pt operação de carga parcial *(f)*

fonte *(f) n* Pt
De Quelle *(f)*
En source
Es fuente *(f)*
Fr source *(f)*
It sorgente *(f)*

fonte de neutrões *(f)* Pt
De Neutronenquelle *(f)*
En neutron source
Es fuente de neutrones *(f)*
Fr source de neutrons *(f)*
It fonte di neutroni *(f)*

fonte di neutroni *(f)* It
De Neutronenquelle *(f)*
En neutron source
Es fuente de neutrones *(f)*
Fr source de neutrons *(f)*
Pt fonte de neutrões *(f)*

fontes de água quente *(f pl)* Pt
De heiβe Quellen *(pl)*
En hot springs *(pl)*
Es termas *(f pl)*
Fr sources chaudes *(f pl)*
It sorgenti calde *(f pl)*

fora da praia Pt
De auf See
En offshore
Es marino
Fr au large
It al largo

forage *(m) n* Fr
De Bohren *(n)*
En drilling
Es perforación *(f)*
It trivellazione *(f)*
Pt perfuração *(f)*

forage à l'air comprimé *(m)* Fr
De Druckluftbohren *(n)*
En air drilling
Es perforación por aire comprimido *(f)*
It trapanatura ad aria compressa *(f)*
Pt perfuração pneumática *(f)*

forage au câble *(m)* Fr
De Seilbohren *(n)*
En cable-tool drilling
Es perforación a cable *(f)*
It trivellazione con utensile a fune *(f)*
Pt perfuração a cabo *(f)*

forage dirigé *(m)* Fr
De gerichtetes Bohren *(n)*
En directional drilling
Es perforación direccional *(f)*
It trivellazione direzionale *(f)*
Pt perfuração direccional *(f)*

forage rotary *(m)* Fr
De Drehbohren *(n)*
En rotary drilling
Es perforación rotativa *(f)*
It sondaggio a rotazione *(m)*
Pt perfuração rotativa *(f)*

força *(f)* *n* Pt
De Kraft *(f)*
En force
Es fuerza *(f)*
Fr force *(f)*
It forza *(f)*

força centrífuga *(f)* Pt
De Zentrifugalkraft *(f)*
En centrifugal force
Es fuerza contrífuga *(f)*
Fr force centrifuge *(f)*
It forza centrifuga *(f)*

força centrípeda *(f)* Pt
De Zentripetalkraft *(f)*
En centripetal force
Es fuerza centrípeta *(f)*
Fr force centripète *(f)*
It forza centripeta *(f)*

força contraelectromotriz *(f)* Pt
De Gegenelektromotorische Kraft *(f)*
En back electromotive force
Es fuerza contraelectromotriz *(f)*
Fr force contreélectromotrice *(f)*
It forza controelettromotrice *(f)*

força de gravidade *(f)* Pt
De Erdanziehungskraft *(f)*
En gravitational force
Es fuerza gravitatoria *(f)*
Fr force de gravitation *(f)*
It forza di gravità *(f)*

força electromotora (f.e.m.) *(f)* Pt
De elektromotorische Kraft (EMK) *(f)*
En electromotive force (e.m.f.)
Es fuerza electromotriz (f.e.m.) *(f)*
Fr force électromotrice (f.é.m.) *(f)*
It forza elettromotrice (f.e.m.) *(f)*

força electromotora induzida *(f)* Pt
De induzierte elektromotorische Kraft *(f)*
En induced electromotive force
Es fuerza electromotriz inducida *(f)*
Fr force électromotrice induite *(f)*
It forza elettromotrice indotta *(f)*

força magnética *(f)* Pt
De Magnetkraft *(n)*
En magnetic force
Es fuerza magnética *(f)*
Fr force magnétique *(f)*
It forza magnetica *(f)*

force *n* En; Fr *(f)*
De Kraft *(f)*
Es fuerza *(f)*
It forza *(f)*
Pt força *(f)*

force centrifuge *(f)* Fr
De Zentrifugalkraft *(f)*
En centrifugal force
Es fuerza contrífuga *(f)*
It forza centrifuga *(f)*
Pt força centrífuga *(f)*

force centripète *(f)* Fr
De Zentripetalkraft *(f)*
En centripetal force
Es fuerza centrípeta *(f)*
It forza centripeta *(f)*
Pt força centrípeda *(f)*

force contreélectromotrice *(f)* Fr
De Gegenelektromotorische Kraft *(f)*
En back electromotive force
Es fuerza contraelectromotriz *(f)*
It forza controelettromotrice *(f)*
Pt força contraelectromotriz *(f)*

forced convection En
De Gebläsekonvektion *(f)*
Es convección forzada *(f)*
Fr convection forcée *(f)*
It convezione forzata *(f)*
Pt convecção forçada *(f)*

force de gravitation *(f)* Fr
De Erdanziehungskraft *(f)*
En gravitational force
Es fuerza gravitatoria *(f)*
It forza di gravità *(f)*
Pt força de gravidade *(f)*

force électromotrice (f.é.m.) *(f)* Fr
De elektromotorische Kraft (EMK) *(f)*
En electromotive force (e.m.f.)
Es fuerza electromotriz (f.e.m.) *(f)*
It forza elettromotrice (f.e.m.) *(f)*
Pt força electromotora (f.e.m.) *(f)*

force électromotrice induite *(f)* Fr
De induzierte elektromotorische Kraft *(f)*
En induced electromotive force
Es fuerza electromotriz inducida *(f)*
It forza elettromotrice indotta *(f)*
Pt força electromotora induzida *(f)*

force magnétique *(f)* Fr
De Magnetkraft *(n)*
En magnetic force
Es fuerza magnética *(f)*
It forza magnetica *(f)*
Pt força magnética *(f)*

formación de turba *(f)* Es
De Vertorfung *(f)*
En peatification
Fr formation de la tourbe *(f)*
It produzione di torba *(f)*
Pt turfização *(f)*

formación hullera *(f)* Es
De Kohlenlager *(n)*
En coal measure
Fr couche de houille *(f)*
It giacimento carbonifero *(m)*
Pt medida de carvão *(f)*

forma de onda *(f)* Es, Pt
De Wellenform *(f)*
En waveform
Fr forme d´onde *(f)*
It forma d´onda *(f)*

forma d'onda *(f)* It
De Wellenform *(f)*
En waveform
Es forma de onda *(f)*
Fr forme d´onde *(f)*
Pt forma de onda *(f)*

formation de la tourbe *(f)* Fr
De Vertorfung *(f)*
En peatification
Es formación de turba *(f)*
It produzione di torba *(f)*
Pt turfização *(f)*

formed coke En
De geformter Koks *(m)*
Es coque conformado *(m)*
Fr coke formé *(m)*
It coke formato *(m)*
Pt coque formado *(m)*

forme d'onde *(f)* Fr
De Wellenform *(f)*
En waveform
Es forma de onda *(f)*
It forma d´onda *(f)*
Pt forma de onda *(f)*

fornalha *(f) n* Pt
De Feuerkiste *(f)*
En firebox
Es hogar *(m)*
Fr foyer *(m)*
It focolaio *(m)*

fornello solare *(m)* It
De Solarherd *(m)*
En solar cooker
Es cocina solar *(f)*
Fr fourneau solaire *(m)*
Pt fervedor solar *(m)*

forno *(m) n* It
De Brennofen; Ofen *(m)*
En furnace; kiln; oven
Es estufa *(f)*; horno *(m)*
Fr four *(m)*
Pt forno; forno de requeima *(m)*

forno *(m) n* Pt
De Ofen *(m)*
En furnace; oven
Es estufa *(f)*; horno *(m)*
Fr four *(m)*
It forno *(m)*

forno a arco voltáico *(m)* Pt
De Lichtbogenofen *(m)*
En arc furnace
Es horno de arco *(m)*
Fr four à arc *(m)*
It forno ad arco *(m)*

forno a ciclone *(m)* It
De Wirbelofen *(m)*
En cyclone furnace
Es hogar de turbulencia *(m)*
Fr four à cyclone *(m)*
Pt forno de ciclone *(m)*

forno a crogiolo *(m)* It
De Tiegelofen *(m)*
En crucible furnace
Es horno de crisol *(m)*
Fr fourneau à creuset *(m)*
Pt forno de cadinho *(m)*

forno ad arco *(m)* It
De Lichtbogenofen *(m)*
En arc furnace
Es horno de arco *(m)*
Fr four à arc *(m)*
Pt forno a arco voltáico *(m)*

forno ad arco a depressione *(m)* It
De Vakuum-Lichtbogenofen *(m)*
En vacuum arc furnace
Es horno de arco eléctrico en vacío *(m)*
Fr four à arc dans le vide *(m)*
Pt forno de arco voltáico em vácuo *(m)*

forno ad induzione a canale *(m)* It
De Kanalinduktionsofen *(m)*
En channel induction furnace
Es horno de inducción de canal *(m)*
Fr four à induction à chenal *(m)*
Pt forno de indução em canal *(m)*

forno ad induzione depressione *(m)* It
De Vakuuminduktions-ofen *(m)*
En vacuum induction furnace
Es horno de inducción en vacío *(m)*
Fr four à induction à vide *(m)*
Pt forno de indução em vácuo *(m)*

forno a muffola *(m)* It
De Muffelofen *(m)*
En muffle furnace
Es horno de mufla *(m)*
Fr four à moufle *(m)*
Pt forno com câmara *(m)*

forno a riscaldamento radiante *(m)* It
De Strahlungsheizofen *(m)*
En radiant-heating furnace
Es horno de calentamiento por radiación *(m)*
Fr four à chauffage radiant *(m)*
Pt forno de aquecimento por radiação *(m)*

forno a rubo radiante *(m)* It
De Strahlungsröhren-ofen *(m)*
En radiant-tube furnace
Es horno de tubo radiante *(m)*
Fr four à tube radiant *(m)*
Pt forno de tubo de radiação *(m)*

forno a suola *(m)* It
De Herdofen *(m)*
En hearth furnace
Es horno de forja *(m)*
Fr four à sole *(m)*
Pt forno de fornalha *(m)*

forno a tino *(m)* It
De Schachtofen *(m)*
En shaft furnace
Es horno de cuba *(m)*
Fr four à cuve *(m)*
Pt forno de chaminé *(m)*

forno com câmara *(m)* Pt
De Muffelofen *(m)*
En muffle furnace
Es horno de mufla *(m)*
Fr four à moufle *(m)*
It forno a muffola *(m)*

forno de aquecimento por radiação *(m)* Pt
De Strahlungsheizofen *(m)*
En radiant-heating furnace
Es horno de calentamiento por radiación *(m)*
Fr four à chauffage radiant *(m)*
It forno a riscaldamento radiante *(m)*

forno de arco voltáico em vácuo *(m)* Pt
De Vakuum-Lichtbogenofen *(m)*
En vacuum arc furnace
Es horno de arco eléctrico en vacío *(m)*
Fr four à arc dans le vide *(m)*
It forno ad arco a depressione *(m)*

forno de cadinho *(m)* Pt
De Tiegelofen *(m)*
En crucible furnace
Es horno de crisol *(m)*
Fr fourneau à creuset *(m)*
It forno a crogiolo *(m)*

forno de chaminé *(m)* Pt
De Schachtofen *(m)*
En shaft furnace
Es horno de cuba *(m)*
Fr four à cuve *(m)*
It forno a tino *(m)*

forno de ciclone *(m)* Pt
De Wirbelofen *(m)*
En cyclone furnace
Es hogar de turbulencia *(m)*
Fr four à cyclone *(m)*
It forno a ciclone *(m)*

forno de cinze seca *(m)* Pt
De Trockenasche-Ofen *(m)*
En dry-ash furnace
Es hogar de cenizas pulverulentas *(m)*
Fr four à cendres sèches *(m)*
It forno di cenere secca *(m)*

forno de fornalha *(m)* Pt
De Herdofen *(m)*
En hearth furnace
Es horno de forja *(m)*
Fr four à sole *(m)*
It forno a suola *(m)*

forno de fornalha aberta *(m)* Pt
De Siemens-Martinofen *(m)*
En open-hearth furnace
Es horno Martin-Siemens *(m)*
Fr four Martin *(m)*
It forno Martin *(m)*

forno de indução em canal *(m)* Pt
De Kanalinduktionsofen *(m)*
En channel induction furnace
Es horno de inducción de canal *(m)*
Fr four à induction à chenal *(m)*

It forno ad induzione a canale *(m)*

forno de indução em vácuo *(m)* Pt
De Vakuuminduktionsofen *(m)*
En vacuum induction furnace
Es horno de inducción en vacío *(m)*
Fr four à induction à vide *(m)*
It forno ad induzione depressione *(m)*

forno de lingotes *(m)* Pt
De Blockofen *(m)*
En billet furnace
Es horno de palanquillas *(m)*
Fr four à billettes *(m)*
It forno per billette *(m)*

forno de requeima *(m)* Pt
De Brennofen *(m)*
En kiln
Es estufa *(f)*
Fr four *(m)*
It forno *(m)*

forno de tubo de radiação *(m)* Pt
De Strahlungsröhrenofen *(m)*
En radiant-tube furnace
Es horno de tubo radiante *(m)*
Fr four à tube radiant *(m)*
It forno a rubo radiante *(m)*

forno di cenere secca *(m)* It
De Trockenasche-Ofen *(m)*
En dry-ash furnace
Es hogar de cenizas pulverulentas *(m)*
Fr four à cendres sèches *(m)*
Pt forno de cinze seca *(m)*

forno di riscaldo *(m)* It
De Nachbrennofen *(m)*
En reheating furnace
Es horno de recalentamiento *(m)*
Fr four à réchauffer *(m)*
Pt forno reaquecedor *(m)*

forno eléctrico *(m)* Pt
De Elektroofen *(m)*
En electric furnace
Es horno eléctrico *(m)*
Fr four électrique *(m)*
It forno elettrico *(m)*

forno eléctrico em vácuo *(m)* Pt
De Vakuum-Elektroofen *(m)*
En vacuum electric furnace
Es horno eléctrico en vacío *(m)*
Fr four électrique à vide *(m)*
It forno elettrico a depressione *(m)*

forno elettrico *(m)* It
De Elektroofen *(m)*
En electric furnace
Es horno eléctrico *(m)*
Fr four électrique *(m)*
Pt forno eléctrico *(m)*

forno elettrico a depressione *(m)* It
De Vakuum-Elektroofen *(m)*
En vacuum electric furnace
Es horno eléctrico en vacío *(m)*
Fr four électrique à vide *(m)*
Pt forno eléctrico em vácuo *(m)*

forno Martin *(m)* It
De Siemens-Martinofen *(m)*
En open-hearth furnace
Es horno Martin-Siemens *(m)*
Fr four Martin *(m)*
Pt forno de fornalha aberta *(m)*

forno per billette *(m)* It
De Blockofen *(m)*
En billet furnace
Es horno de palanquillas *(m)*
Fr four à billettes *(m)*
Pt forno de lingotes *(m)*

forno reaquecedor *(m)* Pt
De Nachbrennofen *(m)*
En reheating furnace
Es horno de recalentamiento *(m)*
Fr four à réchauffer *(m)*
It forno di riscaldo *(m)*

forno solar *(m)* Pt
De Sonnenofen *(m)*
En solar furnace
Es horno solar *(m)*
Fr four solaire *(m)*
It forno solare *(m)*

forno solare *(m)* It
De Sonnenofen *(m)*
En solar furnace
Es horno solar *(m)*
Fr four solaire *(m)*
Pt forno solar *(m)*

fortgeschrittener gasgekühlter Reaktor *(m)* De
En advanced gas-cooled reactor (AGR)
Es reactor avanzado refrigerado por gas *(m)*
Fr reacteur avancé à refroidissement au gaz *(m)*
It reattore avanzato raffreddato a gas *(m)*
Pt reactor avançado arrefecido a gás *(m)*

forza *(f)* *n* It
De Kraft *(f)*
En force
Es fuerza *(f)*
Fr force *(f)*
Pt força *(f)*

forza centrifuga *(f)* It
De Zentrifugalkraft *(f)*
En centrifugal force
Es fuerza contrífuga *(f)*
Fr force centrifuge *(f)*
Pt força centrífuga *(f)*

forza centripeta *(f)* It
De Zentripetalkraft *(f)*
En centripetal force
Es fuerza centrípeta *(f)*
Fr force centripète *(f)*
Pt força centrípeda *(f)*

forza controelettromotrice *(f)* It
De Gegenelektromotorische Kraft *(f)*
En back electromotive force
Es fuerza contraelectromotriz *(f)*
Fr force contreélectromotrice *(f)*
Pt força contraelectromotriz *(f)*

forza di gravità *(f)* It
De Erdanziehungskraft *(f)*
En gravitational force
Es fuerza gravitatoria *(f)*
Fr force de gravitation *(f)*
Pt força de gravidade *(f)*

forza elettromotrice (f.e.m.) *(f)* It
De elektromotorische Kraft (EMK) *(f)*
En electromotive force (e.m.f.)
Es fuerza electromotriz (f.e.m.) *(f)*
Fr force électromotrice (f.é.m.) *(f)*
Pt força electromotora (f.e.m.) *(f)*

forza elettromotrice indotta *(f)* It
De induzierte elektromotorische Kraft *(f)*
En induced electromotive force
Es fuerza electromotriz inducida *(f)*
Fr force électromotrice induite *(f)*
Pt força electromotora induzida *(f)*

forza magnetica *(f)* It
De Magnetkraft *(n)*
En magnetic force
Es fuerza magnética *(f)*
Fr force magnétique *(f)*
Pt força magnética *(f)*

fossiler Brennstoff *(m)* De
En fossil fuel

Es combustible fósil *(m)*
Fr combustible fossile *(m)*
It combustibile fossile *(m)*
Pt combustível fóssil *(m)*

fossil fuel En
De fossiler Brennstoff *(m)*
Es combustible fósil *(m)*
Fr combustible fossile *(m)*
It combustibile fossile *(m)*
Pt combustível fóssil *(m)*

fotão *(m) n* Pt
De Photon *(n)*
En photon
Es fotón *(m)*
Fr photon *(m)*
It fotone *(m)*

fotocélula *(f) n* Es, Pt
De Photozelle *(f)*
En photocell
Fr cellule photoélectrique *(f)*
It cellula fotoelettrica *(f)*

fotoconductor *adj* Es
De photoleitend
En photoconductive
Fr photoconducteur
It fotoconduttivo
Pt fotocondutor

fotocondutor *adj* Pt
De photoleitend
En photoconductive
Es fotoconductor
Fr photoconducteur
It fotoconduttivo

fotoconduttivo *adj* It
De photoleitend
En photoconductive
Es fotoconductor
Fr photoconducteur
Pt fotocondutor

fotoeléctrico *adj* Es, Pt
De photoelektrisch
En photoelectric
Fr photoélectrique
It fotoelettrico

fotoelettrico *adj* It
De photoelektrisch
En photoelectric
Es fotoeléctrico
Fr photoélectrique
Pt fotoeléctrico

fotón *(m) n* Es
De Photon *(n)*
En photon
Fr photon *(m)*
It fotone *(m)*
Pt fotão *(m)*

fotone *(m) n* It
De Photon *(n)*
En photon
Es fotón *(m)*
Fr photon *(m)*
Pt fotão *(m)*

fotosíntese *(m) n* Pt
De Photosynthese *(f)*
En photosynthesis
Es fotosíntesis *(f)*
Fr photosynthèse *(f)*
It fotosintesi *(f)*

fotosintesi *(f) n* It
De Photosynthese *(f)*
En photosynthesis
Es fotosíntesis *(f)*
Fr photosynthèse *(f)*
Pt fotosíntese *(m)*

fotosíntesis *(f) n* Es
De Photosynthese *(f)*
En photosynthesis
Fr photosynthèse *(f)*
It fotosintesi *(f)*
Pt fotosíntese *(m)*

fotovoltaico *adj* Es, It
De photovoltaisch
En photovoltaic
Fr photovoltaïque
Pt fotovoltáico

fotovoltáico *adj* Pt
De photovoltaisch
En photovoltaic
Es fotovoltaico
Fr photovoltaïque
It fotovoltaico

fouling *n* En
De Verstopfung *(f)*
Es incrustación *(f)*
Fr encrassement *(m)*
It incrostazione *(f)*
Pt incrustação *(m)*

four *(m) n* Fr
De Brennofen; Ofen *(m)*
En furnace; kiln; oven
Es estufa *(f)*; horno *(m)*
It forno *(m)*
Pt forno; forno de requeima *(m)*

four à arc *(m)* Fr
De Lichtbogenofen *(m)*
En arc furnace
Es horno de arco *(m)*
It forno ad arco *(m)*
Pt forno a arco voltáico *(m)*

four à arc dans le vide *(m)* Fr
De Vakuum-Lichtbogenofen *(m)*
En vacuum arc furnace
Es horno de arco eléctrico en vacío *(m)*
It forno ad arco a depressione *(m)*
Pt forno de arco voltáico em vácuo *(m)*

four à billettes *(m)* Fr
De Blockofen *(m)*
En billet furnace
Es horno de palanquillas *(m)*
It forno per billette *(m)*
Pt forno de lingotes *(m)*

four à cendres sèches *(m)* Fr
De Trockenasche-Ofen *(m)*
En dry-ash furnace
Es hogar de cenizas pulverulentas *(m)*
It forno di cenere secca *(m)*
Pt forno de cinze seca *(m)*

four à chauffage radiant *(m)* Fr
De Strahlungsheizofen *(m)*
En radiant-heating furnace
Es horno de calentamiento por radiación *(m)*
It forno a riscaldamento radiante *(m)*
Pt forno de aquecimento por radiação *(m)*

four à cuve *(m)* Fr
De Schachtofen *(m)*
En shaft furnace
Es horno de cuba *(m)*
It forno a tino *(m)*
Pt forno de chaminé *(m)*

four à cyclone *(m)* Fr
De Wirbelofen *(m)*
En cyclone furnace
Es hogar de turbulencia *(m)*
It forno a ciclone *(m)*
Pt forno de ciclone *(m)*

four à induction à chenal *(m)* Fr
De Kanalinduktionsofen *(m)*
En channel induction furnace
Es horno de inducción de canal *(m)*
It forno ad induzione a canale *(m)*
Pt forno de indução em canal *(m)*

four à induction à vide *(m)* Fr
De Vakuuminduktions-ofen *(m)*
En vacuum induction furnace
Es horno de inducción en vacío *(m)*
It forno ad induzione depressione *(m)*
Pt forno de indução em vácuo *(m)*

four à moufle *(m)* Fr
De Muffelofen *(m)*
En muffle furnace
Es horno de mufla *(m)*
It forno a muffola *(m)*
Pt forno com câmara *(m)*

four à réchauffer *(m)* Fr
De Nachbrennofen *(m)*
En reheating furnace
Es horno de recalentamiento *(m)*
It forno di riscaldo *(m)*

Pt forno reaquecedor *(m)*

four à sole *(m)* Fr
De Herdofen *(m)*
En hearth furnace
Es horno de forja *(m)*
It forno a suola *(m)*
Pt forno de fornalha *(m)*

four à tube radiant *(m)* Fr
De Strahlungsröhrenofen *(m)*
En radiant-tube furnace
Es horno de tubo radiante *(m)*
It forno a rubo radiante *(m)*
Pt forno de tubo de radiação *(m)*

four électrique *(m)* Fr
De Elektroofen *(m)*
En electric furnace
Es horno eléctrico *(m)*
It forno elettrico *(m)*
Pt forno eléctrico *(m)*

four électrique à vide *(m)* Fr
De Vakuum-Elektroofen *(m)*
En vacuum electric furnace
Es horno eléctrico en vacío *(m)*
It forno elettrico a depressione *(m)*
Pt forno eléctrico em vácuo *(m)*

four Martin *(m)* Fr
De Siemens-Martinofen *(m)*
En open-hearth furnace
Es horno Martin-Siemens *(m)*
It forno Martin *(m)*
Pt forno de fornalha aberta *(m)*

fourneau à creuset *(m)* Fr
De Tiegelofen *(m)*
En crucible furnace
Es horno de crisol *(m)*
It forno a crogiolo *(m)*
Pt forno de cadinho *(m)*

fourneau solaire *(m)* Fr
De Solarherd *(m)*
En solar cooker
Es cocina solar *(f)*
It fornello solare *(m)*
Pt fervedor solar *(m)*

four solaire *(m)* Fr
De Sonnenofen *(m)*
En solar furnace
Es horno solar *(m)*
It forno solare *(m)*
Pt forno solar *(m)*

four-stroke engine En
De Viertaktmotor *(m)*
Es motor de cuatro tiempos *(m)*
Fr moteur à quatre temps *(m)*
It motore a quattro tempi *(m)*
Pt motor a quatro tempos *(m)*

foyer *(m) n* Fr
De Feuerkiste *(f)*
En firebox
Es hogar *(m)*
It focolaio *(m)*
Pt fornalha *(f)*

fracção *(f) n* Pt
De Fraktion *(f)*
En fraction
Es fracción *(f)*
Fr fraction *(f)*
It frazione *(f)*

fracción *(f) n* Es
De Fraktion *(f)*
En fraction
Fr fraction *(f)*
It frazione *(f)*
Pt fracção *(f)*

fracciónes de bajo punto de ebullición *(f pl)* Es
De Tiefsiedepunkt-fraktionen *(pl)*
En low-boiling fractions *(pl)*
Fr fractions à bas point d'ébullition *(f pl)*
It frazioni a basso punto di ebollizione *(f pl)*
Pt fracções de baixo ponto de ebulição *(f pl)*

fracções de baixo ponto de ebulição *(f pl)* Pt
De Tiefsiedepunkt-fraktionen *(pl)*
En low-boiling fractions *(pl)*
Es fracciónes de bajo punto de ebullición *(f pl)*
Fr fractions à bas point d'ébullition *(f pl)*
It frazioni a basso punto di ebollizione *(f pl)*

fraction *n* En; Fr *(f)*
De Fraktion *(f)*
Es fracción *(f)*
It frazione *(f)*
Pt fracção *(f)*

fractional distillation En
De fraktionierte Destillation *(f)*
Es destilación fraccionada *(f)*
Fr distillation fractionnée *(f)*
It distillazione a frazione *(f)*
Pt destilação fraccional *(f)*

fractions à bas point d'ébullition *(f pl)* Fr
De Tiefsiedepunkt-fraktionen *(pl)*
En low-boiling fractions *(pl)*
Es fracciónes de bajo punto de ebullición *(f pl)*
It frazioni a basso punto di ebollizione *(f pl)*
Pt fracções de baixo ponto de ebulição *(f pl)*

fracturação hidráulica *(f)* Pt
De hydraulische Rißbildung *(f)*
En hydraulic fracturing
Es agrietamiento hidráulico *(m)*
Fr fracturation hydraulique *(f)*
It fratturazione idraulica *(f)*

fracturation hydraulique *(f)* Fr
De hydraulische Rißbildung *(f)*
En hydraulic fracturing
Es agrietamiento hidráulico *(m)*
It fratturazione idraulica *(f)*
Pt fracturação hidráulica *(f)*

frais de distribution *(m)* Fr
De Vertriebskosten *(pl)*
En distribution cost
Es coste de distribución *(m)*
It costo di distribuzione *(m)*
Pt custo de distribuição *(m)*

frais d'exploitation *(m pl)* Fr
De Betriebskosten *(pl)*
En running costs
Es costes de explotación *(m pl)*
It costi di esercizio *(m pl)*
Pt custo de exploração *(m)*

Fraktion *(f) n* De
En fraction
Es fracción *(f)*
Fr fraction *(f)*
It frazione *(f)*
Pt fracção *(f)*

fraktionierte Destillation *(f)* De
En fractional distillation
Es destilación fraccionada *(f)*
Fr distillation fractionnée *(f)*
It distillazione a frazione *(f)*
Pt destilação fraccional *(f)*

fratturazione idraulica *(f)* It
De hydraulische Rißbildung *(f)*
En hydraulic fracturing
Es agrietamiento hidráulico *(m)*
Fr fracturation hydraulique *(f)*

Pt fracturação hidráulica *(f)*

frazione *(f) n* It
De Fraktion *(f)*
En fraction
Es fracción *(f)*
Fr fraction *(f)*
Pt fracção *(f)*

frazioni a basso punto di ebollizione *(f pl)* It
De Tiefsiedepunktfraktionen *(pl)*
En low-boiling fractions *(pl)*
Es fracciónes de bajo punto de ebullición *(f pl)*
Fr fractions à bas point d'ébullition *(f pl)*
Pt fracções de baixo ponto de ebulição *(f pl)*

frecuencia *(f) n* Es
De Frequenz *(f)*
En frequency
Fr fréquence *(f)*
It frequenza *(f)*
Pt frequência *(f)*

frecuencia ultraelevada *(f)* Es
De Ultrahochfrequenz *(f)*
En ultrahigh frequency (UHF)
Fr hyperfréquence *(f)*
It frequenza ultraelevata *(f)*
Pt frequência ultra-elevada *(f)*

free energy En
De freigesetzte Energie *(f)*
Es energía libre *(f)*
Fr énergie libre *(f)*
It energia libera *(f)*
Pt energia livre *(f)*

freeze drying En
De Gefriertrocknung *(f)*
Es liofilización *(f)*
Fr lyophilisation *(f)*
It liofilizzazione *(f)*
Pt secagem por congelação *(f)*

freeze protection En
De Gefrierschutz *(m)*
Es protección contra la congelación *(f)*
Fr protection contre le gel *(f)*
It protezione dal congelamento *(f)*
Pt protecção antigelo *(f)*

freezing point En
De Gefrierpunkt *(m)*
Es punto de congelación *(m)*
Fr point de congélation *(m)*
It punto di congelamento *(m)*
Pt ponto de congelação *(m)*

freigesetzte Energie *(f)* De
En free energy
Es energía libre *(f)*
Fr énergie libre *(f)*
It energia libera *(f)*
Pt energia livre *(f)*

freinage à récupération *(m)* Fr
De Rückstromgewinnbremsung *(f)*
En regenerative braking
Es frenado de recuperación *(m)*
It frenatura rigenerativo *(m)*
Pt quebra por regeneração *(f)*

freinage par résistance *(m)* Fr
De Widerstandsbremsung *(f)*
En rheostatic braking
Es frenado reostático *(m)*
It frenatura reostatica *(f)*
Pt quebra reostática *(f)*

frenado de recuperación *(m)* Es
De Rückstromgewinnbremsung *(f)*
En regenerative braking
Fr freinage à récupération *(m)*
It frenatura rigenerativo *(m)*
Pt quebra por regeneração *(f)*

frenado reostático *(m)* Es
De Widerstandsbremsung *(f)*
En rheostatic braking
Fr freinage par résistance *(m)*
It frenatura reostatica *(f)*
Pt quebra reostática *(f)*

frenatura reostatica *(f)* It
De Widerstandsbremsung *(f)*
En rheostatic braking
Es frenado reostático *(m)*
Fr freinage par résistance *(m)*
Pt quebra reostática *(f)*

frenatura rigenerativo *(m)* It
De *(f)*
En regenerative braking
Es frenado de recuperación *(m)*
Fr freinage à récupération *(m)*
Pt quebra por regeneração *(f)*

frente de onda *(m)* Es
De Wellenfront *(f)*
En wavefront
Fr front d'onde *(m)*
It fronte d'onda *(f)*
Pt frente ondulatória *(f)*

frente ondulatória *(f)* Pt
De Wellenfront *(f)*
En wavefront
Es frente de onda *(m)*
Fr front d'onde *(m)*
It fronte d'onda *(f)*

freon *n* En; It, Pt *(m)*
De Freon *(n)*
Es freón *(m)*
Fr fréon *(m)*

Freon *(n) n* De
En freon
Es freón *(m)*
Fr fréon *(m)*
It freon *(m)*
Pt freon *(m)*

freón *(m) n* Es
De Freon *(n)*
En freon
Fr fréon *(m)*
It freon *(m)*
Pt freon *(m)*

fréon *(m) n* Fr
De Freon *(n)*
En freon
Es freón *(m)*
It freon *(m)*
Pt freon *(m)*

fréquence *(f) n* Fr
De Frequenz *(f)*
En frequency
Es frecuencia *(f)*
It frequenza *(f)*
Pt frequência *(f)*

frequência *(f) n* Pt
De Frequenz *(f)*
En frequency
Es frecuencia *(f)*
Fr fréquence *(f)*
It frequenza *(f)*

frequência rádio *(f)* Pt
De Radiofrequenz *(f)*
En radiofrequency
Es radiofrecuencia *(f)*
Fr radiofréquence *(f)*
It radiofrequenza *(f)*

frequência ultra-elevada *(f)* Pt
De Ultrahochfrequenz *(f)*
En ultrahigh frequency (UHF)
Es frecuencia ultraelevada *(f)*
Fr hyperfréquence *(f)*
It frequenza ultraelevata *(f)*

frequency *n* En
De Frequenz *(f)*
Es frecuencia *(f)*
Fr fréquence *(f)*
It frequenza *(f)*
Pt frequência *(f)*

Frequenz *(f) n* De
En frequency
Es frecuencia *(f)*
Fr fréquence *(f)*

It frequenza *(f)*
Pt frequência *(f)*

frequenza *(f) n* It
De Frequenz *(f)*
En frequency
Es frecuencia *(f)*
Fr fréquence *(f)*
Pt frequência *(f)*

frequenza ultraelevata *(f)* It
De Ultrahochfrequenz *(f)*
En ultrahigh frequency (UHF)
Es frecuencia ultraelevada *(f)*
Fr hyperfréquence *(f)*
Pt frequência ultra-elevada *(f)*

frittage *(m) n* Fr
De Sintern *(n)*
En sintering
Es sinterización *(f)*
It sinterizzazione *(f)*
Pt concreção *(f)*

front d'onde *(m)* Fr
De Wellenfront *(f)*
En wavefront
Es frente de onda *(m)*
It fronte d'onda *(f)*
Pt frente ondulatória *(f)*

fronte d'onda *(f)* It
De Wellenfront *(f)*
En wavefront
Es frente de onda *(m)*
Fr front d'onde *(m)*
Pt frente ondulatória *(f)*

fuel *n* En
De Brennstoff *(m)*
Es combustible *(m)*
Fr combustible *(m)*
It combustibile *(m)*
Pt combustível *(m)*

fuel cell En
De Brennstoffzelle *(f)*
Es célula de combustible *(f)*
Fr pile à combustible *(f)*
It pila a combustibile *(f)*
Pt célula de combustível *(f)*

fuel channel En
De Brennstoffkanal *(m)*
Es canal de combustible *(m)*
Fr canal de combustible *(m)*
It canale del combustibile *(m)*
Pt canal de combustível *(m)*

fuel consumption En
De Brennstoffverbrauch *(m)*
Es consumo de combustible *(m)*
Fr consommation de combustible *(f)*
It consumo di combustibile *(m)*
Pt consumo de combustível *(m)*

fuel cycle En
De Brennstoffzyklus *(m)*
Es ciclo de los combustibles *(m)*
Fr cycle du combustible *(m)*
It ciclo del combustibile *(m)*
Pt ciclo de combustível *(m)*

fuel element En
De Brennelement *(n)*
Es elemento combustible *(m)*
Fr élément combustible *(m)*
It elemento combustibile *(m)*
Pt elemento combustível *(m)*

fuel oil En; Es, Pt *(m)*
De Treiböl *(n)*
Fr mazout *(m)*
It nafta *(f)*

fuel-oil de destilados *(m)* Pt
De Destillatöl *(n)*
En distillate fuel oil
Es fuel-oil destilado *(m)*
Fr fuel-oil distillé *(m)*
It nafta distillata *(f)*

fuel-oil destilado *(m)* Es
De Destillatöl *(n)*
En distillate fuel oil
Fr fuel-oil distillé *(m)*
It nafta distillata *(f)*
Pt fuel-oil de destilados *(m)*

fuel-oil distillé *(m)* Fr
De Destillatöl *(n)*
En distillate fuel oil
Es fuel-oil destilado *(m)*
It nafta distillata *(f)*
Pt fuel-oil de destilados *(m)*

fuel-oil para queimadores *(m)* Pt
De Brenner-Treiböl *(n)*
En burner fuel-oil
Es fuel-oil para quemador *(m)*
Fr pétrole lampant *(m)*
It nafta per bruciatori *(f)*

fuel-oil para quemador *(m)* Es
De Brenner-Treiböl *(n)*
En burner fuel-oil
Fr pétrole lampant *(m)*
It nafta per bruciatori *(f)*
Pt fuel-oil para queimadores *(m)*

fuel résiduel *(m)* Fr
De Rückstandsöl *(n)*
En residual fuels *(pl)*
Es combustibles residuales *(m pl)*
It combustibili residui *(m pl)*
Pt combustíveis residuais *(m pl)*

fuel rod En
De Brennstoffstab *(m)*
Es barra de combustible *(f)*
Fr barre de combustible *(f)*
It barra combustibile *(f)*
Pt vara de combustível *(f)*

fuente *(f) n* Es
De Quelle *(f)*
En source
Fr source *(f)*
It sorgente *(f)*
Pt fonte *(f)*

fuente de neutrones *(f)* Es
De Neutronenquelle *(f)*
En neutron source
Fr source de neutrons *(f)*
It fonte di neutroni *(f)*
Pt fonte de neutrões *(f)*

fuerza *(f) n* Es
De Kraft *(f)*
En force
Fr force *(f)*
It forza *(f)*
Pt força *(f)*

fuerza centrípeta *(f)* Es
De Zentripetalkraft *(f)*
En centripetal force
Fr force centripète *(f)*
It forza centripeta *(f)*
Pt força centrípeda *(f)*

fuerza contraelectromotriz *(f)* Es
De Gegenelektromotorische Kraft *(f)*
En back electromotive force
Fr force contreélectromotrice *(f)*
It forza controelettromotrice *(f)*
Pt força contraelectromotriz *(f)*

fuerza contrífuga *(f)* Es
De Zentrifugalkraft *(f)*
En centrifugal force
Fr force centrifuge *(f)*
It forza centrifuga *(f)*
Pt força centrífuga *(f)*

fuerza electromotriz (f.e.m.) *(f)* Es
De elektromotorische Kraft (EMK) *(f)*
En electromotive force (e.m.f.)
Fr force électromotrice (f.é.m.) *(f)*
It forza elettromotrice (f.e.m.) *(f)*
Pt força electromotora (f.e.m.) *(f)*

fuerza electromotriz inducida *(f)* Es
De induzierte elektromotorische Kraft *(f)*
En induced electromotive force

Fr force électromotrice induite *(f)*
It forza elettromotrice indotta *(f)*
Pt força electromotora induzida *(f)*

fuerza gravitatoria *(f)* Es
De Erdanziehungskraft *(f)*
En gravitational force
Fr force de gravitation *(f)*
It forza di gravità *(f)*
Pt força de gravidade *(f)*

fuerza magnética *(f)* Es
De Magnetkraft *(n)*
En magnetic force
Fr force magnétique *(f)*
It forza magnetica *(f)*
Pt força magnética *(f)*

fuga *(f) n* Es, Pt
De Leck *(n)*
En leakage
Fr fuite *(f)*
It perdita *(f)*

fuga d'aria *(f)* It
De Luftleck *(n)*
En air leakage
Es fuga de aire *(f)*
Fr fuite d'air *(f)*
Pt fuga de ar *(f)*

fuga de aire *(f)* Es
De Luftleck *(n)*
En air leakage
Fr fuite d'air *(f)*
It fuga d'aria *(f)*
Pt fuga de ar *(f)*

fuga de ar *(f)* Pt
De Luftleck *(n)*
En air leakage
Es fuga de aire *(f)*
Fr fuite d'air *(f)*
It fuga d'aria *(f)*

fuga de gas *(f)* Es
De Gasaustritt *(m)*
En gas escape
Fr échappement de gaz *(m)*
It fuga di gas *(f)*
Pt escape de gás *(f)*

fuga di gas *(f)* It
De Gasaustritt *(m)*
En gas escape
Es fuga de gas *(f)*
Fr échappement de gaz *(m)*
Pt escape de gás *(f)*

fuite *(f) n* Fr
De Leck *(n)*
En leakage
Es fuga *(f)*
It perdita *(f)*
Pt fuga *(f)*

fuite d'air *(f)* Fr
De Luftleck *(n)*
En air leakage
Es fuga de aire *(f)*
It fuga d'aria *(f)*
Pt fuga de ar *(f)*

fulígem *(f) n* Pt
De Ruß *(m)*
En soot
Es hollín *(m)*
Fr suie *(f)*
It nerofumo *(m)*

Füllstandsregler *(m) n* De
En level control
Es control de nivel *(m)*
Fr commande de niveau *(f)*
It regolatore di livello *(m)*
Pt controle de nível *(m)*

full-wave rectifier En
De Ganzwellengleich-richter *(m)*
Es rectificador de onda completa *(m)*
Fr redresseur biphasé *(m)*
It raddrizzatore di onda intera *(m)*
Pt rectificador de onda completa *(m)*

fumée *(f) n* Fr
De Rauch *(m)*
En smoke
Es humo *(m)*
It fumo *(m)*
Pt fumo *(m)*

fumeiro *(m) n* Pt
De Abzug *(m)*
En flue
Es humero *(m)*
Fr carneau *(m)*
It condotto *(m)*

fumeiros múltiplos *(m pl)* Pt
De Sammel-Abgasleitungen *(pl)*
En multiple flues *(pl)*
Es conductos de humo múltiples *(m pl)*
Fr carneaux multiples *(m pl)*
It condotti multipli *(m pl)*

fumo *(m) n* It, Pt
De Rauch *(m)*
En smoke
Es humo *(m)*
Fr fumée *(f)*

função de trabalho *(f)* Pt
De Arbeitsfunktion *(f)*
En work function
Es función de trabajo *(f)*
Fr travail de sortie *(m)*
It funzione di lavoro *(f)*

funcionamiento con carga reducida *(m)* Es
De Teillastbetrieb *(m)*
En part-load operation
Fr fonctionnement sous charge partielle *(m)*
It funzionamento a carico parziale *(m)*
Pt operação de carga parcial *(f)*

función de trabajo *(f)* Es
De Arbeitsfunktion *(f)*
En work function
Fr travail de sortie *(m)*
It funzione di lavoro *(f)*
Pt função de trabalho *(f)*

Funkenzündung *(f) n* De
En spark ignition
Es encendido por chispa *(m)*
Fr allumage par étincelle *(m)*
It accensione a scintilla *(f)*
Pt ignição por faísca *(f)*

funzionamento a carico parziale *(m)* It
De Teillastbetrieb *(m)*
En part-load operation
Es funcionamiento con carga reducida *(m)*
Fr fonctionnement sous charge partielle *(m)*
Pt operação de carga parcial *(f)*

funzione di lavoro *(f)* It
De Arbeitsfunktion *(f)*
En work function
Es función de trabajo *(f)*
Fr travail de sortie *(m)*
Pt função de trabalho *(f)*

furnace *n* En
De Ofen *(m)*
Es horno *(m)*
Fr four *(m)*
It forno *(m)*
Pt forno *(m)*

fusão congruente *(f)* Pt
De kongruentes Schmelzen *(n)*
En congruent melting
Es fusión congruente *(f)*
Fr fusion congruente *(f)*
It fusione congruente *(f)*

fusão nuclear *(f)* Pt
De Nuklearfusion *(f)*
En nuclear fusion
Es fusión nuclear *(f)*
Fr fusion nucléaire *(f)*
It fusione nucleare *(f)*

fuse *n* En
De Sicherung *(f)*
Es fusible *(m)*
Fr fusible *(m)*
It fusibile *(m)*
Pt fusível *(m)*

fusée nucléaire *(f)* Fr
De Kernkraftrakete *(f)*
En nuclear rocket
Es cohete nuclear *(m)*
It razzo nucleare *(m)*
Pt foguetão nuclear *(m)*

fusibile *(m) n* It
De Sicherung *(f)*
En fuse
Es fusible *(m)*
Fr fusible *(m)*
Pt fusível *(m)*

fusible *(m) n* Es, Fr
De Sicherung *(f)*
En fuse
It fusibile *(m)*
Pt fusível *(m)*

fusion congruente *(f)* Fr
De kongruentes Schmelzen *(n)*
En congruent melting
Es fusión congruente *(f)*
It fusione congruente *(f)*
Pt fusão congruente *(f)*

fusión congruente *(f)* Es
De kongruentes Schmelzen *(n)*
En congruent melting
Fr fusion congruente *(f)*
It fusione congruente *(f)*
Pt fusão congruente *(f)*

fusione congruente *(f)* It
De kongruentes Schmelzen *(n)*
En congruent melting
Es fusión congruente *(f)*
Fr fusion congruente *(f)*
Pt fusão congruente *(f)*

fusione nucleare *(f)* It
De Nuklearfusion *(f)*
En nuclear fusion
Es fusión nuclear *(f)*
Fr fusion nucléaire *(f)*
Pt fusão nuclear *(f)*

fusion nucléaire *(f)* Fr
De Nuklearfusion *(f)*
En nuclear fusion
Es fusión nuclear *(f)*
It fusione nucleare *(f)*
Pt fusão nuclear *(f)*

fusión nuclear *(f)* Es
De Nuklearfusion *(f)*
En nuclear fusion
Fr fusion nucléaire *(f)*
It fusione nucleare *(f)*
Pt fusão nuclear *(f)*

fusível *(m) n* Pt
De Sicherung *(f)*
En fuse
Es fusible *(m)*
Fr fusible *(m)*
It fusibile *(m)*

G

gainage *(m) n* Fr
De Umhüllung *(f)*
En cladding
Es encamisado *(m)*
It incamiciatura *(f)*
Pt revestimento *(m)*

gain de chaleur incident *(m)* Fr
De zufällige Wärmezunahme *(f)*
En incidental heat gain
Es ganancia de calor incidental *(f)*
It guadagno di calore incidentale *(m)*
Pt ganho de calor acessório *(m)*

gain solaire *(m)* Fr
De Solarverstärkung *(f)*
En solar gain
Es ganancia solar *(f)*
It guadagno solare *(m)*
Pt ganho solar *(m)*

gallium arsenide solar cell En
De Galliumarsenid-Solarzelle *(f)*
Es célula solar de arseniuro de galio *(f)*
Fr cellule solaire à l'arséniure de gallium *(f)*
It cellula solare all'arseniuro di gallio *(f)*
Pt célula solar de arseneto de gálio *(f)*

Galliumarsenid-Solarzelle *(f) n* De
En gallium arsenide solar cell
Es célula solar de arseniuro de galio *(f)*
Fr cellule solaire à l'arséniure de gallium *(f)*
It cellula solare all'arseniuro di gallio *(f)*
Pt célula solar de arseneto de gálio *(f)*

galvanic *adj* En
De galvanisch
Es galvánico
Fr galvanique
It galvanico
Pt galvânico

galvanic cell En
De galvanische Zelle *(f)*
Es elemento galvánico *(m)*
Fr pile galvanique *(f)*
It cellula galvanica *(f)*
Pt célula galvânica *(f)*

galvanico *adj* It
De galvanisch
En galvanic
Es galvánico
Fr galvanique
Pt galvânico

galvánico *adj* Es
De galvanisch
En galvanic
Fr galvanique
It galvanico
Pt galvânico

galvânico *adj* Pt
De galvanisch
En galvanic
Es galvánico
Fr galvanique
It galvanico

galvanique *adj* Fr
De galvanisch
En galvanic
Es galvánico
It galvanico
Pt galvânico

galvanisch *adj* De
En galvanic
Es galvánico
Fr galvanique
It galvanico
Pt galvânico

galvanische Zelle *(f)* De
En galvanic cell
Es elemento galvánico *(m)*
Fr pile galvanique *(f)*
It cellula galvanica *(f)*
Pt célula galvânica *(f)*

gamma-rays *(pl) n* En
De Gammastrahlen *(pl)*
Es rayos gamma *(m pl)*
Fr rayons gamma *(m pl)*
It raggi gamma *(m pl)*
Pt raios gama *(m pl)*

Gammastrahlen *(pl) n* De
En gamma-rays *(pl)*
Es rayos gamma *(m pl)*
Fr rayons gamma *(m pl)*
It raggi gamma *(m pl)*
Pt raios gama *(m pl)*

ganancia de calor incidental *(f)* Es
De zufällige Wärmezunahme *(f)*
En incidental heat gain
Fr gain de chaleur incident *(m)*
It guadagno di calore incidentale *(m)*
Pt ganho de calor acessório *(m)*

ganancia solar *(f)* Es
De Solarverstärkung *(f)*
En solar gain
Fr gain solaire *(m)*
It guadagno solare *(m)*
Pt ganho solar *(m)*

ganho de calor acessório *(m)* Pt
De zufällige Wärmezunahme *(f)*
En incidental heat gain
Es ganancia de calor incidental *(f)*
Fr gain de chaleur incident *(m)*
It guadagno di calore incidentale *(m)*

ganho solar *(m)* Pt
De Solarverstärkung *(f)*
En solar gain
Es ganancia solar *(f)*
Fr gain solaire *(m)*
It guadagno solare *(m)*

Ganzwellengleichrichter *(m) n* De
En full-wave rectifier
Es rectificador de onda completa *(m)*
Fr redresseur biphasé *(m)*
It raddrizzatore di onda intera *(m)*

Pt rectificador de onda completa *(m)*

garrafa magnética *(f)* Pt
De magnetische Flasche *(f)*
En magnetic bottle
Es botella magnética *(f)*
Fr bouteille magnétique *(f)*
It bottiglia magnetica *(f)*

Gärung *(f) n* De
En fermentation
Es fermentación *(f)*
Fr fermentation *(f)*
It fermentazione *(f)*
Pt fermentação *(f)*

gas *n* En; Es, It *(m)*
De Gas *(n)*
Fr gaz *(m)*
Pt gás *(m)*

Gas *(n) n* De
En gas
Es gas *(m)*
Fr gaz *(m)*
It gas *(m)*
Pt gás *(m)*

gás *(m) n* Pt
De Gas *(n)*
En gas
Es gas *(m)*
Fr gaz *(m)*
It gas *(m)*

Gasanalyse *(f) n* De
En gas analysis
Es análisis del gas *(m)*
Fr analyse des gaz *(f)*
It analisi del gas *(f)*
Pt análise de gás *(f)*

gas analysis En
De Gasanalyse *(f)*
Es análisis del gas *(m)*
Fr analyse des gaz *(f)*
It analisi del gas *(f)*
Pt análise de gás *(f)*

Gasaustritt *(m) n* De
En gas escape
Es fuga de gas *(f)*
Fr échappement de gaz *(m)*
It fuga di gas *(f)*
Pt escape de gas *(f)*

Gaschromatographie *(f) n* De
En gas chromatography
Es cromatografía de gases *(f)*
Fr chromatographie gazeuse *(f)*
It cromatografia in fase gassosa *(f)*
Pt cromatografia gasosa *(f)*

gas chromatography En
De Gaschromatographie *(f)*
Es cromatografía de gases *(f)*
Fr chromatographie gazeuse *(f)*
It cromatografia in fase gassosa *(f)*
Pt cromatografia gasosa *(f)*

gas-cooled nuclear reactor En
De gasgekühlter Kernreaktor *(m)*
Es reactor nuclear refrigerado con gas *(m)*
Fr réacteur nucléaire à refroidissement au gaz *(m)*
It reattore nucleare raffreddato a gas *(m)*
Pt reactor nuclear arrefecido a gás *(m)*

gas d'acqua *(m)* It
De Wassergas *(n)*
En water gas
Es gas de agua *(m)*
Fr gaz à l'eau *(m)*
Pt gás de água *(m)*

gas de agua *(m)* Es
De Wassergas *(n)*
En water gas
Fr gaz à l'eau *(m)*
It gas d'acqua *(m)*
Pt gás de água *(m)*

gás de água *(m)* Pt
De Wassergas *(n)*
En water gas
Es gas de agua *(m)*
Fr gaz à l'eau *(m)*
It gas d'acqua *(m)*

gas de carbón *(m)* Es
De Steinkohlengas *(n)*
En coal gas
Fr gaz de houille *(m)*
It gas illuminante *(m)*
Pt gás de carvão *(m)*

gás de carvão *(m)* Pt
De Steinkohlengas *(n)*
En coal gas
Es gas de carbón *(m)*
Fr gaz de houille *(m)*
It gas illuminante *(m)*

gas de chimenea *(m)* Es
De Abgas *(n)*
En flue gas
Fr gaz de carneau *(m)*
It gas della combustione *(m)*
Pt gas de fumeiro *(m)*

gás de cidade *(m)* Pt
De Stadtgas *(n)*
En town gas
Es gas de ciudad *(m)*
Fr gaz de ville *(m)*
It gas per uso domestico *(m)*

gas de ciudad *(m)* Es
De Stadtgas *(n)*
En town gas
Fr gaz de ville *(m)*
It gas per uso domestico *(m)*
Pt gás de cidade *(m)*

gas de fumeiro *(m)* Pt
De Abgas *(n)*
En flue gas
Es gas de chimenea *(m)*
Fr gaz de carneau *(m)*
It gas della combustione *(m)*

gas della combustione *(m)* It
De Abgas *(n)*
En flue gas
Es gas de chimenea *(m)*
Fr gaz de carneau *(m)*
Pt gas de fumeiro *(m)*

gas del Mar del Norte *(m)* Es
De Nordseegas *(n)*
En North Sea gas
Fr gaz de Mer du Nord *(m)*
It gas del Mare del Nord *(m)*
Pt gás do Mar do Norte *(m)*

gas del Mare del Nord *(m)* It
De Nordseegas *(n)*
En North Sea gas
Es gas del Mar del Norte *(m)*
Fr gaz de Mer du Nord *(m)*
Pt gás do Mar do Norte *(m)*

gás de petróleo liquefeito *(m)* Pt
De verflüssigtes Petroleumgas *(n)*
En liquefied petroleum gas (LPG)
Es gas licuado del petróleo (GLP) *(m)*
Fr gaz de pétrole liquéfié (GPL) *(m)*
It gas liquido di petrolio *(m)*

gas dielectric En
De Gas-Dielektrikum *(n)*
Es dieléctrico de gas *(m)*
Fr diélectrique à gaz *(m)*
It dielettrico gas *(m)*
Pt dieléctrico gasoso *(m)*

Gas-Dielektrikum *(n) n* De
En gas dielectric
Es dieléctrico de gas *(m)*
Fr diélectrique à gaz *(m)*
It dielettrico gas *(m)*
Pt dieléctrico gasoso *(m)*

Gasdiffusion *(f) n* De
En gaseous diffusion
Es difusión gaseosa *(f)*
Fr diffusion gazeuse *(f)*
It diffusione gassosa *(f)*
Pt difusão gasosa *(f)*

gas di gassogeno *(m)* It
De Generatorgas *(n)*
En producer gas
Es gas probre *(m)*
Fr gaz de gazogène *(m)*
Pt gás probre *(m)*

gas di scarico *(m pl)* It
De Abgase *(pl)*
En waste gases *(pl)*
Es gases de desecho *(m pl)*
Fr gaz perdus *(m pl)*
Pt gases peridos *(m pl)*

gás do Mar do Norte *(m)* Pt
De Nordseegas *(n)*
En North Sea gas
Es gas del Mar del Norte *(m)*
Fr gaz de Mer du Nord *(m)*
It gas del Mare del Nord *(m)*

gaseificação *(f) n* Pt
De Vergasung *(f)*
En gasification
Es gasificación *(f)*
Fr gazéification *(f)*
It gassificazione *(f)*

gaseificação por oxidação parcial *(f)* Pt
De partielle Oxidationsvergasung *(f)*
En partial oxidation gasification
Es gasificación don oxidación parcial *(f)*
Fr gazéification à oxydation partielle *(f)*
It gassificazione con ossidazione parziale *(f)*

gaseificação subterrânea *(f)* Pt
De unterirdische Gasifizierung *(f)*
En underground gasification
Es gasificación subterránea *(f)*
Fr gazéification souterraine *(f)*
It gassificazione sotterranea *(f)*

gas engine En
De Gasmotor *(m)*
Es motor de gas *(m)*
Fr moteur à gaz *(m)*
It motore a gas *(m)*
Pt motor a gás *(m)*

gaseous diffusion En
De Gasdiffusion *(f)*
Es difusión gaseosa *(f)*
Fr diffusion gazeuse *(f)*
It diffusione gassosa *(f)*
Pt difusão gasosa *(f)*

gas escape En
De Gasaustritt *(m)*
Es fuga de gas *(f)*
Fr échappement de gaz *(m)*
It fuga di gas *(f)*
Pt escape de gás *(f)*

gases de desecho *(m pl)* Es
De Abgase *(pl)*
En waste gases *(pl)*
Fr gaz perdus *(m pl)*
It gas di scarico *(m pl)*
Pt gases peridos *(m pl)*

gases de escape *(m pl)* Es
De Auspuffgas *(n)*
En exhaust
Fr échappement *(m)*
It scarico *(m)*
Pt escape *(m)*

gases peridos *(m pl)* Pt
De Abgase *(pl)*
En waste gases *(pl)*
Es gases de desecho *(m pl)*
Fr gaz perdus *(m pl)*
It gas di scarico *(m pl)*

gas-fired *adj* En
De gasgefeuert
Es alimentado por gas
Fr chauffé au gaz
It alimentato a gas
Pt alimentado a gás

gasgefeuert *adj* De
En gas-fired
Es alimentado por gas
Fr chauffé au gaz
It alimentato a gas
Pt alimentado a gás

gasgekühlter Kernreaktor *(m)* De
En gas-cooled nuclear reactor
Es reactor nuclear refrigerado con gas *(m)*
Fr réacteur nucléaire à refroidissement au gaz *(m)*
It reattore nucleare raffreddato a gas *(m)*
Pt reactor nuclear arrefecido a gás *(m)*

gás hidráulico *(m)* Pt
De Wassergas *(n)*
En water gas
Es gas de agua *(m)*
Fr gaz à l'eau *(m)*
It gas d'acqua *(m)*

gas holder Am
De Gasometer *(m)*
En gasometer
Es gasómetro *(m)*
Fr gazomètre *(m)*
It gasometro *(m)*
Pt gasómetro *(m)*

gás ideal *(m)* Pt
De ideales Gas *(n)*
En ideal gas
Es gas perfecto *(m)*
Fr gaz idéal *(m)*
It gas perfetto *(m)*

gasificación *(f) n* Es
De Vergasung *(f)*
En gasification
Fr gazéification *(f)*
It gassificazione *(f)*
Pt gaseificação *(f)*

gasificación don oxidación parcial *(f)* Es
De partielle Oxidationsvergasung *(f)*
En partial oxidation gasification
Fr gazéification à oxydation partielle *(f)*
It gassificazione con ossidazione parziale *(f)*
Pt gaseificação por oxidação parcial *(f)*

gasificación subterránea *(f)* Es
De unterirdische Gasifizierung *(f)*
En underground gasification
Fr gazéification souterraine *(f)*
It gassificazione sotterranea *(f)*
Pt gaseificação subterrânea *(f)*

gasification *n* En
De Vergasung *(f)*
Es gasificación *(f)*
Fr gazéification *(f)*
It gassificazione *(f)*
Pt gaseificação *(f)*

gas illuminante *(m)* It
De Steinkohlengas *(n)*
En coal gas
Es gas de carbón *(m)*
Fr gaz de houille *(m)*
Pt gás de carvão *(m)*

gas licuado del petróleo (GLP) *(m)* Es
De verflüssigtes Petroleumgas *(n)*
En liquefied petroleum gas (LPG)
Fr gaz de pétrole liquéfié (GPL) *(m)*
It gas liquido di petrolio *(m)*
Pt gás de petróleo liquefeito *(m)*

gas liquido di petrolio *(m)* It
De verflüssigtes Petroleumgas *(n)*
En liquefied petroleum gas (LPG)
Es gas licuado del petróleo (GLP) *(m)*
Fr gaz de pétrole liquéfié (GPL) *(m)*
Pt gás de petróleo liquefeito *(m)*

Gasmotor *(m) n* De
En gas engine
Es motor de gas *(m)*
Fr moteur à gaz *(m)*
It motore a gas *(m)*
Pt motor a gás *(m)*

gas natural *(m)* Es
De Erdgas *(n)*
En natural gas
Fr gaz naturel *(m)*
It gas naturale *(m)*
Pt gás natural *(m)*

gás natural *(m)* Pt
De Erdgas *(n)*
En natural gas
Es gas natural *(m)*

Fr gaz naturel *(m)*
It gas naturale *(m)*

gás natural de substituição *(m)* Pt
De Erdgasersatz *(m)*
En substitute natural gas
Es gas natural de sustitución *(m)*
Fr gaz naturel de remplacement *(m)*
It sostituto di gas naturale *(m)*

gas natural de sustitución *(m)* Es
De Erdgasersatz *(m)*
En substitute natural gas
Fr gaz naturel de remplacement *(m)*
It sostituto di gas naturale *(m)*
Pt gás natural de substituição *(m)*

gas naturale *(m)* It
De Erdgas *(n)*
En natural gas
Es gas natural *(m)*
Fr gaz naturel *(m)*
Pt gás natural *(m)*

gas naturale liquido *(m)* It
De verflüssigtes Erdgas *(n)*
En liquefied natural gas (LNG)
Es gas natural licuado (GNL) *(m)*
Fr gaz naturel liquéfié (GNL) *(m)*
Pt gás natural liquefeito *(m)*

gas naturale non associato *(m)* It
De nicht assoziertes Erdgas *(n)*
En nonassociated natural gas
Es gas natural no asociada *(m)*
Fr gaz naturel non associé *(m)*
Pt gás natural não associada *(m)*

gas naturale sintetico *(m)* It
De synthetisches Erdgas *(n)*
En synthetic natural gas (SNG)
Es gas natural sintético (GNS) *(m)*
Fr gaz naturel de synthèse *(m)*
Pt gás natural sintético *(m)*

gas natural licuado (GNL) *(m)* Es
De verflüssigtes Erdgas *(n)*
En liquefied natural gas (LNG)
Fr gaz naturel liquéfié (GNL) *(m)*
It gas naturale liquido *(m)*
Pt gás natural liquefeito *(m)*

gás natural liquefeito *(m)* Pt
De verflüssigtes Erdgas *(n)*
En liquefied natural gas (LNG)
Es gas natural licuado (GNL) *(m)*
Fr gaz naturel liquéfié (GNL) *(m)*
It gas naturale liquido *(m)*

gás natural não associada *(m)* Pt
De nicht assoziertes Erdgas *(n)*
En nonassociated natural gas
Es gas natural no asociada *(m)*
Fr gaz naturel non associé *(m)*
It gas naturale non associato *(m)*

gas natural no asociada *(m)* Es
De nicht assoziertes Erdgas *(n)*
En nonassociated natural gas
Fr gaz naturel non associé *(m)*
It gas naturale non associato *(m)*
Pt gás natural não associada *(m)*

gás natural sintético *(m)* Pt
De synthetisches Erdgas *(n)*
En synthetic natural gas (SNG)
Es gas natural sintético (GNS) *(m)*
Fr gaz naturel de synthèse *(m)*
It gas naturale sintetico *(m)*

gas natural sintético (GNS) *(m)* Es
De synthetisches Erdgas *(n)*
En synthetic natural gas (SNG)
Fr gaz naturel de synthèse *(m)*
It gas naturale sintetico *(m)*
Pt gás natural sintético *(m)*

gas oil En
De Gasöl *(n)*
Es gas-oil *(m)*
Fr gazole *(f)*
It gasolio *(m)*
Pt gasoil *(m)*

gas-oil *(m)* *n* Es
De Gasöl *(n)*
En gas oil
Fr gazole *(f)*
It gasolio *(m)*
Pt gasoil *(m)*

gasoil *(m)* *n* Pt
De Gasöl *(n)*
En gas oil
Es gas-oil *(m)*
Fr gazole *(f)*
It gasolio *(m)*

Gasöl *(n)* *n* De
En gas oil
Es gas-oil *(m)*
Fr gazole *(f)*
It gasolio *(m)*
Pt gasoil *(m)*

gasolina *(f)* *n* Es, Pt
Am gasoline
De Benzin *(n)*
En petrol
Fr essence *(f)*
It benzina *(f)*

gasolina de aviación *(f)* Es
De Flugkraftstoff *(m)*
En aviation fuel
Fr carburation aviation *(m)*
It combustibile per aviazione *(m)*
Pt combustível de aviação *(m)*

gasoline *n* Am
De Benzin *(n)*
En petrol
Es gasolina *(f)*
Fr essence *(f)*
It benzina *(f)*
Pt gasolina *(f)*

gasolio *(m)* *n* It
De Gasöl *(n)*
En gas oil
Es gas-oil *(m)*
Fr gazole *(f)*
Pt gasoil *(m)*

gasometer *n* En
Am gas holder
De Gasometer *(m)*
Es gasómetro *(m)*
Fr gazomètre *(m)*
It gasometro *(m)*
Pt gasómetro *(m)*

Gasometer *(m)* *n* De
Am gas holder
En gasometer
Es gasómetro *(m)*
Fr gazomètre *(m)*
It gasometro *(m)*
Pt gasómetro *(m)*

gasometro *(m)* *n* It
Am gas holder
De Gasometer *(m)*
En gasometer
Es gasómetro *(m)*
Fr gazomètre *(m)*
Pt gasómetro *(m)*

gasómetro *(m)* *n* Es, Pt
Am gas holder
De Gasometer *(m)*
En gasometer
Fr gazomètre *(m)*
It gasometro *(m)*

gas perfecto *(m)* Es
De ideales Gas *(n)*
En ideal gas
Fr gaz idéal *(m)*

It gas perfetto *(m)*
Pt gás ideal *(m)*

gas perfetto *(m)* It
De ideales Gas *(n)*
En ideal gas
Es gas perfecto *(m)*
Fr gaz idéal *(m)*
Pt gás ideal *(m)*

gas per uso domestico *(m)* It
De Stadtgas *(n)*
En town gas
Es gas de ciudad *(m)*
Fr gaz de ville *(m)*
Pt gás de cidade *(m)*

gas probre *(m)* Es
De Generatorgas *(n)*
En producer gas
Fr gaz de gazogène *(m)*
It gas di gassogeno *(m)*
Pt gás probre *(m)*

gás probre *(m)* Pt
De Generatorgas *(n)*
En producer gas
Es gas probre *(m)*
Fr gaz de gazogène *(m)*
It gas di gassogeno *(m)*

Gasschiff *(n) n* De
En gas ship
Es barco de gas *(m)*
Fr navire méthanier *(m)*
It nave del gas *(f)*
Pt barco de gás *(m)*

gas ship En
De Gasschiff *(n)*
Es barco de gas *(m)*
Fr navire méthanier *(m)*
It nave del gas *(f)*
Pt barco de gás *(m)*

gassificazione *(f) n* It
De Vergasung *(f)*
En gasification
Es gasificación *(f)*
Fr gazéification *(f)*
Pt gaseificação *(f)*

gassificazione con ossidazione parziale *(f)* It
De partielle Oxidationsvergasung *(f)*
En partial oxidation gasification
Es gasificación don oxidación parcial *(f)*
Fr gazéification à oxydation partielle *(f)*
Pt gaseificação por oxidação parcial *(f)*

gassificazione sotterranea *(f)* It
De unterirdische Gasifizierung *(f)*
En underground gasification
Es gasificación subterránea *(f)*
Fr gazéification souterraine *(f)*
Pt gaseificação subterrânea *(f)*

gas thermometer En
De Gasthermometer *(n)*
Es termómetro de gas *(m)*
Fr thermomètre à gaz *(m)*
It termometro a gas *(m)*
Pt termómetro de gás *(m)*

Gasthermometer *(n) n* De
En gas thermometer
Es termómetro de gas *(m)*
Fr thermomètre à gaz *(m)*
It termometro a gas *(m)*
Pt termómetro de gás *(m)*

gas turbine En
De Gasturbine *(f)*
Es turbina de gas *(f)*
Fr turbine à gaz *(f)*
It turbina a gas *(f)*
Pt turbina a gás *(f)*

Gasturbine *(f) n* De
En gas turbine
Es turbina de gas *(f)*
Fr turbine à gaz *(f)*
It turbina a gas *(f)*
Pt turbina a gás *(f)*

gaz *(m) n* Fr
De Gas *(n)*
En gas
Es gas *(m)*
It gas *(m)*
Pt gás *(m)*

gaz à l'eau *(m)* Fr
De Wassergas *(n)*
En water gas
Es gas de agua *(m)*
It gas d'acqua *(m)*
Pt gás de água *(m)*

gaz de carneau *(m)* Fr
De Abgas *(n)*
En flue gas
Es gas de chimenea *(m)*
It gas della combustione *(m)*
Pt gas de fumeiro *(m)*

gaz de gazogène *(m)* Fr
De Generatorgas *(n)*
En producer gas
Es gas probre *(m)*
It gas di gassogeno *(m)*
Pt gás probre *(m)*

gaz de houille *(m)* Fr
De Steinkohlengas *(n)*
En coal gas
Es gas de carbón *(m)*
It gas illuminante *(m)*
Pt gás de carvão *(m)*

gaz de Mer du Nord *(m)* Fr
De Nordseegas *(n)*
En North Sea gas
Es gas del Mar del Norte *(m)*
It gas del Mare del Nord *(m)*
Pt gás do Mar do Norte *(m)*

gaz de pétrole liquéfié (GPL) *(m)* Fr
De verflüssigtes Petroleumgas *(n)*
En liquefied petroleum gas (LPG)
Es gas licuado del petróleo (GLP) *(m)*
It gas liquido di petrolio *(m)*
Pt gás de petróleo liquefeito *(m)*

gaz de ville *(m)* Fr
De Stadtgas *(n)*
En town gas
Es gas de ciudad *(m)*
It gas per uso domestico *(m)*
Pt gás de cidade *(m)*

gazéification *(f) n* Fr
De Vergasung *(f)*
En gasification (of coal)
Es gasificación *(f)*
It gassificazione *(f)*
Pt gaseificação *(f)*

gazéification à oxydation partielle *(f)* Fr
De partielle Oxidationsvergasung *(f)*
En partial oxidation gasification
Es gasificación don oxidación parcial *(f)*
It gassificazione con ossidazione parziale *(f)*
Pt gaseificação por oxidação parcial *(f)*

gazéification souterraine *(f)* Fr
De unterirdische Gasifizierung *(f)*
En underground gasification
Es gasificación subterránea *(f)*
It gassificazione sotterranea *(f)*
Pt gaseificação subterrânea *(f)*

gaz idéal *(m)* Fr
De ideales Gas *(n)*
En ideal gas
Es gas perfecto *(m)*
It gas perfetto *(m)*
Pt gás ideal *(m)*

gaz naturel *(m)* Fr
De Erdgas *(n)*
En natural gas
Es gas natural *(m)*
It gas naturale *(m)*
Pt gás natural *(m)*

gaz naturel de remplacement *(m)* Fr
De Erdgasersatz *(m)*
En substitute natural gas
Es gas natural de sustitución *(m)*

It sostituto di gas naturale *(m)*
Pt gás natural de substituição *(m)*

gaz naturel de synthèse *(m)* Fr
De synthetisches Erdgas *(n)*
En synthetic natural gas (SNG)
Es gas natural sintético (GNS) *(m)*
It gas naturale sintetico *(m)*
Pt gás natural sintético *(m)*

gaz naturel liquéfié (GNL) *(m)* Fr
De verflüssigtes Erdgas *(n)*
En liquefied natural gas (LNG)
Es gas natural licuado (GNL) *(m)*
It gas naturale liquido *(m)*
Pt gás natural liquefeito *(m)*

gaz naturel non associé *(m)* Fr
De nicht assoziertes Erdgas *(n)*
En nonassociated natural gas
Es gas natural no asociada *(m)*
It gas naturale non associato *(m)*
Pt gás natural não associada *(m)*

gazole *(f) n* Fr
De Gasöl *(n)*
En gas oil
Es gas-oil *(m)*
It gasolio *(m)*
Pt gasoil *(m)*

gazomètre *(m) n* Fr
Am gas holder
De Gasometer *(m)*
En gasometer
Es gasómetro *(m)*
It gasometro *(m)*
Pt gasómetro *(m)*

gaz perdus *(m pl)* Fr
De Abgase *(pl)*
En waste gases *(pl)*
Es gases de desecho *(m pl)*
It gas di scarico *(m pl)*
Pt gases peridos *(m pl)*

Gebläse *(n) n* De
En fan
Es ventilador *(m)*
Fr ventilateur *(m)*
It ventilatore *(m)*
Pt ventoinha *(f)*

Gebläsekonvektion *(f) n* De
En forced convection
Es convección forzada *(f)*
Fr convection forcée *(f)*
It convezione forzata *(f)*
Pt convecção forçada *(f)*

gebundene Wärme *(f)* De
En latent heat
Es calor latente *(m)*
Fr chaleur latente *(f)*
It calore latente *(m)*
Pt calor latente *(m)*

geformter Koks *(m)* De
En formed coke
Es coque conformado *(m)*
Fr coke formé *(m)*
It coke formato *(m)*
Pt coque formado *(m)*

Gefrierpunkt *(m) n* De
En freezing point
Es punto de congelación *(m)*
Fr point de congélation *(m)*
It punto di congelamento *(m)*
Pt ponto de congelação *(m)*

Gefrierschutz *(m) n* De
En freeze protection
Es protección contra la congelación *(f)*
Fr protection contre le gel *(f)*
It protezione dal congelamento *(f)*
Pt protecção antigelo *(f)*

Gefriertrocknung *(f) n* De
En freeze drying
Es liofilización *(f)*
Fr lyophilisation *(f)*
It liofilizzazione *(f)*
Pt secagem por congelação *(f)*

Gegendruckturbine *(f) n* De
En back-pressure turbine
Es turbina de contrapresión *(f)*
Fr turbine à contre-pression *(f)*
It turbina a contropressione *(f)*
Pt turbina de contrapressão *(f)*

Gegenelektro-motorische Kraft *(f)* De
En back electromotive force
Es fuerza contraelectromotriz *(f)*
Fr force contreélectromotrice *(f)*
It forza controelettromotrice *(f)*
Pt força contraelectromotriz *(f)*

Gegeninduktivität *(f) n* De
En mutual inductance
Es inductancia mutua *(f)*
Fr inductance mutuelle *(f)*
It induttanza mutua *(f)*
Pt indutância mútua *(f)*

Geiger counter En
De Geiger-Zähler *(m)*
Es contador de Geiger *(m)*
Fr compteur Geiger *(m)*
It contatore Geiger *(m)*
Pt contador Geiger *(m)*

Geiger-Zähler *(m) n* De
En Geiger counter
Es contador de Geiger *(m)*
Fr compteur Geiger *(m)*
It contatore Geiger *(m)*
Pt contador Geiger *(m)*

geiser *(m) n* Pt
De Geysir *(m)*
En geyser
Es géiser *(m)*
Fr geyser *(m)*
It geyser *(m)*

géiser *(m) n* Es
De Geysir *(m)*
En geyser
Fr geyser *(m)*
It geyser *(m)*
Pt geiser *(m)*

generación de electricidad *(f)* Es
De Stromerzeugung *(f)*
En electricity generation
Fr production d'électricité *(f)*
It generazione elettrica *(f)*
Pt geração de electricidade *(f)*

generador *(m) n* Es
De Generator *(m)*
En generator
Fr générateur *(m)*
It generatore *(m)*
Pt gerador *(m)*

generador de impulsos *(m)* Es
De Impulsgenerator *(m)*
En pulse generator
Fr générateur d'impulsions *(m)*
It generatore d'impulsi *(m)*
Pt gerador de impulsos *(m)*

generador de ondas *(m)* Es
De Wellengenerator *(m)*
En wave generator
Fr générateur d'ondes *(f)*
It generatore d'onda *(m)*
Pt gerador de ondas *(m)*

generador eólico *(m)* Es
De windgetriebener Generator *(m)*
En wind-driven generator
Fr générateur éolien *(m)*
It generatore a vento *(m)*

Pt gerador accionado pelo vento *(m)*

generador magnetohidrodinámico *(m)* Es
De MagnetohydrodynamikGenerator *(m)*
En magnetohydrodynamic generator
Fr générateur magnétohydrodynamique *(m)*
It generatore magnetoidrodinamico *(m)*
Pt gerador magnetohidrodinámico *(m)*

générateur *(m)* *n* Fr
De Generator *(m)*
En generator
Es generador *(m)*
It generatore *(m)*
Pt gerador *(m)*

générateur d'impulsions *(m)* Fr
De Impulsgenerator *(m)*
En pulse generator
Es generador de impulsos *(m)*
It generatore d'impulsi *(m)*
Pt gerador de impulsos *(m)*

générateur d'ondes *(f)* Fr
De Wellengenerator *(m)*
En wave generator
Es generador de ondas *(m)*
It generatore d'onda *(m)*
Pt gerador de ondas *(m)*

générateur éolien *(m)* Fr
De windgetriebener Generator *(m)*
En wind-driven generator
Es generador eólico *(m)*
It generatore a vento *(m)*
Pt gerador accionado pelo vento *(m)*

générateur magnétohydrodynamique *(m)* Fr
De MagnetohydrodynamikGenerator *(m)*
En magnetohydrodynamic generator
Es generador magnetohidrodinámico *(m)*
It generatore magnetoidrodinamico *(m)*
Pt gerador magnetohidrodinámico *(m)*

generating capacity En
De Erzeugungskapazität *(f)*
Es capacidad generadora *(f)*
Fr capacité génératrice *(f)*
It capacità di generazione *(f)*
Pt capacidade geradora *(f)*

generating efficiency En
De Erzeugungsleistung *(f)*
Es rendimiento de generación *(m)*
Fr rendement générateur *(m)*
It efficienza di generazione *(f)*
Pt eficiência geradora *(f)*

generator *n* En
De Generator *(m)*
Es generador *(m)*
Fr générateur *(m)*
It generatore *(m)*
Pt gerador *(m)*

Generator *(m)* *n* De
En generator
Es generador *(m)*
Fr générateur *(m)*
It generatore *(m)*
Pt gerador *(m)*

generatore *(m)* *n* It
De Generator *(m)*
En generator
Es generador *(m)*
Fr générateur *(m)*
Pt gerador *(m)*

generatore a vento *(m)* It
De windgetriebener Generator *(m)*
En wind-driven generator
Es generador eólico *(m)*
Fr générateur éolien *(m)*
Pt gerador accionado pelo vento *(m)*

generatore d'impulsi *(m)* It
De Impulsgenerator *(m)*
En pulse generator
Es generador de impulsos *(m)*
Fr générateur d'impulsions *(m)*
Pt gerador de impulsos *(m)*

generatore d'onda *(m)* It
De Wellengenerator *(m)*
En wave generator
Es generador de ondas *(m)*
Fr générateur d'ondes *(f)*
Pt gerador de ondas *(m)*

generatore magnetoidrodinamico *(m)* It
De MagnetohydrodynamikGenerator *(m)*
En magnetohydrodynamic generator
Es generador magnetohidrodinámico *(m)*
Fr générateur magnétohydrodynamique *(m)*
Pt gerador magnetohidrodinámico *(m)*

Generatorgas *(n)* *n* De
En producer gas
Es gas probre *(m)*
Fr gaz de gazogène *(m)*
It gas di gassogeno *(m)*
Pt gás probre *(m)*

generazione elettrica *(f)* It
De Stromerzeugung *(f)*
En electricity generation
Es generación de electricidad *(f)*
Fr production d'électricité *(f)*
Pt geração de electricidade *(f)*

geochemical exploration En
De geochemische Untersuchung *(f)*
Es exploración geoquímica *(f)*
Fr exploration géochimique *(f)*
It esplorazione geochimica *(f)*
Pt exploração geoquímica *(f)*

geochemical prospecting En
De geochemisches Schürfen *(n)*
Es prospección geoquímica *(f)*
Fr prospection géochimique *(f)*
It prospezione geochimica *(f)*
Pt prospecção geoquímica *(f)*

geochemisches Schürfen *(n)* De
En geochemical prospecting
Es prospección geoquímica *(f)*
Fr prospection géochimique *(f)*
It prospezione geochimica *(f)*
Pt prospecção geoquímica *(f)*

geochemische Untersuchung *(f)* De
En geochemical exploration
Es exploración geoquímica *(f)*
Fr exploration géochimique *(f)*
It esplorazione geochimica *(f)*
Pt exploração geoquímica *(f)*

geofisica *(f)* *n* It
De Geophysik *(f)*
En geophysics
Es geofísica *(f)*

Fr géophysique *(f)*
Pt geofísica *(f)*

geofísica *(f) n* Es, Pt
De Geophysik *(f)*
En geophysics
Fr géophysique *(f)*
It geofisica *(f)*

geophysical exploration En
De geophysische Untersuchung *(f)*
Es exploración geofísica *(f)*
Fr exploration géophysique *(f)*
It esplorazione geofisica *(f)*
Pt exploração geofísica *(f)*

geophysical prospecting En
De geophysisches Schürfen *(n)*
Es prospección geofísica *(f)*
Fr prospection géophysique *(f)*
It prospezione geofisica *(f)*
Pt prospecção geofísica *(f)*

geophysics *n* En
De Geophysik *(f)*
Es geofísica *(f)*
Fr géophysique *(f)*
It geofisica *(f)*
Pt geofísica *(f)*

Geophysik *(f) n* De
En geophysics
Es geofísica *(f)*
Fr géophysique *(f)*
It geofisica *(f)*
Pt geofísica *(f)*

géophysique *(f) n* Fr
De Geophysik *(f)*
En geophysics
Es geofísica *(f)*
It geofisica *(f)*
Pt geofísica *(f)*

geophysisches Schürfen *(n)* De
En geophysical prospecting
Es prospección geofísica *(f)*
Fr prospection géophysique *(f)*
It prospezione geofisica *(f)*
Pt prospecção geofísica *(f)*

geophysische Untersuchung *(f)* De
En geophysical exploration
Es exploración geofísica *(f)*
Fr exploration géophysique *(f)*
It esplorazione geofisica *(f)*
Pt exploração geofísica *(f)*

geothermal brine En
De geothermische Sole *(f)*
Es salmuera geotérmica *(f)*
Fr saumure géothermique *(f)*
It acqua salmastra geotermica *(f)*
Pt salmoura geotérmica *(f)*

geothermal energy En
De geothermische Energie *(f)*
Es energía geotérmica *(f)*
Fr énergie géothermique *(f)*
It energia geotermica *(f)*
Pt energia geotérmica *(f)*

geothermal gradient En
De geothermisches Gefälle *(n)*
Es gradiente geotérmico *(m)*
Fr gradient géothermique *(m)*
It gradiente geotermico *(m)*
Pt gradiente geotérmico *(m)*

geothermische Energie *(f)* De
En geothermal energy
Es energía geotérmica *(f)*
Fr énergie géothermique *(f)*
It energia geotermica *(f)*
Pt energia geotérmica *(f)*

geothermisches Gefälle *(n)* De
En geothermal gradient
Es gradiente geotérmico *(m)*
Fr gradient géothermique *(m)*
It gradiente geotermico *(m)*
Pt gradiente geotérmico *(m)*

geothermische Sole *(f)* De
En geothermal brine
Es salmuera geotérmica *(f)*
Fr saumure géothermique *(f)*
It acqua salmastra geotermica *(f)*
Pt salmoura geotérmica *(f)*

gepumpte Speicherung *(f)* De
En pumped storage
Es almacenamiento bombeado *(m)*
Fr accumulation pompée *(f)*
It immagazzinaggi a pompa *(m)*
Pt armazenagem realizada à bomba *(f)*

geração de electricidade *(f)* Pt
De Stromerzeugung *(f)*
En electricity generation
Es generación de electricidad *(f)*
Fr production d'électricité *(f)*
It generazione elettrica *(f)*

gerador *(m) n* Pt
De Generator *(m)*
En generator
Es generador *(m)*
Fr générateur *(m)*
It generatore *(m)*

gerador accionado pelo vento *(m)* Pt
De windgetriebener Generator *(m)*
En wind-driven generator
Es generador eólico *(m)*
Fr générateur éolien *(m)*
It generatore a vento *(m)*

gerador de impulsos *(m)* Pt
De Impulsgenerator *(m)*
En pulse generator
Es generador de impulsos *(m)*
Fr générateur d'impulsions *(m)*
It generatore d'impulsi *(m)*

gerador de ondas *(m)* Pt
De Wellengenerator *(m)*
En wave generator
Es generador de ondas *(m)*
Fr générateur d'ondes *(f)*
It generatore d'onda *(m)*

gerador magnetohidro-dinámico *(m)* Pt
De Magneto-hydrodynamik-Generator *(m)*
En magneto-hydrodynamic generator
Es generador magnetohidro-dinámico *(m)*
Fr générateur magnétohydro-dynamique *(m)*
It generatore magnetoidro-dinamico *(m)*

gerichtetes Bohren *(n)* De
En directional drilling
Es perforación direccional *(f)*
Fr forage dirigé *(m)*
It trivellazione direzionale *(f)*
Pt perfuração direccional *(f)*

Gesamtenergie *(f) n* De
En total energy
Es energía total *(f)*
Fr énergie totale *(f)*
It energia totale *(f)*
Pt energia total *(f)*

Gesamterzeugungskapazität *(f) n* De
En overall generating capacity
Es capacidad generadora total *(f)*
Fr capacité génératrice globale *(f)*
It capacità totale di generazione *(f)*
Pt capacidade geradora global *(f)*

Gesamtleistungsfähigkeit *(f) n* De
En overall efficiency
Es rendimiento global *(m)*
Fr rendement global *(m)*
It rendimento totale *(m)*
Pt eficiência global *(f)*

gesättigter Luft *(f)* De
En saturated air
Es aire saturado *(m)*
Fr air saturé *(m)*
It aria satura *(f)*
Pt ar saturado *(m)*

gesättigter Dampf *(m)* De
En saturated steam
Es vapor saturado *(m)*
Fr vapeur saturée *(f)*
It vapore saturo *(m)*
Pt vapor saturado *(m)*

gesättigter Dampfdruck *(m)* De
Am saturated vapor pressure
En saturated vapour pressure
Es presión de vapor saturado *(f)*
Fr tension de vapeur saturée *(f)*
It pressione del vapore saturo *(m)*
Pt pressão de vapor saturado *(f)*

geschäumte Polystyrolkugel *(f)* De
En expanded polystyrene pellet
Es gránulo de poliestireno expandido *(m)*
Fr granule de polystyrène expansé *(m)*
It palline di polistirolo espanso *(f)*
Pt grânulo de polistireno expandido *(m)*

geschlossener Kreislauf *(m)* De
En closed cycle
Es ciclo cerrado *(m)*
Fr circuit fermé *(m)*
It ciclo chiuso *(m)*
Pt ciclo fechado *(m)*

geschlossener Regelkreis *(m)* De
En closed loop
Es bucle cerrado *(m)*
Fr boucle de retour *(f)*
It circuito chiuso *(m)*
Pt circuito fechado *(m)*

Geschwindigkeit *(f) n* De
En speed; velocity
Es velocidad *(f)*
Fr vitesse *(f)*
It velocità *(f)*
Pt velocidade *(f)*

Gesteinskunde *(f) n* De
En lithology
Es litología *(f)*
Fr lithologie *(f)*
It litologia *(f)*
Pt litologia *(f)*

getto *(m) n* It
De Düse *(f)*
En jet
Es chorro *(m)*
Fr jet *(m)*
Pt jacto *(m)*

Gewicht *(n) n* De
En weight
Es peso *(m)*
Fr poids *(m)*
It peso *(m)*
Pt peso *(m)*

gewinnbare Reserven *(pl)* De
En recoverable reserves
Es reservas recuperables *(f pl)*
Fr réserves récupérables *(f pl)*
It riserve recuperabili *(f pl)*
Pt reservas recuperáveis *(f pl)*

gewogener Durchschnitt *(m)* De
En weighted mean
Es media compensada *(f)*
Fr moyenne pondérée *(f)*
It media ponderata *(f)*
Pt média ponderada *(f)*

geyser *n* En; Fr, It *(m)*
De Geysir *(m)*
Es géiser *(m)*
Pt geiser *(m)*

Geysir *(m) n* De
En geyser
Es géiser *(m)*
Fr geyser *(m)*
It geyser *(m)*
Pt geiser *(m)*

Gezeitendamm *(m) n* De
En tidal barrage
Es presa de marea *(f)*
Fr barrage marémoteur *(m)*
It sbarramento di marea *(m)*
Pt barragem de marés *(f)*

Gezeitenkraft *(f) n* De
En tidal power
Es potencia mareal *(f)*
Fr puissance marémotrice *(f)*
It potenza della marea *(f)*
Pt potência das marés *(f)*

Gezeitenkraft in Flußmündungen *(f)* De
En estuary tidal power
Es potencia mareal de estuario *(f)*
Fr puissance marémotrice d'estuaire *(f)*
It potenza della marea di estuario *(f)*
Pt potência de marés de estuário *(f)*

giacimento carbonifero *(m)* It
De Kohlenlager *(n)*
En coal measure
Es formación hullera *(f)*
Fr couche de houille *(f)*
Pt medida de carvão *(f)*

giacimento petrolifero *(m)* It
De Ölfeld *(n)*
En oilfield
Es yacimiento de petróleo *(m)*
Fr gisement pétrolifère *(m)*
Pt jazigo de petróleo *(m)*

gisement pétrolifère *(m)* Fr
De Ölfeld *(n)*
En oilfield
Es yacimiento de petróleo *(m)*
It giacimento petrolifero *(m)*
Pt jazigo de petróleo *(m)*

Gitter *(n) n* De
En grid (electronics)
Es rejilla *(f)*
Fr grille *(f)*
It griglia *(f)*
Pt grade *(f)*

Glasfaserisolierung *(f) n* De
Am glass-fiber insulation
En glass-fibre insulation
Es aislamiento con fibra de vidrio *(m)*
Fr isolement par fibre de verre *(m)*
It isolamento con fibra di vetro *(m)*
Pt isolamento com fibra de vidro *(m)*

glass-fiber insulation Am
De Glasfaserisolierung *(f)*
En glass-fibre insulation
Es aislamiento con fibra de vidrio *(m)*

Fr isolement par fibre de verre *(m)*
It isolamento con fibra di vetro *(m)*
Pt isolamento com fibra de vidro *(m)*

glass-fibre insulation En
Am glass-fiber insulation
De Glasfaserisolierung *(f)*
Es aislamiento con fibra de vidrio *(m)*
Fr isolement par fibre de verre *(m)*
It isolamento con fibra di vetro *(m)*
Pt isolamento com fibra de vidro *(m)*

glass heat exchanger En
De Glaswärmeaustauscher *(m)*
Es intercambiador de calor de vidrio *(m)*
Fr échangeur de chaleur en verre *(m)*
It scambiatore di calore di vetro *(m)*
Pt termo-permutador de vidro *(m)*

glassified *adj* En
De verglast
Es vitrificado
Fr vitrifié
It vetrificato
Pt vitrificado

Glaswärmeaustauscher *(m) n* De
En glass heat exchanger
Es intercambiador de calor de vidrio *(m)*
Fr échangeur de chaleur en verre *(m)*
It scambiatore di calore di vetro *(m)*
Pt termo-permutador de vidro *(m)*

Glattrohr-Kollektor *(m) n* De
En bare collector
Es colector desnudo *(m)*
Fr collecteur nu *(m)*
It collettore nudo *(m)*
Pt colector descoberto *(m)*

Gleichgewicht *(n) n* De
En equilibrium
Es equilibrio *(m)*
Fr équilibre *(m)*
It equilibrio *(m)*
Pt equilíbrio *(m)*

Gleichrichter (elektrischer) *(m) n* De
En rectifier (electrical)
Es rectificador (eléctrico) *(m)*
Fr redresseur (électrique) *(m)*
It raddrizzatore (elettrico) *(m)*
Pt rectificador (eléctrico) *(m)*

Gleichstrom *(m) n* De
En direct current (d.c.)
Es corriente continua (c.c.) *(f)*
Fr courant continu (c.c.) *(m)*
It corrente continua (c.c.) *(f)*
Pt corrente directa (c.d.) *(f)*

glicole etilenico *(m)* It
De Äthylenglykol *(n)*
En ethylene glycol
Es glicoletileno *(m)*
Fr éthylène glycol *(m)*
Pt glicol etilénico *(m)*

glicol etilénico *(m)* Pt
De Äthylenglykol *(n)*
En ethylene glycol
Es glicoletileno *(m)*
Fr éthylène glycol *(m)*
It glicole etilenico *(m)*

glicoletileno *(m) n* Es
De Äthylenglykol *(n)*
En ethylene glycol
Fr éthylène glycol *(m)*
It glicole etilenico *(m)*
Pt glicol etilénico *(m)*

Glühfaden *(m) n* De
En filament
Es filamento *(m)*
Fr filament *(m)*
It filamento *(m)*
Pt filamento *(m)*

goudron *(m) n* Fr
De Koksofenteer *(m)*
En coke-oven tar
Es alquitrán de horno de coquización *(m)*
It catrame di forno da coke *(m)*
Pt alcatrão de forno de coque *(m)*

goudron à basse température *(m)* Fr
De Tieftemperaturteer *(m)*
En low-temperature tar
Es alquitrán de baja temperatura *(m)*
It catrame a bassa temperatura *(m)*
Pt alcatrão de baixa temperatura *(m)*

goudron de houille *(m)* Fr
De Steinkohlenteer *(m)*
En coal tar
Es alquitrán de hulla *(m)*
It catrame di carbon fossile *(m)*
Pt alcatrão de hulha *(m)*

grade *(f) n* Pt
De Gitter *(n)*
En grid (electronics)
Es rejilla *(f)*
Fr grille *(f)*
It griglia *(f)*

gradient de température *(m)* Fr
De Temperaturgefälle *(n)*
En temperature gradient
Es gradiente de temperatura *(m)*
It gradiente di temperatura *(m)*
Pt gradiente de temperatura *(m)*

gradiente de temperatura *(m)* Es, Pt
De Temperaturgefälle *(n)*
En temperature gradient
Fr gradient de température *(m)*
It gradiente di temperatura *(m)*

gradiente di temperatura *(m)* It
De Temperaturgefälle *(n)*
En temperature gradient
Es gradiente de temperatura *(m)*
Fr gradient de température *(m)*
Pt gradiente de temperatura *(m)*

gradiente geotermico *(m)* It
De geothermisches Gefälle *(n)*
En geothermal gradient
Es gradiente geotérmico *(m)*
Fr gradient géothermique *(m)*
Pt gradiente geotérmico *(m)*

gradiente geotérmico *(m)* Es, Pt
De geothermisches Gefälle *(n)*
En geothermal gradient
Fr gradient géothermique *(m)*
It gradiente geotermico *(m)*

gradiente termico oceanico *(m)* It
De Meereswärmegefälle *(n)*
En ocean thermal gradient
Es gradiente térmico oceánico *(m)*
Fr gradient thermique des océans *(m)*
Pt gradiente térmico oceânico *(m)*

gradiente térmico oceánico *(m)* Es
De Meereswärmegefälle *(n)*
En ocean thermal gradient
Fr gradient thermique des océans *(m)*
It gradiente termico oceanico *(m)*
Pt gradiente térmico oceânico *(m)*

gradiente térmico oceânico *(m)* Pt
De Meereswärmegefälle *(n)*
En ocean thermal gradient
Es gradiente térmico oceánico *(m)*
Fr gradient thermique des océans *(m)*

It gradiente termico oceanico *(m)*

gradient géothermique *(m)* Fr
De geothermisches Gefälle *(n)*
En geothermal gradient
Es gradiente geotérmico *(m)*
It gradiente geotermico *(m)*
Pt gradiente geotérmico *(m)*

gradient thermique des océans *(m)* Fr
De Meereswärmegefälle *(n)*
En ocean thermal gradient
Es gradiente térmico oceánico *(m)*
It gradiente termico oceanico *(m)*
Pt gradiente térmico oceânico *(m)*

grading *n* En
De Klassierung *(f)*
Es clasificación *(f)*
Fr criblage *(m)*
It cernita *(f)*
Pt granulometria *(f)*

gráfica *(f) n* Es
De Diagramm *(n)*
En graph
Fr graphique *(m)*
It grafico *(m)*
Pt gráfico *(m)*

grafico *(m) n* It
De Diagramm *(n)*
En graph
Es gráfica *(f)*
Fr graphique *(m)*
Pt gráfico *(m)*

gráfico *(m) n* Pt
De Diagramm *(n)*
En graph
Es gráfica *(f)*
Fr graphique *(m)*
It grafico *(m)*

grafite *(f) n* It, Pt
De Graphit *(n)*
En graphite
Es grafito *(m)*
Fr graphite *(m)*

grafito *(m) n* Es
De Graphit *(n)*
En graphite
Fr graphite *(m)*
It grafite *(f)*
Pt grafite *(f)*

grandezza scalare *(f)* It
De Skalarquantität *(f)*
En scalar quantity
Es cantidad escalar *(f)*
Fr quantité scalaire *(f)*
Pt quantidade escalar *(f)*

granja solar *(f)* Es
De Solarfarm *(f)*
En solar farm
Fr ferme solaire *(f)*
It podere solare *(m)*
Pt centro de produção solar *(f)*

granule de polystyrène expansé *(m)* Fr
De geschäumte Polystyrolkugel *(f)*
En expanded polystyrene pellet
Es gránulo de poliestireno expandido *(m)*
It palline di polistirolo espanso *(f)*
Pt grânulo de polistireno expandido *(m)*

gránulo de poliestireno expandido *(m)* Es
De geschäumte Polystyrolkugel *(f)*
En expanded polystyrene pellet
Fr granule de polystyrène expansé *(m)*
It palline di polistirolo espanso *(f)*
Pt grânulo de polistireno expandido *(m)*

grânulo de polistireno expandido *(m)* Pt
De geschäumte Polystyrolkugel *(f)*
En expanded polystyrene pellet
Es gránulo de poliestireno expandido *(m)*
Fr granule de polystyrène expansé *(m)*
It palline di polistirolo espanso *(f)*

granulometria *(f) n* Pt
De Klassierung *(f)*
En grading
Es clasificación *(f)*
Fr criblage *(m)*
It cernita *(f)*

graph *n* En
De Diagramm *(n)*
Es gráfica *(f)*
Fr graphique *(m)*
It grafico *(m)*
Pt gráfico *(m)*

graphique *(m) n* Fr
De Diagramm *(n)*
En graph
Es gráfica *(f)*
It grafico *(m)*
Pt gráfico *(m)*

Graphit *(n) n* De
En graphite
Es grafito *(m)*
Fr graphite *(m)*
It grafite *(f)*
Pt grafite *(f)*

graphite *n* En, Fr *(m)*
De Graphit *(n)*
Es grafito *(m)*
It grafite *(f)*
Pt grafite *(f)*

graphite-moderated *adj* En
De graphitmoderiert
Es moderado por grafito
Fr modéré au graphite
It moderato a grafite
Pt moderado com grafite

graphitmoderiert *adj* De
En graphite-moderated
Es moderado por grafito
Fr modéré au graphite
It moderato a grafite
Pt moderado com grafite

gravedad *(f) n* Es
De Schwerkraft *(f)*
En gravity
Fr gravité *(f)*
It gravità *(f)*
Pt gravidade *(f)*

gravidade *(f) n* Pt
De Schwerkraft *(f)*
En gravity
Es gravedad *(f)*
Fr gravité *(f)*
It gravità *(f)*

gravità *(f) n* It
De Schwerkraft *(f)*
En gravity
Es gravedad *(f)*
Fr gravité *(f)*
Pt gravidade *(f)*

gravitational constant En
De Gravitations-konstante *(f)*
Es constante de gravitación *(f)*
Fr constante de gravitation *(f)*
It costante gravitazionale *(f)*
Pt constante gravitacional *(f)*

gravitational field En
De Gravitationsfeld *(n)*
Es campo gravitacional *(m)*
Fr champ de gravitation *(m)*
It campo gravitazionale *(m)*
Pt campo de gravidade *(m)*

gravitational force En
De Erdanziehungskraft *(f)*
Es fuerza gravitatoria *(f)*
Fr force de gravitation *(f)*
It forza di gravità *(f)*
Pt força de gravidade *(f)*

Gravitationsfeld *(n) n* De
En gravitational field
Es campo gravitacional *(m)*
Fr champ de gravitation *(m)*
It campo gravitazionale *(m)*
Pt campo de gravidade *(m)*

Gravitationskonstante *(f) n* De
En gravitational constant
Es constante de gravitación *(f)*
Fr constante de gravitation *(f)*
It costante gravitazionale *(f)*
Pt constante gravitacional *(f)*

gravité *(f) n* Fr
De Schwerkraft *(f)*
En gravity
Es gravedad *(f)*
It gravità *(f)*
Pt gravidade *(f)*

gravity *n* En
De Schwerkraft *(f)*
Es gravedad *(f)*
Fr gravité *(f)*
It gravità *(f)*
Pt gravidade *(f)*

gravity-feed *n* En
De Schwerkraftspeisung *(f)*
Es alimentación por gravedad *(f)*
Fr alimentation par gravité *(f)*
It alimentazione a gravità *(f)*
Pt alimentação por gravidade

grenzflächenaktiver Stoff *(m)* De
En surfactant
Es surfactante *(m)*
Fr agent tensio-actif *(m)*
It agente tensioattivo *(m)*
Pt agente tenso-activo *(m)*

grid (electronics) *n* En
De Gitter *(n)*
Es rejilla *(f)*
Fr grille *(f)*
It griglia *(f)*
Pt grade *(f)*

grid (electrical power) *n* En
De Überlandleitungsnetz *(n)*
Es red nacional de energía eléctrica *(f)*
Fr réseau électrique national *(m)*
It rete nazionale *(f)*
Pt rede eléctrica *(f)*

griglia *(f) n* It
De Gitter *(n)*
En grid (electronics)
Es rejilla *(f)*
Fr grille *(f)*
Pt grade *(f)*

grille *(f) n* Fr
De Gitter *(n)*
En grid (electronics)
Es rejilla *(f)*
It griglia *(f)*
Pt grade *(f)*

gross *adj* En
De brutto
Es bruto
Fr brut
It lordo
Pt bruto

ground *n* Am
De Erde *(f)*
En earth
Es tierra *(f)*
Fr terre *(f)*
It terra *(f)*
Pt massa *(f)*

ground-scattered *adj* En
De bodengestreut
Es dispersado en tierra
Fr diffusé au sol
It diffuso sul suolo
Pt disperso pelo solo

Grubenbetrieb *(m) n* De
En underground mining
Es minería subterránea *(f)*
Fr exploitation minière souterraine *(f)*
It scavo sotteraneo *(m)*
Pt mineração subterrânea *(f)*

Grundheizung *(f) n* De
En background heat
Es calor de fondo *(m)*
Fr chaleur ambiante *(f)*
It calore di fondo *(m)*
Pt calor de fundo *(m)*

Grundkraft *(f) n* De
En base load
Es carga base *(f)*
Fr charge minimale *(f)*
It carico base *(m)*
Pt carga fundamental *(f)*

Grundumsatz *(m) n* De
En basal metabolic rate
Es índice metabólico basal *(m)*
Fr métabolisme basal *(m)*
It tasso metabolico basale *(m)*
Pt índice de metabolismo basal *(m)*

gruppo motore-dinamo *(m)* It
De Motor-Generator *(m)*
En motor-generator
Es motor-generador *(m)*
Fr convertisseur *(m)*
Pt moto-gerador *(m)*

guadagno di calore incidentale *(m)* It
De zufällige Wärmezunahme *(f)*
En incidental heat gain
Es ganancia de calor incidental *(f)*
Fr gain de chaleur incident *(m)*
Pt ganho de calor acessório *(m)*

guadagno solare *(m)* It
De Solarverstärkung *(f)*
En solar gain
Es ganancia solar *(f)*
Fr gain solaire *(m)*
Pt ganho solar *(m)*

H

Hahn *(m) n* De
En cock
Es espita *(f)*
Fr robinet *(m)*
It rubinetto *(m)*
Pt torneira *(f)*

halbbituminöse Kohle *(f)* De
En semibituminous coal
Es carbón semibituminoso *(m)*
Fr houille demi-grasse *(f)*
It carbone semibituminoso *(m)*
Pt carvão semibetuminoso *(m)*

Halbleiter *(m) n* De
En semiconductor
Es semiconductor *(m)*
Fr semiconducteur *(m)*
It semiconduttore *(m)*
Pt semicondutor *(m)*

Halbwertszeit *(f) n* De
En half-life
Es media vida *(f)*
Fr demi-vie *(f)*
It periodo di dimezzamento *(m)*
Pt meia vida *(f)*

Haldenabfall *(m) n* De
En tailings *(pl)*
Es desechos *(m pl)*
Fr produits de queue *(m pl)*
It residui di scarto *(m pl)*
Pt colas *(f pl)*

half-life *n* En
De Halbwertszeit *(f)*
Es media vida *(f)*
Fr demi-vie *(f)*
It periodo di dimezzamento *(m)*
Pt meia vida *(f)*

hard coal En
De Steinkohle *(f)*
Es carbón duro *(m)*
Fr charbon dure *(m)*
It carbone duro *(m)*
Pt carvão duro *(m)*

hard radiation En
De harte Strahlung *(f)*
Es radiación dura *(f)*
Fr radiation dure *(f)*
It radiazione dura *(f)*
Pt radiação dura *(f)*

Harnstoff-Formaldehydschaumstoff *(m) n* De

En urea-formaldehyde foam
Es espuma de urea-formaldehido *(f)*
Fr mousse urée et formaldéhyde *(f)*
It espanso di urea-formaldeide *(m)*
Pt espuma de ureia-formaldeído *(f)*

harte Strahlung *(f)* De
En hard radiation
Es radiación dura *(f)*
Fr radiation dure *(f)*
It radiazione dura *(f)*
Pt radiação dura *(f)*

haste de controle *(f)* Pt
De Kontrollstab *(m)*
En control rod
Es varilla de control *(f)*
Fr barre de commande *(f)*
It barra di controllo *(f)*

Haushalt-Energie-verbrauch *(m) n* De
En domestic energy consumption
Es consumo doméstico de energía *(m)*
Fr consommation domestique d'énergie *(f)*
It consumo di energia per usi domestici *(m)*
Pt consumo doméstico de energia *(m)*

haute tension *(f)* Fr
De Hochspannung *(f)*
En high tension (HT)
Es alta tensión *(f)*
It alta tensione (AT) *(f)*
Pt alta tensão *(f)*

hauteur de refoulement *(f)* Fr
De Druckhöhe *(f)*
En pressure head
Es altura manométrica *(f)*
It altezza manometrica *(f)*
Pt altura manométrica *(f)*

haut fourneau *(m)* Fr
De Hochofen *(m)*
En blast furnace
Es alto horno *(m)*
It alto forno *(m)*
Pt alto-forno *(m)*

hearth furnace En
De Herdofen *(m)*
Es horno de forja *(m)*
Fr four à sole *(m)*
It forno a suola *(m)*
Pt forno de fornalha *(m)*

heat *n* En
De Wärme *(f)*
Es calor *(m)*
Fr chaleur *(f)*
It calore *(m)*
Pt calor *(m)*

heat balance En
De Wärmeausgleich *(m)*
Es equilibrio térmico *(m)*
Fr bilan calorifique *(m)*
It bilancio termico *(m)*
Pt equilíbrio térmico *(m)*

heat capacity En
De Wärmekapazität *(f)*
Es capacidad térmica *(f)*
Fr capacité calorifique *(f)*
It capacità termica *(f)*
Pt capacidade térmica *(f)*

heat energy En
De Wärmeenergie *(f)*
Es energía térmica *(f)*
Fr énergie calorifique; énergie thermique *(f)*
It energia termica *(f)*
Pt energia térmica *(f)*

heat engine En
De Wärmekraftmaschine *(f)*
Es motor térmico *(m)*
Fr moteur thermique *(m)*
It motore termico *(m)*
Pt motor térmico *(m)*

heater *n* En
De Heizelement *(n)*
Es calentador *(m)*
Fr réchauffeur *(m)*
It riscaldatore *(m)*
Pt aquecedor *(m)*

heater coil En
De Heizschlange *(f)*
Es serpentín de calentamiento *(m)*
Fr serpentin de chauffage *(m)*
It serpentina di riscaldamento *(f)*
Pt serpentina de aquecimento *(f)*

heat exchanger En
De Wärmeaustauscher *(m)*
Es intercambiador de calor *(m)*
Fr échangeur de chaleur *(m)*
It scambiatore di calore *(m)*
Pt termo-permutador *(m)*

heating *n* En
De Heizung *(f)*
Es calefacción *(f)*
Fr chauffage *(m)*
It riscaldamento *(m)*
Pt aquecimento *(m)*

heating oil En
De Heizöl *(n)*
Es petróleo de calefacción *(m)*
Fr huile de chauffe *(f)*
It olio per riscaldamento *(m)*
Pt óleo para aquecimento *(m)*

heat loss En
De Wärmeverlust *(m)*
Es pérdida de calor *(f)*
Fr perte de chaleur *(f)*
It perdita di calore *(f)*
Pt perda de calor *(f)*

heat of combustion En
De Verbrennungswärme *(f)*
Es calor de combustión *(m)*
Fr chaleur de combustion *(f)*
It calore di combustione *(m)*
Pt calor de combustão *(m)*

heat of reaction En
De Reaktionswärme *(f)*
Es calor de reacción *(m)*
Fr chaleur de réaction *(f)*
It calore di reazione *(m)*
Pt calor de reacção *(m)*

heat pipe En
De Wärmeübertragungs-rohr *(n)*
Es tubo calefactor *(m)*
Fr caloduc *(m)*
It tubo di calore *(m)*
Pt tubo de calor *(m)*

heat pump En
De Wärmepumpe *(f)*
Es bomba de calor *(f)*
Fr thermopompe *(f)*
It pompa di calore *(f)*
Pt bomba térmica *(f)*

heat recovery En
De Wärmerück-gewinnung *(f)*
Es recuperación del calor *(f)*
Fr récupération de chaleur *(f)*
It recupero di calore *(m)*
Pt recuperação térmica *(f)*

heat reservoir En
De Wärmespeicher *(m)*
Es depósito de calor *(m)*
Fr réservoir de chaleur *(m)*
It serbatoio di calore *(m)*
Pt reservatório de calor *(m)*

heat shield En
De Wärmeschutz *(m)*
Es pantalla térmica *(f)*
Fr bouclier thermique *(m)*
It schermo termico *(m)*
Pt protector térmico *(m)*

heat storage En
De Wärmespeicherung *(f)*
Es almacenamiento de calor *(m)*
Fr accumulation de chaleur *(f)*
It immagazzinaggio di calore *(m)*
Pt armazenagem de calor *(f)*

heat transfer En
De Wärmeübertragung *(f)*
Es transferencia térmica *(f)*
Fr transfert de chaleur *(m)*
It scambio di calore *(m)*
Pt transferência térmica *(f)*

heat-transfer coefficient En
De Wärmeübertragungszahl *(f)*
Es coeficiente de transferencia térmica *(m)*
Fr coefficient de transfert de chaleur *(m)*
It coefficiente di scambio di calore *(m)*
Pt coeficiente de transferência térmica *(m)*

heat-transfer fluid En
De Wärmeübertragungsflüssigkeit *(f)*
Es flúido de transferencia térmica *(m)*
Fr fluide de transfert de chaleur *(m)*
It fluido di scambio del calore *(m)*
Pt fluido de transferência térmica *(m)*

heat treatment En
De Warmbehandlung *(f)*
Es tratamiento térmico *(m)*
Fr traitement thermique *(m)*
It trattamento termico *(m)*
Pt tratamento térmico *(m)*

heat-work cycle En
De Wärmearbeitszyklus *(m)*
Es ciclo de calor-trabajo *(m)*
Fr cycle de chaleur-travail *(m)*
It ciclo calore-lavoro *(m)*
Pt ciclo de calor-trabalho *(m)*

heavy-fuel engines *(pl)* En
De Masutmotoren *(pl)*
Es motores de combustible pesado *(m pl)*
Fr moteurs au fuel lourd *(m pl)*
It motori a combustibile pesante *(m pl)*
Pt motores a óleos pesados *(m pl)*

heavy hydrogen En
De schwerer Wasserstoff *(m)*
Es hidrógeno pesado *(m)*
Fr hydrogène lourd *(m)*
It idrogeno pesante *(m)*
Pt hidrogénio pesado *(m)*

heavy water En
De Schwerwasser *(n)*
Es agua pesada *(f)*
Fr eau lourde *(f)*
It acqua pesante *(f)*
Pt água pesada *(f)*

heavy-water reactor En
De Schwerwasserreaktor *(m)*
Es reactor de agua pesada *(m)*
Fr réacteur à eau lourde *(m)*
It reattore ad acqua pesante *(m)*
Pt reactor de água pesada *(m)*

heiße Quellen *(pl)* De
En hot springs *(pl)*
Es termas *(f pl)*
Fr sources chaudes *(f pl)*
It sorgenti calde *(f pl)*
Pt fontes de água quente *(f pl)*

Heißluftmotor *(m) n* De
En hot-air engine
Es motor de aire caliente *(m)*
Fr moteur à air chaud *(m)*
It motore ad aria calda *(m)*
Pt motor a ar quente *(m)*

Heißluftspeicherung *(f) n* De
En hot-air storage
Es almacenamiento de aire caliente *(m)*
Fr accumulation d'air chaud *(f)*
It immagazzinaggio di aria calda *(m)*
Pt armazenagem de ar quente *(f)*

Heißwasserkühlung *(f) n* De
En ebullient cooling
Es enfriamiento desde la ebullición *(m)*
Fr refroidissement par ébullition *(m)*
It raffreddamento a ebollizione *(m)*
Pt arrefecimento ebuliente *(m)*

Heizelement *(n) n* De
En heater
Es calentador *(m)*
Fr réchauffeur *(m)*
It riscaldatore *(m)*
Pt aquecedor *(m)*

Heizelement mit Gebläse *(n)* De
En fan-assisted heater
Es calentador auxiliado por ventilador *(m)*
Fr radiateur soufflant *(m)*
It riscaldatore a ventilatore *(m)*
Pt aquecedor auxiliado por ventoinha *(m)*

Heizelement mit offenem Abzug *(n)* De
En open-flue heater
Es calentador de chimenea abierta *(m)*
Fr appareil de chauffage à conduit ouvert *(m)*
It riscaldatore a condotto aperto *(m)*
Pt aquecedor de fumeiro aberto *(m)*

Heizfaden *(m) n* De
En hot-wire filament
Es filamento de hilo caliente *(m)*
Fr filament thermique *(m)*
It filamento a filo caldo *(m)*
Pt filamento de fio quente *(m)*

Heizkörper *(m) n* De
En radiator
Es radiador *(m)*
Fr radiateur *(m)*
It radiatore *(m)*
Pt radiador *(m)*

Heizkraft *(f) n* De
En thermal power
Es potencia térmica *(f)*
Fr puissance thermique *(f)*
It potenza termica *(f)*
Pt potência térmica *(f)*

Heizöl *(n) n* De
En heating oil
Es petróleo de calefacción *(m)*
Fr huile de chauffe *(f)*
It olio per riscaldamento *(m)*
Pt óleo para aquecimento *(m)*

Heizrohrkessel *(m) n* De
En fire-tube boiler
Es caldera pirotubular *(f)*
Fr chaudière à tube de fumée *(f)*
It caldaia a tubo di fumo *(f)*
Pt caldeira de tubo de fumaça *(f)*

Heizschlange *(f) n* De
En calorifier; heater coil
Es calorífero; serpentín de calentamiento *(m)*
Fr calorifiant; serpentin de chauffage *(m)*
It calorifero *(m)*; serpentina di riscaldamento *(f)*
Pt calorificador *(m)*; serpentina de aquecimento *(f)*

Heizung *(f) n* De
En heating
Es calefacción *(f)*
Fr chauffage *(m)*
It riscaldamento *(m)*
Pt aquecimento *(m)*

Heizwert *(m) n* De
En calorific value
Es potencia calorífica *(f)*
Fr valeur calorifique *(f)*
It potere calorifico *(m)*
Pt valor calorífico *(m)*

helio *(m) n* Es
De Helium *(n)*
En helium
Fr hélium *(m)*
It elio *(m)*
Pt hélio *(m)*

hélio *(m) n* Pt
De Helium *(n)*
En helium
Es helio *(m)*
Fr hélium *(m)*
It elio *(m)*

heliostat *n* En
De Heliostat *(m)*
Es heliostato *(m)*
Fr héliostat *(m)*
It eliostato *(m)*
Pt heliostato *(m)*

Heliostat *(m) n* De
En heliostat
Es heliostato *(m)*
Fr héliostat *(m)*
It eliostato *(m)*
Pt heliostato *(m)*

héliostat *(m) n* Fr
De Heliostat *(m)*
En heliostat
Es heliostato *(m)*
It eliostato *(m)*
Pt heliostato *(m)*

heliostato *(m) n* Es, Pt
De Heliostat *(m)*
En heliostat
Fr héliostat *(m)*
It eliostato *(m)*

helium *n* En
De Helium *(n)*
Es helio *(m)*
Fr hélium *(m)*
It elio *(m)*
Pt hélio *(m)*

Helium *(n) n* De
En helium
Es helio *(m)*
Fr hélium *(m)*
It elio *(m)*
Pt hélio *(m)*

hélium *(m) n* Fr
De Helium *(n)*
En helium
Es helio *(m)*
It elio *(m)*
Pt hélio *(m)*

Helligkeit *(f) n* De
En brightness
Es brillo *(m)*
Fr brillance *(f)*
It luminosità *(f)*
Pt brilho *(m)*

Herdofen *(m) n* De
En hearth furnace
Es horno de forja *(m)*
Fr four à sole *(m)*
It forno a suola *(m)*
Pt forno de fornalha *(m)*

Hexan *(n) n* De
En hexane
Es hexano *(m)*
Fr hexane *(m)*
It esano *(m)*
Pt hexano *(m)*

hexane *n* En; Fr *(m)*
De Hexan *(n)*
Es hexano *(m)*
It esano *(m)*
Pt hexano *(m)*

hexano *(m) n* Es, Pt
De Hexan *(n)*
En hexane
Fr hexane *(m)*
It esano *(m)*

hidracina *(f) n* Es
De Hydrazin *(n)*
En hydrazine
Fr hydrazine *(f)*
It idrazina *(f)*
Pt hidrazina *(f)*

hidrano *(m) n* Es, Pt
De Hydran *(n)*
En hydrane
Fr hydrane *(m)*
It idrano *(m)*

hidráulico *adj* Es, Pt
De hydraulisch
En hydraulic
Fr hydraulique
It idraulico

hidrazina *(f) n* Pt
De Hydrazin *(n)*
En hydrazine
Es hidracina *(f)*
Fr hydrazine *(f)*
It idrazina *(f)*

hidroalmacenamiento *(m) n* Es
De Wasserspeicherung *(f)*
En hydrostorage
Fr hydrostockage *(m)*
It idroimmagazzinaggio *(m)*
Pt hidroarmazenamento *(m)*

hidroarmazenamento *(m) n* Pt
De Wasserspeicherung *(f)*
En hydrostorage
Es hidroalmacen-amiento *(m)*
Fr hydrostockage *(m)*
It idroimmagazzinaggio *(m)*

hidrocarboneto *(m) n* Pt
De Kohlenwasserstoff *(m)*
En hydrocarbon
Es hidrocarburo *(m)*
Fr hydrocarbure *(m)*
It idrocarburo *(m)*

hidrocarbonetos líquidos *(m pl)* Pt
De flüssige Kohlenwasserstoffe *(pl)*
En liquid hydrocarbons
Es hidrocarburos líquidos *(m pl)*
Fr hydrocarbures liquides *(m pl)*
It idrocarburi liquidi *(m pl)*

hidrocarburo *(m) n* Es
De Kohlenwasserstoff *(m)*
En hydrocarbon
Fr hydrocarbure *(m)*
It idrocarburo *(m)*
Pt hidrocarboneto *(m)*

hidrocarburos líquidos *(m pl)* Es
De flüssige Kohlenwasserstoffe *(pl)*
En liquid hydrocarbons
Fr hydrocarbures liquides *(m pl)*
It idrocarburi liquidi *(m pl)*
Pt hidrocarbonetos líquidos *(m pl)*

hidro-cracking *(m) n* Pt
De Hydrokracken *(n)*
En hydrocracking
Es hidrocrácking *(m)*
Fr hydrocraquage *(m)*
It piroscissione idraulica *(f)*

hidrocrácking *(m) n* Es
De Hydrokracken *(n)*
En hydrocracking
Fr hydrocraquage *(m)*
It piroscissione idraulica *(f)*
Pt hidro-cracking *(m)*

hidroelectricidad *(f) n* Es
De Hydroelektrizität *(f)*
En hydroelectricity
Fr hydroélectricité *(f)*
It idroelettricità *(f)*
Pt hidroelectricidade *(f)*

hidroelectricidade *(f) n* Pt
De Hydroelektrizität *(f)*
En hydroelectricity
Es hidroelectricidad *(f)*
Fr hydroélectricité *(f)*
It idroelettricità *(f)*

hidrogaseificação *(f) n* Pt
De Hydro-Vergasung *(f)*
En hydrogasification
Es hidrogasificación *(f)*
Fr hydrogazéification *(f)*
It idrogassificazione *(f)*

hidrogasificación *(f) n* Es
De Hydro-Vergasung *(f)*
En hydrogasification
Fr hydrogazéification *(f)*
It idrogassificazione *(f)*
Pt hidrogaseificação *(f)*

hidrogenação *(f)* *n* Pt
De Hydrierung *(f)*
En hydrogenation
Es hidrogenación *(f)*
Fr hydrogénation *(f)*
It idrogenazione *(f)*

hidrogenación *(f)* *n* Es
De Hydrierung *(f)*
En hydrogenation
Fr hydrogénation *(f)*
It idrogenazione *(f)*
Pt hidrogenação *(f)*

hidrogénio *(m)* *n* Pt
De Wasserstoff *(m)*
En hydrogen
Es hidrógeno *(m)*
Fr hydrogène *(m)*
It idrogeno *(m)*

hidrogénio pesado *(m)* Pt
De schwerer Wasserstoff *(m)*
En heavy hydrogen
Es hidrógeno pesado *(m)*
Fr hydrogène lourd *(m)*
It idrogeno pesante *(m)*

hidrógeno *(m)* *n* Es
De Wasserstoff *(m)*
En hydrogen
Fr hydrogène *(m)*
It idrogeno *(m)*
Pt hidrogénio *(m)*

hidrógeno pesado *(m)* Es
De schwerer Wasserstoff *(m)*
En heavy hydrogen
Fr hydrogène lourd *(m)*
It idrogeno pesante *(m)*
Pt hidrogénio pesado *(m)*

hidrólise *(f)* *n* Pt
De Hydrolyse *(f)*
En hydrolysis
Es hidrólisis *(f)*
Fr hydrolyse *(f)*
It idrolisi *(f)*

hidrologia *(f)* *n* Pt
De Hydrologie *(f)*
En hydrology
Es hidrología *(f)*
Fr hydrologie *(f)*
It idrologia *(f)*

hidrología *(f)* *n* Es
De Hydrologie *(f)*
En hydrology
Fr hydrologie *(f)*
It idrologia *(f)*
Pt hidrologia *(f)*

hidroseparador *(m)* *n* Es, Pt
De Unterwasser-Stromapparat *(m)*
En hydroseparator
Fr hydroséparateur *(m)*
It idroseparatore *(m)*

hidrotérmico *adj* Es, Pt
De hydrothermal
En hydrothermal
Fr hydrothermique
It idrotermico

high-frequency dryer En
De Hochfrequenz-Trockner *(m)*
Es secador de alta frecuencia *(m)*
Fr sécheur haute fréquence *(m)*
It essiccatore ad alta frequenza *(m)*
Pt secador de alta frequência *(m)*

high-temperature reactor (HTR) En
De Hochtemperatur-reaktor *(m)*
Es reactor de alta temperatura *(m)*
Fr réacteur à haute température *(m)*
It reattore ad alta temperatura *(m)*
Pt reactor de alta temperatura *(m)*

high tension (HT) En
De Hochspannung *(f)*
Es alta tensión *(f)*
Fr haute tension *(f)*
It alta tensione (AT) *(f)*
Pt alta tensão *(f)*

hilada hidrófuga *(f)* Es
De Feuchtigkeits-sperrschicht *(f)*
En damp-proof course
Fr couche hydrofuge *(f)*
It trattamento impermeabilizzante *(m)*
Pt curso a prova de imfietrações *(m)*

hilos explosivos *(m pl)* Es
De Sprengdrähte *(pl)*
En exploding wires
Fr fils explosifs *(m pl)*
It fili esplosivi *(m pl)*
Pt fios de explosão *(m pl)*

histerese *(f)* *n* Pt
De Hysterese *(f)*
En hysteresis
Es histéresis *(f)*
Fr hystérésis *(f)*
It isteresi *(f)*

histéresis *(f)* *n* Es
De Hysterese *(f)*
En hysteresis
Fr hystérésis *(f)*
It isteresi *(f)*
Pt histerese *(f)*

Hochfrequenz-Trockner *(m)* *n* De
En high-frequency dryer
Es secador de alta frecuencia *(m)*
Fr sécheur haute fréquence *(m)*
It essiccatore ad alta frequenza *(m)*
Pt secador de alta frequência *(m)*

Hochofen *(m)* *n* De
En blast furnace
Es alto horno *(m)*
Fr haut fourneau *(m)*
It alto forno *(m)*
Pt alto-forno *(m)*

Hochspannung *(f)* *n* De
En high tension (HT)
Es alta tensión *(f)*
Fr haute tension *(f)*
It alta tensione (AT) *(f)*
Pt alta tensão *(f)*

Höchstspannungs-transmission *(f)* *n* De
En extra-high voltage transmission
Es transmisión con tensión muy alta *(f)*
Fr transmission des ultra-hautes tensions *(f)*
It trasmissione a tensione extra elevata *(f)*
Pt transmissão de voltagem extra-alta *(f)*

Hochtemperaturreaktor *(m)* *n* De
En high-temperature reactor (HTR)
Es reactor de alta temperatura *(m)*
Fr réacteur à haute température *(m)*
It reattore ad alta temperatura *(m)*
Pt reactor de alta temperatura *(m)*

hogar *(m)* *n* Es
De Feuerkiste *(f)*
En firebox
Fr foyer *(m)*
It focolaio *(m)*
Pt fornalha *(f)*

hogar de cenizas pulverulentas *(m)* Es
De Trockenasche-Ofen *(m)*
En dry-ash furnace
Fr four à cendres sèches *(m)*
It forno di cenere secca *(m)*
Pt forno de cinze seca *(m)*

hogar de turbulencia *(m)* Es
De Wirbelofen *(m)*
En cyclone furnace
Fr four à cyclone *(m)*
It forno a ciclone *(m)*
Pt forno de ciclone *(m)*

Höhe *(f)* *n* De
En altitude
Es altitud *(f)*
Fr altitude *(f)*
It altitudine *(f)*
Pt altitude *(f)*

hole conduction En
De Defektleitung *(f)*
Es conducción por lagunas *(f)*
Fr conduction par les trous *(f)*

It conduzione dei buchi *(f)*
Pt condução por furos *(f)*

hollín *(m) n* Es
De Ruβ *(m)*
En soot
Fr suie *(f)*
It nerofumo *(m)*
Pt fulígem *(f)*

Holzkohle *(f) n* De
En charcoal
Es carbón vegetal *(m)*
Fr charbon de bois *(m)*
It carbone dolce *(m)*
Pt carvão vegital *(m)*

honeycomb structure En
De Wabenstruktur *(f)*
Es estructura alveolar *(f)*
Fr structure en nids d'abeilles *(f)*
It structura a nido d'ape *(f)*
Pt estrutura alveolar *(f)*

horizontaler Sonnenfluβ *(m)* De
En horizontal solar flux
Es flujo solar horizontal *(m)*
Fr flux solaire horizontal *(m)*
It flusso solare orizzontale *(m)*
Pt fluxo solar horizontal *(m)*

horizontal solar flux En
De horizontaler Sonnenfluβ *(m)*
Es flujo solar horizontal *(m)*
Fr flux solaire horizontal *(m)*
It flusso solare orizzontale *(m)*
Pt fluxo solar horizontal *(m)*

hormigón cargado *(m)* Es
De beschwerter Beton *(m)*
En loaded concrete
Fr béton chargé *(m)*
It cemento pesante *(m)*
Pt betão carregado *(m)*

horno *(m) n* Es
De Ofen *(m)*
En furnace
Fr four *(m)*
It forno *(m)*
Pt forno *(m)*

horno de arco *(m)* Es
De Lichtbogenofen *(m)*
En arc furnace
Fr four à arc *(m)*
It forno ad arco *(m)*
Pt forno a arco voltáico *(m)*

horno de arco eléctrico en vacío *(m)* Es
De Vakuum-Lichtbogenofen *(m)*
En vacuum arc furnace
Fr four à arc dans le vide *(m)*
It forno ad arco a depressione *(m)*
Pt forno de arco voltáico em vácuo *(m)*

horno de calentamiento por radiación *(m)* Es
De Strahlungsheizofen *(m)*
En radiant-heating furnace
Fr four à chauffage radiant *(m)*
It forno a riscaldamento radiante *(m)*
Pt forno de aquecimento por radiação *(m)*

horno de crisol *(m)* Es
De Tiegelofen *(m)*
En crucible furnace
Fr fourneau à creuset *(m)*
It forno a crogiolo *(m)*
Pt forno de cadinho *(m)*

horno de cuba *(m)* Es
De Schachtofen *(m)*
En shaft furnace
Fr four à cuve *(m)*
It forno a tino *(m)*
Pt forno de chaminé *(m)*

horno de forja *(m)* Es
De Herdofen *(m)*
En hearth furnace
Fr four à sole *(m)*
It forno a suola *(m)*
Pt forno de fornalha *(m)*

horno de inducción de canal *(m)* Es
De Kanalinduktionsofen *(m)*
En channel induction furnace
Fr four à induction à chenal *(m)*
It forno ad induzione a canale *(m)*
Pt forno de indução em canal *(m)*

horno de inducción en vacío *(m)* Es
De Vakuuminduktions-ofen *(m)*
En vacuum induction furnace
Fr four à induction à vide *(m)*
It forno ad induzione depressione *(m)*
Pt forno de indução em vácuo *(m)*

horno de mufla *(m)* Es
De Muffelofen *(m)*
En muffle furnace
Fr four à moufle *(m)*
It forno a muffola *(m)*
Pt forno com câmara *(m)*

horno de palanquillas *(m)* Es
De Blockofen *(m)*
En billet furnace
Fr four à billettes *(m)*
It forno per billette *(m)*
Pt forno de lingotes *(m)*

horno de recalentamiento *(m)* Es
De Nachbrennofen *(m)*
En reheating furnace
Fr four à réchauffer *(m)*
It forno di riscaldo *(m)*
Pt forno reaquecedor *(m)*

horno de tubo radiante *(m)* Es
De Strahlungsröhren-ofen *(m)*
En radiant-tube furnace
Fr four à tube radiant *(m)*
It forno a rubo radiante *(m)*
Pt forno de tubo de radiação *(m)*

horno eléctrico *(m)* Es
De Elektroofen *(m)*
En electric furnace
Fr four électrique *(m)*
It forno elettrico *(m)*
Pt forno eléctrico *(m)*

horno eléctrico en vacío *(m)* Es
De Vakuum-Elektroofen *(m)*
En vacuum electric furnace
Fr four électrique à vide *(m)*
It forno elettrico a depressione *(m)*
Pt forno eléctrico em vácuo *(m)*

horno Martin-Siemens *(m)* Es
De Siemens-Martinofen *(m)*
En open-hearth furnace
Fr four Martin *(m)*
It forno Martin *(m)*
Pt forno de fornalha aberta *(m)*

horno solar *(m)* Es
De Sonnenofen *(m)*
En solar furnace
Fr four solaire *(m)*
It forno solare *(m)*
Pt forno solar *(m)*

hot-air engine En
De Heiβluftmotor *(m)*
Es motor de aire caliente *(m)*
Fr moteur à air chaud *(m)*
It motore ad aria calda *(m)*
Pt motor a ar quente *(m)*

hot-air storage En
De Heiβluftspeicherung *(f)*
Es almacenamiento de aire caliente *(m)*
Fr accumulation d'air chaud *(f)*
It immagazzinaggio di aria calda *(m)*

Pt armazenagem de ar quente *(f)*

hot springs *(pl)* En
De heiβe Quellen *(pl)*
Es termas *(f pl)*
Fr sources chaudes *(f pl)*
It sorgenti calde *(f pl)*
Pt fontes de água quente *(f pl)*

hot water En
De Warmwasser *(n)*
Es agua caliente *(f)*
Fr eau chaude *(f)*
It acqua calda *(f)*
Pt água quente *(f)*

hot-water storage En
De Warmwasser-speicherung *(f)*
Es almacenamiento de agua caliente *(m)*
Fr accumulation d'eau chaude *(f)*
It immagazzinaggio di acqua calda *(m)*
Pt armazenagem de água quente *(f)*

hot-wire filament En
De Heizfaden *(m)*
Es filamento de hilo caliente *(m)*
Fr filament thermique *(m)*
It filamento a filo caldo *(m)*
Pt filamento de fio quente *(m)*

houille *(f) n* Fr
De Schwarzkohle *(f)*
En black coal
Es carbón negro *(m)*
It carbone nero *(m)*
Pt carvão negro *(m)*

houille collante *(f)* Fr
De Backkohle *(f)*
En caking coal
Es hulla grasa *(f)*
It carbone agglutinante *(m)*
Pt hulha gorda *(f)*

houille demi-grasse *(f)* Fr
De halbbituminöse Kohle *(f)*
En semibituminous coal
Es carbón semibituminoso *(m)*
It carbone semibituminoso *(m)*
Pt carvão semibetuminoso *(m)*

houille grasse *(f)* Fr
De Mattkohle *(f)*
En cannel coal
Es canel *(m)*
It carbone a lunga fiamma *(m)*
Pt carvão cannel *(m)*

houille maigre *(f)* Fr
De nicht anbackende Kohle *(f)*
En noncaking coal
Es carbón inaglutinable *(m)*
It carbone non agglutinante *(m)*
Pt carvaõ não aglutinante *(m)*

H-S-Diagramm *(n)* De
En Mollier steam diagram
Es diagrama de vapor de Mollier *(m)*
Fr diagramme de Mollier *(m)*
It diagramma di Mollier *(m)*
Pt diagrama de vapor de Mollier *(m)*

huile *(f) n* Fr
De Öl *(n)*
En oil
Es petróleo *(m)*
It olio *(m)*
Pt óleo *(m)*

huile cokéfiante *(f)* Fr
De Koksöl *(n)*
En coking oil
Es petróleo de coquización *(m)*
It olio cokificante *(m)*
Pt petróleo de coqueificação *(m)*

huile de chauffe *(f)* Fr
De Heizöl *(n)*
En heating oil
Es petróleo de calefacción *(m)*
It olio per riscaldamento *(m)*
Pt óleo para aquecimento *(m)*

huile de graissage *(f)* Fr
De Schmieröl *(n)*
En lubricating oil
Es aceite lubricante *(m)*
It olio lubrificante *(m)*
Pt óleo de lubrificação *(m)*

huile de schiste *(f)* Fr
De Schieferöl *(n)*
En shale oil
Es aceite de lutita *(m)*
It olio di schisto *(m)*
Pt petróleo xistoso *(m)*

huile légère *(f)* Fr
De leichtes Öl *(n)*
En light oil
Es petróleo ligero *(m)*
It olio leggero *(m)*
Pt óleo leve *(m)*

hulha gorda *(f)* Pt
De Backkohle *(f)*
En caking coal
Es hulla grasa *(f)*
Fr houille collante *(f)*
It carbone agglutinante *(m)*

hulla grasa *(f)* Es
De Backkohle *(f)*
En caking coal
Fr houille collante *(f)*
It carbone agglutinante *(m)*
Pt hulha gorda *(f)*

humedad *(f) n* Es
De Feuchtigkeit *(f)*
En humidity
Fr humidité *(f)*
It umidità *(f)*
Pt humidade *(f)*

humedad absoluta *(f)* Es
De absolute Feuchtigkeit *(f)*
En absolute humidity
Fr humidité absolue *(f)*
It umidità assoluta *(f)*
Pt humidade absoluta *(f)*

humedad relativa *(f)* Es
De relative Feuchtigkeit *(f)*
En relative humidity
Fr humidité relative *(f)*
It umidità relativa *(f)*
Pt humidade relativa *(f)*

humero *(m) n* Es
De Abzug *(m)*
En flue
Fr carneau *(m)*
It condotto *(m)*
Pt fumeiro *(m)*

humic coal En
De Humuskohle *(f)*
Es carbón húmico *(m)*
Fr charbon humique *(m)*
It carbone umico *(m)*
Pt carvão humico *(m)*

humidade *(f) n* Pt
De Feuchtigkeit *(f)*
En humidity
Es humedad *(f)*
Fr humidité *(f)*
It umidità *(f)*

humidade absoluta *(f)* Pt
De absolute Feuchtigkeit *(f)*
En absolute humidity
Es humedad absoluta *(f)*
Fr humidité absolue *(f)*
It umidità assoluta *(f)*

humidade relativa *(f)* Pt
De relative Feuchtigkeit *(f)*
En relative humidity
Es humedad relativa *(f)*
Fr humidité relative *(f)*
It umidità relativa *(f)*

humidité *(f) n* Fr
De Feuchtigkeit *(f)*
En humidity
Es humedad *(f)*
It umidità *(f)*
Pt humidade *(f)*

humidité absolue *(f)* Fr
De absolute Feuchtigkeit *(f)*
En absolute humidity
Es humedad absoluta *(f)*
It umidità assoluta *(f)*
Pt humidade absoluta *(f)*

humidité relative *(f)* Fr
De relative Feuchtigkeit *(f)*
En relative humidity
Es humedad relativa *(f)*
It umidità relativa *(f)*
Pt humidade relativa *(f)*

humidity *n* En
De Feuchtigkeit *(f)*
Es humedad *(f)*
Fr humidité *(f)*
It umidità *(f)*
Pt humidade *(f)*

humo *(m) n* Es
De Rauch *(m)*
En smoke
Fr fumée *(f)*
It fumo *(m)*
Pt fumo *(m)*

Humuskohle *(f) n* De
En humic coal
Es carbón húmico *(m)*
Fr charbon humique *(m)*
It carbone umico *(m)*
Pt carvão humico *(m)*

Hybrid-Raketen-treibstoff *(m) n* De
En hybrid rocket propellant
Es propulsantede cohete híbrido *(m)*
Fr propergol hybride *(m)*
It propellente ibrido per missili *(m)*
Pt combustível propulsor de foguetão híbrido *(m)*

hybrid rocket propellant En
De Hybrid-Raketen-treibstoff *(m)*
Es propulsantede cohete híbrido *(m)*
Fr propergol hybride *(m)*
It propellente ibrido per missili *(m)*
Pt combustível propulsor de foguetão híbrido *(m)*

Hybrid-Solaranlage *(f) n* De
En hybrid solar plant
Es planta solar híbrida *(f)*
Fr centrale solaire hybride *(f)*
It impianto solare ibrido *(m)*
Pt central de energia solar híbrida *(f)*

hybrid solar plant En
De Hybrid-Solaranlage *(f)*
Es planta solar híbrida *(f)*
Fr centrale solaire hybride *(f)*
It impianto solare ibrido *(m)*
Pt central de energia solar híbrida *(f)*

Hydran *(n) n* De
En hydrane
Es hidrano *(m)*
Fr hydrane *(m)*
It idrano *(m)*
Pt hidrano *(m)*

hydrane *n* En; Fr *(m)*
De Hydran *(n)*
Es hidrano *(m)*
It idrano *(m)*
Pt hidrano *(m)*

hydraulic *adj* En
De hydraulisch
Es hidráulico
Fr hydraulique
It idraulico
Pt hidráulico

hydraulic fracturing En
De hydraulische Rißbildung *(f)*
Es agrietamiento hidráulico *(m)*
Fr fracturation hydraulique *(f)*
It fratturazione idraulica *(f)*
Pt fracturação hidráulica *(f)*

hydraulique *adj* Fr
De hydraulisch
En hydraulic
Es hidráulico
It idraulico
Pt hidráulico

hydraulisch *adj* De
En hydraulic
Es hidráulico
Fr hydraulique
It idraulico
Pt hidráulico

hydraulische Rißbildung *(f)* De
En hydraulic fracturing
Es agrietamiento hidráulico *(m)*
Fr fracturation hydraulique *(f)*
It fratturazione idraulica *(f)*
Pt fracturação hidráulica *(f)*

Hydrazin *(n) n* De
En hydrazine
Es hidracina *(f)*
Fr hydrazine *(f)*
It idrazina *(f)*
Pt hidrazina *(f)*

hydrazine *n* En; Fr *(f)*
De Hydrazin *(n)*
Es hidracina *(f)*
It idrazina *(f)*
Pt hidrazina *(f)*

Hydrierung *(f) n* De
En hydrogenation
Es hidrogenación *(f)*
Fr hydrogénation *(f)*
It idrogenazione *(f)*
Pt hidrogenação *(f)*

hydrocarbon *n* En
De Kohlenwasserstoff *(m)*
Es hidrocarburo *(m)*
Fr hydrocarbure *(m)*
It idrocarburo *(m)*
Pt hidrocarboneto *(m)*

hydrocarbure *(m) n* Fr
De Kohlenwasserstoff *(m)*
En hydrocarbon
Es hidrocarburo *(m)*
It idrocarburo *(m)*
Pt hidrocarboneto *(m)*

hydrocarbures liquides *(m pl)* Fr
De flüssige Kohlenwasserstoffe *(pl)*
En liquid hydrocarbons
Es hidrocarburos líquidos *(m pl)*
It idrocarburi liquidi *(m pl)*
Pt hidrocarbonetos líquidos *(m pl)*

hydrocracking *n* En
De Hydrokracken *(n)*
Es hidrocrácking *(m)*
Fr hydrocraquage *(m)*
It piroscissione idraulica *(f)*
Pt hidro-cracking *(m)*

hydrocraquage *(m) n* Fr
De Hydrokracken *(n)*
En hydrocracking
Es hidrocrácking *(m)*
It piroscissione idraulica *(f)*
Pt hidro-cracking *(m)*

hydroelectric energy En
De hydroelektrische Energie *(f)*
Es energía hidroeléctrica *(f)*
Fr énergie hydroélectrique *(f)*
It energia idroelettrica *(f)*
Pt energia hidroeléctrica *(f)*

hydroélectricité *(f) n* Fr
De Hydroelektrizität *(f)*
En hydroelectricity
Es hidroelectricidad *(f)*
It idroelettricità *(f)*
Pt hidroelectricidade *(f)*

hydroelectricity *n* En
De Hydroelektrizität *(f)*
Es hidroelectricidad *(f)*
Fr hydroélectricité *(f)*
It idroelettricità *(f)*
Pt hidroelectricidade *(f)*

hydroelectric power station En
De hydroelektrisches Kraftwerk *(n)*
Es central hidroeléctrica *(f)*
Fr centrale hydroélectrique *(f)*
It centrale idroelettrica *(f)*
Pt central hidroeléctrica *(f)*

hydroelektrische Energie *(f)* De
En hydroelectric energy
Es energía hidroeléctrica *(f)*
Fr énergie hydroélectrique *(f)*

It energia idroelettrica (f)
Pt energia hidroeléctrica (f)

hydroelektrisches Kraftwerk *(n)* De
En hydroelectric power station
Es central hidroeléctrica (f)
Fr centrale hydroélectrique (f)
It centrale idoelettrica (f)
Pt central hidroeléctrica (f)

Hydroelektrizität *(f) n* De
En hydroelectricity
Es hidroelectricidad (f)
Fr hydroélectricité (f)
It idroelettricità (f)
Pt hidroelectricidade (f)

hydrogasification *n* En
De Hydro-Vergasung (f)
Es hidrogasificación (f)
Fr hydrogazéification (f)
It idrogassificazione (f)
Pt hidrogaseificação (f)

hydrogazéification *(f) n* Fr
De Hydro-Vergasung (f)
En hydrogasification
Es hidrogasificación (f)
It idrogassificazione (f)
Pt hidrogaseificação (f)

hydrogen *n* En
De Wasserstoff (m)
Es hidrógeno (m)
Fr hydrogène (m)
It idrogeno (m)
Pt hidrogénio (m)

hydrogen-air fuel cell En
De Wasserstoff-Luft-brennstoffzelle (f)
Es célula de combustible de hidrógeno-aire (f)
Fr pile à combustible hydrogène-air (f)
It cellula combustibile idrogeno-aria (f)
Pt célula de combustível hidrogénio-ar (f)

hydrogenation *n* En
De Hydrierung (f)
Es hidrogenación (f)
Fr hydrogénation (f)
It idrogenazione (f)
Pt hidrogenação (f)

hydrogénation *(f) n* Fr
De Hydrierung (f)
En hydrogenation
Es hidrogenación (f)
It idrogenazione (f)
Pt hidrogenação (f)

hydrogen bomb En
De Wasserstoffbombe (f)
Es bomba de hidrógeno (f)
Fr bombe à hydrogène (f)
It bomba all'idrogeno (f)
Pt bomba de hidrogénio (f)

hydrogen-cooled *adj* En
De wasserstoffgekühlt
Es enfriada con hidrógeno
Fr refroidi par l'hydrogène
It raffreddata ad idrogeno
Pt arrefecido a hidrogénio

hydrogène *(m) n* Fr
De Wasserstoff (m)
En hydrogen
Es hidrógeno (m)
It idrogeno (m)
Pt hidrogénio (m)

hydrogen economy En
De Wasserstoffwirt-schaft (f)
Es economía de hidrógeno (f)
Fr économie d'hydrogène (f)
It economia dell'idrogeno (f)
Pt economia de hidrogénio (f)

hydrogène lourd *(m)* Fr
De schwerer Wasserstoff (m)
En heavy hydrogen
Es hidrógeno pesado (m)
It idrogeno pesante (m)
Pt hidrogénio pesado (m)

hydrogen-oil ratio En
De Wasserstoff-Ölverhältnis (n)
Es relación hidrógeno-aceite (f)
Fr rapport hydrogène-huile (m)
It rapporto idrogeno-olio (m)
Pt razão hidrogénio-óleo (f)

hydrogravitational energy En
De Wasserschwerkraft-Energie (f)
Es energía hidrogravitacional (f)
Fr énergie d'hydrogravitation (f)
It energia idrogravitazionale (f)
Pt energia hidrogravitacional (f)

Hydrokracken *(n) n* De
En hydrocracking
Es hidrocrácking (m)
Fr hydrocraquage (m)
It piroscissione idraulica (f)
Pt hidro-cracking (m)

hydrólisis *(f) n* Es
De Hydrolyse (f)
En hydrolysis
Fr hydrolyse (f)
It idrolisi (f)
Pt hidrólise (f)

hydrologie *(f) n* Fr
De Hydrologie (f)
En hydrology
Es hidrología (f)
It idrologia (f)
Pt hidrologia (f)

Hydrologie *(f) n* De
En hydrology
Es hidrología (f)
Fr hydrologie (f)
It idrologia (f)
Pt hidrologia (f)

hydrology *n* En
De Hydrologie (f)
Es hidrología (f)
Fr hydrologie (f)
It idrologia (f)
Pt hidrologia (f)

hydrolyse *(f) n* Fr
De Hydrolyse (f)
En hydrolysis
Es hydrólisis (f)
It idrolisi (f)
Pt hidrólise (f)

Hydrolyse *(f) n* De
En hydrolysis
Es hydrólisis (f)
Fr hydrolyse (f)
It idrolisi (f)
Pt hidrólise (f)

hydrolysis *n* En
De Hydrolyse (f)
Es hydrólisis (f)
Fr hydrolyse (f)
It idrolisi (f)
Pt hidrólise (f)

hydroséparateur *(m) n* Fr
De Unterwasser-Stromapparat (m)
En hydroseparator
Es hidroseparador (m)
It idroseparatore (m)
Pt hidroseparador (m)

hydroseparator *n* En
De Unterwasser-Stromapparat (m)
Es hidroseparador (m)
Fr hydroséparateur (m)
It idroseparatore (m)
Pt hidroseparador (m)

hydrostockage *(m) n* Fr
De Wasserspeicherung (f)
En hydrostorage
Es hidroalmacen-amiento (m)
It idroimmagazzinaggio (m)
Pt hidroarmazenamento (m)

hydrostorage *n* En
De Wasserspeicherung (f)
Es hidroalmacen-amiento (m)
Fr hydrostockage (m)
It idroimmagazzinaggio (m)
Pt hidroarmazen-amento (m)

hydrothermal *adj* De, En
Es hidrotérmico
Fr hydrothermique
It idrotermico
Pt hidrotérmico

hydrothermique *adj* Fr
De hydrothermal
En hydrothermal
Es hidrotérmico
It idrotermico
Pt hidrotérmico

Hydro-Vergasung *(f) n* De
En hydrogasification
Es hidrogasificación *(f)*
Fr hydrogazéification *(f)*
It idrogassificazione *(f)*
Pt hidrogaseificação *(f)*

hyperfréquence *(f)* Fr
De Ultrahochfrequenz *(f)*
En ultrahigh frequency (UHF)
Es frecuencia ultraelevada *(f)*
It frequenza ultraelevata *(f)*
Pt frequência ultra-elevada *(f)*

Hysterese *(f) n* De
En hysteresis
Es histéresis *(f)*
Fr hystérésis *(f)*
It isteresi *(f)*
Pt histerese *(f)*

Hystereseverlust *(m) n* De
En hysteresis loss
Es pérdida por histéresis *(f)*
Fr perte par hystérésis *(f)*
It perdita per isteresi *(f)*
Pt perda por histerese *(f)*

hysteresis *n* En
De Hysterese *(f)*
Es histéresis *(f)*
Fr hystérésis *(f)*
It isteresi *(f)*
Pt histerese *(f)*

hystérésis *(f) n* Fr
De Hysterese *(f)*
En hysteresis
Es histéresis *(f)*
It isteresi *(f)*
Pt histerese *(f)*

hysteresis loss En
De Hystereseverlust *(m)*
Es pérdida por histéresis *(f)*
Fr perte par hystérésis *(f)*
It perdita per isteresi *(f)*
Pt perda por histerese *(f)*

I

ião *(m) n* Pt
De Ion *(n)*
En ion
Es ion *(m)*
Fr ion *(m)*
It ione *(m)*

ideales Gas *(n)* De
En ideal gas
Es gas perfecto *(m)*
Fr gaz idéal *(m)*
It gas perfetto *(m)*
Pt gás ideal *(m)*

ideal gas En
De ideales Gas *(n)*
Es gas perfecto *(m)*
Fr gaz idéal *(m)*
It gas perfetto *(m)*
Pt gás ideal *(m)*

idrano *(m) n* It
De Hydran *(n)*
En hydrane
Es hidrano *(m)*
Fr hydrane *(m)*
Pt hidrano *(m)*

idraulico *adj* It
De hydraulisch
En hydraulic
Es hidráulico
Fr hydraulique
Pt hidráulico

idrazina *(f) n* It
De Hydrazin *(n)*
En hydrazine
Es hidracina *(f)*
Fr hydrazine *(f)*
Pt hidrazina *(f)*

idrocarburi liquidi *(m pl)* It
De flüssige Kohlenwasserstoffe *(pl)*
En liquid hydrocarbons
Es hidrocarburos líquidos *(m pl)*
Fr hydrocarbures liquides *(m pl)*
Pt hidrocarbonetos líquidos *(m pl)*

idrocarburo *(m) n* It
De Kohlenwasserstoff *(m)*
En hydrocarbon
Es hidrocarburo *(m)*
Fr hydrocarbure *(m)*
Pt hidrocarboneto *(m)*

idroelettricità *(f) n* It
De Hydroelektrizität *(f)*
En hydroelectricity
Es hidroelectricidad *(f)*
Fr hydroélectricité *(f)*
Pt hidroelectricidade *(f)*

idrogassificazione *(f) n* It
De Hydro-Vergasung *(f)*
En hydrogasification
Es hidrogasificación *(f)*
Fr hydrogazéification *(f)*
Pt hidrogaseificação *(f)*

idrogenazione *(f) n* It
De Hydrierung *(f)*
En hydrogenation
Es hidrogenación *(f)*
Fr hydrogénation *(f)*
Pt hidrogenação *(f)*

idrogeno *(m) n* It
De Wasserstoff *(m)*
En hydrogen
Es hidrógeno *(m)*
Fr hydrogène *(m)*
Pt hidrogénio *(m)*

idrogeno pesante *(m)* It
De schwerer Wasserstoff *(m)*
En heavy hydrogen
Es hidrógeno pesado *(m)*
Fr hydrogène lourd *(m)*
Pt hidrogénio pesado *(m)*

idroimmagazzinaggio *(m) n* It
De Wasserspeicherung *(f)*
En hydrostorage
Es hidroalmacenamiento *(m)*
Fr hydrostockage *(m)*
Pt hidroarmazenamento *(m)*

idrolisi *(f) n* It
De Hydrolyse *(f)*
En hydrolysis
Es hydrólisis *(f)*
Fr hydrolyse *(f)*
Pt hidrólise *(f)*

idrologia *(f) n* It
De Hydrologie *(f,*
En hydrology
Es hidrología *(f)*
Fr hydrologie *(f)*
Pt hidrologia *(f)*

idroseparatore *(m) n* It
De Unterwasser-Stromapparat *(m)*
En hydroseparator
Es hidroseparador *(m)*
Fr hydroséparateur *(m)*
Pt hidroseparador *(m)*

idrotermico *adj* It
De hydrothermal
En hydrothermal
Es hidrotérmico
Fr hydrothermique
Pt hidrotérmico

ignição *(f) n* Pt
De Zündung *(f)*
En ignition
Es encendido *(m)*
Fr allumage *(m)*
It accensione *(f)*

ignição dupla *(f)* Pt
De Doppelzündung *(f)*
En dual ignition
Es doble encendido *(m)*
Fr double allumage *(m)*
It accensione doppia *(f)*

ignição espontânea *(f)* Pt
De spontane Zündung *(f)*
En spontaneous ignition
Es encendido espontáneo *(m)*
Fr allumage spontané *(m)*
It autoaccensione *(f)*

ignição por faísca *(f)* Pt
De Funkenzündung *(f)*
En spark ignition
Es encendido por chispa *(m)*
Fr allumage par étincelle *(m)*
It accensione a scintilla *(f)*

ignição por impulso electrónico *(f)* Pt
De elektronische Impulszündung *(f)*
En electronic pulse ignition
Es encendido por impulsos electrónicos *(m)*
Fr allumage électronique à impulsions *(m)*
It accensione ad impulso elettronico *(f)*

ignition *n* En
De Zündung *(f)*
Es encendido *(m)*
Fr allumage *(m)*
It accensione *(f)*
Pt ignição *(f)*

ignition temperature En
De Zündtemperatur *(f)*
Es temperatura de encendido *(f)*
Fr température d'allumage *(f)*
It temperatura di accensione *(f)*
Pt temperatura de ignição *(f)*

illumination *n* En
De Beleuchtung *(f)*
Es iluminación *(f)*
Fr éclairage *(m)*
It illuminazione *(f)*
Pt iluminação *(f)*

illuminazione *(f) n* It
De Beleuchtung *(f)*
En illumination
Es iluminación *(f)*
Fr éclairage *(m)*
Pt iluminação *(f)*

iluminação *(f) n* Pt
De Beleuchtung *(f)*
En illumination
Es iluminación *(f)*
Fr éclairage *(m)*
It illuminazione *(f)*

iluminación *(f) n* Es
De Beleuchtung *(f)*
En illumination
Fr éclairage *(m)*
It illuminazione *(f)*
Pt iluminação *(f)*

ímã *(m) n* Pt
De Magnet *(m)*
En magnet
Es imán *(m)*
Fr aimant *(m)*
It magmete *(m)*

ímã móvel Pt
De Dreheisen-
En moving-iron
Es núcleo giratorio
Fr fer mobile
It ferro rotante

imán *(m) n* Es
De Magnet *(m)*
En magnet
Fr aimant *(m)*
It magmete *(m)*
Pt ímã *(m)*

imán superconductor *(m)* Es
De supraleitfähiger Magnet *(m)*
En superconducting magnet
Fr aimant superconducteur *(m)*
It magnete superconduttore *(m)*
Pt ímã supercondutor *(m)*

ímã supercondutor *(m)* Pt
De supraleitfähiger Magnet *(m)*
En superconducting magnet
Es imán superconductor *(m)*
Fr aimant superconducteur *(m)*
It magnete superconduttore *(m)*

immagazzinaggi a pompa *(m)* It
De gepumpte Speicherung *(f)*
En pumped storage
Es almacenamiento bombeado *(m)*
Fr accumulation pompée *(f)*
Pt armazenagem realizada à bomba *(f)*

immagazzinaggio a transizione di fase *(m)* It
De Phasenumwandlungs-speicherung *(f)*
En phase-transition storage
Es almacenamiento de transición de fase *(m)*
Fr accumulation à transition de phase *(f)*
Pt armazenagem de transição de fase *(f)*

immagazzinaggio di acqua calda *(m)* It
De Warmwasser-speicherung *(f)*
En hot-water storage
Es almacenamiento de agua caliente *(m)*
Fr accumulation d'eau chaude *(f)*
Pt armazenagem de água quente *(f)*

immagazzinaggio di aria calda *(m)* It
De Heiβluftspeicherung *(f)*
En hot-air storage
Es almacenamiento de aire caliente *(m)*
Fr accumulation d'air chaud *(f)*
Pt armazenagem de ar quente *(f)*

immagazzinaggio di aria compressa *(m)* It
De Druckluftspeicherung *(f)*
En compressed-air storage
Es almacenamiento de aire comprimido *(m)*
Fr stockage de l'air comprimé *(m)*
Pt armazenagem de ar comprimido *(f)*

immagazzinaggio di calore *(m)* It
De Wärmespeicherung *(f)*
En heat storage
Es almacenamiento de calor *(m)*
Fr accumulation de chaleur *(f)*
Pt armazenagem de calor *(f)*

immagazzinaggio di calore diurno *(m)* It
De tägliche Warmespeicherung *(f)*
En diurnal heat storage
Es almacenamiento diurno del calor *(m)*
Fr accumulation de chaleur diurne *(f)*
Pt armazenagem diurna de calor *(f)*

immagazzinaggio di energia termica *(m)* It
De Wärmeenergie-speicherung *(f)*
En thermal-energy storage
Es almacenamiento de energía térmica *(m)*
Fr accumulation d'énergie thermique *(n)*
Pt armazenagem de energia térmica *(f)*

immagazzinaggio elettrochimico *(m)* It
De elektrochemische Speicherung *(f)*
En electrochemical storage
Es almacenamiento electroquímico *(m)*
Fr accumulation électrochimique *(f)*
Pt armazenagem electroquímica *(f)*

immagazzinaggio magnetico *(m)* It
De Magnetspeicher *(m)*
En magnetic storage
Es almacenamiento magnético *(m)*
Fr accumulation magnétique *(f)*
Pt armazenagem magnética *(f)*

immagazzinaggio meccanico *(m)* It
De mechanische Speicherung *(f)*
En mechanical storage
Es almacenamiento mecánico *(m)*
Fr accumulation mécanique *(f)*
Pt armazenagem mecânica *(f)*

immersion heater En
De Tauchsieder *(m)*
Es calentador de inmersión *(m)*
Fr thermoplongeur *(m)*
It riscaldatore ad immersione *(m)*
Pt aquecedor por imersão *(m)*

impermeabilización *(f) n* Es
De Wetterschutz *(m)*
En weatherproofing
Fr protection contre les intempéries *(f)*
It resistenza alle intemperie *(f)*
Pt protecção contra a intempérie *(f)*

impianto ausiliario caldaia *(m)* It
De Kesselhilfsanlage *(f)*
En boiler auxiliary plant
Es planta auxiliar de calderas *(f)*
Fr chaufferie auxiliaire *(f)*
Pt instalação auxiliar de caldeira *(f)*

impianto di trivellazione per petrolio *(m)* It
De Ölbohrturm *(m)*
En oilrig
Es tren de perforación de petróleo *(m)*
Fr installation de forage *(f)*
Pt plataforma petroleira *(f)*

impianto pilota *(m)* It
De Versuchsanlage *(f)*
En pilot plant
Es planta piloto *(f)*
Fr installation pilote *(f)*
Pt fábrica piloto *(f)*

impianto solare ibrido *(m)* It
De Hybrid-Solaranlage *(f)*
En hybrid solar plant
Es planta solar híbrida *(f)*
Fr centrale solaire hybride *(f)*
Pt central de energia solar híbrida *(f)*

Impuls *(m) n* De
En impulse
Es impulso *(m)*
Fr impulsion *(f)*
It impulso *(m)*
Pt impulso *(m)*

impulse *n* En
De Impuls *(m)*
Es impulso *(m)*
Fr impulsion *(f)*
It impulso *(m)*
Pt impulso *(m)*

Impulsgenerator *(m) n* De
En pulse generator
Es generador de impulsos *(m)*
Fr générateur d'impulsions *(m)*
It generatore d'impulsi *(m)*
Pt gerador de impulsos *(m)*

impulsion *(f) n* Fr
De Impuls *(m)*
En impulse
Es impulso *(m)*
It impulso *(m)*
Pt impulso *(m)*

impulsion spécifique *(f)* Fr
De spezifischer Impuls *(m)*
En specific impulse
Es impulso específico *(m)*
It impulso specifico *(m)*
Pt impulso específico *(m)*

impulso *(m) n* Es, It, Pt
De Impuls *(m)*
En impulse
Fr impulsion *(f)*

impulso específico *(m)* Es, Pt
De spezifischer Impuls *(m)*
En specific impulse
Fr impulsion spécifique *(f)*
It impulso specifico *(m)*

impulso specifico *(m)* It
De spezifischer Impuls *(m)*
En specific impulse
Es impulso específico *(m)*
Fr impulsion spécifique *(f)*
Pt impulso específico *(m)*

incamiciatura *(f) n* It
De Umhüllung *(f)*
En cladding
Es encamisado *(m)*
Fr gainage *(m)*
Pt revestimento *(m)*

incandescent *adj* En, Fr
De weißglühend
Es incandescente
It incandescente
Pt incandescente

incandescente *adj* Es, It, Pt
De weißglühend
En incandescent
Fr incandescent

incenerimento catalitico *(m)* It
De katalytische Veraschung *(f)*
En catalytic incineration
Es incineración catalítica *(f)*
Fr incinération catalytique *(f)*
Pt incineração catalítica *(f)*

incenerimento dei rifiuti *(m)* It
De Abfallverbrennung *(f)*
En refuse incineration
Es incineración de basuras *(f)*
Fr incinération de déchets *(f)*
Pt incineração de refugo *(f)*

incidental heat gain En
De zufällige Wärmezunahme *(f)*
Es ganancia de calor incidental *(f)*
Fr gain de chaleur incident *(m)*
It guadagno di calore incidentale *(m)*
Pt ganho de calor acessório *(m)*

incidente di perdita del refrigerante *(m)* It
De Unfall durch Kühlmittelverlust *(m)*
En loss of coolant accident (LOCA)
Es accidente por pérdida de refrigerante *(m)*
Fr accident par perte de fluide réfrigérant *(m)*
Pt acidente por perda de refrigerante *(m)*

incineração catalítica *(f)* Pt
De katalytische Veraschung *(f)*
En catalytic incineration
Es incineración catalítica *(f)*
Fr incinération catalytique *(f)*
It incenerimento catalitico *(m)*

incineração de refugo *(f)* Pt
De Abfallverbrennung *(f)*
En refuse incineration
Es incineración de basuras *(f)*
Fr incinération de déchets *(f)*
It incenerimento dei rifiuti *(m)*

incineración catalítica *(f)* Es
- De katalytische Veraschung *(f)*
- En catalytic incineration
- Fr incinération catalytique *(f)*
- It incenerimento catalitico *(m)*
- Pt incineração catalítica *(f)*

incineración de basuras *(f)* Es
- De Abfallverbrennung *(f)*
- En refuse incineration
- Fr incinération de déchets *(f)*
- It incenerimento dei rifiuti *(m)*
- Pt incineração de refugo *(f)*

incinération catalytique *(f)* Fr
- De katalytische Veraschung *(f)*
- En catalytic incineration
- Es incineración catalítica *(f)*
- It incenerimento catalitico *(m)*
- Pt incineração catalítica *(f)*

incinération de déchets *(f)* Fr
- De Abfallverbrennung *(f)*
- En refuse incineration
- Es incineración de basuras *(f)*
- It incenerimento dei rifiuti *(m)*
- Pt incineração de refugo *(f)*

incrostazione *(f)* *n* It
- De Verstopfung *(f)*
- En fouling (boiler)
- Es incrustación *(f)*
- Fr encrassement *(m)*
- Pt incrustação *(m)*

incrostazione marina *(f)* It
- De Seeverschmutzung *(f)*
- En marine fouling
- Es suciedad por depósitos marinos *(f)*
- Fr accumulation de dépôts marins *(f)*
- Pt depósitos marítimos *(m)*

incrustação *(m)* *n* Pt
- De Verstopfung *(f)*
- En fouling (boiler)
- Es incrustación *(f)*
- Fr encrassement *(m)*
- It incrostazione *(f)*

incrustación *(f)* *n* Es
- De Verstopfung *(f)*
- En fouling (boiler)
- Fr encrassement *(m)*
- It incrostazione *(f)*
- Pt incrustação *(m)*

indicator diagram En
- De Indikatordiagramm *(n)*
- Es diagrama indicador *(m)*
- Fr schéma indicateur *(m)*
- It diagramma del ciclo indicato *(m)*
- Pt diagrama indicador *(m)*

indice de cétane *(m)* Fr
- De Cetanzahl *(f)*
- En cetane number
- Es número cetano *(m)*
- It numero di cetano *(m)*
- Pt número cetânico *(m)*

índice de metabolismo basal *(m)* Pt
- De Grundumsatz *(m)*
- En basal metabolic rate
- Es índice metabólico basal *(m)*
- Fr métabolisme basal *(m)*
- It tasso metabolico basale *(m)*

índice de octanas *(m)* Pt
- De Oktanzahl *(f)*
- En octane number
- Es número octano *(m)*
- Fr indice d'octane *(m)*
- It numero di ottano *(m)*

índice de refracção *(m)* Pt
- De Brechungsindex *(m)*
- En refractive index
- Es índice de refracción *(m)*
- Fr indice de réfraction *(m)*
- It indice di rifrazione *(m)*

índice de refracción *(m)* Es
- De Brechungsindex *(m)*
- En refractive index
- Fr indice de réfraction *(m)*
- It indice di rifrazione *(m)*
- Pt índice de refracção *(m)*

indice de réfraction *(m)* Fr
- De Brechungsindex *(m)*
- En refractive index
- Es índice de refracción *(m)*
- It indice di rifrazione *(m)*
- Pt índice de refracção *(m)*

indice di rifrazione *(m)* It
- De Brechungsindex *(m)*
- En refractive index
- Es índice de refracción *(m)*
- Fr indice de réfraction *(m)*
- Pt índice de refracção *(m)*

indice d'octane *(m)* Fr
- De Oktanzahl *(f)*
- En octane number
- Es número octano *(m)*
- It numero di ottano *(m)*
- Pt índice de octanas *(m)*

índice metabólico basal *(m)* Es
- De Grundumsatz *(m)*
- En basal metabolic rate
- Fr métabolisme basal *(m)*
- It tasso metabolico basale *(m)*
- Pt índice de metabolismo basal *(m)*

Indikatordiagramm *(n)* *n* De
- En indicator diagram
- Es diagrama indicador *(m)*
- Fr schéma indicateur *(m)*
- It diagramma del ciclo indicato *(m)*
- Pt diagrama indicador *(m)*

indirect heating En
- De indirekte Heizung *(f)*
- Es calefacción indirecto *(f)*
- Fr chauffage indirect *(m)*
- It riscaldamento indiretto *(m)*
- Pt aquecimento indirecto *(m)*

indirekte Heizung *(f)* De
- En indirect heating
- Es calefacción indirecto *(f)*
- Fr chauffage indirect *(m)*
- It riscaldamento indiretto *(m)*
- Pt aquecimento indirecto *(m)*

indução *(f)* *n* Pt
- De Induktion *(f)*
- En induction
- Es inducción *(f)*
- Fr induction *(f)*
- It induzione *(f)*

inducción *(f)* *n* Es
- De Induktion *(f)*
- En induction
- Fr induction *(f)*
- It induzione *(f)*
- Pt indução *(f)*

induced current En
- De induzierter Strom *(m)*
- Es corriente inducida *(f)*
- Fr courant induit *(m)*
- It corrente indotta *(f)*
- Pt corrente induzida *(f)*

induced electromotive force En
- De induzierte elektromotorische Kraft *(f)*
- Es fuerza electromotriz inducida *(f)*
- Fr force électromotrice induite *(f)*
- It forza elettromotrice indotta *(f)*

Pt força electromotora induzida *(f)*

induced radioactivity En
De induzierte Radioaktivität *(f)*
Es radiactividad inducida *(f)*
Fr radioactivité induite *(f)*
It radioattività indotta *(f)*
Pt radioactividade induzida *(f)*

inducido *(m) n* Es
De Anker *(m)*
En armature
Fr induit *(m)*
It armatura *(f)*
Pt armadura *(f)*

inductance mutuelle *(f)* Fr
De Gegeninduktivität *(f)*
En mutual inductance
Es inductancia mutua *(f)*
It induttanza mutua *(f)*
Pt indutância mútua *(f)*

inductancia mutua *(f)* Es
De Gegeninduktivität *(f)*
En mutual inductance
Fr inductance mutuelle *(f)*
It induttanza mutua *(f)*
Pt indutância mútua *(f)*

induction *n* En; Fr *(f)*
De Induktion *(f)*
Es inducción *(f)*
It induzione *(f)*
Pt indução *(f)*

induction coil En
De Induktionsspule *(f)*
Es bobina de inducción *(f)*
Fr bobine d'induction *(f)*
It rocchetto d'induzione *(m)*
Pt bobina de indução *(f)*

induction heating En
De Induktionsheizung *(f)*
Es caldeo por inducción *(m)*
Fr chauffage par induction *(m)*
It riscaldamento ad induzione *(m)*
Pt aquecimento por indução *(m)*

induction motor En
De Induktionsmotor *(m)*
Es motor de inducción *(m)*
Fr moteur d'induction *(m)*
It motore ad induzione *(m)*
Pt motor de indução *(m)*

induit *(m) n* Fr
De Anker *(m)*
En armature
Es inducido *(m)*
It armatura *(f)*
Pt armadura *(f)*

Induktion *(f) n* De
En induction
Es inducción *(f)*
Fr induction *(f)*
It induzione *(f)*
Pt indução *(f)*

Induktionsheizung *(f) n* De
En induction heating
Es caldeo por inducción *(m)*
Fr chauffage par induction *(m)*
It riscaldamento ad induzione *(m)*
Pt aquecimento por indução *(m)*

Induktionsmotor *(m) n* De
En induction motor
Es motor de inducción *(m)*
Fr moteur d'induction *(m)*
It motore ad induzione *(m)*
Pt motor de indução *(m)*

Induktionsspule *(f) n* De
En induction coil
Es bobina de inducción *(f)*
Fr bobine d'induction *(f)*
It rocchetto d'induzione *(m)*
Pt bobina de indução *(f)*

industrial energy consumption En
De Energieverbrauch der Industrie *(m)*
Es consumo industrial de energía *(m)*
Fr consommation industrielle d'énergie *(f)*
It consumo industriale di energia *(m)*
Pt consumo industrial de energia *(m)*

indutância mútua *(f)* Pt
De Gegeninduktivität *(f)*
En mutual inductance
Es inductancia mutua *(f)*
Fr inductance mutuelle *(f)*
It induttanza mutua *(f)*

induttanza mutua *(f)* It
De Gegeninduktivität *(f)*
En mutual inductance
Es inductancia mutua *(f)*
Fr inductance mutuelle *(f)*
Pt indutância mútua *(f)*

induzierte elektromotorische Kraft *(f)* De
En induced electromotive force
Es fuerza electromotriz inducida *(f)*
Fr force électromotrice induite *(f)*
It forza elettromotrice indotta *(f)*
Pt força electromotora induzida *(f)*

induzierte Radioaktivität *(f)* De
En induced radioactivity
Es radiactividad inducida *(f)*
Fr radioactivité induite *(f)*
It radioattività indotta *(f)*
Pt radioactividade induzida *(f)*

induzierter Strom *(m)* De
En induced current
Es corriente inducida *(f)*
Fr courant induit *(m)*
It corrente indotta *(f)*
Pt corrente induzida *(f)*

induzione *(f) n* It
De Induktion *(f)*
En induction
Es inducción *(f)*
Fr induction *(f)*
Pt indução *(f)*

inercia *(f) n* Es
De Trägheitsmoment *(m)*
En inertia
Fr inertie *(f)*
It inerzia *(f)*
Pt inércia *(f)*

inércia *(f) n* Pt
De Trägheitsmoment *(m)*
En inertia
Es inercia *(f)*
Fr inertie *(f)*
It inerzia *(f)*

inercia térmica *(f)* Es
De Wärmeträgheit *(f)*
En thermal inertia
Fr inertie thermique *(f)*
It inerzia termica *(f)*
Pt inércia térmica *(f)*

inércia térmica *(f)* Pt
De Wärmeträgheit *(f)*
En thermal inertia
Es inercia térmica *(f)*
Fr inertie thermique *(f)*
It inerzia termica *(f)*

inert *adj* En, De
Es inerte
Fr inerte
It inerte
Pt inerte

inerte *adj* Es, Fr, It, Pt
De inert
En inert

inertia *n* En
De Trägheitsmoment *(m)*
Es inercia *(f)*
Fr inertie *(f)*
It inerzia *(f)*
Pt inércia *(f)*

inertie *(f) n* Fr
De Trägheitsmoment *(m)*
En inertia
Es inercia *(f)*

It inerzia *(f)*
Pt inércia *(f)*

inertie thermique *(f)* Fr
De Wärmeträgheit *(f)*
En thermal inertia
Es inercia térmica *(f)*
It inerzia termica *(f)*
Pt inércia térmica *(f)*

inerzia *(f) n* It
De Trägheitsmoment *(m)*
En inertia
Es inercia *(f)*
Fr inertie *(f)*
Pt inércia *(f)*

inerzia termica *(f)* It
De Wärmeträgheit *(f)*
En thermal inertia
Es inercia térmica *(f)*
Fr inertie thermique *(f)*
Pt inércia térmica *(f)*

infiammabile *adj* It
De feuergefährlich
En flammable
Es inflamable
Fr inflammable
Pt inflamável

inflamable *adj* Es
De feuergefährlich
En flammable
Fr inflammable
It infiammabile
Pt inflamável

inflamável *adj* Pt
De feuergefährlich
En flammable
Es inflamable
Fr inflammable
It infiammabile

inflammable *adj* Fr
De feuergefährlich
En flammable
Es inflamable
It infiammabile
Pt inflamável

infraestructura *(f) n* Es
De Infrastruktur *(f)*
En infrastructure
Fr infrastructure *(f)*
It infrastruttura *(f)*
Pt infraestrutura *(f)*

infraestrutura *(f) n* Pt
De Infrastruktur *(f)*
En infrastructure
Es infraestructura *(f)*
Fr infrastructure *(f)*
It infrastruttura *(f)*

infrared radiation (IR) En
De Infrarotstrahlung *(f)*
Es radiación infrarroja *(f)*
Fr radiation infrarouge *(f)*
It radiazione infrarossa *(f)*
Pt radiação infravermelha *(f)*

Infrarotstrahlung *(f) n* De
En infrared radiation (IR)
Es radiación infrarroja *(f)*
Fr radiation infrarouge *(f)*
It radiazione infrarossa *(f)*
Pt radiação infravermelha *(f)*

Infraschall *(m) n* De
En infrasound
Es infrasonido *(m)*
Fr infrason *(m)*
It infrasuono *(m)*
Pt infrasom *(m)*

infrasom *(m) n* Pt
De Infraschall *(m)*
En infrasound
Es infrasonido *(m)*
Fr infrason *(m)*
It infrasuono *(m)*

infrason *(m) n* Fr
De Infraschall *(m)*
En infrasound
Es infrasonido *(m)*
It infrasuono *(m)*
Pt infrasom *(m)*

infrasonido *(m) n* Es
De Infraschall *(m)*
En infrasound
Fr infrason *(m)*
It infrasuono *(m)*
Pt infrasom *(m)*

infrasound *n* En
De Infraschall *(m)*
Es infrasonido *(m)*
Fr infrason *(m)*
It infrasuono *(m)*
Pt infrasom *(m)*

infrastructure *n* En; Fr *(f)*
De Infrastruktur *(f)*
Es infraestructura *(f)*
It infrastruttura *(f)*
Pt infraestrutura *(f)*

Infrastruktur *(f) n* De
En infrastructure
Es infraestructura *(f)*
Fr infrastructure *(f)*
It infrastruttura *(f)*
Pt infraestrutura *(f)*

infrastruttura *(f) n* It
De Infrastruktur *(f)*
En infrastructure
Es infraestructura *(f)*
Fr infrastructure *(f)*
Pt infraestrutura *(f)*

infrasuono *(m) n* It
De Infraschall *(m)*
En infrasound
Es infrasonido *(m)*
Fr infrason *(m)*
Pt infrasom *(m)*

injection well En
De Einpreßbohrung *(f)*
Es pozo de inyección *(m)*
Fr puits d'injection *(m)*
It pozzo petrolifero sotto pressione *(m)*
Pt poço de injecção *(m)*

Innenarbeit *(f) n* De
En internal work
Es trabajo interno *(m)*
Fr travail interne *(m)*
It lavoro interno *(m)*
Pt trabalho interno *(m)*

innere Energie *(f)* De
En internal energy
Es energía interna *(f)*
Fr énergie interne *(f)*
It energia interna *(f)*
Pt energia interna *(f)*

input energy En
De Eingabeenergie *(f)*
Es energía de entrada *(f)*
Fr énergie d'entrée *(f)*
It energia immessa *(f)*
Pt energia de entrada *(f)*

inquinamento *(m) n* It
De Verschmutzung *(f)*
En pollution
Es contaminación *(f)*
Fr pollution *(f)*
Pt poluição *(f)*

inquinamento atmosferico *(m)* It
De Luftverschmutzung *(f)*
En air pollution
Es contaminación del aire *(f)*
Fr pollution de l'air *(f)*
Pt poluição do ar *(f)*

insolação *(f) n* Pt
De Sonnenbestrahlung *(f)*
En insolation
Es insolación *(f)*
Fr insolation *(f)*
It insolazione *(f)*

insolación *(f) n* Es
De Sonnenbestrahlung *(f)*
En insolation
Fr insolation *(f)*
It insolazione *(f)*
Pt insolação *(f)*

insolation *n* En; Fr *(f)*
De Sonnenbestrahlung *(f)*
Es insolación *(f)*
It insolazione *(f)*
Pt insolação *(f)*

insolazione *(f) n* It
De Sonnenbestrahlung *(f)*
En insolation
Es insolación *(f)*
Fr insolation *(f)*
Pt insolação *(f)*

instalação auxiliar de caldeira *(f)* Pt
De Kesselhilfsanlage *(f)*
En boiler auxiliary plant
Es planta auxiliar de calderas *(f)*
Fr chaufferie auxiliaire *(f)*
It impianto ausiliario caldaia *(m)*

installation de forage *(f)* Fr
De Ölbohrturm *(m)*
En oilrig
Es tren de perforación de petróleo *(m)*
It impianto di trivellazione per petrolio *(m)*
Pt plataforma petroleira *(f)*

installation pilote *(f)* Fr
De Versuchsanlage *(f)*
En pilot plant
Es planta piloto *(f)*
It impianto pilota *(m)*
Pt fábrica piloto *(f)*

instantaneous water heater En
De Wasser-Schneller-hitzer *(m)*
Es calentador de agua instantáneo *(m)*
Fr chauffe-eau instantané *(m)*
It riscaldatore istantaneo di acqua *(m)*
Pt aquecedor de água instantâneo *(m)*

insulation *n* En
De Isolierung *(f)*
Es aislamiento *(m)*
Fr isolement *(m)*
It isolamento *(m)*
Pt isolamento *(m)*

integrated environmental design (IED) En
De integrierte Umweltplanung *(f)*
Es diseño ambiental integrado *(m)*
Fr étude de l'environment integreé *(f)*
It design ambientale integrato *(m)*
Pt projecto de ambiente integrado *(m)*

integrierte Umweltplanung *(f)* De
En integrated environmental design (IED)
Es diseño ambiental integrado *(m)*
Fr étude de l'environment integreé *(f)*
It design ambientale integrato *(m)*
Pt projecto de ambiente integrado *(m)*

intensidad de campo *(f)* Es
De Feldstärke *(f)*
En field strength
Fr intensité de champ *(f)*
It intensità di campo *(f)*
Pt intensidade de campo *(f)*

intensidade de campo *(f)* Pt
De Feldstärke *(f)*
En field strength
Es intensidad de campo *(f)*
Fr intensité de champ *(f)*
It intensità di campo *(f)*

intensità di campo *(f)* It
De Feldstärke *(f)*
En field strength
Es intensidad de campo *(f)*
Fr intensité de champ *(f)*
Pt intensidade de campo *(f)*

intensité de champ *(f)* Fr
De Feldstärke *(f)*
En field strength
Es intensidad de campo *(f)*
It intensità di campo *(f)*
Pt intensidade de campo *(f)*

inter-arrefecedor *(m) n* Pt
De Zwischenkühler *(m)*
En intercooler
Es enfriador intermedio *(m)*
Fr refroidisseur intermédiaire *(m)*
It refrigeratore intermedio *(m)*

intercambiador *(m) n* Es
De Austauscher *(m)*
En exchanger
Fr échangeur *(m)*
It scambiatore *(m)*
Pt permutador *(m)*

intercambiador de calor *(m)* Es
De Wärmeaustauscher *(m)*
En heat exchanger
Fr échangeur de chaleur *(m)*
It scambiatore di calore *(m)*
Pt termo-permutador *(m)*

intercambiador de calor con serpentín en baño *(m)* Es
De Wärmeaustauscher bestehend aus Rohrschlangen im Bad *(m)*
En coil-in-bath heat exchanger
Fr échangeur de chaleur à serpentin immergé *(m)*
It scambiatore di calore a serpentino in bagno *(m)*
Pt termo-permutador de bobina em banho *(m)*

intercambiador de calor de coraza y tubos *(m)* Es
De Röhrenwärmeaus-tauscher *(m)*
En shell-and-tube heat exchanger
Fr échangeur de chaleur à calandre *(m)*
It scambiatore di calore a cilindro e tubi *(m)*
Pt termo-permutador de camisa e tubo *(m)*

intercambiador de calor de placas *(m)* Es
De Plattenwärmeaus-tauscher *(m)*
En plate heat-exchanger
Fr échangeur de chaleur à plaque *(m)*
It scambiatore di calore a piastra *(m)*
Pt termo-permutador de placa *(m)*

intercambiador de calor de tubos concéntricos *(m)* Es
De Wärmeaustauscher mit konzentrischen Röhren *(m)*
En concentric-tube heat exchanger
Fr échangeur de chaleur à tubes concentriques (m)
It scambiatore di calore a tubo concentrico *(m)*
Pt termo-permutador de tubo concêntrico *(m)*

intercambiador de calor de tubos en espiral *(m)* Es
De Spiralröhren-Wärmeaustauscher *(m)*
En spiral-tube heat exchanger
Fr échangeur de chaleur à serpentin *(m)*
It scambiatore di calore con tubo a spirale *(m)*
Pt termo-permutador de tubo em espiral *(m)*

intercambiador de calor de vidrio *(m)* Es
De Glaswärmeaus-tauscher *(m)*
En glass heat exchanger
Fr échangeur de chaleur en verre *(m)*
It scambiatore di calore di vetro *(m)*
Pt termo-permutador de vidro *(m)*

intercambio de iones *(m)* Es
De Ionenaustausch *(m)*
En ion exchange
Fr échange d'ions *(m)*
It scambio ionico *(m)*
Pt permuta de ioẽs *(f)*

intercooler *n* En
De Zwischenkühler *(m)*
Es enfriador intermedio *(m)*
Fr refroidisseur intermédiaire *(m)*
It refrigeratore intermedio *(m)*
Pt inter-arrefecedor *(m)*

internal-combustion engine En
De Verbrennungsmotor *(m)*
Es motor de combustión interna *(m)*
Fr moteur à combustion interne *(m)*
It motore a scoppio *(m)*
Pt motor de combustão interna *(m)*

internal energy En
De innere Energie *(f)*
Es energía interna *(f)*
Fr énergie interne *(f)*
It energia interna *(f)*
Pt energia interna *(f)*

internal work En
De Innenarbeit *(f)*
Es trabajo interno *(m)*
Fr travail interne *(m)*
It lavoro interno *(m)*
Pt trabalho interno *(m)*

interruttore a tempo *(m)* It
De Zeitschalter *(m)*
En time switch
Es temporizador *(m)*
Fr minuterie *(f)*
Pt comutador cronométrico *(m)*

interruttore automatico *(m)* It
De Leitungsschalter *(m)*
En circuit breaker
Es disyuntor *(m)*
Fr disjoncteur *(m)*
Pt corta-circuito *(m)*

invecchiamento termico *(m)* It
De thermische Alterung *(f)*
En thermal ageing
Es envejecimiento térmico *(m)*
Fr vieillissement thermique *(m)*
Pt envelhecimento térmico *(m)*

inverseur *(m) n* Fr
De Wechselrichter *(m)*
En inverter
Es inversor *(m)*
It invertitore *(m)*
Pt inversor *(m)*

inversor *(m) n* Es, Pt
De Wechselrichter *(m)*
En inverter
Fr inverseur *(m)*
It invertitore *(m)*

inverter *n* En
De Wechselrichter *(m)*
Es inversor *(m)*
Fr inverseur *(m)*
It invertitore *(m)*
Pt inversor *(m)*

invertitore *(m) n* It
De Wechselrichter *(m)*
En inverter
Es inversor *(m)*
Fr inverseur *(m)*
Pt inversor *(m)*

ion *n* En; Es, Fr *(m)*
De Ion *(n)*
It ione *(m)*
Pt ião *(m)*

Ion *(n) n* De
En ion
Es ion *(m)*
Fr ion *(m)*
It ione *(m)*
Pt ião *(m)*

ione *(m) n* It
De Ion *(n)*
En ion
Es ion *(m)*
Fr ion *(m)*
Pt ião *(m)*

Ionenaustausch *(m) n* De
En ion exchange
Es intercambio de iones *(m)*
Fr échange d'ions *(m)*
It scambio ionico *(m)*
Pt permuta de ioẽs *(f)*

ion engine En
De Ionentriebwerk *(n)*
Es motor de iones *(m)*
Fr moteur ionique *(m)*
It motore a ioni *(m)*
Pt motor a iões *(m)*

Ionentriebwerk *(n) n* De
En ion engine
Es motor de iones *(m)*
Fr moteur ionique *(m)*
It motore a ioni *(m)*
Pt motor a iões *(m)*

ion exchange En
De Ionenaustausch *(m)*
Es intercambio de iones *(m)*
Fr échange d'ions *(m)*
It scambio ionico *(m)*
Pt permuta de ioẽs *(f)*

ionisation *(f) n* Fr
De Ionisation *(f)*
En ionization
Es ionización *(f)*
It ionizzazione *(f)*
Pt ionização *(f)*

Ionisation *(f) n* De
En ionization
Es ionización *(f)*
Fr ionisation *(f)*
It ionizzazione *(f)*
Pt ionização *(f)*

Ionisierstrahlung *(f) n* De
En ionizing radiation
Es radiación ionizante *(f)*
Fr rayonnement d'ionisation *(m)*
It radiazione ionizzante *(f)*
Pt radiação ionizante *(f)*

Ionisierungsenergie *(f) n* De
En ionization potential
Es potencial de ionización *(m)*
Fr potentiel d'ionisation *(m)*
It potenziale di ionizzazione *(m)*
Pt potencial de ionização *(m)*

ionização *(f) n* Pt
De Ionisation *(f)*
En ionization
Es ionización *(f)*
Fr ionisation *(f)*
It ionizzazione *(f)*

ionización *(f) n* Es
De Ionisation *(f)*
En ionization
Fr ionisation *(f)*
It ionizzazione *(f)*
Pt ionização *(f)*

ionization *n* En
De Ionisation *(f)*
Es ionización *(f)*
Fr ionisation *(f)*
It ionizzazione *(f)*
Pt ionização *(f)*

ionization potential En
De Ionisierungsenergie *(f)*
Es potencial de ionización *(m)*
Fr potentiel d'ionisation *(m)*
It potenziale di ionizzazione *(m)*
Pt potencial de ionização *(m)*

ionizing radiation En
De Ionisierstrahlung *(f)*
Es radiación ionizante *(f)*
Fr rayonnement d'ionisation *(m)*
It radiazione ionizzante *(f)*
Pt radiação ionizante *(f)*

ionizzazione *(f) n* It
De Ionisation *(f)*
En ionization
Es ionización *(f)*
Fr ionisation *(f)*
Pt ionização *(f)*

ionosfera *(f) n* Es, It, Pt
De Ionosphäre *(f)*
En ionosphere
Fr ionosphère *(f)*

Ionosphäre *(f) n* De
En ionosphere
Es ionosfera *(f)*
Fr ionosphère *(f)*
It ionosfera *(f)*
Pt ionosfera *(f)*

ionosphere *n* En
De Ionosphäre *(f)*
Es ionosfera *(f)*
Fr ionosphère *(f)*
It ionosfera *(f)*
Pt ionosfera *(f)*

ionosphère *(f) n* Fr
De Ionosphäre *(f)*
En ionosphere
Es ionosfera *(f)*
It ionosfera *(f)*
Pt ionosfera *(f)*

irradiação *(f) n* Pt
De Bestrahlung *(f)*
En irradiation
Es irradiación *(f)*
Fr irradiation *(f)*
It irradiazione *(f)*

irradiación *(f) n* Es
De Bestrahlung *(f)*
En irradiation
Fr irradiation *(f)*
It irradiazione *(f)*
Pt irradiação *(f)*

irradiado *adj* Es, Pt
De bestrahlt
En irradiated
Fr irradié
It irradiato

irradiated *adj* En
De bestrahlt
Es irradiado
Fr irradié
It irradiato
Pt irradiado

irradiation *n* En; Fr *(f)*
De Bestrahlung *(f)*
Es irradiación *(f)*
It irradiazione *(f)*
Pt irradiação *(f)*

irradiato *adj* It
De bestrahlt
En irradiated
Es irradiado
Fr irradié
Pt irradiado

irradiazione *(f) n* It
De Bestrahlung *(f)*
En irradiation
Es irradiación *(f)*
Fr irradiation *(f)*
Pt irradiação *(f)*

irradié *adj* Fr
De bestrahlt
En irradiated
Es irradiado
It irradiato
Pt irradiado

irreversibile *adj* It
De nicht umkehrbar
En irreversible
Es irreversible
Fr irréversible
Pt irrevesível

irreversible *adj* En, Es
De nicht umkehrbar
Fr irréversible
It irreversibile
Pt irrevesível

irréversible *adj* Fr
De nicht umkehrbar
En irreversible
Es irreversible
It irreversibile
Pt irrevesível

irrevesível *adj* Pt
De nicht umkehrbar
En irreversible
Es irreversible
Fr irréversible
It irreversibile

isentalpico *adj* It
De isenthalpisch
En isenthalpic
Es isentálpico
Fr isenthalpique
Pt isentálpico

isentálpico *adj* Es, Pt
De isenthalpisch
En isenthalpic
Fr isenthalpique
It isentalpico

isenthalpic *adj* En
De isenthalpisch
Es isentálpico
Fr isenthalpique
It isentalpico
Pt isentálpico

isenthalpique *adj* Fr
De isenthalpisch
En isenthalpic
Es isentálpico
It isentalpico
Pt isentálpico

isenthalpisch *adj* De
En isenthalpic
Es isentálpico
Fr isenthalpique
It isentalpico
Pt isentálpico

isentropic *adj* En
De isentropisch
Es isentrópico
Fr isentropique
It isentropico
Pt isentrópico

isentropico *adj* It
De isentropisch
En isentropic
Es isentrópico
Fr isentropique
Pt isentrópico

isentrópico *adj* Es, Pt
De isentropisch
En isentropic
Fr isentropique
It isentropico

isentropique *adj* Fr
De isentropisch
En isentropic
Es isentrópico
It isentropico
Pt isentrópico

isentropisch *adj* De
En isentropic
Es isentrópico
Fr isentropique
It isentropico
Pt isentrópico

Isobutan *(n) n* De
En isobutane
Es isobutano *(m)*
Fr isobutane *(m)*
It isobutano *(m)*
Pt isobutano *(m)*

isobutane *n* En; Fr *(m)*
De Isobutan *(n)*
Es isobutano *(m)*
It isobutano *(m)*
Pt isobutano *(m)*

isobutano *(m) n* Es, It, Pt
De Isobutan *(n)*
En isobutane
Fr isobutane *(m)*

isolamento *(m) n* It, Pt
De Isolierung *(f)*
En insulation
Es aislamiento *(m)*
Fr isolement *(m)*

isolamento com fibra de vidro *(m)* Pt
Am glass-fiber insulation
De Glasfaserisolierung *(f)*
En glass-fibre insulation
Es aislamiento con fibra de vidrio *(m)*
Fr isolement par fibre de verre *(m)*
It isolamento con fibra di vetro *(m)*

isolamento com fibras *(m)* Pt
Am fiber insulation
De Faserisolierung *(f)*
En fibre insulation
Es aislamiento con fibra *(m)*
Fr isolement par fibre *(m)*
It isolamento con fibra *(m)*

isolamento com fibras de lã mineral *(m)* Pt
Am mineral-wool fiber insulation
De Steinwollefaser-isolierung *(f)*
En mineral-wool fibre insulation
Es aislamiento con fibra de lana mineral *(m)*
Fr isolement par fibre de laine minérale *(m)*
It isolamento con fibra di lana minerale *(m)*

isolamento con fibra di lana minerale *(m)* It
Am mineral-wool fiber insulation
De Steinwollefaser-isolierung *(f)*
En mineral-wool fibre insulation
Es aislamiento con fibra de lana mineral *(m)*
Fr isolement par fibre de laine minérale *(m)*
Pt isolamento com fibras de lã mineral *(m)*

isolamento con fibra *(m)* It
Am fiber insulation
De Faserisolierung *(f)*
En fibre insulation
Es aislamiento con fibra *(m)*
Fr isolement par fibre *(m)*
Pt isolamento com fibras *(m)*

isolamento con fibra di vetro *(m)* It
Am glass-fiber insulation
De Glasfaserisolierung *(f)*

En glass-fibre insulation
Es aislamiento de fibra de vidrio *(m)*
Fr isolement par fibre de verre *(m)*
Pt isolamento com fibra de vidro *(m)*

isolamento con poliuretano *(m)* It
De Polyurethanisolierung *(f)*
En polyurethane insulation
Es aislamiento de poliuretano *(m)*
Fr isolement au polyuréthane *(m)*
Pt isolamento de poliuretano *(m)*

isolamento da cobertura *(m)* Pt
De Dachisolierung *(f)*
En roof insulation
Es aislamiento de techos *(m)*
Fr isolement des toits *(m)*
It isolamento del tetto *(m)*

isolamento de janela *(m)* Pt
De Fensterisolierung *(f)*
En window insulation
Es aislamiento de ventanas *(m)*
Fr isolement des fenêtres *(m)*
It isolamento delle finestre *(m)*

isolamento della soffitta *(m)* It
De Dachbodenisolierung *(f)*
En loft insulation
Es aislamiento de galerías *(m)*
Fr isolement des combles *(m)*
Pt isolamento do sotão *(m)*

isolamento delle finestre *(m)* It
De Fensterisolierung *(f)*
En window insulation
Es aislamiento de ventanas *(m)*
Fr isolement des fenêtres *(m)*
Pt isolamento de janela *(m)*

isolamento delle pareti *(m)* It
De Wandisolierung *(f)*
En wall insulation
Es aislamiento de paredes *(m)*
Fr isolement des murs *(m)*
Pt isolamento de paredes *(m)*

isolamento del pavimento *(m)* It
De Bodenisolierung *(f)*
En floor insulation
Es aislamiento de suelos *(m)*
Fr isolement du plancher *(m)*
Pt isolamento de pavimento *(m)*

isolamento del tetto *(m)* It
De Dachisolierung *(f)*
En roof insulation
Es aislamiento de techos *(m)*
Fr isolement des toits *(m)*
Pt isolamento da cobertura *(m)*

isolamento de paredes *(m)* Pt
De Wandisolierung *(f)*
En wall insulation
Es aislamiento de paredes *(m)*
Fr isolement des murs *(m)*
It isolamento delle pareti *(m)*

isolamento de paredes de cavidade *(m)* Pt
De Isolierung von Hohlziegelmauerwerk *(f)*
En cavity wall insulation
Es aislamiento de pared de cavidad *(m)*
Fr isolement des murs doubles *(m)*
It isolamento di muro a cassavuota *(m)*

isolamento de pavimento *(m)* Pt
De Bodenisolierung *(f)*
En floor insulation
Es aislamiento de suelos *(m)*
Fr isolement du plancher *(m)*
It isolamento del pavimento *(m)*

isolamento de poliuretano *(m)* Pt
De Polyurethanisolierung *(f)*
En polyurethane insulation
Es aislamiento de poliuretano *(m)*
Fr isolement au polyuréthane *(m)*
It isolamento con poliuretano *(m)*

isolamento di muro a cassavuota *(m)* It
De Isolierung von Hohlziegelmauerwerk *(f)*
En cavity wall insulation
Es aislamiento de pared de cavidad *(m)*
Fr isolement des murs doubles *(m)*
Pt isolamento de paredes de cavidade *(m)*

isolamento do sotão *(m)* Pt
De Dachbodenisolierung *(f)*
En loft insulation
Es aislamiento de galerías *(m)*
Fr isolement des combles *(m)*
It isolamento della soffitta *(m)*

isolement *(m)* *n* Fr
De Isolierung *(f)*
En insulation
Es aislamiento *(m)*
It isolamento *(m)*
Pt isolamento *(m)*

isolement au polyuréthane *(m)* Fr
De Polyurethanisolierung *(f)*
En polyurethane insulation
Es aislamiento de poliuretano *(m)*
It isolamento con poliuretano *(m)*
Pt isolamento de poliuretano *(m)*

isolement des combles *(m)* Fr
De Dachbodenisolierung *(f)*
En loft insulation
Es aislamiento de galerías *(m)*
It isolamento della soffitta *(m)*
Pt isolamento do sotão *(m)*

isolement des fenêtres *(m)* Fr
De Fensterisolierung *(f)*
En window insulation
Es aislamiento de ventanas *(m)*
It isolamento delle finestre *(m)*
Pt isolamento de janela *(m)*

isolement des murs *(m)* Fr
De Wandisolierung *(f)*
En wall insulation
Es aislamiento de paredes *(m)*
It isolamento delle pareti *(m)*
Pt isolamento de paredes *(m)*

isolement des murs doubles *(m)* Fr
De Isolierung von Hohlziegelmauerwerk *(f)*
En cavity wall insulation
Es aislamiento de pared de cavidad *(m)*
It isolamento di muro a cassavuota *(m)*
Pt isolamento de paredes de cavidade *(m)*

isolement des toits *(m)* Fr
De Dachisolierung *(f)*
En roof insulation

Es aislamiento de techos *(m)*
It isolamento del tetto *(m)*
Pt isolamento da cobertura *(m)*

isolement du plancher *(m)* Fr
De Bodenisolierung *(f)*
En floor insulation
Es aislamiento de suelos *(m)*
It isolamento del pavimento *(m)*
Pt isolamento de pavimento *(m)*

isolement par fibre *(m)* Fr
Am fiber insulation
De Faserisolierung *(f)*
En fibre insulation
Es aislamiento con fibra *(m)*
It isolamento con fibra *(m)*
Pt isolamento com fibras *(m)*

isolement par fibre de laine minérale *(m)* Fr
Am mineral-wool fiber insulation
De Steinwollefaser-isolierung *(f)*
En mineral-wool fibre insulation
Es aislamiento con fibra de lana mineral *(m)*
It isolamento con fibra di lana minerale *(m)*
Pt isolamento com fibras de lã mineral *(m)*

isolement par fibre de verre *(m)* Fr
Am glass-fiber insulation
De Glasfaserisolierung *(f)*
En glass-fibre insulation
Es aislamiento de fibra de vidrio *(m)*
It isolamento di fibra di vetro *(m)*
Pt isolamento com fibra de vidro *(m)*

Isolierung *(f)* *n* De
En insulation
Es aislamiento *(m)*
Fr isolement *(m)*
It isolamento *(m)*
Pt isolamento *(m)*

Isolierung von Hohlziegelmauerwerk *(f)* De
En cavity wall insulation
Es aislamiento de pared de cavidad *(m)*
Fr isolement des murs doubles *(m)*
It isolamento di muro a cassavuota *(m)*
Pt isolamento de paredes de cavidade *(m)*

isomer *n* En
De Isomer *(n)*
Es isómero *(m)*
Fr isomère *(m)*
It isomero *(m)*
Pt isómero *(m)*

Isomer *(n)* *n* De
En isomer
Es isómero *(m)*
Fr isomère *(m)*
It isomero *(m)*
Pt isómero *(m)*

isomère *(m)* *n* Fr
De Isomer *(n)*
En isomer
Es isómero *(m)*
It isomero *(m)*
Pt isómero *(m)*

isomérisation *(f)* *n* Fr
De Isomerisierungs-verfahren *(n)*
En isomerization
Es isomerización *(f)*
It isomerizzazione *(f)*
Pt isomerização *(f)*

Isomerisierungs-verfahren *(n)* *n* De
En isomerization
Es isomerización *(f)*
Fr isomérisation *(f)*
It isomerizzazione *(f)*
Pt isomerização *(f)*

isomerização *(f)* *n* Pt
De Isomerisierungs-verfahren *(n)*
En isomerization
Es isomerización *(f)*
Fr isomérisation *(f)*
It isomerizzazione *(f)*

isomerización *(f)* *n* Es
De Isomerisierungs-verfahren *(n)*
En isomerization
Fr isomérisation *(f)*
It isomerizzazione *(f)*
Pt isomerização *(f)*

isomerization *n* En
De Isomerisierungs-verfahren *(n)*
Es isomerización *(f)*
Fr isomérisation *(f)*
It isomerizzazione *(f)*
Pt isomerização *(f)*

isomerizzazione *(f)* *n* It
De Isomerisierungs-verfahren *(n)*
En isomerization
Es isomerización *(f)*
Fr isomérisation *(f)*
Pt isomerização *(f)*

isomero *(m)* *n* It
De Isomer *(n)*
En isomer
Es isómero *(m)*
Fr isomère *(m)*
Pt isómero *(m)*

isómero *(m)* *n* Es, Pt
De Isomer *(n)*
En isomer
Fr isomère *(m)*
It isomero *(m)*

Isopentan *(n)* *n* De
En isopentane
Es isopentano *(m)*
Fr isopentane *(m)*
It isopentano *(m)*
Pt isopentano *(m)*

isopentane *n* En; Fr *(m)*
De Isopentan *(n)*
Es isopentano *(m)*
It isopentano *(m)*
Pt isopentano *(m)*

isopentano *(m)* *n* Es, It, Pt
De Isopentan *(n)*
En isopentane
Fr isopentane *(m)*

isopropanol *n* En; Es, Fr, Pt *(m)*
De Isopropanol *(n)*
It isopropanolo *(m)*

Isopropanol *(n)* *n* De
En isopropanol
Es isopropanol *(m)*
Fr isopropanol *(m)*
It isopropanolo *(m)*
Pt isopropanol *(m)*

isopropanolo *(m)* *n* It
De Isopropanol *(n)*
En isopropanol
Es isopropanol *(m)*
Fr isopropanol *(m)*
Pt isopropanol *(m)*

isoterma *(f)* *n* Es, It, Pt
De Isotherme *(f)*
En isotherm
Fr isotherme *(f)*

isotermico *adj* It
De isothermisch
En isothermal
Es isotérmico
Fr isothermique
Pt isotérmico

isotérmico *adj* Es, Pt
De isothermisch
En isothermal
Fr isothermique
It isotermico

isotherm *n* En
De Isotherme *(f)*
Es isoterma *(f)*
Fr isotherme *(f)*
It isoterma *(f)*
Pt isoterma *(f)*

isothermal *adj* En
De isothermisch
Es isotérmico
Fr isothermique
It isotermico
Pt isotérmico

isotherme *(f)* *n* Fr
De Isotherme *(f)*
En isotherm
Es isoterma *(f)*
It isoterma *(f)*
Pt isoterma *(f)*

Isotherme *(f) n* De
En isotherm
Es isoterma *(f)*
Fr isotherme *(f)*
It isoterma *(f)*
Pt isoterma *(f)*

isothermique *adj* Fr
De isothermisch
En isothermal
Es isotérmico
It isotermico
Pt isotérmico

isothermisch *adj* De
En isothermal
Es isotérmico
Fr isothermique
It isotermico
Pt isotérmico

Isotop *(n) n* De
En isotope
Es isótopo *(m)*
Fr isotope *(m)*
It isotopo *(m)*
Pt isótopo *(m)*

isotope *n* En; Fr *(m)*
De Isotop *(n)*
Es isótopo *(m)*
It isotopo *(m)*
Pt isótopo *(m)*

Isotopentrennung *(f) n* De
En isotope separation
Es separación de isótopos *(f)*
Fr séparation isotopique *(f)*
It separazione isotopica *(f)*
Pt separação de isótopos *(f)*

isotope separation En
De Isotopentrennung *(f)*
Es separación de isótopos *(f)*
Fr séparation isotopique *(f)*
It separazione isotopica *(f)*
Pt separação de isótopos *(f)*

isotopo *(m) n* It
De Isotop *(n)*
En isotope
Es isótopo *(m)*
Fr isotope *(m)*
Pt isótopo *(m)*

isótopo *(m) n* Es, Pt
De Isotop *(n)*
En isotope
Fr isotope *(m)*
It isotopo *(m)*

isteresi *(f) n* It
De Hysterese *(f)*
En hysteresis
Es histéresis *(f)*
Fr hystérésis *(f)*
Pt histerese *(f)*

J

jacto *(m) n* Pt
De Düse *(f)*
En jet
Es chorro *(m)*
Fr jet *(m)*
It getto *(m)*

janela de absorção *(f)* Pt
De Absorptionsfenster *(n)*
En absorption window
Es ventanilla de absorción *(f)*
Fr fenêtre d'absorption *(f)*
It finestra di assorbimento *(f)*

jazigo de petróleo *(m)* Pt
De Ölfeld *(n)*
En oilfield
Es yacimiento de petróleo *(m)*
Fr gisement pétrolifère *(m)*
It giacimento petrolifero *(m)*

jet *n* En; Fr *(m)*
De Düse *(f)*
Es chorro *(m)*
It getto *(m)*
Pt jacto *(m)*

jet burner En
De Düsenbrenner *(m)*
Es quemador de chorro *(m)*
Fr brûleur-jet *(m)*
It bruciatore a getto *(m)*
Pt queimador de jacto *(m)*

jet fuel En
De Düsentreibstoff *(m)*
Es combustible de propulsión a chorro *(m)*
Fr carburéacteur *(m)*
It combustibile per aviogetti *(m)*
Pt combustível de jactos *(m)*

jet propulsion En
De Strahlantrieb *(m)*
Es propulsión a chorro *(f)*
Fr propulsion par réaction *(f)*
It propulsione a getto *(m)*
Pt propulsão a jacto *(f)*

joint *(m) n* Fr
De Abdichtung *(f)*
En seal (boiler)
Es cierre *(m)*
It dispositivo di tenuta *(m)*
Pt vedação *(f)*

K

Kadmiumbatterie *(f) n* De
En cadmium battery
Es batería de cadmio *(f)*
Fr pile au cadmium *(f)*
It batteria al cadmio *(f)*
Pt bateria de cádmio *(f)*

Kalorimeter *(n) n* De
En calorimeter
Es calorímetro *(m)*
Fr calorimètre *(m)*
It calorimetro *(m)*
Pt calorímetro *(m)*

Kaltstart *(m) n* De
En cold start
Es arranque en frío *(m)*
Fr démarrage à froid *(m)*
It avviamento a freddo *(m)*
Pt arranque a frio *(m)*

Kanalinduktionsofen *(m) n* De
En channel induction furnace
Es horno de inducción de canal *(m)*
Fr four à induction à chenal *(m)*
It forno ad induzione a canale *(m)*
Pt forno de indução em canal *(m)*

Kapazitanz *(f) n* De
En capacitance
Es capacitancia *(f)*
Fr capacitance *(f)*
It capacitanza *(f)*
Pt capacitância *(f)*

kapazitive Hochfrequenzerwärmung *(f) n* De
En dielectric heating
Es caldeo dieléctrico *(m)*
Fr chauffage par pertes diélectriques *(m)*
It riscaldamento dielettrico *(m)*
Pt aquecimento dieléctrico *(m)*

Kapitalausstattung *(f) n* De
En capitalization
Es capitalización *(f)*
Fr capitalisation *(f)*
It capitalizzazione *(f)*
Pt capitalização *(f)*

Karbonisation *(f) n* De
En carbonization
Es carbonización *(f)*
Fr carbonisation *(f)*
It carbonizzazione *(f)*
Pt carbonização *(f)*

Kaskadenprozeβ *(m) n* De
En cascade cycle
Es ciclo en cascada *(m)*

Fr cycle en cascade *(m)*
It ciclo di cascata *(m)*
Pt ciclo de cascada *(m)*

Kassamarkt *(m) n* De
En spot market
Es mercado de disponibles *(m)*
Fr marché disponible *(m)*
It mercato del disponibile *(m)*
Pt mercado de ocasião *(m)*

Katalysator *(m) n* De
En catalyst
Es catalizador *(m)*
Fr catalyseur *(m)*
It catalizzatore *(m)*
Pt catalisador *(m)*

Katalyse *(f) n* De
En catalysis
Es catálisis *(f)*
Fr catalyse *(f)*
It catalisi *(f)*
Pt catálise *(f)*

katalytische Reformierung *(f)* De
En catalytic reforming
Es reformación catalítica *(f)*
Fr reformage catalytique *(m)*
It riforma catalitica *(f)*
Pt reforma catalítica *(f)*

katalytisches Kracken *(n)* De
En catalytic cracking
Es cráking catalítico *(m)*
Fr craquage catalytique *(m)*
It piroscissione per catalisi *(f)*
Pt cracking catalítico *(m)*

katalytische Veraschung *(f)* De
En catalytic incineration
Es incineración catalítica *(f)*
Fr incinération catalytique *(f)*
It incenerimento catalitico *(m)*
Pt incineração catalítica *(f)*

katalytische Verbrennung *(f)* De
En catalytic combustion
Es combustión catalítica *(f)*
Fr combustion catalytique *(f)*
It combustione per catalisi *(f)*
Pt combustão catalítica *(f)*

Kathode *(f) n* De
En cathode
Es cátodo *(m)*
Fr cathode *(f)*
It catodo *(m)*
Pt cátodo *(m)*

Kathodenstrahlröhre *(f) n* De
En cathode-ray tube
Es tubo de rayos catódicos *(m)*
Fr tube à rayons cathodiques *(m)*
It tubo a raggi catodici *(m)*
Pt válvula de raios catódicos *(f)*

Kennkraft-U-Boot *(n) n* De
En nuclear submarine
Es submarino nuclear *(m)*
Fr sous-marin nucléaire *(m)*
It sottomarino nucleare *(m)*
Pt submarino nuclear *(m)*

Kern (atomischer) *(m) n* De
En nucleus (atomic)
Es núcleo (atómico) *(m)*
Fr noyau (atomique) *(m)*
It nucleo (atomico) *(m)*
Pt núcleo (atómico) *(m)*

Kern (magnetischer) *(m)* De
En core (magnetic)
Es alma (magnética) *(f)*
Fr noyau (magnétique) *(m)*
It nucleo (magnetico) *(m)*
Pt núcleo (magnético) *(m)*

Kernkraftrakete *(f) n* De
En nuclear rocket
Es cohete nuclear *(m)*
Fr fusée nucléaire *(f)*
It razzo nucleare *(m)*
Pt foguetão nuclear *(m)*

Kernkraftschiff *(n) n* De
En nuclear ship
Es barco nuclear *(m)*
Fr navire nucléaire *(m)*
It nave nucleare *(f)*
Pt barco nuclear *(m)*

Kernkraftwerk *(n) n* De
En nuclear power station
Es central nuclear *(f)*
Fr centrale nucléaire *(f)*
It centrale nucleare *(f)*
Pt central de energia nuclear *(f)*

Kernreaktor *(m) n* De
En nuclear reactor
Es reactor nuclear *(m)*
Fr réacteur nucléaire *(m)*
It reattore nucleare *(m)*
Pt reactor nuclear *(m)*

kerogen *n* En
De Kerogen *(n)*
Es kerógeno *(m)*
Fr kérogène *(m)*
It kerogene *(m)*
Pt querogéneo *(m)*

Kerogen *(n) n* De
En kerogen
Es kerógeno *(m)*
Fr kérogène *(m)*
It kerogene *(m)*
Pt querogéneo *(m)*

kerogene *(m) n* It
De Kerogen *(n)*
En kerogen
Es kerógeno *(m)*
Fr kérogène *(m)*
Pt querogéneo *(m)*

kérogène *(m) n* Fr
De Kerogen *(n)*
En kerogen
Es kerógeno *(m)*
It kerogene *(m)*
Pt querogéneo *(m)*

kerógeno *(m) n* Es
De Kerogen *(n)*
En kerogen
Fr kérogène *(m)*
It kerogene *(m)*
Pt querogéneo *(m)*

Kerosen *(n) n* De
En kerosene
Es keroseno *(m)*
Fr kérosène *(m)*
It kerosene *(m)*
Pt queroseno *(m)*

kerosene *n* En
De Kerosen *(n)*
Es keroseno *(m)*
Fr kérosène *(m)*
It kerosene *(m)*
Pt queroseno *(m)*

kerosene *(m) n* It
De Kerosen *(n)*
En kerosene
Es keroseno *(m)*
Fr kérosène *(m)*
Pt queroseno *(m)*

kérosène *(m) n* Fr
De Kerosen *(n)*
En kerosene
Es keroseno *(m)*
It kerosene *(m)*
Pt queroseno *(m)*

keroseno *(m) n* Es
De Kerosen *(n)*
En kerosene
Fr kérosène *(m)*
It kerosene *(m)*
Pt queroseno *(m)*

Kessel *(m) n* De
En boiler
Es caldera *(f)*
Fr chaudière *(f)*
It caldaia *(f)*
Pt caldeira *(f)*

Kesselhilfsanlage *(f) n* De
En boiler auxiliary plant
Es planta auxiliar de calderas *(f)*
Fr chaufferie auxiliaire *(f)*
It impianto ausiliario caldaia *(m)*
Pt instalação auxiliar de caldeira *(f)*

Kesselrippenrohr *(n) n* De
En ribbed boiler tube
Es tubo de aletas de caldera *(m)*
Fr tube de chaudière aileté *(m)*
It tubo nervato per caldaia *(m)*
Pt tubo de caldeira com nervuras *(m)*

Kettenreaktion *(f) n* De
En chain reaction
Es reacción en cadena *(f)*
Fr réaction en chaîne *(f)*
It reazione a catena *(f)*
Pt reacção em cadeia *(f)*

kiln *n* En
De Brennofen *(m)*
Es estufa *(f)*
Fr four *(m)*
It forno *(m)*
Pt forno de requeima *(m)*

kinematics *n* En
De Kinematik *(f)*
Es cinemática *(f)*
Fr cinématique *(f)*
It cinematica *(f)*
Pt cinemática *(f)*

Kinematik *(f) n* De
En kinematics
Es cinemática *(f)*
Fr cinématique *(f)*
It cinematica *(f)*
Pt cinemática *(f)*

kinetic energy En
De kinetische Energie *(f)*
Es energía cinética *(f)*
Fr énergie cinétique *(f)*
It energia cinetica *(f)*
Pt energia cinética *(f)*

kinetische Energie *(f)* De
En kinetic energy
Es energía cinética *(f)*
Fr énergie cinétique *(f)*
It energia cinetica *(f)*
Pt energia cinética *(f)*

Kippen *(n) n* De
En dumping
Es vaciamiento *(m)*
Fr déversement *(m)*
It pressatura *(f)*
Pt esvaziamento *(m)*

Klassierung *(f) n* De
En grading
Es clasificación *(f)*
Fr criblage *(m)*
It cernita *(f)*
Pt granulometria *(f)*

klastische Gesteine *(pl)* De
En conglomerate rocks *(pl)*
Es rocas conglomeradas *(f pl)*
Fr roches conglomérés *(f pl)*
It rocce conglomerate *(f pl)*
Pt rochas conglomeradas *(f pl)*

Klemme (elektrische) *(f) n* De
En terminal (electric)
Es terminal (eléctrico) *(m)*
Fr borne (électrique) *(f)*
It terminale (elettrico) *(m)*
Pt borne (eléctrico) *(m)*

Klima *(n) n* De
En climate
Es clima *(m)*
Fr climat *(m)*
It clima *(m)*
Pt clima *(m)*

Klimaanlage *(f) n* De
En air conditioning
Es acondicionamiento de aire *(m)*
Fr climatisation *(f)*
It condizionamento dell'aria *(m)*
Pt ar condicionado *(m)*

klopffest *adj* De
En antiknock
Es antidetonante
Fr antidétonant
It antidetonante
Pt antidetonante

Kohle *(f) n* De
En coal
Es carbón *(m)*
Fr charbon *(m)*
It carbone *(m)*
Pt carvão *(m)*

kohlegefeuertes Kraftwerk *(n)* De
En coal-fired power station
Es central eléctrica de carbón *(f)*
Fr centrale électrique au charbon *(f)*
It centrale elettrica a carbone *(f)*
Pt central de energia a carvão *(f)*

Kohlendioxid *(n) n* De
En carbon dioxide
Es anhídrido carbónico *(m)*
Fr anhydride carbonique *(m)*
It anidride carbonica *(f)*
Pt anidrido carbónico *(m)*

Kohlenersatz *(m) n* De
En coal equivalent
Es carbón equivalente *(m)*
Fr équivalent charbon *(m)*
It equivalente del carbone *(m)*
Pt equivalente carbónico *(m)*

Kohlenflöz *(n) n* De
En coal seam
Es capa de carbón *(f)*
Fr filon houiller *(m)*
It filone di carbone *(m)*
Pt camada de carvão *(f)*

kohlenhaltig *adj* De
En carboniferous
Es carbonífero
Fr carbonifère
It carbonifero
Pt carbonífero

Kohlenlager *(n) n* De
En coal measure
Es formación hullera *(f)*
Fr couche de houille *(f)*
It giacimento carbonifero *(m)*
Pt medida de carvão *(f)*

Kohlenmonoxid *(n) n* De
En carbon monoxide
Es óxido de carbono *(m)*
Fr oxyde de carbone *(m)*
It ossido di carbonio *(m)*
Pt óxido de carbono *(m)*

Kohlenstoffgehalt *(m) n* De
En carbon content
Es contenido de carbono *(m)*
Fr teneur en carbone *(f)*
It contenuto di carbonio *(m)*
Pt teor de carbono *(m)*

kohlenstoffhaltiger Brennstoff *(m)* De
En carbonaceous fuel
Es combustible carbonoso *(m)*
Fr combustible charbonneux *(m)*
It combustibile carbonioso *(m)*
Pt combustível carbonoso *(m)*

kohlenstoffhaltiger feuerfester Stoff *(m)* De
En carbon refractory
Es refractario de carbón *(m)*
Fr réfractaire au carbone *(m)*
It refrattario al carbonio *(m)*
Pt refractário ao carbono *(m)*

kohlenverbrennend *adj* De
En coal-burning
Es caldeado con carbón
Fr chauffé au charbon
It scaldato a carbone
Pt aquecido a carvão

Kohlenwasserstoff *(m) n* De
En hydrocarbon
Es hidrocarburo *(m)*
Fr hydrocarbure *(m)*
It idrocarburo *(m)*
Pt hidrocarboneto *(m)*

Kohlereserven *(pl) n* De
En coal reserves *(pl)*

Es reservas carboníferas *(f pl)*
Fr réserves de houille *(f pl)*
It riserves di carbone *(f pl)*
Pt reservas de carvão *(f pl)*

Koks *(m) n* De
En coke
Es coque *(m)*
Fr coke *(m)*
It coke *(m)*
Pt coque *(m)*

Kokskohle *(f) n* De
En coking coal
Es carbón de coquización *(m)*
Fr charbon cokéfiant *(m)*
It carbone cokificante *(m)*
Pt carvão de coqueificação *(m)*

Koksofenteer *(m) n* De
En coke-oven tar
Es alquitrán de horno de coquización *(m)*
Fr goudron *(m)*
It catrame di forno da coke *(m)*
Pt alcatrão de forno de coque *(m)*

Koksöl *(n) n* De
En coking oil
Es petróleo de coquización *(m)*
Fr huile cokéfiante *(f)*
It olio cokificante *(m)*
Pt petróleo de coqueificação *(m)*

Koksrückstand *(m) n* De
En carbon residue
Es residuo de carbón *(m)*
Fr résidu de carbone *(m)*
It residuo di carbonio *(m)*
Pt resíduo de carbono *(m)*

Kolbenmaschine *(f) n* De
En reciprocating engine
Es motor alternativo *(m)*
Fr moteur alternatif *(m)*
It motore a movimento alternativo *(m)*
Pt motor alternativo *(m)*

Kombinationsbetrieb *(m) n* De
En combined cycle
Es ciclo combinado *(m)*
Fr cycle mixte *(m)*
It ciclo combinato *(m)*
Pt ciclo combinado *(m)*

kombinierte Wärme und Energie *(f)* De
En combined heat and power
Es calor y potencia combinados *(m)*
Fr chaleur et puissance combinées *(f)*
It potenza e calore combinati *(f)*
Pt calor e potência combinados *(m)*

kommerzieller Schnellreaktor *(m)* De
En commercial fast reactor (CFR)
Es reactor rápido comercial *(m)*
Fr réacteur rapide industriel *(m)*
It reattore rapido commerciale *(m)*
Pt reactor rápido comercial *(m)*

Kommutator *(m) n* De
En commutator
Es colector *(m)*
Fr collecteur *(m)*
It commutatore *(m)*
Pt comutador *(m)*

Kompressionsverhältnis *(n) n* De
En compression ratio
Es relación de compresión *(f)*
Fr taux de compression *(m)*
It rapporto di compressione *(m)*
Pt taxa de compressão *(f)*

Kondensation *(f) n* De
En condensation
Es condensación *(f)*
Fr condensation *(f)*
It condensazione *(f)*
Pt condensação *(f)*

Kondensator (elektrischer) *(m) n* De
En capacitor
Es condensador (eléctrico) *(m)*
Fr condensateur *(m)*
It condensatore (elettrico) *(m)*
Pt capacitor *(m)*

Kondensator (chemischer) *(m) n* De
En condenser (chemical)
Es condensador (químico) *(m)*
Fr condenseur *(m)*
It condensatore (chimico) *(m)*
Pt condensador *(m)*

Kondensator-Speisewasser *(n) n* De
En condenser feedwater
Es agua de alimentación del condensador *(f)*
Fr eau d'alimentation de condenseur *(f)*
It acqua di alimentazione del condensatore *(f)*
Pt água de alimentação de condensador *(f)*

kongruentes Schmelzen *(n)* De
En congruent melting
Es fusión congruente *(f)*
Fr fusion congruente *(f)*
It fusione congruente *(f)*
Pt fusão congruente *(f)*

Konsolverfahren *(n) n* De
En consol process
Es proceso consol *(m)*
Fr procédé consol *(m)*
It processo consol *(m)*
Pt processo consol *(m)*

Kontaktspannung *(f) n* De
En contact potential
Es potencial de contacto *(m)*
Fr potentiel de contact *(m)*
It potenziale di contatto *(m)*
Pt potencial de contacto *(m)*

kontaminierter Abfall *(m)* De
En contaminated waste
Es residuo contaminado *(m)*
Fr déchets contaminés *(m pl)*
It rifiuti contaminati *(m pl)*
Pt desperdício contaminado *(m)*

Kontrollstab *(m) n* De
En control rod
Es varilla de control *(f)*
Fr barre de commande *(f)*
It barra di controllo *(f)*
Pt haste de controle *(f)*

Konturenabbau *(m) n* De
En contour mining
Es minería siguiendo el perfil del terreno *(f)*
Fr exploitation minière des contours *(f)*
It scavo a contorno *(m)*
Pt mineração de curvas de nível *(f)*

Konvektion *(f) n* De
En convection
Es convección *(f)*
Fr convection *(f)*
It convezione *(f)*
Pt convecção *(f)*

Konvektionsrekuperator *(m) n* De
En convection recuperator
Es recuperador de convección *(m)*
Fr récupérateur à convection *(m)*
It recuperatore di convezione *(m)*
Pt recuperador por convecção *(m)*

Konvektionsstrom *(m) n* De
En convection current

Es corriente de convección *(f)*
Fr courant de convection *(m)*
It corrente di convezione *(f)*
Pt corrente de convecção *(f)*

Konvektor mit entlastetem Abzug *(m)* De
En balanced-flue convector
Es convector de chimenea equilibrada *(m)*
Fr convecteur à carneau équilibré *(m)*
It convettore a focolare equilibrato *(m)*
Pt convector de fluxo equilibrado *(m)*

Konvektor ohne Abzug *(m)* De
En flueless convector
Es convector sin tubo *(m)*
Fr convecteur sans carneau *(m)*
It convettore senza condotto del fumo *(m)*
Pt convector sem fumeiro *(m)*

Konverter *(m) n* De
En converter reactor
Es reactor convertidor *(m)*
Fr réacteur convertisseur *(m)*
It reattore convertitore *(m)*
Pt reactor de conversão *(m)*

Kracken *(n) n* De
En cracking
Es cráсking *(m)*
Fr craquage *(m)*
It piroscissione *(f)*
Pt cracking *(m)*

Kraft *(f) n* De
En force
Es fuerza *(f)*
Fr force *(f)*
It forza *(f)*
Pt força *(f)*

Kraftlinie *(f) n* De
En line of force
Es línea de fuerza *(f)*
Fr ligne de force *(f)*
It linea di forza *(f)*
Pt linha de força *(f)*

Kraftwerk *(n) n* De
En power station
Es central eléctrica *(f)*
Fr centrale électrique *(f)*
It centrale elettrica *(f)*
Pt central energética *(f)*

Kreisbrenner *(m) n* De
En circular burner
Es quemador circular *(m)*
Fr brûleur circulaire *(m)*
It bruciatore circolare *(m)*
Pt queimador circular *(m)*

Kresol *(n) n* De
En cresol
Es cresol *(m)*
Fr crésol *(m)*
It cresolo *(m)*
Pt cresol *(m)*

kritisch *adj* De
En critical
Es crítico
Fr critique
It critico
Pt crítico

Kryostat *(m) n* De
En cryostat
Es crióstato *(m)*
Fr cryostat *(m)*
It criostato *(m)*
Pt criostato *(m)*

Kühlapparat *(m) n* De
En refrigerator
Es refrigerador *(m)*
Fr réfrigérateur *(m)*
It refrigeratore *(m)*
Pt refrigerador *(m)*

Kühlmittel *(n) n* De
En coolant; refrigerant
Es refrigerante *(m)*
Fr réfrigérant *(m)*
It refrigerante *(m)*
Pt refrigerante *(m)*

Kühlmittelkreislauf *(m) n* De
En cooling circuit
Es circuito de refrigeración *(m)*
Fr circuit de refroidissement *(m)*
It circuito di raffreddamento *(m)*
Pt circuito refrigerante *(m)*

Kühlteich *(m) n* De
En cooling pond
Es piscina de refrigeración *(f)*
Fr piscine de refroidissement *(f)*
It bacino di raffreddamento *(m)*
Pt tanque refrigerante *(m)*

Kühlwasser *(n) n* De
En cooling water
Es agua de refrigeración *(f)*
Fr eau de refroidissement *(f)*
It acqua di raffreddamento *(f)*
Pt água de refrigeração *(f)*

Kupferverlust *(m) n* De
En copper loss
Es pérdida en el cobre *(f)*
Fr perte dans le cuivre *(f)*
It perdita nel rame *(f)*
Pt perda no cobre *(f)*

Kuppel *(f) n* De
En cupola
Es cúpula *(f)*
Fr cubilot *(m)*
It cupola *(f)*
Pt cúpula *(f)*

Küstenlinie *(f) n* De
En shoreline
Es litoral *(m)*
Fr ligne de rivage *(f)*
It litorale *(m)*
Pt linha de costa *(f)*

K-value En
De K-Wert *(m)*
Es valor K *(m)*
Fr valeur K *(f)*
It valore K *(m)*
Pt valor de K *(m)*

K-Wert *(m) n* De
En K-value
Es valor K *(m)*
Fr valeur K *(f)*
It valore K *(m)*
Pt valor de K *(m)*

L

ladrillo refractario *(m)* Es
De feuerfester Stein *(m)*
En firebrick
Fr brique réfractaire *(f)*
It mattone refrattario *(m)*
Pt tijolo refractário *(m)*

Ladung (elektrische) *(f) n* De
En charge (electrical)
Es carga (eléctrica) *(f)*
Fr charge (électrique) *(f)*
It carica (elettrica) *(f)*
Pt carga (eléctrica) *(f)*

Lagerstättengestein *(n) n* De
En source rock
Es roca madre *(f)*
Fr roche mère *(f)*
It roccia madre *(f)*
Pt rocha mãe *(f)*

lagging *n* En
De Ummantellung *(f)*
Es revestimiento *(m)*
Fr revêtement calorifuge *(m)*
It rivestimento isolante *(m)*
Pt revestimento *(m)*

laghetto solare salino *(m)* It
De Solar-Salzteich *(m)*
En saline solar pond
Es piscina solar salina *(f)*
Fr bassin solaire salant *(m)*
Pt lago solar de sal *(m)*

lago solar de sal *(m)* Pt
De Solar-Salzteich *(m)*
En saline solar pond
Es piscina solar salina *(f)*
Fr bassin solaire salant *(m)*
It laghetto solare salino *(m)*

laitier *(m) n* Fr
De Schlacke *(f)*
En slag
Es escoria *(f)*
It scoria *(f)*
Pt escória *(f)*

Langmodul *(n) n* De
En long module
Es módulo largo *(m)*
Fr module long *(m)*
It modulo lungo *(m)*
Pt módulo comprido *(m)*

langsames Neutron *(n)* De
En slow neutron
Es neutrón lento *(m)*
Fr neutron lent *(m)*
It neutrone lento *(m)*
Pt neutrão lento *(m)*

laser *n* En; Fr, It, Pt *(m)*
De Laser *(m)*
Es láser *(m)*

Laser *(m) n* De
En laser
Es láser *(m)*
Fr laser *(m)*
It laser *(m)*
Pt laser *(m)*

láser *(m) n* Es
De Laser *(m)*
En laser
Fr laser *(m)*
It laser *(m)*
Pt laser *(m)*

Last (mechanische) *(f) n* De
En load (mechanical)
Es carga (mecánica) *(f)*
Fr charge (méchanique) *(f)*
It carico (meccanico) *(m)*
Pt carga (mecânica) *(f)*

Lastabgabe *(f) n* De
En load-shedding
Es restricción de la carga *(f)*
Fr délestage *(m)*
It spargimento di carico *(m)*
Pt restrição de carga *(f)*

Lastfaktor *(m) n* De
En load factor
Es factor de carga *(m)*
Fr facteur de charge *(m)*
It fattore di carico *(m)*
Pt factor de carga *(m)*

lastre limpio *(m)* Es
De sauberer Ballast *(m)*
En clean ballast
Fr ballast propre *(m)*
It zavorra pulita *(f)*
Pt balastro puro *(m)*

latent heat En
De gebundene Wärme *(f)*
Es calor latente *(m)*
Fr chaleur latente *(f)*
It calore latente *(m)*
Pt calor latente *(m)*

latitude effect En
De Breiteneffekt *(m)*
Es efecto de latitud *(m)*
Fr effet de la latitude *(m)*
It effetto latitudine *(m)*
Pt efeito da latitude *(m)*

lattice *n* En
De Spaltstoffgitter *(n)*
Es celosía *(f)*
Fr treillis *(m)*
It traliccio *(m)*
Pt treliça *(f)*

Läufer *(m) n* De
En rotor
Es rotor *(m)*
Fr rotor *(m)*
It rotore *(m)*
Pt rotor *(m)*

Läuterung *(f) n* De
En purging
Es purga *(f)*
Fr purge *(f)*
It spurgo *(m)*
Pt purga *(f)*

lavado cáustico *(m)* Es
De Neutralisieren *(n)*
En caustic washing
Fr lavage à la soude caustique *(m)*
It lavaggio caustico *(m)*
Pt lavagem caústica *(f)*

lavado húmedo *(m)* Es
De Naβreinigen *(n)*
En wet scrubbing
Fr épuration humide *(f)*
It lavaggio *(m)*
Pt limpeza com lavagem *(f)*

lavage à la soude caustique *(m)* Fr
De Neutralisieren *(n)*
En caustic washing
Es lavado cáustico *(m)*
It lavaggio caustico *(m)*
Pt lavagem caústica *(f)*

lavagem caústica *(f)* Pt
De Neutralisieren *(n)*
En caustic washing
Es lavado cáustico *(m)*
Fr lavage à la soude caustique *(m)*
It lavaggio caustico *(m)*

lavaggio *(m) n* It
De Naβreinigen *(n)*
En wet scrubbing
Es lavado húmedo *(m)*
Fr épuration humide *(f)*
Pt limpeza com lavagem *(f)*

lavaggio caustico *(m)* It
De Neutralisieren *(n)*
En caustic washing
Es lavado cáustico *(m)*
Fr lavage à la soude caustique *(m)*
Pt lavagem caústica *(f)*

lavoro *(m) n* It
De Arbeit *(f)*
En work
Es trabajo *(m)*
Fr travail *(m)*
Pt trabalho *(m)*

lavoro esterno *(m)* It
De Auβenarbeit *(f)*
En external work
Es trabajo externo *(m)*
Fr travail extérieur *(m)*
Pt trabalho externo *(m)*

lavoro interno *(m)* It
De Innenarbeit *(f)*
En internal work
Es trabajo interno *(m)*
Fr travail interne *(m)*
Pt trabalho interno *(m)*

lavoro virtuale *(m)* It
De virtuelle Arbeit *(f)*
En virtual work
Es trabajo virtual *(m)*
Fr travail virtuel *(m)*
Pt trabalho virtual *(m)*

lead-acid battery En
De Blei-Säure- Batterie *(f)*
Es batería de ácido-plomo *(f)*
Fr batterie au plomb *(f)*
It batteria acida al piombo *(f)*
Pt bateria de chumbo-ácido *(f)*

lead content En
De Bleigehalt *(m)*
Es contenido de plomo *(m)*
Fr teneur en plomb *(f)*
It contenuto di piombo *(m)*
Pt teor de chumbo *(m)*

leakage *n* En
De Leck *(n)*
Es fuga *(f)*
Fr fuite *(f)*
It perdita *(f)*
Pt fuga *(f)*

Lebendmasse *(f) n* De
En biomass
Es biomasa *(f)*
Fr biomasse *(f)*
It biomassa *(f)*
Pt biomassa *(f)*

Leck *(n) n* De
En leakage
Es fuga *(f)*
Fr fuite *(f)*
It perdita *(f)*
Pt fuga *(f)*

lega *(f) n* It
De Legierung *(f)*
En alloy
Es aleación *(f)*
Fr alliage *(m)*
Pt liga *(f)*

lega di zirconio *(f)* It
De Zirkonlegierung *(f)*
En zircalloy
Es aleación de zirconio *(f)*
Fr alliage de zirconium *(m)*
Pt liga de zircónio *(f)*

Legierung *(f) n* De
En alloy
Es aleación *(f)*
Fr alliage *(m)*
It lega *(f)*
Pt liga *(f)*

Lehmbehandlung *(f) n* De
En clay treatment
Es tratamiento con arcilla *(m)*
Fr traitement à l'argile *(m)*
It trattamento all'argilla *(m)*
Pt tratamento de argila *(m)*

leichtes Öl *(n)* De
En light oil
Es petróleo ligero *(m)*
Fr huile légère *(f)*
It olio leggero *(m)*
Pt óleo leve *(m)*

leichtes Rohöl *(n)* De
En light crude oil
Es crudo ligero *(m)*
Fr brut léger *(m)*
It petrolio greggio leggero *(m)*
Pt petróleo crú leve *(m)*

Leichtwasserreaktor *(m) n* De
En light-water reactor
Es reactor de agua ligera *(m)*
Fr réacteur à eau légère *(m)*
It reattore ad acqua naturale *(m)*
Pt reactor de água leve *(m)*

Leistung *(f) n* De
En power
Es potencia *(f)*
Fr puissance *(f)*
It potenza *(f)*
Pt potência *(f)*

Leistungsdichte *(f) n* De
En power density
Es densidad de potencia *(f)*
Fr puissance volumique *(f)*
It densità di potenza *(f)*
Pt densidade de potência *(f)*

Leistungsfähigkeit *(f) n* De
En efficiency
Es eficiencia *(f)*
Fr rendement *(m)*
It efficienza *(f)*
Pt eficiência *(f)*

Leistungsfaktor *(m) n* De
En power factor
Es factor de potencia *(m)*
Fr facteur de puissance *(m)*
It fattore di potenza *(m)*
Pt factor de potência *(m)*

Leitung *(f) n* De
En conduction
Es conducción *(f)*
Fr conduction *(f)*
It conduzione *(f)*
Pt condução *(f)*

Leitungsschalter *(m) n* De
En circuit breaker
Es disyuntor *(m)*
Fr disjoncteur *(m)*
It interruttore automatico *(m)*
Pt corta-circuito *(m)*

Leitungsschiene *(f) n* De
En busbar
Es barra colectora *(f)*
Fr barre omnibus *(f)*
It sbarra *(f)*
Pt barra colectora *(f)*

lenhite *(f) n* Pt
De Braunkohle *(f)*; Lignit *(n)*
En brown coal; lignite
Es lignito *(m)*
Fr lignite *(m)*
It lignite *(f)*

lens *n* En
De Objektiv *(n)*
Es lente *(f)*
Fr lentille *(f)*
It lente *(f)*
Pt lente *(f)*

lente *(f) n* Es, It, Pt
De Objektiv *(n)*
En lens
Fr lentille *(f)*

lenticular coal deposit En
De linsenförmiges Kohlenlager *(n)*
Es depósito de carbón lenticular *(m)*
Fr dépôt de charbon lenticulaire *(m)*
It deposito di carbone lenticolare *(m)*
Pt depósito de carvão lenticular *(m)*

lentille *(f) n* Fr
De Objektiv *(n)*
En lens
Es lente *(f)*
It lente *(f)*
Pt lente *(f)*

lethal dose En
De tödliche Dosis *(f)*
Es dosis letal *(f)*
Fr dose mortelle *(f)*
It dose letale *(f)*
Pt dose letal *(f)*

Leuchtdichte *(f) n* De
En luminance
Es luminancia *(f)*
Fr luminance *(f)*
It luminanza *(f)*
Pt luminância *(f)*

Leuchtenergie *(f) n* De
En luminous energy
Es energía luminosa *(f)*
Fr énergie lumineuse *(f)*
It energia luminosa *(f)*
Pt energia luminosa *(f)*

levantamento magnético *(m)* Pt
De Magnetaufnahme *(f)*
En magnetic survey
Es levantamiento magnético *(m)*
Fr prospection magnétique *(f)*
It perizia magnetica *(f)*

levantamiento magnético *(m)* Es
De Magnetaufnahme *(f)*
En magnetic survey
Fr prospection magnétique *(f)*
It perizia magnetica *(f)*
Pt levantamento magnético *(m)*

level control En
De Füllstandsregler *(m)*
Es control de nivel *(m)*
Fr commande de niveau *(f)*
It regolatore di livello *(m)*
Pt controle de nível *(m)*

Lichtbogen-Heizelement *(m) n* De
En arc heater
Es calentador de arco *(m)*
Fr appareil de chauffage à arc *(m)*
It riscaldatore ad arco *(m)*
Pt aquecedor a arco voltáico *(m)*

Lichtbogenofen *(m) n* De
En arc furnace
Es horno de arco *(m)*
Fr four à arc *(m)*
It forno ad arco *(m)*
Pt forno a arco voltáico *(m)*

licuación *(f) n* Es
De Verflüssigung *(f)*
En liquefaction
Fr liquéfaction *(f)*
It liquefazione *(f)*
Pt liquefacção *(f)*

liga *(f) n* Pt
De Legierung *(f)*
En alloy
Es aleación *(f)*
Fr alliage *(m)*
It lega *(f)*

ligação à terra múltipla protectora *(f)* Pt
De Mehrfach-Schutzerdung *(f)*
En protective multiple earthing
Es toma de tierra múltiple protectora *(f)*
Fr mise à la terre multiple de protection *(f)*
It messa a terra multipla protettiva *(f)*

liga de zircónio *(f)* Pt
De Zirkonlegierung *(f)*
En zircalloy
Es aleación de zirconio *(f)*
Fr alliage de zirconium *(m)*
It lega di zirconio *(f)*

light crude oil En
De leichtes Rohöl *(n)*
Es crudo ligero *(m)*
Fr brut léger *(m)*
It petrolio greggio leggero *(m)*
Pt petróleo crú leve *(m)*

light oil En
De leichtes Öl *(n)*
Es petróleo ligero *(m)*
Fr huile légère *(f)*
It olio leggero *(m)*
Pt óleo leve *(m)*

light-water reactor En
De Leichtwasserreaktor *(m)*
Es reactor de agua ligera *(m)*
Fr réacteur à eau légère *(m)*
It reattore ad acqua naturale *(m)*
Pt reactor de água leve *(m)*

ligne de force *(f)* Fr
De Kraftlinie *(f)*
En line of force
Es línea de fuerza *(f)*
It linea di forza *(f)*
Pt linha de força *(f)*

ligne de rivage *(f)* Fr
De Küstenlinie *(f)*
En shoreline
Es litoral *(m)*
It litorale *(m)*
Pt linha de costa *(f)*

ligne de transmission *(f)* Fr
De Transmissionsleitung *(f)*
En transmission line
Es línea de transmisión *(f)*
It linea di trasmissione *(f)*
Pt linha de transmissão *(f)*

Lignit *(n)* *n* De
En brown coal; lignite
Es lignito *(m)*
Fr lignite *(m)*
It lignite *(f)*
Pt carvão castanho *(m)*; lenhite *(f)*

lignite *n* En; Fr *(m)*; It *(f)*
De Braunkohle *(f)*; Lignit *(n)*
Es lignito *(m)*
Fr lignite *(m)*
It lignite *(f)*
Pt carvão castanho *(m)*; lenhite *(f)*

lignito *(m)* *n* Es
De Braunkohle *(f)*; Lignit *(n)*
En brown coal; lignite
Fr lignite *(m)*
It lignite *(m)*
Pt carvão castanho *(m)*; lenhite *(f)*

limpeza com lavagem *(f)* Pt
De Naßreinigen *(n)*
En wet scrubbing
Es lavado húmedo *(m)*
Fr épuration humide *(f)*
It lavaggio *(m)*

línea de fuerza *(f)* Es
De Kraftlinie *(f)*
En line of force
Fr ligne de force *(f)*
It linea di forza *(f)*
Pt linha de força *(f)*

línea de transmisión *(f)* Es
De Transmissionsleitung *(f)*
En transmission line
Fr ligne de transmission *(f)*
It linea di trasmissione *(f)*
Pt linha de transmissão *(f)*

linea di forza *(f)* It
De Kraftlinie *(f)*
En line of force
Es línea de fuerza *(f)*
Fr ligne de force *(f)*
Pt linha de força *(f)*

linea di trasmissione *(f)* It
De Transmissionsleitung *(f)*
En transmission line
Es línea de transmisión *(f)*
Fr ligne de transmission *(f)*
Pt linha de transmissão *(f)*

line of force En
De Kraftlinie *(f)*
Es línea de fuerza *(f)*
Fr ligne de force *(f)*
It linea di forza *(f)*
Pt linha de força *(f)*

linha de costa *(f)* Pt
De Küstenlinie *(f)*
En shoreline
Es litoral *(m)*
Fr ligne de rivage *(f)*
It litorale *(m)*

linha de força *(f)* Pt
De Kraftlinie *(f)*
En line of force
Es línea de fuerza *(f)*
Fr ligne de force *(f)*
It linea di forza *(f)*

linha de transmissão *(f)* Pt
De Transmissionsleitung *(f)*
En transmission line
Es línea de transmisión *(f)*
Fr ligne de transmission *(f)*
It linea di trasmissione *(f)*

linsenförmiges Kohlenlager *(n)* De
En lenticular coal deposit
Es depósito de carbón lenticular *(m)*
Fr dépôt de charbon lenticulaire *(m)*
It deposito di carbone lenticolare *(m)*
Pt depósito de carvão lenticular *(m)*

liofilización *(f)* *n* Es
De Gefriertrocknung *(f)*
En freeze drying
Fr lyophilisation *(f)*
It liofilizzazione *(f)*
Pt secagem por congelação *(f)*

liofilizzazione *(f)* *n* It
De Gefriertrocknung *(f)*
En freeze drying
Es liofilización *(f)*
Fr lyophilisation *(f)*
Pt secagem por congelação *(f)*

liquefacção *(f)* *n* Pt
De Verflüssigung *(f)*
En liquefaction
Es licuación *(f)*
Fr liquéfaction *(f)*
It liquefazione *(f)*

liquefaction *n* En
De Verflüssigung *(f)*
Es licuación *(f)*
Fr liquéfaction *(f)*
It liquefazione *(f)*
Pt liquefacção *(f)*

liquéfaction *(f)* *n* Fr
De Verflüssigung *(f)*
En liquefaction
Es licuación *(f)*
It liquefazione *(f)*
Pt liquefacção *(f)*

liquefazione *(f)* *n* It
De Verflüssigung *(f)*
En liquefaction
Es licuación *(f)*
Fr liquéfaction *(f)*
Pt liquefacção *(f)*

liquefied natural gas (LNG) En
De verflüssigtes Erdgas *(n)*

Es gas natural licuado (GNL) *(m)*
Fr gaz naturel liquéfié (GNL) *(m)*
It gas naturale liquido *(m)*
Pt gás natural liquefeito *(m)*

liquefied petroleum gas (LPG) En
De verflüssigtes Petroleumgas *(n)*
Es gas licuado del petróleo (GLP) *(m)*
Fr gaz de pétrole liquéfié (GPL) *(m)*
It gas liquido di petrolio *(m)*
Pt gás de petróleo liquefeito *(m)*

liquid hydrocarbons En
De flüssige Kohlenwasserstoffe *(pl)*
Es hidrocarburos líquidos *(m pl)*
Fr hydrocarbures liquides *(m pl)*
It idrocarburi liquidi *(m pl)*
Pt hidrocarbonetos líquidos *(m pl)*

liquid-metal fast breeder reactor (LMFBR) En
De Flüssigmetall-Schnellbrüter *(m)*
Es reactor reproductor rápido de metal líquido *(m)*
Fr réacteur surrégénérateur rapide à métal liquide *(m)*
It reattore veloce autofissilizzante a metallo liquido *(m)*
Pt reactor reproductor rápido de metal líquido *(m)*

líquido *adj* Pt
De netto
En net
Es neto
Fr net
It netto

liquid propellants *(pl)* En
De flüssige Treibstoffe *(pl)*
Es propulsantes líquidos *(m pl)*
Fr propergols liquides *(m pl)*
It propellenti liquidi *(m pl)*
Pt combustíveis propulsores líquidos *(m pl)*

lithologie *(f) n* Fr
De Gesteinskunde *(f)*
En lithology
Es litología *(f)*
It litologia *(f)*
Pt litologia *(f)*

lithology *n* En
De Gesteinskunde *(f)*
Es litología *(f)*
Fr lithologie *(f)*
It litologia *(f)*
Pt litologia *(f)*

Lithosphäre *(f) n* De
En lithosphere
Es litosfera *(f)*
Fr lithosphère *(f)*
It litosfera *(f)*
Pt litosfera *(f)*

lithosphere *n* En
De Lithosphäre *(f)*
Es litosfera *(f)*
Fr lithosphère *(f)*
It litosfera *(f)*
Pt litosfera *(f)*

lithosphère *(f) n* Fr
De Lithosphäre *(f)*
En lithosphere
Es litosfera *(f)*
It litosfera *(f)*
Pt litosfera *(f)*

litologia *(f) n* It, Pt
De Gesteinskunde *(f)*
En lithology
Es litología *(f)*
Fr lithologie *(f)*

litología *(f) n* Es
De Gesteinskunde *(f)*
En lithology
Fr lithologie *(f)*
It litologia *(f)*
Pt litologia *(f)*

litoral *(m) n* Es
De Küstenlinie *(f)*
En shoreline
Fr ligne de rivage *(f)*
It litorale *(m)*
Pt linha de costa *(f)*

litorale *(m) n* It
De Küstenlinie *(f)*
En shoreline
Es litoral *(m)*
Fr ligne de rivage *(f)*
Pt linha de costa *(f)*

litosfera *(f) n* Es, It, Pt
De Lithosphäre *(f)*
En lithosphere
Fr lithosphère *(f)*

llama *(f) n* Es
De Flamme *(f)*
En flame
Fr flamme *(f)*
It fiamma *(f)*
Pt chama *(f)*

llama aireada *(f)* Es
De belüftete Flamme *(f)*
En aerated flame
Fr flamme aérée *(f)*
It fiamma aerata *(f)*
Pt chama arejada *(f)*

llama no aireada *(f)* Es
De unbelüftete Flamme *(f)*
En nonaerated flame
Fr flamme non aérée *(f)*
It fiamma non aerata *(f)*
Pt chama não arejada *(f)*

llama piloto *(f)* Es
De Zündflamme *(f)*
En pilot flame
Fr veilleuse *(f)*
It semprevivo *(m)*
Pt chama piloto *(f)*

load (mechanical) *n* En
De Last (mechanische) *(f)*
Es carga (mecánica) *(f)*
Fr charge (méchanique) *(f)*
It carico (meccanico) *(m)*
Pt carga (mecânica) *(f)*

loaded concrete En
De beschwerter Beton *(m)*
Es hormigón cargado *(m)*
Fr béton chargé *(m)*
It cemento pesante *(m)*
Pt betão carregado *(m)*

load factor En
De Lastfaktor *(m)*
Es factor de carga *(m)*
Fr facteur de charge *(m)*
It fattore di carico *(m)*
Pt factor de carga *(m)*

load-shedding *n* En
De Lastabgabe *(f)*
Es restricción de la carga *(f)*
Fr délestage *(m)*
It spargimento di carico *(m)*
Pt restrição de carga *(f)*

lodo de perforación *(m)* Es
De Bohrschlamm *(m)*
En drilling mud
Fr boue de forage *(f)*
It fango di trivellazione *(m)*
Pt lodo de perfuração *(m)*

lodo de perfuração *(m) n* Pt
De Bohrschlamm *(m)*
En drilling mud
Es lodo de perforación *(m)*
Fr boue de forage *(f)*
It fango di trivellazione *(m)*

loft insulation En
De Dachbodenisolierung *(f)*
Es aislamiento de galerías *(m)*
Fr isolement des combles *(m)*
It isolamento della soffitta *(m)*
Pt isolamento do sotão *(m)*

longitud de onda *(f)* Es
De Wellenlänge *(f)*
En wavelength
Fr longueur d'onde *(f)*
It lunghezza d'onda *(f)*

Pt comprimento de onda *(m)*

long module En
De Langmodul *(n)*
Es módulo largo *(m)*
Fr module long *(m)*
It modulo lungo *(m)*
Pt módulo comprido *(m)*

longueur d'onde *(f)* Fr
De Wellenlänge *(f)*
En wavelength
Es longitud de onda *(f)*
It lunghezza d'onda *(f)*
Pt comprimento de onda *(m)*

lordo *adj* It
De brutto
En gross
Es bruto
Fr brut
Pt bruto

loss of coolant accident (LOCA) En
De Unfall durch Kühlmittelverlust *(m)*
Es accidente por pérdida de refrigerante *(m)*
Fr accident par perte de fluide réfrigérant *(m)*
It incidente di perdita del refrigerante *(m)*
Pt acidente por perda de refrigerante *(m)*

Lösungsmittel *(n) n* De
En solvent
Es disolvente *(m)*
Fr solvant *(m)*
It solvente *(m)*
Pt dissolvente *(m)*

low-boiling fractions *(pl)* En
De Tiefsiedepunkt-fraktionen *(pl)*
Es fracciónes de bajo punto de ebullición *(f pl)*
Fr fractions à bas point d'ébullition *(f pl)*
It frazioni a basso punto di ebollizione *(f pl)*
Pt fracções de baixo ponto de ebulição *(f pl)*

low-temperature tar En
De Tieftemperaturteer *(m)*
Es alquitrán de baja temperatura *(m)*
Fr goudron à basse température *(m)*
It catrame a bassa temperatura *(m)*
Pt alcatrão de baixa temperatura *(m)*

Luft *(f) n* De
En air
Es aire *(m)*
Fr air *(m)*
It aria *(f)*
Pt ar *(m)*

lubricating oil En
De Schmieröl *(n)*
Es aceite lubricante *(m)*
Fr huile de graissage *(f)*
It olio lubrificante *(m)*
Pt óleo de lubrificação *(m)*

Luftabsorption *(f) n* De
En atmospheric absorption
Es absorción atmosférica *(f)*
Fr absorption atmosphérique *(f)*
It assorbimento atmosferico *(m)*
Pt absorção atmosférica *(f)*

luftdichte Doppelverglasungs-elemente *(pl)* De
En sealed-unit double glazing
Es doble encristalado de unidad sellada *(m)*
Fr double vitrage à éléments scellés *(m)*
It doppia vetrata sigillata *(f)*
Pt vitrificação dupla de unidade vedada *(f)*

Luftdruck *(m) n* De
En atmospheric pressure
Es presión atmosférica *(f)*
Fr pression atmosphérique *(f)*
It pressione atmosferica *(f)*
Pt pressão atmosférica *(f)*

Lufteinlaß *(m) n* De
En air inlet
Es entrada de aire *(f)*
Fr admission d'air *(f)*
It presa d'aria *(f)*
Pt admissão de ar *(f)*

Lufterhitzer *(m) n* De
En air heater
Es calentador de aire *(m)*
Fr réchauffeur d'air *(m)*
It riscaldatore ad aria *(m)*
Pt aquecedor de ar *(m)*

luftgekühlt *adj* De
En air-cooled
Es refrigerado por aire
Fr refroidi par air
It raffreddato ad aria
Pt arrefecido a ar

Luftgenerator *(m) n* De
En aerogenerator
Es aerogenerador *(m)*
Fr aérogénérateur *(m)*
It aerogeneratore *(m)*
Pt aerogerador *(m)*

Luftkompressor *(m) n* De
En air compressor
Es compresor de aire *(m)*
Fr compresseur d'air *(m)*
It compressore d'aria *(m)*
Pt compressor de ar *(m)*

Luftleck *(n) n* De
En air leakage
Es fuga de aire *(f)*
Fr fuite d'air *(f)*
It fuga d'aria *(f)*
Pt fuga de ar *(f)*

Luftreibungsverlust *(m) n* De
En windage loss
Es pérdida por resistencia aerodinámica *(f)*
Fr perte due au jeu *(f)*
It perdita per resistenza aerodinamica *(f)*
Pt perda por fricção do vento *(f)*

Lufttrockner *(m) n* De
En air dryer
Es secador de aire *(m)*
Fr sécheur d'air *(m)*
It essiccatore ad aria *(m)*
Pt secador por ar *(m)*

Luftüberschuß *(m) n* De
En excess air
Es aire sobrante *(m)*
Fr excès d'air *(m)*
It eccesso d'aria *(m)*
Pt ar em excesso *(m)*

Lüftung *(f) n* De
En aeration
Es aireación *(f)*
Fr aération *(f)*
It aerazione *(f)*
Pt arejamento *(m)*

Luftverschmutzung *(f) n* De
En air pollution
Es contaminación del aire *(f)*
Fr pollution de l'air *(f)*
It inquinamento atmosferico *(m)*
Pt poluição do ar *(f)*

Luftzerstäuber *(m) n* De
En air atomizer
Es atomizador de aire *(m)*
Fr atomiseur d'air *(m)*
It nebulizzatore d'aria *(m)*
Pt atomizador de ar *(m)*

Luftzugablenkung *(f) n* De
Am draft diverter
En draught diverter
Es desviador de corrientes *(m)*
Fr déflecteur de courant d'air *(m)*
It deviatore d'aria *(m)*
Pt deflector de corrente de ar *(m)*

Luftzugschutz *(m) n* De
Am draft excluder
En draught excluder

Es eliminador de corrientes *(m)*
Fr brise-bise *(m)*
It esclusore d'aria *(m)*
Pt eliminador de corrente de ar *(m)*

luminance *n* En; Fr *(f)*
De Leuchtdichte *(f)*
Es luminancia *(f)*
It luminanza *(f)*
Pt luminância *(f)*

luminancia *(f) n* Es
De Leuchtdichte *(f)*
En luminance
Fr luminance *(f)*
It luminanza *(f)*
Pt luminância *(f)*

luminância *(f) n* Pt
De Leuchtdichte *(f)*
En luminance
Es luminancia *(f)*
Fr luminance *(f)*
It luminanza *(f)*

luminanza *(f) n* It
De Leuchtdichte *(f)*
En luminance
Es luminancia *(f)*
Fr luminance *(f)*
Pt luminância *(f)*

luminescent *adj* En, Fr
De lumineszierend
Es luminiscente
It luminescente
Pt luminescente

luminescente *adj* It, Pt
De lumineszierend
En luminescent
Es luminiscente
Fr luminescent

lumineszierend *adj* De
En luminescent
Es luminiscente
Fr luminescent
It luminescente
Pt luminescente

luminiscente *adj* Es
De lumineszierend
En luminescent
Fr luminescent
It luminescente
Pt luminescente

luminosità *(f) n* It
De Helligkeit *(f)*
En brightness
Es brillo *(m)*
Fr brillance *(f)*
Pt brilho *(m)*

luminosità effettiva *(f)* It
De effektive Helligkeit *(f)*
En effective brightness
Es brillo efectivo *(m)*
Fr brillance efficace *(f)*
Pt brilho efectivo *(m)*

luminous energy En
De Leuchtenergie *(f)*
Es energía luminosa *(f)*
Fr énergie lumineuse *(f)*
It energia luminosa *(f)*
Pt energia luminosa *(f)*

lunghezza d'onda *(f)* It
De Wellenlänge *(f)*
En wavelength
Es longitud de onda *(f)*
Fr longueur d'onde *(f)*
Pt comprimento de onda *(m)*

lyophilisation *(f) n* Fr
De Gefriertrocknung *(f)*
En freeze drying
Es liofilización *(f)*
It liofilizzazione *(f)*
Pt secagem por congelação *(f)*

M

maçarico de plasma *(m)* Pt
De Plasmabrenner *(m)*
En plasma torch
Es soplete para plasma *(m)*
Fr chalumeau à plasma *(m)*
It torcia a plasma *(f)*

macchina a vapore *(f)* It
De Dampfmaschine *(f)*
En steam engine
Es motor de vapor *(m)*
Fr machine à vapeur *(f)*
Pt motor a vapor *(m)*

macchina eccitata in derivazione *(f)* It
De Nebenschlußmaschine *(f)*
En shunt-wound machine
Es máquina excitada en derivación *(f)*
Fr machine shunt *(f)*
Pt máquina enrolada em shunt *(f)*

macchina eccitata in serie *(f)* It
De seriengewickelte Maschine *(f)*
En series-wound machine
Es maquina excitada en serie *(f)*
Fr machine série *(f)*
Pt máquina enrolada em série *(f)*

machine à vapeur *(f)* Fr
De Dampfmaschine *(f)*
En steam engine
Es motor de vapor *(m)*
It macchina a vapore *(f)*
Pt motor a vapor *(m)*

machine série *(f)* Fr
De seriengewickelte Maschine *(f)*
En series-wound machine
Es maquina excitada en serie *(f)*
It macchina eccitata in serie *(f)*
Pt máquina enrolada em série *(f)*

machine shunt *(f)* Fr
De Nebenschlußmaschine *(f)*
En shunt-wound machine
Es máquina excitada en derivación *(f)*
It macchina eccitata in derivazione *(f)*
Pt máquina enrolada em shunt *(f)*

Magmadampf *(f) n* De
En magmatic steam
Es vapor magmático *(m)*
Fr vapeur magmatique *(f)*
It vapore magmatico *(m)*
Pt vapor magmático *(m)*

magmatic steam En
De Magmadampf *(f)*
Es vapor magmático *(m)*
Fr vapeur magmatique *(f)*
It vapore magmatico *(m)*
Pt vapor magmático *(m)*

magmete *(m) n* It
De Magnet *(m)*
En magnet
Es imán *(m)*
Fr aimant *(m)*
Pt ímã *(m)*

magnet *n* En
De Magnet *(m)*
Es imán *(m)*
Fr aimant *(m)*
It magmete *(m)*
Pt ímã *(m)*

Magnet *(m) n* De
En magnet
Es imán *(m)*
Fr aimant *(m)*
It magmete *(m)*
Pt ímã *(m)*

Magnetaufnahme *(f) n* De
En magnetic survey
Es levantamiento magnético *(m)*
Fr prospection magnétique *(f)*
It perizia magnetica *(f)*
Pt levantamento magnético *(m)*

magnete *(m) n* It
De Magnetzünder *(m)*
En magneto
Es magneto *(f)*
Fr magnéto *(f)*
Pt magneto *(m)*

magnete superconduttore *(m)* It
De supraleitfähiger Magnet *(m)*
En superconducting magnet

Es imán superconductor (m)
Fr aimant superconducteur (m)
Pt ímã supercondutor (m)

Magnetfeld *(n) n* De
En magnetic field
Es campo magnético *(m)*
Fr champ magnétique *(m)*
It campo magnetico *(m)*
Pt campo magnético *(m)*

magnetic bottle En
De magnetische Flasche *(f)*
Es botella magnética *(f)*
Fr bouteille magnétique *(f)*
It bottiglia magnetica *(f)*
Pt garrafa magnética *(f)*

magnetic constant En
De Magnetkonstante *(f)*
Es constante magnética *(f)*
Fr constante magnétique *(f)*
It costante magnetica *(f)*
Pt constante magnética *(f)*

magnetic field En
De Magnetfeld *(n)*
Es campo magnético *(m)*
Fr champ magnétique *(m)*
It campo magnetico *(m)*
Pt campo magnético *(m)*

magnetic force En
De Magnetkraft *(n)*
Es fuerza magnética *(f)*
Fr force magnétique *(f)*
It forza magnetica *(f)*
Pt força magnética *(f)*

magnetic storage En
De Magnetspeicher *(m)*
Es almacenamiento magnético *(m)*
Fr accumulation magnétique *(f)*
It immagazzinaggio magnetico *(m)*
Pt armazenagem magnética *(f)*

magnetic survey En
De Magnetaufnahme *(f)*
Es levantamiento magnético *(m)*
Fr prospection magnétique *(f)*
It perizia magnetica *(f)*
Pt levantamento magnético *(m)*

magnétisation *(f) n* Fr
De Magnetisierung *(f)*
En magnetization
Es magnetización *(f)*
It magnetizzazione *(f)*
Pt magnetização *(m)*

magnetische Flasche *(f)* De
En magnetic bottle
Es botella magnética *(f)*
Fr bouteille magnétique *(f)*
It bottiglia magnetica *(f)*
Pt garrafa magnética *(f)*

Magnetisierung *(f) n* De
En magnetization
Es magnetización *(f)*
Fr magnétisation *(f)*
It magnetizzazione *(f)*
Pt magnetização *(m)*

magnetism *n* En
De Magnetismus *(m)*
Es magnetismo *(m)*
Fr magnétisme *(m)*
It magnetismo *(m)*
Pt magnetismo *(m)*

magnétisme *(m) n* Fr
De Magnetismus *(m)*
En magnetism
Es magnetismo *(m)*
It magnetismo *(m)*
Pt magnetismo *(m)*

magnetismo *(m) n* Es, It, Pt
De Magnetismus *(m)*
En magnetism
Fr magnétisme *(m)*

Magnetismus *(m) n* De
En magnetism
Es magnetismo *(m)*
Fr magnétisme *(m)*
It magnetismo *(m)*
Pt magnetismo *(m)*

magnetização *(m) n* Pt
De Magnetisierung *(f)*
En magnetization
Es magnetización *(f)*
Fr magnétisation *(f)*
It magnetizzazione *(f)*

magnetización *(f) n* Es
De Magnetisierung *(f)*
En magnetization
Fr magnétisation *(f)*
It magnetizzazione *(f)*
Pt magnetização *(m)*

magnetization *n* En
De Magnetisierung *(f)*
Es magnetización *(f)*
Fr magnétisation *(f)*
It magnetizzazione *(f)*
Pt magnetização *(m)*

magnetizzazione *(f) n* It
De Magnetisierung *(f)*
En magnetization
Es magnetización *(f)*
Fr magnétisation *(f)*
Pt magnetização *(m)*

Magnetkonstante *(f) n* De
En magnetic constant
Es constante magnética *(f)*
Fr constante magnétique *(f)*
It costante magnetica *(f)*
Pt constante magnética *(f)*

Magnetkraft *(n) n* De
En magnetic force
Es fuerza magnética *(f)*
Fr force magnétique *(f)*
It forza magnetica *(f)*
Pt força magnética *(f)*

magneto *n* En; Es *(f)*; Pt *(m)*
De Magnetzünder *(m)*
Fr magnéto *(f)*
It magnete *(m)*

magnéto *(f) n* Fr
De Magnetzünder *(m)*
En magneto
Es magneto *(f)*
It magnete *(m)*
Pt magneto *(m)*

magnetohidrodinámica *(f) n* Es
De Magnetohydrodynamik *(f)*
En magnetohydrodynamics
Fr magnétohydrodynamique *(f)*
It magnetoidrodinamica *(f)*
Pt magnetohidrodinâmica *(f)*

magnetohidrodinâmica *(f) n* Pt
De Magnetohydrodynamik *(f)*
En magnetohydrodynamics
Es magnetohidrodinámica *(f)*
Fr magnétohydrodynamique *(f)*
It magnetoidrodinamica *(f)*

magnetohydrodynamic generator En
De Magneto-hydrodynamik-Generator *(m)*
Es generador magnetohidrodinámico *(m)*
Fr générateur magnétohydrodynamique *(m)*
It generatore magnetoidrodinamico *(m)*
Pt gerador magnetohidrodinámico *(m)*

magnetohydrodynamics *n* En
De Magnetohydrodynamik *(f)*
Es magnetohidrodinámica *(f)*
Fr magnétohydrodynamique *(f)*
It magnetoidrodinamica *(f)*
Pt magnetohidrodinâmica *(f)*

Magnetohydrodynamik *(f) n* De
En magnetohydrodynamics
Es magnetohidrodinámica *(f)*
Fr magnétohydrodynamique *(f)*
It magnetoidrodinamica *(f)*
Pt magnetohidrodinâmica *(f)*

Magnetohydrodynamik-Generator *(m) n* De
En magnetohydrodynamic generator
Es generador magnetohidrodinámico *(m)*
Fr générateur magnétohydrodynamique *(m)*
It generatore magnetoidrodinamico *(m)*
Pt gerador magnetohidrodinámico *(m)*

magnétohydrodynamique *(f) n* Fr
De Magnetohydrodynamik *(f)*
En magnetohydrodynamics
Es magnetohidrodinámica *(f)*
It magnetoidrodinamica *(f)*
Pt magnetohidrodinâmica *(f)*

magnetoidrodinamica *(f) n* It
De Magnetohydrodynamik *(f)*
En magnetohydrodynamics
Es magnetohidrodinámica *(f)*
Fr magnétohydrodynamique *(f)*
Pt magnetohidrodinâmica *(f)*

Magnetspeicher *(m) n* De
En magnetic storage
Es almacenamiento magnético *(m)*
Fr accumulation magnétique *(f)*
It immagazzinaggio magnetico *(m)*
Pt armazenagem magnética *(f)*

mains supply En
De Netzstrom *(m)*
Es alimentación principal *(f)*
Fr alimentation secteur *(f)*
It alimentazione di rete *(f)*
Pt distribuição pela linha principal *(f)*

mancha de óleo *(f)* Pt
De Ölschlamm *(m)*
En oil slick
Es mancha de petróleo flotante *(f)*
Fr nappe d'huile *(f)*
It scia d'olio grezzo in mare *(f)*

mancha de petróleo flotante *(f)* Es
De Ölschlamm *(m)*
En oil slick
Fr nappe d'huile *(f)*
It scia d'olio grezzo in mare *(f)*
Pt mancha de óleo *(f)*

manned platform En
De bemannte Ölbohrinsel *(f)*
Es plataforma atendida por personal *(f)*
Fr plate-forme occupée *(f)*
It piattaforma abitata *(f)*
Pt plataforma com tripulação *(f)*

manometer *n* En
De Manometer *(n)*
Es manómetro *(m)*
Fr manomètre *(m)*
It manometro *(m)*
Pt manómetro *(m)*

Manometer *(n) n* De
En manometer
Es manómetro *(m)*
Fr manomètre *(m)*
It manometro *(m)*
Pt manómetro *(m)*

manomètre *(m) n* Fr
De Manometer *(n)*
En manometer
Es manómetro *(m)*
It manometro *(m)*
Pt manómetro *(m)*

manometro *(m) n* It
De Manometer *(n)*
En manometer
Es manómetro *(m)*
Fr manomètre *(m)*
Pt manómetro *(m)*

manómetro *(m) n* Es, Pt
De Manometer *(n)*
En manometer
Fr manomètre *(m)*
It manometro *(m)*

manometro Bourdon *(m)* It
De Bourdonmanometer *(n)*
En Bourdon gauge
Es manómetro de Bourdon *(m)*
Fr manomètre de Bourdon *(m)*
Pt manómetro de Bourdon *(m)*

manómetro de Bourdon *(m)* Es, Pt
De Bourdonmanometer *(n)*
En Bourdon gauge
Fr manomètre de Bourdon *(m)*
It manometro Bourdon *(m)*

manomètre de Bourdon *(m)* Fr
De Bourdonmanometer *(n)*
En Bourdon gauge
Es manómetro de Bourdon *(m)*
It manometro Bourdon *(m)*
Pt manómetro de Bourdon *(m)*

manteau terrestre *(m)* Fr
De Unterboden *(m)*
En mantle (geology)
Es manto *(m)*
It manto *(m)*
Pt manto *(m)*

mantle (geology) *n* En
De Unterboden *(m)*
Es manto *(m)*
Fr manteau terrestre *(m)*
It manto *(m)*
Pt manto *(m)*

manto *(m) n* Es, It, Pt
De Unterboden *(m)*
En mantle (geology)
Fr manteau terrestre *(m)*

manual control En
De manuelle Steuerung *(f)*
Es control manual *(m)*
Fr commande manuelle *(f)*
It comando manuale *(m)*
Pt comando manual *(m)*

manuelle Steuerung *(f)* De
En manual control
Es control manual *(m)*
Fr commande manuelle *(f)*
It comando manuale *(m)*
Pt comando manual *(m)*

máquina enrolada em série *(f)* Pt
De seriengewickelte Maschine *(f)*
En series-wound machine
Es maquina excitada en serie *(f)*
Fr machine série *(f)*
It macchina eccitata in serie *(f)*

máquina enrolada em shunt *(f)* Pt
De Nebenschlußmaschine *(f)*
En shunt-wound machine
Es máquina excitada en derivación *(f)*
Fr machine shunt *(f)*
It macchina eccitata in derivazione *(f)*

máquina excitada en derivación *(f)* Es
De Nebenschlußmaschine *(f)*

En shunt-wound machine
Fr machine shunt *(f)*
It macchina eccitata in derivazione *(f)*
Pt máquina enrolada em shunt *(f)*

máquina excitada en serie *(f)* Es
De seriengewickelte Maschine *(f)*
En series-wound machine
Fr machine série *(f)*
It macchina eccitata in serie *(f)*
Pt máquina enrolada em série *(f)*

marché disponible *(m)* Fr
De Kassamarkt *(m)*
En spot market
Es mercado de disponibles *(m)*
It mercato del disponibile *(m)*
Pt mercado de ocasião *(m)*

marine diesel-fuel En
De Schiffsdieseltreibstoff *(m)*
Es combustible diesel para marina *(m)*
Fr combustible diesel marin *(m)*
It combustibile diesel per motori marini *(m)*
Pt diesel-fuel marítimo *(m)*

marine fouling En
De Seeverschmutzung *(f)*
Es suciedad por depósitos marinos *(f)*
Fr accumulation de dépôts marins *(f)*
It incrostazione marina *(f)*
Pt depósitos marítimos *(m)*

marine pipeline En
De Seerohrleitung *(f)*
Es conducción submarina *(f)*
Fr pipeline marin *(m)*
It condotto marino *(m)*
Pt oleoduto marítimo *(m)*

marino *adj* Es
De auf See
En offshore
Fr au large
It al largo
Pt fora da praia

masa atómica *(f)* Es
De Atommasse *(f)*
En atomic mass
Fr masse atomique *(f)*
It massa dell'atomo *(f)*
Pt massa atómica *(f)*

maser *n* En; Fr, It, Pt *(m)*
De Maser *(m)*
Es máser *(m)*

Maser *(m) n* De
En maser
Es máser *(m)*
Fr maser *(m)*
It maser *(m)*
Pt maser *(m)*

máser *(m) n* Es
De Maser *(m)*
En maser
Fr maser *(m)*
It maser *(m)*
Pt maser *(m)*

máser de masa *(m)* Es
De Massenmaser *(m)*
En mass maser
Fr maser de masse *(m)*
It maser di massa *(m)*
Pt maser de massa *(m)*

maser de massa *(m)* Pt
De Massenmaser *(m)*
En mass maser
Es máser de masa *(m)*
Fr maser de masse *(m)*
It maser di massa *(m)*

maser de masse *(m)* Fr
De Massenmaser *(m)*
En mass maser
Es máser de masa *(m)*
It maser di massa *(m)*
Pt maser de massa *(m)*

maser di massa *(m)* It
De Massenmaser *(m)*
En mass maser
Es máser de masa *(m)*
Fr maser de masse *(m)*
Pt maser de massa *(m)*

massa *(f) n* Pt
Am ground
De Erde *(f)*
En earth
Es tierra *(f)*
Fr terre *(f)*
It terra *(f)*

massa atómica *(f)* Pt
De Atommasse *(f)*
En atomic mass
Es masa atómica *(f)*
Fr masse atomique *(f)*
It massa dell'atomo *(f)*

massa dell'atomo *(f)* It
De Atommasse *(f)*
En atomic mass
Es masa atómica *(f)*
Fr masse atomique *(f)*
Pt massa atómica *(f)*

mass balance En
De Massenausgleich *(m)*
Es equilibrio de masa *(m)*
Fr équilibre de masse *(m)*
It bilancio di massa *(m)*
Pt equilíbrio de massa *(m)*

mass defect En
De Massendefekt *(m)*
Es defecto de masa *(m)*
Fr défaut de masse *(m)*
It difetto di massa *(m)*
Pt defeito de massa *(m)*

masse atomique *(f)* Fr
De Atommasse *(f)*
En atomic mass
Es masa atómica *(f)*
It massa dell'atomo *(f)*
Pt massa atómica *(f)*

Massenausgleich *(m) n* De
En mass balance
Es equilibrio de masa *(m)*
Fr équilibre de masse *(m)*
It bilancio di massa *(m)*
Pt equilíbrio de massa *(m)*

Massendefekt *(m) n* De
En mass defect
Es defecto de masa *(m)*
Fr défaut de masse *(m)*
It difetto di massa *(m)*
Pt defeito de massa *(m)*

Massenenergie-Erhaltung *(f) n* De
En mass-energy conservation
Es conservación masa-energía *(f)*
Fr conservation d'énergie-masse *(f)*
It conservazione dell'energia-massa *(f)*
Pt conservação de energia-massa *(f)*

mass-energy conservation En
De Massenenergie-Erhaltung *(f)*
Es conservación masa-energía *(f)*
Fr conservation d'énergie-masse *(f)*
It conservazione dell'energia-massa *(f)*
Pt conservação de energia-massa *(f)*

Massenmaser *(m) n* De
En mass maser
Es máser de masa *(m)*
Fr maser de masse *(m)*
It maser di massa *(m)*
Pt maser de massa *(m)*

Massenspektroskopie *(f) n* De
En mass spectroscopy
Es espectroscopia de masa *(f)*
Fr spectroscopie de masse *(f)*
It spettroscopia di massa *(f)*
Pt espectroscopia de massa *(f)*

mass maser En
De Massenmaser *(m)*
Es máser de masa *(m)*
Fr maser de masse *(m)*
It maser di massa *(m)*
Pt maser de massa *(m)*

mass spectroscopy En
De Massenspektro-skopie *(f)*

Es espectroscopia de masa *(f)*
Fr spectroscopie de masse *(f)*
It spettroscopia di massa *(f)*
Pt espectroscopia de massa *(f)*

Masutmotoren *(pl) n* De
En heavy-fuel engines *(pl)*
Es motores de combustible pesado *(m pl)*
Fr moteurs au fuel lourd *(m pl)*
It motori a combustibile pesante *(m pl)*
Pt motores a óleos pesados *(m pl)*

material and energy balance En
De Stoff- und Energiebilanz *(f)*
Es equilibrio materia-energía *(m)*
Fr balance de matériel et énergie *(f)*
It bilancio materiale-energia *(m)*
Pt equilíbrio de material e energia *(m)*

material de carga *(m)* Es
De Stangenmaterial *(n)*
En feedstock
Fr stock d'alimentation *(m)*
It materiale di alimentazione *(m)*
Pt stock de abastecimento *(m)*

materiale di alimentazione *(m)* It
De Stangenmaterial *(n)*
En feedstock
Es material de carga *(m)*
Fr stock d'alimentation *(m)*
Pt stock de abastecimento *(m)*

materiale di alimentazione chimico *(m)* It
De chemisches Stangenmaterial *(n)*
En chemical feedstock
Es material químico de carga *(m)*
Fr stock d'alimentation chimique *(m)*
Pt stock de abastecimento químico *(m)*

materiale di sterro *(m)* It
De Abraum *(m)*
En spoil (mining)
Es estériles *(m pl)*
Fr déblai *(m)*
Pt entulho *(m)*

materiale di sterro di roccia *(m)* It
De Felsabraum *(m)*
En rock spoils *(pl)*
Es desperdicios de roca *(m pl)*
Fr déblais de roche *(m pl)*
Pt detritos de rocha *(m pl)*

material químico de carga *(m)* Es
De chemisches Stangenmaterial *(n)*
En chemical feedstock
Fr stock d'alimentation chimique *(m)*
It materiale di alimentazione chimico *(m)*
Pt stock de abastecimento químico *(m)*

materia volátil *(f)* Es
De flüchtiger Bestandteil *(m)*
En volatile matter
Fr matière volatile *(f)*
It sostanza volatile *(f)*
Pt matéria volátil *(f)*

matéria volátil *(f)* Pt
De flüchtiger Bestandteil *(m)*
En volatile matter
Es materia volátil *(f)*
Fr matière volatile *(f)*
It sostanza volatile *(f)*

matière volatile *(f)* Fr
De flüchtiger Bestandteil *(m)*
En volatile matter
Es materia volátil *(f)*
It sostanza volatile *(f)*
Pt matéria volátil *(f)*

Mattkohle *(f) n* De
En cannel coal
Es canel *(m)*
Fr houille grasse *(f)*
It carbone a lunga fiamma *(m)*
Pt carvão cannel *(m)*

mattone refrattario *(m)* It
De feuerfester Stein *(m)*
En firebrick
Es ladrillo refractario *(m)*
Fr brique réfractaire *(f)*
Pt tijolo refractário *(m)*

mazout *(m) n* Fr
De Treiböl *(n)*
En fuel oil
Es fuel oil *(m)*
It nafta *(f)*
Pt fuel oil *(m)*

mean *n* En
De Durchschnitt *(m)*
Es media *(f)*
Fr moyen *(m)*
It medio *(f)*
Pt médio *(m)*

mean deviation En
De durchschnittliche Abweichung *(f)*
Es desviación media *(f)*
Fr déviation moyenne *(f)*
It deviazione media *(f)*
Pt desvio médio *(m)*

mecánica *(f) n* Es
De Mechanik *(f)*
En mechanics
Fr mécanique *(f)*
It meccanica *(f)*
Pt mecânica *(f)*

mecânica *(f) n* Pt
De Mechanik *(f)*
En mechanics
Es mecánica *(f)*
Fr mécanique *(f)*
It meccanica *(f)*

mecánica cuántica *(f)* Es
De Quantenmechanik *(f)*
En quantum mechanics
Fr mécanique quantique *(f)*
It meccanica dei quanti *(f)*
Pt mecânica quântica *(f)*

mecânica quântica *(f)* Pt
De Quantenmechanik *(f)*
En quantum mechanics
Es mecánica cuántica *(f)*
Fr mécanique quantique *(f)*
It meccanica dei quanti *(f)*

mecánico *adj* Es
De mechanisch
En mechanical
Fr mécanique
It meccanico
Pt mecânico

mecânico *adj* Pt
De mechanisch
En mechanical
Es mecánico
Fr mécanique
It meccanico

mécanique *adj* Fr
De mechanisch
En mechanical
Es mecánico
It meccanico
Pt mecânico

mécanique *(f) n* Fr
De Mechanik *(f)*
En mechanics
Es mecánica *(f)*
It meccanica *(f)*
Pt mecânica *(f)*

mécanique quantique *(f)* Fr
De Quantenmechanik *(f)*
En quantum mechanics
Es mecánica cuántica *(f)*
It meccanica dei quanti *(f)*
Pt mecânica quântica *(f)*

meccanica *(f) n* It
De Mechanik *(f)*
En mechanics
Es mecánica *(f)*
Fr mécanique *(f)*
Pt mecânica *(f)*

meccanica dei quanti *(f)* It
De Quantenmechanik *(f)*

En quantum mechanics
Es mecánica cuántica *(f)*
Fr mécanique quantique *(f)*
Pt mecânica quântica *(f)*

meccanico *adj* It
De mechanisch
En mechanical
Es mecánico
Fr mécanique
Pt mecânico

mechanical *adj* En
De mechanisch
Es mecánico
Fr mécanique
It meccanico
Pt mecânico

mechanical efficiency En
De mechanischer Wirkungsgrad *(m)*
Es rendimiento mecánico *(m)*
Fr rendement mécanique *(m)*
It resa meccanica *(f)*
Pt eficiência mecânica *(f)*

mechanical energy En
De mechanische Energie *(f)*
Es energía mecánica *(f)*
Fr énergie mécanique *(f)*
It energia meccanica *(f)*
Pt energia mecânica *(f)*

mechanical storage En
De mechanische Speicherung *(f)*
Es almacenamiento mecánico *(m)*
Fr accumulation mécanique *(f)*
It immagazzinaggio meccanico *(m)*
Pt armazenagem mecânica *(f)*

mechanics *n* En
De Mechanik *(f)*
Es mecánica *(f)*
Fr mécanique *(f)*
It meccanica *(f)*
Pt mecânica *(f)*

Mechanik *(f) n* De
En mechanics
Es mecánica *(f)*
Fr mécanique *(f)*
It meccanica *(f)*
Pt mecânica *(f)*

mechanisch *adj* De
En mechanical
Es mecánico
Fr mécanique
It meccanico
Pt mecânico

mechanische Energie *(f)* De
En mechanical energy
Es energía mecánica *(f)*
Fr énergie mécanique *(f)*
It energia meccanica *(f)*
Pt energia mecânica *(f)*

mechanischer Wirkungsgrad *(m)* De
En mechanical efficiency
Es rendimiento mecánico *(m)*
Fr rendement mécanique *(m)*
It resa meccanica *(f)*
Pt eficiência mecânica *(f)*

mechanische Speicherung *(f)* De
En mechanical storage
Es almacenamiento mecánico *(m)*
Fr accumulation mécanique *(f)*
It immagazzinaggio meccanico *(m)*
Pt armazenagem mecânica *(f)*

media *(f) n* Es
De Durchschnitt *(m)*
En mean
Fr moyen *(m)*
It medio *(f)*
Pt médio *(m)*

media compensada *(f)* Es
De gewogener Durchschnitt *(m)*
En weighted mean
Fr moyenne pondérée *(f)*
It media ponderata *(f)*
Pt média ponderada *(f)*

média ponderada *(f)* Pt
De gewogener Durchschnitt *(m)*
En weighted mean
Es media compensada *(f)*
Fr moyenne pondérée *(f)*
It media ponderata *(f)*

media ponderata *(f)* It
De gewogener Durchschnitt *(m)*
En weighted mean
Es media compensada *(f)*
Fr moyenne pondérée *(f)*
Pt média ponderada *(f)*

media vida *(f)* Es
De Halbwertszeit *(f)*
En half-life
Fr demi-vie *(f)*
It periodo di dimezzamento *(m)*
Pt meia vida *(f)*

medida de carvão *(f)* Pt
De Kohlenlager *(n)*
En coal measure
Es formación hullera *(f)*
Fr couche de houille *(f)*
It giacimento carbonifero *(m)*

medidor *(m) n* Es
De Meßgerät *(n)*
En meter
Fr compteur *(m)*
It metro *(m)*
Pt contador *(m)*

medidor de flujo *(m)* Es
De Durchflußmesser *(m)*
En flowmeter
Fr débitmètre *(m)*
It flussometro *(m)*
Pt fluxómetro *(m)*

medidor de humidade *(m)* Pt
De Naßmesser *(m)*
En wet meter
Es medidor húmedo *(m)*
Fr compteur humide *(m)*
It contatore a liquido *(m)*

medidor de seco *(m)* Pt
De Trockenmeßgerät *(n)*
En dry meter
Es medidor seco *(m)*
Fr dessicomètre *(m)*
It contatore a secco *(m)*

medidor de watt *(m)* Pt
De Wattmesser *(m)*
En wattmeter
Es watímetro *(m)*
Fr wattmètre *(m)*
It wattometro *(m)*

medidor húmedo *(m)* Es
De Naßmesser *(m)*
En wet meter
Fr compteur humide *(m)*
It contatore a liquido *(m)*
Pt medidor de humidade *(m)*

medidor seco *(m)* Es
De Trockenmeßgerät *(n)*
En dry meter
Fr dessicomètre *(m)*
It contatore a secco *(m)*
Pt medidor de seco *(m)*

medidor Venturi *(m)* Es
De Venturimesser *(m)*
En Venturi meter
Fr compteur Venturi *(m)*
It venturimetro *(m)*
Pt contador de Venturi *(m)*

medio *(f) n* It
De Durchschnitt *(m)*
En mean
Es media *(f)*
Fr moyen *(m)*
Pt médio *(m)*

médio *(m) n* Pt
De Durchschnitt *(m)*
En mean
Es media *(f)*
Fr moyen *(m)*
It medio *(f)*

Meereswärmeenergie *(f)* De
En ocean thermal energy
Es energía térmica oceánica *(f)*
Fr énergie thermique des océans *(f)*
It energia termica oceanica *(f)*
Pt energia térmica oceânica *(f)*

Meereswärmegefälle *(n) n* De
En ocean thermal gradient
Es gradiente térmico oceánico *(m)*
Fr gradient thermique des océans *(m)*
It gradiente termico oceanico *(m)*
Pt gradiente térmico oceânico *(m)*

Meerwasserentsalzung *(f) n* De
En desalination
Es desalación *(f)*
Fr dessalage *(m)*
It desalinizzazione *(f)*
Pt dessalinização *(f)*

Mehrfacheffekt-Destillation *(f)* De
En multiple-effect distillation
Es destilación de efecto múltiple *(f)*
Fr distillation à effet multiple *(f)*
It distillazione ad effetto multiplo *(f)*
Pt destilação de efeito múltiplo *(f)*

Mehrfach-Schutzerdung *(f) n* De
En protective multiple earthing
Es toma de tierra múltiple protectora *(f)*
Fr mise à la terre multiple de protection *(f)*
It messa a terra multipla protettiva *(f)*
Pt ligação à terra múltipla protectora *(f)*

mehrphasig *adj* De
En polyphase
Es polifase
Fr polyphase
It polifase
Pt polifase

meia vida *(f)* Pt
De Halbwertszeit *(f)*
En half-life
Es media vida *(f)*
Fr demi-vie *(f)*
It periodo di dimezzamento *(m)*

mejoramiento *(m) n* Es
De Aufkonzentrierung *(f)*
En upgrading
Fr amélioration *(f)*
It promovimento a grado superiore *(m)*
Pt melhoramento *(m)*

melhoramento *(m) n* Pt
De Aufkonzentrierung *(f)*
En upgrading
Es mejoramiento *(m)*
Fr amélioration *(f)*
It promovimento a grado superiore *(m)*

melting point En
De Schmelzpunkt *(m)*
Es punto de fusión *(f)*
Fr point de fusion *(f)*
It punto di fusione *(m)*
Pt ponto de fusão *(m)*

mercado de disponibles *(m)* Es
De Kassamarkt *(m)*
En spot market
Fr marché disponible *(m)*
It mercato del disponibile *(m)*
Pt mercado de ocasião *(m)*

mercado de ocasião *(m)* Pt
De Kassamarkt *(m)*
En spot market
Es mercado de disponibles *(m)*
Fr marché disponible *(m)*
It mercato del disponibile *(m)*

mercato del disponibile *(m)* It
De Kassamarkt *(m)*
En spot market
Es mercado de disponibles *(m)*
Fr marché disponible *(m)*
Pt mercado de ocasião *(m)*

meson *n* En; Pt *(m)*
De Meson *(n)*
Es mesón *(m)*
Fr méson *(m)*
It mesone *(m)*

Meson *(n) n* De
En meson
Es mesón *(m)*
Fr méson *(m)*
It mesone *(m)*
Pt meson *(m)*

mesón *(m) n* Es
De Meson *(n)*
En meson
Fr méson *(m)*
It mesone *(m)*
Pt meson *(m)*

méson *(m) n* Fr
De Meson *(n)*
En meson
Es mesón *(m)*
It mesone *(m)*
Pt meson *(m)*

mesone *(m) n* It
De Meson *(n)*
En meson
Es mesón *(m)*
Fr méson *(m)*
Pt meson *(m)*

messa a terra multipla protettiva *(f)* It
De Mehrfach-Schutzerdung *(f)*
En protective multiple earthing
Es toma de tierra múltiple protectora *(f)*
Fr mise à la terre multiple de protection *(f)*
Pt ligação à terra múltipla protectora *(f)*

Meßgerät *(n) n* De
En meter
Es medidor *(m)*
Fr compteur *(m)*
It metro *(m)*
Pt contador *(m)*

Meßwandler *(m) n* De
En transducer
Es transductor *(m)*
Fr transducteur *(m)*
It trasduttore *(m)*
Pt transdutor *(m)*

metabituminöse Kohle *(f)* De
En metabituminous coal
Es carbón metabituminoso *(m)*
Fr charbon métabitumineux *(m)*
It carbone metabituminoso *(m)*
Pt carvão metabetuminoso *(m)*

metabituminous coal En
De metabituminöse Kohle *(f)*
Es carbón metabituminoso *(m)*
Fr charbon métabitumineux *(m)*
It carbone metabituminoso *(m)*
Pt carvão metabetuminoso *(m)*

métabolisme basal *(m)* Fr
De Grundumsatz *(m)*
En basal metabolic rate
Es índice metabólico basal *(m)*
It tasso metabolico basale *(m)*
Pt índice de metabolismo basal *(m)*

metaestable *adj* Es
De metastabil
En metastable
Fr métastable
It metastabile
Pt metaestável

metaestável *adj* Pt
De metastabil
En metastable
Es metaestable

Fr métastable
It metastabile

metal-air cell En
De Metall-Luftzelle *(f)*
Es célula de metal-aire *(f)*
Fr pile métal-air *(f)*
It cellula metallo-aria *(f)*
Pt célula de metal-ar *(f)*

metal deactivators *(pl)* En
De Metall-Deaktivierungsmittel *(pl)*
Es desactivadores de metales *(m pl)*
Fr désactiveurs de métaux *(m pl)*
It disattivatori di metallo *(m pl)*
Pt desactivadores metálicos *(m pl)*

Metall-Deaktivierungsmittel *(pl) n* De
En metal deactivators *(pl)*
Es desactivadores de metales *(m pl)*
Fr désactiveurs de métaux *(m pl)*
It disattivatori di metallo *(m pl)*
Pt desactivadores metálicos *(m pl)*

Metall-Luftzelle *(f) n* De
En metal-air cell
Es célula de metal-aire *(f)*
Fr pile métal-air *(f)*
It cellula metallo-aria *(f)*
Pt célula de metal-ar *(f)*

metallurgical coke En
De metallurgischer Koks *(m)*
Es coque metalúrgico *(m)*
Fr coke métallurgique *(m)*
It coke metallurgico *(m)*
Pt coque metalúrgico *(m)*

metallurgischer Koks *(m)* De
En metallurgical coke
Es coque metalúrgico *(m)*
Fr coke métallurgique *(m)*
It coke metallurgico *(m)*
Pt coque metalúrgico *(m)*

metamorphic rock En
De metamorphosisches Gestein *(n)*
Es roca metamórfica *(f)*
Fr roche métamorphique *(f)*
It roccia metamorfica *(f)*
Pt rocha metamórfica *(f)*

metamorphosisches Gestein *(n)* De
En metamorphic rock
Es roca metamórfica *(f)*
Fr roche métamorphique *(f)*
It roccia metamorfica *(f)*
Pt rocha metamórfica *(f)*

metanação *(f) n* Pt
De Methanisierung *(f)*
En methanation
Es metanación *(f)*
Fr méthanisation *(f)*
It metanizzazione *(f)*

metanación *(f) n* Es
De Methanisierung *(f)*
En methanation
Fr méthanisation *(f)*
It metanizzazione *(f)*
Pt metanação *(f)*

metanizzazione *(f) n* It
De Methanisierung *(f)*
En methanation
Es metanación *(f)*
Fr méthanisation *(f)*
Pt metanação *(f)*

metano *(m) n* Es, It, Pt
De Methan *(n)*
En methane
Fr méthane *(m)*

metanol *(m) n* Es, Pt
De Methanol *(n)*
En methanol
Fr méthanol *(m)*
It metanolo *(m)*

metanolo *(m) n* It
De Methanol *(n)*
En methanol
Es metanol *(m)*
Fr méthanol *(m)*
Pt metanol *(m)*

metastabil *adj* De
En metastable
Es metaestable
Fr métastable
It metastabile
Pt metaestável

metastabile *adj* It
De metastabil
En metastable
Es metaestable
Fr métastable
Pt metaestável

metastable *adj* En
De metastabil
Es metaestable
Fr métastable
It metastabile
Pt metaestável

métastable *adj* Fr
De metastabil
En metastable
Es metaestable
It metastabile
Pt metaestável

meteoritic steam En
De meteoritischer Dampf *(m)*
Es vapor meteorítico *(m)*
Fr vapeur météorique *(f)*
It vapore meteoritico *(m)*
Pt vapor meteórico *(m)*

meteoritischer Dampf *(m)* De
En meteoritic steam
Es vapor meteorítico *(m)*
Fr vapeur météorique *(f)*
It vapore meteoritico *(m)*
Pt vapor meteórico *(m)*

meter *n* En
De Meßgerät *(n)*
Es medidor *(m)*
Fr compteur *(m)*
It metro *(m)*
Pt contador *(m)*

Methan *(n) n* De
En methane
Es metano *(m)*
Fr méthane *(m)*
It metano *(m)*
Pt metano *(m)*

methanation *n* En
De Methanisierung *(f)*
Es metanación *(f)*
Fr méthanisation *(f)*
It metanizzazione *(f)*
Pt metanação *(f)*

methane *n* En
De Methan *(n)*
Es metano *(m)*
Fr méthane *(m)*
It metano *(m)*
Pt metano *(m)*

méthane *(m) n* Fr
De Methan *(n)*
En methane
Es metano *(m)*
It metano *(m)*
Pt metano *(m)*

méthanisation *(f) n* Fr
De Methanisierung *(f)*
En methanation
Es metanación *(f)*
It metanizzazione *(f)*
Pt metanação *(f)*

Methanisierung *(f) n* De
En methanation
Es metanación *(f)*
Fr méthanisation *(f)*
It metanizzazione *(f)*
Pt metanação *(f)*

methanol *n* En
De Methanol *(n)*
Es metanol *(m)*
Fr méthanol *(m)*
It metanolo *(m)*
Pt metanol *(m)*

Methanol *(n) n* De
En methanol
Es metanol *(m)*
Fr méthanol *(m)*
It metanolo *(m)*
Pt metanol *(m)*

méthanol *(m) n* Fr
De Methanol *(n)*
En methanol
Es metanol *(m)*

It metanolo *(m)*
Pt metanol *(m)*

méthode d'Orsat *(m)* Fr
De Orsat-Verfahren *(n)*
En Orsat method
Es método de Orsat *(m)*
It metodi di Orsat *(m)*
Pt método de Orsat *(m)*

metodi di Orsat *(m)* It
De Orsat-Verfahren *(n)*
En Orsat method
Es método de Orsat *(m)*
Fr méthode d'Orsat *(m)*
Pt método de Orsat *(m)*

método de Orsat *(m)* Es, Pt
De Orsat-Verfahren *(n)*
En Orsat method
Fr méthode d'Orsat *(m)*
It metodi di Orsat *(m)*

metro *(m) n* It
De Meßgerät *(n)*
En meter
Es medidor *(m)*
Fr compteur *(m)*
Pt contador *(m)*

metro di spostamento rotante *(m)* It
De Drehverschiebungs-messer *(m)*
En rotary-displacement meter
Es contador de desplazamiento rotativo *(m)*
Fr compteur à piston rotatif *(m)*
Pt contador de deslocação rotativa *(m)*

metrologia *(f) n* It, Pt
De Metrologie *(f)*
En metrology
Es metrología *(f)*
Fr métrologie *(f)*

metrología *(f) n* Es
De Metrologie *(f)*
En metrology
Fr métrologie *(f)*
It metrologia *(f)*
Pt metrologia *(f)*

Metrologie *(f) n* De
En metrology
Es metrología *(f)*
Fr métrologie *(f)*
It metrologia *(f)*
Pt metrologia *(f)*

métrologie *(f) n* Fr
De Metrologie *(f)*
En metrology
Es metrología *(f)*
It metrologia *(f)*
Pt metrologia *(f)*

metrology *n* En
De Metrologie *(f)*
Es metrología *(f)*
Fr métrologie *(f)*
It metrologia *(f)*
Pt metrologia *(f)*

microcomputador *(m) n* Pt
De Mikrocomputer *(n)*
En microcomputer
Es microordenador *(m)*
Fr micro-ordinateur *(m)*
It microelaboratore *(m)*

microcomputer *n* En
De Mikrocomputer *(n)*
Es microordenador *(m)*
Fr micro-ordinateur *(m)*
It microelaboratore *(m)*
Pt microcomputador *(m)*

microelaboratore *(m) n* It
De Mikrocomputer *(n)*
En microcomputer
Es microordenador *(m)*
Fr micro-ordinateur *(m)*
Pt microcomputador *(m)*

microonda *(f) n* Es
De Mikrowelle *(f)*
En microwave
Fr micro-onde *(f)*
It microonda *(f)*
Pt microonda *(f)*

microonda *(f) n* It, Pt
De Mikrowelle *(f)*
En microwave
Es microonda *(f)*
Fr micro-onde *(f)*

micro-onde *(f) n* Fr
De Mikrowelle *(f)*
En microwave
Es microonda *(f)*
It microonda *(f)*
Pt microonda *(f)*

microordenador *(m) n* Es
De Mikrocomputer *(n)*
En microcomputer
Fr micro-ordinateur *(m)*
It microelaboratore *(m)*
Pt microcomputador *(m)*

micro-ordinateur *(m) n* Fr
De Mikrocomputer *(n)*
En microcomputer
Es microordenador *(m)*
It microelaboratore *(m)*
Pt microcomputador *(m)*

microwave *n* En
De Mikrowelle *(f)*
Es microonda *(f)*
Fr micro-onde *(f)*
It microonda *(f)*
Pt microonda *(f)*

Mikrocomputer *(n) n* De
En microcomputer
Es microordenador *(m)*
Fr micro-ordinateur *(m)*
It microelaboratore *(m)*
Pt microcomputador *(m)*

Mikrowelle *(f) n* De
En microwave
Es microonda *(f)*
Fr micro-onde *(f)*
It microonda *(f)*
Pt microonda *(f)*

mina *(f) n* Es, Pt
De Bergwerk *(n)*
En mine
Fr mine *(f)*
It miniera *(f)*

mina aberta *(f)* Pt
De Tagebau *(m)*
En open-cast mine
Es mina a cielo abierto *(f)*
Fr mine à ciel ouvert *(f)*
It miniera a cielo aperto *(f)*

mina a cielo abierto *(f)* Es
De Tagebau *(m)*
En open-cast mine
Fr mine à ciel ouvert *(f)*
It miniera a cielo aperto *(f)*
Pt mina aberta *(f)*

mine *n* En; Fr *(f)*
De Bergwerk *(n)*
Es mina *(f)*
It miniera *(f)*
Pt mina *(f)*

mine à ciel ouvert *(f)* Fr
De Tagebau *(m)*
En open-cast mine
Es mina a cielo abierto *(f)*
It miniera a cielo aperto *(f)*
Pt mina aberta *(f)*

mineração *(f) n* Pt
De Bergbau *(m)*
En mining
Es minería *(f)*
Fr exploitation minière *(f)*
It estrazione di minerali *(f)*

mineração de curvas de nível *(f)* Pt
De Konturenabbau *(m)*
En contour mining
Es minería siguiendo el perfil del terreno *(f)*
Fr exploitation minière des contours *(f)*
It scavo a contorno *(m)*

mineração profunda *(f)* Pt
De Tiefbau *(m)*
En deep mining
Es explotación en profundidad *(f)*
Fr exploitation des niveaux inférieurs *(f)*
It scavo profondo *(m)*

mineração subterrânea *(f)* Pt
De Grubenbetrieb *(m)*
En underground mining
Es minería subterránea *(f)*
Fr exploitation minière souterraine *(f)*
It scavo sotteraneo *(m)*

mineral wax En
De Ozokerit *(m)*
Es cera mineral *(f)*
Fr cire minérale *(f)*
It cera minerale *(f)*
Pt cera mineral *(f)*

mineral-wool fiber insulation Am
De Steinwollefaser-isolierung *(f)*
En mineral-wool fibre insulation
Es aislamiento con fibra de lana mineral *(m)*
Fr isolement par fibre de laine minérale *(m)*
It isolamento con fibra di lana minerale *(m)*
Pt isolamento com fibras de lã mineral *(m)*

mineral-wool fibre insulation En
Am mineral-wool fiber insulation
De Steinwollefaser-isolierung *(f)*
Es aislamiento con fibra de lana mineral *(m)*
Fr isolement par fibre de laine minérale *(m)*
It isolamento con fibra di lana minerale *(m)*
Pt isolamento com fibras de lã mineral *(m)*

minería *(f) n* Es
De Bergbau *(m)*
En mining
Fr exploitation minière *(f)*
It estrazione di minerali *(f)*
Pt mineração *(f)*

minería siguiendo el perfil del terreno *(f)* Es
De Konturenabbau *(m)*
En contour mining
Fr exploitation minière des contours *(f)*
It scavo a contorno *(m)*
Pt mineração de curvas de nível *(f)*

minería subterránea *(f)* Es
De Grubenbetrieb *(m)*
En underground mining
Fr exploitation minière souterraine *(f)*
It scavo sotteraneo *(m)*
Pt mineração subterrânea *(f)*

minicomputador *(m) n* Pt
De Minicomputer *(m)*
En minicomputer
Es miniordenador *(m)*
Fr mini-ordinateur *(m)*
It minielaboratore *(m)*

minicomputer *n* En
De Minicomputer *(m)*
Es miniordenador *(m)*
Fr mini-ordinateur *(m)*
It minielaboratore *(m)*
Pt minicomputador *(m)*

Minicomputer *(m) n* De
En minicomputer
Es miniordenador *(m)*
Fr mini-ordinateur *(m)*
It minielaboratore *(m)*
Pt minicomputador *(m)*

minielaboratore *(m) n* It
De Minicomputer *(m)*
En minicomputer
Es miniordenador *(m)*
Fr mini-ordinateur *(m)*
Pt minicomputador *(m)*

miniera *(f) n* It
De Bergwerk *(n)*
En mine
Es mina *(f)*
Fr mine *(f)*
Pt mina *(f)*

miniera a cielo aperto *(f)* It
De Tagebau *(m)*
En open-cast mine
Es mina a cielo abierto *(f)*
Fr mine à ciel ouvert *(f)*
Pt mina aberta *(f)*

mining *n* En
De Bergbau *(m)*
Es minería *(f)*
Fr exploitation minière *(f)*
It estrazione di minerali *(f)*
Pt mineração *(f)*

miniordenador *(m) n* Es
De Minicomputer *(m)*
En minicomputer
Fr mini-ordinateur *(m)*
It minielaboratore *(m)*
Pt minicomputador *(m)*

mini-ordinateur *(m) n* Fr
De Minicomputer *(m)*
En minicomputer
Es miniordenador *(m)*
It minielaboratore *(m)*
Pt minicomputador *(m)*

minuterie *(f) n* Fr
De Zeitschalter *(m)*
En time switch
Es temporizador *(m)*
It interruttore a tempo *(m)*
Pt comutador cronométrico *(m)*

mise à la terre multiple de protection *(f)* Fr
De Mehrfach-Schutzerdung *(f)*
En protective multiple earthing
Es toma de tierra múltiple protectora *(f)*
It messa a terra multipla protettiva *(f)*
Pt ligação à terra múltipla protectora *(f)*

moderado com grafite Pt
De graphitmoderiert
En graphite-moderated
Es moderado por grafito
Fr modéré au graphite
It moderato a grafite

moderado por grafito Es
De graphitmoderiert
En graphite-moderated
Fr modéré au graphite
It moderato a grafite
Pt moderado com grafite

moderador *(m) n* Es, Pt
De Moderator *(m)*
En moderator
Fr modérateur *(m)*
It moderatore *(m)*

modérateur *(m) n* Fr
De Moderator *(m)*
En moderator
Es moderador *(m)*
It moderatore *(m)*
Pt moderador *(m)*

moderato a grafite It
De graphitmoderiert
En graphite-moderated
Es moderado por grafito
Fr modéré au graphite
Pt moderado com grafite

moderator *n* En
De Moderator *(m)*
Es moderador *(m)*
Fr modérateur *(m)*
It moderatore *(m)*
Pt moderador *(m)*

Moderator *(m) n* De
En moderator
Es moderador *(m)*
Fr modérateur *(m)*
It moderatore *(m)*
Pt moderador *(m)*

moderatore *(m) n* It
De Moderator *(m)*
En moderator
Es moderador *(m)*
Fr modérateur *(m)*
Pt moderador *(m)*

modéré au graphite Fr
De graphitmoderiert
En graphite-moderated
Es moderado por grafito
It moderato a grafite
Pt moderado com grafite

module long *(m)* Fr
De Langmodul *(n)*
En long module
Es módulo largo *(m)*
It modulo lungo *(m)*
Pt módulo comprido *(m)*

módulo comprido *(m)* Pt
De Langmodul *(n)*
En long module
Es módulo largo *(m)*
Fr module long *(m)*
It modulo lungo *(m)*

módulo largo *(m)* Es
De Langmodul *(n)*
En long module
Fr module long *(m)*

It modulo lungo *(m)*
Pt módulo comprido *(m)*

modulo lungo *(m)* It
De Langmodul *(n)*
En long module
Es módulo largo *(m)*
Fr module long *(m)*
Pt módulo comprido *(m)*

moínho de vento *(m)* Pt
De Windmühle *(f)*
En windmill
Es molino de viento *(m)*
Fr moulin à vent *(m)*
It mulino a vento *(m)*

moínho de vento de eixo vertical *(m)* Pt
De Windmühle mit Hochachse *(f)*
En vertical-axis windmill
Es molino de viento de eje vertical *(m)*
Fr moulin à vent à axe vertical *(m)*
It mulino a vento a asse verticale *(m)*

moínho hidráulico *(m)* Pt
De Wassermühle *(f)*
En watermill
Es molino hidráulico *(m)*
Fr moulin à eau *(m)*
It mulino ad acqua *(m)*

molaire *adj* Fr
De molar
En molar
Es molar
It molare
Pt molar

molar *adj* De, En, Es, Pt
Fr molaire
It molare

molare *adj* It
De molar
En molar
Es molar
Fr molaire
Pt molar

molare gebundene Wärme *(f)* De
En molar latent heat
Es calor latente molar *(m)*
Fr chaleur latente molaire *(f)*
It calore latente molare *(m)*
Pt calor latente molar *(m)*

molare Wärmekapazität *(f)* De
En molar heat capacity
Es capacidad de calor molar *(f)*
Fr capacité de chaleur molaire *(f)*
It capacità termica molare *(f)*
Pt termo-capacidade molar *(f)*

molar heat capacity En
De molare Wärmekapazität *(f)*
Es capacidad de calor molar *(f)*
Fr capacité de chaleur molaire *(f)*
It capacità termica molare *(f)*
Pt termo-capacidade molar *(f)*

molar latent heat En
De molare gebundene Wärme *(f)*
Es calor latente molar *(m)*
Fr chaleur latente molaire *(f)*
It calore latente molare *(m)*
Pt calor latente molar *(m)*

molecular absorption En
De molekulare Absorption *(f)*
Es absorción molecular *(f)*
Fr absorption moléculaire *(f)*
It assorbimento molecolare *(m)*
Pt absorção molecular *(f)*

molekulare Absorption *(f)* De
En molecular absorption
Es absorción molecular *(f)*
Fr absorption moléculaire *(f)*
It assorbimento molecolare *(m)*
Pt absorção molecular *(f)*

molino de viento *(m)* Es
De Windmühle *(f)*
En windmill
Fr moulin à vent *(m)*
It mulino a vento *(m)*
Pt moínho de vento *(m)*

molino de viento de eje vertical *(m)* Es
De Windmühle mit Hochachse *(f)*
En vertical-axis windmill
Fr moulin à vent à axe vertical *(m)*
It mulino a vento a asse verticale *(m)*
Pt moínho de vento de eixo vertical *(m)*

molino hidráulico *(m)* Es
De Wassermühle *(f)*
En watermill
Fr moulin à eau *(m)*
It mulino ad acqua *(m)*
Pt moínho hidráulico *(m)*

Mollier steam diagram En
De H-S-Diagramm *(n)*
Es diagrama de vapor de Mollier *(m)*
Fr diagramme de Mollier *(m)*
It diagramma di Mollier *(m)*
Pt diagrama de vapor de Mollier *(m)*

moment *n* En; Fr *(m)*
De Moment *(n)*
Es momento *(m)*
It momento *(m)*
Pt momento *(m)*

Moment *(n) n* De
En moment
Es momento *(m)*
Fr moment *(m)*
It momento *(m)*
Pt momento *(m)*

momento *(m) n* Es, It, Pt
De Moment *(n)*
En moment
Fr moment *(m)*

moteur *(m) n* Fr
De Motor *(m)*
En engine; motor
Es motor *(m)*
It motore *(m)*
Pt motor *(m)*

moteur à air chaud *(m)* Fr
De Heiβluftmotor *(m)*
En hot-air engine
Es motor de aire caliente *(m)*
It motore ad aria calda *(m)*
Pt motor a ar quente *(m)*

moteur à combustion interne *(m)* Fr
De Verbrennungsmotor *(m)*
En internal-combustion engine
Es motor de combustión interna *(m)*
It motore a scoppio *(m)*
Pt motor de combustão interna *(m)*

moteur à deux temps *(m)* Fr
De Zweitaktmotor *(m)*
En two-stroke engine
Es motor de dos tiempos *(m)*
It motore a due tempi *(m)*
Pt motor de dois tempos *(m)*

moteur à essence *(m)* Fr
De Benzinmotor *(m)*
En petrol engine
Es motor de gasolina *(m)*
It motore a benzina *(m)*
Pt motor a gasolina *(m)*

moteur à gaz *(m)* Fr
De Gasmotor *(m)*
En gas engine
Es motor de gas *(m)*
It motore a gas *(m)*
Pt motor a gás *(m)*

moteur alternatif *(m)* Fr
De Kolbenmaschine *(f)*
En reciprocating engine
Es motor alternativo *(m)*

It motore a movimento alternativo *(m)*
Pt motor alternativo *(m)*

moteur à quatre temps *(m)* Fr
De Viertaktmotor *(m)*
En four-stroke engine
Es motor de cuatro tiempos *(m)*
It motore a quattro tempi *(m)*
Pt motor a quatro tempos *(m)*

moteur diesel *(m)* Fr
De Dieselmotor *(m)*
En diesel engine
Es motor diesel *(m)*
It motore diesel *(m)*
Pt motor diesel *(m)*

moteur d'induction *(m)* Fr
De Induktionsmotor *(m)*
En induction motor
Es motor de inducción *(m)*
It motore ad induzione *(m)*
Pt motor de indução *(m)*

moteur électrique *(m)* Fr
De Elektromotor *(m)*
En electric motor
Es motor eléctrico *(m)*
It motore elettrico *(m)*
Pt motor eléctrico *(m)*

moteur ionique *(m)* Fr
De Ionentriebwerk *(n)*
En ion engine
Es motor de iones *(m)*
It motore a ioni *(m)*
Pt motor a iões *(m)*

moteur polycarburant *(m)* Fr
De Doppeltreibstoff-motor *(m)*
En dual-fuel engine
Es motor de dos combustibles *(m)*
It motore a due combustibili *(m)*
Pt motor de dois combustíveis *(m)*

moteurs au fuel lourd *(m pl)* Fr
De Masut-Motoren *(pl)*
En heavy-fuel engines *(pl)*
Es motores de combustible pesado *(m pl)*
It motori a combustibile pesante *(m pl)*
Pt motores a óleos pesados *(m pl)*

moteur synchrone *(m)* Fr
De Synchronomotor *(m)*
En synchronous motor
Es motor síncrono *(m)*
It motore sincrono *(m)*
Pt motor sincronizado *(m)*

moteur thermique *(m)* Fr
De Wärmekraftmaschine *(f)*
En heat engine
Es motor térmico *(m)*
It motore termico *(m)*
Pt motor térmico *(m)*

moteur universel *(m)* Fr
De Universalmotor *(m)*
En universal motor
Es motor universal *(m)*
It motore universale *(m)*
Pt motor universal *(m)*

moteur Wankel *(m)* Fr
De Wankel-Motor *(m)*
En Wankel engine
Es motor Wankel *(m)*
It motore di Wankel *(m)*
Pt motor Wankel *(m)*

moto-gerador *(m) n* Pt
De Motor-Generator *(m)*
En motor-generator
Es motor-generador *(m)*
Fr convertisseur *(m)*
It gruppo motore-dinamo *(m)*

motor *(m) n* Es, Pt
De Motor *(m)*
En engine; motor
Fr moteur *(m)*
It motore *(m)*

motor *n* En
De Motor *(m)*
Es motor *(m)*
Fr moteur *(m)*
It motore *(m)*
Pt motor *(m)*

Motor *(m) n* De
En engine; motor
Es motor *(m)*
Fr moteur *(m)*
It motore *(m)*
Pt motor *(m)*

motor a ar quente *(m)* Pt
De Heiβluftmotor *(m)*
En hot-air engine
Es motor de aire caliente *(m)*
Fr moteur à air chaud *(m)*
It motore ad aria calda *(m)*

motor a gás *(m)* Pt
De Gasmotor *(m)*
En gas engine
Es motor de gas *(m)*
Fr moteur à gaz *(m)*
It motore a gas *(m)*

motor a gasolina *(m)* Pt
De Benzinmotor *(m)*
En petrol engine
Es motor de gasolina *(m)*
Fr moteur à essence *(m)*
It motore a benzina *(m)*

motor a iões *(m)* Pt
De Ionentriebwerk *(n)*
En ion engine
Es motor de iones *(m)*
Fr moteur ionique *(m)*
It motore a ioni *(m)*

motor alternativo *(m)* Es, Pt
De Kolbenmaschine *(f)*
En reciprocating engine
Fr moteur alternatif *(m)*
It motore a movimento alternativo *(m)*

motor a quatro tempos *(m)* Pt
De Viertaktmotor *(m)*
En four-stroke engine
Es motor de cuatro tiempos *(m)*
Fr moteur à quatre temps *(m)*
It motore a quattro tempi *(m)*

motor a vapor *(m)* Pt
De Dampfmaschine *(f)*
En steam engine
Es motor de vapor *(m)*
Fr machine à vapeur *(f)*
It macchina a vapore *(f)*

motor de aire caliente *(m)* Es
De Heiβluftmotor *(m)*
En hot-air engine
Fr moteur à air chaud *(m)*
It motore ad aria calda *(m)*
Pt motor a ar quente *(m)*

motor de arranque *(m)* Es, Pt
De Starter *(m)*
En starter motor
Fr démarreur *(m)*
It avviatore *(m)*

motor de combustão interna *(m)* Pt
De Verbrennungsmotor *(m)*
En internal-combustion engine
Es motor de combustión interna *(m)*
Fr moteur à combustion interne *(m)*
It motore a scoppio *(m)*

motor de combustión interna *(m)* Es
De Verbrennungsmotor *(m)*
En internal-combustion engine
Fr moteur à combustion interne *(m)*
It motore a scoppio *(m)*
Pt motor de combustão interna *(m)*

motor de cuatro tiempos *(m)* Es
De Viertaktmotor *(m)*
En four-stroke engine
Fr moteur à quatre temps *(m)*
It motore a quattro tempi *(m)*
Pt motor a quatro tempos *(m)*

motor de dois combustíveis *(m)* Pt
De Doppeltreibstoff-motor *(m)*
En dual-fuel engine
Es motor de dos combustibles *(m)*
Fr moteur polycarburant *(m)*
It motore a due combustibili *(m)*

motor de dois tempos *(m)* Pt
De Zweitaktmotor *(m)*
En two-stroke engine
Es motor de dos tiempos *(m)*
Fr moteur à deux temps *(m)*
It motore a due tempi *(m)*

motor de dos combustibles *(m)* Es
De Doppeltreibstoff-motor *(m)*
En dual-fuel engine
Fr moteur polycarburant *(m)*
It motore a due combustibili *(m)*
Pt motor de dois combustíveis *(m)*

motor de dos tiempos *(m)* Es
De Zweitaktmotor *(m)*
En two-stroke engine
Fr moteur à deux temps *(m)*
It motore a due tempi *(m)*
Pt motor de dois tempos *(m)*

motor de gas *(m)* Es
De Gasmotor *(m)*
En gas engine
Fr moteur à gaz *(m)*
It motore a gas *(m)*
Pt motor a gás *(m)*

motor de gasolina *(m)* Es
De Benzinmotor *(m)*
En petrol engine
Fr moteur à essence *(m)*
It motore a benzina *(m)*
Pt motor a gasolina *(m)*

motor de indução *(m)* Pt
De Induktionsmotor *(m)*
En induction motor
Es motor de inducción *(m)*
Fr moteur d'induction *(m)*
It motore ad induzione *(m)*

motor de inducción *(m)* Es
De Induktionsmotor *(m)*
En induction motor
Fr moteur d'induction *(m)*
It motore ad induzione *(m)*
Pt motor de indução *(m)*

motor de iones *(m)* Es
De Ionentriebwerk *(n)*
En ion engine
Fr moteur ionique *(m)*
It motore a ioni *(m)*
Pt motor a iões *(m)*

motor de vapor *(m)* Es
De Dampfmaschine *(f)*
En steam engine
Fr machine à vapeur *(f)*
It macchina a vapore *(f)*
Pt motor a vapor *(m)*

motor diesel *(m)* Es
De Dieselmotor *(m)*
En diesel engine
Fr moteur diesel *(m)*
It motore diesel *(m)*
Pt motor diesel *(m)*

motor diesel *(m)* Pt
De Dieselmotor *(m)*
En diesel engine
Es motor diesel *(m)*
Fr moteur diesel *(m)*
It motore diesel *(m)*

motore *(m)* *n* It
De Motor *(m)*
En engine; motor
Es motor *(m)*
Fr moteur *(m)*
Pt motor *(m)*

motore a benzina *(m)* It
De Benzinmotor *(m)*
En petrol engine
Es motor de gasolina *(m)*
Fr moteur à essence *(m)*
Pt motor a gasolina *(m)*

motore ad aria calda *(m)* It
De Heißluftmotor *(m)*
En hot-air engine
Es motor de aire caliente *(m)*
Fr moteur à air chaud *(m)*
Pt motor a ar quente *(m)*

motore ad induzione *(m)* It
De Induktionsmotor *(m)*
En induction motor
Es motor de inducción *(m)*
Fr moteur d'induction *(m)*
Pt motor de indução *(m)*

motore a due combustibili *(m)* It
De Doppeltreibstoff-motor *(m)*
En dual-fuel engine
Es motor de dos combustibles *(m)*
Fr moteur polycarburant *(m)*
Pt motor de dois combustíveis *(m)*

motore a due tempi *(m)* It
De Zweitaktmotor *(m)*
En two-stroke engine
Es motor de dos tiempos *(m)*
Fr moteur à deux temps *(m)*
Pt motor de dois tempos *(m)*

motore a gas *(m)* It
De Gasmotor *(m)*
En gas engine
Es motor de gas *(m)*
Fr moteur à gaz *(m)*
Pt motor a gás *(m)*

motore a ioni *(m)* It
De Ionentriebwerk *(n)*
En ion engine
Es motor de iones *(m)*
Fr moteur ionique *(m)*
Pt motor a iões *(m)*

motore a movimento alternativo *(m)* It
De Kolbenmaschine *(f)*
En reciprocating engine
Es motor alternativo *(m)*
Fr moteur alternatif *(m)*
Pt motor alternativo *(m)*

motore a quattro tempi *(m)* It
De Viertaktmotor *(m)*
En four-stroke engine
Es motor de cuatro tiempos *(m)*
Fr moteur à quatre temps *(m)*
Pt motor a quatro tempos *(m)*

motore a scoppio *(m)* It
De Verbrennungsmotor *(m)*
En internal-combustion engine
Es motor de combustión interna *(m)*
Fr moteur à combustion interne *(m)*
Pt motor de combustão interna *(m)*

motore diesel *(m)* It
De Dieselmotor *(m)*
En diesel engine
Es motor diesel *(m)*
Fr moteur diesel *(m)*
Pt motor diesel *(m)*

motore di Wankel *(m)* It
De Wankel-Motor *(m)*
En Wankel engine
Es motor Wankel *(m)*
Fr moteur Wankel *(m)*
Pt motor Wankel *(m)*

motore elettrico *(m)* It
De Elektromotor *(m)*
En electric motor
Es motor eléctrico *(m)*
Fr moteur électrique *(m)*
Pt motor eléctrico *(m)*

motor eléctrico *(m)* Es, Pt
De Elektromotor *(m)*
En electric motor
Fr moteur électrique *(m)*
It motore elettrico *(m)*

motores a óleos pesados *(m pl)* Pt
De Masutmotoren *(pl)*
En heavy-fuel engines *(pl)*
Es motores de combustible pesado *(m pl)*
Fr moteurs au fuel lourd *(m pl)*
It motori a combustibile pesante *(m pl)*

motores de combustible pesado *(m pl)* Es
De Masutmotoren *(pl)*
En heavy-fuel engines *(pl)*
Fr moteurs au fuel lourd *(m pl)*
It motori a combustibile pesante *(m pl)*
Pt motores a óleos pesados *(m pl)*

motore sincrono *(m)* It
De Synchronomotor *(m)*
En synchronous motor
Es motor síncrono *(m)*
Fr moteur synchrone *(m)*
Pt motor sincronizado *(m)*

motore termico *(m)* It
De Wärmekraftmaschine *(f)*
En heat engine
Es motor térmico *(m)*
Fr moteur thermique *(m)*
Pt motor térmico *(m)*

motore universale *(m)* It
De Universalmotor *(m)*
En universal motor
Es motor universal *(m)*
Fr moteur universel *(m)*
Pt motor universal *(m)*

motor-generador *(m) n* Es
De Motor-Generator *(m)*
En motor-generator
Fr convertisseur *(m)*
It gruppo motore-dinamo *(m)*
Pt moto-gerador *(m)*

motor-generator *n* En
De Motor-Generator *(m)*
Es motor-generador *(m)*
Fr convertisseur *(m)*
It gruppo motore-dinamo *(m)*
Pt moto-gerador *(m)*

Motor-Generator *(m) n* De
En motor-generator
Es motor-generador *(m)*
Fr convertisseur *(m)*
It gruppo motore-dinamo *(m)*
Pt moto-gerador *(m)*

motori a combustibile pesante *(m pl)* It
De Masutmotoren *(pl)*
En heavy-fuel engines *(pl)*
Es motores de combustibile pesante *(m pl)*
Fr moteurs au fuel lourd *(m pl)*
Pt motores a óleos pesados *(m pl)*

motor sincronizado *(m)* Pt
De Synchronomotor *(m)*
En synchronous motor
Es motor síncrono *(m)*
Fr moteur synchrone *(m)*
It motore sincrono *(m)*

motor síncrono *(m)* Es
De Synchronomotor *(m)*
En synchronous motor
Fr moteur synchrone *(m)*
It motore sincrono *(m)*
Pt motor sincronizado *(m)*

motor térmico *(m)* Es, Pt
De Wärmekraftmaschine *(f)*
En heat engine
Fr moteur thermique *(m)*
It motore termico *(m)*

motor universal *(m)* Es, Pt
De Universalmotor *(m)*
En universal motor
Fr moteur universel *(m)*
It motore universale *(m)*

motor Wankel *(m)* Es, Pt
De Wankel-Motor *(m)*
En Wankel engine
Fr moteur Wankel *(m)*
It motore di Wankel *(m)*

moulin à eau *(m)* Fr
De Wassermühle *(f)*
En watermill
Es molino hidráulico *(m)*
It mulino ad acqua *(m)*
Pt moínho hidráulico *(m)*

moulin à vent *(m)* Fr
De Windmühle *(f)*
En windmill
Es molino de viento *(m)*
It mulino a vento *(m)*
Pt moínho de vento *(m)*

moulin à vent à axe vertical *(m)* Fr
De Windmühle mit Hochachse *(f)*
En vertical-axis windmill
Es molino de viento de eje vertical *(m)*
It mulino a vento a asse verticale *(m)*
Pt moínho de vento de eixo vertical *(m)*

mousse urée et formaldéhyde *(f)* Fr
De Harnstoff-Formaldehydschaumstoff *(m)*
En urea-formaldehyde foam
Es espuma de urea-formaldehido *(f)*
It espanso di urea-formaldeide *(m)*
Pt espuma de ureia-formaldeído *(f)*

moving-coil En
De Drehspule-
Es bobina móvil
Fr bobine mobile
It bobina mobile
Pt bobina móvel

moving-iron En
De Dreheisen-
Es núcleo giratorio
Fr fer mobile
It ferro rotante
Pt ímã móvel

moyen *(m) n* Fr
De Durchschnitt *(m)*
En mean
Es media *(f)*
It medio *(f)*
Pt médio *(m)*

moyenne pondérée *(f)* Fr
De gewogener Durchschnitt *(m)*
En weighted mean
Es media compensada *(f)*
It media ponderata *(f)*
Pt média ponderada *(f)*

Muffelofen *(m) n* De
En muffle furnace
Es horno de mufla *(m)*
Fr four à moufle *(m)*
It forno a muffola *(m)*
Pt forno com câmara *(m)*

muffle furnace En
De Muffelofen *(m)*
Es horno de mufla *(m)*
Fr four à moufle *(m)*
It forno a muffola *(m)*
Pt forno com câmara *(m)*

mulino ad acqua *(m)* It
De Wassermühle *(f)*
En watermill
Es molino hidráulico *(m)*
Fr moulin à eau *(m)*
Pt moínho hidráulico *(m)*

mulino a vento *(m)* It
De Windmühle *(f)*
En windmill
Es molino de viento *(m)*
Fr moulin à vent *(m)*
Pt moínho de vento *(m)*

mulino a vento a asse verticale *(m)* It
De Windmühle mit Hochachse *(f)*
En vertical-axis windmill
Es molino de viento de eje vertical *(m)*
Fr moulin à vent à axe vertical *(m)*
Pt moínho de vento de eixo vertical *(m)*

multiple-effect distillation En
De Mehrfacheffekt-Destillation *(f)*
Es destilación de efecto múltiple *(f)*
Fr distillation à effet multiple *(f)*
It distillazione ad effetto multiplo *(f)*
Pt destilação de efeito múltiplo *(f)*

multiple flues *(pl)* En
De Sammel-Abgasleitungen *(pl)*
Es conductos de humo múltiples *(m pl)*
Fr carneaux multiples *(m pl)*
It condotti multipli *(m pl)*
Pt fumeiros múltiplos *(m pl)*

Mündungsplatte *(f) n* De
En orifice plate
Es placa de orificio *(f)*
Fr plaque perforée *(f)*
It placca di orifizio *(f)*
Pt placa de orifício *(f)*

muon *n* En; Fr, It *(m)*
De Muon *(n)*
Es muón *(m)*
Pt múon *(m)*

Muon *(n) n* De
En muon
Es muón *(m)*
Fr muon *(m)*
It muon *(m)*
Pt múon *(m)*

muón *(m) n* Es
De Muon *(n)*
En muon
Fr muon *(m)*
It muon *(m)*
Pt múon *(m)*

múon *(m) n* Pt
De Muon *(n)*
En muon
Es muón *(m)*
Fr muon *(m)*
It muon *(m)*

mutual inductance En
De Gegeninduktivität *(f)*
Es inductancia mutua *(f)*
Fr inductance mutuelle *(f)*
It induttanza mutua *(f)*
Pt indutância mútua *(f)*

N

nachbelüfteter Brenner *(m)* De
En post-aerated burner
Es quemador posaireado *(m)*
Fr brûleur à post-aération *(m)*
It bruciatore post-aerato *(m)*
Pt queimador pós-arejado *(m)*

Nachbrenner *(m) n* De
En reheater
Es recalentador *(m)*
Fr réchauffeur *(m)*
It ricombustore *(m)*
Pt reaquecedor *(m)*

Nachbrennofen *(m) n* De
En reheating furnace
Es horno de recalentamiento *(m)*
Fr four à réchauffer *(m)*
It forno di riscaldo *(m)*
Pt forno reaquecedor *(m)*

Nachtstrahlungs-kühlung *(f) n* De
En night-radiation cooling
Es refrigeración de radiación nocturna *(f)*
Fr refroidissement par rayonnement nocturne *(f)*
It raffreddamento a radiazioni notturne *(m)*
Pt arrefecimento por radiação nocturna *(m)*

Nachtstrom *(m) n* De
En night-generated electricity
Es electricidad de generación nocturna *(f)*
Fr électricité produite de nuit *(f)*
It elettricità generata di notte *(f)*
Pt electricidade de geração nocturna *(f)*

nafta *(f) n* It
De Dieselöl; Treiböl *(n)*
En diesel oil; fuel oil
Es aceite diesel; fuel oil *(m)*
Fr combustible diesel; mazout *(m)*
Pt fuel oil; óleo diesel *(m)*

nafta *(f) n* Es, It, Pt
De Naphtha *(n)*
En naphtha
Fr naphta *(m)*

nafta distillata *(f)* It
De Destillatöl *(n)*
En distillate fuel oil
Es fuel-oil destilado *(m)*
Fr fuel-oil distillé *(m)*
Pt fuel-oil de destilados *(m)*

naftaleno *(m) n* Es, Pt
De Naphtalen *(n)*
En naphthalene
Fr naphtalène *(m)*
It naftalina *(f)*

naftalina *(f) n* It
De Naphthalen *(n)*
En naphthalene
Es naftaleno *(m)*
Fr naphtalène *(m)*
Pt naftaleno *(m)*

nafta per bruciatori *(f)* It
De Brenner-Treiböl *(n)*
En burner fuel-oil
Es fuel-oil para quemador *(m)*
Fr pétrole lampant *(m)*
Pt fuel-oil para queimadores *(m)*

naphta *(m) n* Fr
De Naphtha *(n)*
En naphtha
Es nafta *(f)*
It nafta *(f)*
Pt nafta *(f)*

naphtalène *(m) n* Fr
De Naphthalen *(n)*
En naphthalene
Es naftaleno *(m)*
It naftalina *(f)*
Pt naftaleno *(m)*

naphtha *n* En
De Naphtha *(n)*
Es nafta *(f)*
Fr naphta *(m)*
It nafta *(f)*
Pt nafta *(f)*

Naphtha *(n) n* De
En naphtha
Es nafta *(f)*
Fr naphta *(m)*
It nafta *(f)*
Pt nafta *(f)*

Naphthalen *(n) n* De
En naphthalene
Es naftaleno *(m)*
Fr naphtalène *(m)*
It naftalina *(f)*
Pt naftaleno *(m)*

naphthalene *n* En
De Naphthalen *(n)*
Es naftaleno *(m)*
Fr naphtalène *(m)*
It naftalina *(f)*
Pt naftaleno *(m)*

naphthenic crude oil En
De naphthenisches Rohöl *(n)*
Es crudo nafténico *(m)*
Fr brut à base naphténique *(m)*
It petrolio greggio naftenico *(m)*
Pt petróleo crú nafténico *(m)*

naphthenisches Rohöl *(n)* De
En naphthenic crude oil
Es crudo nafténico *(m)*
Fr brut à base naphténique *(m)*
It petrolio greggio naftenico *(m)*
Pt petróleo crü nafténico *(m)*

nappe d'huile *(f)* Fr
De Ölschlamm *(m)*
En oil slick

Es mancha de petróleo flotante *(f)*
It scia d'olio grezzo in mare *(f)*
Pt mancha de óleo *(f)*

Naβmesser *(m) n* De
En wet meter
Es medidor húmedo *(m)*
Fr compteur humide *(m)*
It contatore a liquido *(m)*
Pt medidor de humidade *(m)*

Naβreinigen *(n) n* De
En wet scrubbing
Es lavado húmedo *(m)*
Fr épuration humide *(f)*
It lavaggio *(m)*
Pt limpeza com lavagem *(f)*

nastro bimetallico *(m)* It
De Bimetallstreifen *(m)*
En bimetallic strip
Es tira bimetálica *(f)*
Fr bilame *(m)*
Pt tira bimetálica *(f)*

natural abundance En
De natürliche Isotopenhäufigkeit *(f)*
Es abundancia natural *(f)*
Fr abondance naturelle *(f)*
It abbondanza naturale *(f)*
Pt abundância natural *(f)*

natural convection En
De natürliche Konvektion *(f)*
Es convección natural *(f)*
Fr convection naturelle *(f)*
It convezione naturale *(f)*
Pt convecção natural *(f)*

natural-draft burner Am
De natürlicher Luftstrombrenner *(m)*
En natural-draught burner
Es quemador de tiro natural *(m)*
Fr brûleur à tirage naturel *(m)*
It bruciatore a corrente d'aria naturale *(m)*
Pt queimador de corrente de ar natural *(m)*

natural-draught burner En
Am natural-draft burner
De natürlicher Luftstrombrenner *(m)*
Es quemador de tiro natural *(m)*
Fr brûleur à tirage naturel *(m)*
It bruciatore a corrente d'aria naturale *(m)*
Pt queimador de corrente de ar natural *(m)*

natural gas En
De Erdgas *(n)*
Es gas natural *(m)*
Fr gaz naturel *(m)*
It gas naturale *(m)*
Pt gás natural *(m)*

natürliche Isotopenhäufigkeit *(f)* De
En natural abundance
Es abundancia natural *(f)*
Fr abondance naturelle *(f)*
It abbondanza naturale *(f)*
Pt abundância natural *(f)*

natürliche Konvektion *(f)* De
En natural convection
Es convección natural *(f)*
Fr convection naturelle *(f)*
It convezione naturale *(f)*
Pt convecção natural *(f)*

natürlicher Luftstrombrenner *(m)* De
Am natural-draft burner
En natural-draught burner
Es quemador de tiro natural *(m)*
Fr brûleur à tirage naturel *(m)*
It bruciatore a corrente d'aria naturale *(m)*
Pt queimador de corrente de ar natural *(m)*

nave del gas *(f)* It
De Gasschiff *(n)*
En gas ship
Es barco de gas *(m)*
Fr navire méthanier *(m)*
Pt barco de gás *(m)*

nave nucleare *(f)* It
De Kernkraftschiff *(n)*
En nuclear ship
Es barco nuclear *(m)*
Fr navire nucléaire *(m)*
Pt barco nuclear *(m)*

navire méthanier *(m)* Fr
De Gasschiff *(n)*
En gas ship
Es barco de gas *(m)*
It nave del gas *(f)*
Pt barco de gás *(m)*

navire nucléaire *(m)* Fr
De Kernkraftschiff *(n)*
En nuclear ship
Es barco nuclear *(m)*
It nave nucleare *(f)*
Pt barco nuclear *(m)*

Nebenschluβmaschine *(f) n* De
En shunt-wound machine
Es máquina excitada en derivación *(f)*
Fr machine shunt *(f)*
It macchina eccitata in derivazione *(f)*
Pt máquina enrolada em shunt *(f)*

nebulizzatore *(m) n* It
De Zerstäuber *(m)*
En atomizer
Es atomizador *(m)*
Fr atomiseur *(m)*
Pt atomizador *(m)*

nebulizzatore d'aria *(m)* It
De Luftzerstäuber *(m)*
En air atomizer
Es atomizador de aire *(m)*
Fr atomiseur d'air *(m)*
Pt atomizador de ar *(m)*

negative feedback En
De negative Rückkoppelung *(f)*
Es realimentación negativa *(f)*
Fr contre-réaction *(f)*
It controreazione *(f)*
Pt retôrno negativo *(m)*

negative Rückkoppelung *(f)* De
En negative feedback
Es realimentación negativa *(f)*
Fr contre-réaction *(f)*
It controreazione *(f)*
Pt retôrno negativo *(m)*

negro de carbón *(m)* Es
De Ruβschwarz *(n)*
En carbon black
Fr noir de carbone *(m)*
It nerofumo di gas *(m)*
Pt negro de carvão *(m)*

negro de carvão *(m)* Pt
De Ruβschwarz *(n)*
En carbon black
Es negro de carbón *(m)*
Fr noir de carbone *(m)*
It nerofumo di gas *(m)*

Neopentan *(n) n* De
En neopentane
Es neopentano *(m)*
Fr néopentane *(m)*
It neopentano *(m)*
Pt neopentano *(m)*

neopentane *n* En
De Neopentan *(n)*
Es neopentano *(m)*
Fr néopentane *(m)*
It neopentano *(m)*
Pt neopentano *(m)*

néopentane *(m) n* Fr
De Neopentan *(n)*
En neopentane
Es neopentano *(m)*
It neopentano *(m)*
Pt neopentano *(m)*

neopentano *(m) n* Es, It, Pt
De Neopentan *(n)*
En neopentane
Fr néopentane *(m)*

neptunio *(m)* *n* Es
De Neptunium *(n)*
En neptunium
Fr neptunium *(m)*
It nettunio *(m)*
Pt neptúnio *(m)*

neptúnio *(m)* *n* Pt
De Neptunium *(n)*
En neptunium
Es neptunio *(m)*
Fr neptunium *(m)*
It nettunio *(m)*

neptunium *n* En; Fr *(m)*
De Neptunium *(n)*
Es neptunio *(m)*
It nettunio *(m)*
Pt neptúnio *(m)*

Neptunium *(n)* *n* De
En neptunium
Es neptunio *(m)*
Fr neptunium *(m)*
It nettunio *(m)*
Pt neptúnio *(m)*

nerofumo *(m)* *n* It
De Ruβ *(m)*
En soot
Es hollín *(m)*
Fr suie *(f)*
Pt fulígem *(f)*

nerofumo di gas *(m)* It
De Ruβschwarz *(n)*
En carbon black
Es negro de carbón *(m)*
Fr noir de carbone *(m)*
Pt negro de carvão *(m)*

net *adj* En, Fr
De netto
Es neto
It netto
Pt líquido

neto *adj* Es
De netto
En net
Fr net
It netto
Pt líquido

netto *adj* De, It
En net
Es neto
Fr net
Pt líquido

nettunio *(m)* *n* It
De Neptunium *(n)*
En neptunium
Es neptunio *(m)*
Fr neptunium *(m)*
Pt neptúnio *(m)*

network *n* En
De Netz *(n)*
Es red *(f)*
Fr réseau *(m)*
It rete *(f)*
Pt rede *(f)*

Netz *(n)* *n* De
En network
Es red *(f)*
Fr réseau *(m)*
It rete *(f)*
Pt rede *(f)*

Netzstrom *(m)* *n* De
En mains supply
Es alimentación principal *(f)*
Fr alimentation secteur *(f)*
It alimentazione di rete *(f)*
Pt distribuição pela linha principal *(f)*

nuclear reactor En
De Kernreaktor *(m)*
Es reactor nuclear *(m)*
Fr réacteur nucléaire *(m)*
It reattore nucleare *(m)*
Pt reactor nuclear *(m)*

neumoconiosis *(f)* *n* Es
De Staublunge *(f)*
En pneumoconiosis
Fr pneumoconiose *(f)*
It pneumoconiosi *(f)*
Pt pneumoconiose *(f)*

Neutralisieren *(n)* *n* De
En caustic washing
Es lavado cáustico *(m)*
Fr lavage à la soude caustique *(m)*
It lavaggio caustico *(m)*
Pt lavagem caústica *(f)*

neutrão *(m)* Pt
De Neutron *(n)*
En neutron
Es neutrón *(m)*
Fr neutron *(m)*
It neutrone *(m)*

neutrão lento *(m)* Pt
De langsames Neutron *(n)*
En slow neutron
Es neutrón lento *(m)*
Fr neutron lent *(m)*
It neutrone lento *(m)*

neutrão rápido *(m)* Pt
De schnelles Neutron *(n)*
En fast neutron
Es neutrón rápido *(m)*
Fr neutron rapide *(m)*
It neutrone veloce *(m)*

neutrão retardado *(m)* Pt
De verzögertes Neutron *(n)*
En delayed neutron
Es neutrón diferido *(m)*
Fr neutron retardé *(m)*
It neutrone ritardato *(m)*

neutrão térmico *(m)* Pt
De thermisches Neutron *(n)*
En thermal neutron
Es neutrón térmico *(m)*
Fr neutron thermique *(m)*
It neutrone termico *(m)*

neutrino *n* En; Es, Fr, It, Pt *(m)*
De Neutrino *(m)*

Neutrino *(m)* *n* De
En neutrino
Es neutrino *(m)*
Fr neutrino *(m)*
It neutrino *(m)*
Pt neutrino *(m)*

neutron En; Fr *(m)*
De Neutron *(n)*
Es neutrón *(m)*
It neutrone *(m)*
Pt neutrão *(m)*

Neutron *(n)* De
En neutron
Es neutrón *(m)*
Fr neutron *(m)*
It neutrone *(m)*
Pt neutrão *(m)*

neutrón *(m)* Es
De Neutron *(n)*
En neutron
Fr neutron *(m)*
It neutrone *(m)*
Pt neutrão *(m)*

neutrón diferido *(m)* Es
De verzögertes Neutron *(n)*
En delayed neutron
Fr neutron retardé *(m)*
It neutrone ritardato *(m)*
Pt neutrão retardado *(m)*

neutrone *(m)* It
De Neutron *(n)*
En neutron
Es neutrón *(m)*
Fr neutron *(m)*
Pt neutrão *(m)*

neutrone lento *(m)* It
De langsames Neutron *(n)*
En slow neutron
Es neutrón lento *(m)*
Fr neutron lent *(m)*
Pt neutrão lento *(m)*

Neutronenabbremsung *(f)* *n* De
En thermalization
Es termalización *(f)*
Fr thermalisation *(f)*
It termalizzazione *(f)*
Pt termalização *(f)*

Neutronenfluβ *(m)* *n* De
En neutron flux
Es flujo de neutrones *(m)*
Fr flux de neutrons *(m)*
It flusso di neutroni *(m)*
Pt fluxo de neutrões *(m)*

Neutronenlebensdauer *(f)* *n* De
En neutron lifetime
Es vida de neutrón *(f)*
Fr durée de vie des neutrons *(f)*
It vita dei neutroni *(f)*
Pt vida de neutrão *(f)*

Neutronenquelle *(f)* *n* De
En neutron source
Es fuente de neutrones *(f)*
Fr source de neutrons *(f)*

It fonte di neutroni *(f)*
Pt fonte de neutrões *(f)*

Neutronentemperatur *(f) n* De
En neutron temperature
Es temperatura neutrónica *(f)*
Fr température neutronique *(f)*
It temperatura dei neutroni *(f)*
Pt temperatura de neutrão *(f)*

neutrone ritardato *(m)* It
De verzögertes Neutron *(n)*
En delayed neutron
Es neutrón diferido *(m)*
Fr neutron retardé *(m)*
Pt neutrão retardado *(m)*

neutrone termico *(m)* It
De thermisches Neutron *(n)*
En thermal neutron
Es neutrón térmico *(m)*
Fr neutron thermique *(m)*
Pt neutrão térmico *(m)*

neutrone veloce *(m)* It
De schnelles Neutron *(n)*
En fast neutron
Es neutrón rápido *(m)*
Fr neutron rapide *(m)*
Pt neutrão rápido *(m)*

neutron flux En
De Neutronenfluß *(m)*
Es flujo de neutrones *(m)*
Fr flux de neutrons *(m)*
It flusso di neutroni *(m)*
Pt fluxo de neutrões *(m)*

neutron lent *(m)* Fr
De langsames Neutron *(n)*
En slow neutron
Es neutrón lento *(m)*
It neutrone lento *(m)*
Pt neutrão lento *(m)*

neutrón lento *(m)* Es
De langsames Neutron *(n)*
En slow neutron
Fr neutron lent *(m)*
It neutrone lento *(m)*
Pt neutrão lento *(m)*

neutron lifetime En
De Neutronenlebensdauer *(f)*
Es vida de neutrón *(f)*
Fr durée de vie des neutrons *(f)*
It vita dei neutroni *(f)*
Pt vida de neutrão *(f)*

neutron rapide *(m)* Fr
De schnelles Neutron *(n)*
En fast neutron
Es neutrón rápido *(m)*
It neutrone veloce *(m)*
Pt neutrão rápido *(m)*

neutrón rápido *(m)* Es
De schnelles Neutron *(n)*
En fast neutron
Fr neutron rapide *(m)*
It neutrone veloce *(m)*
Pt neutrão rápido *(m)*

neutron retardé *(m)* Fr
De verzögertes Neutron *(n)*
En delayed neutron
Es neutrón diferido *(m)*
It neutrone ritardato *(m)*
Pt neutrão retardado *(m)*

neutron source En
De Neutronenquelle *(f)*
Es fuente de neutrones *(f)*
Fr source de neutrons *(f)*
It fonte di neutroni *(f)*
Pt fonte de neutrões *(f)*

neutron temperature En
De Neutronentemperatur *(f)*
Es temperatura neutrónica *(f)*
Fr température neutronique *(f)*
It temperatura dei neutroni *(f)*
Pt temperatura de neutrão *(f)*

neutrón térmico *(m)* Es
De thermisches Neutron *(n)*
En thermal neutron
Fr neutron thermique *(m)*
It neutrone termico *(m)*
Pt neutrão térmico *(m)*

neutron thermique *(m)* Fr
De thermisches Neutron *(n)*
En thermal neutron
Es neutrón térmico *(m)*
It neutrone termico *(m)*
Pt neutrão térmico *(m)*

nicht anbackende Kohle *(f)* De
En noncaking coal
Es carbón inaglutinable *(m)*
Fr houille maigre *(f)*
It carbone non agglutinante *(m)*
Pt carvaõ não aglutinante *(m)*

nicht assoziiertes Erdgas *(n)* De
En nonassociated natural gas
Es gas natural no asociada *(m)*
Fr gaz naturel non associé *(m)*
It gas naturale non associato *(m)*
Pt gás natural não associada *(m)*

nicht erneuerbare Energie *(f)* De
En nonrenewable energy
Es energía no recuperable *(f)*
Fr énergie non renouvelable *(f)*
It energia non rinnovabile *(f)*
Pt energia não renovável *(f)*

nicht flüchtiger Kohlenstoff *(m)* De
En fixed carbon
Es carbón fijo *(m)*
Fr carbone fixe *(m)*
It carbone fisso *(m)*
Pt carvão fixo *(m)*

nicht klumpende Kohlen *(pl)* De
En nonagglomerating coals
Es carbones no aglomerantes *(m pl)*
Fr charbons non-agglomérants *(m pl)*
It carboni non agglomeranti *(m pl)*
Pt carvões não-aglomerantes *(m pl)*

nicht umkehrbar De
En irreversible
Es irreversible
Fr irréversible
It irreversibile
Pt irrevesível

nicht verfügbare Wärme *(f)* De
En unavailable heat
Es calor no disponible *(m)*
Fr chaleur non disponible *(f)*
It calore non disponibile *(m)*
Pt calor não disponível *(m)*

nickel-cadmium cell En
De Nickel-Kadmiumzelle *(f)*
Es pila de níquel-cadmio *(f)*
Fr pile au nickel-cadmium *(f)*
It cellula al nichel-cadmio *(f)*
Pt célula de níquel-cádmio *(f)*

Nickel-Eisenzelle *(f) n* De
En nickel-iron cell
Es pila de níquel-hierro *(f)*
Fr pile au nickel-fer *(f)*
It cellula al nichel-ferro *(f)*
Pt célula de níquel-ferro *(f)*

nickel-iron cell En
De Nickel-Eisenzelle *(f)*
Es pila de níquel-hierro *(f)*
Fr pile au nickel-fer *(f)*
It cellula al nichel-ferro *(f)*

Pt célula de níquel-ferro *(f)*

Nickel-Kadmiumzelle *(f)* *n* De
En nickel-cadmium cell
Es pila de níquel-cadmio *(f)*
Fr pile au nickel-cadmium *(f)*
It cellula al nichel-cadmio *(f)*
Pt célula de níquel-cádmio *(f)*

night-generated electricity En
De Nachtstrom *(m)*
Es electricidad de generación nocturna *(f)*
Fr électricité produite de nuit *(f)*
It elettricità generata di notte *(f)*
Pt electricidade de geração nocturna *(f)*

night-radiation cooling En
De Nachtstrahlungs-kühlung *(f)*
Es refrigeración de radiación nocturna *(f)*
Fr refroidissement par rayonnement nocturne *(f)*
It raffreddamento a radiazioni notturne *(m)*
Pt arrefecimento por radiação nocturna *(m)*

nitrogen *n* En
De Stickstoff *(m)*
Es nitrógeno *(m)*
Fr azote *(m)*
It azoto *(m)*
Pt azoto *(m)*

nitrógeno *(m)* *n* Es
De Stickstoff *(m)*
En nitrogen
Fr azote *(m)*
It azoto *(m)*
Pt azoto *(m)*

nitrogen oxides *(pl)* En
De Stickstoffoxide *(pl)*
Es óxidos de nitrógeno *(m pl)*
Fr oxydes d'azote *(m pl)*
It ossidi d'azoto *(m pl)*
Pt óxidos de azoto *(m pl)*

nitroglicerina *(f)* *n* Es, It, Pt
De Nitroglyzerin *(n)*
En nitroglycerine
Fr nitroglycérine *(f)*

nitroglycerine *n* En
De Nitroglyzerin *(n)*
Es nitroglicerina *(f)*
Fr nitroglycérine *(f)*
It nitroglicerina *(f)*
Pt nitroglicerina *(f)*

nitroglycérine *(f)* *n* Fr
De Nitroglyzerin *(n)*
En nitroglycerine
Es nitroglicerina *(f)*
It nitroglicerina *(f)*
Pt nitroglicerina *(f)*

Nitroglyzerin *(n)* *n* De
En nitroglycerine
Es nitroglicerina *(f)*
Fr nitroglycérine *(f)*
It nitroglicerina *(f)*
Pt nitroglicerina *(f)*

noir de carbone *(m)* Fr
De Rußschwarz *(n)*
En carbon black
Es negro de carbón *(m)*
It nerofumo di gas *(m)*
Pt negro de carvão *(m)*

noix de charbon *(f)* Fr
De Nußkohle *(f)*
En nut coal
Es almendrilla *(f)*
It carbone di pezzatura noce *(m)*
Pt carvão em nozes *(m)*

nombre atomique *(m)* Fr
De Atomnummer *(f)*
En atomic number
Es número atómico *(m)*
It numero atomico *(m)*
Pt número atómico *(m)*

nombre de Wobbe *(m)* Fr
De Wobbe-Zahl *(f)*
En Wobbe number
Es número Wobbe *(m)*
It numero di Wobbe *(m)*
Pt número de Wobbe *(m)*

nombre quantique *(m)* Fr
De Quantenzahl *(f)*
En quantum number
Es número cuántico *(m)*
It numero quantico *(m)*
Pt número quântico *(m)*

nomográfico *(m)* *n* Pt
De Nomograph *(m)*
En nomograph
Es nomograma *(m)*
Fr nomogramme *(m)*
It nomogramma *(m)*

nomograma *(m)* *n* Es
De Nomograph *(m)*
En nomograph
Fr nomogramme *(m)*
It nomogramma *(m)*
Pt nomográfico *(m)*

nomogramma *(m)* *n* It
De Nomograph *(m)*
En nomograph
Es nomograma *(m)*
Fr nomogramme *(m)*
Pt nomográfico *(m)*

nomogramme *(m)* *n* Fr
De Nomograph *(m)*
En nomograph
Es nomograma *(m)*
It nomogramma *(m)*
Pt nomográfico *(m)*

nomograph *n* En
De Nomograph *(m)*
Es nomograma *(m)*
Fr nomogramme *(m)*
It nomogramma *(m)*
Pt nomográfico *(m)*

Nomograph *(m)* *n* De
En nomograph
Es nomograma *(m)*
Fr nomogramme *(m)*
It nomogramma *(m)*
Pt nomográfico *(m)*

nonaerated flame En
De unbelüftete Flamme *(f)*
Es llama no aireada *(f)*
Fr flamme non aérée *(f)*
It fiamma non aerata *(f)*
Pt chama não arejada *(f)*

nonagglomerating coals En
De nicht klumpende Kohlen *(pl)*
Es carbones no aglomerantes *(m pl)*
Fr charbons non-agglomérants *(m pl)*
It carboni non agglomeranti *(m pl)*
Pt carvões não-aglomerantes *(m pl)*

nonassociated natural gas En
De nicht assoziiertes Erdgas *(n)*
Es gas natural no asociada *(m)*
Fr gaz naturel non associé *(m)*
It gas naturale non associato *(m)*
Pt gás natural não associada *(m)*

noncaking coal En
De nicht anbackende Kohle *(f)*
Es carbón inaglutinable *(m)*
Fr houille maigre *(f)*
It carbone non agglutinante *(m)*
Pt carvaõ não aglutinante *(m)*

nonrenewable energy En
De nicht erneuerbare Energie *(f)*
Es energía no recuperable *(f)*
Fr énergie non renouvelable *(f)*
It energia non rinnovabile *(f)*
Pt energia não renovável *(f)*

Nordseegas *(n)* *n* De
En North Sea gas
Es gas del Mar del Norte *(m)*
Fr gaz de Mer du Nord *(m)*
It gas del Mare del Nord *(m)*
Pt gás do Mar do Norte *(m)*

Nordseeöl *(n) n* De
En North Sea oil
Es petróleo del Mar del Norte *(m)*
Fr pétrole de Mer du Nord *(m)*
It petrolio del Mare del Nord *(m)*
Pt petróleo do Mar do Norte *(m)*

Normalkohle *(f)* De
En run-of-mine coal
Es carbón todouno *(m)*
Fr charbon tout-venant *(m)*
It carbone tout-venant *(m)*
Pt carvão tal como sai da mina *(m)*

North Sea gas En
De Nordseegas *(n)*
Es gas del Mar del Norte *(m)*
Fr gaz de Mer du Nord *(m)*
It gas del Mare del Nord *(m)*
Pt gás do Mar do Norte *(m)*

North Sea oil En
De Nordseeöl *(n)*
Es petróleo del Mar del Norte *(m)*
Fr pétrole de Mer du Nord *(m)*
It petrolio del Mare del Nord *(m)*
Pt petróleo do Mar do Norte *(m)*

noyau (atomique) *(m) n* Fr
De Kern (atomischer) *(m)*
En nucleus (atomic)
Es núcleo (atómico) *(m)*
It nucleo (atomico) *(m)*
Pt núcleo (atómico) *(m)*

noyau (magnétique) *(m) n* Fr
De Kern (magnetischer) *(m)*
En core (magnetic)
Es alma (magnética) *(f)*
It nucleo (magnetico) *(m)*
Pt núcleo (magnético) *(m)*

nuclear energy En
De Nuklearenergie *(f)*
Es energía nuclear *(f)*
Fr énergie nucléaire *(f)*
It energia nucleare *(f)*
Pt energia nuclear *(f)*

nuclear fission En
De Nuklearspaltung *(f)*
Es fisión nuclear *(f)*
Fr fission nucléaire *(f)*
It fissione nucleare *(f)*
Pt desintegração nuclear *(f)*

nuclear fusion En
De Nuklearfusion *(f)*
Es fusión nuclear *(f)*
Fr fusion nucléaire *(f)*
It fusione nucleare *(f)*
Pt fusão nuclear *(f)*

nuclear power station En
De Kernkraftwerk *(n)*
Es central nuclear *(f)*
Fr centrale nucléaire *(f)*
It centrale nucleare *(f)*
Pt central de energia nuclear *(f)*

nuclear reactor En
De Kernkraftreaktor *(m)*
Es reactor nuclear *(m)*
Fr réacteur nucléaire *(m)*
It reattore nucleare *(m)*
Pt reactor nuclear *(m)*

nuclear rocket En
De Kernkraftrakete *(f)*
Es cohete nuclear *(m)*
Fr fusée nucléaire *(f)*
It razzo nucleare *(m)*
Pt foguetão nuclear *(m)*

nuclear ship En
De Kernkraftschiff *(n)*
Es barco nuclear *(m)*
Fr navire nucléaire *(m)*
It nave nucleare *(f)*
Pt barco nuclear *(m)*

nuclear submarine En
De Kennkraft-U-Boot *(n)*
Es submarino nuclear *(m)*
Fr sous-marin nucléaire *(m)*
It sottomarino nucleare *(m)*
Pt submarino nuclear *(m)*

nucleido *(m) n* Es
De Nuklid *(n)*
En nuclide
Fr nuclide *(m)*
It nuclide *(m)*
Pt nucleto *(m)*

nucleo (atomico) *(m) n* It
De Kern (atomischer) *(m)*
En nucleus (atomic)
Es núcleo (atómico) *(m)*
Fr noyau (atomique) *(m)*
Pt núcleo (atómico) *(m)*

núcleo (atómico) *(m) n* Es, Pt
De Kern (atomischer) *(m)*
En nucleus (atomic)
Fr noyau (atomique) *(m)*
It nucleo (atomico) *(m)*

nucleo (magnetico) *(m) n* It
De Kern (magnetischer) *(m)*
En core (magnetic)
Es alma (magnética) *(f)*
Fr noyau (magnétique) *(m)*
Pt núcleo (magnético) *(m)*

núcleo (magnético) *(m) n* Pt
De Kern (magnetischer) *(m)*
En core (magnetic)
Es alma (magnética) *(f)*
Fr noyau (magnétique) *(m)*
It nucleo (magnetico) *(m)*

núcleo (reactor nuclear) *(m) n* Es, Pt
De Spaltzone (Kernkraftreaktor) *(f)*
En core (nuclear reactor)
Fr coeur (réacteur nucléaire) *(m)*
It cuore (reattore nucléaire) *(m)*

núcleo giratorio Es
De Dreheisen-
En moving-iron
Fr fer mobile
It ferro rotante
Pt ímã móvel

nucleto *(m) n* Pt
De Nuklid *(n)*
En nuclide
Es nucleido *(m)*
Fr nuclide *(m)*
It nuclide *(m)*

nucleus (atomic) *n* En
De Kern (atomischer) *(m)*
Es núcleo (atómico) *(m)*
Fr noyau (atomique) *(m)*
It nucleo (atomico) *(m)*
Pt núcleo (atómico) *(m)*

nuclide *n* En; Fr, It *(m)*
De Nuklid *(n)*
Es nucleido *(m)*
Pt nucleto *(m)*

Nuklearenergie *(f) n* De
En nuclear energy
Es energía nuclear *(f)*
Fr énergie nucléaire *(f)*
It energia nucleare *(f)*
Pt energia nuclear *(f)*

Nuklearfusion *(f) n* De
En nuclear fusion
Es fusión nuclear *(f)*
Fr fusion nucléaire *(f)*
It fusione nucleare *(f)*
Pt fusão nuclear *(f)*

Nuklearspaltung *(f) n* De
En nuclear fission
Es fisión nuclear *(f)*
Fr fission nucléaire *(f)*
It fissione nucleare *(f)*
Pt desintegração nuclear *(f)*

Nuklid *(n) n* De
En nuclide
Es nucleido *(m)*
Fr nuclide *(m)*
It nuclide *(m)*
Pt nucleto *(m)*

Nulleffektstrahlung *(f) n* De
En background radiation
Es radiación ambiente *(f)*
Fr rayonnement provoquant le mouvement propre *(m)*

It effetto di fondo *(m)*
Pt radiação de fundo *(f)*

numero atomico *(m)* It
De Atomnummer *(f)*
En atomic number
Es número atómico *(m)*
Fr nombre atomique *(m)*
Pt número atómico *(m)*

número atómico *(m)* Es, Pt
De Atomnummer *(f)*
En atomic number
Fr nombre atomique *(m)*
It numero atomico *(m)*

número cetânico *(m)* Pt
De Cetanzahl *(f)*
En cetane number
Es número cetano *(m)*
Fr indice de cétane *(m)*
It numero di cetano *(m)*

número cetano *(m)* Es
De Cetanzahl *(f)*
En cetane number
Fr indice de cétane *(m)*
It numero di cetano *(m)*
Pt número cetânico *(m)*

número cuántico *(m)* Es
De Quantenzahl *(f)*
En quantum number
Fr nombre quantique *(m)*
It numero quantico *(m)*
Pt número quântico *(m)*

número de Wobbe *(m)* Pt
De Wobbe-Zahl *(f)*
En Wobbe number
Es número Wobbe *(m)*
Fr nombre de Wobbe *(m)*
It numero di Wobbe *(m)*

numero di cetano *(m)* It
De Cetanzahl *(f)*
En cetane number
Es número cetano *(m)*
Fr indice de cétane *(m)*
Pt número cetânico *(m)*

numero di ottano *(m)* It
De Oktanzahl *(f)*
En octane number
Es número octano *(m)*
Fr indice d'octane *(m)*
Pt índice de octanas *(m)*

numero di Wobbe *(m)* It
De Wobbe-Zahl *(f)*
En Wobbe number
Es número Wobbe *(m)*
Fr nombre de Wobbe *(m)*
Pt número de Wobbe *(m)*

número octano *(m)* Es
De Oktanzahl *(f)*
En octane number
Fr indice d'octane *(m)*
It numero di ottano *(m)*
Pt índice de octanas *(m)*

numero quantico *(m)* It
De Quantenzahl *(f)*
En quantum number
Es número cuántico *(m)*
Fr nombre quantique *(m)*
Pt número quântico *(m)*

número quântico *(m)* Pt
De Quantenzahl *(f)*
En quantum number
Es número cuántico *(m)*
Fr nombre quantique *(m)*
It numero quantico *(m)*

número Wobbe *(m)* Es
De Wobbe-Zahl *(f)*
En Wobbe number
Fr nombre de Wobbe *(m)*
It numero di Wobbe *(m)*
Pt número de Wobbe *(m)*

Nußkohle *(f)* *n* De
En nut coal
Es almendrilla *(f)*
Fr noix de charbon *(f)*
It carbone di pezzatura noce *(m)*
Pt carvão em nozes *(m)*

nut coal En
De Nußkohle *(f)*
Es almendrilla *(f)*
Fr noix de charbon *(f)*
It carbone di pezzatura noce *(m)*
Pt carvão em nozes *(m)*

O

Objektiv *(n)* *n* De
En lens
Es lente *(f)*
Fr lentille *(f)*
It lente *(f)*
Pt lente *(f)*

ocean thermal energy En
De Meereswärme-energie *(f)*
Es energía térmica oceánica *(f)*
Fr énergie thermique des océans *(f)*
It energia termica oceanica *(f)*
Pt energia térmica oceânica *(f)*

ocean thermal gradient En
De Meereswärmegefälle *(n)*
Es gradiente térmico oceánico *(m)*
Fr gradient thermique des océans *(m)*
It gradiente termico oceanico *(m)*
Pt gradiente térmico oceânico *(m)*

octane number En
De Oktanzahl *(f)*
Es número octano *(m)*
Fr indice d'octane *(m)*
It numero di ottano *(m)*
Pt índice de octanas *(m)*

Ofen *(m)* *n* De
En furnace; oven
Es estufa *(f)*; horno *(m)*
Fr four *(m)*
It forno *(m)*
Pt forno *(m)*

offener Kreislauf *(m)* De
En open loop
Es bucle abierto *(m)*
Fr boucle ouverte *(f)*
It circuito aperto *(m)*
Pt circuito aberto *(m)*

offener Zyklus *(m)* De
En open cycle
Es ciclo abierto *(m)*
Fr circuit ouvert *(m)*
It ciclo aperto *(m)*
Pt ciclo aberto *(m)*

off-peak electricity En
De Strom aus der Schwachlastzeit *(m)*
Es electricidad fuera de horas punta *(f)*
Fr électricité hors-pointe *(f)*
It elettricità fuori di ore di punta *(f)*
Pt electricidade fora de ponta *(f)*

off-peak period En
De Schwachlastzeit *(f)*
Es período fuera de horas punta *(m)*
Fr période hors-pointe *(f)*
It periodo fuori di ore di punta *(m)*
Pt período fora de ponta *(m)*

offshore *adj* En
De auf See
Es marino
Fr au large
It al largo
Pt fora da praia

oil *n* En
De Öl *(n)*
Es petróleo *(m)*
Fr huile *(f)*
It olio *(m)*
Pt óleo *(m)*

oil ash En
De Ölasche *(f)*
Es escoria vanadosa de petróleo *(f)*
Fr cendre de pétrole *(f)*

It cenere di petrolio *(f)*
Pt cinza de óleo *(f)*

oil cartel En
De Ölkartell *(n)*
Es cartel de petróleo *(m)*
Fr cartel pétrolier *(m)*
It cartello del petrolio *(m)*
Pt cartel de petróleo *(m)*

oil-consuming country En
De ölverbrauchendes Land *(n)*
Es país consumidor de petróleo *(m)*
Fr pays consommateur de pétrole *(m)*
It paese consumatore di petrolio *(m)*
Pt país consumidor de petróleo *(m)*

oil consumption En
De Ölverbrauch *(m)*
Es consumo de petróleo *(m)*
Fr consommation de pétrole *(f)*
It consumo di petrolio *(m)*
Pt consumo de petróleo *(m)*

oil equivalent En
De Ölersatz *(m)*
Es petróleo equivalente *(m)*
Fr équivalent pétrole *(m)*
It equivalente del petrolio *(m)*
Pt equivalente de petróleo *(m)*

oilfield *n* En
De Ölfeld *(n)*
Es yacimiento de petróleo *(m)*
Fr gisement pétrolifère *(m)*
It giacimento petrolifero *(m)*
Pt jazigo de petróleo *(m)*

oil-filled radiator En
De ölgefüllter Heizkörper *(m)*
Es radiador relleno de petróleo *(m)*
Fr radiateur à circulation d'huile *(m)*
It radiatore riempito d'olio *(m)*
Pt radiador de óleo *(m)*

oil-fired *adj* En
De ölgefeuert
Es alimentado con petróleo
Fr chauffé au mazout
It alimentato a nafta
Pt alimentado a óleo

oil-fired central heating En
De Zentralheizung mit Ölfeuerung *(f)*
Es calefacción central con petróleo *(f)*
Fr chauffage central au mazout *(m)*
It riscaldamento centrale a nafta *(m)*
Pt aquecimento central a óleo *(m)*

oil-importing country En
De ölimportierendes Land *(n)*
Es país importador de petróleo *(m)*
Fr pays importateur d'huile *(m)*
It paese importatore di petrolio *(m)*
Pt país importador de petróleo *(m)*

oil-producing country En
De ölerzeugendes Land *(n)*
Es país productor de petróleo *(m)*
Fr pays producteur de pétrole *(m)*
It paese produttore di petrolio *(m)*
Pt país productor de petróleo *(m)*

oil refinery En
De Ölraffinerie *(f)*
Es refinería de petróleo *(f)*
Fr raffinerie de pétrole *(f)*
It raffineria di petrolio *(f)*
Pt refinaria de petróleo *(f)*

oilrig *n* En
De Ölbohrturm *(m)*
Es tren de perforación de petróleo *(m)*
Fr installation de forage *(f)*
It impianto di trivellazione per petrolio *(m)*
Pt plataforma petroleira *(f)*

oil shales *(pl)* En
De Ölschiefer *(pl)*
Es pizarras bituminosas *(f pl)*
Fr schistes bitumineux *(m pl)*
It schisti bituminosi *(m pl)*
Pt xistos betuminosos *(m pl)*

oil slick En
De Ölschlamm *(m)*
Es mancha de petróleo flotante *(f)*
Fr nappe d'huile *(f)*
It scia d'olio grezzo in mare *(f)*
Pt mancha de óleo *(f)*

oil tanker En
De Öltanker *(m)*
Es petrolero *(m)*
Fr pétrolier *(m)*
It petroliera *(f)*
Pt petroleiro *(m)*

oil well En
De Ölbohrung *(f)*
Es pozo petrolífero *(m)*
Fr puits de pétrole *(m)*
It pozzo petrolifero *(m)*
Pt poço de petróleo *(m)*

Oktanzahl *(f) n* De
En octane number
Es número octano *(m)*
Fr indice d'octane *(m)*
It numero di ottano *(m)*
Pt índice de octanas *(m)*

Öl *(n) n* De
En oil
Es petróleo *(m)*
Fr huile *(f)*
It olio *(m)*
Pt óleo *(m)*

Ölasche *(f) n* De
En oil ash
Es escoria vanadosa de petróleo *(f)*
Fr cendre de pétrole *(f)*
It cenere di petrolio *(f)*
Pt cinza de óleo *(f)*

Ölbohrturm *(m) n* De
En oilrig
Es tren de perforación de petróleo *(m)*
Fr installation de forage *(f)*
It impianto di trivellazione per petrolio *(m)*
Pt plataforma petroleira *(f)*

Ölbohrung *(f) n* De
En oil well
Es pozo petrolífero *(m)*
Fr puits de pétrole *(m)*
It pozzo petrolifero *(m)*
Pt poço de petróleo *(m)*

olefinas *(f pl)* Es, Pt
De Olefine *(pl)*
En olefins *n pl*
Fr oléfines *(m pl)*
It olefine *(f pl)*

olefine *(f pl)* It
De Olefine *(pl)*
En olefins *n pl*
Es olefinas *(f pl)*
Fr oléfines *(m pl)*
Pt olefinas *(f pl)*

Olefine *(pl)* De
En olefins *n pl*
Es olefinas *(f pl)*
Fr oléfines *(m pl)*
It olefine *(f pl)*
Pt olefinas *(f pl)*

oléfines *(m pl)* Fr
De Olefine *(pl)*
En olefins *n pl*
Es olefinas *(f pl)*
It olefine *(f pl)*
Pt olefinas *(f pl)*

olefins *(pl)* En
De Olefine *(pl)*
Es olefinas *(f pl)*
Fr oléfines *(m pl)*
It olefine *(f pl)*
Pt olefinas *(f pl)*

óleo *(m) n* Pt
De Öl *(n)*
En oil
Es petróleo *(m)*
Fr huile *(f)*
It olio *(m)*

óleo aromático crú *(m)* Pt
De aromatisches Rohöl *(n)*
En aromatic crude oil
Es crudo aromático *(m)*
Fr pétrole brut aromatique *(m)*
It petrolio greggio aromatico *(m)*

óleo crú *(m)* Pt
De Rohöl *(n)*
En crude oil
Es petróleo crudo *(m)*
Fr pétrole brut *(m)*
It petrolio grezzo *(m)*

óleo crú parafínico *(m)* Pt
De paraffinisches Rohöl *(n)*
En paraffinic crude oil
Es crudo parafínico *(m)*
Fr pétrole brut paraffinique *(m)*
It petrolio grezzo paraffinico *(m)*

óleo de lubrificação *(m)* Pt
De Schmieröl *(n)*
En lubricating oil
Es aceite lubricante *(m)*
Fr huile de graissage *(f)*
It olio lubrificante *(m)*

óleo diesel *(m)* Pt
De Dieselöl *(n)*
En diesel oil
Es aceite diesel *(m)*
Fr combustible diesel *(m)*
It nafta *(f)*

oléoduc *(m) n* Fr
De Rohrleitung *(f)*
En pipeline
Es oleoducto *(m)*
It tubazione *(f)*
Pt oleoduto *(m)*

oleoducto *(m) n* Es
De Rohrleitung *(f)*
En pipeline
Fr oléoduc *(m)*
It tubazione *(f)*
Pt oleoduto *(m)*

oleoduto *(m) n* Pt
De Rohrleitung *(f)*
En pipeline
Es oleoducto *(m)*
Fr oléoduc *(m)*
It tubazione *(f)*

oleoduto marítimo *(m)* Pt
De Seerohrleitung *(f)*
En marine pipeline
Es conducción submarina *(f)*
Fr pipeline marin *(m)*
It condotto marino *(m)*

óleo leve *(m)* Pt
De leichtes Öl *(n)*
En light oil
Es petróleo ligero *(m)*
Fr huile légère *(f)*
It olio leggero *(m)*

óleo para aquecimento *(m)* Pt
De Heizöl *(n)*
En heating oil
Es petróleo de calefacción *(m)*
Fr huile de chauffe *(f)*
It olio per riscaldamento *(m)*

Ölersatz *(m) n* De
En oil equivalent
Es petróleo equivalente *(m)*
Fr équivalent pétrole *(m)*
It equivalente del petrolio *(m)*
Pt equivalente de petróleo *(m)*

ölerzeugendes Land *(n)* De
En oil-producing country
Es país productor de petróleo *(m)*
Fr pays producteur de pétrole *(m)*
It paese produttore di petrolio *(m)*
Pt país productor de petróleo *(m)*

Ölfeld *(n) n* De
En oilfield
Es yacimiento de petróleo *(m)*
Fr gisement pétrolifère *(m)*
It giacimento petrolifero *(m)*
Pt jazigo de petróleo *(m)*

ölgefeuert *adj* De
En oil-fired
Es alimentado con petróleo
Fr chauffé au mazout
It alimentato a nafta
Pt alimentado a óleo

ölgefüllter Heizkörper *(m)* De
En oil-filled radiator
Es radiador relleno de petróleo *(m)*
Fr radiateur à circulation d'huile *(m)*
It radiatore riempito d'olio *(m)*
Pt radiador de óleo *(m)*

ölimportierendes Land *(n)* De
En oil-importing country
Es país importador de petróleo *(m)*
Fr pays importateur d'huile *(m)*
It paese importatore di petrolio *(m)*
Pt país importador de petróleo *(m)*

olio *(m) n* It
De Öl *(n)*
En oil
Es petróleo *(m)*
Fr huile *(f)*
Pt óleo *(m)*

olio cokificante *(m)* It
De Koksöl *(n)*
En coking oil
Es petróleo de coquización *(m)*
Fr huile cokéfiante *(f)*
Pt petróleo de coqueificação *(m)*

olio di schisto *(m)* It
De Schieferöl *(n)*
En shale oil
Es aceite de lutita *(m)*
Fr huile de schiste *(f)*
Pt petróleo xistoso *(m)*

olio leggero *(m)* It
De leichtes Öl *(n)*
En light oil
Es petróleo ligero *(m)*
Fr huile légère *(f)*
Pt óleo leve *(m)*

olio lubrificante *(m)* It
De Schmieröl *(n)*
En lubricating oil
Es aceite lubricante *(m)*
Fr huile de graissage *(f)*
Pt óleo de lubrificação *(m)*

olio per riscaldamento *(m)* It
De Heizöl *(n)*
En heating oil
Es petróleo de calefacción *(m)*
Fr huile de chauffe *(f)*
Pt óleo para aquecimento *(m)*

Ölkartell *(n) n* De
En oil cartel
Es cartel de petróleo *(m)*
Fr cartel pétrolier *(m)*
It cartello del petrolio *(m)*
Pt cartel de petróleo *(m)*

Ölraffinerie *(f) n* De
En oil refinery
Es refinería de petróleo *(f)*
Fr raffinerie de pétrole *(f)*
It raffineria di petrolio *(f)*
Pt refinaria de petróleo *(f)*

Ölsammelschiff *(n) n* De
En bulk oil-carrier
Es petrolero a granel *(m)*
Fr pétrolier vraquier *(m)*
It petroliera all'ingrosso *(f)*
Pt petroleiro a granel *(m)*

Ölsande *(pl)* De
En tar sands
Es arenas impregnadas de brea *(f pl)*

Fr sables asphaltiques (m pl)
It sabbie impregnate di catrame (f pl)
Pt areias com alcatrão (f pl)

Ölschiefer *(pl)* De
En oil shales *(pl)*
Es pizarras bituminosas *(f pl)*
Fr schistes bitumineux *(m pl)*
It schisti bituminosi *(m pl)*
Pt xistos betuminosos *(m pl)*

Ölschlamm *(m) n* De
En oil slick
Es mancha de petróleo flotante *(f)*
Fr nappe d'huile *(f)*
It scia d'olio grezzo in mare *(f)*
Pt mancha de óleo *(f)*

Öltanker *(m) n* De
En oil tanker
Es petrolero *(m)*
Fr pétrolier *(m)*
It petroliera *(f)*
Pt petroleiro *(m)*

Ölverbrauch *(m) n* De
En oil consumption
Es consumo de petróleo *(m)*
Fr consommation de pétrole *(f)*
It consumo di petrolio *(m)*
Pt consumo de petróleo *(m)*

ölverbrauchendes Land *(n)* De
En oil-consuming country
Es país consumidor de petróleo *(m)*
Fr pays consommateur de pétrole *(m)*
It paese consumatore di petrolio *(m)*
Pt país consumidor de petróleo *(m)*

on-board energy conversion En
De Energieumwandlung an Bord *(f)*
Es conversión de energía a bordo *(f)*
Fr transformation d'énergie à bord *(f)*
It conversione di energia a bordo *(f)*
Pt conversão de energia a bordo *(m)*

once-through boiler En
De Zwangsdurchlaufkessel *(m)*
Es caldera de proceso directo *(f)*
Fr chaudière sans recyclage *(f)*
It caldaia a processo diretto *(f)*
Pt caldeira de uma só passagem *(f)*

onda *(f) n* Es, It, Pt
De Welle *(f)*
En wave
Fr onde *(f)*

onda sinusoidal *(f)* Es, Pt
De Sinuswelle *(f)*
En sine wave
Fr onde sinusoïdale *(f)*
It sinusoide *(m)*

onde *(f) n* Fr
De Welle *(f)*
En wave
Es onda *(f)*
It onda *(f)*
Pt onda *(f)*

onde sinusoïdale *(f)* Fr
De Sinuswelle *(f)*
En sine wave
Es onda sinusoidal *(f)*
It sinusoide *(m)*
Pt onda sinusoidal *(f)*

open-cast mine En
De Tagebau *(m)*
Es mina a cielo abierto *(f)*
Fr mine à ciel ouvert *(f)*
It miniera a cielo aperto *(f)*
Pt mina aberta *(f)*

open cycle En
De offener Zyklus *(m)*
Es ciclo abierto *(m)*
Fr circuit ouvert *(m)*
It ciclo aperto *(m)*
Pt ciclo aberto *(m)*

open-flue heater En
De Heizelement mit offenem Abzug *(n)*
Es calentador de chimenea abierta *(m)*
Fr appareil de chauffage à conduit ouvert *(m)*
It riscaldatore a condotto aperto *(m)*
Pt aquecedor de fumeiro aberto *(m)*

open-hearth furnace En
De Siemens-Martinofen *(m)*
Es horno Martin-Siemens *(m)*
Fr four Martin *(m)*
It forno Martin *(m)*
Pt forno de fornalha aberta *(m)*

open loop En
De offener Kreislauf *(m)*
Es bucle abierto *(m)*
Fr boucle ouverte *(f)*
It circuito aperto *(m)*
Pt circuito aberto *(m)*

operação de carga parcial *(f)* Pt
De Teillastbetrieb *(m)*
En part-load operation
Es funcionamiento con carga reducida *(m)*
Fr fonctionnement sous charge partielle *(m)*
It funzionamento a carico parziale *(m)*

óptica de fibras *(f)* Pt
Am fiber optics
De Faseroptik *(f)*
En fibre optics
Es óptica de las fibras *(f)*
Fr optique des fibres *(f)*
It fibra ottica *(f)*

óptica de las fibras *(f)* Es
Am fiber optics
De Faseroptik *(f)*
En fibre optics
Fr optique des fibres *(f)*
It fibra ottica *(f)*
Pt óptica de fibras *(f)*

optical pyrometer En
De Strahlungspyrometer *(n)*
Es pirómetro óptico *(m)*
Fr pyromètre optique *(m)*
It pirometro ottico *(m)*
Pt pirómetro óptico *(m)*

Optimierung *(f) n* De
En optimization
Es optimización *(f)*
Fr optimisation *(f)*
It ottimizzazione *(f)*
Pt optimização *(f)*

optimisation *(f) n* Fr
De Optimierung *(f)*
En optimization
Es optimización *(f)*
It ottimizzazione *(f)*
Pt optimização *(f)*

optimização *(f) n* Pt
De Optimierung *(f)*
En optimization
Es optimización *(f)*
Fr optimisation *(f)*
It ottimizzazione *(f)*

optimización *(f) n* Es
De Optimierung *(f)*
En optimization
Fr optimisation *(f)*
It ottimizzazione *(f)*
Pt optimização *(f)*

optimization *n* En
De Optimierung *(f)*
Es optimización *(f)*
Fr optimisation *(f)*
It ottimizzazione *(f)*
Pt optimização *(f)*

optique des fibres *(f)* Fr
Am fiber optics
De Faseroptik *(f)*
En fibre optics
Es óptica de las fibras *(f)*
It fibra ottica *(f)*
Pt óptica de fibras *(f)*

orifice plate En
De Mündungsplatte *(f)*
Es placa de orificio *(f)*
Fr plaque perforée *(f)*
It placca di orifizio *(f)*
Pt placa de orifício *(f)*

Orsat method En
De Orsat-Verfahren *(n)*
Es método de Orsat *(m)*
Fr méthode d'Orsat *(m)*
It metodi di Orsat *(m)*
Pt método de Orsat *(m)*

Orsat-Verfahren *(n)* *n* De
En Orsat method
Es método de Orsat *(m)*
Fr méthode d'Orsat *(m)*
It metodi di Orsat *(m)*
Pt método de Orsat *(m)*

oscilloscope *n* En; Fr *(m)*
De Oszilloskop *(n)*
Es osciloscopio *(m)*
It oscilloscopio *(m)*
Pt osciloscópio *(m)*

oscilloscopio *(m)* *n* It
De Oszilloskop *(n)*
En oscilloscope
Es osciloscopio *(m)*
Fr oscilloscope *(m)*
Pt osciloscópio *(m)*

osciloscopio *(m)* *n* Es
De Oszilloskop *(n)*
En oscilloscope
Fr oscilloscope *(m)*
It oscilloscopio *(m)*
Pt osciloscópio *(m)*

osciloscópio *(m)* *n* Pt
De Oszilloskop *(n)*
En oscilloscope
Es osciloscopio *(m)*
Fr oscilloscope *(m)*
It oscilloscopio *(m)*

osmose *(f)* *n* Fr, Pt
De Osmose *(f)*
En osmosis
Es ósmosis *(f)*
It osmosi *(f)*

Osmose *(f)* *n* De
En osmosis
Es ósmosis *(f)*
Fr osmose *(f)*
It osmosi *(f)*
Pt osmose *(f)*

osmose inversa *(f)* Pt
De umgekehrte Osmose *(f)*
En reverse osmosis
Es ósmosis inversa *(f)*
Fr osmose inverse *(f)*
It osmosi inversa *(f)*

osmose inverse *(f)* Fr
De umgekehrte Osmose *(f)*
En reverse osmosis
Es ósmosis inversa *(f)*
It osmosi inversa *(f)*
Pt osmose inversa *(f)*

osmosi *(f)* *n* It
De Osmose *(f)*
En osmosis
Es ósmosis *(f)*
Fr osmose *(f)*
Pt osmose *(f)*

osmosi inversa *(f)* It
De umgekehrte Osmose *(f)*
En reverse osmosis
Es ósmosis inversa *(f)*
Fr osmose inverse *(f)*
Pt osmose inversa *(f)*

osmosis *n* En
De Osmose *(f)*
Es ósmosis *(f)*
Fr osmose *(f)*
It osmosi *(f)*
Pt osmose *(f)*

ósmosis *(f)* *n* Es
De Osmose *(f)*
En osmosis
Fr osmose *(f)*
It osmosi *(f)*
Pt osmose *(f)*

ósmosis inversa *(f)* Es
De umgekehrte Osmose *(f)*
En reverse osmosis
Fr osmose inverse *(f)*
It osmosi inversa *(f)*
Pt osmose inversa *(f)*

osmotic pressure En
De osmotischer Druck *(m)*
Es presión osmótica *(f)*
Fr pression osmotique *(f)*
It pressione osmotica *(f)*
Pt pressão osmótica *(f)*

osmotischer Druck *(m)* De
En osmotic pressure
Es presión osmótica *(f)*
Fr pression osmotique *(f)*
It pressione osmotica *(f)*
Pt pressão osmótica *(f)*

ossidante *(m)* *n* It
De Oxidationsmittel *(n)*
En oxidant
Es oxidante *(m)*
Fr oxydant *(m)*
Pt oxidante *(m)*

ossidazione *(f)* *n* It
De Oxidation *(f)*
En oxidation
Es oxidación *(f)*
Fr oxydation *(f)*
Pt oxidação *(f)*

ossidi d'azoto *(m pl)* It
De Stickstoffoxide *(pl)*
En nitrogen oxides
Es óxidos de nitrógeno *(m pl)*
Fr oxydes d'azote *(m pl)*
Pt óxidos de azoto *(m pl)*

ossido *(m)* *n* It
De Oxid *(n)*
En oxide
Es óxido *(m)*
Fr oxyde *(m)*
Pt óxido *(m)*

ossido di carbonio *(m)* It
De Kohlenmonoxid *(n)*
En carbon monoxide
Es óxido de carbono *(m)*
Fr oxyde de carbone *(m)*
Pt óxido de carbono *(m)*

ossigeno *(m)* *n* It
De Sauerstoff *(m)*
En oxygen
Es oxígeno *(m)*
Fr oxygène *(m)*
Pt oxigénio *(m)*

Oszilloskop *(n)* *n* De
En oscilloscope
Es osciloscopio *(m)*
Fr oscilloscope *(m)*
It oscilloscopio *(m)*
Pt osciloscópio *(m)*

ottimizzazione *(f)* *n* It
De Optimierung *(f)*
En optimization
Es optimización *(f)*
Fr optimisation *(f)*
Pt optimização *(f)*

Otto cycle En
De Otto-Zyklus *(m)*
Es ciclo de Otto *(m)*
Fr cycle d'Otto *(m)*
It ciclo di Otto *(m)*
Pt ciclo de Otto *(m)*

Otto-Zyklus *(m)* *n* De
En Otto cycle
Es ciclo de Otto *(m)*
Fr cycle d'Otto *(m)*
It ciclo di Otto *(m)*
Pt ciclo de Otto *(m)*

output *n* En
De Produktion *(f)*
Es salida *(f)*
Fr production *(f)*
It produzione *(f)*
Pt saída *(f)*

output energy En
De abgegebene Energie *(f)*
Es energía de salida *(f)*
Fr énergie produite *(f)*
It energia erogata *(f)*
Pt energia de saída *(f)*

oven *n* En
De Ofen *(m)*
Es estufa *(f)*
Fr four *(m)*
It forno *(m)*
Pt forno *(m)*

overall efficiency En
De Gesamtleistungs-fähigkeit *(f)*
Es rendimiento global *(m)*
Fr rendement global *(m)*
It rendimento totale *(m)*
Pt eficiência global *(f)*

overall generating capacity En
De Gesamterzeuger-ungskapazität *(f)*
Es capacidad generadora total *(f)*
Fr capacité génératrice globale *(f)*
It capacità totale di generazione *(f)*
Pt capacidade geradora global *(f)*

overload *n* En
De Überlastung *(f)*
Es sobrecarga *(f)*
Fr sucharge *(f)*

It sovraccarico *(m)*
Pt sobrecarga *(f)*

overvoltage *n* En
De Überspannung *(f)*
Es sobretensión *(f)*
Fr surtension *(f)*
It sovratensione *(f)*
Pt sobrevoltagem *(f)*

Oxid *(n) n* De
En oxide
Es óxido *(m)*
Fr oxyde *(m)*
It ossido *(m)*
Pt óxido *(m)*

oxidação *(f) n* Pt
De Oxidation *(f)*
En oxidation
Es oxidación *(f)*
Fr oxydation *(f)*
It ossidazione *(f)*

oxidación *(f) n* Es
De Oxidation *(f)*
En oxidation
Fr oxydation *(f)*
It ossidazione *(f)*
Pt oxidação *(f)*

oxidant *n* En
De Oxidationsmittel *(n)*
Es oxidante *(m)*
Fr oxydant *(m)*
It ossidante *(m)*
Pt oxidante *(m)*

oxidante *(m) n* Es, Pt
De Oxidationsmittel *(n)*
En oxidant
Fr oxydant *(m)*
It ossidante *(m)*

oxidation *n* En
De Oxidation *(f)*
Es oxidación *(f)*
Fr oxydation *(f)*
It ossidazione *(f)*
Pt oxidação *(f)*

Oxidation *(f) n* De
En oxidation
Es oxidación *(f)*
Fr oxydation *(f)*
It ossidazione *(f)*
Pt oxidação *(f)*

Oxidationsmittel *(n) n* De
En oxidant
Es oxidante *(m)*
Fr oxydant *(m)*
It ossidante *(m)*
Pt oxidante *(m)*

oxide *n* En
De Oxid *(n)*
Es óxido *(m)*
Fr oxyde *(m)*
It ossido *(m)*
Pt óxido *(m)*

óxido *(m) n* Es, Pt
De Oxid *(n)*
En oxide
Fr oxyde *(m)*
It ossido *(m)*

óxido de carbono *(m)* Es, Pt
De Kohlenmonoxid *(n)*
En carbon monoxide
Fr oxyde de carbone *(m)*
It ossido di carbonio *(m)*

óxidos de azoto *(m pl)* Pt
De Stickstoffoxide *(pl)*
En nitrogen oxides
Es óxidos de nitrógeno *(m pl)*
Fr oxydes d'azote *(m pl)*
It ossidi d'azoto *(m pl)*

óxidos de nitrógeno *(m pl)* Es
De Stickstoffoxide *(pl)*
En nitrogen oxides
Fr oxydes d'azote *(m pl)*
It ossidi d'azoto *(m pl)*
Pt óxidos de azoto *(m pl)*

oxigénio *(m) n* Pt
De Sauerstoff *(m)*
En oxygen
Es oxígeno *(m)*
Fr oxygène *(m)*
It ossigeno *(m)*

oxígeno *(m) n* Es
De Sauerstoff *(m)*
En oxygen
Fr oxygène *(m)*
It ossigeno *(m)*
Pt oxigénio *(m)*

oxydant *(m) n* Fr
De Oxidationsmittel *(n)*
En oxidant
Es oxidante *(m)*
It ossidante *(m)*
Pt oxidante *(m)*

oxydation *(f) n* Fr
De Oxidation *(f)*
En oxidation
Es oxidación *(f)*
It ossidazione *(f)*
Pt oxidação *(f)*

oxyde *(m) n* Fr
De Oxid *(n)*
En oxide
Es óxido *(m)*
It ossido *(m)*
Pt óxido *(m)*

oxyde de carbone *(m)* Fr
De Kohlenmonoxid *(n)*
En carbon monoxide
Es óxido de carbono *(m)*
It ossido di carbonio *(m)*
Pt óxido de carbono *(m)*

oxydes d'azote *(m pl)* Fr
De Stickstoffoxide *(pl)*
En nitrogen oxides
Es óxidos de nitrógeno *(m pl)*
It ossidi d'azoto *(m pl)*
Pt óxidos de azoto *(m pl)*

oxygen *n* En
De Sauerstoff *(m)*
Es oxígeno *(m)*
Fr oxygène *(m)*
It ossigeno *(m)*
Pt oxigénio *(m)*

oxygène *(m) n* Fr
De Sauerstoff *(m)*
En oxygen
Es oxígeno *(m)*
It ossigeno *(m)*
Pt oxigénio *(m)*

Ozokerit *(m) n* De
En mineral wax
Es cera mineral *(f)*
Fr cire minérale *(f)*
It cera minerale *(f)*
Pt cera mineral *(f)*

Ozon *(m) n* De
En ozone
Es ozono *(m)*
Fr ozone *(m)*
It ozono *(m)*
Pt ozónio *(m)*

ozone *n* En; Fr *(m)*
De Ozon *(m)*
Es ozono *(m)*
It ozono *(m)*
Pt ozónio *(m)*

ozónio *(m) n* Pt
De Ozon *(m)*
En ozone
Es ozono *(m)*
Fr ozone *(m)*
It ozono *(m)*

ozono *(m) n* Es, It
De Ozon *(m)*
En ozone
Fr ozone *(m)*
Pt ozónio *(m)*

ozonosfera *(f) n* Es, It, Pt
De Ozonschicht *(f)*
En ozonosphere
Fr ozonosphère *(f)*

ozonosphere *n* En
De Ozonschicht *(f)*
Es ozonosfera *(f)*
Fr ozonosphère *(f)*
It ozonosfera *(f)*
Pt ozonosfera *(f)*

ozonosphère *(f) n* Fr
De Ozonschicht *(f)*
En ozonosphere
Es ozonosfera *(f)*
It ozonosfera *(f)*
Pt ozonosfera *(f)*

Ozonschicht *(f) n* De
En ozonosphere
Es ozonosfera *(f)*
Fr ozonosphère *(f)*
It ozonosfera *(f)*
Pt ozonosfera *(f)*

P

paarweise Produktion *(f)* De
- En pair production
- Es producción por pares *(f)*
- Fr production en paire *(f)*
- It produzione di coppie *(f)*
- Pt produção a par *(f)*

paese consumatore di petrolio *(m)* It
- De ölverbrauchendes Land *(n)*
- En oil-consuming country
- Es país consumidor de petróleo *(m)*
- Fr pays consommateur de pétrole *(m)*
- Pt país consumidor de petróleo *(m)*

paese importatore di petrolio *(m)* It
- De ölimportierendes Land *(n)*
- En oil-importing country
- Es país importador de petróleo *(m)*
- Fr pays importateur d'huile *(m)*
- Pt país importador de petróleo *(m)*

paese produttore di petrolio *(m)* It
- De ölerzeugendes Land *(n)*
- En oil-producing country
- Es país productor de petróleo *(m)*
- Fr pays producteur de pétrole *(m)*
- Pt país productor de petróleo *(m)*

painel solar *(m)* Pt
- De Solarplatte *(f)*
- En solar panel
- Es panel solar *(m)*
- Fr panneau solaire *(m)*
- It pannello solare *(m)*

pair production En
- De paarweise Produktion *(f)*
- Es producción por pares *(f)*
- Fr production en paire *(f)*
- It produzione di coppie *(f)*
- Pt produção a par *(f)*

país consumidor de petróleo *(m)* Es
- De ölverbrauchendes Land *(n)*
- En oil-consuming country
- Fr pays consommateur de pétrole *(m)*
- It paese consumatore di petrolio *(m)*
- Pt país consumidor de petróleo *(m)*

país consumidor de petróleo *(m)* Pt
- De ölverbrauchendes Land *(n)*
- En oil-consuming country
- Es país consumidor de petróleo *(m)*
- Fr pays consommateur de pétrole *(m)*
- It paese consumatore di petrolio *(m)*

país importador de petróleo *(m)* Es
- De ölimportierendes Land *(n)*
- En oil-importing country
- Fr pays importateur d'huile *(m)*
- It paese importatore di petrolio *(m)*
- Pt país importador de petróleo *(m)*

país importador de petróleo *(m)* Pt
- De ölimportierendes Land *(n)*
- En oil-importing country
- Es país importador de petróleo *(m)*
- Fr pays importateur d'huile *(m)*
- It paese importatore di petrolio *(m)*

país productor de petróleo *(m)* Es, Pt
- De ölerzeugendes Land *(n)*
- En oil-producing country
- Fr pays producteur de pétrole *(m)*
- It paese produttore di petrolio *(m)*

palline di polistirolo espanso *(f)* It
- De geschäumte Polystyrolkugel *(f)*
- En expanded polystyrene pellet
- Es gránulo de poliestireno expandido *(m)*
- Fr granule de polystyrène expansé *(m)*
- Pt grânulo de polistireno expandido *(m)*

panel solar *(m)* Es
- De Solarplatte *(f)*
- En solar panel
- Fr panneau solaire *(m)*
- It pannello solare *(m)*
- Pt painel solar *(m)*

panneau solaire *(m)* Fr
- De Solarplatte *(f)*
- En solar panel
- Es panel solar *(m)*
- It pannello solare *(m)*
- Pt painel solar *(m)*

pannello solare *(m)* It
- De Solarplatte *(f)*
- En solar panel
- Es panel solar *(m)*
- Fr panneau solaire *(m)*
- Pt painel solar *(m)*

pantalla térmica *(f)* Es
- De Wärmeschutz *(m)*
- En heat shield; thermal shield
- Fr bouclier thermique *(m)*
- It schermo termico *(m)*
- Pt protector térmico *(m)*

parada *(f)* *n* Es
- De Abschalten *(n)*
- En shutdown
- Fr arrêt *(m)*
- It arresto *(m)*
- Pt paralização *(f)*

parada de emergencia *(f)* Es
- De Abschaltung *(f)*
- En scram (nuclear reactor)
- Fr arrêt d'urgence *(m)*
- It spegnimento immediato *(m)*
- Pt paragem de emergência *(f)*

paraffin *n* En
- De Paraffin *(n)*
- Es parafina *(f)*
- Fr paraffine *(f)*
- It paraffina *(f)*
- Pt parafina *(f)*

Paraffin *(n)* *n* De
- En paraffin
- Es parafina *(f)*
- Fr paraffine *(f)*
- It paraffina *(f)*
- Pt parafina *(f)*

paraffina *(f)* *n* It
- De Paraffin *(n)*
- En paraffin
- Es parafina *(f)*
- Fr paraffine *(f)*
- Pt parafina *(f)*

Paraffinausscheidungspunkt *(m)* *n* De
- En cloud point
- Es punto de opacidad *(m)*
- Fr point de trouble *(m)*
- It punto di intorbidimento *(m)*
- Pt ponto de révoa *(m)*

paraffine *(f)* *n* Fr
- De Paraffin *(n)*
- En paraffin
- Es parafina *(f)*
- It paraffina *(f)*
- Pt parafina *(f)*

paraffinic crude oil En
- De paraffinisches Rohöl *(n)*
- Es crudo parafínico *(m)*
- Fr pétrole brut paraffinique *(m)*
- It petrolio grezzo paraffinico *(m)*
- Pt óleo crú parafínico *(m)*

paraffinisches Rohöl *(n)* De
En paraffinic crude oil
Es crudo parafínico *(m)*
Fr pétrole brut paraffinique *(m)*
It petrolio grezzo paraffinico *(m)*
Pt óleo crú parafínico *(m)*

parafiamma *(m) n* It
De Prallplatte *(f)*
En baffle
Es deflector *(m)*
Fr déflecteur *(m)*
Pt deflector *(m)*

parafina *(f) n* Es, Pt
De Paraffin *(n)*
En paraffin
Fr paraffine *(f)*
It paraffina *(f)*

paragem de emergência *(f)* Pt
De Abschaltung *(f)*
En scram (nuclear reactor)
Es parada de emergencia *(f)*
Fr arrêt d'urgence *(m)*
It spegnimento immediato *(m)*

paralização *(f) n* Pt
De Abschalten *(n)*
En shutdown
Es parada *(f)*
Fr arrêt *(m)*
It arresto *(m)*

partial oxidation gasification En
De partielle Oxidationsvergasung *(f)*
Es gasificación don oxidación parcial *(f)*
Fr gazéification à oxydation partielle *(f)*
It gassificazione con ossidazione parziale *(f)*
Pt gaseificação por oxidação parcial *(f)*

particella alfa *(f)* It
De Alpha-Teilchen *(n)*
En alpha particle
Es partícula alfa *(f)*
Fr particule alpha *(f)*
Pt partícula alfa *(f)*

particella beta *(f)* It
De Beta-Partikel *(f)*
En beta particle
Es partícula beta *(f)*
Fr particule bêta *(f)*
Pt partícula beta *(f)*

partícula alfa *(f)* Es, Pt
De Alpha-Teilchen *(n)*
En alpha particle
Fr particule alpha *(f)*
It particella alfa *(f)*

partícula beta *(f)* Es, Pt
De Beta-Partikel *(f)*
En beta particle
Fr particule bêta *(f)*
It particella beta *(f)*

particulaire *adj* Fr
De Partikel-
En particulate
Es de macropartículas
It delle macroparticelle
Pt em partículas

particulate *adj* En
De Partikel-
Es de macropartículas
Fr particulaire
It delle macroparticelle
Pt em partículas

particulate filter En
De Partikelfilter *(m)*
Es filtro de macropartículas *(m)*
Fr filtre particulaire *(m)*
It filtro delle macroparticelle *(m)*
Pt filtro de partículas *(m)*

particule alpha *(f)* Fr
De Alpha-Teilchen *(n)*
En alpha particle
Es partícula alfa *(f)*
It particella alfa *(f)*
Pt partícula alfa *(f)*

particule bêta *(f)* Fr
De Beta-Partikel *(f)*
En beta particle
Es partícula beta *(f)*
It particella beta *(f)*
Pt partícula beta *(f)*

partielle Oxidationsvergasung *(f)* De
En partial oxidation gasification
Es gasificación don oxidación parcial *(f)*
Fr gazéification à oxydation partielle *(f)*
It gassificazione con ossidazione parziale *(f)*
Pt gaseificação por oxidação parcial *(f)*

Partikel- De
En particulate
Es de macropartículas
Fr particulaire
It delle macroparticelle
Pt em partículas

Partikelfilter *(m) n* De
En particulate filter
Es filtro de macropartículas *(m)*
Fr filtre particulaire *(m)*
It filtro delle macroparticelle *(m)*
Pt filtro de partículas *(m)*

partitore di tensione *(m)* It
De Spannungsteiler *(m)*
En potential divider; voltage divider
Es divisor de potencial; divisor de tensión *(m)*
Fr diviseur de tension *(m)*
Pt divisor de potencial; divisor de tensão *(m)*

part-load operation En
De Teillastbetrieb *(m)*
Es funcionamiento con carga reducida *(m)*
Fr fonctionnement sous charge partielle *(m)*
It funzionamento a carico parziale *(m)*
Pt operação de carga parcial *(f)*

passive sunlight reflector En
De Sonnenlicht-Umlenkspiegel *(m)*
Es reflector solar pasivo *(m)*
Fr réflecteur solaire passif *(m)*
It riflettore passivo della luce del sole *(m)*
Pt reflector de raios solares passivo *(m)*

payback period En
De Rückzahldauer *(f)*
Es período de reembolso *(m)*
Fr durée de remboursement *(f)*
It periodo di remborso *(m)*
Pt período de reembolso *(m)*

pays consommateur de pétrole *(m)* Fr
De ölverbrauchendes Land *(n)*
En oil-consuming country
Es país consumidor de petróleo *(m)*
It paese consumatore di petrolio *(m)*
Pt país consumidor de petróleo *(m)*

pays importateur d'huile *(m)* Fr
De ölimportierendes Land *(n)*
En oil-importing country
Es país importador de petróleo *(m)*
It paese importatore di petrolio *(m)*
Pt país importador de petróleo *(m)*

pays producteur de pétrole *(m)* Fr
De ölerzeugendes Land *(n)*
En oil-producing country
Es país productor de petróleo *(m)*
It paese produttore di petrolio *(m)*
Pt país productor de petróleo *(m)*

peak demand En
De Spitzenbedarf *(m)*
Es demanda punta *(f)*
Fr consommation de pointe *(f)*
It domanda di punta *(f)*
Pt procura de ponta *(f)*

peaking power En
De Spitzenkraft *(f)*
Es potencia máxima *(f)*
Fr puissance de crête *(f)*
It potenza massima *(f)*
Pt potência de ponta *(f)*

peak load En
De Spitzenbelastung *(f)*
Es carga de punta *(f)*
Fr charge de pointe *(f)*
It carico di punta *(m)*
Pt carga de ponta *(f)*

peat *n* En
De Torf *(m)*
Es turba *(f)*
Fr tourbe *(f)*
It torba *(f)*
Pt turfa *(f)*

peatification *n* En
De Vertorfung *(f)*
Es formación de turba *(f)*
Fr formation de la tourbe *(f)*
It produzione di torba *(f)*
Pt turfização *(f)*

pece *(f) n* It
De Pech *(n)*
En pitch
Es brea *(f)*
Fr brai *(m)*
Pt pez *(f)*

Pech *(n) n* De
En pitch
Es brea *(f)*
Fr brai *(m)*
It pece *(f)*
Pt pez *(f)*

Pechkoks *(m) n* De
En pitch coke
Es coque con ligante *(m)*
Fr coke de brai *(m)*
It coke di pece *(m)*
Pt coque de pez *(m)*

Pentan *(n) n* De
En pentane
Es pentano *(m)*
Fr pentane *(m)*
It pentano *(m)*
Pt pentano *(m)*

pentane *n* En; Fr *(m)*
De Pentan *(n)*
Es pentano *(m)*
It pentano *(m)*
Pt pentano *(m)*

pentano *(m) n* Es, It, Pt
De Pentan *(n)*
En pentane
Fr pentane *(m)*

per capita energy consumption En
De Stromverbrauch pro Kopf *(m)*
Es consumo de energía per capita *(m)*
Fr consommation d'énergie par personne *(f)*
It consumo di energia pro capite *(m)*
Pt consumo de energia per capita *(m)*

per capita energy production En
De Stromerzeugung pro Kopf *(f)*
Es producción de energía per capita *(f)*
Fr production d'énergie par personne *(f)*
It produzione di energia pro capite *(f)*
Pt produção de energia per capita *(f)*

perda de calor *(f)* Pt
De Wärmeverlust *(m)*
En heat loss
Es pérdida de calor *(f)*
Fr perte de chaleur *(f)*
It perdita di calore *(f)*

perda dieléctrica *(f)* Pt
De dielektrischer Verlust *(m)*
En dielectric loss
Es pérdida dieléctrica *(f)*
Fr perte diélectrique *(f)*
It perdita dielettrica *(f)*

perda no cobre *(f)* Pt
De Kupferverlust *(m)*
En copper loss
Es pérdida en el cobre *(f)*
Fr perte dans le cuivre *(f)*
It perdita nel rame *(f)*

perda por fricção do vento *(f)* Pt
De Luftreibungsverlust *(m)*
En windage loss
Es pérdida por resistencia aerodinámica *(f)*
Fr perte due au jeu *(f)*
It perdita per resistenza aerodinamica *(f)*

perda por histerese *(f)* Pt
De Hystereseverlust *(m)*
En hysteresis loss
Es pérdida por histéresis *(f)*
Fr perte par hystérésis *(f)*
It perdita per isteresi *(f)*

perda por transmissão *(f)* Pt
De Durchgangsdämpfung *(f)*
En transmission loss
Es pérdida por transmisión *(f)*
Fr perte par transmission *(f)*
It perdita per trasmissione *(f)*

pérdida de calor *(f)* Es
De Wärmeverlust *(m)*
En heat loss
Fr perte de chaleur *(f)*
It perdita di calore *(f)*
Pt perda de calor *(f)*

pérdida dieléctrica *(f)* Es
De dielektrischer Verlust *(m)*
En dielectric loss
Fr perte diélectrique *(f)*
It perdita dielettrica *(f)*
Pt perda dieléctrica *(f)*

pérdida en el cobre *(f)* Es
De Kupferverlust *(m)*
En copper loss
Fr perte dans le cuivre *(f)*
It perdita nel rame *(f)*
Pt perda no cobre *(f)*

pérdida por histéresis *(f)* Es
De Hystereseverlust *(m)*
En hysteresis loss
Fr perte par hystérésis *(f)*
It perdita per isteresi *(f)*
Pt perda por histerese *(f)*

pérdida por resistencia aerodinámica *(f)* Es
De Luftreibungsverlust *(m)*
En windage loss
Fr perte due au jeu *(f)*
It perdita per resistenza aerodinamica *(f)*
Pt perda por fricção do vento *(f)*

pérdida por transmisión *(f)* Es
De Durchgangsdämpfung *(f)*
En transmission loss
Fr perte par transmission *(f)*
It perdita per trasmissione *(f)*
Pt perda por transmissão *(f)*

perdita *(f) n* It
De Leck *(n)*
En leakage
Es fuga *(f)*
Fr fuite *(f)*
Pt fuga *(f)*

perdita di calore *(f)* It
De Wärmeverlust *(m)*
En heat loss
Es pérdida de calor *(f)*
Fr perte de chaleur *(f)*
Pt perda de calor *(f)*

perdita dielettrica *(f)* It
De dielektrischer Verlust *(m)*
En dielectric loss
Es pérdida dieléctrica *(f)*
Fr perte diélectrique *(f)*
Pt perda dieléctrica *(f)*

perdita nel rame *(f)* It
De Kupferverlust *(m)*
En copper loss
Es pérdida en el cobre *(f)*
Fr perte dans le cuivre *(f)*
Pt perda no cobre *(f)*

perdita per isteresi *(f)* It
De Hystereseverlust *(m)*
En hysteresis loss
Es pérdida por histéresis *(f)*
Fr perte par hystérésis *(f)*
Pt perda por histerese *(f)*

perdita per resistenza aerodinamica *(f)* It
De Luftreibungsverlust *(m)*
En windage loss
Es pérdida por resistencia aerodinámica *(f)*
Fr perte due au jeu *(f)*
Pt perda por fricção do vento *(f)*

perdita per trasmissione *(f)* It
De Durchgangsdämpfung *(f)*
En transmission loss
Es pérdida por transmisión *(f)*
Fr perte par transmission *(f)*
Pt perda por transmissão *(f)*

perforación *(f)* *n* Es
De Bohren *(n)*
En drilling
Fr forage *(m)*
It trivellazione *(f)*
Pt perfuração *(f)*

perforación a cable *(f)* Es
De Seilbohren *(n)*
En cable-tool drilling
Fr forage au câble *(m)*
It trivellazione con utensile a fune *(f)*
Pt perfuração a cabo *(f)*

perforación direccional *(f)* Es
De gerichtetes Bohren *(n)*
En directional drilling
Fr forage dirigé *(m)*
It trivellazione direzionale *(f)*
Pt perfuração direccional *(f)*

perforación por aire comprimido *(f)* Es
De Druckluftbohren *(n)*
En air drilling
Fr forage à l'air comprimé *(m)*
It trapanatura ad aria compressa *(f)*
Pt perfuração pneumática *(f)*

perforación rotativa *(f)* Es
De Drehbohren *(n)*
En rotary drilling
Fr forage rotary *(m)*
It sondaggio a rotazione *(m)*
Pt perfuração rotativa *(f)*

perfuração *(f)* *n* Pt
De Bohren *(n)*
En drilling
Es perforación *(f)*
Fr forage *(m)*
It trivellazione *(f)*

perfuração a cabo *(f)* Pt
De Seilbohren *(n)*
En cable-tool drilling
Es perforación a cable *(f)*
Fr forage au câble *(m)*
It trivellazione con utensile a fune *(f)*

perfuração direccional *(f)* Pt
De gerichtetes Bohren *(n)*
En directional drilling
Es perforación direccional *(f)*
Fr forage dirigé *(m)*
It trivellazione direzionale *(f)*

perfuração pneumática *(f)* Pt
De Druckluftbohren *(n)*
En air drilling
Es perforación por aire comprimido *(f)*
Fr forage à l'air comprimé *(m)*
It trapanatura ad aria compressa *(f)*

perfuração rotativa *(f)* Pt
De Drehbohren *(n)*
En rotary drilling
Es perforación rotativa *(f)*
Fr forage rotary *(m)*
It sondaggio a rotazione *(m)*

pericolo di incendio *(m)* It
De Feuergefahr *(f)*
En fire hazard
Es riesgo de incendio *(m)*
Fr danger d'incendie *(m)*
Pt perigo de incêndio *(m)*

pericolo di radiazioni *(m)* It
De Strahlungsgefahr *(f)*
En radiation hazard
Es riesgo de radiación *(m)*
Fr danger de radiation *(m)*
Pt perigo de radiação *(m)*

perigo de incêndio *(m)* Pt
De Feuergefahr *(f)*
En fire hazard
Es riesgo de incendio *(m)*
Fr danger d'incendie *(m)*
It pericolo di incendio *(m)*

perigo de radiação *(m)* Pt
De Strahlungsgefahr *(f)*
En radiation hazard
Es riesgo de radiación *(m)*
Fr danger de radiation *(m)*
It pericolo di radiazioni *(m)*

period *n* En
De Periode *(f)*
Es período *(m)*
Fr période *(f)*
It periodo *(m)*
Pt período *(m)*

Periode *(f)* *n* De
En period
Es período *(m)*
Fr période *(f)*
It periodo *(m)*
Pt período *(m)*

période *(f)* *n* Fr
De Periode *(f)*
En period
Es período *(m)*
It periodo *(m)*
Pt período *(m)*

période hors-pointe *(f)* Fr
De Schwachlastzeit *(f)*
En off-peak period
Es período fuera de horas punta *(m)*
It periodo fuori di ore di punta *(m)*
Pt período fora de ponta *(m)*

periodo *(m)* *n* It
De Periode *(f)*
En period
Es período *(m)*
Fr période *(f)*
Pt período *(m)*

período *(m)* *n* Es, Pt
De Periode *(f)*
En period
Fr période *(f)*
It periodo *(m)*

período de reembolso *(m)* Es, Pt
De Rückzahldauer *(f)*
En payback period
Fr durée de remboursement *(f)*
It periodo di remborso *(m)*

periodo di dimezzamento *(m)* It
De Halbwertszeit *(f)*
En half-life
Es media vida *(f)*
Fr demi-vie *(f)*
Pt meia vida *(f)*

periodo di remborso *(m)* It
De Rückzahldauer *(f)*
En payback period
Es período de reembolso *(m)*
Fr durée de remboursement *(f)*
Pt período de reembolso *(m)*

período fora de ponta *(m)* Pt
De Schwachlastzeit *(f)*
En off-peak period
Es período fuera de horas punta *(m)*
Fr période hors-pointe *(f)*
It periodo fuori di ore di punta *(m)*

período fuera de horas punta *(m)* Es
De Schwachlastzeit *(f)*
En off-peak period
Fr période hors-pointe *(f)*
It periodo fuori di ore di punta *(m)*
Pt período fora de ponta *(m)*

periodo fuori di ore di punta *(m)* It
De Schwachlastzeit *(f)*
En off-peak period
Es período fuera de horas punta *(m)*
Fr période hors-pointe *(f)*
Pt período fora de ponta *(m)*

perizia magnetica *(f)* It
De Magnetaufnahme *(f)*
En magnetic survey
Es levantamiento magnético *(m)*
Fr prospection magnétique *(f)*
Pt levantamento magnético *(m)*

permeabilidad *(f) n* Es
De Permeabilität *(f)*
En permeability
Fr perméabilité *(f)*
It permeabilità *(f)*
Pt permeabilidade *(f)*

permeabilidade *(f) n* Pt
De Permeabilität *(f)*
En permeability
Es permeabilidad *(f)*
Fr perméabilité *(f)*
It permeabilità *(f)*

permeabilità *(f) n* It
De Permeabilität *(f)*
En permeability
Es permeabilidad *(f)*
Fr perméabilité *(f)*
Pt permeabilidade *(f)*

Permeabilität *(f) n* De
En permeability
Es permeabilidad *(f)*
Fr perméabilité *(f)*
It permeabilità *(f)*
Pt permeabilidade *(f)*

perméabilité *(f) n* Fr
De Permeabilität *(f)*
En permeability
Es permeabilidad *(f)*
It permeabilità *(f)*
Pt permeabilidade *(f)*

permeability *n* En
De Permeabilität *(f)*
Es permeabilidad *(f)*
Fr perméabilité *(f)*
It permeabilità *(f)*
Pt permeabilidade *(f)*

permitividad *(f) n* Es
De Permitivität *(f)*
En permittivity
Fr permittivité *(f)*
It permittività *(f)*
Pt permitividade *(f)*

permitividade *(f) n* Pt
De Permitivität *(f)*
En permittivity
Es permitividad *(f)*
Fr permittivité *(f)*
It permittività *(f)*

Permitivität *(f) n* De
En permittivity
Es permitividad *(f)*
Fr permittivité *(f)*
It permittività *(f)*
Pt permitividade *(f)*

permittività *(f) n* It
De Permitivität *(f)*
En permittivity
Es permitividad *(f)*
Fr permittivité *(f)*
Pt permitividade *(f)*

permittivité *(f) n* Fr
De Permitivität *(f)*
En permittivity
Es permitividad *(f)*
It permittività *(f)*
Pt permitividade *(f)*

permittivity *n* En
De Permitivität *(f)*
Es permitividad *(f)*
Fr permittivité *(f)*
It permittività *(f)*
Pt permitividade *(f)*

permuta de ioēs *(f)* Pt
De Ionenaustausch *(m)*
En ion exchange
Es intercambio de iones *(m)*
Fr échange d'ions *(m)*
It scambio ionico *(m)*

permutador *(m) n* Pt
De Austauscher *(m)*
En exchanger
Es intercambiador *(m)*
Fr échangeur *(m)*
It scambiatore *(m)*

perte dans le cuivre *(f)* Fr
De Kupferverlust *(m)*
En copper loss
Es pérdida en el cobre *(f)*
It perdita nel rame *(f)*
Pt perda no cobre *(f)*

perte de chaleur *(f)* Fr
De Wärmeverlust *(m)*
En heat loss
Es pérdida de calor *(f)*
It perdita di calore *(f)*
Pt perda de calor *(f)*

perte diélectrique *(f)* Fr
De dielektrischer Verlust *(m)*
En dielectric loss
Es pérdida dieléctrica *(f)*
It perdita dielettrica *(f)*
Pt perda dieléctrica *(f)*

perte due au jeu *(f)* Fr
De Luftreibungsverlust *(m)*
En windage loss
Es pérdida por resistencia aerodinámica *(f)*
It perdita per resistenza aerodinamica *(f)*
Pt perda por fricção do vento *(f)*

perte par hystérésis *(f)* Fr
De Hystereseverlust *(m)*
En hysteresis loss
Es pérdida por histéresis *(f)*
It perdita per isteresi *(f)*
Pt perda por histerese *(f)*

perte par transmission *(f)* Fr
De Durchgangsdämpfung *(f)*
En transmission loss
Es pérdida por transmisión *(f)*
It perdita per trasmissione *(f)*
Pt perda por transmissão *(f)*

peso *(m) n* Es, It, Pt
De Gewicht *(n)*
En weight
Fr poids *(m)*

peso atomico *(m)* It
De Atomgewicht *(n)*
En atomic weight
Es peso atómico *(m)*
Fr poids atomique *(m)*
Pt peso atómico *(m)*

peso atómico *(m)* Es, Pt
De Atomgewicht *(n)*
En atomic weight
Fr poids atomique *(m)*
It peso atomico *(m)*

petrochemicals *pl n* En
De petrochemische Produkte *(pl)*
Es productos petroquímicos *(m pl)*
Fr produits pétrochimiques *(m pl)*
It prodotti petrolchimici *(m pl)*
Pt produtos petroquímicos *(m pl)*

petrochemische Produkte *(pl)* De
En petrochemicals
Es productos petroquímicos *(m pl)*
Fr produits pétrochimiques *(m pl)*
It prodotti petrolchimici *(m pl)*
Pt produtos petroquímicos *(m pl)*

petrografia *(f) n* It, Pt
De Petrographie *(f)*
En petrography
Es petrografía *(f)*
Fr pétrographie *(f)*

petrografía *(f) n* Es
De Petrographie *(f)*
En petrography
Fr pétrographie *(f)*
It petrografia *(f)*
Pt petrografia *(f)*

Petrographie *(f) n* De
En petrography
Es petrografía *(f)*
Fr pétrographie *(f)*
It petrografia *(f)*
Pt petrografia *(f)*

pétrographie *(f) n* Fr
De Petrographie *(f)*
En petrography
Es petrografía *(f)*
It petrografia *(f)*
Pt petrografia *(f)*

petrography *n* En
De Petrographie *(f)*
Es petrografía *(f)*
Fr pétrographie *(f)*
It petrografia *(f)*
Pt petrografia *(f)*

petrol *n* En
Am gasoline
De Benzin *(n)*
Es gasolina *(f)*
Fr essence *(f)*
It benzina *(f)*
Pt gasolina *(f)*

pétrole *(m) n* Fr
De Erdöl; Petroleum *(n)*
En petroleum; rock oil
Es petróleo *(m)*
It petrolio *(m)*
Pt petróleo *(m)*

pétrole brut *(m)* Fr
De Rohöl *(n)*
En crude oil
Es petróleo crudo *(m)*
It petrolio grezzo *(m)*
Pt óleo crú *(m)*

pétrole brut aromatique *(m)* Fr
De aromatisches Rohöl *(n)*
En aromatic crude oil
Es crudo aromático *(m)*
It petrolio greggio aromatico *(m)*
Pt óleo aromático crú *(m)*

pétrole brut paraffinique *(m)* Fr
De paraffinisches Rohöl *(n)*
En paraffinic crude oil
Es crudo parafínico *(m)*
It petrolio grezzo paraffinico *(m)*
Pt óleo crú parafínico *(m)*

pétrole de Mer du Nord *(m)* Fr
De Nordseeöl *(n)*
En North Sea oil
Es petróleo del Mar del Norte *(m)*
It petrolio del Mare del Nord *(m)*
Pt petróleo do Mar do Norte *(m)*

petroleiro *(m) n* Pt
De Öltanker *(m)*
En oil tanker
Es petrolero *(m)*
Fr pétrolier *(m)*
It petroliera *(f)*

petroleiro a granel *(m)* Pt
De Ölsammelschiff *(n)*
En bulk oil-carrier
Es petrolero a granel *(m)*
Fr pétrolier vraquier *(m)*
It petroliera all'ingrosso *(f)*

pétrole lampant *(m)* Fr
De Brenner-Treiböl *(n)*
En burner fuel-oil
Es fuel-oil para quemador *(m)*
It nafta per bruciatori *(f)*
Pt fuel-oil para queimadores *(m)*

petrol engine En
De Benzinmotor *(m)*
Es motor de gasolina *(m)*
Fr moteur à essence *(m)*
It motore a benzina *(m)*
Pt motor a gasolina *(m)*

petróleo *(m) n* Es
De Öl *(n)*
En oil
Fr huile *(f)*
It olio *(m)*
Pt óleo *(m)*

petróleo *(m) n* Es, Pt
De Erdöl; Petroleum *(n)*
En petroleum; rock oil
Fr pétrole *(m)*
It petrolio *(m)*

petróleo crudo *(m)* Es
De Rohöl *(n)*
En crude oil
Fr pétrole brut *(m)*
It petrolio grezzo *(m)*
Pt óleo crú *(m)*

petróleo crú leve *(m)* Pt
De leichtes Rohöl *(n)*
En light crude oil
Es crudo ligero *(m)*
Fr brut léger *(m)*
It petrolio greggio leggero *(m)*

petróleo crú nafténico *(m)* Pt
De naphthenisches Rohöl *(n)*
En naphthenic crude oil
Es crudo nafténico *(m)*
Fr brut à base naphténique *(m)*
It petrolio greggio naftenico *(m)*

petróleo crú sintético *(m)* Pt
De synthetisches Rohöl *(n)*
En synthetic crude oil (SCO)
Es crudo sintético *(m)*
Fr brut de synthèse *(m)*
It petrolio greggio sintetico *(m)*

petróleo de calefacción *(m)* Es
De Heizöl *(n)*
En heating oil
Fr huile de chauffe *(f)*
It olio per riscaldamento *(m)*
Pt óleo para aquecimento *(m)*

petróleo de coqueificação *(m)* Pt
De Koksöl *(n)*
En coking oil
Es petróleo de coquización *(m)*
Fr huile cokéfiant *(f)*
It olio cokificante *(m)*

petróleo de coquización *(m)* Es
De Koksöl *(n)*
En coking oil
Fr huile cokéfiante *(f)*
It olio cokificante *(m)*
Pt petróleo de coqueificação *(m)*

petróleo del Mar del Norte *(m)* Es
De Nordseeöl *(n)*
En North Sea oil
Fr pétrole de Mer du Nord *(m)*
It petrolio del Mare del Nord *(m)*
Pt petróleo do Mar do Norte *(m)*

petróleo do Mar do Norte *(m)* Pt
De Nordseeöl *(n)*
En North Sea oil
Es petróleo del Mar del Norte *(m)*
Fr pétrole de Mer du Nord *(m)*
It petrolio del Mare del Nord *(m)*

petróleo equivalente *(m)* Es
De Ölersatz *(m)*
En oil equivalent
Fr équivalent pétrole *(m)*
It equivalente del petrolio *(m)*
Pt equivalente de petróleo *(m)*

petróleo ligero *(m)* Es
De leichtes Öl *(n)*
En light oil
Fr huile légère *(f)*
It olio leggero *(m)*
Pt óleo leve *(m)*

petróleo xistoso *(m)* Pt
De Schieferöl *(n)*
En shale oil

Es aceite de lutita *(m)*
Fr huile de schiste *(f)*
It olio di schisto *(m)*

petrolero *(m) n* Es
De Öltanker *(m)*
En oil tanker
Fr pétrolier *(m)*
It petroliera *(f)*
Pt petroleiro *(m)*

petrolero a granel *(m)* Es
De Ölsammelschiff *(n)*
En bulk oil-carrier
Fr pétrolier vraquier *(m)*
It petroliera all'ingrosso *(f)*
Pt petroleiro a granel *(m)*

petroleum *n* En
De Erdöl; Petroleum *(n)*
Es petróleo *(m)*
Fr pétrole *(m)*
It petrolio *(m)*
Pt petróleo *(m)*

Petroleum *(n) n* De
En petroleum; rock oil
Es petróleo *(m)*
Fr pétrole *(m)*
It petrolio *(m)*
Pt petróleo *(m)*

pétrolier *(m) n* Fr
De Öltanker *(m)*
En oil tanker
Es petrolero *(m)*
It petroliera *(f)*
Pt petroleiro *(m)*

petroliera *(f) n* It
De Öltanker *(m)*
En oil tanker
Es petrolero *(m)*
Fr pétrolier *(m)*
Pt petroleiro *(m)*

petroliera all'ingrosso *(f)* It
De Ölsammelschiff *(n)*
En bulk oil-carrier
Es petrolero a granel *(m)*
Fr pétrolier vraquier *(m)*
Pt petroleiro a granel *(m)*

pétrolier géant *(m)* Fr
De Supertanker *(m)*
En supertanker
Es superpetrolero *(m)*
It superpetroliera *(f)*
Pt supernaviocisterna *(m)*

pétrolier vraquier *(m)* Fr
De Ölsammelschiff *(n)*
En bulk oil-carrier
Es petrolero a granel *(m)*
It petroliera all'ingrosso *(f)*
Pt petroleiro a granel *(m)*

petrolio *(m) n* It
De Erdöl; Petroleum *(n)*
En petroleum; rock oil
Es petróleo *(m)*
Fr pétrole *(m)*
Pt petróleo *(m)*

petrolio del Mare del Nord *(m)* It
De Nordseeöl *(n)*
En North Sea oil
Es petróleo del Mar del Norte *(m)*
Fr pétrole de Mer du Nord *(m)*
Pt petróleo do Mar do Norte *(m)*

petrolio greggio aromatico *(m)* It
De aromatisches Rohöl *(n)*
En aromatic crude oil
Es crudo aromático *(m)*
Fr pétrole brut aromatique *(m)*
Pt óleo aromático crú *(m)*

petrolio greggio leggero *(m)* It
De leichtes Rohöl *(n)*
En light crude oil
Es crudo ligero *(m)*
Fr brut léger *(m)*
Pt petróleo crú leve *(m)*

petrolio greggio naftenico *(m)* It
De naphthenisches Rohöl *(n)*
En naphthenic crude oil
Es crudo nafténico *(m)*
Fr brut à base naphténique *(m)*
Pt petróleo crú nafténico *(m)*

petrolio greggio sintetico *(m)* It
De synthetisches Rohöl *(n)*
En synthetic crude oil (SCO)
Es crudo sintético *(m)*
Fr brut de synthèse *(m)*
Pt petróleo crú sintético *(m)*

petrolio grezzo *(m)* It
De Rohöl *(n)*
En crude oil
Es petróleo crudo *(m)*
Fr pétrole brut *(m)*
Pt óleo crú *(m)*

petrolio grezzo paraffinico *(m)* It
De paraffinisches Rohöl *(n)*
En paraffinic crude oil
Es crudo parafínico *(m)*
Fr pétrole brut paraffinique *(m)*
Pt óleo crú parafínico *(m)*

pez *(f) n* Pt
De Pech *(n)*
En pitch
Es brea *(f)*
Fr brai *(m)*
It pece *(f)*

phase *n* En; Fr *(f)*
De Phase *(f)*
Es fase *(f)*
It fase *(f)*
Pt fase *(f)*

Phase *(f) n* De
En phase
Es fase *(f)*
Fr phase *(f)*
It fase *(f)*
Pt fase *(f)*

phase-angle *n* En
De Phasenwinkel *(m)*
Es ángulo de fase *(m)*
Fr angle de phase *(m)*
It angolo di fase *(m)*
Pt ângulo de fase *(m)*

Phasenumwandlungs-speicherung *(f) n* De
En phase-transition storage
Es almacenamiento de transición de fase *(m)*
Fr accumulation à transition de phase *(f)*
It immagazzinaggio a transizione di fase *(m)*
Pt armazenagem de transição de fase *(f)*

Phasenwinkel *(m) n* De
En phase-angle
Es ángulo de fase *(m)*
Fr angle de phase *(m)*
It angolo di fase *(m)*
Pt ângulo de fase *(m)*

phase-transition storage En
De Phasenum-wandlungs-speicherung *(f)*
Es almacenamiento de transición de fase *(m)*
Fr accumulation à transition de phase *(f)*
It immagazzinaggio a transizione di fase *(m)*
Pt armazenagem de transição de fase *(f)*

photocell *n* En
De Photozelle *(f)*
Es fotocélula *(f)*
Fr cellule photoélectrique *(f)*
It cellula fotoelettrica *(f)*
Pt fotocélula *(f)*

photoconducteur *adj* Fr
De photoleitend
En photoconductive
Es fotoconductor
It fotoconduttivo
Pt fotocondutor

photoconductive *adj* En
De photoleitend
Es fotoconductor
Fr photoconducteur
It fotoconduttivo
Pt fotocondutor

photoelectric *adj* En
De photoelektrisch
Es fotoeléctrico
Fr photoélectrique
It fotoelettrico
Pt fotoeléctrico

photoélectrique *adj* Fr
De photoelektrisch
En photoelectric
Es fotoeléctrico
It fotoelettrico
Pt fotoeléctrico

photoelektrisch *adj* De
En photoelectric
Es fotoeléctrico
Fr photoélectrique
It fotoelettrico
Pt fotoeléctrico

photoleitend *adj* De
En photoconductive
Es fotoconductor
Fr photoconducteur
It fotoconduttivo
Pt fotocondutor

photon *n* En; Fr *(m)*
De Photon *(n)*
Es fotón *(m)*
It fotone *(m)*
Pt fotão *(m)*

Photon *(n) n* De
En photon
Es fotón *(m)*
Fr photon *(m)*
It fotone *(m)*
Pt fotão *(m)*

Photosynthese *(f) n* De
En photosynthesis
Es fotosíntesis *(f)*
Fr photosynthèse *(f)*
It fotosintesi *(f)*
Pt fotosíntese *(m)*

photosynthèse *(f) n* Fr
De Photosynthese *(f)*
En photosynthesis
Es fotosíntesis *(f)*
It fotosintesi *(f)*
Pt fotosíntese *(m)*

photosynthesis *n* En
De Photosynthese *(f)*
Es fotosíntesis *(f)*
Fr photosynthèse *(f)*
It fotosintesi *(f)*
Pt fotosíntese *(m)*

photovoltaic *adj* En
De photovoltaisch
Es fotovoltaico
Fr photovoltaïque
It fotovoltaico
Pt fotovoltáico

photovoltaïque *adj* Fr
De photovoltaisch
En photovoltaic
Es fotovoltaico
It fotovoltaico
Pt fotovoltáico

photovoltaisch *adj* De
En photovoltaic
Es fotovoltaico
Fr photovoltaïque
It fotovoltaico
Pt fotovoltáico

Photozelle *(f) n* De
En photocell
Es fotocélula *(f)*
Fr cellule photoélectrique *(f)*
It cellula fotoelettrica *(f)*
Pt fotocélula *(f)*

piattaforma abitata *(f)* It
De bemannte Ölbohrinsel *(f)*
En manned platform
Es plataforma atendida por personal *(f)*
Fr plate-forme occupée *(f)*
Pt plataforma com tripulação *(f)*

piezoelectric *adj* En
De piezoelektrisch
Es piezoeléctrico
Fr piézoélectrique
It piezoelettrico
Pt piezoeléctrico

piezoeléctrico *adj* Es, Pt
De piezoelektrisch
En piezoelectric
Fr piézoélectrique
It piezoelettrico

piézoélectrique *adj* Fr
De piezoelektrisch
En piezoelectric
Es piezoeléctrico
It piezoelettrico
Pt piezoeléctrico

piezoelektrisch *adj* De
En piezoelectric
Es piezoeléctrico
Fr piézoélectrique
It piezoelettrico
Pt piezoeléctrico

piezoelettrico *adj* It
De piezoelektrisch
En piezoelectric
Es piezoeléctrico
Fr piézoélectrique
Pt piezoeléctrico

pila a combustibile *(f)* It
De Brennstoffzelle *(f)*
En fuel cell
Es célula de combustible *(f)*
Fr pile à combustible *(f)*
Pt célula de combustível *(f)*

pila de níquel-cadmio *(f)* Es
De Nickel-Kadmiumzelle *(f)*
En nickel-cadmium cell
Fr pile au nickel-cadmium *(f)*
It cellula al nichel-cadmio *(f)*
Pt célula de níquel-cádmio *(f)*

pila de níquel-hierro *(f)* Es
De Nickel-Eisenzelle *(f)*
En nickel-iron cell
Fr pile au nickel-fer *(f)*
It cellula al nichel-ferro *(f)*
Pt célula de níquel-ferro *(f)*

pila de plata-cadmio *(f)* Es
De Silber-Kadmiumzelle *(f)*
En silver-cadmium cell
Fr pile à l'argent-cadmium *(f)*
It cellula di argento-cadmio *(f)*
Pt célula de prata-cádmio *(f)*

pila de plata-zinc *(f)* Es
De Silber-Zinkzelle *(f)*
En silver-zinc cell
Fr pile à l'argent-zinc *(f)*
It cellula di argento-zinco *(f)*
Pt célula de prata-zinco *(f)*

pile à combustible *(f)* Fr
De Brennstoffzelle *(f)*
En fuel cell
Es célula de combustible *(f)*
It pila a combustibile *(f)*
Pt célula de combustível *(f)*

pile à combustible biochimique *(f)* Fr
De biochemische Brennstoffzelle *(f)*
En biochemical fuel cell
Es célula de combustible bioquímico *(f)*
It cellula combustibile biochimica *(f)*
Pt célula de combustível bioquímico *(f)*

pile à combustible hydrogène-air *(f)* Fr
De Wasserstoff-Luft-brennstoffzelle *(f)*
En hydrogen-air fuel cell
Es célula de combustible de hidrógeno-aire *(f)*
It cellula combustibile idrogeno-aria *(f)*
Pt célula de combustível hidrogénio-ar *(f)*

pile à l'argent-cadmium *(f)* Fr
De Silber-Kadmiumzelle *(f)*
En silver-cadmium cell
Es pila de plata-cadmio *(f)*
It cellula di argento-cadmio *(f)*
Pt célula de prata-cádmio *(f)*

pile à l'argent-zinc *(f)* Fr
De Silber-Zinkzelle *(f)*
En silver-zinc cell
Es pila de plata-zinc *(f)*
It cellula di argento-zinco *(f)*
Pt célula de prata-zinco *(f)*

pile au cadmium *(f)* Fr
De Kadmiumbatterie *(f)*
En cadmium battery
Es batería de cadmio *(f)*
It batteria al cadmio *(f)*
Pt bateria de cádmio *(f)*

pile au nickel-cadmium *(f)* Fr
De Nickel-Kadmiumzelle *(f)*
En nickel-cadmium cell
Es pila de níquel-cadmio *(f)*
It cellula al nichel-cadmio *(f)*
Pt célula de níquel-cádmio *(f)*

pile au nickel-fer *(f)* Fr
De Nickel-Eisenzelle *(f)*
En nickel-iron cell
Es pila de níquel-hierro *(f)*
It cellula al nichel-ferro *(f)*
Pt célula de níquel-ferro *(f)*

pile au silicium *(f)* Fr
De Silikonzelle *(f)*
En silicon cell
Es célula de silicio *(f)*
It cellula di silicone *(f)*
Pt célula de silíco *(f)*

pile galvanique *(f)* Fr
De galvanische Zelle *(f)*
En galvanic cell
Es elemento galvánico *(m)*
It cellula galvanica *(f)*
Pt célula galvânica *(f)*

pile métal-air *(f)* Fr
De Metall-Luftzelle *(f)*
En metal-air cell
Es célula de metal-aire *(f)*
It cellula metallo-aria *(f)*
Pt célula de metal-ar *(f)*

pilha de armazenagem *(f)* Pt
De Vorrat *(m)*
En stockpile
Es existencia para emergencia *(f)*
Fr stock de réserves *(m)*
It riserva *(f)*

pilha térmica *(f)* Pt
De Thermosäule *(f)*
En thermopile
Es termopila *(f)*
Fr thermopile *(f)*
It termopila *(f)*

pilot flame En
De Zündflamme *(f)*
Es llama piloto *(f)*
Fr veilleuse *(f)*
It semprevivo *(m)*
Pt chama piloto *(f)*

pilot plant En
De Versuchsanlage *(f)*
Es planta piloto *(f)*
Fr installation pilote *(f)*
It impianto pilota *(m)*
Pt fábrica piloto *(f)*

pinch effect En
De Schnüreffekt *(m)*
Es efecto de estricción *(m)*
Fr effet de pincement *(m)*
It effetto di strizione *(m)*
Pt efeito de aperto *(m)*

piombo tetraetile *(m)* It
De Tetraäthylblei *(n)*
En tetraethyl lead
Es plomo de tetraetilo *(m)*
Fr plomb tétraéthyle *(m)*
Pt tetraetilato de chumbo *(m)*

pipe lagging En
De Rohrummantellung *(f)*
Es revestimiento de tuberías *(m)*
Fr calorifugeage de tuyaux *(m)*
It rivestimento isolante di tubazioni *(m)*
Pt revestimento de canos *(m)*

pipeline *n* En
De Rohrleitung *(f)*
Es oleoducto *(m)*
Fr oléoduc *(m)*
It tubazione *(f)*
Pt oleoduto *(m)*

pipeline marin *(m)* Fr
De Seerohrleitung *(f)*
En marine pipeline
Es conducción submarina *(f)*
It condotto marino *(m)*
Pt oleoduto marítimo *(m)*

piranometro *(m) n* It
De Pyranometer *(n)*
En pyranometer
Es piranómetro *(m)*
Fr pyranomètre *(m)*
Pt piranómetro *(m)*

piranómetro *(m) n* Es, Pt
De Pyranometer *(n)*
En pyranometer
Fr pyranomètre *(m)*
It piranometro *(m)*

pireliometro *(m) n* It
De Pyrheliometer *(n)*
En pyrheliometer
Es pirheliómetro *(m)*
Fr pyrhéliomètre *(m)*
Pt pirheliómetro *(m)*

pirheliómetro *(m) n* Es, Pt
De Pyrheliometer *(n)*
En pyrheliometer
Fr pyrhéliomètre *(m)*
It pireliometro *(m)*

pirólise *(f) n* Pt
De Pyrolyse *(f)*
En pyrolysis
Es pirólisis *(f)*
Fr pyrolyse *(f)*
It pirolisi *(f)*

pirolisi *(f) n* It
De Pyrolyse *(f)*
En pyrolysis
Es pirólisis *(f)*
Fr pyrolyse *(f)*
Pt pirólise *(f)*

pirólisis *(f) n* Es
De Pyrolyse *(f)*
En pyrolysis
Fr pyrolyse *(f)*
It pirolisi *(f)*
Pt pirólise *(f)*

pirometro *(m) n* It
De Pyrometer *(n)*
En pyrometer
Es pirómetro *(m)*
Fr pyromètre *(m)*
Pt pirómetro *(m)*

pirómetro *(m) n* Es, Pt
De Pyrometer *(n)*
En pyrometer
Fr pyromètre *(m)*
It pirometro *(m)*

pirometro ad aspirazione *(m)* It
De Saugpyrometer *(n)*
En suction pyrometer
Es pirómetro de succión *(m)*
Fr pyromètre aspirant *(m)*
Pt pirómetro de sucção *(m)*

pirómetro de sucção *(m)* Pt
De Saugpyrometer *(n)*
En suction pyrometer
Es pirómetro de succión *(m)*
Fr pyromètre aspirant *(m)*
It pirometro ad aspirazione *(m)*

pirómetro de succión *(m)* Es
De Saugpyrometer *(n)*
En suction pyrometer
Fr pyromètre aspirant *(m)*
It pirometro ad aspirazione *(m)*
Pt pirómetro de sucção *(m)*

pirómetro óptico *(m)* Es, Pt
De Strahlungspyrometer *(n)*
En optical pyrometer
Fr pyromètre optique *(m)*
It pirometro ottico *(m)*

pirometro ottico *(m)* It
De Strahlungspyrometer *(n)*
En optical pyrometer
Es pirómetro óptico *(m)*
Fr pyromètre optique *(m)*
Pt pirómetro óptico *(m)*

piroscissione *(f) n* It
De Kracken *(n)*
En cracking
Es cráking *(m)*
Fr craquage *(m)*
Pt cracking *(m)*

piroscissione idraulica *(f)* It
De Hydrokracken *(n)*
En hydrocracking

Es hidrocrácking *(m)*
Fr hydrocraquage *(m)*
Pt hidro-cracking *(m)*

piroscissione per catalisi *(f)* It
De katalytisches Kracken *(n)*
En catalytic cracking
Es crácking catalítico *(m)*
Fr craquage catalytique *(m)*
Pt cracking catalítico *(m)*

piroscissione termica *(f)* It
De thermisches Kracken *(n)*
En thermal cracking
Es crácking térmico *(m)*
Fr craquage thermique *(m)*
Pt cracking térmico *(m)*

piscina de refrigeración *(f)* Es
De Kühlteich *(m)*
En cooling pond
Fr piscine de refroidissement *(f)*
It bacino di raffreddamento *(m)*
Pt tanque refrigerante *(m)*

piscina solar salina *(f)* Es
De Solar-Salzteich *(m)*
En saline solar pond
Fr bassin solaire salant *(m)*
It laghetto solare salino *(m)*
Pt lago solar de sal *(m)*

piscine de refroidissement *(f)* Fr
De Kühlteich *(m)*
En cooling pond
Es piscina de refrigeración *(f)*
It bacino di raffreddamento *(m)*
Pt tanque refrigerante *(m)*

pitch *n* En
De Pech *(n)*
Es brea *(f)*
Fr brai *(m)*
It pece *(f)*
Pt pez *(f)*

pitch coke En
De Pechkoks *(m)*
Es coque con ligante *(m)*
Fr coke de brai *(m)*
It coke di pece *(m)*
Pt coque de pez *(m)*

pizarras bituminosas *(f pl)* Es
De Ölschiefer *(pl)*
En oil shales
Fr schistes bitumineux *(m pl)*
It schisti bituminosi *(m pl)*
Pt xistos betuminosos *(m pl)*

placa de orificio *(f)* Es
De Mündungsplatte *(f)*
En orifice plate
Fr plaque perforée *(f)*
It placca di orifizio *(f)*
Pt placa de orifício *(f)*

placa de orifício *(f)* Pt
De Mündungsplatte *(f)*
En orifice plate
Es placa de orificio *(f)*
Fr plaque perforée *(f)*
It placca di orifizio *(f)*

placca di orifizio *(f)* It
De Mündungsplatte *(f)*
En orifice plate
Es placa de orificio *(f)*
Fr plaque perforée *(f)*
Pt placa de orifício *(f)*

Planckscher Strahler *(m)* De
En black body
Es cuerpo negro *(m)*
Fr corps noir *(m)*
It corpo nero *(m)*
Pt corpo negro *(m)*

Plancksche Strahlung *(f)* De
En black-body radiation
Es radiación de cuerpo negro *(f)*
Fr rayonnement du corps noir *(m)*
It radiazione corpo nero *(f)*
Pt radiação de corpo negro *(f)*

planta auxiliar de calderas *(f)* Es
De Kesselhilfsanlage *(f)*
En boiler auxiliary plant
Fr chaufferie auxiliaire *(f)*
It impianto ausiliario caldaia *(m)*
Pt instalação auxiliar de caldeira *(f)*

planta piloto *(f)* Es
De Versuchsanlage *(f)*
En pilot plant
Fr installation pilote *(f)*
It impianto pilota *(m)*
Pt fábrica piloto *(f)*

planta solar híbrida *(f)* Es
De Hybrid-Solaranlage *(f)*
En hybrid solar plant
Fr centrale solaire hybride *(f)*
It impianto solare ibrido *(m)*
Pt central de energia solar híbrida *(f)*

plaque perforée *(f)* Fr
De Mündungsplatte *(f)*
En orifice plate
Es placa de orificio *(f)*
It placca di orifizio *(f)*
Pt placa de orifício *(f)*

plasma *n* En; Es, Fr, It, Pt *(m)*
De Plasma *(n)*

Plasma *(n) n* De
En plasma
Es plasma *(m)*
Fr plasma *(m)*
It plasma *(m)*
Pt plasma *(m)*

Plasmabrenner *(m) n* De
En plasma torch
Es soplete para plasma *(m)*
Fr chalumeau à plasma *(m)*
It torcia a plasma *(f)*
Pt maçarico de plasma *(m)*

plasma torch En
De Plasmabrenner *(m)*
Es soplete para plasma *(m)*
Fr chalumeau à plasma *(m)*
It torcia a plasma *(f)*
Pt maçarico de plasma *(m)*

plataforma atendida por personal *(f)* Es
De bemannte Ölbohrinsel *(f)*
En manned platform
Fr plate-forme occupée *(f)*
It piattaforma abitata *(f)*
Pt plataforma com tripulação *(f)*

plataforma com tripulação *(f)* Pt
De bemannte Ölbohrinsel *(f)*
En manned platform
Es plataforma atendida por personal *(f)*
Fr plate-forme occupée *(f)*
It piattaforma abitata *(f)*

plataforma petroleira *(f)* Pt
De Ölbohrturm *(m)*
En oilrig
Es tren de perforación de petróleo *(m)*
Fr installation de forage *(f)*
It impianto di trivellazione per petrolio *(m)*

plate-forme occupée *(f)* Fr
De bemannte Ölbohrinsel *(f)*
En manned platform
Es plataforma atendida por personal *(f)*
It piattaforma abitata *(f)*
Pt plataforma com tripulação *(f)*

plate heat-exchanger En
De Plattenwärmeaustauscher *(m)*
Es intercambiador de calor de placas *(m)*
Fr échangeur de chaleur à plaque *(m)*
It scambiatore di calore a piastra *(m)*
Pt termo-permutador de placa *(m)*

Plattenwärmeaustauscher *(m)* De
En plate heat-exchanger
Es intercambiador de calor de placas *(m)*
Fr échangeur de chaleur à plaque *(m)*
It scambiatore di calore a piastra *(m)*
Pt termo-permutador de placa *(m)*

plomb tétraéthyle *(m)* Fr
De Tetraäthylblei *(n)*
En tetraethyl lead
Es plomo de tetraetilo *(m)*
It piombo tetraetile *(m)*
Pt tetraetilato de chumbo *(m)*

plomo de tetraetilo *(m)* Es
De Tetraäthylblei *(n)*
En tetraethyl lead
Fr plomb tétraéthyle *(m)*
It piombo tetraetile *(m)*
Pt tetraetilato de chumbo *(m)*

pneumoconiose *(f) n* Fr, Pt
De Staublunge *(f)*
En pneumoconiosis
Es neumoconiosis *(f)*
It pneumoconiosi *(f)*

pneumoconiosi *(f) n* It
De Staublunge *(f)*
En pneumoconiosis
Es neumoconiosis *(f)*
Fr pneumoconiose *(f)*
Pt pneumoconiose *(f)*

pneumoconiosis *n* En
De Staublunge *(f)*
Es neumoconiosis *(f)*
Fr pneumoconiose *(f)*
It pneumoconiosi *(f)*
Pt pneumoconiose *(f)*

pó *(m) n* Pt
De Staub *(m)*
En dust
Es polvo *(m)*
Fr poussière *(f)*
It polvere *(f)*

poço *(m) n* Pt
De Bohrung *(f)*
En well
Es pozo *(m)*
Fr puits *(m)*
It pozzo *(m)*

poço de injecção *(m)* Pt
De Einpreßbohrung *(f)*
En injection well
Es pozo de inyección *(m)*
Fr puits d'injection *(m)*
It pozzo petrolifero sotto pressione *(m)*

poço de petróleo *(m)* Pt
De Ölbohrung *(f)*
En oil well
Es pozo petrolífero *(m)*
Fr puits de pétrole *(m)*
It pozzo petrolifero *(m)*

podere solare *(m)* It
De Solarfarm *(f)*
En solar farm
Es granja solar *(f)*
Fr ferme solaire *(f)*
Pt centro de produção solar *(f)*

poids *(m) n* Fr
De Gewicht *(n)*
En weight
Es peso *(m)*
It peso *(m)*
Pt peso *(m)*

poids atomique *(m)* Fr
De Atomgewicht *(n)*
En atomic weight
Es peso atómico *(m)*
It peso atomico *(m)*
Pt peso atómico *(m)*

point d'aniline *(m)* Fr
De Anilinpunkt *(m)*
En aniline point
Es punto de anilina *(m)*
It punto di anilina *(m)*
Pt ponto de anilina *(m)*

point d'ébullition *(f)* Fr
De Siedepunkt *(m)*
En boiling point
Es punto de ebullición *(f)*
It punto di ebollizione *(m)*
Pt ponto de ebulição *(m)*

point d'éclair *(m)* Fr
De Flammpunkt *(m)*
En flash point
Es punto de deflagración *(m)*
It punto di infiammabilità *(m)*
Pt ponto de fulgor *(m)*

point de congélation *(m)* Fr
De Gefrierpunkt *(m)*
En freezing point
Es punto de congelación *(m)*
It punto di congelamento *(m)*
Pt ponto de congelação *(m)*

point de fumée *(m)* Fr
De Rauchpunkt *(m)*
En smoke point
Es temperatura de formación de humo *(f)*
It punto di fumo *(m)*
Pt ponto de fumo *(m)*

point de fusion *(f)* Fr
De Schmelzpunkt *(m)*
En melting point
Es punto de fusión *(f)*
It punto di fusione *(m)*
Pt ponto de fusão *(m)*

point de rosée *(m)* Fr
De Taupunkt *(m)*
En dew point
Es punto de condensación *(m)*
It punto di rugiada *(m)*
Pt ponto de orvalho *(m)*

point de trouble *(m)* Fr
De Paraffinausscheidungspunkt *(m)*
En cloud point
Es punto de opacidad *(m)*
It punto di intorbidimento *(m)*
Pt ponto de névoa *(m)*

point mort *(m)* Fr
De Rentabilitätsgrenze *(f)*
En break-even point
Es punto comparativo *(m)*
It punto di pareggio *(m)*
Pt ponto de equilíbrio *(m)*

polarisation *(f) n* Fr
De Polarisation *(f)*
En polarization
Es polarización *(f)*
It polarizzazione *(f)*
Pt polarização *(f)*

Polarisation *(f) n* De
En polarization
Es polarización *(f)*
Fr polarisation *(f)*
It polarizzazione *(f)*
Pt polarização *(f)*

polarização *(f) n* Pt
De Polarisation *(f)*
En polarization
Es polarización *(f)*
Fr polarisation *(f)*
It polarizzazione *(f)*

polarización *(f) n* Es
De Polarisation *(f)*
En polarization
Fr polarisation *(f)*
It polarizzazione *(f)*
Pt polarização *(f)*

polarization *n* En
De Polarisation *(f)*
Es polarización *(f)*
Fr polarisation *(f)*
It polarizzazione *(f)*
Pt polarização *(f)*

polarizzazione *(f) n* It
De Polarisation *(f)*
En polarization
Es polarización *(f)*
Fr polarisation *(f)*
Pt polarização *(f)*

polifase *adj* Es, It, Pt
De mehrphasig
En polyphase
Fr polyphase

polimerização *(f) n* Pt
De Polymerisation *(f)*
En polymerization
Es polimerización *(f)*
Fr polymérisation *(f)*
It polimerizzazione *(f)*

polimerización *(f) n* Es
De Polymerisation *(f)*
En polymerization
Fr polymérisation *(f)*
It polimerizzazione *(f)*
Pt polimerização *(f)*

polimerizzazione *(f)* *n* It
De Polymerisation *(f)*
En polymerization
Es polimerización *(f)*
Fr polymérisation *(f)*
Pt polimerização *(f)*

polimero *(m)* *n* It
De Polymer *(n)*
En polymer
Es polímero *(m)*
Fr polymère *(m)*
Pt polímero *(m)*

polímero *(m)* *n* Es, Pt
De Polymer *(n)*
En polymer
Fr polymère *(m)*
It polimero *(m)*

politica energetica *(f)* It
De Energiepolitik *(f)*
En energy policy
Es política energética *(f)*
Fr politique de l'énergie *(f)*
Pt política energética *(f)*

política energética *(f)* Es, Pt
De Energiepolitik *(f)*
En energy policy
Fr politique de l'énergie *(f)*
It politica energetica *(f)*

politique de l'énergie *(f)* Fr
De Energiepolitik *(f)*
En energy policy
Es política energética *(f)*
It politica energetica *(f)*
Pt política energética *(f)*

pollution *n* En; Fr *(f)*
De Verschmutzung *(f)*
Es contaminación *(f)*
It inquinamento *(m)*
Pt poluição *(f)*

pollution de l'air *(f)* Fr
De Luftverschmutzung *(f)*
En air pollution
Es contaminación del aire *(f)*
It inquinamento atmosferico *(m)*
Pt poluição do ar *(f)*

poluição *(f)* *n* Pt
De Verschmutzung *(f)*
En pollution
Es contaminación *(f)*
Fr pollution *(f)*
It inquinamento *(m)*

poluição do ar *(f)* Pt
De Luftverschmutzung *(f)*
En air pollution
Es contaminación del aire *(f)*
Fr pollution de l'air *(f)*
It inquinamento atmosferico *(m)*

polvere *(f)* *n* It
De Staub *(m)*
En dust
Es polvo *(m)*
Fr poussière *(f)*
Pt pó *(m)*

polvo *(m)* *n* Es
De Staub *(m)*
En dust
Fr poussière *(f)*
It polvere *(f)*
Pt pó *(m)*

polymer *n* En
De Polymer *(n)*
Es polímero *(m)*
Fr polymère *(m)*
It polimero *(m)*
Pt polímero *(m)*

Polymer *(n)* *n* De
En polymer
Es polímero *(m)*
Fr polymère *(m)*
It polimero *(m)*
Pt polímero *(m)*

polymère *(m)* *n* Fr
De Polymer *(n)*
En polymer
Es polímero *(m)*
It polimero *(m)*
Pt polímero *(m)*

Polymerisation *(f)* *n* De
En polymerization
Es polimerización *(f)*
Fr polymérisation *(f)*
It polimerizzazione *(f)*
Pt polimerização *(f)*

polymérisation *(f)* *n* Fr
De Polymerisation *(f)*
En polymerization
Es polimerización *(f)*
It polimerizzazione *(f)*
Pt polimerização *(f)*

polymerization *n* En
De Polymerisation *(f)*
Es polimerización *(f)*
Fr polymérisation *(f)*
It polimerizzazione *(f)*
Pt polimerização *(f)*

polyphase *adj* En, Fr
De mehrphasig
Es polifase
It polifase
Pt polifase

polyurethane insulation En
De Polyurethanisolierung *(f)*
Es aislamiento de poliuretano *(m)*
Fr isolement au polyuréthane *(m)*
It isolamento con poliuretano *(m)*
Pt isolamento de poliuretano *(m)*

Polyurethanisolierung *(f)* *n* De
En polyurethane insulation
Es aislamiento de poliuretano *(m)*
Fr isolement au polyuréthane *(m)*
It isolamento con poliuretano *(m)*
Pt isolamento de poliuretano *(m)*

pompa *(f)* *n* It
De Pumpe *(f)*
En pump
Es bomba *(f)*
Fr pompe *(f)*
Pt bomba *(f)*

pompa di alimentazione *(f)* It
De Speisepumpe *(f)*
En feed pump
Es bomba de alimentación *(f)*
Fr pompe d'alimentation *(f)*
Pt bomba de alimentação *(f)*

pompa di calore *(f)* It
De Wärmepumpe *(f)*
En heat pump
Es bomba de calor *(f)*
Fr thermopompe *(f)*
Pt bomba térmica *(f)*

pompe *(f)* *n* Fr
De Pumpe *(f)*
En pump
Es bomba *(f)*
It pompa *(f)*
Pt bomba *(f)*

pompe à vide *(f)* Fr
De Vakuumpumpe *(f)*
En vacuum pump
Es bomba de vacío *(f)*
It depressore *(m)*
Pt bomba de vácuo *(f)*

pompe d'alimentation *(f)* Fr
De Speisepumpe *(f)*
En feed pump
Es bomba de alimentación *(f)*
It pompa di alimentazione *(f)*
Pt bomba de alimentação *(f)*

pompe-turbine *(f)* *n* Fr
De Pumpenturbine *(f)*
En pump-turbine
Es turbina-bomba *(f)*
It turbopompa *(f)*
Pt turbobomba *(f)*

ponto de anilina *(m)* Pt
De Anilinpunkt *(m)*
En aniline point
Es punto de anilina *(m)*
Fr point d'aniline *(m)*
It punto di anilina *(m)*

ponto de congelação *(m)* Pt
De Gefrierpunkt *(m)*
En freezing point
Es punto de congelación *(m)*
Fr point de congélation *(m)*
It punto di congelamento *(m)*

ponto de ebulição *(m)* Pt
De Siedepunkt *(m)*
En boiling point
Es punto de ebullición *(f)*
Fr point d'ébullition *(f)*
It punto di ebollizione *(m)*

ponto de equilíbrio *(m)* Pt
De Rentabilitätsgrenze *(f)*
En break-even point
Es punto comparativo *(m)*
Fr point mort *(m)*
It punto di pareggio *(m)*

ponto de fulgor *(m)* Pt
De Flammpunkt *(m)*
En flash point
Es punto de deflagración *(m)*
Fr point d'éclair *(m)*
It punto di infiammabilità *(m)*

ponto de fumo *(m)* Pt
De Rauchpunkt *(m)*
En smoke point
Es temperatura de formación de humo *(f)*
Fr point de fumée *(m)*
It punto di fumo *(m)*

ponto de fusão *(m)* Pt
De Schmelzpunkt *(m)*
En melting point
Es punto de fusión *(f)*
Fr point de fusion *(f)*
It punto di fusione *(m)*

ponto de névoa *(m)* Pt
De Paraffinausscheid-ungspunkt *(m)*
En cloud point
Es punto de opacidad *(m)*
Fr point de trouble *(m)*
It punto di intorbidimento *(m)*

ponto de orvalho *(m)* Pt
De Taupunkt *(m)*
En dew point
Es punto de condensación *(m)*
Fr point de rosée *(m)*
It punto di rugiada *(m)*

positrão *(m) n* Pt
De Positron *(n)*
En positron
Es positrón *(m)*
Fr positron *(m)*
It positrone *(m)*

positron *n* En; Fr *(m)*
De Positron *(n)*
Es positrón *(m)*
It positrone *(m)*
Pt positrão *(m)*

Positron *(n) n* De
En positron
Es positrón *(m)*
Fr positron *(m)*
It positrone *(m)*
Pt positrão *(m)*

positrón *(m) n* Es
De Positron *(n)*
En positron
Fr positron *(m)*
It positrone *(m)*
Pt positrão *(m)*

positrone *(m) n* It
De Positron *(n)*
En positron
Es positrón *(m)*
Fr positron *(m)*
Pt positrão *(m)*

post-aerated burner En
De nachbelüfteter Brenner *(m)*
Es quemador posaireado *(m)*
Fr brûleur à post-aération *(m)*
It bruciatore post-aerato *(m)*
Pt queimador pós-arejado *(m)*

potencia *(f) n* Es
De Leistung *(f)*
En power
Fr puissance *(f)*
It potenza *(f)*
Pt potência *(f)*

potência *(f) n* Pt
De Leistung *(f)*
En power
Es potencia *(f)*
Fr puissance *(f)*
It potenza *(f)*

potencia calorífica *(f)* Es
De Heizwert *(m)*
En calorific value
Fr valeur calorifique *(f)*
It potere calorifico *(m)*
Pt valor calorífico *(m)*

potência das marés *(f)* Pt
De Gezeitenkraft *(f)*
En tidal power
Es potencia mareal *(f)*
Fr puissance marémotrice *(f)*
It potenza della marea *(f)*

potencia de la onda *(f)* Es
De Wellenkraft *(f)*
En wavepower
Fr puissance ondulatoire *(f)*
It potenza d'onda *(f)*
Pt potência ondulatória *(f)*

potencia del viento *(f)* Es
De Windkraft *(f)*
En windpower
Fr puissance éolienne *(f)*
It potenza del vento *(f)*
Pt potência eólica *(f)*

potência de marés di estuário *(f)* Pt
De Gezeitenkraft in Flußmündungen *(f)*
En estuary tidal power
Es potencia mareal de estuario *(f)*
Fr puissance marémotrice d'estuaire *(f)*
It potenza della marea di estuario *(f)*

potência de ponta *(f)* Pt
De Spitzenkraft *(f)*
En peaking power
Es potencia màxima *(f)*
Fr puissance de crête *(f)*
It potenza massima *(f)*

potência eólica *(f)* Pt
De Windkraft *(f)*
En windpower
Es potencia del viento *(f)*
Fr puissance éolienne *(f)*
It potenza del vento *(f)*

potencial *(m) n* Es, Pt
De Potential *(n)*
En potential
Fr potentiel *(m)*
It potenziale *(m)*

potencial de contacto *(m)* Es, Pt
De Kontaktspannung *(f)*
En contact potential
Fr potentiel de contact *(m)*
It potenziale di contatto *(m)*

potencial de ionização *(m)* Pt
De Ionisierungsenergie *(f)*
En ionization potential
Es potencial de ionización *(m)*
Fr potentiel d'ionisation *(m)*
It potenziale di ionizzazione *(m)*

potencial de ionización *(m)* Es
De Ionisierungsenergie *(f)*
En ionization potential
Fr potentiel d'ionisation *(m)*
It potenziale di ionizzazione *(m)*
Pt potencial de ionização *(m)*

potencia mareal *(f)* Es
De Gezeitenkraft *(f)*
En tidal power
Fr puissance marémotrice *(f)*
It potenza della marea *(f)*
Pt potência das marés *(f)*

potencia mareal de estuario *(f)* Es
De Gezeitenkraft in Flußmündungen *(f)*
En estuary tidal power
Fr puissance marémotrice d'estuaire *(f)*
It potenza della marea di estuario *(f)*
Pt potência de marés de estuário *(f)*

potencia máxima *(f)* Es
De Spitzenkraft *(f)*
En peaking power
Fr puissance de crête *(f)*
It potenza massima *(f)*
Pt potência de ponta *(f)*

potência ondulatória *(f)* Pt
De Wellenkraft *(f)*
En wavepower
Es potencia de la onda *(f)*
Fr puissance ondulatoire *(f)*
It potenza d'onda *(f)*

potencia térmica *(f)* Es
De Heizkraft *(f)*
En thermal power
Fr puissance thermique *(f)*
It potenza termica *(f)*
Pt potência térmica *(f)*

potência térmica *(f)* Pt
De Heizkraft *(f)*
En thermal power
Es potencia térmica *(f)*
Fr puissance thermique *(f)*
It potenza termica *(f)*

potenciómetro *(m) n* Es, Pt
De Potentiometer *(n)*
En potentiometer
Fr potentiomètre *(m)*
It potenziometro *(m)*

potential *n* En
De Potential *(n)*
Es potencial *(m)*
Fr potentiel *(m)*
It potenziale *(m)*
Pt potencial *(m)*

Potential *(n) n* De
En potential
Es potencial *(m)*
Fr potentiel *(m)*
It potenziale *(m)*
Pt potencial *(m)*

potential converter En
De Spannungswandler *(m)*
Es convertidor de potencial *(m)*
Fr convertisseur de tension *(m)*
It convertitore di tensione *(m)*
Pt conversor de potencial *(m)*

potential difference En
De Spannungsunterschied *(m)*
Es diferencia de potencial *(f)*
Fr différence de potentiel *(f)*
It caduta di tensione *(f)*
Pt diferença de potencial *(f)*

potential divider En
De Spannungsteiler *(m)*
Es divisor de potencial; divisor de tensión *(m)*
Fr diviseur de tension *(m)*
It partitore di tensione *(m)*
Pt divisor de potencial; divisor de tensão *(m)*

potential energy En
De potentielle Energie *(f)*
Es energía potencial *(f)*
Fr énergie potentielle *(f)*
It energia potenziale *(f)*
Pt energia potencial *(f)*

potentiel *(m) n* Fr
De Potential *(n)*
En potential
Es potencial *(m)*
It potenziale *(m)*
Pt potencial *(m)*

potentiel de contact *(m)* Fr
De Kontaktspannung *(f)*
En contact potential
Es potencial de contacto *(m)*
It potenziale di contatto *(m)*
Pt potencial de contacto *(m)*

potentiel d'ionisation *(m)* Fr
De Ionisierungsenergie *(f)*
En ionization potential
Es potencial de ionización *(m)*
It potenziale di ionizzazione *(m)*
Pt potencial de ionização *(m)*

potentielle Energie *(f)* De
En potential energy
Es energía potencial *(f)*
Fr énergie potentielle *(f)*
It energia potenziale *(f)*
Pt energia potencial *(f)*

Potentiometer *(n) n* De
En potentiometer
Es potenciómetro *(m)*
Fr potentiomètre *(m)*
It potenziometro *(m)*
Pt potenciómetro *(m)*

potentiometer *n* En
De Potentiometer *(n)*
Es potenciómetro *(m)*
Fr potentiomètre *(m)*
It potenziometro *(m)*
Pt potenciómetro *(m)*

potentiomètre *(m) n* Fr
De Potentiometer *(n)*
En potentiometer
Es potenciómetro *(m)*
It potenziometro *(m)*
Pt potenciómetro *(m)*

potenza *(f) n* It
De Leistung *(f)*
En power
Es potencia *(f)*
Fr puissance *(f)*
Pt potência *(f)*

potenza della marea *(f)* It
De Gezeitenkraft *(f)*
En tidal power
Es potencia mareal *(f)*
Fr puissance marémotrice *(f)*
Pt potência das marés *(f)*

potenza della marea di estuario *(f)* It
De Gezeitenkraft in Flußmündungen *(f)*
En estuary tidal power
Es potencia mareal de estuario *(f)*
Fr puissance marémotrice d'estuaire *(f)*
Pt potência de marés de estuário *(f)*

potenza del vento *(f)* It
De Windkraft *(f)*
En windpower
Es potencia del viento *(f)*
Fr puissance éolienne *(f)*
Pt potência eólica *(f)*

potenza d'onda *(f)* It
De Wellenkraft *(f)*
En wavepower
Es potencia de la onda *(f)*
Fr puissance ondulatoire *(f)*
Pt potência ondulatória *(f)*

potenza e calore combinati *(f)* It
De kombinierte Wärme und Energie *(f)*
En combined heat and power
Es calor y potencia combinados *(m)*
Fr chaleur et puissance combinées *(f)*
Pt calor e potência combinados *(m)*

potenza massima *(f)* It
De Spitzenkraft *(f)*
En peaking power
Es potencia máxima *(f)*
Fr puissance de crête *(f)*
Pt potência de ponta *(f)*

potenza termica *(f)* It
De Heizkraft *(f)*
En thermal power
Es potencia térmica *(f)*
Fr puissance thermique *(f)*
Pt potência térmica *(f)*

potenziale *(m) n* It
De Potential *(n)*
En potential
Es potencial *(m)*
Fr potentiel *(m)*
Pt potencial *(m)*

potenziale di contatto *(m)* It
De Kontaktspannung *(f)*
En contact potential
Es potencial de contacto *(m)*
Fr potentiel de contact *(m)*
Pt potencial de contacto *(m)*

potenziale di ionizzazione *(m)* It
De Ionisierungsenergie *(f)*
En ionization potential
Es potencial de ionización *(m)*
Fr potentiel d'ionisation *(m)*
Pt potencial de ionização *(m)*

potenziometro *(m) n* It
De Potentiometer *(n)*
En potentiometer
Es potenciómetro *(m)*
Fr potentiomètre *(m)*
Pt potenciómetro *(m)*

potere calorifico *(m)* It
De Heizwert *(m)*
En calorific value
Es potencia calorífica *(f)*
Fr valeur calorifique *(f)*
Pt valor calorífico *(m)*

poupança de energia *(f)* Pt
De Energiesparen *(n)*
En energy saving
Es ahorro energético *(m)*
Fr économie d'énergie *(f)*
It risparmio di energia *(m)*

poussière *(f) n* Fr
De Staub *(m)*
En dust
Es polvo *(m)*
It polvere *(f)*
Pt pó *(m)*

pouvoir émissif *(m)* Fr
De Emissionsvermögen *(n)*
En emissivity
Es emisividad *(f)*
It emissività *(f)*
Pt emissividade *(f)*

power *n* En
De Leistung *(f)*
Es potencia *(f)*
Fr puissance *(f)*
It potenza *(f)*
Pt potência *(f)*

power density En
De Leistungsdichte *(f)*
Es densidad de potencia *(f)*
Fr puissance volumique *(f)*
It densità di potenza *(f)*
Pt densidade de potência *(f)*

power factor En
De Leistungsfaktor *(m)*
Es factor de potencia *(m)*
Fr facteur de puissance *(m)*
It fattore di potenza *(m)*
Pt factor de potência *(m)*

power station En
De Kraftwerk *(n)*
Es central eléctrica *(f)*
Fr centrale électrique *(f)*
It centrale elettrica *(f)*
Pt central energética *(f)*

pozo *(m) n* Es
De Bohrung *(f)*
En well
Fr puits *(m)*
It pozzo *(m)*
Pt poço *(m)*

pozo de inyección *(m)* Es
De Einpreßbohrung *(f)*
En injection well
Fr puits d'injection *(m)*
It pozzo petrolifero sotto pressione *(m)*
Pt poço de injecção *(m)*

pozo petrolífero *(m)* Es
De Ölbohrung *(f)*
En oil well
Fr puits de pétrole *(m)*
It pozzo petrolifero *(m)*
Pt poço de petróleo *(m)*

pozzo *(m) n* It
De Bohrung *(f)*
En well
Es pozo *(m)*
Fr puits *(m)*
Pt poço *(m)*

pozzo petrolifero *(m)* It
De Ölbohrung *(f)*
En oil well
Es pozo petrolífero *(m)*
Fr puits de pétrole *(m)*
Pt poço de petróleo *(m)*

pozzo petrolifero sotto pressione *(m)* It
De Einpreßbohrung *(f)*
En injection well
Es pozo de inyección *(m)*
Fr puits d'injection *(m)*
Pt poço de injecção *(m)*

Prallplatte *(f) ι.* De
En baffle
Es deflector *(m)*
Fr déflecteur *(m)*
It parafiamma *(m)*
Pt deflector *(m)*

pre-aerated burner En
De vorbelüfteter Brenner *(m)*
Es quemador preaireado *(m)*
Fr brûleur à pré-aération *(m)*
It bruciatore pre-aerato *(m)*
Pt queimador pré-arejado *(m)*

pré-aquecedor *(m) n* Pt
De Vorheizer *(m)*
En preheater
Es precalentador *(m)*
Fr pré-chauffeur *(m)*
It preriscaldatore *(m)*

precalentador *(m) n* Es
De Vorheizer *(m)*
En preheater
Fr pré-chauffeur *(m)*
It preriscaldatore *(m)*
Pt pré-aquecedor *(m)*

pré-chauffeur *(m) n* Fr
De Vorheizer *(m)*
En preheater
Es precalentador *(m)*
It preriscaldatore *(m)*
Pt pré-aquecedor *(m)*

precipitação radioactiva *(f)* Pt
De radioaktiver Niederschlag *(m)*
En fall-out
Es precipitación radiactiva *(f)*
Fr retombée radioactive *(f)*
It precipitazione radioattiva *(f)*

precipitación radiactiva *(f)* Es
De radioaktiver Niederschlag *(m)*
En fall-out
Fr retombée radioactive *(f)*
It precipitazione radioattiva *(f)*
Pt precipitação radioactiva *(f)*

precipitazione radioattiva *(f)* It
De radioaktiver Niederschlag *(m)*
En fall-out
Es precipitación radiactiva *(f)*
Fr retombée radioactive *(f)*
Pt precipitação radioactiva *(f)*

preheater *n* En
De Vorheizer *(m)*
Es precalentador *(m)*
Fr pré-chauffeur *(m)*
It preriscaldatore *(m)*
Pt pré-aquecedor *(m)*

preriscaldatore *(m) n* It
De Vorheizer *(m)*
En preheater
Es precalentador *(m)*
Fr pré-chauffeur *(m)*
Pt pré-aquecedor *(m)*

presa *(f) n* Es
De Damm *(m)*
En dam
Fr barrage *(m)*
It diga *(f)*
Pt barragem *(f)*

presa d'aria *(f)* It
De Lufteinlaß *(m)*
En air inlet
Es entrada de aire *(f)*
Fr admission d'air *(f)*
Pt admissão de ar *(f)*

presa de marea *(f)* Es
De Gezeitendamm *(m)*
En tidal barrage
Fr barrage marémoteur *(m)*
It sbarramento di marea *(m)*
Pt barragem de marés *(f)*

presión *(f)* *n* Es
De Druck *(m)*
En pressure
Fr pression *(f)*
It pressione *(f)*
Pt pressão *(f)*

presión atmosférica *(f)* Es
De Luftdruck *(m)*
En atmospheric pressure
Fr pression atmosphérique *(f)*
It pressione atmosferica *(f)*
Pt pressão atmosférica *(f)*

presión del vapor *(f)* Es
Am vapor pressure
De Dampfdruck *(m)*
En vapour pressure
Fr tension de vapeur *(f)*
It tensione di vapore *(f)*
Pt pressão de vapor *(f)*

presión de vapor saturado *(f)* Es
Am saturated vapor pressure
De gesättigter Dampfdruck *(m)*
En saturated vapour pressure
Fr tension de vapeur saturée *(f)*
It pressione del vapore saturo *(m)*
Pt pressão de vapor saturado *(f)*

presión osmótica *(f)* Es
De osmotischer Druck *(m)*
En osmotic pressure
Fr pression osmotique *(f)*
It pressione osmotica *(f)*
Pt pressão osmótica *(f)*

pressão *(f)* *n* Pt
De Druck *(m)*
En pressure
Es presión *(f)*
Fr pression *(f)*
It pressione *(f)*

pressão atmosférica *(f)* Pt
De Luftdruck *(m)*
En atmospheric pressure
Es presión atmosférica *(f)*
Fr pression atmosphérique *(f)*
It pressione atmosferica *(f)*

pressão de vapor *(f)* Pt
Am vapor pressure
De Dampfdruck *(m)*
En vapour pressure
Es presión del vapor *(f)*
Fr tension de vapeur *(f)*
It tensione di vapore *(f)*

pressão de vapor saturado *(f)* Pt
Am saturated vapor pressure
De gesättigter Dampfdruck *(m)*
En saturated vapour pressure
Es presión de vapor saturado *(f)*
Fr tension de vapeur saturée *(f)*
It pressione del vapore saturo *(m)*

pressão osmótica *(f)* Pt
De osmotischer Druck *(m)*
En osmotic pressure
Es presión osmótica *(f)*
Fr pression osmotique *(f)*
It pressione osmotica *(f)*

pressatura *(f)* *n* It
De Kippen *(n)*
En dumping
Es vaciamiento *(m)*
Fr déversement *(m)*
Pt esvaziamento *(m)*

pression *(f)* *n* Fr
De Druck *(m)*
En pressure
Es presión *(f)*
It pressione *(f)*
Pt pressão *(f)*

pression atmosphérique *(f)* Fr
De Luftdruck *(m)*
En atmospheric pressure
Es presión atmosférica *(f)*
It pressione atmosferica *(f)*
Pt pressão atmosférica *(f)*

pressione *(f)* *n* It
De Druck *(m)*
En pressure
Es presión *(f)*
Fr pression *(f)*
Pt pressão *(f)*

pressione atmosferica *(f)* It
De Luftdruck *(m)*
En atmospheric pressure
Es presión atmosférica *(f)*
Fr pression atmosphérique *(f)*
Pt pressão atmosférica *(f)*

pressione del vapore saturo *(m)* It
Am saturated vapor pressure
De gesättigter Dampfdruck *(m)*
En saturated vapour pressure
Es presión de vapor saturado *(f)*
Fr tension de vapeur saturée *(f)*
Pt pressão de vapor saturado *(f)*

pressione osmotica *(f)* It
De osmotischer Druck *(m)*
En osmotic pressure
Es presión osmótica *(f)*
Fr pression osmotique *(f)*
Pt pressão osmótica *(f)*

pression osmotique *(f)* Fr
De osmotischer Druck *(m)*
En osmotic pressure
Es presión osmótica *(f)*
It pressione osmotica *(f)*
Pt pressão osmótica *(f)*

pressure *n* En
De Druck *(m)*
Es presión *(f)*
Fr pression *(f)*
It pressione *(f)*
Pt pressão *(f)*

pressure drop En
De Druckabfall *(m)*
Es caída de presión *(f)*
Fr chute de pression *(f)*
It caduta di pressione *(f)*
Pt queda de pressão *(f)*

pressure head En
De Druckhöhe *(f)*
Es altura manométrica *(f)*
Fr hauteur de refoulement *(f)*
It altezza manometrica *(f)*
Pt altura manométrica *(f)*

pressure-tube reactor En
De Druckröhrenreaktor *(m)*
Es reactor de tubos de presión *(m)*
Fr réacteur à tube de force *(m)*
It reattore con tubo di pressione *(m)*
Pt reactor de tubo de pressão *(m)*

pressurized-water reactor (PWR) En
De Druckwasserreaktor *(m)*
Es reactor de agua a presión *(m)*
Fr réacteur à eau sous pression *(m)*
It reattore ad acqua pressurizzata *(m)*
Pt reactor de água pressurizada *(m)*

pretreatment *n* En
De Vorbehandlung *(f)*
Es tratamiento previo *(m)*
Fr traitement préalable *(m)*
It trattamento previo *(m)*
Pt tratamento prévio *(m)*

primäre Energie *(f)* De
- En primary energy
- Es energía primaria *(f)*
- Fr énergie primaire *(f)*
- It energia primaria *(f)*
- Pt energia primária *(f)*

primäre Luft *(f)* De
- En primary air
- Es aire primario *(m)*
- Fr air primaire *(m)*
- It aria primaria *(f)*
- Pt ar primário *(m)*

primäres Kühlmittel *(n)* De
- En primary coolant
- Es refrigerante primario *(m)*
- Fr réfrigérant primaire *(m)*
- It refrigerante primario *(m)*
- Pt refrigerante primário *(m)*

primary air En
- De primäre Luft *(f)*
- Es aire primario *(m)*
- Fr air primaire *(m)*
- It aria primaria *(f)*
- Pt ar primário *(m)*

primary coolant En
- De primäres Kühlmittel *(n)*
- Es refrigerante primario *(m)*
- Fr réfrigérant primaire *(m)*
- It refrigerante primario *(m)*
- Pt refrigerante primário *(m)*

primary energy En
- De primäre Energie *(f)*
- Es energía primaria *(f)*
- Fr énergie primaire *(f)*
- It energia primaria *(f)*
- Pt energia primária *(f)*

probabilidad *(f) n* Es
- De Wahrscheinlichkeit *(f)*
- En probability
- Fr probabilité *(f)*
- It probabilità *(f)*
- Pt probabilidade *(f)*

probabilidade *(f) n* Pt
- De Wahrscheinlichkeit *(f)*
- En probability
- Es probabilidad *(f)*
- Fr probabilité *(f)*
- It probabilità *(f)*

probabilità *(f) n* It
- De Wahrscheinlichkeit *(f)*
- En probability
- Es probabilidad *(f)*
- Fr probabilité *(f)*
- Pt probabilidade *(f)*

probabilité *(f) n* Fr
- De Wahrscheinlichkeit *(f)*
- En probability
- Es probabilidad *(f)*
- It probabilità *(f)*
- Pt probabilidade *(f)*

probability *n* En
- De Wahrscheinlichkeit *(f)*
- Es probabilidad *(f)*
- Fr probabilité *(f)*
- It probabilità *(f)*
- Pt probabilidade *(f)*

probable reserves En
- De wahrscheinliche Reserven *(pl)*
- Es reservas probables *(f pl)*
- Fr réserves probables *(f pl)*
- It riserve probabili *(f pl)*
- Pt reservas prováveis *(f pl)*

procédé consol *(m)* Fr
- De Konsolverfahren *(n)*
- En consol process
- Es proceso consol *(m)*
- It processo consol *(m)*
- Pt processo consol *(m)*

proceso consol *(m)* Es
- De Konsolverfahren *(n)*
- En consol process
- Fr procédé consol *(m)*
- It processo consol *(m)*
- Pt processo consol *(m)*

processo consol *(m)* It
- De Konsolverfahren *(n)*
- En consol process
- Es proceso consol *(m)*
- Fr procédé consol *(m)*
- Pt processo consol *(m)*

processo consol *(m)* Pt
- De Konsolverfahren *(n)*
- En consol process
- Es proceso consol *(m)*
- Fr procédé consol *(m)*
- It processo consol *(m)*

procura de ponta *(f)* Pt
- De Spitzenbedarf *(m)*
- En peak demand
- Es demanda punta *(f)*
- Fr consommation de pointe *(f)*
- It domanda di punta *(f)*

prodotti petrolchimici *(m pl)* It
- De petrochemische Produkte *(pl)*
- En petrochemicals
- Es productos petroquímicos *(m pl)*
- Fr produits pétrochimiques *(m pl)*
- Pt produtos petroquímicos *(m pl)*

prodotto di fissione *(m)* It
- De Spaltprodukt *(n)*
- En fission product
- Es producto de fisión *(m)*
- Fr produit de la fission *(m)*
- Pt produto de desintegração *(m)*

produção a par *(f)* Pt
- De paarweise Produktion *(f)*
- En pair production
- Es producción por pares *(f)*
- Fr production en paire *(f)*
- It produzione di coppie *(f)*

produção de energia per capita *(f)* Pt
- De Stromerzeugung pro Kopf *(f)*
- En per capita energy production
- Es producción de energía per capita *(f)*
- Fr production d'énergie par personne *(f)*
- It produzione di energia pro capite *(f)*

producción de energía per capita *(f)* Es
- De Stromerzeugung pro Kopf *(f)*
- En per capita energy production
- Fr production d'énergie par personne *(f)*
- It produzione di energia pro capite *(f)*
- Pt produção de energia per capita *(f)*

producción final *(f)* Es
- De Ausbeutefaktor *(m)*
- En ultimate recovery
- Fr récupération finale *(f)*
- It recupero ultimo *(m)*
- Pt recuperação final *(f)*

producción por pares *(f)* Es
- De paarweise Produktion *(f)*
- En pair production
- Fr production en paire *(f)*
- It produzione di coppie *(f)*
- Pt produção a par *(f)*

producer gas En
- De Generatorgas *(n)*
- Es gas probre *(m)*
- Fr gaz de gazogène *(m)*
- It gas di gassogeno *(m)*
- Pt gás probre *(m)*

production *(f) n* Fr
- De Produktion *(f)*
- En output
- Es salida *(f)*
- It produzione *(f)*
- Pt saída *(f)*

production d'électricité *(f)* Fr
- De Stromerzeugung *(f)*
- En electricity generation
- Es generación de electricidad *(f)*
- It generazione elettrica *(f)*
- Pt geração de electricidade *(f)*

production d'énergie par personne *(f)* Fr
- De Stromerzeugung pro Kopf *(f)*
- En per capita energy production

Es producción de energía per capita *(f)*
It produzione di energia pro capite *(f)*
Pt produção de energia per capita *(f)*

production en paire *(f)* Fr
De paarweise Produktion *(f)*
En pair production
Es producción por pares *(f)*
It produzione di coppie *(f)*
Pt produção a par *(f)*

producto de fisión *(m)* Es
De Spaltprodukt *(n)*
En fission product
Fr produit de la fission *(m)*
It prodotto di fissione *(m)*
Pt produto de desintegração *(m)*

productos petroquímicos *(m pl)* Es
De petrochemische Produkte *(pl)*
En petrochemicals
Fr produits pétrochimiques *(m pl)*
It prodotti petrolchimici *(m pl)*
Pt produtos petroquímicos *(m pl)*

produit de la fission *(m)* Fr
De Spaltprodukt *(n)*
En fission product
Es producto de fisión *(m)*
It prodotto di fissione *(m)*
Pt produto de desintegração *(m)*

produits de queue *(m pl)* Fr
De Haldenabfall *(m)*
En tailings *(pl)*
Es desechos *(m pl)*
It residui di scarto *(m pl)*
Pt colas *(f pl)*

produits pétrochimiques *(m pl)* Fr
De petrochemische Produkte *(pl)*
En petrochemicals
Es productos petroquímicos *(m pl)*
It prodotti petrolchimici *(m pl)*
Pt produtos petroquímicos *(m pl)*

Produktion *(f) n* De
En output
Es salida *(f)*
Fr production *(f)*
It produzione *(f)*
Pt saída *(f)*

produto de desintegração *(m)* Pt
De Spaltprodukt *(n)*
En fission product
Es producto de fisión *(m)*
Fr produit de la fission *(m)*
It prodotto di fissione *(m)*

produtos petroquímicos *(m pl)* Pt
De petrochemische Produkte *(pl)*
En petrochemicals
Es productos petroquímicos *(m pl)*
Fr produits pétrochimiques *(m pl)*
It prodotti petrolchimici *(m pl)*

produzione *(f) n* It
De Produktion *(f)*
En output
Es salida *(f)*
Fr production *(f)*
Pt saída *(f)*

produzione catrame *(f)* It
De Teergiebigkeit *(f)*
En tar yield
Es riqueza de alquitrán *(f)*
Fr rendement en goudron *(m)*
Pt rendimento de alcatrão *(m)*

produzione di coppie *(f)* It
De paarweise Produktion *(f)*
En pair production
Es producción por pares *(f)*
Fr production en paire *(f)*
Pt produção a par *(f)*

produzione di energia pro capite *(f)* It
De Stromerzeugung pro Kopf *(f)*
En per capita power production
Es producción de energía per capita *(f)*
Fr production d'énergie par personne *(f)*
Pt produção de energia per capita *(f)*

produzione di torba *(f)* It
De Vertorfung *(f)*
En peatification
Es formación de turba *(f)*
Fr formation de la tourbe *(f)*
Pt turfização *(f)*

projecto de ambiente integrado *(m)* Pt
De integrierte Umweltplanung *(f)*
En integrated environmental design (IED)
Es diseño ambienta integrado *(m)*
Fr étude de l'environment integreé *(f)*
It design ambientale integrato *(m)*

promovimento a grado superiore *(m)* It
De Aufkonzentrierung *(f)*
En upgrading
Es mejoramiento *(m)*
Fr amélioration *(f)*
Pt melhoramento *(m)*

Propan *(n) n* De
En propane
Es propano *(m)*
Fr propane *(m)*
It propano *(m)*
Pt propano *(m)*

propane *n* En; Fr *(m)*
De Propan *(n)*
Es propano *(m)*
It propano *(m)*
Pt propano *(m)*

propano *(m) n* Es, It, Pt
De Propan *(n)*
En propane
Fr propane *(m)*

propanol *n* En; Es, Fr, Pt *(m)*
De Propanol *(n)*
It propanolo *(m)*

Propanol *(n) n* De
En propanol
Es propanol *(m)*
Fr propanol *(m)*
It propanolo *(m)*
Pt propanol *(m)*

propanolo *(m) n* It
De Propanol *(n)*
En propanol
Es propanol *(m)*
Fr propanol *(m)*
Pt propanol *(m)*

propellant *n* En
De Treibstoff *(m)*
Es propulsante *(m)*
Fr propergol *(m)*
It propellente *(m)*
Pt combustível propulsor *(m)*

propellente *(m) n* It
De Treibstoff *(m)*
En propellant
Es propulsante *(m)*
Fr propergol *(m)*
Pt combustível propulsor *(m)*

propellente ibrido per missili *(m)* It
De Hybrid-Raketen-treibstoff *(m)*
En hybrid rocket propellant
Es propulsantede cohete híbrido *(m)*
Fr propergol hybride *(m)*
Pt combustível propulsor de foguetão híbrido *(m)*

propellenti liquidi *(m pl)* It
De flüssige Treibstoffe *(pl)*
En liquid propellants *(pl)*
Es propulsantes líquidos *(m pl)*
Fr propergols liquides *(m pl)*
Pt combustíveis propulsores líquidos *(m pl)*

propergol *(m) n* Fr
De Treibstoff *(m)*
En propellant
Es propulsante *(m)*
It propellente *(m)*
Pt combustível propulsor *(m)*

propergol hybride *(m)* Fr
De Hybrid-Raketen-treibstoff *(m)*
En hybrid rocket propellant
Es propulsantede cohete híbrido *(m)*
It propellente ibrido per missili *(m)*
Pt combustível propulsor de foguetão híbrido *(m)*

propergols liquides *(m pl)* Fr
De flüssige Treibstoffe *(pl)*
En liquid propellants *(pl)*
Es propulsantes líquidos *(m pl)*
It propellenti liquidi *(m pl)*
Pt combustíveis propulsores líquidos *(m pl)*

propilene *(m) n* It
De Propylen *(n)*
En propylene
Es propileno *(m)*
Fr propylène *(m)*
Pt propileno *(m)*

propileno *(m) n* Es, Pt
De Propylen *(n)*
En propylene
Fr propylène *(m)*
It propilene *(m)*

propulsante *(m) n* Es
De Treibstoff *(m)*
En propellant
Fr propergol *(m)*
It propellente *(m)*
Pt combustível propulsor *(m)*

propulsantede cohete híbrido *(m)* Es
De Hybrid-Raketen-treibstoff *(m)*
En hybrid rocket propellant
Fr propergol hybride *(m)*
It propellente ibrido per missili *(m)*
Pt combustível propulsor de foguetão híbrido *(m)*

propulsantes líquidos *(m pl)* Es
De flüssige Treibstoffe *(pl)*
En liquid propellants *(pl)*
Fr propergols liquides *(m pl)*
It propellenti liquidi *(m pl)*
Pt combustíveis propulsores líquidos *(m pl)*

propulsão a jacto *(f)* Pt
De Strahlantrieb *(m)*
En jet propulsion
Es propulsión a chorro *(f)*
Fr propulsion par réaction *(f)*
It propulsione a getto *(m)*

propulsión a chorro *(f)* Es
De Strahlantrieb *(m)*
En jet propulsion
Fr propulsion par réaction *(f)*
It propulsione a getto *(m)*
Pt propulsão a jacto *(f)*

propulsione a getto *(m)* It
De Strahlantrieb *(m)*
En jet propulsion
Es propulsión a chorro *(f)*
Fr propulsion par réaction *(f)*
Pt propulsão a jacto *(f)*

propulsion par réaction *(f)* Fr
De Strahlantrieb *(m)*
En jet propulsion
Es propulsión a chorro *(f)*
It propulsione a getto *(m)*
Pt propulsão a jacto *(f)*

Propylen *(n) n* De
En propylene
Es propileno *(m)*
Fr propylène *(m)*
It propilene *(m)*
Pt propileno *(m)*

propylene *n* En
De Propylen *(n)*
Es propileno *(m)*
Fr propylène *(m)*
It propilene *(m)*
Pt propileno *(m)*

propylène *(m) n* Fr
De Propylen *(n)*
En propylene
Es propileno *(m)*
It propilene *(m)*
Pt propileno *(m)*

prospecção geofísica *(f)* Pt
De geophysisches Schürfen *(n)*
En geophysical prospecting
Es prospección geofísica *(f)*
Fr prospection géophysique *(f)*
It prospezione geofisica *(f)*

prospecção geoquímica *(f)* Pt
De geochemisches Schürfen *(n)*
En geochemical prospecting
Es prospección geoquímica *(f)*
Fr prospection géochimique *(f)*
It prospezione geochimica *(f)*

prospección geofísica *(f)* Es
De geophysisches Schürfen *(n)*
En geophysical prospecting
Fr prospection géophysique *(f)*
It prospezione geofisica *(f)*
Pt prospecção geofísica *(f)*

prospección geoquímica *(f)* Es
De geochemisches Schürfen *(n)*
En geochemical prospecting
Fr prospection géochimique *(f)*
It prospezione geochimica *(f)*
Pt prospecção geoquímica *(f)*

prospection géochimique *(f)* Fr
De geochemisches Schürfen *(n)*
En geochemical prospecting
Es prospección geoquímica *(f)*
It prospezione geochimica *(f)*
Pt prospecção geoquímica *(f)*

prospection géophysique *(f)* Fr
De geophysisches Schürfen *(n)*
En geophysical prospecting
Es prospección geofísica *(f)*
It prospezione geofisica *(f)*
Pt prospecção geofísica *(f)*

prospection magnétique *(f)* Fr
De Magnetaufnahme *(f)*
En magnetic survey
Es levantamiento magnético *(m)*
It perizia magnetica *(f)*
Pt levantamento magnético *(m)*

prospezione geochimica *(f)* It
De geochemisches Schürfen *(n)*
En geochemical prospecting
Es prospección geoquímica *(f)*
Fr prospection géochimique *(f)*
Pt prospecção geoquímica *(f)*

prospezione geofisica *(f)* It
De geophysisches Schürfen *(n)*
En geophysical prospecting
Es prospección geofísica *(f)*
Fr prospection géophysique *(f)*
Pt prospecção geofísica *(f)*

protecção antigelo *(f)* Pt
De Gefrierschutz *(m)*
En freeze protection
Es protección contra la congelación *(f)*
Fr protection contre le gel *(f)*
It protezione dal congelamento *(f)*

protecção contra a intempérie *(f)* Pt
De Wetterschutz *(m)*
En weatherproofing
Es impermeabilización *(f)*
Fr protection contre les intempéries *(f)*
It resistenza alle intemperie *(f)*

protección contra la congelación *(f)* Es
De Gefrierschutz *(m)*
En freeze protection
Fr protection contre le gel *(f)*
It protezione dal congelamento *(f)*
Pt protecção antigelo *(f)*

protection contre le gel *(f)* Fr
De Gefrierschutz *(m)*
En freeze protection
Es protección contra la congelación *(f)*
It protezione dal congelamento *(f)*
Pt protecção antigelo *(f)*

protection contre les intempéries *(f)* Fr
De Wetterschutz *(m)*
En weatherproofing
Es impermeabilización *(f)*
It resistenza alle intemperie *(f)*
Pt protecção contra a intempérie *(f)*

protection system En
De Schutzsystem *(n)*
Es sistema de protección *(m)*
Fr système de protection *(m)*
It sistema di protezione *(m)*
Pt sistema de protecção *(m)*

protective multiple earthing *n* En
De Mehrfach-Schutzerdung *(f)*
Es toma de tierra múltiple protectora *(f)*
Fr mise à la terre multiple de protection *(f)*
It messa a terra multipla protettiva *(f)*
Pt ligação à terra múltipla protectora *(f)*

protector *(m) n* Pt
De Abschirmung *(f)*
En shield
Es blindaje *(m)*
Fr bouclier *(m)*
It schermo *(m)*

protector biológico *(m)* Pt
De biologischer Schild *(m)*
En biological shield
Es blindaje biológico *(m)*
Fr bouclier biologique *(m)*
It schermo biologico *(m)*

protector térmico *(m)* Pt
De Wärmeschutz *(m)*
En heat shield; thermal shield
Es pantalla térmica *(f)*
Fr bouclier thermique *(m)*
It schermo termico *(m)*

protezione dal congelamento *(f)* It
De Gefrierschutz *(m)*
En freeze protection
Es protección contra la congelación *(f)*
Fr protection contre le gel *(f)*
Pt protecção antigelo *(f)*

prototipo di reattore veloce *(m)* It
De Prototypen-Schnellreaktor *(m)*
En prototype fast reactor (PFR)
Es reactor rápido prototipo *(m)*
Fr réacteur rapide prototype *(m)*
Pt reactor rápido prototipo *(m)*

prototype fast reactor (PFR) En
De Prototypen-Schnellreaktor *(m)*
Es reactor rápido prototipo *(m)*
Fr réacteur rapide prototype *(m)*
It prototipo di reattore veloce *(m)*
Pt reactor rápido prototipo *(m)*

Prototypen-Schnellreaktor *(m) n* De
En prototype fast reactor (PFR)
Es reactor rápido prototipo *(m)*
Fr réacteur rapide prototype *(m)*
It prototipo di reattore veloce *(m)*
Pt reactor rápido prototipo *(m)*

prova di frangibilità *(f)* It
De Sturzfestigkeits-prüfung *(f)*
En shatter test
Es prueba de cohesión *(f)*
Fr essai de résistance au choc *(m)*
Pt teste de fragmentação *(m)*

prova di ringonfiamento *(f)* It
De Quellversuch *(m)*
En swelling test
Es prueba de esponjamiento *(f)*
Fr essai de gonflement *(m)*
Pt teste de inchamento *(m)*

prova di sedimentazione *(f)* It
De Rückstandsprüfung *(f)*
En sediment test
Es prueba de sedimentos *(f)*
Fr essai de sédimentation *(m)*
Pt teste de sedimentos *(m)*

proven reserves En
De erwiesene Reserven *(pl)*
Es reservas probadas *(f pl)*
Fr réserves prouvées *(f pl)*
It riserve dimostrate *(f pl)*
Pt reservas comprovadas *(f pl)*

prueba de cohesión *(f)* Es
De Sturzfestigkeits-prüfung *(f)*
En shatter test
Fr essai de résistance au choc *(m)*
It prova di frangibilità *(f)*
Pt teste de fragmentação *(m)*

prueba de esponjamiento *(f)* Es
De Quellversuch *(m)*
En swelling test
Fr essai de gonflement *(m)*

It prova di ringonfiamento *(f)*
Pt teste de inchamento *(m)*

prueba de sedimentos *(f)* Es
De Rückstandsprüfung *(f)*
En sediment test
Fr essai de sédimentation *(m)*
It prova di sedimentazione *(f)*
Pt teste de sedimentos *(m)*

PTL-Triebwerk *(n)* *n* De
En turboprop
Es turbohélice *(f)*
Fr turbopropulseur *(m)*
It turboelica *(f)*
Pt turbo-hélice *(f)*

puissance *(f)* *n* Fr
De Leistung *(f)*
En power
Es potencia *(f)*
It potenza *(f)*
Pt potência *(f)*

puissance de crête *(f)* Fr
De Spitzenkraft *(f)*
En peaking power
Es potencia máxima *(f)*
It potenza massima *(f)*
Pt potência de ponta *(f)*

puissance éolienne *(f)* Fr
De Windkraft *(f)*
En windpower
Es potencia del viento *(f)*
It potenza del vento *(f)*
Pt potência eólica *(f)*

puissance marémotrice *(f)* Fr
De Gezeitenkraft *(f)*
En tidal power
Es potencia mareal *(f)*
It potenza della marea *(f)*
Pt potência das marés *(f)*

puissance marémotrice d'estuaire *(f)* Fr
De Gezeitenkraft in Flußmündungen *(f)*
En estuary tidal power
Es potencia mareal de estuario *(f)*
It potenza della marea di estuario *(f)*
Pt potência das marés de estuário *(f)*

puissance ondulatoire *(f)* Fr
De Wellenkraft *(f)*
En wavepower
Es potencia de la onda *(f)*
It potenza d'onda *(f)*
Pt potência ondulatória *(f)*

puissance thermique *(f)* Fr
De Heizkraft *(f)*
En thermal power
Es potencia térmica *(f)*
It potenza termica *(f)*
Pt potência térmica *(f)*

puissance volumique *(f)* Fr
De Leistungsdichte *(f)*
En power density
Es densidad de potencia *(f)*
It densità di potenza *(f)*
Pt densidade de potência *(f)*

puits *(m)* *n* Fr
De Bohrung *(f)*
En well
Es pozo *(m)*
It pozzo *(m)*
Pt poço *(m)*

puits de pétrole *(m)* Fr
De Ölbohrung *(f)*
En oil well
Es pozo petrolífero *(m)*
It pozzo petrolifero *(m)*
Pt poço de petróleo *(m)*

puits d'injection *(m)* Fr
De Einpreßbohrung *(f)*
En injection well
Es pozo de inyección *(m)*
It pozzo petrolifero sotto pressione *(m)*
Pt poço de injecção *(m)*

pulsação *(f)* *n* Pt
De Pulsierung *(f)*
En pulsation
Es pulsación *(f)*
Fr pulsation *(f)*
It pulsazione *(f)*

pulsación *(f)* *n* Es
De Pulsierung *(f)*
En pulsation
Fr pulsation *(f)*
It pulsazione *(f)*
Pt pulsação *(f)*

pulsation *n* En; Fr *(f)*
De Pulsierung *(f)*
Es pulsación *(f)*
It pulsazione *(f)*
Pt pulsação *(f)*

pulsazione *(f)* *n* It
De Pulsierung *(f)*
En pulsation
Es pulsación *(f)*
Fr pulsation *(f)*
Pt pulsação *(f)*

pulse generator En
De Impulsgenerator *(m)*
Es generador de impulsos *(m)*
Fr générateur d'impulsions *(m)*
It generatore d'impulsi *(m)*
Pt gerador de impulsos *(m)*

Pulsierung *(f)* *n* De
En pulsation
Es pulsación *(f)*
Fr pulsation *(f)*
It pulsazione *(f)*
Pt pulsação *(f)*

pulverisierte Kohle *(f)* De
En pulverized coal
Es carbón pulverizado *(m)*
Fr charbon pulvérisé *(m)*
It carbone polverizzato *(m)*
Pt carvão pulverizado *(m)*

pulverized coal En
De pulverisierte Kohle *(f)*
Es carbón pulverizado *(m)*
Fr charbon pulvérisé *(m)*
It carbone polverizzato *(m)*
Pt carvão pulverizado *(m)*

pump *n* En
De Pumpe *(f)*
Es bomba *(f)*
Fr pompe *(f)*
It pompa *(f)*
Pt bomba *(f)*

Pumpe *(f)* *n* De
En pump
Es bomba *(f)*
Fr pompe *(f)*
It pompa *(f)*
Pt bomba *(f)*

pumped storage En
De gepumpte Speicherung *(f)*
Es almacenamiento bombeado *(m)*
Fr accumulation pompée *(f)*
It immagazzinaggi a pompa *(m)*
Pt armazenagem realizada à bomba *(f)*

Pumpenturbine *(f)* *n* De
En pump-turbine
Es turbina-bomba *(f)*
Fr pompe-turbine *(f)*
It turbopompa *(f)*
Pt turbobomba *(f)*

pump-turbine *n* En
De Pumpenturbine *(f)*
Es turbina-bomba *(f)*
Fr pompe-turbine *(f)*
It turbopompa *(f)*
Pt turbobomba *(f)*

punta *(f)* *n* It
De Bohrmeißel *(m)*
En bit
Es broca *(f)*
Fr trépan *(m)*
Pt broca *(f)*

punto comparativo *(m)* Es
De Rentabilitätsgrenze *(f)*
En break-even point
Fr point mort *(m)*
It punto di pareggio *(m)*
Pt ponto de equilíbrio *(m)*

punto de anilina *(m)* Es
De Anilinpunkt *(m)*
En aniline point
Fr point d'aniline *(m)*

It punto di anilina *(m)*
Pt ponto de anilina *(m)*

punto de condensación *(m)* Es
De Taupunkt *(m)*
En dew point
Fr point de rosée *(m)*
It punto di rugiada *(m)*
Pt ponto de orvalho *(m)*

punto de congelación *(m)* Es
De Gefrierpunkt *(m)*
En freezing point
Fr point de congélation *(m)*
It punto di congelamento *(m)*
Pt ponto de congelação *(m)*

punto de deflagración *(m)* Es
De Flammpunkt *(m)*
En flash point
Fr point d'éclair *(m)*
It punto di infiammabilità *(m)*
Pt ponto de fulgor *(m)*

punto de ebullición *(f)* Es
De Siedepunkt *(m)*
En boiling point
Fr point d'ébullition *(f)*
It punto di ebollizione *(m)*
Pt ponto de ebulição *(m)*

punto de fusión *(f)* Es
De Schmelzpunkt *(m)*
En melting point
Fr point de fusion *(f)*
It punto di fusione *(m)*
Pt ponto de fusão *(m)*

punto de opacidad *(m)* Es
De Paraffinausscheidungspunkt *(m)*
En cloud point
Fr point de trouble *(m)*
It punto di intorbidimento *(m)*
Pt ponto de névoa *(m)*

punto di anilina *(m)* It
De Anilinpunkt *(m)*
En aniline point
Es punto de anilina *(m)*
Fr point d'aniline *(m)*
Pt ponto de anilina *(m)*

punto di congelamento *(m)* It
De Gefrierpunkt *(m)*
En freezing point
Es punto de congelación *(m)*
Fr point de congélation *(m)*
Pt ponto de congelação *(m)*

punto di ebollizione *(m)* It
De Siedepunkt *(m)*
En boiling point
Es punto de ebullición *(f)*
Fr point d'ébullition *(f)*
Pt ponto de ebulição *(m)*

punto di fumo *(m)* It
De Rauchpunkt *(m)*
En smoke point
Es temperatura de formación de humo *(f)*
Fr point de fumée *(m)*
Pt ponto de fumo *(m)*

punto di fusione *(m)* It
De Schmelzpunkt *(m)*
En melting point
Es punto de fusión *(f)*
Fr point de fusion *(f)*
Pt ponto de fusão *(m)*

punto di infiammabilità *(m)* It
De Flammpunkt *(m)*
En flash point
Es punto de deflagración *(m)*
Fr point d'éclair *(m)*
Pt ponto de fulgor *(m)*

punto di intorbidimento *(m)* It
De Paraffinausscheidungspunkt *(m)*
En cloud point
Es punto de opacidad *(m)*
Fr point de trouble *(m)*
Pt ponto de névoa *(m)*

punto di pareggio *(m)* It
De Rentabilitätsgrenze *(f)*
En break-even point
Es punto comparativo *(m)*
Fr point mort *(m)*
Pt ponto de equilíbrio *(m)*

punto di rugiada *(m)* It
De Taupunkt *(m)*
En dew point
Es punto de condensación *(m)*
Fr point de rosée *(m)*
Pt ponto de orvalho *(m)*

pure-coal basis En
De Reinkohlengrundlage *(f)*
Es base de carbón puro *(f)*
Fr base de charbon pur *(f)*
It base di carbone puro *(f)*
Pt base de carvão puro *(f)*

pureté *(f)* *n* Fr
De Reinheit *(f)*
En purity
Es pureza *(f)*
It purezza *(f)*
Pt pureza *(f)*

pureza *(f)* *n* Es, Pt
De Reinheit *(f)*
En purity
Fr pureté *(f)*
It purezza *(f)*

purezza *(f)* *n* It
De Reinheit *(f)*
En purity
Es pureza *(f)*
Fr pureté *(f)*
Pt pureza *(f)*

purga *(f)* *n* Es, Pt
De Läuterung *(f)*
En purging
Fr purge *(f)*
It spurgo *(m)*

purge *(f)* *n* Fr
De Läuterung *(f)*
En purging
Es purga *(f)*
It spurgo *(m)*
Pt purga *(f)*

purging *n* En
De Läuterung *(f)*
Es purga *(f)*
Fr purge *(f)*
It spurgo *(m)*
Pt purga *(f)*

purity *n* En
De Reinheit *(f)*
Es pureza *(f)*
Fr pureté *(f)*
It purezza *(f)*
Pt pureza *(f)*

pyranometer *n* En
De Pyranometer *(n)*
Es piranómetro *(m)*
Fr pyranomètre *(m)*
It piranometro *(m)*
Pt piranómetro *(m)*

Pyranometer *(n)* *n* De
En pyranometer
Es piranómetro *(m)*
Fr pyranomètre *(m)*
It piranometro *(m)*
Pt piranómetro *(m)*

pyranomètre *(m)* *n* Fr
De Pyranometer *(n)*
En pyranometer
Es piranómetro *(m)*
It piranometro *(m)*
Pt piranómetro *(m)*

pyrheliometer *n* En
De Pyrheliometer *(n)*
Es pirheliómetro *(m)*
Fr pyrhéliomètre *(m)*
It pireliometro *(m)*
Pt pirheliómetro *(m)*

Pyrheliometer *(n)* *n* De
En pyrheliometer
Es pirheliómetro *(m)*
Fr pyrhéliomètre *(m)*
It pireliometro *(m)*
Pt pirheliómetro *(m)*

pyrhéliomètre *(m)* *n* Fr
De Pyrheliometer *(n)*
En pyrheliometer
Es pirheliómetro *(m)*
It pireliometro *(m)*
Pt pirheliómetro *(m)*

pyrolyse *(f)* *n* Fr
De Pyrolyse *(f)*
En pyrolysis
Es pirólisis *(f)*

It pirolisi *(f)*
Pt pirólise *(f)*

Pyrolyse *(f) n* De
En pyrolysis
Es pirólisis *(f)*
Fr pyrolyse *(f)*
It pirolisi *(f)*
Pt pirólise *(f)*

pyrolysis *n* En
De Pyrolyse *(f)*
Es pirólisis *(f)*
Fr pyrolyse *(f)*
It pirolisi *(f)*
Pt pirólise *(f)*

pyrometer *n* En
De Pyrometer *(n)*
Es pirómetro *(m)*
Fr pyromètre *(m)*
It pirometro *(m)*
Pt pirómetro *(m)*

Pyrometer *(n) n* De
En pyrometer
Es pirómetro *(m)*
Fr pyromètre *(m)*
It pirometro *(m)*
Pt pirómetro *(m)*

pyromètre *(m) n* Fr
De Pyrometer *(n)*
En pyrometer
Es pirómetro *(m)*
It pirometro *(m)*
Pt pirómetro *(m)*

pyromètre aspirant *(m)* Fr
De Saugpyrometer *(n)*
En suction pyrometer
Es pirómetro de succión *(m)*
It pirometro ad aspirazione *(m)*
Pt pirómetro de sucção *(m)*

pyromètre optique *(m)* Fr
De Strahlungspyrometer *(n)*
En optical pyrometer
Es pirómetro óptico *(m)*
It pirometro ottico *(m)*
Pt pirómetro óptico *(m)*

Q

quadratura *(f) n* It, Pt
De Flächeninhaltsbestimmung *(f)*
En quadrature
Es cuadratura *(f)*
Fr quadrature *(f)*

quadrature *n* En; Fr *(f)*
De Flächeninhaltsbestimmung *(f)*
Es cuadratura *(f)*
It quadratura *(f)*
Pt quadratura *(f)*

qualidade *(f) n* Pt
De Qualität *(f)*
En quality
Es calidad *(f)*
Fr qualité *(f)*
It qualità *(f)*

qualidade do vapor *(f)* Pt
De Dampfqualität *(f)*
En steam quality
Es calidad del vapor *(f)*
Fr qualité de vapeur *(f)*
It qualità del vapore *(f)*

qualità *(f) n* It
De Qualität *(f)*
En quality
Es calidad *(f)*
Fr qualité *(f)*
Pt qualidade *(f)*

qualità del vapore *(f)* It
De Dampfqualität *(f)*
En steam quality
Es calidad del vapor *(f)*
Fr qualité de vapeur *(f)*
Pt qualidade do vapor *(f)*

Qualität *(f) n* De
En quality
Es calidad *(f)*
Fr qualité *(f)*
It qualità *(f)*
Pt qualidade *(f)*

qualité *(f) n* Fr
De Qualität *(f)*
En quality
Es calidad *(f)*
It qualità *(f)*
Pt qualidade *(f)*

qualité de vapeur *(f)* Fr
De Dampfqualität *(f)*
En steam quality
Es calidad del vapor *(f)*
It qualità del vapore *(f)*
Pt qualidade do vapor *(f)*

quality *n* En
De Qualität *(f)*
Es calidad *(f)*
Fr qualité *(f)*
It qualità *(f)*
Pt qualidade *(f)*

Quantenelektrodynamik *(f) n* De
En quantum electrodynamics
Es electrodinámica cuántica *(f)*
Fr électrodynamique quantique *(f)*
It elettrodinamica dei quanti *(f)*
Pt electrodinâmica quântica *(f)*

Quantenmechanik *(f) n* De
En quantum mechanics
Es mecánica cuántica *(f)*
Fr mécanique quantique *(f)*
It meccanica dei quanti *(f)*
Pt mecânica quântica *(f)*

Quantentheorie *(f) n* De
En quantum theory
Es teoría de los cuantos *(f)*
Fr théorie des quanta *(f)*
It teoria dei quanti *(f)*
Pt teoría quântica *(f)*

Quantenzahl *(f) n* De
En quantum number
Es número cuántico *(m)*
Fr nombre quantique *(m)*
It numero quantico *(m)*
Pt número quântico *(m)*

quantidade de vector *(f)* Pt
De Vektorquantität *(f)*
En vector quantity
Es cantidad vectorial *(f)*
Fr quantité vectorielle *(f)*
It quantità di vettore *(f)*

quantidade escalar *(f)* Pt
De Skalarquantität *(f)*
En scalar quantity
Es cantidad escalar *(f)*
Fr quantité scalaire *(f)*
It grandezza scalare *(f)*

quantità di vettore *(f)* It
De Vektorquantität *(f)*
En vector quantity
Es cantidad vectorial *(f)*
Fr quantité vectorielle *(f)*
Pt quantidade de vector *(f)*

quantité scalaire *(f)* Fr
De Skalarquantität *(f)*
En scalar quantity
Es cantidad escalar *(f)*
It grandezza scalare *(f)*
Pt quantidade escalar *(f)*

quantité vectorielle *(f)* Fr
De Vektorquantität *(f)*
En vector quantity
Es cantidad vectorial *(f)*
It quantità di vettore *(f)*
Pt quantidade de vector *(f)*

quantum *n* En; Fr, It, Pt *(m)*
De Quantum *(n)*
Es cuando *(m)*

Quantum *(n) n* De
En quantum
Es cuando *(m)*
Fr quantum *(m)*
It quantum *(m)*
Pt quantum *(m)*

quantum electrodynamics En
De Quantenelektrodynamik *(f)*
Es electrodinámica cuántica *(f)*
Fr électrodynamique quantique *(f)*
It elettrodinamica dei quanti *(f)*
Pt electrodinâmica quântica *(f)*

quantum mechanics En
De Quantenmechanik *(f)*
Es mecánica cuántica *(f)*

Fr mécanique quantique *(f)*
It meccanica dei quanti *(f)*
Pt mecânica quântica *(f)*

quantum number En
De Quantenzahl *(f)*
Es número cuántico *(m)*
Fr nombre quantique *(m)*
It numero quantico *(m)*
Pt número quântico *(m)*

quantum theory En
De Quantentheorie *(f)*
Es teoría de los cuantos *(f)*
Fr théorie des quanta *(f)*
It teoria dei quanti *(f)*
Pt teoría quântica *(f)*

quebra por regeneração *(f)* Pt
De Rückstromgewinn-bremsung *(f)*
En regenerative braking
Es frenado de recuperación *(m)*
Fr freinage à récupération *(m)*
It frenatura rigenerativo *(m)*

quebra reostática *(f)* Pt
De Widerstands-bremsung *(f)*
En rheostatic braking
Es frenado reostático *(m)*
Fr freinage par résistance *(m)*
It frenatura reostatica *(f)*

queda de pressão *(f)* Pt
De Druckabfall *(m)*
En pressure drop
Es caída de presión *(f)*
Fr chute de pression *(f)*
It caduta di pressione *(f)*

queimador *(m)* *n* Pt
De Brenner *(m)*
En burner
Es quemador *(m)*
Fr brûleur *(m)*
It bruciatore *(m)*

queimador circular *(m)* Pt
De Kreisbrenner *(m)*
En circular burner
Es quemador circular *(m)*
Fr brûleur circulaire *(m)*
It bruciatore circolare *(m)*

queimador de chama direccional *(m)* Pt
De flammengerichteter Brenner *(m)*
En directional flame burner
Es quemador de llama direccional *(m)*
Fr brûleur à flamme dirigée *(m)*
It bruciatore di fiamma direzionale *(m)*

queimador de corrente de ar *(m)* Pt
De Windbrenner *(m)*
En air-blast burner
Es quemador de chorro de aire *(m)*
Fr brûleur à air soufflé *(m)*
It bruciatore a ventilazione forzata *(m)*

queimador de corrente de ar natural *(m)* Pt
Am natural-draft burner
De natürlicher Luftstrombrenner *(m)*
En natural-draught burner
Es quemador de tiro natural *(m)*
Fr brûleur à tirage naturel *(m)*
It bruciatore a corrente d'aria naturale *(m)*

queimador de jacto *(m)* Pt
De Düsenbrenner *(m)*
En jet burner
Es quemador de chorro *(m)*
Fr brûleur-jet *(m)*
It bruciatore a getto *(m)*

queimador misturador retardado *(m)* Pt
De Brenner mit verzögerter Luftmischung *(m)*
En delayed-mixing burner
Es quemador de mezcla retardada *(m)*
Fr brûleur a mélange différé *(m)*
It bruciatore di mescolata ritardato *(m)*

queimador pós-arejado *(m)* Pt
De nachbelüfteter Brenner *(m)*
En post-aerated burner
Es quemador posaireado *(m)*
Fr brûleur à post-aération *(m)*
It bruciatore post-aerato *(m)*

queimador pré-arejado *(m)* Pt
De vorbelüfteter Brenner *(m)*
En pre-aerated burner
Es quemador preaireado *(m)*
Fr brûleur à pré-aération *(m)*
It bruciatore pre-aerato *(m)*

Quelle *(f)* *n* De
En source
Es fuente *(f)*
Fr source *(f)*
It sorgente *(f)*
Pt fonte *(f)*

Quellversuch *(m)* *n* De
En swelling test
Es prueba de esponjamiento *(f)*
Fr essai de gonflement *(m)*
It prova di ringonfiamento *(f)*
Pt teste de inchamento *(m)*

quemador *(m)* *n* Es
De Brenner *(m)*
En burner
Fr brûleur *(m)*
It bruciatore *(m)*
Pt queimador *(m)*

quemador circular *(m)* Es
De Kreisbrenner *(m)*
En circular burner
Fr brûleur circulaire *(m)*
It bruciatore circolare *(m)*
Pt queimador circular *(m)*

quemador de chorro *(m)* Es
De Düsenbrenner *(m)*
En jet burner
Fr brûleur-jet *(m)*
It bruciatore a getto *(m)*
Pt queimador de jacto *(m)*

quemador de chorro de aire *(m)* Es
De Windbrenner *(m)*
En air-blast burner
Fr brûleur à air soufflé *(m)*
It bruciatore a ventilazione forzata *(m)*
Pt queimador de corrente de ar *(m)*

quemador de llama direccional *(m)* Es
De flammengerichteter Brenner *(m)*
En directional flame burner
Fr brûleur à flamme dirigée *(m)*
It bruciatore di fiamma direzionale *(m)*
Pt queimador de chama direccional *(m)*

quemador de mezcla retardada *(m)* Es
De Brenner mit verzögerter Luftmischung *(m)*
En delayed-mixing burner
Fr brûleur a mélange différé *(m)*
It bruciatore di mescolata ritardato *(m)*
Pt queimador misturador retardado *(m)*

quemador de tiro natural *(m)* Es
Am natural-draft burner
De natürlicher Luftstrombrenner *(m)*
En natural-draught burner
Fr brûleur à tirage naturel *(m)*
It bruciatore a corrente d'aria naturale *(m)*
Pt queimador de corrente de ar natural *(m)*

quemador posaireado *(m)* Es
De nachbelüfteter Brenner *(m)*
En post-aerated burner
Fr brûleur à post-aération *(m)*
It bruciatore post-aerato *(m)*
Pt queimador pós-arejado *(m)*

quemador preaireado *(m)* Es
De vorbelüfteter Brenner *(m)*
En pre-aerated burner
Fr brûleur à pré-aération *(m)*
It bruciatore pre-aerato *(m)*
Pt queimador pré-arejado *(m)*

querogéneo *(m) n* Pt
De Kerogen *(n)*
En kerogen
Es kerógeno *(m)*
Fr kérogène *(m)*
It kerogene *(m)*

queroseno *(m) n* Pt
De Naphta *(n)*
En kerosene
Es keroseno *(m)*
Fr kérosène *(m)*
It kerosene *(m)*

química das radiações *(f)* Pt
De Strahlungschemie *(f)*
En radiation chemistry
Es química de la radiación *(f)*
Fr radiochimie *(f)*
It chimica delle radiazioni *(f)*

química de la radiación *(f)* Es
De Strahlungschemie *(f)*
En radiation chemistry
Fr radiochimie *(f)*
It chimica delle radiazioni *(f)*
Pt química das radiações *(f)*

R

raccogliatore a finestra unica *(m)* It
De Einzelfenster-Kollektor *(m)*
En single-window collector
Es colector de ventana única *(m)*
Fr collecteur à simple fenêtre *(m)*
Pt colector de janela única *(m)*

raddrizzatore di onda intera *(m)* It
De Ganzwellengleich-richter *(m)*
En full-wave rectifier
Es rectificador de onda completa *(m)*
Fr redresseur biphasé *(m)*
Pt rectificador de onda completa *(m)*

raddrizzatore (elettrico) *(m)* It
De Gleichrichter (elektrischer) *(m)*
En rectifier (electrical)
Es rectificador (eléctrico) *(m)*
Fr redresseur (électrique) *(m)*
Pt rectificador (eléctrico) *(m)*

radiação *(f) n* Pt
De Strahlung *(f)*
En radiation
Es radiación *(f)*
Fr radiation *(f)*
It radiazione *(f)*

radiação de corpo negro *(f)* Pt
De Plancksche Strahlung *(f)*
En black-body radiation
Es radiación de cuerpo negro *(f)*
Fr rayonnement du corps noir *(m)*
It radiazione corpo nero *(f)*

radiação de fundo *(f)* Pt
De Nulleffektstrahlung *(f)*
En background radiation
Es radiación ambiente *(f)*
Fr rayonnement provoquant le mouvement propre *(m)*
It effetto di fondo *(m)*

radiação difusa *(f)* Pt
De Streustrahlung *(f)*
En diffuse radiation
Es radiación difusa *(f)*
Fr rayonnement diffus *(m)*
It radiazione diffusa *(f)*

radiação dura *(f)* Pt
De harte Strahlung *(f)*
En hard radiation
Es radiación dura *(f)*
Fr radiation dure *(f)*
It radiazione dura *(f)*

radiação electromagnética *(f)* Pt
De elektromagnetische Strahlung *(f)*
En electromagnetic radiation
Es radiación electromagnética *(f)*
Fr rayonnement électromagnétique *(m)*
It radiazione elettromagnetica *(f)*

radiação infravermelha *(f)* Pt
De Infrarotstrahlung *(f)*
En infrared radiation (IR)
Es radiación infrarroja *(f)*
Fr radiation infrarouge *(f)*
It radiazione infrarossa *(f)*

radiação ionizante *(f)* Pt
De Ionisierstrahlung *(f)*
En ionizing radiation
Es radiación ionizante *(f)*
Fr rayonnement d'ionisation *(m)*
It radiazione ionizzante *(f)*

radiação solar *(f)* Pt
De Sonnenstrahlung *(f)*
En solar radiation
Es radiación solar *(f)*
Fr rayonnement solaire *(m)*
It radiazione solare *(f)*

radiação suave *(f)* Pt
De Weichstrahlung *(f)*
En soft radiation
Es radiación blanda *(f)*
Fr radiation molle *(f)*
It radiazione soffice *(f)*

radiação térmica *(f)* Pt
De Wärmestrahlung *(f)*
En thermal radiation
Es radiación térmica *(f)*
Fr rayonnement thermique *(f)*
It radiazione termica *(f)*

radiação ultravioleta *(f)* Pt
De Ultraviolettstrahlung *(f)*
En ultraviolet radiation (UV)
Es radiación ultravioleta *(f)*
Fr radiation ultraviolette *(f)*
It radiazione ultravioletta *(f)*

radiación *(f) n* Es
De Strahlung *(f)*
En radiation
Fr radiation *(f)*
It radiazione *(f)*
Pt radiação *(f)*

radiación ambiente *(f)* Es
De Nulleffektstrahlung *(f)*
En background radiation
Fr rayonnement provoquant le mouvement propre *(m)*
It effetto di fondo *(m)*
Pt radiação de fundo *(f)*

radiación blanda *(f)* Es
De Weichstrahlung *(f)*
En soft radiation
Fr radiation molle *(f)*
It radiazione soffice *(f)*
Pt radiação suave *(f)*

radiación de cuerpo negro *(f)* Es
De Plancksche Strahlung *(f)*
En black-body radiation
Fr rayonnement du corps noir *(m)*
It radiazione corpo nero *(f)*
Pt radiação de corpo negro *(f)*

radiación difusa *(f)* Es
De Streustrahlung *(f)*
En diffuse radiation
Fr rayonnement diffus *(m)*
It radiazione diffusa *(f)*
Pt radiação difusa *(f)*

radiación dura *(f)* Es
De harte Strahlung *(f)*
En hard radiation
Fr radiation dure *(f)*
It radiazione dura *(f)*
Pt radiação dura *(f)*

radiación electromagnética *(f)* Es
De elektromagnetische Strahlung *(f)*
En electromagnetic radiation
Fr rayonnement électromagnétique *(m)*
It radiazione elettromagnetica *(f)*
Pt radiação electromagnética *(f)*

radiación infrarroja *(f)* Es
De Infrarotstrahlung *(f)*
En infrared radiation (IR)
Fr radiation infrarouge *(f)*
It radiazione infrarossa *(f)*
Pt radiação infravermelha *(f)*

radiación ionizante *(f)* Es
De Ionisierstrahlung *(f)*
En ionizing radiation
Fr rayonnement d'ionisation *(m)*
It radiazione ionizzante *(f)*
Pt radiação ionizante *(f)*

radiación solar *(f)* Es
De Sonnenstrahlung *(f)*
En solar radiation
Fr rayonnement solaire *(m)*
It radiazione solare *(f)*
Pt radiação solar *(f)*

radiación térmica *(f)* Es
De Wärmestrahlung *(f)*
En thermal radiation
Fr rayonnement thermique *(f)*
It radiazione termica *(f)*
Pt radiação térmica *(f)*

radiación ultravioleta *(f)* Es
De Ultraviolettstrahlung *(f)*
En ultraviolet radiation (UV)
Fr radiation ultraviolette *(f)*
It radiazione ultravioletta *(f)*
Pt radiação ultravioleta *(f)*

radiactividad *(f)* *n* Es
De Radioaktivität *(f)*
En radioactivity
Fr radioactivité *(f)*
It radioattività *(f)*
Pt radioactividade *(f)*

radiactividad inducida *(f)* Es
De induzierte Radioaktivität *(f)*
En induced radioactivity
Fr radioactivité induite *(f)*
It radioattività indotta *(f)*
Pt radioactividade induzida *(f)*

radiador *(m)* *n* Es, Pt
De Heizkörper *(m)*
En radiator
Fr radiateur *(m)*
It radiatore *(m)*

radiador de óleo *(m)* Pt
De ölgefüllter Heizkörper *(m)*
En oil-filled radiator
Es radiador relleno de petróleo *(m)*
Fr radiateur à circulation d'huile *(m)*
It radiatore riempito d'olio *(m)*

radiador relleno de petróleo *(m)* Es
De ölgefüllter Heizkörper *(m)*
En oil-filled radiator
Fr radiateur à circulation d'huile *(m)*
It radiatore riempito d'olio *(m)*
Pt radiador de óleo *(m)*

radialer Turboverdichter *(m)* De
En centrifugal compressor
Es compresor centrífugo *(m)*
Fr compresseur centrifuge *(m)*
It compressore centrifugo *(m)*
Pt compressor centrífugo *(m)*

radiant boiler En
De Strahlungsheizkessel *(m)*
Es caldera radiante *(f)*
Fr chaudière radiante *(f)*
It caldaia radiante *(f)*
Pt convector de radiação *(m)*

radiant convector En
De Strahlungskonvektor *(m)*
Es convector radiante *(m)*
Fr convecteur radiant *(m)*
It convettore radiante *(m)*
Pt convector de radiação *(m)*

radiant-heating furnace En
De Strahlungsheizofen *(m)*
Es horno de calentamiento por radiación *(m)*
Fr four à chauffage radiant *(m)*
It forno a riscaldamento radiante *(m)*
Pt forno de aquecimento por radiação *(m)*

radiant-tube furnace En
De Strahlungsröhrenofen *(m)*
Es horno de tubo radiante *(m)*
Fr four à tube radiant *(m)*
It forno a rubo radiante *(m)*
Pt forno de tubo de radiação *(m)*

radiateur *(m)* *n* Fr
De Heizkörper *(m)*
En radiator
Es radiador *(m)*
It radiatore *(m)*
Pt radiador *(m)*

radiateur à accumulation *(m)* Fr
De Speicherheizung *(f)*
En storage heater
Es calentador para almacenamiento térmico *(m)*
It riscaldatore a conservazione *(m)*
Pt aquecedor de armazenagem *(m)*

radiateur à circulation d'huile *(m)* Fr
De ölgefüllter Heizkörper *(m)*
En oil-filled radiator
Es radiador relleno de petróleo *(m)*
It radiatore riempito d'olio *(m)*
Pt radiador de óleo *(m)*

radiateur soufflant *(m)* Fr
De Heizelement mit Gebläse *(n)*
En fan-assisted heater

Es calentador auxiliado por ventilador *(m)*
It riscaldatore a ventilatore *(m)*
Pt aquecedor auxiliado por ventoinha *(m)*

radiation *n* En; Fr *(f)*
De Strahlung *(f)*
Es radiación *(f)*
It radiazione *(f)*
Pt radiação *(f)*

radiation chemistry En
De Strahlungschemie *(f)*
Es química de la radiación *(f)*
Fr radiochimie *(f)*
It chimica delle radiazioni *(f)*
Pt química das radiações *(f)*

radiation dure *(f)* Fr
De harte Strahlung *(f)*
En hard radiation
Es radiación dura *(f)*
It radiazione dura *(f)*
Pt radiação dura *(f)*

radiation hazard En
De Strahlungsgefahr *(f)*
Es riesgo de radiación *(m)*
Fr danger de radiation *(m)*
It pericolo di radiazioni *(m)*
Pt perigo de radiação *(m)*

radiation infrarouge *(f)* Fr
De Infrarotstrahlung *(f)*
En infrared radiation (IR)
Es radiación infrarroja *(f)*
It radiazione infrarossa *(f)*
Pt radiação infravermelha *(f)*

radiation molle *(f)* Fr
De Weichstrahlung *(f)*
En soft radiation
Es radiación blanda *(f)*
It radiazione soffice *(f)*
Pt radiação suave *(f)*

radiation suppression En
De Strahlungsunterdrückung *(f)*
Es supresión de la radiación *(f)*
Fr suppression des radiations *(f)*
It soppressione delle radiazioni *(f)*
Pt supressão de radiações *(f)*

radiation ultraviolette *(f)* Fr
De Ultraviolettstrahlung *(f)*
En ultraviolet radiation (UV)
Es radiación ultravioleta *(f)*
It radiazione ultravioletta *(f)*
Pt radiação ultravioleta *(f)*

radiator *n* En
De Heizkörper *(m)*
Es radiador *(m)*
Fr radiateur *(m)*
It radiatore *(m)*
Pt radiador *(m)*

radiatore *(m) n* It
De Heizkörper *(m)*
En radiator
Es radiador *(m)*
Fr radiateur *(m)*
Pt radiador *(m)*

radiatore riempito d'olio *(m)* It
De ölgefüllter Heizkörper *(m)*
En oil-filled radiator
Es radiador relleno de petróleo *(m)*
Fr radiateur à circulation d'huile *(m)*
Pt radiador de óleo *(m)*

radiazione *(f) n* It
De Strahlung *(f)*
En radiation
Es radiación *(f)*
Fr radiation *(f)*
Pt radiação *(f)*

radiazione corpo nero *(f)* It
De Plancksche Strahlung *(f)*
En black-body radiation
Es radiación de cuerpo negro *(f)*
Fr rayonnement du corps noir *(m)*
Pt radiação de corpo negro *(f)*

radiazione diffusa *(f)* It
De Streustrahlung *(f)*
En diffuse radiation
Es radiación difusa *(f)*
Fr rayonnement diffus *(m)*
Pt radiação difusa *(f)*

radiazione dura *(f)* It
De harte Strahlung *(f)*
En hard radiation
Es radiación dura *(f)*
Fr radiation dure *(f)*
Pt radiação dura *(f)*

radiazione elettromagnetica *(f)* It
De elektromagnetische Strahlung *(f)*
En electromagnetic radiation
Es radiación electromagnética *(f)*
Fr rayonnement électromagnétique *(m)*
Pt radiação electromagnética *(f)*

radiazione infrarossa *(f)* It
De Infrarotstrahlung *(f)*
En infrared radiation (IR)
Es radiación infrarroja *(f)*
Fr radiation infrarouge *(f)*
Pt radiação infravermelha *(f)*

radiazione ionizzante *(f)* It
De Ionisierstrahlung *(f)*
En ionizing radiation
Es radiación ionizante *(f)*
Fr rayonnement d'ionisation *(m)*
Pt radiação ionizante *(f)*

radiazione soffice *(f)* It
De Weichstrahlung *(f)*
En soft radiation
Es radiación blanda *(f)*
Fr radiation molle *(f)*
Pt radiação suave *(f)*

radiazione solare *(f)* It
De Sonnenstrahlung *(f)*
En solar radiation
Es radiación solar *(f)*
Fr rayonnement solaire *(m)*
Pt radiação solar *(f)*

radiazione termica *(f)* It
De Wärmestrahlung *(f)*
En thermal radiation
Es radiación térmica *(f)*
Fr rayonnement thermique *(f)*
Pt radiação térmica *(f)*

radiazione ultravioletta *(f)* It
De Ultraviolettstrahlung *(f)*
En ultraviolet radiation (UV)
Es radiación ultravioleta *(f)*
Fr radiation ultraviolette *(f)*
Pt radiação ultravioleta *(f)*

radioactif *adj* Fr
De radioaktiv
En radioactive
Es radioactivo
It radioattivo
Pt radioactivo

radioactive *adj* En
De radioaktiv
Es radioactivo
Fr radioactif
It radioattivo
Pt radioactivo

radioactive series En
De radioaktive Serie *(f)*
Es serie radiactiva *(f)*
Fr série radioactive *(f)*
It serie radioattiva *(f)*
Pt série radioactiva *(f)*

radioactividade *(f) n* Pt
De Radioaktivität *(f)*
En radioactivity
Es radiactividad *(f)*
Fr radioactivité *(f)*
It radioattività *(f)*

radioactividade induzida *(f)* Pt
De induzierte Radioaktivität *(f)*

En induced radioactivity
Es radiactividad inducida *(f)*
Fr radioactivité induite *(f)*
It radioattività indotta *(f)*

radioactivité *(f) n* Fr
De Radioaktivität *(f)*
En radioactivity
Es radiactividad *(f)*
It radioattività *(f)*
Pt radioactividade *(f)*

radioactivité induite *(f)* Fr
De induzierte Radioaktivität *(f)*
En induced radioactivity
Es radiactividad inducida *(f)*
It radioattività indotta *(f)*
Pt radioactividade induzida *(f)*

radioactivity *n* En
De Radioaktivität *(f)*
Es radiactividad *(f)*
Fr radioactivité *(f)*
It radioattività *(f)*
Pt radioactividade *(f)*

radioactivo *adj* Es, Pt
De radioaktiv
En radioactive
Fr radioactif
It radioattivo

radioaktiv *adj* De
En radioactive
Es radioactivo
Fr radioactif
It radioattivo
Pt radioactivo

radioaktive Abfallstoffe *(pl)* De
En active waste
Es residuo activo *(m)*
Fr déchets actifs *(m pl)*
It rifiuti attivi *(m pl)*
Pt desperdícios activos *(m pl)*

radioaktiver Niederschlag *(m)* De
En fall-out
Es precipitación radiactiva *(f)*
Fr retombée radioactive *(f)*
It precipitazione radioattiva *(f)*
Pt precipitação radioactiva *(f)*

radioaktive Serie *(f)* De
En radioactive series
Es serie radiactiva *(f)*
Fr série radioactive *(f)*
It serie radioatttiva *(f)*
Pt série radioactiva *(f)*

Radioaktivität *(f) n* De
En radioactivity
Es radiactividad *(f)*
Fr radioactivité *(f)*
It radioattività *(f)*
Pt radioactividade *(f)*

radioattività *(f) n* It
De Radioaktivität *(f)*
En radioactivity
Es radiactividad *(f)*
Fr radioactivité *(f)*
Pt radioactividade *(f)*

radioattività indotta *(f)* It
De induzierte Radioaktivität *(f)*
En induced radioactivity
Es radiactividad inducida *(f)*
Fr radioactivité induite *(f)*
Pt radioactividade induzida *(f)*

radioattivo *adj* It
De radioaktiv
En radioactive
Es radioactivo
Fr radioactif
Pt radioactivo

radiochimie *(f) n* Fr
De Strahlungschemie *(f)*
En radiation chemistry
Es química de la radiación *(f)*
It chimica delle radiazioni *(f)*
Pt química das radiações *(f)*

radiofrecuencia *(f) n* Es
De Radiofrequenz *(f)*
En radiofrequency
Fr radiofréquence *(f)*
It radiofrequenza *(f)*
Pt frequência rádio *(f)*

radiofréquence *(f) n* Fr
De Radiofrequenz *(f)*
En radiofrequency
Es radiofrecuencia *(f)*
It radiofrequenza *(f)*
Pt frequência rádio *(f)*

radiofrequency *n* En
De Radiofrequenz *(f)*
Es radiofrecuencia *(f)*
Fr radiofréquence *(f)*
It radiofrequenza *(f)*
Pt frequência rádio *(f)*

radiofrequency heating En
De Radiofrequenz-heizung *(f)*
Es calentamiento por radiofrecuencia *(m)*
Fr chauffage à radiofréquence *(m)*
It riscaldamento a radiofrequenza *(m)*
Pt aquecimento por frequência rádio *(m)*

Radiofrequenz *(f) n* De
En radiofrequency
Es radiofrequencia *(f)*
Fr radiofréquence *(f)*
It radiofrequenza *(f)*
Pt frequência rádio *(f)*

radiofrequenza *(f) n* It
De Radiofrequenz *(f)*
En radiofrequency
Es radiofrecuencia *(f)*
Fr radiofréquence *(f)*
Pt frequência rádio *(f)*

Radiofrequenzheizung *(f) n* De
En radiofrequency heating
Es calentamiento por radiofrequencia *(m)*
Fr chauffage à radiofréquence *(m)*
It riscaldamento a radiofrequenza *(m)*
Pt aquecimento por frequência rádio *(m)*

Radioisotop *(n) n* De
En radioisotope
Es radioisótopo *(m)*
Fr radioisotope *(m)*
It radioisotopo *(m)*
Pt radioisótopo *(m)*

radioisotope *n* En; Fr *(m)*
De Radioisotop *(n)*
Es radioisótopo *(m)*
It radioisotopo *(m)*
Pt radioisótopo *(m)*

radioisotopo *(m) n* It
De Radioisotop *(n)*
En radioisotope
Es radioisótopo *(m)*
Fr radioisotope *(m)*
Pt radioisótopo *(m)*

radioisótopo *(m) n* Es, Pt
De Radioisotop *(n)*
En radioisotope
Fr radioisotope *(m)*
It radioisotopo *(m)*

radiometer *n* En
De Radiometer *(n)*
Es radiómetro *(m)*
Fr radiomètre *(m)*
It radiometro *(m)*
Pt radiómetro *(m)*

Radiometer *(n) n* De
En radiometer
Es radiómetro *(m)*
Fr radiomètre *(m)*
It radiometro *(m)*
Pt radiómetro *(m)*

radiomètre *(m) n* Fr
De Radiometer *(n)*
En radiometer
Es radiómetro *(m)*
It radiometro *(m)*
Pt radiómetro *(m)*

radiometro *(m) n* It
De Radiometer *(n)*
En radiometer
Es radiómetro *(m)*
Fr radiomètre *(m)*
Pt radiómetro *(m)*

radiómetro *(m) n* Es, Pt
De Radiometer *(n)*
En radiometer
Fr radiomètre *(m)*
It radiometro *(m)*

radiotossicità *(f) n* It
De Radiotoxizität *(f)*
En radiotoxicity

Es radiotoxicidad *(f)*
Fr radiotoxicité *(f)*
Pt radiotoxicidade *(f)*

radiotoxicidad *(f) n* Es
De Radiotoxizität *(f)*
En radiotoxicity
Fr radiotoxicité *(f)*
It radiotossicità *(f)*
Pt radiotoxicidade *(f)*

radiotoxicidade *(f) n* Pt
De Radiotoxizität *(f)*
En radiotoxicity
Es radiotoxicidad *(f)*
Fr radiotoxicité *(f)*
It radiotossicità *(f)*

radiotoxicité *(f) n* Fr
De Radiotoxizität *(f)*
En radiotoxicity
Es radiotoxicidad *(f)*
It radiotossicità *(f)*
Pt radiotoxicidade *(f)*

radiotoxicity *n* En
De Radiotoxizität *(f)*
Es radiotoxicidad *(f)*
Fr radiotoxicité *(f)*
It radiotossicità *(f)*
Pt radiotoxicidade *(f)*

Radiotoxizität *(f) n* De
En radiotoxicity
Es radiotoxicidad *(f)*
Fr radiotoxicité *(f)*
It radiotossicità *(f)*
Pt radiotoxicidade *(f)*

raffineria *(f) n* It
De Raffinerie *(f)*
En refinery
Es refinería *(f)*
Fr raffinerie *(f)*
Pt refinaria *(f)*

raffineria di petrolio *(f)* It
De Ölraffinerie *(f)*
En oil refinery
Es refinería de petróleo *(f)*
Fr raffinerie de pétrole *(f)*
Pt refinaria de petróleo *(f)*

raffinerie *(f) n* Fr
De Raffinerie *(f)*
En refinery
Es refinería *(f)*
It raffineria *(f)*
Pt refinaria *(f)*

Raffinerie *(f) n* De
En refinery
Es refinería *(f)*
Fr raffinerie *(f)*
It raffineria *(f)*
Pt refinaria *(f)*

raffinerie de pétrole *(f)* Fr
De Ölraffinerie *(f)*
En oil refinery
Es refinería de petróleo *(f)*
It raffineria di petrolio *(f)*
Pt refinaria de petróleo *(f)*

raffreddamento a ebollizione *(m)* It
De Heiswasserkühlung *(f)*
En ebullient cooling
Es enfriamiento desde la ebullición *(m)*
Fr refroidissement par ébullition *(m)*
Pt arrefecimento ebuliente *(m)*

raffreddamento a radiazioni notturne *(m)* It
De Nachtstrahlungs-kühlung *(f)*
En night-radiation cooling
Es refrigeración de radiación nocturna *(f)*
Fr refroidissement par rayonnement nocturne *(f)*
Pt arrefecimento por radiação nocturna *(m)*

raffreddata ad idrogeno It
De wasserstoffgekühlt
En hydrogen-cooled
Es enfriada con hidrógeno
Fr refroidi par l'hydrogène
Pt arrefecido a hidrogénio

raffreddato ad aria It
De luftgekühlt
En air-cooled
Es refrigerado por aire
Fr refroidi par air
Pt arrefecido a ar

raggi gamma *(m pl)* It
De Gammastrahlen *(pl)*
En gamma-rays *(pl)*
Es rayos gamma *(m pl)*
Fr rayons gamma *(m pl)*
Pt raios gama *(m pl)*

raggi X *(m pl)* It
De Röntgen-Strahlen *(pl)*
En X-rays *(pl)*
Es rayos X *(m pl)*
Fr rayons X *(m pl)*
Pt raios X *(m pl)*

rail de contact *(m)* Fr
De Stromschiene *(f)*
En third rail
Es riel conductor *(m)*
It terza rotaia *(f)*
Pt terceiro carril *(m)*

raios gama *(m pl)* Pt
De Gammastrahlen *(pl)*
En gamma-rays *(pl)*
Es rayos gamma *(m pl)*
Fr rayons gamma *(m pl)*
It raggi gamma *(m pl)*

raios X *(m pl)* Pt
De Röntgen-Strahlen *(pl)*
En X-rays *(pl)*
Es rayos X *(m pl)*
Fr rayons X *(m pl)*
It raggi X *(m pl)*

Raketentreibstoff *(m) n* De
En rocket fuel
Es combustible para cohetes *(m)*
Fr combustible pour fusées *(m)*
It carburante per missili *(m)*
Pt combustível para foguetões *(m)*

Rankine cycle En
De Clausius-Rankine-Prozeß *(m)*
Es ciclo de Rankine *(m)*
Fr cycle de Rankine *(m)*
It ciclo di Rankine *(m)*
Pt ciclo ranquinizado *(m)*

rapport hydrogène-huile *(m)* Fr
De Wasserstoff-Ölverhältnis *(n)*
En hydrogen-oil ratio
Es relación hidrógeno-aceite *(f)*
It rapporto idrogeno-olio *(m)*
Pt razão hidrogénio-óleo *(f)*

rapporto di compressione *(m)* It
De Kompressions-verhältnis *(n)*
En compression ratio
Es relación de compresión *(f)*
Fr taux de compression *(m)*
Pt taxa de compressão *(f)*

rapporto idrogeno-olio *(m)* It
De Wasserstoff-Ölverhältnis *(n)*
En hydrogen-oil ratio
Es relación hidrógeno-aceite *(f)*
Fr rapport hydrogène-huile *(m)*
Pt razão hidrogénio-óleo *(f)*

Rauch *(m) n* De
En smoke
Es humo *(m)*
Fr fumée *(f)*
It fumo *(m)*
Pt fumo *(m)*

rauchlose Kohle *(f)* De
En smokeless coal
Es carbón fumífugo *(m)*
Fr charbon sans fumée *(m)*
It carbone senza fumo *(m)*
Pt carvão ardendo sem produzir fumo *(m)*

Rauchpunkt *(m) n* De
En smoke point
Es temperatura de

formación de humo (f)
Fr point de fumée (m)
It punto di fumo (m)
Pt ponto de fumo (m)

Raumheizung (f) n De
En space heating
Es calefacción de espacios (f)
Fr chauffage de chambres (m)
It riscaldamento locale (m)
Pt aquecimento espacial (m)

rayonnement diffus (m) Fr
De Streustrahlung (f)
En diffuse radiation
Es radiación difusa (f)
It radiazione diffusa (f)
Pt radiação difusa (f)

rayonnement d'ionisation (m) Fr
De Ionisierstrahlung (f)
En ionizing radiation
Es radiación ionizante (f)
It radiazione ionizzante (f)
Pt radiação ionizante (f)

rayonnement du corps noir (m) Fr
De Plancksche Strahlung (f)
En black-body radiation
Es radiación de cuerpo negro (f)
It radiazione corpo nero (f)
Pt radiação de corpo negro (f)

rayonnement électromagnétique (m) Fr
De elektromagnetische Strahlung (f)
En electromagnetic radiation
Es radiación electromagnética (f)
It radiazione elettromagnetica (f)
Pt radiação electromagnética (f)

rayonnement provoquant le mouvement propre (m) Fr
De Nulleffektstrahlung (f)
En background radiation
Es radiación ambiente (f)
It effetto di fondo (m)
Pt radiação de fundo (f)

rayonnement solaire (m) Fr
De Sonnenstrahlung (f)
En solar radiation
Es radiación solar (f)
It radiazione solare (f)
Pt radiação solar (f)

rayonnement thermique (f) Fr
De Wärmestrahlung (f)
En thermal radiation
Es radiación térmica (f)
It radiazione termica (f)
Pt radiação térmica (f)

rayons gamma (m pl) Fr
De Gammastrahlen (pl)
En gamma-rays (pl)
Es rayos gamma (m pl)
It raggi gamma (m pl)
Pt raios gama (m pl)

rayons X (m pl) Fr
De Röntgen-Strahlen (pl)
En X-rays (pl)
Es rayos X (m pl)
It raggi X (m pl)
Pt raios X (m pl)

rayos gamma (m pl) Es
De Gammastrahlen (pl)
En gamma-rays (pl)
Fr rayons gamma (m pl)
It raggi gamma (m pl)
Pt raios gama (m pl)

rayos X (m pl) Es
De Röntgen-Strahlen (pl)
En X-rays (pl)
Fr rayons X (m pl)
It raggi X (m pl)
Pt raios X (m pl)

razão hidrogénio-óleo (f) Pt
De Wasserstoff-Ölverhältnis (n)
En hydrogen-oil ratio
Es relación hidrógeno-aceite (f)
Fr rapport hydrogène-huile (m)
It rapporto idrogeno-olio (m)

razzo nucleare (m) It
De Kernkraftrakete (f)
En nuclear rocket
Es cohete nuclear (m)
Fr fusée nucléaire (f)
Pt foguetão nuclear (m)

reacção (f) n Pt
De Reaktion (f)
En reaction
Es reacción (f)
Fr réaction (f)
It reazione (f)

reacção em cadeia (f) Pt
De Kettenreaktion (f)
En chain reaction
Es reacción en cadena (f)
Fr réaction en chaîne (f)
It reazione a catena (f)

reacción (f) n Es
De Reaktion (f)
En reaction
Fr réaction (f)
It reazione (f)
Pt reacção (f)

reacción en cadena (f) Es
De Kettenreaktion (f)
En chain reaction
Fr réaction en chaîne (f)
It reazione a catena (f)
Pt reacção em cadeia (f)

réacteur (m) n Fr
De Reaktor (m)
En reactor
Es reactor (m)
It reattore (m)
Pt reactor (m)

réacteur à eau bouillante (m) Fr
De Siedereaktor (m)
En boiling-water reactor (BWR)
Es reactor de agua en ebullición (m)
It reattore ad acqua bollente (m)
Pt reactor de água em ebulição (m)

réacteur à eau légère (m) Fr
De Leichtwasserreaktor (m)
En light-water reactor
Es reactor de agua ligera (m)
It reattore ad acqua naturale (m)
Pt reactor de água leve (m)

réacteur à eau lourde (m) Fr
De Schwerwasser-reaktor (m)
En heavy-water reactor
Es reactor de agua pesada (m)
It reattore ad acqua pesante (m)
Pt reactor de água pesada (m)

réacteur à eau sous pression (m) Fr
De Druckwasserreaktor (m)
En pressurized-water reactor (PWR)
Es reactor de agua a presión (m)
It reattore ad acqua pressurizzata (m)
Pt reactor de água pressurizada (m)

réacteur à haute température (m) Fr
De Hochtemperatur-reaktor (m)
En high-temperature reactor (HTR)
Es reactor de alta temperatura (m)
It reattore ad alta temperatura (m)
Pt reactor de alta temperatura (m)

réacteur à refroidissement par eau (m) Fr
De wassergekühlter Reaktor (m)
En water-cooled reactor
Es reactor refrigerado con agua (m)
It reattore raffreddato ad acqua (m)
Pt reactor arrefecido por água (m)

réacteur à tube de force *(m)* Fr
De Druckröhrenreaktor *(m)*
En pressure-tube reactor
Es reactor de tubos de presión *(m)*
It reattore con tubo di pressione *(m)*
Pt reactor de tubo de pressão *(m)*

reacteur avancé à refroidissement au gaz *(m)* Fr
De fortgeschrittener gasgekühlter Reaktor *(m)*
En advanced gas-cooled reactor (AGR)
Es reactor avanzado refrigerado por gas *(m)*
It reattore avanzato raffreddato a gas *(m)*
Pt reactor avançado arrefecido a gás *(m)*

réacteur convertisseur *(m)* Fr
De Konverter *(m)*
En converter reactor
Es reactor convertidor *(m)*
It reattore convertitore *(m)*
Pt reactor de conversão *(m)*

réacteur nucléaire *(m)* Fr
De Kernreaktor *(m)*
En nuclear reactor
Es reactor nuclear *(m)*
It reattore nucleare *(m)*
Pt reactor nuclear *(m)*

réacteur nucléaire à refroidissement au gaz *(m)* Fr
De gasgekühlter Kernreaktor *(m)*
En gas-cooled nuclear reactor
Es reactor nuclear refrigerado con gas *(m)*
It reattore nucleare raffreddato a gas *(m)*
Pt reactor nuclear arrefecido a gás *(m)*

réacteur rapide *(m)* Fr
De Schnellreaktor *(m)*
En fast reactor
Es reactor rápido *(m)*
It reattore veloce *(m)*
Pt reactor rápido *(m)*

réacteur rapide industriel *(m)* Fr
De kommerzieller Schnellreaktor *(m)*
En commercial fast reactor (CFR)
Es reactor rápido comercial *(m)*
It reattore rapido commerciale *(m)*
Pt reactor rápido comercial *(m)*

réacteur rapide prototype *(m)* Fr
De Prototypen-Schnellreaktor *(m)*
En prototype fast reactor (PFR)
Es reactor rápido prototipo *(m)*
It prototipo di reattore veloce *(m)*
Pt reactor rápido prototipo *(m)*

réacteur surrégénérateur *(m)* Fr
De Brutreaktor *(m)*
En breeder reactor
Es reactor reproductor *(m)*
It reattore autofissilizzante *(m)*
Pt reactor reproductor *(m)*

réacteur surrégénérateur rapide *(m)* Fr
De Schnellbrüter *(m)*
En fast breeder reactor
Es reactor reproductor rápido *(m)*
It reattore veloce autofissilizzante *(m)*
Pt reactor reproductor rápida *(m)*

réacteur surrégénérateur rapide à métal liquide *(m)* Fr
De Flüssigmetall-Schnellbrüter *(m)*
En liquid-metal fast breeder reactor (LMFBR)
Es reactor reproductor rápido de metal líquido *(m)*
It reattore veloce autofissilizzante a metallo liquido *(m)*
Pt reactor reproductor rápido de metal líquido *(m)*

réacteur thermique *(m)* Fr
De thermischer Reaktor *(m)*
En thermal reactor
Es reactor térmico *(m)*
It reattore termico *(m)*
Pt reactor térmico *(m)*

reaction *n* En
De Reaktion *(f)*
Es reacción *(f)*
Fr réaction *(f)*
It reazione *(f)*
Pt reacção *(f)*

réaction *(f) n* Fr
De Reaktion *(f)*
En reaction
Es reacción *(f)*
It reazione *(f)*
Pt reacção *(f)*

réaction en chaîne *(f)* Fr
De Kettenreaktion *(f)*
En chain reaction
Es reacción en cadena *(f)*
It reazione a catena *(f)*
Pt reacção em cadeia *(f)*

reactor *n* En; Es, Pt *(m)*
De Reaktor *(m)*
Fr réacteur *(m)*
It reattore *(m)*

reactor arrefecido por água *(m)* Pt
De wassergekühlter Reaktor *(m)*
En water-cooled reactor
Es reactor refrigerado con agua *(m)*
Fr réacteur à refroidissement par eau *(m)*
It reattore raffreddato ad acqua *(m)*

reactor avançado arrefecido a gás *(m)* Pt
De fortgeschrittener gasgekühlter Reaktor *(m)*
En advanced gas-cooled reactor (AGR)
Es reactor avanzado refrigerado por gas *(m)*
Fr reacteur avancé à refroidissement au gaz *(m)*
It reattore avanzato raffreddato a gas *(m)*

reactor avanzado refrigerado por gas *(m)* Es
De fortgeschrittener gasgekühlter Reaktor *(m)*
En advanced gas-cooled reactor (AGR)
Fr reacteur avancé à refroidissement au gaz *(m)*
It reattore avanzato raffreddato a gas *(m)*
Pt reactor avançado arrefecido a gás *(m)*

reactor convertidor *(m)* Es
De Konverter *(m)*
En converter reactor
Fr réacteur convertisseur *(m)*
It reattore convertitore *(m)*
Pt reactor de conversão *(m)*

reactor de agua a presión *(m)* Es
De Druckwasserreaktor *(m)*
En pressurized-water reactor (PWR)
Fr réacteur à eau sous pression *(m)*
It reattore ad acqua pressurizzata *(m)*
Pt reactor de água pressurizada *(m)*

reactor de água em ebulição *(m)* Pt
De Siedereaktor *(m)*
En boiling-water reactor (BWR)
Es reactor de agua en ebullición *(m)*
Fr réacteur à eau bouillante *(m)*
It reattore ad acqua bollente *(m)*

reactor de aqua en ebullición *(m)* Es
De Siedereaktor *(m)*
En boiling-water reactor (BWR)
Fr réacteur à eau bouillante *(m)*
It reattore ad acqua bollente *(m)*
Pt reactor de água em ebulição *(m)*

reactor de água leve *(m)* Pt
De Leichtwasserreaktor *(m)*
En light-water reactor
Es reactor de agua ligera *(m)*
Fr réacteur à eau légère *(m)*
It reattore ad acqua naturale *(m)*

reactor de agua ligera *(m)* Es
De Leichtwasserreaktor *(m)*
En light-water reactor
Fr réacteur à eau légère *(m)*
It reattore ad acqua naturale *(m)*
Pt reactor de água leve *(m)*

reactor de agua pesada *(m)* Es
De Schwerwasser-reaktor *(m)*
En heavy-water reactor
Fr réacteur à eau lourde *(m)*
It reattore ad acqua pesante *(m)*
Pt reactor de água pesada *(m)*

reactor de água pesada *(m)* Pt
De Schwerwasser-reaktor *(m)*
En heavy-water reactor
Es reactor de agua pesada *(m)*
Fr réacteur à eau lourde *(m)*
It reattore ad acqua pesante *(m)*

reactor de áqua pressurizada *(m)* Pt
De Druckwasserreaktor *(m)*
En pressurized-water reactor (PWR)
Es reactor de aqua a presión *(m)*
Fr réacteur à eau sous pression *(m)*
It reattore ad acqua pressurizzata *(m)*

reactor de alta temperatura *(m)* Es, Pt
De Hochtemperatur-reaktor *(m)*
En high-temperature reactor (HTR)
Fr réacteur à haute température *(m)*
It reattore ad alta temperatura *(m)*

reactor de conversão *(m)* Pt
De Konverter *(m)*
En converter reactor
Es reactor convertidor *(m)*
Fr réacteur convertisseur *(m)*
It reattore convertitore *(m)*

reactor de tubo de pressão *(m)* Pt
De Druckröhrenreaktor *(m)*
En pressure-tube reactor
Es reactor de tubos de presión *(m)*
Fr réacteur à tube de force *(m)*
It reattore con tubo di pressione *(m)*

reactor de tubos de presión *(m)* Es
De Druckröhrenreaktor *(m)*
En pressure-tube reactor
Fr réacteur à tube de force *(m)*
It reattore con tubo di pressione *(m)*
Pt reactor de tubo de pressão *(m)*

reactor nuclear *(m)* Es, Pt
De Kernreaktor *(m)*
En nuclear reactor
Fr réacteur nucléaire *(m)*
It reattore nucleare *(m)*

reactor nuclear arrefecido a gás *(m)* Pt
De gasgekühlter Kernreaktor *(m)*
En gas-cooled nuclear reactor
Es reactor nuclear refrigerado con gas *(m)*
Fr réacteur nucléaire à refroidissement au gaz *(m)*
It reattore nucleare raffreddato a gas *(m)*

reactor nuclear refrigerado con gas *(m)* Es
De gasgekühlter Kernreaktor *(m)*
En gas-cooled nuclear reactor
Fr réacteur nucléaire à refroidissement au gaz *(m)*
It reattore nucleare raffreddato a gas *(m)*
Pt reactor nuclear arrefecido a gás *(m)*

reactor rápido *(m)* Es, Pt
De Schnellreaktor *(m)*
En fast reactor
Fr réacteur rapide *(m)*
It reattore veloce *(m)*

reactor rápido comercial *(m)* Es, Pt
De kommerzieller Schnellreaktor *(m)*
En commercial fast reactor (CFR)
Fr réacteur rapide industriel *(m)*
It reattore rapido commerciale *(m)*

reactor rápido prototipo *(m)* Es, Pt
De Prototypen-Schnellreaktor *(m)*
En prototype fast reactor (PFR)
Fr réacteur rapide prototype *(m)*
It prototipo di reattore veloce *(m)*

reactor refrigerado con agua *(m)* Es
De wassergekühlter Reaktor *(m)*
En water-cooled reactor
Fr réacteur à refroidissement par eau *(m)*
It reattore raffreddato ad acqua *(m)*
Pt reactor arrefecido por água *(m)*

reactor reproductor *(m)* Es, Pt
De Brutreaktor *(m)*
En breeder reactor
Fr réacteur surrégénérateur *(m)*
It reattore autofissilizzante *(m)*

reactor reproductor rápido *(m)* Es, Pt
De Schnellbrüter *(m)*
En fast breeder reactor
Fr réacteur surrégénérateur rapide *(m)*
It reattore veloce autofissilizzante *(m)*

reactor reproductor rápido de metal líquido *(m)* Es, Pt
De Flüssigmetall-Schnellbrüter *(m)*
En liquid-metal fast breeder reactor (LMFBR)
Fr réacteur surrégénérateur rapide à métal liquide *(m)*

It reattore veloce autofissilizzante a metallo liquido *(m)*
Pt reactor reproductor rápido de metal líquido *(m)*

reactor térmico *(m)* Es, Pt
De thermischer Reaktor *(m)*
En thermal reactor
Fr réacteur thermique *(m)*
It reattore termico *(m)*

Reaktion *(f) n* De
En reaction
Es reacción *(f)*
Fr réaction *(f)*
It reazione *(f)*
Pt reacção *(f)*

Reaktionswärme *(f) n* De
En heat of reaction
Es calor de reacción *(m)*
Fr chaleur de réaction *(f)*
It calore di reazione *(m)*
Pt calor de reacção *(m)*

Reaktor *(m) n* De
En reactor
Es reactor *(m)*
Fr réacteur *(m)*
It reattore *(m)*
Pt reactor *(m)*

realimentación *(f) n* Es
De Rückkoppelung *(f)*
En feedback
Fr rétroaction *(f)*
It retroazione *(f)*
Pt retôrno *(m)*

realimentación negativa *(f)* Es
De negative Rückkoppelung *(f)*
En negative feedback
Fr contre-réaction *(f)*
It controreazione *(f)*
Pt retôrno negativo *(m)*

reaquecedor *(m) n* Pt
De Nachbrenner *(m)*
En reheater
Es recalentador *(m)*
Fr réchauffeur *(m)*
It ricombustore *(m)*

reattore *(m) n* It
De Reaktor *(m)*
En reactor
Es reactor *(m)*
Fr réacteur *(m)*
Pt reactor *(m)*

reattore ad acqua bollente *(m)* It
De Siedereaktor *(m)*
En boiling-water reactor (BWR)
Es reactor de agua en ebullición *(m)*
Fr réacteur à eau bouillante *(m)*
Pt reactor de água em ebulição *(m)*

reattore ad acqua naturale *(m)* It
De Leichtwasserreaktor *(m)*
En light-water reactor
Es reactor de agua ligera *(m)*
Fr réacteur à eau légère *(m)*
Pt reactor de água leve *(m)*

reattore ad acqua pesante *(m)* It
De Schwerwasser-reaktor *(m)*
En heavy-water reactor
Es reactor de agua pesada *(m)*
Fr réacteur à eau lourde *(m)*
Pt reactor de água pesada *(m)*

reattore ad acqua pressurizzata *(m)* It
De Druckwasserreaktor *(m)*
En pressurized-water reactor (PWR)
Es reactor de agua a presión *(m)*
Fr réacteur à eau sous pression *(m)*
Pt reactor de água pressurizada *(m)*

reattore ad alta temperatura *(m)* It
De Hochtemperatur-reaktor *(m)*
En high-temperature reactor (HTR)
Es reactor de alta temperatura *(m)*
Fr réacteur à haute température *(m)*
Pt reactor de alta temperatura *(m)*

reattore autofissilizzante *(m)* It
De Brutreaktor *(m)*
En breeder reactor
Es reactor reproductor *(m)*
Fr réacteur surrégénérateur *(m)*
Pt reactor reproductor *(m)*

reattore avanzato raffreddato a gas *(m)* It
De fortgeschrittener gasgekühlter Reaktor *(m)*
En advanced gas-cooled reactor (AGR)
Es reactor avanzado refrigerado por gas *(m)*
Fr reacteur avancé à refroidissement au gaz *(m)*
Pt reactor avançado arrefecido a gás *(m)*

reattore con tubo di pressione *(m)* It
De Druckröhrenreaktor *(m)*
En pressure-tube reactor
Es reactor de tubos de presión *(m)*
Fr réacteur à tube de force *(m)*
Pt reactor de tubo de pressão *(m)*

reattore convertitore *(m)* It
De Konverter *(m)*
En converter reactor
Es reactor convertidor *(m)*
Fr réacteur convertisseur *(m)*
Pt reactor de conversão *(m)*

reattore nucleare *(m)* It
De Kernreaktor *(m)*
En nuclear reactor
Es reactor nuclear *(m)*
Fr réacteur nucléaire *(m)*
Pt reactor nuclear *(m)*

reattore nucleare raffreddato a gas *(m)* It
De gasgekühlter Kernreaktor *(m)*
En gas-cooled nuclear reactor
Es reactor nuclear refrigerado con gas *(m)*
Fr réacteur nucléaire à refroidissement au gaz *(m)*
Pt reactor nuclear arrefecido a gás *(m)*

reattore raffreddato ad acqua *(m)* It
De wassergekühlter Reaktor *(m)*
En water-cooled reactor
Es reactor refrigerado con agua *(m)*
Fr réacteur à refroidissement par eau *(m)*
Pt reactor arrefecido por água *(m)*

reattore rapido commerciale *(m)* It
De kommerzieller Schnellreaktor *(m)*
En commercial fast reactor (CFR)
Es reactor rápido comercial *(m)*
Fr réacteur rapide industriel *(m)*
Pt reactor rápido comercial *(m)*

reattore termico *(m)* It
De thermischer Reaktor *(m)*
En thermal reactor
Es reactor térmico *(m)*
Fr réacteur thermique *(m)*
Pt reactor térmico *(m)*

reattore veloce *(m)* It
De Schnellreaktor *(m)*
En fast reactor
Es reactor rápido *(m)*
Fr réacteur rapide *(m)*
Pt reactor rápido *(m)*

reattore veloce autofissilizzante *(m)* It
De Schnellbrüter *(m)*
En fast breeder reactor
Es reactor reproductor rápido *(m)*
Fr réacteur surrégénérateur rapide *(m)*
Pt reactor reproductor rápido *(m)*

reattore veloce autofissilizzante a metallo liquido *(m)* It
De Flüssigmetall-Schnellbrüter *(m)*
En liquid-metal fast breeder reactor (LMFBR)
Es reactor reproductor rápido de metal líquido *(m)*
Fr réacteur surrégénérateur rapide à métal liquide *(m)*
Pt reactor reproductor rápido de metal líquido *(m)*

reazione *(f) n* It
De Rückkoppelung *(f)*
En feedback
Es realimentación *(f)*
Fr réaction *(f)*
Pt retôrno *(m)*

reazione *(f) n* It
De Reaktion *(f)*
En reaction
Es reacción *(f)*
Fr réaction *(f)*
Pt reacção *(f)*

reazione a catena *(f)* It
De Kettenreaktion *(f)*
En chain reaction
Es reacción en cadena *(f)*
Fr réaction en chaîne *(f)*
Pt reacção em cadeia *(f)*

reboiler *n* En
De Auskocher *(m)*
Es recaldera *(f)*
Fr rebouilleur *(m)*
It recaldaia *(f)*
Pt recaldeira *(f)*

rebouilleur *(m) n* Fr
De Auskocher *(m)*
En reboiler
Es recaldera *(f)*
It recaldaia *(f)*
Pt recaldeira *(f)*

recaldaia *(f) n* It
De Auskocher *(m)*
En reboiler
Es recaldera *(f)*
Fr rebouilleur *(m)*
Pt recaldeira *(f)*

recaldeira *(f) n* Pt
De Auskocher *(m)*
En reboiler
Es recaldera *(f)*
Fr rebouilleur *(m)*
It recaldaia *(f)*

recaldera *(f) n* Es
De Auskocher *(m)*
En reboiler
Fr rebouilleur *(m)*
It recaldaia *(f)*
Pt recaldeira *(f)*

recalentador *(m) n* Es
De Nachbrenner *(m)*
En reheater
Fr réchauffeur *(m)*
It ricombustore *(m)*
Pt reaquecedor *(m)*

réchauffeur *(m) n* Fr
De Heizelement *(n)*; Nachbrenner *(m)*
En heater; reheater
Es calentador; recalentador *(m)*
It ricombustore; riscaldatore *(m)*
Pt aquecedor; reaquecedor *(m)*

réchauffeur d'air *(m)* Fr
De Lufterhitzer *(m)*
En air heater
Es calentador de aire *(m)*
It riscaldatore ad aria *(m)*
Pt aquecedor de ar *(m)*

reciclagem *(f) n* Pt
De Rückführung in den Kreislauf *(f)*
En recycling
Es reciclajo *(m)*
Fr recyclage *(m)*
It riciclaggio *(m)*

reciclajo *(m) n* Es
De Rückführung in den Kreislauf *(f)*
En recycling
Fr recyclage *(m)*
It riciclaggio *(m)*
Pt reciclagem *(f)*

récipient de pression en béton *(m)* Fr
De Betondruckbehälter *(m)*
En concrete pressure vessel
Es recipiente de presión de hormigón *(m)*
It recipiente a pressione di betone *(m)*
Pt recipiento de pressão de betão *(m)*

recipiente a pressione di betone *(m)* It
De Betondruckbehälter *(m)*
En concrete pressure vessel
Es recipiente de presión de hormigón *(m)*
Fr récipient de pression en béton *(m)*
Pt recipiento de pressão de betão *(m)*

recipiente de presión de hormigón *(m)* Es
De Betondruckbehälter *(m)*
En concrete pressure vessel
Fr récipient de pression en béton *(m)*
It recipiente a pressione di betone *(m)*
Pt recipiento de pressão de betão *(m)*

recipiento de pressão de betão *(m)* Pt
De Betondruckbehälter *(m)*
En concrete pressure vessel
Es recipiente de presión de hormigón *(m)*
Fr récipient de pression en béton *(m)*
It recipiente a pressione di betone *(m)*

reciprocating engine En
De Kolbenmaschine *(f)*
Es motor alternativo *(m)*
Fr moteur alternatif *(m)*
It motore a movimento alternativo *(m)*
Pt motor alternativo *(m)*

recoverable reserves *(pl)* En
De gewinnbare Reserven *(pl)*
Es reservas recuperables *(f pl)*
Fr réserves récupérables *(f pl)*
It riserve recuperabili *(f pl)*
Pt reservas recuperáveis *(f pl)*

rectificador (eléctrico) *(m) n* Es, Pt
De Gleichrichter (elektrischer) *(m)*
En rectifier (electrical)
Fr redresseur (électrique) *(m)*
It raddrizzatore (elettrico) *(m)*

rectificador (químico) *(m) n* Es, Pt
De Rektifikator (chemischer) *(m)*
En rectifier (chemical)
Fr rectificateur (chimique) *(m)*
It rettificatore (chimico) *(m)*

rectificador de onda completa *(m)* Es, Pt
De Ganzwellengleichrichter *(m)*
En full-wave rectifier
Fr redresseur biphasé *(m)*
It raddrizzatore di onda intera *(m)*

rectificateur (chimique) *(m) n* Fr
De Rektifikator (chemischer) *(m)*
En rectifier (chemical)
Es rectificador (químico) *(m)*

It rettificatore (chimico) *(m)*
Pt rectificador (químico) *(m)*

rectified spirit En
De rektifizierter Alkohol *(m)*
Es alcohol rectificado *(m)*
Fr alcool rectifié *(m)*
It spirito rettificato *(m)*
Pt álcool rectificado *(m)*

rectifier (electrical) *n* En
De Gleichrichter (elektrischer) *(m)*
Es rectificador (eléctrico) *(m)*
Fr redresseur (électrique) *(m)*
It raddrizzatore (elettrico) *(m)*
Pt rectificador (eléctrico) *(m)*

rectifier (chemical) *n* En
De Rektifikator (chemischer) *(m)*
Es rectificador (químico) *(m)*
Fr rectificateur (chimique) *(m)*
It rettificatore (chimico) *(m)*
Pt rectificador (químico) *(m)*

recuperação de ar gasto *(f)* Pt
De Rückgewinnung verbrauchter Luft *(f)*
En spent-air recovery
Es recuperación del aire gastado *(f)*
Fr récupération de l'air dépensé *(f)*
It recupero dell'aria consumata *(m)*

recuperação de calor residual *(f)* Pt
De Abwärmerück-gewinnung *(f)*
En waste-heat recovery
Es recuperación de calor residual *(f)*
Fr récupération des chaleurs perdues *(f)*
It recupero di calore perduto *(m)*

recuperação final *(f)* Pt
De Ausbeutefaktor *(m)*
En ultimate recovery
Es producción final *(f)*
Fr récupération finale *(f)*
It recupero ultimo *(m)*

recuperação secundária *(f)* Pt
De sekundäre Rückgewinnung *(f)*
En secondary recovery
Es recuperación secundaria *(f)*
Fr récupération secondaire *(f)*
It recupero secondario *(m)*

recuperação térmica *(f)* Pt
De Wärmerück-gewinnung *(f)*
En heat recovery
Es recuperación del calor *(f)*
Fr récupération de chaleur *(f)*
It recupero di calore *(m)*

recuperación de calor residual *(f)* Es
De Abwärmerück-gewinnung *(f)*
En waste-heat recovery
Fr récupération des chaleurs perdues *(f)*
It recupero di calore perduto *(m)*
Pt recuperação de calor residual *(f)*

recuperación del aire gastado *(f)* Es
De Rückgewinnung verbrauchter Luft *(f)*
En spent-air recovery
Fr récupération de l'air dépensé *(f)*
It recupero dell'aria consumata *(m)*
Pt recuperação de ar gasto *(f)*

recuperación del calor *(f)* Es
De Wärmerück-gewinnung *(f)*
En heat recovery
Fr récupération de chaleur *(f)*
It recupero di calore *(m)*
Pt recuperação térmica *(f)*

recuperación secundaria *(f)* Es
De sekundäre Rückgewinnung *(f)*
En secondary recovery
Fr récupération secondaire *(f)*
It recupero secondario *(m)*
Pt recuperação secundária *(f)*

recuperador *(m) n* Es, Pt
De Rekuperator *(m)*
En recuperator
Fr récupérateur *(m)*
It ricuperatore *(m)*

recuperador de convección *(m)* Es
De Konvektions-rekuperator *(m)*
En convection recuperator
Fr récupérateur à convection *(m)*
It recuperatore di convezione *(m)*
Pt recuperador por convecção *(m)*

recuperador por convecção *(m)* Pt
De Konvektions-rekuperator *(m)*
En convection recuperator
Es recuperador de convección *(m)*
Fr récupérateur à convection *(m)*
It recuperatore di convezione *(m)*

recuperador tubulare *(m)* Es, Pt
De Röhrenrekuperator *(m)*
En tubular recuperator
Fr récupérateur tubulaire *(m)*
It ricuperatore tubolare *(m)*

récupérateur *(m) n* Fr
De Rekuperator *(m)*
En recuperator
Es recuperador *(m)*
It ricuperatore *(m)*
Pt recuperador *(m)*

récupérateur à convection *(m)* Fr
De Konvektions-rekuperator *(m)*
En convection recuperator
Es recuperador de convección *(m)*
It recuperatore di convezione *(m)*
Pt recuperador por convecção *(m)*

récupérateur tubulaire *(m)* Fr
De Röhrenrekuperator *(m)*
En tubular recuperator
Es recuperador tubulare *(m)*
It ricuperatore tubolare *(m)*
Pt recuperador tubulare *(m)*

récupération de chaleur *(f)* Fr
De Wärmerück-gewinnung *(f)*
En heat recovery
Es recuperación del calor *(f)*
It recupero di calore *(m)*
Pt recuperação térmica *(f)*

récupération de l'air dépensé *(f)* Fr
De Rückgewinnung verbrauchter Luft *(f)*
En spent-air recovery
Es recuperación del aire gastado *(f)*
It recupero dell'aria consumata *(m)*
Pt recuperação de ar gasto *(f)*

récupération des chaleurs perdues *(f)* Fr
De Abwärmerück-gewinnung *(f)*
En waste-heat recovery
Es recuperación de calor residual *(f)*

It recupero di calore perduto *(m)*
Pt recuperação de calor residual *(f)*

récupération finale *(f)* Fr
De Ausbeutefaktor *(m)*
En ultimate recovery
Es producción final *(f)*
It recupero ultimo *(m)*
Pt recuperação final *(f)*

récupération secondaire *(f)* Fr
De sekundäre Rückgewinnung *(f)*
En secondary recovery
Es recuperación secundaria *(f)*
It recupero secondario *(m)*
Pt recuperação secundária *(f)*

recuperator *n* En
De Rekuperator *(m)*
Es recuperador *(m)*
Fr récupérateur *(m)*
It ricuperatore *(m)*
Pt recuperador *(m)*

recuperatore di convezione *(m)* It
De Konvektions-rekuperator *(m)*
En convection recuperator
Es recuperador de convección *(m)*
Fr récupérateur à convection *(m)*
Pt recuperador por convecção *(m)*

recupero dell'aria consumata *(m)* It
De Rückgewinnung verbrauchter Luft *(f)*
En spent-air recovery
Es recuperación del aire gastado *(f)*
Fr récupération de l'air dépensé *(f)*
Pt recuperação de ar gasto *(f)*

recupero di calore *(m)* It
De Wärmerück-gewinnung *(f)*
En heat recovery
Es recuperación del calor *(f)*
Fr récupération de chaleur *(f)*
Pt recuperação térmica *(f)*

recupero di calore perduto *(m)* It
De Abwärmerück-gewinnung *(f)*
En waste-heat recovery
Es recuperación de calor residual *(f)*
Fr récupération des chaleurs perdues *(f)*
Pt recuperação de calor residual *(f)*

recupero secondario *(m)* It
De sekundäre Rückgewinnung *(f)*
En secondary recovery
Es recuperación secundaria *(f)*
Fr récupération secondaire *(f)*
Pt recuperação secundária *(f)*

recupero ultimo *(m)* It
De Ausbeutefaktor *(m)*
En ultimate recovery
Es producción final *(f)*
Fr récupération finale *(f)*
Pt recuperação final *(f)*

recursos *(m pl) n* Es, Pt
De Bodenschätze *(pl)*
En resources *(pl)*
Fr ressources *(f pl)*
It risorse *(f pl)*

recursos energéticos *(m pl)* Es
De Energiequellen *(pl)*
En energy resources *(pl)*
Fr ressources d'énergie *(f pl)*
It risorse energetiche *(f pl)*
Pt recursos energéticos *(m pl)*

recursos energéticos *(m pl)* Pt
De Energiequellen *(pl)*
En energy resources *(pl)*
Es recursos energéticos *(m pl)*
Fr ressources d'énergie *(f pl)*
It risorse energetiche *(f pl)*

recyclage *(m) n* Fr
De Rückführung in den Kreislauf *(f)*
En recycling
Es reciclajo *(m)*
It riciclaggio *(m)*
Pt reciclagem *(f)*

recycling *n* En
De Rückführung in den Kreislauf *(f)*
Es reciclajo *(m)*
Fr recyclage *(m)*
It riciclaggio *(m)*
Pt reciclagem *(f)*

red *(f) n* Es
De Netz *(n)*
En network
Fr réseau *(m)*
It rete *(f)*
Pt rede *(f)*

rede *(f) n* Pt
De Netz *(n)*
En network
Es red *(f)*
Fr réseau *(m)*
It rete *(f)*

rede eléctrica *(f)* Pt
De Überlandleitungsnetz *(n)*
En grid (electrical power)
Es red nacional de energía eléctrica *(f)*
Fr réseau électrique national *(m)*
It rete nazionale *(f)*

red nacional de energía eléctrica *(f)* Es
De Überlandleitungsnetz *(n)*
En grid (electrical power)
Fr réseau électrique national *(m)*
It rete nazionale *(f)*
Pt rede eléctrica *(f)*

redresseur biphasé *(m)* Fr
De Ganzwellengleich-richter *(m)*
En full-wave rectifier
Es rectificador de onda completa *(m)*
It raddrizzatore di onda intera *(m)*
Pt rectificador de onda completa *(m)*

redresseur (électrique) *(m) n* Fr
De Gleichrichter (elektrischer) *(m)*
En rectifier (electrical)
Es rectificador (eléctrico) *(m)*
It raddrizzatore (elettrico) *(m)*
Pt rectificador (eléctrico) *(m)*

reelaboración *(f) n* Es
De Wiederaufbereitung *(f)*
En reprocessing
Fr traitement du combustible irradié *(m)*
It riutilazzione *(f)*
Pt reprocessamento *(m)*

refinaria *(f) n* Pt
De Raffinerie *(f)*
En refinery
Es refinería *(f)*
Fr raffinerie *(f)*
It raffineria *(f)*

refinaria de petróleo *(f)* Pt
De Ölraffinerie *(f)*
En oil refinery
Es refinería de petróleo *(f)*
Fr raffinerie de pétrole *(f)*
It raffineria di petrolio *(f)*

refinería *(f) n* Es
De Raffinerie *(f)*
En refinery
Fr raffinerie *(f)*
It raffineria *(f)*
Pt refinaria *(f)*

refinería de petróleo *(f) n* Es
De Ölraffinerie *(f)*
En oil refinery
Fr raffinerie de pétrole *(f)*
It raffineria di petrolio *(f)*
Pt refinaria de petróleo *(f)*

refinery *n* En
De Raffinerie *(f)*
Es refinería *(f)*
Fr raffinerie *(f)*
It raffineria *(f)*
Pt refinaria *(f)*

réflecteur solaire passif *(m)* Fr
De Sonnenlicht-Umlenkspiegel *(m)*
En passive sunlight reflector
Es reflector solar pasivo *(m)*
It riflettore passivo della luce del sole *(m)*
Pt reflector de raios solares passivo *(m)*

réflectif *adj* Fr
De reflektierend
En reflective
Es reflexivo
It riflettente
Pt reflectivo

reflective *adj* En
De reflektierend
Es reflexivo
Fr réflectif
It riflettente
Pt reflectivo

reflectivo *adj* Pt
De reflektierend
En reflective
Es reflexivo
Fr réflectif
It riflettente

reflector de raios solares passivo *(m)* Pt
De Sonnenlicht-Umlenkspiegel *(m)*
En passive sunlight reflector
Es reflector solar pasivo *(m)*
Fr réflecteur solaire passif *(m)*
It riflettore passivo della luce del sole *(m)*

reflector solar pasivo *(m)* Es
De Sonnenlicht-Umlenkspiegel *(m)*
En passive sunlight reflector
Fr réflecteur solaire passif *(m)*
It riflettore passivo della luce del sole *(m)*
Pt reflector de raios solares passivo *(m)*

reflektierend *adj* De
En reflective
Es reflexivo
Fr réflectif
It riflettente
Pt reflectivo

reflexivo *adj* Es
De reflektierend
En reflective
Fr réflectif
It riflettente
Pt reflectivo

reforma catalítica *(f)* Pt
De katalytische Reformierung *(f)*
En catalytic reforming
Es reformación catalítica *(f)*
Fr reformage catalytique *(m)*
It riforma catalitica *(f)*

reformación catalítica *(f)* Es
De katalytische Reformierung *(f)*
En catalytic reforming
Fr reformage catalytique *(m)*
It riforma catalitica *(f)*
Pt reforma catalítica *(f)*

reformage catalytique *(m)* Fr
De katalytische Reformierung *(f)*
En catalytic reforming
Es reformación catalítica *(f)*
It riforma catalitica *(f)*
Pt reforma catalítica *(f)*

refracção *(f) n* Pt
De Brechung *(f)*
En refraction
Es refracción *(f)*
Fr réfraction *(f)*
It rifrazione *(f)*

refracción *(f) n* Es
De Brechung *(f)*
En refraction
Fr réfraction *(f)*
It rifrazione *(f)*
Pt refracção *(f)*

réfractaire *adj* Fr
De feuerfest
En refractory
Es refractario
It refrattario
Pt refractário

réfractaire au carbone *(m)* Fr
De kohlenstoffhaltiger feuerfester Stoff *(m)*
En carbon refractory
Es refractario de carbón *(m)*
It refrattario al carbonio *(m)*
Pt refractário ao carbono *(m)*

réfractaire aux cermets *(m)* Fr
De feuerfestes Cermet *(n)*
En cermet refractory
Es refractario de cerametal *(m)*
It refrattario al cermete *(m)*
Pt refractário metalocerâmico *(m)*

refractario *adj* Es
De feuerfest
En refractory
Fr réfractaire
It refrattario
Pt refractário

refractário *adj* Pt
De feuerfest
En refractory
Es refractario
Fr réfractaire
It refrattario

refractário ao carbono *(m)* Pt
De kohlenstoffhaltiger feuerfester Stoff *(m)*
En carbon refractory
Es refractario de carbón *(m)*
Fr réfractaire au carbone *(m)*
It refrattario al carbonio *(m)*

refractario de carbón *(m)* Es
De kohlenstoffhaltiger feuerfester Stoff *(m)*
En carbon refractory
Fr réfractaire au carbone *(m)*
It refrattario al carbonio *(m)*
Pt refractário ao carbono *(m)*

refractario de cerametal *(m)* Es
De feuerfestes Cermet *(n)*
En cermet refractory
Fr réfractaire aux cermets *(m)*
It refrattario al cermete *(m)*
Pt refractário metalocerâmico *(m)*

refractário metalocerâmico *(m)* Pt
De feuerfestes Cermet *(n)*
En cermet refractory
Es refractario de cerametal *(m)*
Fr réfractaire aux cermets *(m)*
It refrattario al cermete *(m)*

refraction *n* En
De Brechung *(f)*
Es refracción *(f)*
Fr réfraction *(f)*
It rifrazione *(f)*
Pt refracção *(f)*

réfraction *(f) n* Fr
De Brechung *(f)*
En refraction
Es refracción *(f)*
It rifrazione *(f)*
Pt refracção *(f)*

refractive index En
De Brechungsindex *(m)*
Es índice de refracción *(m)*
Fr indice de réfraction *(m)*
It indice di rifrazione *(m)*
Pt índice de refracção *(m)*

refractory *adj* En
De feuerfest
Es refractario
Fr réfractaire
It refrattario
Pt refractário

refrattario *adj* It
De feuerfest
En refractory
Es refractario
Fr réfractaire
Pt refractário

refrattario al carbonio *(m)* It
De kohlenstoffhaltiger feuerfester Stoff *(m)*
En carbon refractory
Es refractario de carbón *(m)*
Fr réfractaire au carbone *(m)*
Pt refractário ao carbono *(m)*

refrattario al cermete *(m)* It
De feuerfestes Cermet *(n)*
En cermet refractory
Es refractario de cerametal *(m)*
Fr réfractaire aux cermets *(m)*
Pt refractário metalocerâmico *(m)*

refrigeración de radiación nocturna *(f)* Es
De Nachtstrahlungs-kühlung *(f)*
En night-radiation cooling
Fr refroidissement par rayonnement nocturne *(f)*
It raffreddamento a radiazioni notturne *(m)*
Pt arrefecimento por radiação nocturna *(m)*

refrigerado por aire Es
De luftgekühlt
En air-cooled
Fr refroidi par air
It raffreddato ad aria
Pt arrefecido a ar

refrigerador *(m) n* Es, Pt
De Kühlapparat *(m)*
En refrigerator
Fr réfrigérateur *(m)*
It refrigeratore *(m)*

refrigerant *n* En
De Kühlmittel *(n)*
Es refrigerante *(m)*
Fr réfrigérant *(m)*
It refrigerante *(m)*
Pt refrigerante *(m)*

réfrigérant *(m) n* Fr
De Kühlmittel *(n)*
En coolant; refrigerant
Es refrigerante *(m)*
It refrigerante *(m)*
Pt refrigerante *(m)*

refrigerante *(m) n* Es, It, Pt
De Kühlmittel *(n)*
En coolant; refrigerant
Fr réfrigérant *(m)*

refrigerante primario *(m)* Es, It
De primäres Kühlmittel *(n)*
En primary coolant
Fr réfrigérant primaire *(m)*
Pt refrigerante primário *(m)*

refrigerante primário *(m)* Pt
De primäres Kühlmittel *(n)*
En primary coolant
Es refrigerante primario *(m)*
Fr réfrigérant primaire *(m)*
It refrigerante primario *(m)*

réfrigérant primaire *(m)* Fr
De primäres Kühlmittel *(n)*
En primary coolant
Es refrigerante primario *(m)*
It refrigerante primario *(m)*
Pt refrigerante primário *(m)*

réfrigérateur *(m) n* Fr
De Kühlapparat *(m)*
En refrigerator
Es refrigerador *(m)*
It refrigeratore *(m)*
Pt refrigerador *(m)*

refrigerator *n* En
De Kühlapparat *(m)*
Es refrigerador *(m)*
Fr réfrigérateur *(m)*
It refrigeratore *(m)*
Pt refrigerador *(m)*

refrigeratore *(m) n* It
De Kühlapparat *(m)*
En refrigerator
Es refrigerador *(m)*
Fr réfrigérateur *(m)*
Pt refrigerador *(m)*

refrigeratore intermedio *(m)* It
De Zwischenkühler *(m)*
En intercooler
Es enfriador intermedio *(m)*
Fr refroidisseur intermédiaire *(m)*
Pt inter-arrefecedor *(m)*

refroidi par air Fr
De luftgekühlt
En air-cooled
Es refrigerado por aire
It raffreddato ad aria
Pt arrefecido a ar

refroidi par l'hydrogène Fr
De wasserstoffgekühlt
En hydrogen-cooled
Es enfriada con hidrógeno
It raffreddata ad idrogeno
Pt arrefecido a hidrogénio

refroidissement par ébullition *(m)* Fr
De Heißwasserkühlung *(f)*
En ebullient cooling
Es enfriamiento desde la ebullición *(m)*
It raffreddamento a ebollizione *(m)*
Pt arrefecimento ebuliente *(m)*

refroidissement par rayonnement nocturne *(f)* Fr
De Nachtstrahlungs-kühlung *(f)*
En night-radiation cooling
Es refrigeración de radiación nocturna *(f)*
It raffreddamento a radiazioni notturne *(m)*
Pt arrefecimento por radiação nocturna *(m)*

refroidisseur intermédiaire *(m)* Fr
De Zwischenkühler *(m)*
En intercooler
Es enfriador intermedio *(m)*
It refrigeratore intermedio *(m)*
Pt inter-arrefecedor *(m)*

refuse incineration En
De Abfallverbrennung *(f)*
Es incineración de basuras *(f)*
Fr incinération de déchets *(f)*
It incenerimento dei rifiuti *(m)*
Pt incineração de refugo *(f)*

regenerador *(m) n* Es, Pt
De Regenerator *(m)*
En regenerator
Fr régénérateur *(m)*
It ricuperatore di calore *(m)*

régénérateur *adj* Fr
De regenerativ
En regenerative
Es regenerativo
It rigenerativo
Pt regenerativo

régénérateur *(m) n* Fr
De Regenerator *(m)*
En regenerator
Es regenerador *(m)*
It ricuperatore di calore *(m)*
Pt regenerador *(m)*

regenerativ *adj* De
En regenerative
Es regenerativo

Fr régénérateur
It rigenerativo
Pt regenerativo

regenerative *adj* En
De regenerativ
Es regenerativo
Fr régénérateur
It rigenerativo
Pt regenerativo

regenerative braking En
De Rückstromgewinnbremsung *(f)*
Es frenado de recuperación *(m)*
Fr freinage à récupération *(m)*
It frenatura rigenerativo *(m)*
Pt quebra por regeneração *(f)*

regenerativo *adj* Es, Pt
De regenerativ
En regenerative
Fr régénérateur
It rigenerativo

regenerator *n* En
De Regenerator *(m)*
Es regenerador *(m)*
Fr régénérateur *(m)*
It ricuperatore di calore *(m)*
Pt regenerador *(m)*

Regenerator *(m) n* De
En regenerator
Es regenerador *(m)*
Fr régénérateur *(m)*
It ricuperatore di calore *(m)*
Pt regenerador *(m)*

Regler *(m) n* De
En regulator
Es regulador *(m)*
Fr régulateur *(m)*
It regolatore *(m)*
Pt regulador *(m)*

regolatore *(m) n* It
De Regler *(m)*
En regulator
Es regulador *(m)*
Fr régulateur *(m)*
Pt regulador *(m)*

regolatore di livello *(m)* It
De Füllstandsregler *(m)*
En level control
Es control de nivel *(m)*
Fr commande de niveau *(f)*
Pt controle de nível *(m)*

regolatore di tensione *(m)* It
De Spannungsregler *(m)*
En voltage regulator
Es regulador de tensión *(m)*
Fr régulateur de tension *(m)*
Pt regulador de tensão *(m)*

regulador *(m) n* Es, Pt
De Regler *(m)*
En regulator
Fr régulateur *(m)*
It regolatore *(m)*

regulador de tensão *(m)* Pt
De Spannungsregler *(m)*
En voltage regulator
Es regulador de tensión *(m)*
Fr régulateur de tension *(m)*
It regolatore di tensione *(m)*

regulador de tensión *(m)* Es
De Spannungsregler *(m)*
En voltage regulator
Fr régulateur de tension *(m)*
It regolatore di tensione *(m)*
Pt regulador de tensão *(m)*

régulateur *(m) n* Fr
De Regler *(m)*
En regulator
Es regulador *(m)*
It regolatore *(m)*
Pt regulador *(m)*

régulateur de tension *(m)* Fr
De Spannungsregler *(m)*
En voltage regulator
Es regulador de tensión *(m)*
It regolatore di tensione *(m)*
Pt regulador de tensão *(m)*

regulator *n* En
De Regler *(m)*
Es regulador *(m)*
Fr régulateur *(m)*
It regolatore *(m)*
Pt regulador *(m)*

reheater *n* En
De Nachbrenner *(m)*
Es recalentador *(m)*
Fr réchauffeur *(m)*
It ricombustore *(m)*
Pt reaquecedor *(m)*

reheating furnace En
De Nachbrennofen *(m)*
Es horno de recalentamiento *(m)*
Fr four à réchauffer *(m)*
It forno di riscaldo *(m)*
Pt forno reaquecedor *(m)*

Reinkohlengrundlage *(f)* De
En pure-coal basis
Es base de carbón puro *(f)*
Fr base de charbon pur *(f)*
It base di carbone puro *(f)*
Pt base de carvão puro *(f)*

Reinheit *(f) n* De
En purity
Es pureza *(f)*
Fr pureté *(f)*
It purezza *(f)*
Pt pureza *(f)*

reject fuel En
De Ausschußbrennstoff *(m)*
Es combustible recusable *(m)*
Fr combustible de rebut *(m)*
It combustibile di scarto *(m)*
Pt combustível rejeitado *(m)*

rejilla *(f) n* Es
De Gitter *(n)*
En grid (electronics)
Fr grille *(f)*
It griglia *(f)*
Pt grade *(f)*

Rektifikator (chemischer) *(m) n* De
En rectifier (chemical)
Es rectificador (químico) *(m)*
Fr rectificateur (chimique) *(m)*
It rettificatore (chimico) *(m)*
Pt rectificador (químico) *(m)*

rektifizierter Alkohol *(m)* De
En rectified spirit
Es alcohol rectificado *(m)*
Fr alcool rectifié *(m)*
It spirito rettificato *(m)*
Pt álcool rectificado *(m)*

Rekuperator *(m) n* De
En recuperator
Es recuperador *(m)*
Fr récupérateur *(m)*
It ricuperatore *(m)*
Pt recuperador *(m)*

relación de compresión *(f)* Es
De Kompressionsverhältnis *(n)*
En compression ratio
Fr taux de compression *(m)*
It rapporto di compressione *(m)*
Pt taxa de compressão *(f)*

relación hidrógeno-aceite *(f)* Es
De Wasserstoff-Ölverhältnis *(n)*
En hydrogen-oil ratio
Fr rapport hydrogène-huile *(m)*
It rapporto idrogeno-olio *(m)*
Pt razão hidrogénio-óleo *(f)*

relais *(m) n* Fr
De Relais *(n)*
En relay
Es relé *(m)*
It relè *(m)*
Pt relé *(m)*

Relais *(n) n* De
En relay
Es relé *(m)*
Fr relais *(m)*
It relè *(m)*
Pt relé *(m)*

relatif *adj* Fr
De relativ
En relative
Es relativa
It relativo
Pt relativo

relativ *adj* De
En relative
Es relativo
Fr relatif
It relativo
Pt relativo

relative *adj* En
De relativ
Es relativo
Fr relatif
It relativo
Pt relativo

relative Feuchtigkeit *(f)* De
En relative humidity
Es humedad relativa *(f)*
Fr humidité relative *(f)*
It umidità relativa *(f)*
Pt humidade relativa *(f)*

relative humidity En
De relative Feuchtigkeit *(f)*
Es humedad relativa *(f)*
Fr humidité relative *(f)*
It umidità relativa *(f)*
Pt humidade relativa *(f)*

relativo *adj* Es, It, Pt
De relativ
En relative
Fr relatif

relay *n* En
De Relais *(n)*
Es relé *(m)*
Fr relais *(m)*
It relè *(m)*
Pt relé *(m)*

relé *(m) n* Es, Pt
De Relais *(n)*
En relay
Fr relais *(m)*
It relè *(m)*

relè *(m) n* It
De Relais *(n)*
En relay
Es relé *(m)*
Fr relais *(m)*
Pt relé *(m)*

reliability theory En
De Verläßlichkeitstheorie *(f)*
Es teoría de la fiabilidad *(f)*
Fr théorie de fiabilité *(f)*
It teoria dell'affidabilità *(f)*
Pt teoria da segurança *(f)*

rendement *(m) n* Fr
De Ertrag *(m)*; Leistungsfähigkeit *(f)*
En efficiency; yield
Es eficiencia *(f)*; rendimiento *(m)*
It efficienza *(f)*; rendimento *(m)*
Pt eficiência *(f)*; rendimento *(m)*

rendement en goudron *(m)* Fr
De Teergiebigkeit *(f)*
En tar yield
Es riqueza de alquitrán *(f)*
It produzione catrame *(f)*
Pt rendimento de alcatrão *(m)*

rendement générateur *(m)* Fr
De Erzeugungsleistung *(f)*
En generating efficiency
Es rendimiento de generación *(m)*
It efficienza di generazione *(f)*
Pt eficiência geradora *(f)*

rendement global *(m)* Fr
De Gesamtleistungsfähigkeit *(f)*
En overall efficiency
Es rendimiento global *(m)*
It rendimento totale *(m)*
Pt eficiência global *(f)*

rendement mécanique *(m)* Fr
De mechanischer Wirkungsgrad *(m)*
En mechanical efficiency
Es rendimiento mecánico *(m)*
It resa meccanica *(f)*
Pt eficiência mecânica *(f)*

rendement thermique *(m)* Fr
De thermischer Wirkungsgrad *(m)*
En thermal efficiency
Es rendimiento térmico *(m)*
It rendimento termico *(m)*
Pt eficiência térmica *(f)*

rendimento *(m) n* It, Pt
De Ertrag *(m)*
En yield
Es rendimiento *(m)*
Fr rendement *(m)*

rendimento de alcatrão *(m)* Pt
De Teergiebigkeit *(f)*
En tar yield
Es riqueza de alquitrán *(f)*
Fr rendement en goudron *(m)*
It produzione catrame *(f)*

rendimento termico *(m)* It
De thermischer Wirkungsgrad *(m)*
En thermal efficiency
Es rendimiento térmico *(m)*
Fr rendement thermique *(m)*
Pt eficiência térmica *(f)*

rendimento totale *(m)* It
De Gesamtleistungsfähigkeit *(f)*
En overall efficiency
Es rendimiento global *(m)*
Fr rendement global *(m)*
Pt eficiência global *(f)*

rendimiento *(m) n* Es
De Ertrag *(m)*
En yield
Fr rendement *(m)*
It rendimento *(m)*
Pt rendimento *(m)*

rendimiento de generación *(m)* Es
De Erzeugungsleistung *(f)*
En generating efficiency
Fr rendement générateur *(m)*
It efficienza di generazione *(f)*
Pt eficiência geradora *(f)*

rendimiento global *(m)* Es
De Gesamtleistungsfähigkeit *(f)*
En overall efficiency
Fr rendement global *(m)*
It rendimento totale *(m)*
Pt eficiência global *(f)*

rendimiento mecánico *(m)* Es
De mechanischer Wirkungsgrad *(m)*
En mechanical efficiency
Fr rendement mécanique *(m)*
It resa meccanica *(f)*
Pt eficiência mecânica *(f)*

rendimiento térmico *(m)* Es
De thermischer Wirkungsgrad *(m)*
En thermal efficiency
Fr rendement thermique *(m)*
It rendimento termico *(m)*
Pt eficiência térmica *(f)*

renewable energy En
De erneuerbare Energie *(f)*

Es energía renovable *(f)*
Fr énergie renouvelable *(f)*
It energia rinnovabile *(f)*
Pt energia renovável *(f)*

Rentabilitätsgrenze *(f) n* De
En break-even point
Es punto comparativo *(m)*
Fr point mort *(m)*
It punto di pareggio *(m)*
Pt ponto de equilíbrio *(m)*

reostato *(m) n* Es, It
De Rheostat *(m)*
En rheostat
Fr rhéostat *(m)*
Pt reóstato *(m)*

reóstato *(m) n* Pt
De Rheostat *(m)*
En rheostat
Es reostato *(m)*
Fr rhéostat *(m)*
It reostato *(m)*

reprocessamento *(m) n* Pt
De Wiederaufbereitung *(f)*
En reprocessing
Es reelaboración *(f)*
Fr traitement du combustible irradié *(m)*
It riutilazzione *(f)*

reprocessing (nuclear fuel) *n* En
De Wiederaufbereitung *(f)*
Es reelaboración *(f)*
Fr traitement du combustible irradié *(m)*
It riutilazzione *(f)*
Pt reprocessamento *(m)*

resa meccanica *(f)* It
De mechanischer Wirkungsgrad *(m)*
En mechanical efficiency
Es rendimiento mecánico *(m)*
Fr rendement mécanique *(m)*
Pt eficiência mecânica *(f)*

réseau *(m) n* Fr
De Netz *(n)*
En network
Es red *(f)*
It rete *(f)*
Pt rede *(f)*

réseau électrique national *(m)* Fr
De Überlandleitungsnetz *(n)*
En grid (electrical power)
Es red nacional de energía eléctrica *(f)*
It rete nazionale *(f)*
Pt rede eléctrica *(f)*

réseau secteur *(m)* Fr
De Ringleitung *(f)*
En ring main
Es canalización circular *(f)*
It conduttori ad anello *(m)*
Pt condutor em anel fechado *(m)*

reservas carboníferas *(f pl)* Es
De Kohlereserven *(pl)*
En coal reserves *(pl)*
Fr réserves de houille *(f pl)*
It riserves di carbone *(f pl)*
Pt reservas de carvão *(f pl)*

reservas comprovadas *(f pl)* Pt
De erwiesene Reserven *(pl)*
En proven reserves *(pl)*
Es reservas probadas *(f pl)*
Fr réserves prouvées *(f pl)*
It riserve dimostrate *(f pl)*

reservas de carvão *(f pl)* Pt
De Kohlereserven *(pl)*
En coal reserves *(pl)*
Es reservas carboníferas *(f pl)*
Fr réserves de houille *(f pl)*
It riserves di carbone *(f pl)*

reservas de petróleo estratégicas *(f pl)* Pt
De strategische Ölreserven *(pl)*
En strategic oil reserves *(pl)*
Es reservas petrolíferas estratégicas *(f pl)*
Fr réserves de pétrole stratégiques *(f pl)*
It riserve strategiche di petrolio *(f pl)*

reservas energéticas *(f pl)* Es, Pt
De Energiereserven *(pl)*
En energy reserves *(pl)*
Fr réserves d'énergie *(f pl)*
It riserve energetiche *(f pl)*

reservas petrolíferas estratégicas *(f pl)* Es
De strategische Ölreserven *(pl)*
En strategic oil reserves *(pl)*
Fr réserves de pétrole stratégiques *(f pl)*
It riserve strategiche di petrolio *(f pl)*
Pt reservas de petróleo estratégicas *(f pl)*

reservas probables *(f pl)* Es
De wahrscheinliche Reserven *(pl)*
En probable reserves *(pl)*
Fr réserves probables *(f pl)*
It riserve probabili *(f pl)*
Pt reservas prováveis *(f pl)*

reservas probadas *(f pl)* Es
De erwiesene Reserven *(pl)*
En proven reserves *(pl)*
Fr réserves prouvées *(f pl)*
It riserve dimostrate *(f pl)*
Pt reservas comprovadas *(f pl)*

reservas prováveis *(f pl)* Pt
De wahrscheinliche Reserven *(pl)*
En probable reserves *(pl)*
Es reservas probables *(f pl)*
Fr réserves probables *(f pl)*
It riserve probabili *(f pl)*

reservas recuperables *(f pl)* Es
De gewinnbare Reserven *(pl)*
En recoverable reserves *(pl)*
Fr réserves récupérables *(f pl)*
It riserve recuperabili *(f pl)*
Pt reservas recuperáveis *(f pl)*

reservas recuperáveis *(f pl)* Pt
De gewinnbare Reserven *(pl)*
En recoverable reserves *(pl)*
Es reservas recuperables *(f pl)*
Fr réserves récupérables *(f pl)*
It riserve recuperabili *(f pl)*

reservatório *(m) n* Pt
De Sammelbecken *(n)*
En reservoir
Es depósito *(m)*
Fr réservoir *(m)*
It serbatoio *(m)*

reservatório de calor *(m)* Pt
De Wärmespeicher *(m)*
En heat reservoir
Es depósito de calor *(m)*
Fr réservoir de chaleur *(m)*
It serbatoio di calore *(m)*

Reserve - De
En standby
Es de reserva
Fr en attente
It emergenza
Pt de reserva

réserves de houille *(f pl)* Fr
De Kohlereserven *(pl)*
En coal reserves *(pl)*
Es reservas carboníferas *(f pl)*
It riserves di carbone *(f pl)*
Pt reservas de carvão *(f pl)*

réserves d'énergie *(f pl)* Fr
De Energiereserven *(pl)*
En energy reserves *(pl)*
Es reservas energéticas *(f pl)*
It riserve energetiche *(f pl)*
Pt reservas energéticas *(f pl)*

réserves de pétrole stratégiques *(f pl)* Fr
De strategische Ölreserven *(pl)*
En strategic oil reserves *(pl)*
Es reservas petrolíferas estratégicas *(f pl)*
It riserve strategiche di petrolio *(f pl)*
Pt reservas de petróleo estratégicas *(f pl)*

réserves probables *(f pl)* Fr
De wahrscheinliche Reserven *(pl)*
En probable reserves *(pl)*
Es reservas probables *(f pl)*
It riserve probabili *(f pl)*
Pt reservas prováveis *(f pl)*

réserves prouvées *(f pl)* Fr
De erwiesene Reserven *(pl)*
En proven reserves *(pl)*
Es reservas probadas *(f pl)*
It riserve dimostrate *(f pl)*
Pt reservas comprovadas *(f pl)*

réserves récupérables *(f pl)* Fr
De gewinnbare Reserven *(pl)*
En recoverable reserves *(pl)*
Es reservas recuperables *(f pl)*
It riserve recuperabili *(f pl)*
Pt reservas recuperáveis *(f pl)*

reservoir *n* En
De Sammelbecken *(n)*
Es depósito *(m)*
Fr réservoir *(m)*
It serbatoio *(m)*
Pt reservatório *(m)*

réservoir *(m) n* Fr
De Sammelbecken *(n)*
En reservoir
Es depósito *(m)*
It serbatoio *(m)*
Pt reservatório *(m)*

réservoir de chaleur *(m)* Fr
De Wärmespeicher *(m)*
En heat reservoir
Es depósito de calor *(m)*
It serbatoio di calore *(m)*
Pt reservatório de calor *(m)*

résidu *(m) n* Fr
De Rückstand *(m)*
En residuum
Es residuo *(m)*
It residuo *(m)*
Pt resíduo *(m)*

residual fuels *(pl)* En
De Rückstandsöl *(n)*
Es combustibles residuales *(m pl)*
Fr fuel résiduel *(m)*
It combustibili residui *(m pl)*
Pt combustíveis residuais *(m pl)*

résidu de carbone *(m)* Fr
De Koksrückstand *(m)*
En carbon residue
Es residuo de carbón *(m)*
It residuo di carbonio *(m)*
Pt resíduo de carbono *(m)*

residui di scarto *(m pl)* It
De Haldenabfall *(m)*
En tailings *(pl)*
Es desechos *(m pl)*
Fr produits de queue *(m pl)*
Pt colas *(f pl)*

residuo *(m) n* Es, It
De Rückstand *(m)*
En residuum
Es residuo *(m)*
Fr résidu *(m)*
Pt resíduo *(m)*

resíduo *(m) n* Pt
De Rückstand *(m)*
En residuum
Es residuo *(m)*
Fr résidu *(m)*
It residuo *(m)*

residuo activo *(m)* Es
De radioaktive Abfallstoffe *(pl)*
En active waste
Fr déchets actifs *(m pl)*
It rifiuti attivi *(m pl)*
Pt desperdícios activos *(m pl)*

residuo contaminado *(m)* Es
De kontaminierter Abfall *(m)*
En contaminated waste
Fr déchets contaminés *(m pl)*
It rifiuti contaminati *(m pl)*
Pt desperdício contaminado *(m)*

residuo de carbón *(m)* Es
De Koksrückstand *(m)*
En carbon residue
Fr résidu de carbone *(m)*
It residuo di carbonio *(m)*
Pt resíduo de carbono *(m)*

resíduo da carbono *(m)* Pt
De Koksrückstand *(m)*
En carbon residue
Es residuo de carbón *(m)*
Fr résidu de carbone *(m)*
It residuo di carbonio *(m)*

residuo di carbonio *(m)* It
De Koksrückstand *(m)*
En carbon residue
Es residuo de carbón *(m)*
Fr résidu de carbone *(m)*
Pt resíduo de carbono *(m)*

residuum *n* En
De Rückstand *(m)*
Es residuo *(m)*
Fr résidu *(m)*
It residuo *(m)*
Pt resíduo *(m)*

resistance *n* En
De Widerstand *(m)*
Es resistencia *(f)*
Fr résistance *(f)*
It resistenza *(f)*
Pt resistência *(f)*

résistance *(f) n* Fr
De Widerstand *(m)*
En resistance
Es resistencia *(f)*
It resistenza *(f)*
Pt resistência *(f)*

resistencia *(f) n* Es
De Widerstand *(m)*
En resistance
Fr résistance *(f)*
It resistenza *(f)*
Pt resistência *(f)*

resistência *(f) n* Pt
De Widerstand *(m)*
En resistance
Es resistencia *(f)*
Fr résistance *(f)*
It resistenza *(f)*

resistenza *(f) n* It
De Widerstand *(m)*
En resistance
Es resistencia *(f)*
Fr résistance *(f)*
Pt resistência *(f)*

resistenza alle intemperie *(f)* It
De Wetterschutz *(m)*
En weatherproofing

Es impermeabilización (f)
Fr protection contre les intempéries (f)
Pt protecção contra a intempérie (f)

resonance *n* En
De Resonanz *(f)*
Es resonancia *(f)*
Fr résonance *(f)*
It risonanza *(f)*
Pt ressonância *(f)*

résonance *(f) n* Fr
De Resonanz *(f)*
En resonance
Es resonancia *(f)*
It risonanza *(f)*
Pt ressonância *(f)*

resonancia *(f) n* Es
De Resonanz *(f)*
En resonance
Fr résonance *(f)*
It risonanza *(f)*
Pt ressonância *(f)*

Resonanz *(f) n* De
En resonance
Es resonancia *(f)*
Fr résonance *(f)*
It risonanza *(f)*
Pt ressonância *(f)*

resources *(pl n)* En
De Bodenschätze *(pl)*
Es recursos *(m pl)*
Fr ressources *(f pl)*
It risorse *(f pl)*
Pt recursos *(m pl)*

respirador de ar *(m)* Pt
De Entlüftungsschlitz *(m)*
En air vent
Es toma de aire *(f)*
Fr évent *(m)*
It sfiato aria *(m)*

ressonância *(f) n* Pt
De Resonanz *(f)*
En resonance
Es resonancia *(f)*
Fr résonance *(f)*
It risonanza *(f)*

ressources *(f pl) n* Fr
De Bodenschätze *(pl)*
En resources *(pl)*
Es recursos *(m pl)*
It risorse *(f pl)*
Pt recursos *(m pl)*

ressources d'énergie *(f pl)* Fr
De Energiequellen *(pl)*
En energy resources *(pl)*
Es recursos energéticos *(m pl)*
It risorse energetiche *(f pl)*
Pt recursos energéticos *(m pl)*

restrição de carga *(f)* Pt
De Lastabgabe *(f)*
En load-shedding
Es restricción de la carga *(f)*
Fr délestage *(m)*
It spargimento di carico *(m)*

restrição de carga a sobfrequência *(f)* Pt
De Unterfrequenz-Lastabschaltung *(f)*
En underfrequency load-shedding
Es restricción de la carga a baja frecuencia *(f)*
Fr délestage à sous-fréquence *(m)*
It spargimento di carico a sottofrequenza *(m)*

restricción de la carga *(f)* Es
De Lastabgabe *(f)*
En load-shedding
Fr délestage *(m)*
It spargimento di carico *(m)*
Pt restrição de carga *(f)*

restricción de la carga a baja frecuencia *(f)* Es
De Unterfrequenz-Lastabschaltung *(f)*
En underfrequency load-shedding
Fr délestage à sous-fréquence *(m)*
It spargimento di carico a sottofrequenza *(m)*
Pt restrição de carga a sobfrequência *(f)*

rete *(f) n* It
De Netz *(n)*
En network
Es red *(f)*
Fr réseau *(m)*
Pt rede *(f)*

rete nazionale *(f)* It
De Überlandleitungsnetz *(n)*
En grid (electrical power)
Es red nacional de energía eléctrica *(f)*
Fr réseau électrique national *(m)*
Pt rede eléctrica *(f)*

retombée radioactive *(f)* Fr
De radioaktiver Niederschlag *(m)*
En fall-out
Es precipitación radiactiva *(f)*
It precipitazione radioattiva *(f)*
Pt precipitação radioactiva *(f)*

retôrno *(m) n* Pt
De Rückkoppelung *(f)*
En feedback
Es realimentación *(f)*
Fr rétroaction *(f)*
It retroazione *(f)*

retôrno negativo *(m)* Pt
De negative Rückkoppelung *(f)*
En negative feedback
Es realimentación negativa *(f)*
Fr contre-réaction *(f)*
It controreazione *(f)*

rétroaction *(f) n* Fr
De Rückkoppelung *(f)*
En feedback
Es realimentación *(f)*
It retroazione *(f)*
Pt retôrno *(m)*

retroazione *(f) n* It
De Ruckkoppelung *(f)*
En feedback
Es realimentación *(f)*
Fr rétroaction *(f)*
Pt retôrno *(m)*

rettificatore (chimico) *(m) n* It
De Rektifikator (chemischer) *(m)*
En rectifier (chemical)
Es rectificador (químico) *(m)*
Fr rectificateur (chimique) *(m)*
Pt rectificador (químico) *(m)*

reventón *(m) n* Es
De Ausbruch *(m)*
En blowout
Fr éruption *(f)*
It eruzione *(f)*
Pt estouro *(m)*

reverse osmosis En
De umgekehrte Osmose *(f)*
Es ósmosis inversa *(f)*
Fr osmose inverse *(f)*
It osmosi inversa *(f)*
Pt osmose inversa *(f)*

reversible cycle En
De umkehrbarer Zyklus *(m)*
Es ciclo reversible *(m)*
Fr cycle réversible *(m)*
It ciclo reversibile *(m)*
Pt ciclo reversível *(m)*

revêtement calorifuge *(m)* Fr
De Ummantellung *(f)*
En lagging
Es revestimiento *(m)*
It rivestimento isolante *(m)*
Pt revestimento *(m)*

revestimento *(m) n* Pt
De Umhüllung; Ummantellung *(f)*
En cladding; lagging
Es encamisado; revestimiento *(m)*
Fr gainage; revêtement calorifuge *(m)*
It incamiciatura *(f)*; rivestimento isolante *(m)*

revestimento de canos *(m)* Pt
De Rohrummantellung *(f)*
En pipe lagging

Es revestimiento de tuberías *(m)*
Fr calorifugeage de tuyaux *(m)*
It rivestimento isolante di tubazioni *(m)*

revestimiento *(m) n* Es
De Ummantellung *(f)*
En lagging
Fr revêtement calorifuge *(m)*
It rivestimento isolante *(m)*
Pt revestimento *(m)*

revestimiento de tuberías *(m)* Es
De Rohrummantellung *(f)*
En pipe lagging
Fr calorifugeage de tuyaux *(m)*
It rivestimento isolante di tubazioni *(m)*
Pt revestimento de canos *(m)*

rheostat *n* En
De Rheostat *(m)*
Es reostato *(m)*
Fr rhéostat *(m)*
It reostato *(m)*
Pt reóstato *(m)*

Rheostat *(m) n* De
En rheostat
Es reostato *(m)*
Fr rhéostat *(m)*
It reostato *(m)*
Pt reóstato *(m)*

rhéostat *(m) n* Fr
De Rheostat *(m)*
En rheostat
Es reostato *(m)*
It reostato *(m)*
Pt reóstato *(m)*

rheostatic braking En
De Widerstands-bremsung *(f)*
Es frenado reostático *(m)*
Fr freinage par résistance *(m)*
It frenatura reostatica *(f)*
Pt quebra reostática *(f)*

ribbed boiler tube En
De Kesselrippenrohr *(n)*
Es tubo de aletas de caldera *(m)*
Fr tube de chaudière aileté *(m)*
It tubo nervato per caldaia *(m)*
Pt tubo de caldeira com nervuras *(m)*

riciclaggio *(m) n* It
De Rückführung in den Kreislauf *(f)*
En recycling
Es reciclajo *(m)*
Fr recyclage *(m)*
Pt reciclagem *(f)*

ricombustore *(m) n* It
De Nachbrenner *(m)*
En reheater
Es recalentador *(m)*
Fr réchauffeur *(m)*
Pt reaquecedor *(m)*

ricuperatore *(m) n* It
De Rekuperator *(m)*
En recuperator
Es recuperador *(m)*
Fr récupérateur *(m)*
Pt recuperador *(m)*

ricuperatore di calore *(m)* It
De Regenerator *(m)*
En regenerator
Es regenerador *(m)*
Fr régénérateur *(m)*
Pt regenerador *(m)*

ricuperatore tubolare *(m)* It
De Röhrenrekuperator *(m)*
En tubular recuperator
Es recuperador tubulare *(m)*
Fr récupérateur tubulaire *(m)*
Pt recuperador tubulare *(m)*

riel conductor *(m)* Es
De Stromschiene *(f)*
En third rail
Fr rail de contact *(m)*
It terza rotaia *(f)*
Pt terceiro carril *(m)*

riesgo de incendio *(m)* Es
De Feuergefahr *(f)*
En fire hazard
Fr danger d'incendie *(m)*
It pericolo di incendio *(m)*
Pt perigo de incêndio *(m)*

riesgo de radiación *(m)* Es
De Strahlungsgefahr *(f)*
En radiation hazard
Fr danger de radiation *(m)*
It pericolo di radiazioni *(m)*
Pt perigo de radiação *(m)*

rifiuti attivi *(m pl)* It
De radioaktive Abfallstoffe *(pl)*
En active waste
Es residuo activo *(m)*
Fr déchets actifs *(m pl)*
Pt desperdícios activos *(m pl)*

rifiuti contaminati *(m pl)* It
De kontaminierter Abfall *(m)*
En contaminated waste
Es residuo contaminado *(m)*
Fr déchets contaminés *(m pl)*
Pt desperdício contaminado *(m)*

riflettente *adj* It
De reflektierend
En reflective
Es reflexivo
Fr réflectif
Pt reflectivo

riflettore passivo della luce del sole *(m)* It
De Sonnenlicht-Umlenkspiegel *(m)*
En passive sunlight reflector
Es reflector solar pasivo *(m)*
Fr réflecteur solaire passif *(m)*
Pt reflector de raios solares passivo *(m)*

riforma catalitica *(f)* It
De katalytische Reformierung *(f)*
En catalytic reforming
Es reformación catalitica *(f)*
Fr reformage catalytique *(m)*
Pt reforma catalítica *(f)*

rifrazione *(f) n* It
De Brechung *(f)*
En refraction
Es refracción *(f)*
Fr réfraction *(f)*
Pt refracção *(f)*

rigenerativo *adj* It
De regenerativ
En regenerative
Es regenerativo
Fr régénérateur
Pt regenerativo

Ringleitung *(f) n* De
En ring main
Es canalización circular *(f)*
Fr réseau secteur *(m)*
It conduttori ad anello *(m)*
Pt condutor em anel fechado *(m)*

ring main En
De Ringleitung *(f)*
Es canalización circular *(f)*
Fr réseau secteur *(m)*
It conduttori ad anello *(m)*
Pt condutor em anel fechado *(m)*

riqueza de alquitrán *(f)* Es
De Teergiebigkeit *(f)*
En tar yield
Fr rendement en goudron *(m)*
It produzione catrame *(f)*
Pt rendimento de alcatrão *(m)*

riscaldamento *(m) n* It
De Heizung *(f)*
En heating
Es calefacción *(f)*
Fr chauffage *(m)*
Pt aquecimento *(m)*

riscaldamento ad induzione *(m)* It
De Induktionsheizung *(f)*
En induction heating
Es caldeo por inducción *(m)*
Fr chauffage par induction *(m)*
Pt aquecimento por indução *(m)*

riscaldamento a radiofrequenza *(m)* It
De Radiofrequenzheizung *(f)*
En radiofrequency heating
Es calentamiento por radiofrecuencia *(m)*
Fr chauffage à radiofréquence *(m)*
Pt aquecimento por frequência rádio *(m)*

riscaldamento centrale *(m)* It
De Zentralheizung *(f)*
En central heating
Es calefacción central *(f)*
Fr chauffage central *(m)*
Pt aquecimento central *(m)*

riscaldamento centrale a nafta *(m)* It
De Zentralheizung mit Ölfeuerung *(f)*
En oil-fired central heating
Es calefacción central con petróleo *(f)*
Fr chauffage central au mazout *(m)*
Pt aquecimento central a óleo *(m)*

riscaldamento dielettrico *(m)* It
De kapazitive Hochfrequenzerwärmung *(f)*
En dielectric heating
Es caldeo dieléctrico *(m)*
Fr chauffage par pertes diélectriques *(m)*
Pt aquecimento dieléctrico *(m)*

riscaldamento di sezione *(m)* It
De Fernheizung *(f)*
En district heating
Es calefacción de distritos *(f)*
Fr chauffage urbain *(m)*
Pt aquecimento de distrito *(m)*

riscaldamento indiretto *(m)* It
De indirekte Heizung *(f)*
En indirect heating
Es calefacción indirecto *(f)*
Fr chauffage indirect *(m)*
Pt aquecimento indirecto *(m)*

riscaldamento locale *(m)* It
De Raumheizung *(f)*
En space heating
Es calefacción de espacios *(f)*
Fr chauffage de chambres *(m)*
Pt aquecimento espacial *(m)*

riscaldatore *(m) n* It
De Heizelement *(n)*
En heater
Es calentador *(m)*
Fr réchauffeur *(m)*
Pt aquecedor *(m)*

riscaldatore a condotto aperto *(m)* It
De Heizelement mit offenem Abzug *(n)*
En open-flue heater
Es calentador de chimenea abierta *(m)*
Fr appareil de chauffage à conduit ouvert *(m)*
Pt aquecedor de fumeiro aberto *(m)*

riscaldatore a conservazione *(m)* It
De Speicherheizung *(f)*
En storage heater
Es calentador para almacenamiento térmico *(m)*
Fr radiateur à accumulation *(m)*
Pt aquecedor de armazenagem *(m)*

riscaldatore ad arco *(m)* It
De Lichtbogen-Heizelement *(m)*
En arc heater
Es calentador de arco *(m)*
Fr appareil de chauffage à arc *(m)*
Pt aquecedor a arco voltáico *(m)*

riscaldatore ad aria *(m)* It
De Lufterhitzer *(m)*
En air heater
Es calentador de aire *(m)*
Fr réchauffeur d'air *(m)*
Pt aquecedor de ar *(m)*

riscaldatore ad aria tiepida *(m)* It
De Warmluftheizung *(f)*
En warm-air heater
Es calentador con aire caliente *(m)*
Fr chauffage à air chaud *(m)*
Pt aquecedor de ar quente *(m)*

riscaldatore ad immersione *(m)* It
De Tauchsieder *(m)*
En immersion heater
Es calentador de inmersión *(m)*
Fr thermoplongeur *(m)*
Pt aquecedor por imersão *(m)*

riscaldatore a ventilatore *(m)* It
De Heizelement mit Gebläse *(n)*
En fan-assisted heater
Es calentador auxiliado por ventilador *(m)*
Fr radiateur soufflant *(m)*
Pt aquecedor auxiliado por ventoinha *(m)*

riscaldatore di acqua *(m)* It
De Wassererhitzer *(m)*
En water heater
Es calentador de agua *(m)*
Fr chauffe-eau *(m)*
Pt aquecedor hidráulico *(m)*

riscaldatore di acqua a conservazione *(m)* It
De Speicherwasserheizung *(f)*
En storage water-heater
Es calentador de agua de acumulación *(m)*
Fr chauffe-eau à accumulation *(m)*
Pt aquecedor a água de armazenagem *(m)*

riscaldatore di billette *(m)* It
De Blockerwärmer *(m)*
En billet heater
Es calentador de palanquillas *(m)*
Fr chauffe-billettes *(m)*
Pt aquecedor de lingotes *(m)*

riscaldatore istantaneo di acqua *(m)* It
De Wasser-Schnellerhitzer *(m)*
En instantaneous water heater
Es calentador de agua instantáneo *(m)*
Fr chauffe-eau instantané *(m)*
Pt aquecedor de água instantâneo *(m)*

riserva *(f) n* It
De Vorrat *(m)*
En stockpile
Es existencia para emergencia *(f)*
Fr stock de réserves *(m)*
Pt pilha de armazenagem *(f)*

riserve dimostrate *(f pl)* It
De erwiesene Reserven *(pl)*
En proven reserves *(pl)*
Es reservas probadas *(f pl)*
Fr réserves prouvées *(f pl)*
Pt reservas comprovadas *(f pl)*

riserve energetiche *(f pl)* It
De Energiereserven *(pl)*
En energy reserves *(pl)*
Es reservas energéticas *(f pl)*
Fr réserves d'énergie *(f pl)*
Pt reservas energéticas *(f pl)*

riserve probabili *(f pl)* It
De wahrscheinliche Reserven *(pl)*
En probable reserves *(pl)*
Es reservas probables *(f pl)*
Fr réserves probables *(f pl)*
Pt reservas prováveis *(f pl)*

riserve recuperabili *(f pl)* It
De gewinnbare Reserven *(pl)*
En recoverable reserves *(pl)*
Es reservas recuperables *(f pl)*
Fr réserves récupérables *(f pl)*
Pt reservas recuperáveis *(f pl)*

riserves di carbone *(f pl)* It
De Kohlereserven *(pl)*
En coal reserves *(pl)*
Es reservas carboníferas *(f pl)*
Fr réserves de houille *(f pl)*
Pt reservas de carvão *(f pl)*

riserve strategiche di petrolio *(f pl)* It
De strategische Ölreserven *(pl)*
En strategic oil reserves *(pl)*
Es reservas petrolíferas estratégicas *(f pl)*
Fr réserves de pétrole stratégiques *(f pl)*
Pt reservas de petróleo estratégicas *(f pl)*

risonanza *(f) n* It
De Resonanz *(f)*
En resonance
Es resonancia *(f)*
Fr résonance *(f)*
Pt ressonância *(f)*

risorse *(f pl) n* It
De Bodenschätze *(pl)*
En resources *(pl)*
Es recursos *(m pl)*
Fr ressources *(f pl)*
Pt recursos *(m pl)*

risorse energetiche *(f pl)* It
De Energiequellen *(pl)*
En energy resources *(pl)*
Es recursos energéticos *(m pl)*
Fr ressources d'énergie *(f pl)*
Pt recursos energéticos *(m pl)*

risparmio di energia *(m)* It
De Energiesparen *(n)*
En energy saving
Es ahorro energético *(m)*
Fr économie d'énergie *(f)*
Pt poupança de energia *(f)*

riutilazzione *(f) n* It
De Wiederaufbereitung *(f)*
En reprocessing (nuclear fuel)
Es reelaboración *(f)*
Fr traitement du combustible irradié *(m)*
Pt reprocessamento *(m)*

rivestimento isolante *(m)* It
De Ummantellung *(f)*
En lagging
Es revestimiento *(m)*
Fr revêtement calorifuge *(m)*
Pt revestimento *(m)*

rivestimento isolante di tubazioni *(m)* It
De Rohrummantellung *(f)*
En pipe lagging
Es revestimiento de tuberías *(m)*
Fr calorifugeage de tuyaux *(m)*
Pt revestimento de canos *(m)*

robinet *(m) n* Fr
De Hahn *(m)*
En cock (tap)
Es espita *(f)*
It rubinetto *(m)*
Pt torneira *(f)*

roca madre *(f)* Es
De Lagerstättengestein *(n)*
En source rock
Fr roche mère *(f)*
It roccia madre *(f)*
Pt rocha mãe *(f)*

roca metamórfica *(f)* Es
De metamorphosisches Gestein *(n)*
En metamorphic rock
Fr roche métamorphique *(f)*
It roccia metamorfica *(f)*
Pt rocha metamórfica *(f)*

rocas conglomeradas *(f pl)* Es
De klastische Gesteine *(pl)*
En conglomerate rocks *(pl)*
Fr roches conglomérés *(f pl)*
It rocce conglomerate *(f pl)*
Pt rochas conglomeradas *(f pl)*

rocce conglomerate *(f pl)* It
De klastische Gesteine *(pl)*
En conglomerate rocks *(pl)*
Es rocas conglomeradas *(f pl)*
Fr roches conglomérés *(f pl)*
Pt rochas conglomeradas *(f pl)*

rocchetto d'induzione *(m)* It
De Induktionsspule *(f)*
En induction coil
Es bobina de inducción *(f)*
Fr bobine d'induction *(f)*
Pt bobina de indução *(f)*

roccia madre *(f)* It
De Lagerstättengestein *(n)*
En source rock
Es roca madre *(f)*
Fr roche mère *(f)*
Pt rocha mãe *(f)*

roccia metamorfica *(f)* It
De metamorphosisches Gestein *(n)*
En metamorphic rock
Es roca metamórfica *(f)*
Fr roche métamorphique *(f)*
Pt rocha metamórfica *(f)*

rocha encaixante *(f)* Pt
De Deckgestein *(n)*
En cap rock
Es estrato impermeable de cobertura *(m)*
Fr roche couverture *(f)*
It strato impermeabile di copertura *(m)*

rocha mãe *(f)* Pt
De Lagerstättengestein *(n)*
En source rock
Es roca madre *(f)*
Fr roche mère *(f)*
It roccia madre *(f)*

rocha metamórfica *(f)* Pt
De metamorphosisches Gestein *(n)*
En metamorphic rock
Es roca metamórfica *(f)*
Fr roche métamorphique *(f)*
It roccia metamorfica *(f)*

rochas conglomeradas *(f pl)* Pt
De klastische Gesteine *(pl)*
En conglomerate rocks *(pl)*
Es rocas conglomeradas *(f pl)*
Fr roches conglomérés *(f pl)*
It rocce conglomerate *(f pl)*

roche couverture *(f)* Fr
De Deckgestein *(n)*
En cap rock
Es estrato impermeable de cobertura *(m)*

It strato impermeabile di copertura *(m)*
Pt rocha encaixante *(f)*

roche mère *(f)* Fr
De Lagerstättengestein *(n)*
En source rock
Es roca madre *(f)*
It roccia madre *(f)*
Pt rocha mãe *(f)*

roche métamorphique *(f)* Fr
De metamorphosisches Gestein *(n)*
En metamorphic rock
Es roca metamórfica *(f)*
It roccia metamorfica *(f)*
Pt rocha metamórfica *(f)*

roches conglomérés *(f pl)* Fr
De klastische Gesteine *(pl)*
En conglomerate rocks *(pl)*
Es rocas conglomeradas *(f pl)*
It rocce conglomerate *(f pl)*
Pt rochas conglomeradas *(f pl)*

rocket fuel En
De Raketentreibstoff *(m)*
Es combustible para cohetes *(m)*
Fr combustible pour fusées *(m)*
It carburante per missili *(m)*
Pt combustível para foguetões *(m)*

rock oil En
De Erdöl; Petroleum *(n)*
Es petróleo *(m)*
Fr pétrole *(m)*
It petrolio *(m)*
Pt petróleo *(m)*

rock spoils *(pl)* En
De Felsabraum *(m)*
Es desperdicios de roca *(m pl)*
Fr déblais de roche *(m pl)*
It materiale di sterro di roccia *(m)*
Pt detritos de rocha *(m pl)*

roda hidráulica *(f)* Pt
De Wasserrad *(n)*
En waterwheel
Es rueda hidráulica *(f)*
Fr roue à eau *(f)*
It ruota idraulica *(f)*

Rohöl *(n)* *n* De
En crude oil
Es petróleo crudo *(m)*
Fr pétrole brut *(m)*
It petrolio grezzo *(m)*
Pt óleo crú *(m)*

Röhrenrekuperator *(m)* *n* De
En tubular recuperator
Es recuperador tubulare *(m)*
Fr récupérateur tubulaire *(m)*
It ricuperatore tubolare *(m)*
Pt recuperador tubulare *(m)*

Röhrenwärmeaustauscher *(m)* De
En shell-and-tube heat exchanger
Es intercambiador de calor de coraza y tubos *(m)*
Fr échangeur de chaleur à calandre *(m)*
It scambiatore di calore a cilindro e tubi *(m)*
Pt termo-permutador de camisa e tubo *(m)*

Rohrleitung *(f)* *n* De
En pipeline
Es oleoducto *(m)*
Fr oléoduc *(m)*
It tubazione *(f)*
Pt oleoduto *(m)*

Rohrummantellung *(f)* *n* De
En pipe lagging
Es revestimiento de tuberías *(m)*
Fr calorifugeage de tuyaux *(m)*
It rivestimento isolante di tubazioni *(m)*
Pt revestimento de canos *(m)*

Röntgen-Strahlen *(pl)* *n* De
En X-rays *(pl)*
Es rayos X *(m pl)*
Fr rayons X *(m pl)*
It raggi X *(m pl)*
Pt raios X *(m pl)*

roof insulation En
De Dachisolierung *(f)*
Es aislamiento de techos *(m)*
Fr isolement des toits *(m)*
It isolamento del tetto *(m)*
Pt isolamento da cobertura *(m)*

rotary-displacement meter En
De Drehverschiebungsmesser *(m)*
Es contador de desplazamiento rotativo *(m)*
Fr compteur à piston rotatif *(m)*
It metro di spostamento rotante *(m)*
Pt contador de deslocação rotativa *(m)*

rotary drilling En
De Drehbohren *(n)*
Es perforación rotativa *(f)*
Fr forage rotary *(m)*
It sondaggio a rotazione *(m)*
Pt perfuração rotativa *(f)*

rotational energy En
De Rotationsenergie *(f)*
Es energía rotacional *(f)*
Fr énergie de rotation *(f)*
It energia rotazionale *(f)*
Pt energia de rotação *(f)*

Rotationsenergie *(f)* *n* De
En rotational energy
Es energía rotacional *(f)*
Fr énergie de rotation *(f)*
It energia rotazionale *(f)*
Pt energia de rotação *(f)*

rotor *n* En; Es, Fr, Pt *(m)*
De Läufer *(m)*
It rotore *(m)*

rotore *(m)* *n* It
De Läufer *(m)*
En rotor
Es rotor *(m)*
Fr rotor *(m)*
Pt rotor *(m)*

roue à eau *(f)* Fr
De Wasserrad *(n)*
En waterwheel
Es rueda hidráulica *(f)*
It ruota idraulica *(f)*
Pt roda hidráulica *(f)*

rubinetto *(m)* *n* It
De Hahn *(m)*
En cock
Es espita *(f)*
Fr robinet *(m)*
Pt torneira *(f)*

Rückführung in den Kreislauf *(f)* De
En recycling
Es reciclajo *(m)*
Fr recyclage *(m)*
It riciclaggio *(m)*
Pt reciclagem *(f)*

Rückgewinnung verbrauchter Luft *(f)* De
En spent-air recovery
Es recuperación del aire gastado *(f)*
Fr récupération de l'air dépensé *(f)*
It recupero dell'aria consumata *(m)*
Pt recuperação de ar gasto *(f)*

Rückkoppelung *(f)* *n* De
En feedback
Es realimentación *(f)*
Fr rétroaction *(f)*
It retroazione *(f)*
Pt retôrno *(m)*

Rückstand *(m)* *n* De
En residuum; sediment
Es residuo; sedimento *(m)*
Fr résidu; sédiment *(m)*
It residuo; sedimento *(m)*
Pt resíduo; sedimento *(m)*

Rückstandsöl *(n) n* De
En residual fuels *(pl)*
Es combustibles residuales *(m pl)*
Fr fuel résiduel *(m)*
It combustibili residui *(m pl)*
Pt combustíveis residuais *(m pl)*

Rückstandsprüfung *(f) n* De
En sediment test
Es prueba de sedimentos *(f)*
Fr essai de sédimentation *(m)*
It prova di sedimentazione *(f)*
Pt teste de sedimentos *(m)*

Rückstromgewinn-bremsung *(f)* De
En regenerative braking
Es frenado de recuperación *(m)*
Fr freinage à récupération *(m)*
It frenatura rigenerativo *(m)*
Pt quebra por regeneração *(f)*

Rückzahldauer *(f) n* De
En payback period
Es período de reembolso *(m)*
Fr durée de remboursement *(f)*
It periodo di remborso *(m)*
Pt período de reembolso *(m)*

rueda hidráulica *(f)* Es
De Wasserrad *(n)*
En waterwheel
Fr roue à eau *(f)*
It ruota idraulica *(f)*
Pt roda hidráulica *(f)*

running costs *(pl)* En
De Betriebskosten *(pl)*
Es costes de explotación *(m pl)*
Fr frais d'exploitation *(m pl)*
It costi di esercizio *(m pl)*
Pt custo de exploração *(m)*

run-of-mine coal En
De Normalkohle *(f)*
Es carbón todouno *(m)*
Fr charbon tout-venant *(m)*
It carbone tout-venant *(m)*
Pt carvão tal como sai da mina *(m)*

ruota idraulica *(f)* It
De Wasserrad *(n)*
En waterwheel
Es rueda hidráulica *(f)*
Fr roue à eau *(f)*
Pt roda hidráulica *(f)*

Ruß *(m) n* De
En soot
Es hollín *(m)*
Fr suie *(f)*
It nerofumo *(m)*
Pt fulígem *(f)*

Rußschwarz *(n) n* De
En carbon black
Es negro de carbón *(m)*
Fr noir de carbone *(m)*
It nerofumo di gas *(m)*
Pt negro de carvão *(m)*

R-value *n* En
De R-Wert *(m)*
Es valor R *(m)*
Fr valeur R *(f)*
It valore R *(m)*
Pt valor de R *(m)*

R-Wert *(m) n* De
En R-value
Es valor R *(m)*
Fr valeur R *(f)*
It valore R *(m)*
Pt valor de R *(m)*

S

sabbia bituminosa *(f)* It
De bituminöser Sand *(m)*
En bituminous sand
Es arena bituminosa *(f)*
Fr sable bitumineux *(m)*
Pt areia betuminosa *(f)*

sabbie impregnate di catrame *(f pl)* It
De Ölsande *(pl)*
En tar sands *(pl)*
Es arenas impregnadas de brea *(f pl)*
Fr sables asphaltiques *(m pl)*
Pt areias com alcatrão *(f pl)*

sable bitumineux *(m)* Fr
De bituminöser Sand *(m)*
En bituminous sand
Es arena bituminosa *(f)*
It sabbia bituminosa *(f)*
Pt areia betuminosa *(f)*

sables asphaltiques *(m pl)* Fr
De Ölsande *(pl)*
En tar sands *(pl)*
Es arenas impregnadas de brea *(f pl)*
It sabbie impregnate di catrame *(f pl)*
Pt areias com alcatrão *(f pl)*

safety valve En
De Sicherheitsventil *(n)*
Es válvula de seguridad *(f)*
Fr soupape de sûreté *(f)*
It valvola di sicurezza *(f)*
Pt válvula de segurança *(f)*

saída *(f) n* Pt
De Produktion *(f)*
En output
Es salida *(f)*
Fr production *(f)*
It produzione *(f)*

sala de comando *(f)* Pt
De Schaltraum *(m)*
En control room
Es sala de control *(f)*
Fr salle de commande *(f)*
It camera di manovra *(f)*

sala de control *(f)* Es
De Schaltraum *(m)*
En control room
Fr salle de commande *(f)*
It camera di manovra *(f)*
Pt sala de comando *(f)*

salida *(f) n* Es
De Produktion *(f)*
En output
Fr production *(f)*
It produzione *(f)*
Pt saída *(f)*

saline solar pond En
De Solar-Salzteich *(m)*
Es piscina solar salina *(f)*
Fr bassin solaire salant *(m)*
It laghetto solare salino *(m)*
Pt lago solar de sal *(m)*

salle de commande *(f)* Fr
De Schaltraum *(m)*
En control room
Es sala de control *(f)*
It camera di manovra *(f)*
Pt sala de comando *(f)*

salmoura *(f) n* Pt
De Sole *(f)*
En brine
Es salmuera *(f)*
Fr saumure *(f)*
It acqua salmastra *(f)*

salmoura geotérmica *(f)* Pt
De geothermische Sole *(f)*
En geothermal brine
Es salmuera geotérmica *(f)*
Fr saumure géothermique *(f)*
It acqua salmastra geotermica *(f)*

salmuera *(f) n* Es
De Sole *(f)*
En brine
Fr saumure *(f)*
It acqua salmastra *(f)*
Pt salmoura *(f)*

salmuera geotérmica *(f)* Es
De geothermische Sole *(f)*
En geothermal brine
Fr saumure géothermique *(f)*
It acqua salmastra geotermica *(f)*
Pt salmoura geotérmica *(f)*

Sammel-Abgasleitungen *(pl)* De
En multiple flues *(pl)*
Es conductos de humo múltiples *(m pl)*
Fr carneaux multiples *(m pl)*
It condotti multipli *(m pl)*
Pt fumeiros múltiplos *(m pl)*

Sammelbecken *(n) n* De
En reservoir
Es depósito *(m)*
Fr réservoir *(m)*
It serbatoio *(m)*
Pt reservatório *(m)*

Sankey diagram En
De Sankeydiagramm *(n)*
Es diagrama de Sankey *(m)*
Fr diagramme de Sankey *(m)*
It diagramma di Sankey *(m)*
Pt diagrama de Sankey *(m)*

Sankeydiagramm *(n) n* De
En Sankey diagram
Es diagrama de Sankey *(m)*
Fr diagramme de Sankey *(m)*
It diagramma di Sankey *(m)*
Pt diagrama de Sankey *(m)*

sapropelic coal En
De Sapropelkohle *(f)*
Es carbón sapropélico *(m)*
Fr charbon sapropélique *(m)*
It carbone sapropelico *(m)*
Pt carvão sapropélico *(m)*

Sapropelkohle *(f) n* De
En sapropelic coal
Es carbón sapropélico *(m)*
Fr charbon sapropélique *(m)*
It carbone sapropelico *(m)*
Pt carvão sapropélico *(m)*

satellite energy system En
De Satelliten-Energiesystem *(n)*
Es sistema de energía satélite *(m)*
Fr système d'énergie satellite *(m)*
It sistema di energia satellite *(m)*
Pt sistema de energia satélite *(m)*

Satelliten-Energiesystem *(n) n* De
En satellite energy system
Es sistema de energía satélite *(m)*
Fr système d'énergie satellite *(m)*
It sistema di energia satellite *(m)*
Pt sistema de energia satélite *(m)*

saturated air En
De gesättigte Luft *(f)*
Es aire saturado *(m)*
Fr air saturé *(m)*
It aria satura *(f)*
Pt ar saturado *(m)*

saturated steam En
De gesättigter Dampf *(m)*
Es vapor saturado *(m)*
Fr vapeur saturée *(f)*
It vapore saturo *(m)*
Pt vapor saturado *(m)*

saturated vapor pressure Am
De gesättigter Dampfdruck *(m)*
En saturated vapour pressure
Es presión de vapor saturado *(f)*
Fr tension de vapeur saturée *(f)*
It pressione del vapore saturo *(m)*
Pt pressão de vapor saturado *(f)*

saturated vapour pressure En
Am saturated vapor pressure
De gesättigter Dampfdruck *(m)*
Es presión de vapor saturado *(f)*
Fr tension de vapeur saturée *(f)*
It pressione del vapore saturo *(m)*
Pt pressão de vapor saturado *(f)*

sauberer Ballast *(m)* De
En clean ballast
Es lastre limpio *(m)*
Fr ballast propre *(m)*
It zavorra pulita *(f)*
Pt balastro puro *(m)*

Sauberluftzone *(f)* De
En clean-air zone
Es zona de aire limpio *(f)*
Fr zone d'air pur *(f)*
It zona aria pulita *(f)*
Pt zona de ar puro *(f)*

Sauerstoff *(m) n* De
En oxygen
Es oxígeno *(m)*
Fr oxygène *(m)*
It ossigeno *(m)*
Pt oxigénio *(m)*

Säuerung *(f) n* De
En acidification
Es acidulación *(f)*
Fr acidification *(f)*
It acidificazione *(f)*
Pt acidificação *(f)*

Saugpyrometer *(n) n* De
En suction pyrometer
Es pirómetro de succión *(m)*
Fr pyromètre aspirant *(m)*
It pirometro ad aspirazione *(m)*
Pt pirómetro de sucção *(m)*

saumure *(f) n* Fr
De Sole *(f)*
En brine
Es salmuera *(f)*
It acqua salmastra *(f)*
Pt salmoura *(f)*

saumure géothermique *(f)* Fr
De geothermische Sole *(f)*
En geothermal brine
Es salmuera geotérmica *(f)*
It acqua salmastra geotermica *(f)*
Pt salmoura geotérmica *(f)*

sbarra *(f) n* It
De Leitungsschiene *(f)*
En busbar
Es barra colectora *(f)*
Fr barre omnibus *(f)*
Pt barra colectora *(f)*

sbarramento di marea *(m)* It
De Gezeitendamm *(m)*
En tidal barrage
Es presa de marea *(f)*
Fr barrage marémoteur *(m)*
Pt barragem de marés *(f)*

scalar quantity En
De Skalarquantität *(f)*
Es cantidad escalar *(f)*
Fr quantité scalaire *(f)*
It grandezza scalare *(f)*
Pt quantidade escalar *(f)*

scaldabagno *(m) n* It
De Badheizelement *(n)*
En bath heater
Es calentador de baño *(m)*
Fr chauffe-bain *(m)*
Pt aquecedor de banho *(m)*

scaldato a carbone It
De kohlenverbrennend
En coal-burning
Es caldeado con carbón
Fr chauffé au charbon
Pt aquecido a carvão

scambiatore *(m) n* It
De Austauscher *(m)*
En exchanger
Es intercambiador *(m)*
Fr échangeur *(m)*
Pt permutador *(m)*

scambiatore di calore *(m)* It
De Wärmeaustauscher *(m)*
En heat exchanger
Es intercambiador de calor *(m)*
Fr échangeur de chaleur *(m)*
Pt termo-permutador *(m)*

scambiatore di calore a cilindro e tubi *(m)* It
De Röhrenwärmeaustauscher *(m)*
En shell-and-tube heat exchanger
Es intercambiador de calor de coraza y tubos *(m)*
Fr échangeur de chaleur à calandre *(m)*
Pt termo-permutador de camisa e tubo *(m)*

scambiatore di calore a piastra *(m)* It
De Plattenwärme-austauscher *(m)*
En plate heat-exchanger
Es intercambiador de calor de placas *(m)*
Fr échangeur de chaleur à plaque *(m)*
Pt termo-permutador de placa *(m)*

scambiatore di calore a serpentino in bagno *(m)* It
De Wärmeaustauscher bestehend aus Rohrschlangen im Bad *(m)*
En coil-in-bath heat exchanger
Es intercambiador de calor con serpentín en baño *(m)*
Fr échangeur de chaleur à serpentin immergé *(m)*
Pt termo-permutador de bobina em banho *(m)*

scambiatore di calore a tubo concentrico *(m)* It
De Wärmeaustauscher mit konzentrischen Röhren *(m)*
En concentric-tube heat exchanger
Es intercambiador de calor de tubos concéntricos *(m)*
Fr échangeur de chaleur à tubes concentriques (m)
Pt termo-permutador de tubo concêntrico *(m)*

scambiatore di calore con tubo a spirale *(m)* It
De Spiralröhren-Wärmeaustauscher *(m)*
En spiral-tube heat exchanger
Es intercambiador de calor de tubos en espiral *(m)*
Fr échangeur de chaleur à serpentin *(m)*
Pt termo-permutador de tubo em espiral *(m)*

scambiatore di calore di vetro *(m)* It
De Glaswärmeaustauscher *(m)*
En glass heat exchanger
Es intercambiador de calor de vidrio *(m)*
Fr échangeur de chaleur en verre *(m)*
Pt termo-permutador de vidro *(m)*

scambio ionico *(m)* It
De Ionenaustausch *(m)*
En ion exchange
Es intercambio de iones *(m)*
Fr échange d'ions *(m)*
Pt permuta de ioēs *(f)*

scarica *(f)* *n* It
De Entladung *(f)*
En discharge
Es descarga (bateria) *(f)*
Fr décharge *(f)*
Pt descarga (batería) *(f)*

scarico *(m)* *n* It
De Auspuffgas *(n)*
En exhaust
Es gases de escape *(m pl)*
Fr échappement *(m)*
Pt escape *(m)*

scattering *n* En
De Streuung *(f)*
Es dispersión *(f)*
Fr dispersion *(f)*
It dispersione *(f)*
Pt espalhamento *(m)*

scavo a contorno *(m)* It
De Konturenabbau *(m)*
En contour mining
Es minería siguiendo el perfil del terreno *(f)*
Fr exploitation minière des contours *(f)*
Pt mineração de curvas de nível *(f)*

scavo profondo *(m)* It
De Tiefbau *(m)*
En deep mining
Es explotación en profundidad *(f)*
Fr exploitation des niveaux inférieurs *(f)*
Pt mineração profunda *(f)*

scavo sotteraneo *(m)* It
De Grubenbetrieb *(m)*
En underground mining
Es minería subterránea *(f)*
Fr exploitation minière souterraine *(f)*
Pt mineração subterrânea *(f)*

Schachtofen *(m)* *n* De
En shaft furnace
Es horno de cuba *(m)*
Fr four à cuve *(m)*
It forno a tino *(m)*
Pt forno de chaminé *(m)*

Schalenkalorimeter *(n)* *n* De
En bucket calorimeter
Es calorímetro de cubeta *(m)*
Fr calorimètre à godet *(m)*
It calorimetro di secchio *(m)*
Pt calorímetro de balde *(m)*

Schalenkessel *(m)* *n* De
En shell boiler
Es caldera de coraza *(f)*
Fr chaudière à paroi *(f)*
It caldaia cilindrica *(f)*
Pt caldeira de camisa *(f)*

Schalldämpfer *(m)* *n* De
En silencer
Es silenciador *(m)*
Fr silencieux *(m)*
It silenziatore *(m)*
Pt silenciador *(m)*

Schallenergie *(f)* *n* De
En acoustic energy
Es energía acústica *(f)*
Fr énergie acoustique *(f)*
It energia acustica *(f)*
Pt energia acústica *(f)*

Schaltraum *(m)* *n* De
En control room
Es sala de control *(f)*
Fr salle de commande *(f)*
It camera di manovra *(f)*
Pt sala de comando *(f)*

Schaltung *(f)* *n* De
En circuit
Es circuito *(m)*
Fr circuit *(m)*
It circuito *(m)*
Pt circuito *(m)*

schéma de passage de l'énergie *(m)* Fr
De Energieflußdiagramm *(n)*
En energy-flow diagram
Es diagrama de flujo energético *(m)*
It diagramma del flusso energetico *(m)*
Pt diagrama de fluxo de energia *(m)*

schéma indicateur *(m)* Fr
De Indikatordiagramm *(n)*
En indicator diagram
Es diagrama indicador *(m)*
It diagramma del ciclo indicato *(m)*
Pt diagrama indicador *(m)*

schermo *(m)* *n* It
De Abschirmung *(f)*
En shield
Es blindaje *(m)*
Fr bouclier *(m)*
Pt protector *(m)*

schermo biologico *(m)* It
De biologischer Schild *(m)*
En biological shield
Es blindaje biológico *(m)*
Fr bouclier biologique *(m)*
Pt protector biológico *(m)*

schermo per il vapore *(m)* It
Am vapor barrier
De Dampfsperre *(f)*
En vapour barrier
Es barrera del vapor *(f)*
Fr écran d'étanchéité à la vapeur *(m)*
Pt barreira de vapor *(f)*

schermo termico *(m)* It
De Wärmeschutz *(m)*
En heat shield; thermal shield
Es pantalla térmica *(f)*
Fr bouclier thermique *(m)*
Pt protector térmico *(m)*

Schieferöl *(n) n* De
En shale oil
Es aceite de lutita *(m)*
Fr huile de schiste *(f)*
It olio di schisto *(m)*
Pt petróleo xistoso *(m)*

Schiffsdieseltreibstoff *(m) n* De
En marine diesel-fuel
Es combustible diesel para marina *(m)*
Fr combustible diesel marin *(m)*
It combustibile diesel per motori marini *(m)*
Pt diesel-fuel marítimo *(m)*

schistes bitumineux *(m pl)* Fr
De Ölschiefer *(pl)*
En oil shales *(pl)*
Es pizarras bituminosas *(f pl)*
It schisti bituminosi *(m pl)*
Pt xistos betuminosos *(m pl)*

schisti bituminosi *(m pl)* It
De Ölschiefer *(pl)*
En oil shales *(pl)*
Es pizarras bituminosas *(f pl)*
Fr schistes bitumineux *(m pl)*
Pt xistos betuminosos *(m pl)*

Schlacke *(f) n* De
En slag
Es escoria *(f)*
Fr laitier *(m)*
It scoria *(f)*
Pt escória *(f)*

Schmelzpunkt *(m) n* De
En melting point
Es punto de fusión *(f)*
Fr point de fusion *(f)*
It punto di fusione *(m)*
Pt ponto de fusão *(m)*

Schmieröl *(n) n* De
En lubricating oil
Es aceite lubricante *(m)*
Fr huile de graissage *(f)*
It olio lubrificante *(m)*
Pt óleo de lubrificação *(m)*

Schnellbrüter *(m) n* De
En fast breeder reactor
Es reactor reproductor rápido *(m)*
Fr réacteur surrégénérateur rapide *(m)*
It reattore veloce autofissilizzante *(m)*
Pt reactor reproductor rápido *(m)*

schnelles Neutron *(n)* De
En fast neutron
Es neutrón rápido *(m)*
Fr neutron rapide *(m)*
It neutrone veloce *(m)*
Pt neutrão rápido *(m)*

Schnellreaktor *(m) n* De
En fast reactor
Es reactor rápido *(m)*
Fr réacteur rapide *(m)*
It reattore veloce *(m)*
Pt reactor rápido *(m)*

Schnüreffekt *(m) n* De
En pinch effect
Es efecto de estricción *(m)*
Fr effet de pincement *(m)*
It effetto di strizione *(m)*
Pt efeito de aperto *(m)*

Schornstein *(m) n* De
En chimney
Es chimenea *(f)*
Fr cheminée *(f)*
It camino *(m)*
Pt chaminé *(f)*

Schüttdichte *(f) n* De
En bulk density
Es densidad en masa *(f)*
Fr densité apparente *(f)*
It densità apparente *(m)*
Pt densidade em massa *(f)*

Schütz *(m) n* De
En contactor
Es contactor *(m)*
Fr conjuncteur *(m)*
It contattore *(m)*
Pt contactor *(m)*

Schutzsystem *(n) n* De
En protection system
Es sistema de protección *(m)*
Fr système de protection *(m)*
It sistema di protezione *(m)*
Pt sistema de protecção *(m)*

Schwachlastzeit *(f) n* De
En off-peak period
Es período fuera de horas punta *(m)*
Fr période hors-pointe *(f)*
It periodo fuori di ore di punta *(m)*
Pt período fora de ponta *(m)*

Schwarzkohle *(f) n* De
En black coal
Es carbón negro *(m)*
Fr houille *(f)*
It carbone nero *(m)*
Pt carvão negro *(m)*

Schwefel *(m) n* De
Am sulfur
En sulphur
Es azufre *(m)*
Fr soufre *(m)*
It zolfo *(m)*
Pt enxofre *(m)*

Schwellenwert *(m) n* De
En threshold limit value (TLV)
Es valor límite de umbral *(m)*
Fr valeur de seuil *(f)*
It valore del limite di soglia *(m)*
Pt valor de limite de limiar *(m)*

schwerer Wasserstoff *(m)* De
En heavy hydrogen
Es hidrógeno pesado *(m)*
Fr hydrogène lourd *(m)*
It idrogeno pesante *(m)*
Pt hidrogénio pesado *(m)*

Schwerkraft *(f) n* De
En gravity
Es gravedad *(f)*
Fr gravité *(f)*
It gravità *(f)*
Pt gravidade *(f)*

Schwerkraftspeisung *(f)* De
En gravity-feed
Es alimentación por gravedad *(f)*
Fr alimentation par gravité *(f)*
It alimentazione a gravità *(f)*
Pt alimentação por gravidade

Schwerwasser *(n) n* De
En heavy water
Es agua pesada *(f)*
Fr eau lourde *(f)*
It acqua pesante *(f)*
Pt água pesada *(f)*

Schwerwasserreaktor *(m)* De
En heavy-water reactor
Es reactor de agua pesada *(m)*
Fr réacteur à eau lourde *(m)*
It reattore ad acqua pesante *(m)*
Pt reactor de água pesada *(m)*

Schwungrad *(n) n* De
En flywheel
Es volante *(m)*
Fr volant *(m)*
It volano *(m)*
Pt volante de inércia *(m)*

scia d'olio grezzo in mare *(f)* It
De Ölschlamm *(m)*
En oil slick
Es mancha de petróleo flotante *(f)*
Fr nappe d'huile *(f)*
Pt mancha de óleo *(f)*

scoria *(f) n* It
De Schlacke *(f)*
En slag
Es escoria *(f)*
Fr laitier *(m)*
Pt escória *(f)*

scram (nuclear reactor) *n* En
De Abschaltung *(f)*
Es parada de emergencia *(f)*
Fr arrêt d'urgence *(m)*
It spegnimento immediato *(m)*
Pt paragem de emergência *(f)*

seal *n* En
De Abdichtung *(f)*
Es cierre *(m)*
Fr joint *(m)*
It dispositivo di tenuta *(m)*
Pt vedação *(f)*

sealed-unit double glazing En
De luftdichte Doppelverglasungs-elemente *(pl)*
Es doble encristalado de unidad sellada *(m)*
Fr double vitrage à éléments scellés *(m)*
It doppia vetrata sigillata *(f)*
Pt vitrificação dupla de unidade vedada *(f)*

secador *(m) n* Es, Pt
De Trockner *(m)*
En dryer
Fr sécheur *(m)*
It essiccatore *(m)*

secador de aire *(m)* Es
De Lufttrockner *(m)*
En air dryer
Fr sécheur d'air *(m)*
It essiccatore ad aria *(m)*
Pt secador por ar *(m)*

secador de alta frecuencia *(m)* Es
De Hochfrequenz-Trockner *(m)*
En high-frequency dryer
Fr sécheur haute fréquence *(m)*
It essiccatore ad alta frequenza *(m)*
Pt secador de alta frequência *(m)*

secador de alta frequência *(m)* Pt
De Hochfrequenz-Trockner *(m)*
En high-frequency dryer
Es secador de alta frecuencia *(m)*
Fr sécheur haute fréquence *(m)*
It essiccatore ad alta frequenza *(m)*

secador de película *(m)* Es, Pt
De Filmtrockner *(m)*
En film dryer
Fr sécheur de pellicule *(m)*
It essiccatore di pellicola *(m)*

secador por ar *(m)* Pt
De Lufttrockner *(m)*
En air dryer
Es secador de aire *(m)*
Fr sécheur d'air *(m)*
It essiccatore ad aria *(m)*

secagem por congelação *(f)* Pt
De Gefriertrocknung *(f)*
En freeze drying
Es liofilización *(f)*
Fr lyophilisation *(f)*
It liofilizzazione *(f)*

sécheur *(m) n* Fr
De Trockner *(m)*
En dryer
Es secador *(m)*
It essiccatore *(m)*
Pt secador *(m)*

sécheur d'air *(m)* Fr
De Lufttrockner *(m)*
En air dryer
Es secador de aire *(m)*
It essiccatore ad aria *(m)*
Pt secador por ar *(m)*

sécheur de pellicule *(m)* Fr
De Filmtrockner *(m)*
En film dryer
Es secador de película *(m)*
It essiccatore di pellicola *(m)*
Pt secador de película *(m)*

sécheur haute fréquence *(m)* Fr
De Hochfrequenz-Trockner *(m)*
En high-frequency dryer
Es secador de alta frecuencia *(m)*
It essiccatore ad alta frequenza *(m)*
Pt secador de alta frequência *(m)*

secondary air En
De Sekundärluft *(f)*
Es aire secundario *(m)*
Fr air secondaire *(m)*
It aria secondaria *(f)*
Pt ar secundário *(m)*

secondary energy En
De Sekundärenergie *(f)*
Es energía secundaria *(f)*
Fr énergie secondaire *(f)*
It energia secondaria *(f)*
Pt energia secundária *(f)*

secondary recovery En
De sekundäre Rückgewinnung *(f)*
Es recuperación secundaria *(f)*
Fr récupération secondaire *(f)*
It recupero secondario *(m)*
Pt recuperação secundária *(f)*

secondary-window double glazing En
De Doppelverglasung für sekundäre Fenster *(f)*
Es doble encristalado de ventana secundaria *(m)*
Fr double vitrage secondaire *(f)*
It doppia vetrata di finestre secondarie *(f)*
Pt vitrificação dupla de janela secundária *(f)*

sectional boiler En
De Teilkammerkessel *(m)*
Es caldera compartimentada *(f)*
Fr chaudière sectionnelle *(f)*
It caldaia a sezioni *(f)*
Pt caldeira de secções *(f)*

sediment *n* En
De Rückstand *(m)*
Es sedimento *(m)*
Fr sédiment *(m)*
It sedimento *(m)*
Pt sedimento *(m)*

sédiment *(m) n* Fr
De Rückstand *(m)*
En sediment
Es sedimento *(m)*
It sedimento *(m)*
Pt sedimento *(m)*

sedimento *(m) n* Es, It, Pt
De Rückstand *(m)*
En sediment
Fr sédiment *(m)*

sediment test En
De Rückstandsprüfung *(f)*
Es prueba de sedimentos *(f)*
Fr essai de sédimentation *(m)*
It prova di sedimentazione *(f)*
Pt teste de sedimentos *(m)*

Seerohrleitung *(f) n* De
En marine pipeline
Es conducción submarina *(f)*
Fr pipeline marin *(m)*
It condotto marino *(m)*

Pt oleoduto marítimo *(m)*

Seeverschmutzung *(f)* *n* De
En marine fouling
Es suciedad por depósitos marinos *(f)*
Fr accumulation de dépôts marins *(f)*
It incrostazione marina *(f)*
Pt depósitos marítimos *(m)*

Seilbohren *(n)* *n* De
En cable-tool drilling
Es perforación a cable *(f)*
Fr forage au câble *(m)*
It trivellazione con utensile a fune *(f)*
Pt perfuração a cabo *(f)*

sekundäre Alkalizelle *(f)* De
En alkaline secondary cell
Es elemento alcalino de acumulador *(m)*
Fr élément secondaire alcalin *(m)*
It cellula secondaria alcalina *(f)*
Pt célula secundária alcalina *(f)*

Sekundärenergie *(f)* *n* De
En secondary energy
Es energía secundaria *(f)*
Fr énergie secondaire *(f)*
It energia secondaria *(f)*
Pt energia secundária *(f)*

sekundäre Rückgewinnung *(f)* De
En secondary recovery
Es recuperación secundaria *(f)*
Fr récupération secondaire *(f)*
It recupero secondario *(m)*
Pt recuperação secundária *(f)*

Sekundärluft *(f)* *n* De
En secondary air
Es aire secundario *(m)*
Fr air secondaire *(m)*
It aria secondaria *(f)*
Pt ar secundário *(m)*

selectividad *(f)* *n* Es
De Selektivität *(f)*
En selectivity
Fr sélectivité *(f)*
It selettività *(f)*
Pt selectividade *(f)*

selectividade *(f)* *n* Pt
De Selektivität *(f)*
En selectivity
Es selectividad *(f)*
Fr sélectivité *(f)*
It selettività *(f)*

sélectivité *(f)* *n* Fr
De Selektivität *(f)*
En selectivity
Es selectividad *(f)*
It selettività *(f)*
Pt selectividade *(f)*

selectivity *n* En
De Selektivität *(f)*
Es selectividad *(f)*
Fr sélectivité *(f)*
It selettività *(f)*
Pt selectividade *(f)*

Selektivität *(f)* *n* De
En selectivity
Es selectividad *(f)*
Fr sélectivité *(f)*
It selettività *(f)*
Pt selectividade *(f)*

selettività *(f)* *n* It
De Selektivität *(f)*
En selectivity
Es selectividad *(f)*
Fr sélectivité *(f)*
Pt selectividade *(f)*

semibituminous coal En
De halbbituminöse Kohle *(f)*
Es carbón semibituminoso *(m)*
Fr houille demi-grasse *(f)*
It carbone semibituminoso *(m)*
Pt carvão semibetuminoso *(m)*

semiconducteur *(m)* *n* Fr
De Halbleiter *(m)*
En semiconductor
Es semiconductor *(m)*
It semiconduttore *(m)*
Pt semicondutor *(m)*

semiconductor *n* En; Es *(m)*
De Halbleiter *(m)*
Fr semiconducteur *(m)*
It semiconduttore *(m)*
Pt semicondutor *(m)*

semicondutor *(m)* *n* Pt
De Halbleiter *(m)*
En semiconductor
Es semiconductor *(m)*
Fr semiconducteur *(m)*
It semiconduttore *(m)*

semiconduttore *(m)* *n* It
De Halbleiter *(m)*
En semiconductor
Es semiconductor *(m)*
Fr semiconducteur *(m)*
Pt semicondutor *(m)*

semprevivo *(m)* *n* It
De Zündflamme *(f)*
En pilot flame
Es llama piloto *(f)*
Fr veilleuse *(f)*
Pt chama piloto *(f)*

sensible heat En
De spürbare Wärme *(f)*
Es calor sensible *(m)*
Fr chaleur sensible *(f)*
It calore sensibile *(m)*
Pt calor sensível *(m)*

separação de isótopos *(f)* Pt
De Isotopentrennung *(f)*
En isotope separation
Es separación de isótopos *(f)*
Fr séparation isotopique *(f)*
It separazione isotopica *(f)*

separación de isótopos *(f)* Es
De Isotopentrennung *(f)*
En isotope separation
Fr séparation isotopique *(f)*
It separazione isotopica *(f)*
Pt separação de isótopos *(f)*

séparation isotopique *(f)* Fr
De Isotopentrennung *(f)*
En isotope separation
Es separación de isótopos *(f)*
It separazione isotopica *(f)*
Pt separação de isótopos *(f)*

separazione isotopica *(f)* It
De Isotopentrennung *(f)*
En isotope separation
Es separación de isótopos *(f)*
Fr séparation isotopique *(f)*
Pt separação de isótopos *(f)*

serbatoio *(m)* *n* It
De Sammelbecken *(n)*
En reservoir
Es depósito *(m)*
Fr réservoir *(m)*
Pt reservatório *(m)*

serbatoio di calore *(m)* It
De Wärmespeicher *(m)*
En heat reservoir
Es depósito de calor *(m)*
Fr réservoir de chaleur *(m)*
Pt reservatório de calor *(m)*

seriengewickelte Maschine *(f)* De
En series-wound machine
Es maquina excitada en serie *(f)*
Fr machine série *(f)*
It macchina eccitata in serie *(f)*
Pt máquina enrolada em série *(f)*

serie radiactiva *(f)* Es
De radioaktive Serie *(f)*
En radioactive series
Fr série radioactive *(f)*
It serie radioattiva *(f)*
Pt série radioactiva *(f)*

série radioactiva *(f)* Pt
De radioaktive Serie *(f)*
En radioactive series
Es serie radiactiva *(f)*

Fr série radioactive *(f)*
It serie radioattiva *(f)*

série radioactive *(f)* Fr
De radioaktive Serie *(f)*
En radioactive series
Es serie radiactiva *(f)*
It serie radioattive *(f)*
Pt série radioactiva *(f)*

serie radioattiva *(f)* It
De radioaktive Serie *(f)*
En radioactive series
Es serie radiactiva *(f)*
Fr série radioactive *(f)*
Pt série radioactiva *(f)*

series-wound machine En
De seriengewickelte Maschine *(f)*
Es maquina excitada en serie *(f)*
Fr machine série *(f)*
It macchina eccitata in serie *(f)*
Pt máquina enrolada em série *(f)*

serpentina de aquecimento *(f)* Pt
De Heizschlange *(f)*
En heater coil
Es serpentín de calentamiento *(m)*
Fr serpentin de chauffage *(m)*
It serpentina di riscaldamento *(f)*

serpentina di riscaldamento *(f)* It
De Heizschlange *(f)*
En heater coil
Es serpentín de calentamiento *(m)*
Fr serpentin de chauffage *(m)*
Pt serpentina de aquecimento *(f)*

serpentín de calentamiento *(m)* Es
De Heizschlange *(f)*
En heater coil
Fr serpentin de chauffage *(m)*
It serpentina di riscaldamento *(f)*
Pt serpentina de aquecimento *(f)*

serpentin de chauffage *(m)* Fr
De Heizschlange *(f)*
En heater coil
Es serpentín de calentamiento *(m)*
It serpentina di riscaldamento *(f)*
Pt serpentina de aquecimento *(f)*

servomécanisme *(m) n* Fr
De Servomechanismus *(m)*
En servomechanism
Es servomecanismo *(m)*
It servomeccanismo *(m)*
Pt servomecanismo *(m)*

servomecanismo *(m) n* Es, Pt
De Servomechanismus *(m)*
En servomechanism
Fr servomécanisme *(m)*
It servomeccanismo *(m)*

servomeccanismo *(m) n* It
De Servomechanismus *(m)*
En servomechanism
Es servomecanismo *(m)*
Fr servomécanisme *(m)*
Pt servomecanismo *(m)*

servomechanism *n* En
De Servomechanismus *(m)*
Es servomecanismo *(m)*
Fr servomécanisme *(m)*
It servomeccanismo *(m)*
Pt servomecanismo *(m)*

Servomechanismus *(m) n* De
En servomechanism
Es servomecanismo *(m)*
Fr servomécanisme *(m)*
It servomeccanismo *(m)*
Pt servomecanismo *(m)*

sfiato aria *(m)* It
De Entlüftungsschlitz *(m)*
En air vent
Es toma de aire *(f)*
Fr évent *(m)*
Pt respirador de ar *(m)*

shaft furnace En
De Schachtofen *(m)*
Es horno de cuba *(m)*
Fr four à cuve *(m)*
It forno a tino *(m)*
Pt forno de chaminé *(m)*

shale oil En
De Schieferöl *(n)*
Es aceite de lutita *(m)*
Fr huile de schiste *(f)*
It olio di schisto *(m)*
Pt petróleo xistoso *(m)*

shatter test En
De Sturzfestigkeits-prüfung *(f)*
Es prueba de cohesión *(f)*
Fr essai de résistance au choc *(m)*
It prova di frangibilità *(f)*
Pt teste de fragmentação *(m)*

shell-and-tube heat exchanger En
De Röhrenwärmeaus-tauscher *(m)*
Es intercambiador de calor de coraza y tubos *(m)*
Fr échangeur de chaleur à calandre *(m)*
It scambiatore di calore a cilindro e tubi *(m)*
Pt termo-permutador de camisa e tubo *(m)*

shell boiler En
De Schalenkessel *(m)*
Es caldera de coraza *(f)*
Fr chaudière à paroi *(f)*
It caldaia cilindrica *(f)*
Pt caldeira de camisa *(f)*

shield *n* En
De Abschirmung *(f)*
Es blindaje *(m)*
Fr bouclier *(m)*
It schermo *(m)*
Pt protector *(m)*

shoreline *n* En
De Küstenlinie *(f)*
Es litoral *(m)*
Fr ligne de rivage *(f)*
It litorale *(m)*
Pt linha de costa *(f)*

shunt-wound machine En
De Nebenschlußmaschine *(f)*
Es máquina excitada en derivación *(f)*
Fr machine shunt *(f)*
It macchina eccitata in derivazione *(f)*
Pt máquina enrolada em shunt *(f)*

shutdown *n* En
De Abschalten *(n)*
Es parada *(f)*
Fr arrêt *(m)*
It arresto *(m)*
Pt paralização *(f)*

Sicherheitsventil *(n) n* De
En safety valve
Es válvula de seguridad *(f)*
Fr soupape de sûreté *(f)*
It valvola di sicurezza *(f)*
Pt válvula de segurança *(f)*

Sicherung *(f) n* De
En fuse
Es fusible *(m)*
Fr fusible *(m)*
It fusibile *(m)*
Pt fusível *(m)*

sichtbares Spektrum *(n)* De
En visible spectrum
Es espectro visible *(m)*
Fr spectre visible *(m)*
It spettro visibile *(m)*
Pt espectro visível *(m)*

Siedepunkt *(m) n* De
En boiling point
Es punto de ebullición *(f)*
Fr point d'ébullition *(f)*
It punto di ebollizione *(m)*
Pt ponto de ebulição *(m)*

Siedereaktor *(m) n* De
En boiling-water reactor (BWR)
Es reactor de agua en ebullición *(m)*

Fr réacteur à eau bouillante *(m)*
It reattore ad acqua bollente *(m)*
Pt reactor de água em ebulição *(m)*

Siemens-Martinofen *(m) n* De
En open-hearth furnace
Es horno Martin-Siemens *(m)*
Fr four Martin *(m)*
It forno Martin *(m)*
Pt forno de fornalha aberta *(m)*

Silber-Kadmiumzelle *(f) n* De
En silver-cadmium cell
Es pila de plata-cadmio *(f)*
Fr pile à l'argent-cadmium *(f)*
It cellula di argento-cadmio *(f)*
Pt célula de prata-cádmio *(f)*

Silber-Zinkzelle *(f) n* De
En silver-zinc cell
Es pila de plata-zinc *(f)*
Fr pile à l'argent-zinc *(f)*
It cellula di argento-zinco *(f)*
Pt célula de prata-zinco *(f)*

silencer *n* En
De Schalldämpfer *(m)*
Es silenciador *(m)*
Fr silencieux *(m)*
It silenziatore *(m)*
Pt silenciador *(m)*

silenciador *(m) n* Es, Pt
De Schalldämpfer *(m)*
En silencer
Fr silencieux *(m)*
It silenziatore *(m)*

silencieux *(m) n* Fr
De Schalldämpfer *(m)*
En silencer
Es silenciador *(m)*
It silenziatore *(m)*
Pt silenciador *(m)*

silenziatore *(m) n* It
De Schalldämpfer *(m)*
En silencer
Es silenciador *(m)*
Fr silencieux *(m)*
Pt silenciador *(m)*

silica *n* En
De Siliziumoxid *(n)*
Es sílice *(f)*
Fr silice *(f)*
It silice *(f)*
Pt sílica *(f)*

sílica *(f) n* Pt
De Siliziumoxid *(n)*
En silica
Es sílice *(f)*
Fr silice *(f)*
It silice *(f)*

silice *(f) n* Fr, It
De Siliziumoxid *(n)*
En silica
Es sílice *(f)*
Pt sílica *(f)*

sílice *(f) n* Es
De Siliziumoxid *(n)*
En silica
Fr silice *(f)*
It silice *(f)*
Pt sílica *(f)*

silicio *(m) n* Es, It
De Silizium *(n)*
En silicon
Fr silicium *(m)*
Pt silício *(m)*

silício *(m) n* Pt
De Silizium *(n)*
En silicon
Es silicio *(m)*
Fr silicium *(m)*
It silicio *(m)*

silicium *(m) n* Fr
De Silizium *(n)*
En silicon
Es silicio *(m)*
It silicio *(m)*
Pt silício *(m)*

silicon *n* En
De Silizium *(n)*
Es silicio *(m)*
Fr silicium *(m)*
It silicio *(m)*
Pt silício *(m)*

silicon cell En
De Silikonzelle *(f)*
Es célula de silicio *(f)*
Fr pile au silicium *(f)*
It cellula di silicone *(f)*
Pt célula de silíco *(f)*

Silikonzelle *(f) n* De
En silicon cell
Es célula de silicio *(f)*
Fr pile au silicium *(f)*
It cellula di silicone *(f)*
Pt célula de silíco *(f)*

Silizium *(n) n* De
En silicon
Es silicio *(m)*
Fr silicium *(m)*
It silicio *(m)*
Pt silício *(m)*

Siliziumoxid *(n) n* De
En silica
Es sílice *(m)*
Fr silice *(f)*
It silice *(m)*
Pt sílica *(f)*

silver-cadmium cell En
De Silber-Kadmiumzelle *(f)*
Es pila de plata-cadmio *(f)*
Fr pile à l'argent-cadmium *(f)*
It cellula di argento-cadmio *(f)*
Pt célula de prata-cádmio *(f)*

silver-zinc cell En
De Silber-Zinkzelle *(f)*
Es pila de plata-zinc *(f)*
Fr pile à l'argent-zinc *(f)*
It cellula di argento-zinco *(f)*
Pt célula de prata-zinco *(f)*

simple cycle *(m)* Fr
De Einzelgang *(m)*
En single cycle
Es ciclo sencillo *(m)*
It ciclo unico *(m)*
Pt ciclo único *(m)*

sincronização *(f) n* Pt
De Synchronisierung *(f)*
En synchronization
Es sincronización *(f)*
Fr synchronisation *(f)*
It sincronizzazione *(f)*

sincronización *(f) n* Es
De Synchronisierung *(f)*
En synchronization
Fr synchronisation *(f)*
It sincronizzazione *(f)*
Pt sincronização *(f)*

sincronizzazione *(f) n* It
De Synchronisierung *(f)*
En synchronization
Es sincronización *(f)*
Fr synchronisation *(f)*
Pt sincronização *(f)*

sine wave En
De Sinuswelle *(f)*
Es onda sinusoidal *(f)*
Fr onde sinusoïdale *(f)*
It sinusoide *(m)*
Pt onda sinusoidal *(f)*

single cycle *n* En
De Einzelgang *(m)*
Es ciclo sencillo *(m)*
Fr simple cycle *(m)*
It ciclo unico *(m)*
Pt ciclo único *(m)*

single-effect distillation En
De Einzeleffekt-Destillation *(f)*
Es destilación de simple efecto *(f)*
Fr distillation simple effet *(f)*
It distillazione ad effetto unico *(f)*
Pt destilação de efeito simples *(f)*

single-window collector En
De Einzelfenster-Kollektor *(m)*
Es colector de ventana única *(m)*
Fr collecteur à simple fenêtre *(m)*
It raccogliatore a finestra unica *(m)*
Pt colector de janela única *(m)*

sintering *n* En
De Sintern *(n)*
Es sinterización *(f)*
Fr frittage *(m)*

It sinterizzazione *(f)*
Pt concreção *(f)*

sinterización *(f) n* Es
De Sintern *(n)*
En sintering
Fr frittage *(m)*
It sinterizzazione *(f)*
Pt concreção *(f)*

sinterizzazione *(f) n* It
De Sintern *(n)*
En sintering
Es sinterización *(f)*
Fr frittage *(m)*
Pt concreção *(f)*

Sintern *(n) n* De
En sintering
Es sinterización *(f)*
Fr frittage *(m)*
It sinterizzazione *(f)*
Pt concreção *(f)*

sinusförmig *adj* De
En sinusoidal
Es sinusoidal
Fr sinusoïdal
It sinusoidale
Pt sinusoidal

sinusoidal *adj* En, Es, Pt
De sinusförmig
Fr sinusoïdal
It sinusoidale

sinusoïdal *adj* Fr
De sinusförmig
En sinusoidal
Es sinusoidal
It sinusoidale
Pt sinusoidal

sinusoidale *adj* It
De sinusförmig
En sinusoidal
Es sinusoidal
Fr sinusoïdal
Pt sinusoidal

sinusoide *(m) n* It
De Sinuswelle *(f)*
En sine wave
Es onda sinusoidal *(f)*
Fr onde sinusoïdale *(f)*
Pt onda sinusoidal *(f)*

Sinuswelle *(f) n* De
En sine wave
Es onda sinusoidal *(f)*
Fr onde sinusoïdale *(f)*
It sinusoide *(m)*
Pt onda sinusoidal *(f)*

sistema de control *(m)* Es
De Steuersystem *(n)*
En control system
Fr système de commande *(m)*
It sistema di controllo *(m)*
Pt sistema de controle *(m)*

sistema de controle *(m)* Pt
De Steuersystem *(n)*
En control system
Es sistema de control *(m)*
Fr système de commande *(m)*
It sistema di controllo *(m)*

sistema de energía satélite *(m)* Es
De Satelliten-Energiesystem *(n)*
En satellite energy system
Fr système d'énergie satellite *(m)*
It sistema di energia satellite *(m)*
Pt sistema de energia satélite *(m)*

sistema de energia satélite *(m)* Pt
De Satelliten-Energiesystem *(n)*
En satellite energy system
Es sistema de energía satélite *(m)*
Fr système d'énergie satellite *(m)*
It sistema di energia satellite *(m)*

sistema de limpeza *(m)* Pt
De Aufzehrsystem *(n)*
En cleanup system
Es sistema de limpieza *(m)*
Fr système de nettoyage *(m)*
It sistema di depurazione *(m)*

sistema de limpieza *(m)* Es
De Aufzehrsystem *(n)*
En cleanup system
Fr système de nettoyage *(m)*
It sistema di depurazione *(m)*
Pt sistema de limpeza *(m)*

sistema de protecção *(m)* Pt
De Schutzsystem *(n)*
En protection system
Es sistema de protección *(m)*
Fr système de protection *(m)*
It sistema di protezione *(m)*

sistema de protección *(m)* Es
De Schutzsystem *(n)*
En protection system
Fr système de protection *(m)*
It sistema di protezione *(m)*
Pt sistema de protecção *(m)*

sistema di controllo *(m)* It
De Steuersystem *(n)*
En control system
Es sistema de control *(m)*
Fr système de commande *(m)*
Pt sistema de controle *(m)*

sistema di depurazione *(m)* It
De Aufzehrsystem *(n)*
En cleanup system
Es sistema de limpieza *(m)*
Fr système de nettoyage *(m)*
Pt sistema de limpeza *(m)*

sistema di energia satellite *(m)* It
De Satelliten-Energiesystem *(n)*
En satellite energy system
Es sistema de energía satélite *(m)*
Fr système d'énergie satellite *(m)*
Pt sistema de energia satélite *(m)*

sistema di protezione *(m)* It
De Schutzsystem *(n)*
En protection system
Es sistema de protección *(m)*
Fr système de protection *(m)*
Pt sistema de protecção *(m)*

Skalarquantität *(f) n* De
En scalar quantity
Es cantidad escalar *(f)*
Fr quantité scalaire *(f)*
It grandezza scalare *(f)*
Pt quantidade escalar *(f)*

slag *n* En
De Schlacke *(f)*
Es escoria *(f)*
Fr laitier *(m)*
It scoria *(f)*
Pt escória *(f)*

slow neutron En
De langsames Neutron *(n)*
Es neutrón lento *(m)*
Fr neutron lent *(m)*
It neutrone lento *(m)*
Pt neutrão lento *(m)*

smagnetizzazione *(f) n* It
De Entmagnetisierung *(f)*
En demagnetization
Es desmagnetización *(f)*
Fr démagnétisation *(f)*
Pt desmagnetização *(f)*

smoke *n* En
De Rauch *(m)*
Es humo *(m)*
Fr fumée *(f)*
It fumo *(m)*
Pt fumo *(m)*

smokeless coal En
De rauchlose Kohle *(f)*
Es carbón fumífugo *(m)*
Fr charbon sans fumée *(m)*
It carbone senza fumo *(m)*

Pt carvão ardendo sem produzir fumo *(m)*

smoke point En
De Rauchpunkt *(m)*
Es temperatura de formación de humo *(f)*
Fr point de fumée *(m)*
It punto di fumo *(m)*
Pt ponto de fumo *(m)*

sobrecarga *(f) n* Es, Pt
De Überlastung *(f)*
En overload
Fr sucharge *(f)*
It sovraccarico *(m)*

sobretensión *(f) n* Es
De Überspannung *(f)*
En overvoltage
Fr surtension *(f)*
It sovratensione *(f)*
Pt sobrevoltagem *(f)*

sobrevoltagem *(f) n* Pt
De Überspannung *(f)*
En overvoltage
Es sobretensión *(f)*
Fr surtension *(f)*
It sovratensione *(f)*

soft radiation En
De Weichstrahlung *(f)*
Es radiación blanda *(f)*
Fr radiation molle *(f)*
It radiazione soffice *(f)*
Pt radiação suave *(f)*

sol *(m) n* Es, Pt
De Sonne *(f)*
En sun
Fr soleil *(m)*
It sole *(m)*

solaire *adj* Fr
De Solar-
En solar
Es solar
It solare
Pt solar

solar *adj* En, Es, Pt
De Solar-
Fr solaire
It solare

Solar- *adj* De
En solar
Es solar
Fr solaire
It solare
Pt solar

solar absorbtance En
De Solarabsorptionsgrad *(m)*
Es absorsancia solar *(f)*
Fr absorptance solaire *(f)*
It assorbimento solare *(m)*
Pt absorvância solar *(f)*

Solarabsorptionsgrad *(m) n* De
En solar absorbtance
Es absorsancia solar *(f)*
Fr absorptance solaire *(f)*
It assorbimento solare *(m)*
Pt absorvância solar *(f)*

solar availability En
De Solarverfügbarkeit *(f)*
Es disponibilidad solar *(f)*
Fr disponibilité solaire *(f)*
It disponibilità solare *(f)*
Pt disponibilidade solar *(f)*

solar cell En
De Solarzelle *(f)*
Es célula solar *(f)*
Fr cellule solaire *(f)*
It cellula solare *(f)*
Pt célula solar *(f)*

solar constant En
De Solarkonstante *(f)*
Es constante solar *(f)*
Fr constante solaire *(f)*
It costante solare *(f)*
Pt constante solar *(f)*

solar cooker En
De Solarherd *(m)*
Es cocina solar *(f)*
Fr fourneau solaire *(m)*
It fornello solare *(m)*
Pt fervedor solar *(m)*

solare *adj* It
De Solar-
En solar
Es solar
Fr solaire
Pt solar

solar energy En
De Sonnenenergie *(f)*
Es energía solar *(f)*
Fr énergie solaire *(f)*
It energia solare *(f)*
Pt energia solar *(f)*

Solarfarm *(f) n* De
En solar farm
Es granja solar *(f)*
Fr ferme solaire *(f)*
It podere solare *(m)*
Pt centro de produção solar *(f)*

solar farm En
De Solarfarm *(f)*
Es granja solar *(f)*
Fr ferme solaire *(f)*
It podere solare *(m)*
Pt centro de produção solar *(f)*

solar flux En
De Sonnenfluß *(m)*
Es flujo solar *(m)*
Fr flux solaire *(m)*
It flusso solare *(m)*
Pt fluxo solar *(m)*

solar furnace En
De Sonnenofen *(m)*
Es horno solar *(m)*
Fr four solaire *(m)*
It forno solare *(m)*
Pt forno solar *(m)*

solar gain En
De Solarverstärkung *(f)*
Es ganancia solar *(f)*
Fr gain solaire *(m)*
It guadagno solare *(m)*
Pt ganho solar *(m)*

Solarherd *(m) n* De
En solar cooker
Es cocina solar *(f)*
Fr fourneau solaire *(m)*
It fornello solare *(m)*
Pt fervedor solar *(m)*

solarimeter *n* En
De Solarimeter *(n)*
Es solarímetro *(m)*
Fr solarimètre *(m)*
It solarimetro *(m)*
Pt solarímetro *(m)*

Solarimeter *(n) n* De
En solarimeter
Es solarímetro *(m)*
Fr solarimètre *(m)*
It solarimetro *(m)*
Pt solarímetro *(m)*

solarimètre *(m) n* Fr
De Solarimeter *(n)*
En solarimeter
Es solarímetro *(m)*
It solarimetro *(m)*
Pt solarímetro *(m)*

solarimetro *(m) n* It
De Solarimeter *(n)*
En solarimeter
Es solarímetro *(m)*
Fr solarimètre *(m)*
Pt solarímetro *(m)*

solarímetro *(m) n* Es, Pt
De Solarimeter *(n)*
En solarimeter
Fr solarimètre *(m)*
It solarimetro *(m)*

Solarkonstante *(f) n* De
En solar constant
Es constante solar *(f)*
Fr constante solaire *(f)*
It costante solare *(f)*
Pt constante solar *(f)*

solar panel En
De Solarplatte *(f)*
Es panel solar *(m)*
Fr panneau solaire *(m)*
It pannello solare *(m)*
Pt painel solar *(m)*

Solarplatte *(f) n* De
En solar panel
Es panel solar *(m)*
Fr panneau solaire *(m)*
It pannello solare *(m)*
Pt painel solar *(m)*

solar radiation En
De Sonnenstrahlung *(f)*
Es radiación solar *(f)*
Fr rayonnement solaire *(m)*
It radiazione solare *(f)*
Pt radiação solar *(f)*

Solar-Salzteich *(m) n* De
En saline solar pond
Es piscina solar salina *(f)*

Fr bassin solaire salant *(m)*
It laghetto solare salino *(m)*
Pt lago solar de sal *(m)*

Solarverfügbarkeit *(f) n* De
En solar availability
Es disponibilidad solar *(f)*
Fr disponibilité solaire *(f)*
It disponibilità solare *(f)*
Pt disponibilidade solar *(f)*

Solarverstärkung *(f) n* De
En solar gain
Es ganancia solar *(f)*
Fr gain solaire *(m)*
It guadagno solare *(m)*
Pt ganho solar *(m)*

solar wind En
De Sonnenwind *(m)*
Es viento solar *(m)*
Fr vent solaire *(m)*
It vento solare *(m)*
Pt vento solar *(m)*

Solarzelle *(f) n* De
En solar cell
Es célula solar *(f)*
Fr cellule solaire *(f)*
It cellula solare *(f)*
Pt célula solar *(f)*

sole *(m) n* It
De Sonne *(f)*
En sun
Es sol *(m)*
Fr soleil *(m)*
Pt sol *(m)*

Sole *(f) n* De
En brine
Es salmuera *(f)*
Fr saumure *(f)*
It acqua salmastra *(f)*
Pt salmoura *(f)*

soleil *(m) n* Fr
De Sonne *(f)*
En sun
Es sol *(m)*
It sole *(m)*
Pt sol *(m)*

solide fluidifié *(m)* Fr
De verflüssigter Feststoff *(m)*
En fluidized solid
Es sólido fluidizado *(m)*
It solido fluidizzato *(m)*
Pt sólido fluidificado *(m)*

solid fuel En
De Festkraftstoff *(m)*
Es combustible sólido *(m)*
Fr combustible solide *(m)*
It combustibile solido *(m)*
Pt combustível sólido *(m)*

sólido fluidificado *(m)* Pt
De verflüssigter Feststoff *(m)*
En fluidized solid
Es sólido fluidizado *(m)*
Fr solide fluidifié *(m)*
It solido fluidizzato *(m)*

sólido fluidizado *(m)* Es
De verflüssigter Feststoff *(m)*
En fluidized solid
Fr solide fluidifié *(m)*
It solido fluidizzato *(m)*
Pt sólido fluidificado *(m)*

solido fluidizzato *(m)* It
De verflüssigter Feststoff *(m)*
En fluidized solid
Es sólido fluidizado *(m)*
Fr solide fluidifié *(m)*
Pt sólido fluidificado *(m)*

solstice *n* En; Fr
De Sonnenwende *(f)*
Es solsticio *(m)*
It solstizio *(m)*
Pt solstício *(m)*

solsticio *(m) n* Es
De Sonnenwende *(f)*
En solstice
Fr solstice *(m)*
It solstizio *(m)*
Pt solstício *(m)*

solstício *(m) n* Pt
De Sonnenwende *(f)*
En solstice
Es solsticio *(m)*
Fr solstice *(m)*
It solstizio *(m)*

solstizio *(m) n* It
De Sonnenwende *(f)*
En solstice
Es solsticio *(m)*
Fr solstice *(m)*
Pt solstício *(m)*

solvant *(m) n* Fr
De Lösungsmittel *(n)*
En solvent
Es disolvente *(m)*
It solvente *(m)*
Pt dissolvente *(m)*

solvent *n* En
De Lösungsmittel *(n)*
Es disolvente *(m)*
Fr solvant *(m)*
It solvente *(m)*
Pt dissolvente *(m)*

solvente *(m) n* It
De Lösungsmittel *(n)*
En solvent
Es disolvente *(m)*
Fr solvant *(m)*
Pt dissolvente *(m)*

sondaggio a rotazione *(m)* It
De Drehbohren *(n)*
En rotary drilling
Es perforación rotativa *(f)*
Fr forage rotary *(m)*
Pt perfuração rotativa *(f)*

Sonne *(f) n* De
En sun
Es sol *(m)*
Fr soleil *(m)*
It sole *(m)*
Pt sol *(m)*

Sonnenbestrahlung *(f) n* De
En insolation
Es insolación *(f)*
Fr insolation *(f)*
It insolazione *(f)*
Pt insolação *(f)*

Sonnenenergie *(f) n* De
En solar energy
Es energía solar *(f)*
Fr énergie solaire *(f)*
It energia solare *(f)*
Pt energia solar *(f)*

Sonnenfluß *(m) n* De
En solar flux
Es flujo solar *(m)*
Fr flux solaire *(m)*
It flusso solare *(m)*
Pt fluxo solar *(m)*

Sonnenlicht-Umlenkspiegel *(m) n* De
En passive sunlight reflector
Es reflector solar pasivo *(m)*
Fr réflecteur solaire passif *(m)*
It riflettore passivo della luce del sole *(m)*
Pt reflector de raios solares passivo *(m)*

Sonnenofen *(m) n* De
En solar furnace
Es horno solar *(m)*
Fr four solaire *(m)*
It forno solare *(m)*
Pt forno solar *(m)*

Sonnenstrahlung *(f) n* De
En solar radiation
Es radiación solar *(f)*
Fr rayonnement solaire *(m)*
It radiazione solare *(f)*
Pt radiação solar *(f)*

Sonnenwende *(f) n* De
En solstice
Es solsticio *(m)*
Fr solstice *(m)*
It solstizio *(m)*
Pt solstício *(m)*

Sonnenwind *(m) n* De
En solar wind
Es viento solar *(m)*
Fr vent solaire *(m)*
It vento solare *(m)*
Pt vento solar *(m)*

soot *n* En
De Ruß *(m)*
Es hollín *(m)*
Fr suie *(f)*
It nerofumo *(m)*
Pt fulígem *(f)*

soplete para plasma *(m)* Es
De Plasmabrenner *(m)*
En plasma torch
Fr chalumeau à plasma *(m)*
It torcia a plasma *(f)*
Pt maçarico de plasma *(m)*

soppressione delle radiazioni *(f)* It
De Strahlungsunterdrückung *(f)*
En radiation suppression
Es supresión de la radiación *(f)*
Fr suppression des radiations *(f)*
Pt supressão de radiações *(f)*

sorgente *(f) n* It
De Quelle *(f)*
En source
Es fuente *(f)*
Fr source *(f)*
Pt fonte *(f)*

sorgenti calde *(f pl)* It
De heiße Quellen *(pl)*
En hot springs *(pl)*
Es termas *(f pl)*
Fr sources chaudes *(f pl)*
Pt fontes de água quente *(f pl)*

sostanza volatile *(f)* It
De flüchtiger Bestandteil *(m)*
En volatile matter
Es materia volátil *(f)*
Fr matière volatile *(f)*
Pt matéria volátil *(f)*

sostituto di gas naturale *(m)* It
De Erdgasersatz *(m)*
En substitute natural gas
Es gas natural de sustitución *(m)*
Fr gaz naturel de remplacement *(m)*
Pt gás natural de substituição *(m)*

sottomarino nucleare *(m)* It
De Kennkraft-U-Boot *(n)*
En nuclear submarine
Es submarino nuclear *(m)*
Fr sous-marin nucléaire *(m)*
Pt submarino nuclear *(m)*

sottostazione *(f) n* It
De Umspannwerk *(n)*
En substation
Es subestación *(f)*
Fr sous-station *(f)*
Pt subestação *(f)*

soufre *(m) n* Fr
Am sulfur
De Schwefel *(m)*
En sulphur
Es azufre *(m)*
It zolfo *(m)*
Pt enxofre *(m)*

soupape *(f) n* Fr
De Ventil *(n)*
En valve
Es válvula *(f)*
It valvola *(f)*
Pt válvula *(f)*

soupape de sûreté *(f)* Fr
De Sicherheitsventil *(n)*
En safety valve
Es válvula de seguridad *(f)*
It valvola di sicurezza *(f)*
Pt válvula de segurança *(f)*

source *n* En; Fr *(f)*
De Quelle *(f)*
Es fuente *(f)*
It sorgente *(f)*
Pt fonte *(f)*

source de neutrons *(f)* Fr
De Neutronenquelle *(f)*
En neutron source
Es fuente de neutrones *(f)*
It fonte di neutroni *(f)*
Pt fonte de neutrões *(f)*

source rock En
De Lagerstättengestein *(n)*
Es roca madre *(f)*
Fr roche mère *(f)*
It roccia madre *(f)*
Pt rocha mãe *(f)*

sources chaudes *(f pl)* Fr
De heiße Quellen *(pl)*
En hot springs *(pl)*
Es termas *(f pl)*
It sorgenti calde *(f pl)*
Pt fontes de água quente *(f pl)*

sous-marin nucléaire *(m)* Fr
De Kennkraft-U-Boot *(n)*
En nuclear submarine
Es submarino nuclear *(m)*
It sottomarino nucleare *(m)*
Pt submarino nuclear *(m)*

sous-station *(f) n* Fr
De Umspannwerk *(n)*
En substation
Es subestación *(f)*
It sottostazione *(f)*
Pt subestação *(f)*

sovraccarico *(m) n* It
De Überlastung *(f)*
En overload
Es sobrecarga *(f)*
Fr sucharge *(f)*
Pt sobrecarga *(f)*

sovratensione *(f) n* It
De Überspannung *(f)*
En overvoltage
Es sobretensión *(f)*
Fr surtension *(f)*
Pt sobrevoltagem *(f)*

space heating En
De Raumheizung *(f)*
Es calefacción de espacios *(f)*
Fr chauffage de chambres *(m)*
It riscaldamento locale *(m)*
Pt aquecimento espacial *(m)*

spaltbar *adj* De
En fissionable
Es fisionable
Fr fissionable
It fissionabile
Pt desintegrável

Spaltprodukt *(n) n* De
En fission product
Es producto de fisión *(m)*
Fr produit de la fission *(m)*
It prodotto di fissione *(m)*
Pt produto de desintegração *(m)*

Spaltstoffgitter *(n) n* De
En lattice
Es celosía *(f)*
Fr treillis *(m)*
It traliccio *(m)*
Pt treliça *(f)*

Spaltung *(f) n* De
En fission
Es fisión *(f)*
Fr fission *(f)*
It fissione *(f)*
Pt desintegração *(f)*

Spaltzone (Kernkraftreaktor) *(f) n* De
En core (nuclear reactor)
Es núcleo (reactor nuclear) *(m)*
Fr coeur (réacteur nucléaire) *(m)*
It cuore (reattore nucléaire) *(m)*
Pt núcleo (reactor nuclear) *(m)*

Spannung *(f) n* De
En voltage
Es tensión *(f)*
Fr tension *(f)*
It tensione *(f)*
Pt voltagem *(f)*

Spannungsregler *(m) n* De
En voltage regulator
Es regulador de tensión *(m)*
Fr régulateur de tension *(m)*
It regolatore di tensione *(m)*
Pt regulador de tensão *(m)*

Spannungsteiler *(m) n* De
En potential divider; voltage divider
Es divisor de potencial; divisor de tensión *(m)*

Fr diviseur de tension (m)
It partitore di tensione (m)
Pt divisor de potencial; divisor de tensão (m)

Spannungsunterschied (m) n De
En potential difference
Es diferencia de potencial (f)
Fr différence de potentiel (f)
It caduta di tensione (f)
Pt diferença de potencial (f)

Spannungswandler (m) n De
En potential converter
Es convertidor de potencial (m)
Fr convertisseur de tension (m)
It convertitore di tensione (m)
Pt conversor de potencial (m)

spargimento di carico (m) It
De Lastabgabe (f)
En load-shedding
Es restricción de la carga (f)
Fr délestage (m)
Pt restrição de carga (f)

spargimento di carico a sottofrequenza (m) It
De Unterfrequenz-Lastabschaltung (f)
En underfrequency load-shedding
Es restricción de la carga a baja frecuencia (f)
Fr délestage à sous-fréquence (m)
Pt restrição de carga a sobfrequência (f)

spark ignition En
De Funkenzündung (f)
Es encendido por chispa (m)
Fr allumage par étincelle (m)
It accensione a scintilla (f)
Pt ignição por faísca (f)

specific heat capacity En
De spezifische Wärmekapazität (f)
Es capacidad de calor específico (f)
Fr capacité calorifique spécifique (f)
It capacità termica specifica (f)
Pt capacidade de calor específico (f)

specific impulse En
De spezifischer Impuls (m)
Es impulso específico (m)
Fr impulsion spécifique (f)
It impulso specifico (m)
Pt impulso específico (m)

specific latent heat En
De spezifische gebundene Wärme (f)
Es calor latente específico (m)
Fr chaleur latente spécifique (f)
It calore latente specifico (m)
Pt calor latente específico (m)

spectre (m) n Fr
De Spektrum (n)
En spectrum
Es espectro (m)
It spettro (m)
Pt espectro (m)

spectre d'émission (m) Fr
De Emissionsspektrum (n)
En emission spectrum
Es espectro de emisión (m)
It spettro delle emissioni (m)
Pt espectro de emissão (m)

spectre électromagnétique (m) Fr
De elektromagnetisches Spektrum (n)
En electromagnetic spectrum
Es espectro electromagnético (m)
It spettro elettromagnetico (m)
Pt espectro electromagnético (m)

spectre visible (m) Fr
De sichtbares Spektrum (n)
En visible spectrum
Es espectro visible (m)
It spettro visibile (m)
Pt espectro visível (m)

spectrometer n En
De Spektrometer (n)
Es espectrómetro (m)
Fr spectromètre (m)
It spettrometro (m)
Pt espectrómetro (m)

spectromètre (m) n Fr
De Spektrometer (n)
En spectrometer
Es espectrómetro (m)
It spettrometro (m)
Pt espectrómetro (m)

spectroscope n En; Fr (m)
De Spektroskop (m)
Es espectroscopio (m)
It spettroscopio (m)
Pt espectroscópio (m)

spectroscopie (f) n Fr
De Spektroskopie (f)
En spectroscopy
Es espectroscopia (f)
It spettroscopia (f)
Pt espectroscopia (f)

spectroscopie de masse (f) Fr
De Massenspektro-skopie (f)
En mass spectroscopy
Es espectroscopia de masa (f)
It spettroscopia di massa (f)
Pt espectroscopia de massa (f)

spectroscopy n En
De Spektroskopie (f)
Es espectroscopia (f)
Fr spectroscopie (f)
It spettroscopia (f)
Pt espectroscopia (f)

spectrum n En
De Spektrum (n)
Es espectro (m)
Fr spectre (m)
It spettro (m)
Pt espectro (m)

speed n En
De Geschwindigkeit (f)
Es velocidad (f)
Fr vitesse (f)
It velocità (f)
Pt velocidade (f)

spegnimento immediato (m) It
De Abschaltung (f)
En scram (nuclear reactor)
Es parada de emergencia (f)
Fr arrêt d'urgence (m)
Pt paragem de emergência (f)

Speicherheizung (f) n De
En storage heater
Es calentador para almacenamiento térmico (m)
Fr radiateur à accumulation (m)
It riscaldatore a conservazione (m)
Pt aquecedor de armazenagem (m)

Speicherwasser-heizung (f) n De
En storage water-heater
Es calentador de agua de acumulación (m)
Fr chauffe-eau à accumulation (m)
It riscaldatore di acqua a conservazione (m)
Pt aquecedor a água de armazenagem (m)

Speisepumpe (f) n De
En feed pump
Es bomba de alimentación (f)
Fr pompe d'alimentation (f)

It pompa di alimentazione *(f)*
Pt bomba de alimentação *(f)*

Speisewasser *(n) n* De
En feedwater
Es agua de alimentación *(f)*
Fr eau d'alimentation *(f)*
It acqua di alimentazione *(f)*
Pt água de alimentação *(f)*

Spektrometer *(n) n* De
En spectrometer
Es espectrómetro *(m)*
Fr spectromètre *(m)*
It spettrometro *(m)*
Pt espectrómetro *(m)*

Spektroskop *(m) n* De
En spectroscope
Es espectroscopio *(m)*
Fr spectroscope *(m)*
It spettroscopio *(m)*
Pt espectroscópio *(m)*

Spektroskopie *(f) n* De
En spectroscopy
Es espectroscopia *(f)*
Fr spectroscopie *(f)*
It spettroscopia *(f)*
Pt espectroscopia *(f)*

Spektrum *(n) n* De
En spectrum
Es espectro *(m)*
Fr spectre *(m)*
It spettro *(m)*
Pt espectro *(m)*

spent-air recovery En
De Rückgewinnung verbrauchter Luft *(f)*
Es recuperación del aire gastado *(f)*
Fr récupération de l'air dépensé *(f)*
It recupero dell'aria consumata *(m)*
Pt recuperação de ar gasto *(f)*

spettro *(m) n* It
De Spektrum *(n)*
En spectrum
Es espectro *(m)*
Fr spectre *(m)*
Pt espectro *(m)*

spettro delle emissioni *(m)* It
De Emissionsspektrum *(n)*
En emission spectrum
Es espectro de emisión *(m)*
Fr spectre d'émission *(m)*
Pt espectro de emissão *(m)*

spettro elettromagnetico *(m)* It
De elektromagnetisches Spektrum *(n)*
En electromagnetic spectrum
Es espectro electromagnético *(m)*
Fr spectre électromagnétique *(m)*
Pt espectro electromagnético *(m)*

spettrometro *(m) n* It
De Spektrometer *(n)*
En spectrometer
Es espectrómetro *(m)*
Fr spectromètre *(m)*
Pt espectrómetro *(m)*

spettroscopia *(f) n* It
De Spektroskopie *(f)*
En spectroscopy
Es espectroscopia *(f)*
Fr spectroscopie *(f)*
Pt espectroscopia *(f)*

spettroscopia di massa *(f)* It
De Massenspektro-skopie *(f)*
En mass spectroscopy
Es espectroscopia de masa *(f)*
Fr spectroscopie de masse *(f)*
Pt espectroscopia de massa *(f)*

spettroscopio *(m) n* It
De Spektroskop *(m)*
En spectroscope
Es espectroscopio *(m)*
Fr spectroscope *(m)*
Pt espectroscópio *(m)*

spettro visibile *(m)* It
De sichtbares Spektrum *(n)*
En visible spectrum
Es espectro visible *(m)*
Fr spectre visible *(m)*
Pt espectro visível *(m)*

spezifische gebundene Wärme *(f)* De
En specific latent heat
Es calor latente específico *(m)*
Fr chaleur latente spécifique *(f)*
It calore latente specifico *(m)*
Pt calor latente específico *(m)*

spezifischer Impuls *(m)* De
En specific impulse
Es impulso específico *(m)*
Fr impulsion spécifique *(f)*
It impulso specifico *(m)*
Pt impulso específico *(m)*

spezifische Wärmekapazität *(f)* De
En specific heat capacity
Es capacidad de calor específico *(f)*
Fr capacité calorifique spécifique *(f)*
It capacità termica specifica *(f)*
Pt capacidade de calor específico *(f)*

Spiralröhren-Wärmeaus-tauscher *(m) n* De
En spiral-tube heat exchanger
Es intercambiador de calor de tubos en espiral *(m)*
Fr échangeur de chaleur à serpentin *(m)*
It scambiatore di calore con tubo a spirale *(m)*
Pt termo-permutador de tubo em espiral *(m)*

spiral-tube heat exchanger En
De Spiralröhren-Wärmeaustauscher *(m)*
Es intercambiador de calor de tubos en espiral *(m)*
Fr échangeur de chaleur à serpentin *(m)*
It scambiatore di calore con tubo a spirale *(m)*
Pt termo-permutador de tubo em espiral *(m)*

spirito rettificato *(m)* It
De rektifizierter Alkohol *(m)*
En rectified spirit
Es alcohol rectificado *(m)*
Fr alcool rectifié *(m)*
Pt álcool rectificado *(m)*

Spitzenbedarf *(m) n* De
En peak demand
Es demanda punta *(f)*
Fr consommation de pointe *(f)*
It domanda di punta *(f)*
Pt procura de ponta *(f)*

Spitzenbelastung *(f) n* De
En peak load
Es carga de punta *(f)*
Fr charge de pointe *(f)*
It carico di punta *(m)*
Pt carga de ponta *(f)*

Spitzenkraft *(f) n* De
En peaking power
Es potencia máxima *(f)*
Fr puissance de crête *(f)*
It potenza massima *(f)*
Pt potência de ponta *(f)*

spoil (mining) *n* En
De Abraum *(m)*
Es estériles *(m pl)*
Fr déblai *(m)*
It materiale di sterro *(m)*
Pt entulho *(m)*

spontaneous ignition En
De spontane Zündung *(f)*
Es encendido espontáneo *(m)*
Fr allumage spontané *(m)*

It autoaccensione *(f)*
Pt ignição espontânea *(f)*

spontane Zündung *(f)* De
En spontaneous ignition
Es encendido espontáneo *(m)*
Fr allumage spontané *(m)*
It autoaccensione *(f)*
Pt ignição espontânea *(f)*

spot market En
De Kassamarkt *(m)*
Es mercado de disponibles *(m)*
Fr marché disponible *(m)*
It mercato del disponibile *(m)*
Pt mercado de ocasião *(m)*

Sprengdrähte *(pl) n* De
En exploding wires *(pl)*
Es hilos explosivos *(m pl)*
Fr fils explosifs *(m pl)*
It fili esplosivi *(m pl)*
Pt fios de explosão *(m pl)*

Sprengstoffe *(pl) n* De
En explosives *(pl)*
Es explosivos *(m pl)*
Fr explosifs *(m pl)*
It esplosivi *(m pl)*
Pt explosivos *(m pl)*

spürbare Wärme *(f) n* De
En sensible heat
Es calor sensible *(m)*
Fr chaleur sensible *(f)*
It calore sensibile *(m)*
Pt calor sensível *(m)*

spurgo *(m) n* It
De Läuterung *(f)*
En purging
Es purga *(f)*
Fr purge *(f)*
Pt purga *(f)*

Stadtgas *(n) n* De
En town gas
Es gas de ciudad *(m)*
Fr gaz de ville *(m)*
It gas per uso domestico *(m)*
Pt gás de cidade *(m)*

Stadtsonnenfluß *(m) n* De
En urban solar flux
Es flujo solar urbano *(m)*
Fr flux solaire urbain *(m)*
It flusso solare urbano *(m)*
Pt fluxo solar urbano *(m)*

stagnation temperature En
De Starrtemperatur *(f)*
Es temperatura de remanso *(f)*
Fr température de stagnation *(f)*
It temperatura di ristagno *(f)*
Pt temperatura de estagnação *(f)*

standby *adj* En
De Reserve
Es de reserva
Fr en attente
It emergenza
Pt de reserva

Stangenmaterial *(n) n* De
En feedstock
Es material de carga *(m)*
Fr stock d'alimentation *(m)*
It materiale di alimentazione *(m)*
Pt stock de abastecimento *(m)*

Starrtemperatur *(f) n* De
En stagnation temperature
Es temperatura de remanso *(f)*
Fr température de stagnation *(f)*
It temperatura di ristagno *(f)*
Pt temperatura de estagnação *(f)*

Starter *(m) n* De
En starter motor
Es motor de arranque *(m)*
Fr démarreur *(m)*
It avviatore *(m)*
Pt motor de arranque *(m)*

starter motor En
De Starter *(m)*
Es motor de arranque *(m)*
Fr démarreur *(m)*
It avviatore *(m)*
Pt motor de arranque *(m)*

start-up *n* En
De Einschalten *(n)*
Es arranque *(m)*
Fr démarrage *(m)*
It avviamento *(m)*
Pt arranque *(m)*

statistica *(f) n* It
De Statistik *(f)*
En statistics
Es estadística *(f)*
Fr statistiques *(f)*
Pt estatística *(f)*

statistics *n* En
De Statistik *(f)*
Es estadística *(f)*
Fr statistiques *(f)*
It statistica *(f)*
Pt estatística *(f)*

Statistik *(f) n* De
En statistics
Es estadística *(f)*
Fr statistiques *(f)*
It statistica *(f)*
Pt estatística *(f)*

statistiques *(f) n* Fr
De Statistik *(f)*
En statistics
Es estadística *(f)*
It statistica *(f)*
Pt estatística *(f)*

Staub *(m) n* De
En dust
Es polvo *(m)*
Fr poussière *(f)*
It polvere *(f)*
Pt pó *(m)*

Staublunge *(f) n* De
En pneumoconiosis
Es neumoconiosis *(f)*
Fr pneumoconiose *(f)*
It pneumoconiosi *(f)*
Pt pneumoconiose *(f)*

steam *n* En
De Dampf *(m)*
Es vapor *(m)*
Fr vapeur *(f)*
It vapore *(m)*
Pt vapor *(m)*

steam accumulator En
De Dampfspeicher *(m)*
Es acumulador de vapor *(m)*
Fr accumulateur de vapeur *(m)*
It accumulatore di vapore *(m)*
Pt acumulador de vapor *(m)*

steam cycle En
De Dampfzyklus *(m)*
Es ciclo de vapor *(m)*
Fr cycle de la vapeur *(m)*
It ciclo di vapore *(m)*
Pt ciclo de vapor *(m)*

steam engine En
De Dampfmaschine *(f)*
Es motor de vapor *(m)*
Fr machine à vapeur *(f)*
It macchina a vapore *(f)*
Pt motor a vapor *(m)*

steam quality En
De Dampfqualität *(f)*
Es calidad del vapor *(f)*
Fr qualité de vapeur *(f)*
It qualità del vapore *(f)*
Pt qualidade do vapor *(f)*

steam table En
De Dampftafel *(f)*
Es tabla de vapor *(f)*
Fr table à vapeur *(f)*
It tavola di vapore *(f)*
Pt tabela de vapor *(m)*

steam turbine En
De Dampfturbine *(f)*
Es turbina de vapor *(f)*
Fr turbine à vapeur *(f)*
It turbina a vapore *(f)*
Pt turbina a vapor *(f)*

stechiometrico *adj* It
De stöchiometrisch
En stoichiometric
Es estoquiometrico
Fr stoechiométrique
Pt estequiométrico

Steinkohle *(f) n* De
En hard coal
Es carbón duro *(m)*
Fr charbon dure *(m)*
It carbone duro *(m)*
Pt carvão duro *(m)*

Steinkohlenasche *(f) n* De
En coal ash
Es ceniza de carbón *(f)*
Fr cendre de charbon *(f)*
It cenere di carbone *(f)*
Pt cinza de carvão *(f)*

Steinkohlengas *(n) n* De
En coal gas
Es gas de carbón *(m)*
Fr gaz de houille *(m)*
It gas illuminante *(m)*
Pt gás de carvão *(m)*

Steinkohlenteer *(m) n* De
En coal tar
Es alquitrán de hulla *(m)*
Fr goudron de houille *(m)*
It catrame di carbon fossile *(m)*
Pt alcatrão de hulha *(m)*

Steinkohlenteer-Treibstoffe *(pl)* De
En coal-tar fuels *(pl)*
Es combustibles de carbón-alquitrán *(m pl)*
Fr combustibles de gondron de houille *(m pl)*
It combustibili a base di catrame *(m pl)*
Pt combustíveis de alcatrão de hulha *(m pl)*

Steinwollefaser-isolierung *(f)* De
Am mineral-wool fiber insulation
En mineral-wool fibre insulation
Es aislamiento con fibra de lana mineral *(m)*
Fr isolement par fibre de laine minérale *(m)*
It isolamento con fibra di lana minerale *(m)*
Pt isolamento com fibras de lã mineral *(m)*

Steuersystem *(n) n* De
En control system
Es sistema de control *(m)*
Fr système de commande *(m)*
It sistema di controllo *(m)*
Pt sistema de controle *(m)*

Stickstoff *(m) n* De
En nitrogen
Es nitrógeno *(m)*
Fr azote *(m)*
It azoto *(m)*
Pt azoto *(m)*

Stickstoffoxide *(pl) n* De
En nitrogen oxides *(pl)*
Es óxidos de nitrógeno *(m pl)*
Fr oxydes d'azote *(m pl)*
It ossidi d'azoto *(m pl)*
Pt óxidos de azoto *(m pl)*

Stirling cycle En
De Stirling-Zyklus *(m)*
Es ciclo de Stirling *(m)*
Fr cycle de Stirling *(m)*
It ciclo di Stirling *(m)*
Pt ciclo de Stirling *(m)*

Stirling-Zyklus *(m) n* De
En Stirling cycle
Es ciclo de Stirling *(m)*
Fr cycle de Stirling *(m)*
It ciclo di Stirling *(m)*
Pt ciclo de Stirling *(m)*

stöchiometrisch *adj* De
En stoichiometric
Es estoquiometrico
Fr stoechiométrique
It stechiometrico
Pt estequiométrico

stockage de l'air comprimé *(m)* Fr
De Druckluftspeicherung *(f)*
En compressed-air storage
Es almacenamiento de aire comprimido *(m)*
It immagazzinaggio di aria compressa *(m)*
Pt armazenagem de ar comprimido *(f)*

stock d'alimentation *(m)* Fr
De Stangenmaterial *(n)*
En feedstock
Es material de carga *(m)*
It materiale di alimentazione *(m)*
Pt stock de abastecimento *(m)*

stock d'alimentation chimique *(m)* Fr
De chemisches Stangenmaterial *(n)*
En chemical feedstock
Es material químico de carga *(m)*
It materiale di alimentazione chimico *(m)*
Pt stock de abastecimento químico *(m)*

stock de abastecimento *(m)* Pt
De Stangenmaterial *(n)*
En feedstock
Es material de carga *(m)*
Fr stock d'alimentation *(m)*
It materiale di alimentazione *(m)*

stock de abastecimento químico *(m)* Pt
De chemisches Stangenmaterial *(n)*
En chemical feedstock
Es material químico de carga *(m)*
Fr stock d'alimentation chimique *(m)*
It materiale di alimentazione chimico *(m)*

stock de réserves *(m)* Fr
De Vorrat *(m)*
En stockpile
Es existencia para emergencia *(f)*
It riserva *(f)*
Pt pilha de armazenagem *(f)*

stockpile *n* En
De Vorrat *(m)*
Es existencia para emergencia *(f)*
Fr stock de réserves *(m)*
It riserva *(f)*
Pt pilha de armazenagem *(f)*

stoechiométrique *adj* Fr
De stöchiometrisch
En stoichiometric
Es estoquiometrico
It stechiometrico
Pt estequiométrico

Stoff- und Energiebilanz *(f)* De
En material and energy balance
Es equilibrio materia-energía *(m)*
Fr balance de matériel et énergie *(f)*
It bilancio materiale-energia *(m)*
Pt equilíbrio de material e energia *(m)*

stoichiometric *adj* En
De stöchiometrisch
Es estoquiometrico
Fr stoechiométrique
It stechiometrico
Pt estequiométrico

storage heater En
De Speicherheizung *(f)*
Es calentador para almacenamiento térmico *(m)*
Fr radiateur à accumulation *(m)*
It riscaldatore a conservazione *(m)*
Pt aquecedor de armazenagem *(m)*

storage water-heater En
De Speicherwasser-heizung *(f)*
Es calentador de agua de acumulación *(m)*
Fr chauffe-eau à accumulation *(m)*
It riscaldatore di acqua a conservazione *(m)*
Pt aquecedor a água de armazenagem *(m)*

Strahlantrieb *(m) n* De
En jet propulsion
Es propulsión a chorro *(f)*
Fr propulsion par réaction *(f)*

It propulsione a getto *(m)*
Pt propulsão a jacto *(f)*

Strahlung *(f) n* De
En radiation
Es radiación *(f)*
Fr radiation *(f)*
It radiazione *(f)*
Pt radiação *(f)*

Strahlungschemie *(f) n* De
En radiation chemistry
Es química de la radiación *(f)*
Fr radiochimie *(f)*
It chimica delle radiazioni *(f)*
Pt química das radiações *(f)*

Strahlungsgefahr *(f) n* De
En radiation hazard
Es riesgo de radiación *(m)*
Fr danger de radiation *(m)*
It pericolo di radiazioni *(m)*
Pt perigo de radiação *(m)*

Strahlungsheizkessel *(m) n* De
En radiant boiler
Es caldera radiante *(f)*
Fr chaudière radiante *(f)*
It caldaia radiante *(f)*
Pt convector de radiação *(m)*

Strahlungsheizofen *(m) n* De
En radiant-heating furnace
Es horno de calentamiento por radiación *(m)*
Fr four à chauffage radiant *(m)*
It forno a riscaldamento radiante *(m)*
Pt forno de aquecimento por radiação *(m)*

Strahlungskonvektor *(m) n* De
En radiant convector
Es convector radiante *(m)*
Fr convecteur radiant *(m)*
It convettore radiante *(m)*
Pt convector de radiação *(m)*

Strahlungspyrometer *(n) n* De
En optical pyrometer
Es pirómetro óptico *(m)*
Fr pyromètre optique *(m)*
It pirometro ottico *(m)*
Pt pirómetro óptico *(m)*

Strahlungsröhrenofen *(m) n* De
En radiant-tube furnace
Es horno de tubo radiante *(m)*
Fr four à tube radiant *(m)*
It forno a rubo radiante *(m)*
Pt forno de tubo de radiação *(m)*

Strahlungsunterdrückung *(f) n* De
En radiation suppression
Es supresión de la radiación *(f)*
Fr suppression des radiations *(f)*
It soppressione delle radiazioni *(f)*
Pt supressão de radiações *(f)*

Straßenfahrzeug mit Dieselmotor *(n)* De
En diesel engine road vehicle (DERV)
Es vehículo para carretera con motor diesel *(m)*
Fr véhicule diesel routier *(m)*
It automezzo con motore a nafta *(m)*
Pt veículo rodoviário de motor diesel *(m)*

strategic oil reserves *(pl)* En
De strategische Ölreserven *(pl)*
Es reservas petrolíferas estratégicas *(f pl)*
Fr réserves de pétrole stratégiques *(f pl)*
It riserve strategiche di petrolio *(f pl)*
Pt reservas de petróleo estratégicas *(f pl)*

strategische Ölreserven *(pl)* De
En strategic oil reserves *(pl)*
Es reservas petrolíferas estratégicas *(f pl)*
Fr réserves de pétrole stratégiques *(f pl)*
It riserve strategiche di petrolio *(f pl)*
Pt reservas de petróleo estratégicas *(f pl)*

strato impermeabile di copertura *(m)* It
De Deckgestein *(n)*
En cap rock
Es estrato impermeable de cobertura *(m)*
Fr roche couverture *(f)*
Pt rocha encaixante *(f)*

Streustrahlung *(f) n* De
En diffuse radiation
Es radiación difusa *(f)*
Fr rayonnement diffus *(m)*
It radiazione diffusa *(f)*
Pt radiação difusa *(f)*

Streuung *(f) n* De
En diffusion
Es difusión *(f)*
Fr diffusion *(f)*
It diffusione *(f)*
Pt difusão *(f)*

Streuung *(f) n* De
En scattering
Es dispersión *(f)*
Fr dispersion *(f)*
It dispersione *(f)*
Pt espalhamento *(m)*

Strom *(m) n* De
En current
Es corriente *(f)*
Fr courant *(m)*
It corrente *(f)*
Pt corrente *(f)*

Strom aus der Schwachlastzeit *(m)* De
En off-peak electricity
Es electricidad fuera de horas punta *(f)*
Fr électricité hors-pointe *(f)*
It elettricità fuori di ore di punta *(f)*
Pt electricidade fora de ponta *(f)*

Stromerzeugung *(f) n* De
En electricity generation
Es generación de electricidad *(f)*
Fr production d'électricité *(f)*
It generazione elettrica *(f)*
Pt geração de electricidade *(f)*

Stromerzeugung pro Kopf *(f)* De
En per capita energy production
Es producción de energía per capita *(f)*
Fr production d'énergie par personne *(f)*
It produzione di energia pro capite *(f)*
Pt produção de energia per capita *(f)*

Stromschiene *(f) n* De
En third rail
Es riel conductor *(m)*
Fr rail de contact *(m)*
It terza rotaia *(f)*
Pt terceiro carril *(m)*

Stromverbrauch pro Kopf *(m)* De
En per capita energy consumption
Es consumo de energía per capita *(m)*
Fr consommation d'énergie par personne *(f)*
It consumo di energia pro capite *(m)*
Pt consumo de energia per capita *(m)*

structura a nido d'ape *(f)* It
De Wabenstruktur *(f)*
En honeycomb structure
Es estructura alveolar *(f)*

Fr structure en nids d'abeilles *(f)*
Pt estrutura alveolar *(f)*

structure en nids d'abeilles *(f)* Fr
De Wabenstruktur *(f)*
En honeycomb structure
Es estructura alveolar *(f)*
It structura a nido d'ape *(f)*
Pt estrutura alveolar *(f)*

structures de forage *(f pl)* Fr
De Bohrkonstruktionen *(pl)*
En drilling structures *(pl)*
Es estructuras de perforación *(f pl)*
It strutture di trivellazione *(f pl)*
Pt estructuras de perfuração *(f pl)*

strutture di trivellazione *(f pl)* It
De Bohrkonstruktionen *(pl)*
En drilling structures *(pl)*
Es estructuras de perforación *(f pl)*
Fr structures de forage *(f pl)*
Pt estructuras de perfuração *(f pl)*

Sturzfestigkeits-prüfung *(f)* De
En shatter test
Es prueba de cohesión *(f)*
Fr essai de résistance au choc *(m)*
It prova di frangibilità *(f)*
Pt teste de fragmentação *(m)*

subestação *(f)* *n* Pt
De Umspannwerk *(n)*
En substation
Es subestación *(f)*
Fr sous-station *(f)*
It sottostazione *(f)*

subestación *(f)* *n* Es
De Umspannwerk *(n)*
En substation
Fr sous-station *(f)*
It sottostazione *(f)*
Pt subestação *(f)*

submarino nuclear *(m)* Es, Pt
De Kennkraft-U-Boot *(n)*
En nuclear submarine
Fr sous-marin nucléaire *(m)*
It sottomarino nucleare *(m)*

substation *n* En
De Umspannwerk *(n)*
Es subestación *(f)*
Fr sous-station *(f)*
It sottostazione *(f)*
Pt subestação *(f)*

substitute natural gas En
De Erdgasersatz *(m)*
Es gas natural de sustitución *(m)*
Fr gaz naturel de remplacement *(m)*
It sostituto di gas naturale *(m)*
Pt gás natural de substituição *(m)*

sucharge *(f)* *n* Fr
De Überlastung *(f)*
En overload
Es sobrecarga *(f)*
It sovraccarico *(m)*
Pt sobrecarga *(f)*

suciedad por depósitos marinos *(f)* Es
De Seeverschmutzung *(f)*
En marine fouling
Fr accumulation de dépôts marins *(f)*
It incrostazione marina *(f)*
Pt depósitos marítimos *(m)*

suction pyrometer En
De Saugpyrometer *(n)*
Es pirómetro de succión *(m)*
Fr pyromètre aspirant *(m)*
It pirometro ad aspirazione *(m)*
Pt pirómetro de sucção *(m)*

suie *(f)* *n* Fr
De Ruß *(m)*
En soot
Es hollín *(m)*
It nerofumo *(m)*
Pt fulígem *(f)*

sulfur *n* Am
De Schwefel *(m)*
En sulphur
Es azufre *(m)*
Fr soufre *(m)*
It zolfo *(m)*
Pt enxofre *(m)*

sulphur *n* En
Am sulfur
De Schwefel *(m)*
Es azufre *(m)*
Fr soufre *(m)*
It zolfo *(m)*
Pt enxofre *(m)*

sun *n* En
De Sonne *(f)*
Es sol *(m)*
Fr soleil *(m)*
It sole *(m)*
Pt sol *(m)*

superaquecedor *(m)* *n* Pt
De Überhitzer *(m)*
En superheater
Es supercalentador *(m)*
Fr surchauffeur *(m)*
It surriscaldatore *(m)*

supercalentador *(m)* *n* Es
De Überhitzer *(m)*
En superheater
Fr surchauffeur *(m)*
It surriscaldatore *(m)*
Pt superaquecedor *(m)*

superconducteur *(m)* *n* Fr
De Supraleiter *(m)*
En superconductor
Es superconductor *(m)*
It superconduttore *(m)*
Pt supercondutor *(m)*

superconducting magnet En
De supraleitfähiger Magnet *(m)*
Es imán superconductor *(m)*
Fr aimant superconducteur *(m)*
It magnete superconduttore *(m)*
Pt ímã supercondutor *(m)*

superconductor *n* En; Es *(m)*
De Supraleiter *(m)*
Fr superconducteur *(m)*
It superconduttore *(m)*
Pt supercondutor *(m)*

supercondutor *(m)* *n* Pt
De Supraleiter *(m)*
En superconductor
Es superconductor *(m)*
Fr superconducteur *(m)*
It superconduttore *(m)*

superconduttore *(m)* *n* It
De Supraleiter *(m)*
En superconductor
Es superconductor *(m)*
Fr superconducteur *(m)*
Pt supercondutor *(m)*

supercritical *adj* En
De überkritisch
Es supercrítico
Fr supercritique
It supercritico
Pt supercrítico

supercritico *adj* It
De überkritisch
En supercritical
Es supercrítico
Fr supercritique
Pt supercrítico

supercrítico *adj* Es, Pt
De überkritisch
En supercritical
Fr supercritique
It supercritico

supercritique *adj* Fr
De überkritisch
En supercritical
Es supercrítico
It supercritico
Pt supercrítico

superfluid *adj* En
De supraflüssig
Es superflúido
Fr superfluide
It superfluido
Pt superfluido

superfluide *adj* Fr
De supraflüssig
En superfluid
Es superflúido

It superfluido
Pt superfluido

superfluido *adj* It, Pt
De supraflüssig
En superfluid
Es superflúido
Fr superfluide

superflúido *adj* Es
De supraflüssig
En superfluid
Fr superfluide
It superfluido
Pt superfluido

superheated steam En
De überhitzter Dampf *(m)*
Es vapor supercalentado *(m)*
Fr vapeur surchauffée *(f)*
It vapore surriscaldato *(m)*
Pt vapor superaquecido *(m)*

superheater *n* En
De Überhitzer *(m)*
Es supercalentador *(m)*
Fr surchauffeur *(m)*
It surriscaldatore *(m)*
Pt superaquecedor *(m)*

supernaviocisterna *(m)* *n* Pt
De Supertanker *(m)*
En supertanker
Es superpetrolero *(m)*
Fr pétrolier géant *(m)*
It superpetroliera *(f)*

superpetrolero *(m)* *n* Es
De Supertanker *(m)*
En supertanker
Fr pétrolier géant *(m)*
It superpetroliera *(f)*
Pt supernaviocisterna *(m)*

superpetroliera *(f)* *n* It
De Supertanker *(m)*
En supertanker
Es superpetrolero *(m)*
Fr pétrolier géant *(m)*
Pt supernaviocisterna *(m)*

supertanker *n* En
De Supertanker *(m)*
Es superpetrolero *(m)*
Fr pétrolier géant *(m)*
It superpetroliera *(f)*
Pt supernaviocisterna *(m)*

Supertanker *(m)* *n* De
En supertanker
Es superpetrolero *(m)*
Fr pétrolier géant *(m)*
It superpetroliera *(f)*
Pt supernaviocisterna *(m)*

suppression des radiations *(f)* Fr
De Strahlungsunterdrückung *(f)*
En radiation suppression
Es supresión de la radiación *(f)*
It soppressione delle radiazioni *(f)*
Pt supressão de radiações *(f)*

supraflüssig *adj* De
En superfluid
Es superflúido
Fr superfluide
It superfluido
Pt superfluido

Supraleiter *(m)* *n* De
En superconductor
Es superconductor *(m)*
Fr superconducteur *(m)*
It superconduttore *(m)*
Pt supercondutor *(m)*

supraleitfähiger Magnet *(m)* De
En superconducting magnet
Es imán superconductor *(m)*
Fr aimant superconducteur *(m)*
It magnete superconduttore *(m)*
Pt ímã supercondutor *(m)*

supresión de la radiación *(f)* Es
De Strahlungsunterdrückung *(f)*
En radiation suppression
Fr suppression des radiations *(f)*
It soppressione delle radiazioni *(f)*
Pt supressão de radiações *(f)*

supressão de radiações *(f)* Pt
De Strahlungsunterdrückung *(f)*
En radiation suppression
Es supresión de la radiación *(f)*
Fr suppression des radiations *(f)*
It soppressione delle radiazioni *(f)*

surchauffeur *(m)* *n* Fr
De Überhitzer *(m)*
En superheater
Es supercalentador *(m)*
It surriscaldatore *(m)*
Pt superaquecedor *(m)*

surfactant *n* En
De grenzflächenaktiver Stoff *(m)*
Es surfactante *(m)*
Fr agent tensio-actif *(m)*
It agente tensioattivo *(m)*
Pt agente tenso-activo *(m)*

surfactante *(m)* *n* Es
De grenzflächenaktiver Stoff *(m)*
En surfactant
Fr agent tensio-actif *(m)*
It agente tensioattivo *(m)*
Pt agente tenso-activo *(m)*

surriscaldatore *(m)* *n* It
De Überhitzer *(m)*
En superheater
Es supercalentador *(m)*
Fr surchauffeur *(m)*
Pt superaquecedor *(m)*

surtension *(f)* *n* Fr
De Überspannung *(f)*
En overvoltage
Es sobretensión *(f)*
It sovratensione *(f)*
Pt sobrevoltagem *(f)*

swelling test En
De Quellversuch *(m)*
Es prueba de esponjamiento *(f)*
Fr essai de gonflement *(m)*
It prova di ringonfiamento *(f)*
Pt teste de inchamento *(m)*

synchronisation *(f)* *n* Fr
De Synchronisierung *(f)*
En synchronization
Es sincronización *(f)*
It sincronizzazione *(f)*
Pt sincronização *(f)*

Synchronisierung *(f)* *n* De
En synchronization
Es sincronización *(f)*
Fr synchronisation *(f)*
It sincronizzazione *(f)*
Pt sincronização *(f)*

synchronization *n* En
De Synchronisierung *(f)*
Es sincronización *(f)*
Fr synchronisation *(f)*
It sincronizzazione *(f)*
Pt sincronização *(f)*

Synchronomotor *(m)* *n* De
En synchronous motor
Es motor síncrono *(m)*
Fr moteur synchrone *(m)*
It motore sincrono *(m)*
Pt motor sincronizado *(m)*

synchronous motor En
De Synchronomotor *(m)*
Es motor síncrono *(m)*
Fr moteur synchrone *(m)*
It motore sincrono *(m)*
Pt motor sincronizado *(m)*

synthetic crude oil (SCO) En
De synthetisches Rohöl *(n)*
Es crudo sintético *(m)*
Fr brut de synthèse *(m)*
It petrolio greggio sintetico *(m)*
Pt petróleo crú sintético *(m)*

synthetic natural gas (SNG) En
De synthetisches Erdgas *(n)*
Es gas natural sintético (GNS) *(m)*
Fr gaz naturel de synthèse *(m)*
It gas naturale sintetico *(m)*
Pt gás natural sintético *(m)*

synthetisches Erdgas *(n)* De
En synthetic natural gas (SNG)
Es gas natural sintético (GNS) *(m)*
Fr gaz naturel de synthèse *(m)*
It gas naturale sintetico *(m)*
Pt gás natural sintético *(m)*

synthetisches Rohöl *(n)* De
En synthetic crude oil (SCO)
Es crudo sintético *(m)*
Fr brut de synthèse *(m)*
It petrolio greggio sintetico *(m)*
Pt petróleo crú sintético *(m)*

système de commande *(m)* Fr
De Steuersystem *(n)*
En control system
Es sistema de control *(m)*
It sistema di controllo *(m)*
Pt sistema de controle *(m)*

système d'énergie satellite *(m)* Fr
De Satelliten-Energiesystem *(n)*
En satellite energy system
Es sistema de energía satélite *(m)*
It sistema di energia satellite *(m)*
Pt sistema de energia satélite *(m)*

système de nettoyage *(m)* Fr
De Aufzehrsystem *(n)*
En cleanup system
Es sistema de limpieza *(m)*
It sistema di depurazione *(m)*
Pt sistema de limpeza *(m)*

système de protection *(m)* Fr
De Schutzsystem *(n)*
En protection system
Es sistema de protección *(m)*
It sistema di protezione *(m)*
Pt sistema de protecção *(m)*

T

tabela de vapor *(m)* Pt
De Dampftafel *(f)*
En steam table
Es tabla de vapor *(f)*
Fr table à vapeur *(f)*
It tavola di vapore *(f)*

tabla de vapor *(f)* Es
De Dampftafel *(f)*
En steam table
Fr table à vapeur *(f)*
It tavola di vapore *(f)*
Pt tabela de vapor *(m)*

table à vapeur *(f)* Fr
De Dampftafel *(f)*
En steam table
Es tabla de vapor *(f)*
It tavola di vapore *(f)*
Pt tabela de vapor *(m)*

Tagebau *(m)* *n* De
En open-cast mine
Es mina a cielo abierto *(f)*
Fr mine à ciel ouvert *(f)*
It miniera a cielo aperto *(f)*
Pt mina aberta *(f)*

tägliche Wärmespeicherung *(f)* De
En diurnal heat storage
Es almacenamiento diurno del calor *(m)*
Fr accumulation de chaleur diurne *(f)*
It immaganizzinaggio di calore diurno *(m)*
Pt armazenagem diurna de calor *(f)*

tailings *(pl)* *n* En
De Haldenabfall *(m)*
Es desechos *(m pl)*
Fr produits de queue *(m pl)*
It residui di scarto *(m pl)*
Pt colas *(f pl)*

tailrace *n* En
De Ablauf *(m)*
Es canal de descarga *(m)*
Fr bief *(m)*
It canale di scarico *(m)*
Pt água de descarga *(f)*

tanque refrigerante *(m)* Pt
De Kühlteich *(m)*
En cooling pond (nuclear reactor)
Es piscina de refrigeración *(f)*
Fr piscine de refroidissement *(f)*
It bacino di raffreddamento *(m)*

tarif *(m)* *n* Fr
De Tarif *(m)*
En tariff
Es tarifa *(f)*
It tariffa *(f)*
Pt tarifa *(f)*

Tarif *(m)* *n* De
En tariff
Es tarifa *(f)*
Fr tarif *(m)*
It tariffa *(f)*
Pt tarifa *(f)*

tarifa *(f)* *n* Es, Pt
De Tarif *(m)*
En tariff
Fr tarif *(m)*
It tariffa *(f)*

tariff *n* En
De Tarif *(m)*
Es tarifa *(f)*
Fr tarif *(m)*
It tariffa *(f)*
Pt tarifa *(f)*

tariffa *(f)* *n* It
De Tarif *(m)*
En tariff
Es tarifa *(f)*
Fr tarif *(m)*
Pt tarifa *(f)*

tar sands *(pl)* En
De Ölsande *(pl)*
Es arenas impregnadas de brea *(f pl)*
Fr sables asphaltiques *(m pl)*
It sabbie impregnate di catrame *(f p!)*
Pt areias com alcatrão *(f pl)*

tar yield En
De Teergiebigkeit *(f)*
Es riqueza de alquitrán *(f)*
Fr rendement en goudron *(m)*
It produzione catrame *(f)*
Pt rendimento de alcatrão *(m)*

tasso metabolico basale *(m)* It
De Grundumsatz *(m)*
En basal metabolic rate
Es índice metabólico basal *(m)*
Fr métabolisme basal *(m)*
Pt índice de metabolismo basal *(m)*

Tauchsieder *(m)* *n* De
En immersion heater
Es calentador de inmersión *(m)*
Fr thermoplongeur *(m)*
It riscaldatore ad immersione *(m)*
Pt aquecedor por imersão *(m)*

Taupunkt *(m)* *n* De
En dew point
Es punto de condensación *(m)*

Fr point de rosée *(m)*
It punto di rugiada *(m)*
Pt ponto de orvalho *(m)*

taux de compression *(m)* Fr
De Kompressionsverhältnis *(n)*
En compression ratio
Es relación de compresión *(f)*
It rapporto di compressione *(m)*
Pt taxa de compressão *(f)*

tavola di vapore *(f)* It
De Dampftafel *(f)*
En steam table
Es tabla de vapor *(f)*
Fr table à vapeur *(f)*
Pt tabela de vapor *(m)*

taxa de compressão *(f)* Pt
De Kompressionsverhältnis *(n)*
En compression ratio
Es relación de compresión *(f)*
Fr taux de compression *(m)*
It rapporto di compressione *(m)*

Teergiebigkeit *(f) n* De
En tar yield
Es riqueza de alquitrán *(f)*
Fr rendement en goudron *(m)*
It produzione catrame *(f)*
Pt rendimento de alcatrão *(m)*

Teilkammerkessel *(m) n* De
En sectional boiler
Es caldera compartimentada *(f)*
Fr chaudière sectionnelle *(f)*
It caldaia a sezioni *(f)*
Pt caldeira de secções *(f)*

Teillastbetrieb *(m) n* De
En part-load operation
Es funcionamiento con carga reducida *(m)*
Fr fonctionnement sous charge partielle *(m)*
It funzionamento a carico parziale *(m)*
Pt operação de carga parcial *(f)*

Temperatur *(f) n* De
En temperature
Es temperatura *(f)*
Fr température *(f)*
It temperatura *(f)*
Pt temperatura *(f)*

temperatura *(f) n* Es, It, Pt
De Temperatur *(f)*
En temperature
Fr température *(f)*

temperatura absoluta *(f)* Es, Pt
De absolute Temperatur *(f)*
En absolute temperature
Fr température absolue *(f)*
It temperatura assoluta *(f)*

temperatura assoluta *(f)* It
De absolute Temperatur *(f)*
En absolute temperature
Es temperatura absoluta *(f)*
Fr température absolue *(f)*
Pt temperatura absoluta *(f)*

temperatura de encendido *(f)* Es
De Zündtemperatur *(f)*
En ignition temperature
Fr température d'allumage *(f)*
It temperatura di accensione *(f)*
Pt temperatura de ignição *(f)*

temperatura de estagnação *(f)* Pt
De Starrtemperatur *(f)*
En stagnation temperature
Es temperatura de remanso *(f)*
Fr température de stagnation *(f)*
It temperatura di ristagno *(f)*

temperatura de formación de humo *(f)* Es
De Rauchpunkt *(m)*
En smoke point
Fr point de fumée *(m)*
It punto di fumo *(m)*
Pt ponto de fumo *(m)*

temperatura de ignição *(f)* Pt
De Zündtemperatur *(f)*
En ignition temperature
Es temperatura de encendido *(f)*
Fr température d'allumage *(f)*
It temperatura di accensione *(f)*

temperatura dei neutroni *(f)* It
De Neutronentemperatur *(f)*
En neutron temperature
Es temperatura neutrónica *(f)*
Fr température neutronique *(f)*
Pt temperatura de neutrão *(f)*

temperatura de neutrão *(f)* Pt
De Neutronentemperatur *(f)*
En neutron temperature
Es temperatura neutrónica *(f)*
Fr température neutronique *(f)*
It temperatura dei neutroni *(f)*

temperatura de remanso *(f)* Es
De Starrtemperatur *(f)*
En stagnation temperature
Fr température de stagnation *(f)*
It temperatura di ristagno *(f)*
Pt temperatura de estagnação *(f)*

temperatura di accensione *(f)* It
De Zündtemperatur *(f)*
En ignition temperature
Es temperatura de encendido *(f)*
Fr température d'allumage *(f)*
Pt temperatura de ignição *(f)*

temperatura di ristagno *(f)* It
De Starrtemperatur *(f)*
En stagnation temperature
Es temperatura de remanso *(f)*
Fr température de stagnation *(f)*
Pt temperatura de estagnação *(f)*

temperatura neutrónica *(f)* Es
De Neutronentemperatur *(f)*
En neutron temperature
Fr température neutronique *(f)*
It temperatura dei neutroni *(f)*
Pt temperatura de neutrão *(f)*

temperatura termodinamica *(f)* It
De thermodynamische Temperatur *(f)*
En thermodynamic temperature
Es temperatura termodinámica *(f)*
Fr température thermodynamique *(f)*
Pt temperatura termodinâmica *(f)*

temperatura termodinámica *(f)* Es
De thermodynamische Temperatur *(f)*
En thermodynamic temperature
Fr température thermodynamique *(f)*
It temperatura termodinamica *(f)*
Pt temperatura termodinâmica *(f)*

temperatura termodinâmica *(f)* Pt

De thermodynamische Temperatur *(f)*
En thermodynamic temperature
Es temperatura termodinámica *(f)*
Fr température thermodynamique *(f)*
It temperatura termodinamica *(f)*

temperature *n* En
De Temperatur *(f)*
Es temperatura *(f)*
Fr température *(f)*
It temperatura *(f)*
Pt temperatura *(f)*

température *(f) n* Fr
De Temperatur *(f)*
En temperature
Es temperatura *(f)*
It temperatura *(f)*
Pt temperatura *(f)*

température absolue *(f)* Fr
De absolute Temperatur *(f)*
En absolute temperature
Es temperatura absoluta *(f)*
It temperatura assoluta *(f)*
Pt temperatura absoluta *(f)*

température d'allumage *(f)* Fr
De Zündtemperatur *(f)*
En ignition temperature
Es temperatura de encendido *(f)*
It temperatura di accensione *(f)*
Pt temperatura de ignição *(f)*

température de stagnation *(f)* Fr
De Starrtemperatur *(f)*
En stagnation temperature
Es temperatura de remanso *(f)*
It temperatura di ristagno *(f)*
Pt temperatura de estagnação *(f)*

temperature gradient En
De Temperaturgefälle *(n)*
Es gradiente de temperatura *(m)*
Fr gradient de température *(m)*
It gradiente di temperatura *(m)*
Pt gradiente de temperatura *(m)*

température neutronique *(f)* Fr
De Neutronen-temperatur *(f)*
En neutron temperature
Es temperatura neutrónica *(f)*
It temperatura dei neutroni *(f)*
Pt temperatura de neutrão *(f)*

température thermodynamique *(f)* Fr
De thermodynamische Temperatur *(f)*
En thermodynamic temperature
Es temperatura termodinámica *(f)*
It temperatura termodinamica *(f)*
Pt temperatura termodinâmica *(f)*

Temperaturgefälle *(n) n* De
En temperature gradient
Es gradiente de temperatura *(m)*
Fr gradient de température *(m)*
It gradiente di temperatura *(m)*
Pt gradiente de temperatura *(m)*

temporizador *(m) n* Es
De Zeitschalter *(m)*
En time switch
Fr minuterie *(f)*
It interruttore a tempo *(m)*
Pt comutador cronométrico *(m)*

teneur en carbone *(f)* Fr
De Kohlenstoffgehalt *(m)*
En carbon content
Es contenido de carbono *(m)*
It contenuto di carbonio *(m)*
Pt teor de carbono *(m)*

teneur en plomb *(f)* Fr
De Bleigehalt *(m)*
En lead content
Es contenido de plomo *(m)*
It contenuto di piombo *(m)*
Pt teor de chumbo *(m)*

tension *(f) n* Fr
De Spannung *(f)*
En voltage
Es tensión *(f)*
It tensione *(f)*
Pt voltagem *(f)*

tensión *(f) n* Es
De Spannung *(f)*
En voltage
Fr tension *(f)*
It tensione *(f)*
Pt voltagem *(f)*

tension de vapeur *(f)* Fr
Am vapor pressure
De Dampfdruck *(m)*
En vapour pressure
Es presión del vapor *(f)*
It tensione di vapore *(f)*
Pt pressão de vapor *(f)*

tension de vapeur saturée *(f)* Fr
Am saturated vapor pressure
De gesättigter Dampfdruck *(m)*
En saturated vapour pressure
Es presión de vapor saturado *(f)*
It pressione del vapore saturo *(m)*
Pt pressão de vapor saturado *(f)*

tensione *(f) n* It
De Spannung *(f)*
En voltage
Es tensión *(f)*
Fr tension *(f)*
Pt voltagem *(f)*

tensione di vapore *(f)* It
Am vapor pressure
De Dampfdruck *(m)*
En vapour pressure
Es presión del vapor *(f)*
Fr tension de vapeur *(f)*
Pt pressão de vapor *(f)*

tensione ultraelevata *(f)* It
De Ultrahochspannung *(f)*
En ultrahigh voltage (UHV)
Es tensión ultraelevada *(f)*
Fr ultra-haute tension *(f)*
Pt voltagem ultra-elevada *(f)*

tensión ultraelevada *(f)* Es
De Ultrahochspannung *(f)*
En ultrahigh voltage (UHV)
Fr ultra-haute tension *(f)*
It tensione ultraelevata *(f)*
Pt voltagem ultra-elevada *(f)*

teor de carbono *(m)* Pt
De Kohlenstoffgehalt *(m)*
En carbon content
Es contenido de carbono *(m)*
Fr teneur en carbone *(f)*
It contenuto di carbonio *(m)*

teor de chumbo *(m)* Pt
De Bleigehalt *(m)*
En lead content
Es contenido de plomo *(m)*
Fr teneur en plomb *(f)*
It contenuto di piombo *(m)*

teoria da segurança *(f)* Pt
De Verläslichkeitstheorie *(f)*
En reliability theory
Es teoría de la fiabilidad *(f)*
Fr théorie de fiabilité *(f)*
It teoria dell'affidabilità *(f)*

teoria dei quanti *(f)* It
De Quantentheorie *(f)*
En quantum theory
Es teoría de los cuantos *(f)*
Fr théorie des quanta *(f)*
Pt teoría quântica *(f)*

teoría de la fiabilidad *(f)* Es
De Verläslichkeitstheorie *(f)*
En reliability theory
Fr théorie de fiabilité *(f)*
It teoria dell'affidabilità *(f)*
Pt teoria da segurança *(f)*

teoria dell'affidabilità *(f)* It
De Verläslichkeitstheorie *(f)*
En reliability theory
Es teoría de la fiabilidad *(f)*
Fr théorie de fiabilité *(f)*
Pt teoria da segurança *(f)*

teoría de los cuantos *(f)* Es
De Quantentheorie *(f)*
En quantum theory
Fr théorie des quanta *(f)*
It teoria dei quanti *(f)*
Pt teoría quântica *(f)*

teoría quântica *(f)* Pt
De Quantentheorie *(f)*
En quantum theory
Es teoría de los cuantos *(f)*
Fr théorie des quanta *(f)*
It teoria dei quanti *(f)*

terceiro carril *(m)* Pt
De Stromschiene *(f)*
En third rail
Es riel conductor *(m)*
Fr rail de contact *(m)*
It terza rotaia *(f)*

termalização *(f)* *n* Pt
De Neutronen-abbremsung *(f)*
En thermalization
Es termalización *(f)*
Fr thermalisation *(f)*
It termalizzazione *(f)*

termalización *(f)* *n* Es
De Neutronen-abbremsung *(f)*
En thermalization
Fr thermalisation *(f)*
It termalizzazione *(f)*
Pt termalização *(f)*

termalizzazione *(f)* *n* It
De Neutronen-abbremsung *(f)*
En thermalization
Es termalización *(f)*
Fr thermalisation *(f)*
Pt termalização *(f)*

termas *(f pl)* *n* Es
De heiße Quellen *(pl)*
En hot springs *(pl)*
Fr sources chaudes *(f pl)*
It sorgenti calde *(f pl)*
Pt fontes de água quente *(f pl)*

termico *adj* It
De thermisch
En thermal
Es térmico
Fr thermique
Pt térmico

térmico *adj* Es, Pt
De thermisch
En thermal
Fr thermique
It termico

terminal (electric) *n* En; Es *(m)*
De Klemme (elektrische) *(f)*
Fr borne (électrique) *(f)*
It terminale (elettrico) *(m)*
Pt borne (eléctrico) *(m)*

terminal (transport) *n* En; Es, Fr *(m)*; Pt *(f)*
De Terminal (Transport) *(n)*
It terminale (trasporto) *(m)*

Terminal (Transport) *(n)* *n* De
En terminal (transport)
Es terminal (transporte) *(m)*
Fr terminal (transport) *(m)*
It terminale (trasporto) *(m)*
Pt terminal (transporte) *(f)*

terminale (elettrico) *(m)* *n* It
De Klemme (elektrisch) *(f)*
En terminal (electric)
Es terminal (elétrico) *(m)*
Fr borne (electrique) *(f)*
Pt borne (eléctrico) *(m)*

terminale (trasporto) *(m)* *n* It
De Terminal (Transport) *(n)*
En terminal (transport)
Es terminal (transporte) *(m)*
Fr terminal (transport) *(m)*
Pt terminal (transporte) *(f)*

termistor *(m)* *n* Es, Pt
De Thermistor *(m)*
En thermistor
Fr thermistor *(m)*
It termistore *(m)*

termistore *(m)* *n* It
De Thermistor *(m)*
En thermistor
Es termistor *(m)*
Fr thermistor *(m)*
Pt termistor *(m)*

termo-capacidade molar *(f)* Pt
De molare Wärmekapazität *(f)*
En molar heat capacity
Es capacidad de calor molar *(f)*
Fr capacité de chaleur molaire *(f)*
It capacità termica molare *(f)*

termocoppia *(f)* *n* It
De Thermoelement *(n)*
En thermocouple
Es termopar *(m)*
Fr thermocouple *(m)*
Pt termopar *(m)*

termodinamica *(f)* *n* It
De Thermodynamik *(f)*
En thermodynamics
Es termodinámica *(f)*
Fr thermodynamique *(f)*
Pt termodinâmica *(f)*

termodinámica *(f)* *n* Es
De Thermodynamik *(f)*
En thermodynamics
Fr thermodynamique *(f)*
It termodinamica *(f)*
Pt termodinâmica *(f)*

termodinâmica *(f)* *n* Pt
De Thermodynamik *(f)*
En thermodynamics
Es termodinámica *(f)*
Fr thermodynamique *(f)*
It termodinamica *(f)*

termoelectricidad *(f)* *n* Es
De Thermoelektrizität *(f)*
En thermoelectricity
Fr thermoélectricité *(f)*
It termoelettricità *(f)*
Pt termoelectricidade *(f)*

termoelectricidade *(f)* *n* Pt
De Thermoelektrizität *(f)*
En thermoelectricity
Es termoelectricidad *(f)*
Fr thermoélectricité *(f)*
It termoelettricità *(f)*

termoeléctrico *adj* Es, Pt
De thermoelektrisch
En thermoelectric
Fr thermoélectrique
It termoelettrico

termoelettricità *(f)* *n* It
De Thermoelektrizität *(f)*
En thermoelectricity
Es termoelectricidad *(f)*
Fr thermoélectricité *(f)*
Pt termoelectricidade *(f)*

termoelettrico *adj* It
De thermoelektrisch
En thermoelectric
Es termoeléctrico
Fr thermoélectrique
Pt termoeléctrico

termometro *(m)* *n* It
De Thermometer *(n)*
En thermometer
Es termómetro *(m)*
Fr thermomètre *(m)*
Pt termómetro *(m)*

termómetro *(m) n* Es, Pt
De Thermometer *(n)*
En thermometer
Fr thermomètre *(m)*
It termometro *(m)*

termometro a gas *(m)* It
De Gasthermometer *(n)*
En gas thermometer
Es termómetro de gas *(m)*
Fr thermomètre à gaz *(m)*
Pt termómetro de gás *(m)*

termómetro de gas *(m)* Es
De Gasthermometer *(n)*
En gas thermometer
Fr thermomètre à gaz *(m)*
It termometro a gas *(m)*
Pt termómetro de gás *(m)*

termómetro de gás *(m)* Pt
De Gasthermometer *(n)*
En gas thermometer
Es termómetro de gas *(m)*
Fr thermomètre à gaz *(m)*
It termometro a gas *(m)*

termonuclear *adj* Es, Pt
De thermonuklear
En thermonuclear
Fr thermonucléaire
It termonucleare

termonucleare *adj* It
De thermonuklear
En thermonuclear
Es termonuclear
Fr thermonucléaire
Pt termonuclear

termopar *(m) n* Es, Pt
De Thermoelement *(n)*
En thermocouple
Fr thermocouple *(m)*
It termocoppia *(f)*

termo-permutador *(m) n* Pt
De Wärmeaustauscher *(m)*
En heat exchanger
Es intercambiador de calor *(m)*
Fr échangeur de chaleur *(m)*
It scambiatore di calore *(m)*

termo-permutador de bobina em banho *(m)* Pt
De Wärmeaustauscher bestehend aus Rohrschlangen im Bad *(m)*
En coil-in-bath heat exchanger
Es intercambiador de calor con serpentín en baño *(m)*
Fr échangeur de chaleur à serpentin immergé *(m)*
It scambiatore di calore a serpentino in bagno *(m)*

termo-permutador de camisa e tubo *(m)* Pt
De Röhrenwärmeaustauscher *(m)*
En shell-and-tube heat exchanger
Es intercambiador de calor de coraza y tubos *(m)*
Fr échangeur de chaleur à calandre *(m)*
It scambiatore di calore a cilindro e tubi *(m)*

termo-permutador de placa *(m)* Pt
De Plattenwärmeaustauscher *(m)*
En plate heat-exchanger
Es intercambiador de calor de placas *(m)*
Fr échangeur de chaleur à plaque *(m)*
It scambiatore di calore a piastra *(m)*

termo-permutador de tubo concêntrico *(m)* Pt
De Wärmeaustauscher mit konzentrischen Röhren *(m)*
En concentric-tube heat exchanger
Es intercambiador de calor de tubos concéntricos *(m)*
Fr échangeur de chaleur à tubes concentriques (m)
It scambiatore di calore a tubo concentrico *(m)*

termo-permutador de tubo em espiral *(m)* Pt
De Spiralröhren-Wärmeaustauscher *(m)*
En spiral-tube heat exchanger
Es intercambiador de calor de tubos en espiral *(m)*
Fr échangeur de chaleur à serpentin *(m)*
It scambiatore di calore con tubo a spirale *(m)*

termo-permutador de vidro *(m)* Pt
De Glaswärmeaustauscher *(m)*
En glass heat exchanger
Es intercambiador de calor de vidrio *(m)*
Fr échangeur de chaleur en verre *(m)*
It scambiatore di calore di vetro *(m)*

termopila *(f) n* Es, It
De Thermosäule *(f)*
En thermopile
Fr thermopile *(f)*
Pt pilha térmica *(f)*

termosifón *(m) n* Es
De Thermosiphon *(n)*
En thermosiphon
Fr thermosyphon *(m)*
It termosifone *(m)*
Pt termossifão *(m)*

termosifone *(m) n* It
De Thermosiphon *(n)*
En thermosiphon
Es termosifón *(m)*
Fr thermosyphon *(m)*
Pt termossifão *(m)*

termossifão *(m) n* Pt
De Thermosiphon *(n)*
En thermosyphon
Es termosifón *(m)*
Fr thermosyphon *(m)*
It termosifone *(m)*

termostato *(m) n* Es, It, Pt
De Thermostat *(m)*
En thermostat
Fr thermostat *(m)*

termotransferência de dois fluidos *(f)* Pt
De Wärmeübertragung mit zwei Flüssigkeiten *(f)*
En dual-fluid heat transfer
Es transferencia térmica de dos flúidos *(f)*
Fr transfert de chaleur à double fluide *(m)*
It trasferimento di calore a doppio fluido *(m)*

terra *(f) n* It
Am ground
De Erde *(f)*
En earth
Es tierra *(f)*
Fr terre *(f)*
Pt massa *(f)*

terre *(f) n* Fr
Am ground
De Erde *(f)*
En earth
Es tierra *(f)*
It terra *(f)*
Pt massa *(f)*

terrestial heat flow En
De terrestrischer Wärmefluβ *(m)*
Es flujo térmico terrestre *(m)*
Fr courant de chaleur tellurique *(m)*
It flusso termico terrestre *(m)*
Pt fluxo de calor de terra *(m)*

terrestrischer Wärmefluβ *(m)* De
En terrestial heat flow
Es flujo térmico terrestre *(m)*
Fr courant de chaleur tellurique *(m)*
It flusso termico terrestre *(m)*

Pt fluxo de calor de terra *(m)*

terza rotaia *(f)* It
De Stromschiene *(f)*
En third rail
Es riel conductor *(m)*
Fr rail de contact *(m)*
Pt terceiro carril *(m)*

teste de fragmentação *(m)* Pt
De Sturzfestigkeits-prüfung *(f)*
En shatter test
Es prueba de cohesión *(f)*
Fr essai de résistance au choc *(m)*
It prova di frangibilità *(f)*

teste de inchamento *(m)* Pt
De Quellversuch *(m)*
En swelling test
Es prueba de esponjamiento *(f)*
Fr essai de gonflement *(m)*
It prova di ringonfiamento *(f)*

teste de sedimentos *(m)* Pt
De Rückstandsprüfung *(f)*
En sediment test
Es prueba de sedimentos *(f)*
Fr essai de sédimentation *(m)*
It prova di sedimentazione *(f)*

Tetraäthylblei *(n)* *n* De
En tetraethyl lead
Es plomo de tetraetilo *(m)*
Fr plomb tétraéthyle *(m)*
It piombo tetraetile *(m)*
Pt tetraetilato de chumbo *(m)*

tetraethyl lead En
De Tetraäthylblei *(n)*
Es plomo de tetraetilo *(m)*
Fr plomb tétraéthyle *(m)*
It piombo tetraetile *(m)*
Pt tetraetilato de chumbo *(m)*

tetraetilato de chumbo *(m)* Pt
De Tetraäthylblei *(n)*
En tetraethyl lead
Es plomo de tetraetilo *(m)*
Fr plomb tétraéthyle *(m)*
It piombo tetraetile *(m)*

théorie de fiabilité *(f)* Fr
De Verläslichkeitstheorie *(f)*
En reliability theory
Es teoría de la fiabilidad *(f)*
It teoria dell'affidabilità *(f)*
Pt teoria da segurança *(f)*

théorie des quanta *(f)* Fr
De Quantentheorie *(f)*
En quantum theory
Es teoría de los cuantos *(f)*
It teoria dei quanti *(f)*
Pt teoría quântica *(f)*

thermal *adj* En
De thermisch
Es térmico
Fr thermique
It termico
Pt térmico

thermal ageing En
De thermische Alterung *(f)*
Es envejecimiento térmico *(m)*
Fr vieillissement thermique *(m)*
It invecchiamento termico *(m)*
Pt envelhecimento térmico *(m)*

thermal conductivity En
De Wärmeleitfähigkeit *(f)*
Es conductividad térmica *(f)*
Fr conductivité thermique *(f)*
It conduttività termica *(f)*
Pt condutividade térmica *(f)*

thermal cracking En
De thermisches Kracken *(n)*
Es cráking térmico *(m)*
Fr craquage thermique *(m)*
It piroscissione termica *(f)*
Pt cracking térmico *(m)*

thermal diffusion En
De Thermodiffusion *(f)*
Es difusión térmica *(f)*
Fr diffusion thermique *(f)*
It diffusione termica *(f)*
Pt difusão térmica *(f)*

thermal efficiency En
De thermischer Wirkungsgrad *(m)*
Es rendimiento térmico *(m)*
Fr rendement thermique *(m)*
It rendimento termico *(m)*
Pt eficiência térmica *(f)*

thermal energy En
De Wärmeenergie *(f)*
Es energía térmica *(f)*
Fr énergie calorifique; énergie thermique *(f)*
It energia termica *(f)*
Pt energia térmica *(f)*

thermal-energy storage En
De Wärmeenergie-speicherung *(f)*
Es almacenamiento de energía térmica *(m)*
Fr accumulation d'énergie thermique *(n)*
It immagazzinaggio di energia termica *(m)*
Pt armazenagem de energia térmica *(f)*

thermal flux En
De Wärmefluβ *(m)*
Es flujo térmico *(m)*
Fr flux thermique *(m)*
It flusso termico *(m)*
Pt fluxo térmico *(m)*

thermal inertia En
De Wärmeträgheit *(f)*
Es inercia térmica *(f)*
Fr inertie thermique *(f)*
It inerzia termica *(f)*
Pt inércia térmica *(f)*

thermalisation *(f)* *n* Fr
De Neutronen-abbremsung *(f)*
En thermalization
Es termalización *(f)*
It termalizzazione *(f)*
Pt termalização *(f)*

thermalization *n* En
De Neutronen-abbremsung *(f)*
Es termalización *(f)*
Fr thermalisation *(f)*
It termalizzazione *(f)*
Pt termalização *(f)*

thermal neutron En
De thermisches Neutron *(n)*
Es neutrón térmico *(m)*
Fr neutron thermique *(m)*
It neutrone termico *(m)*
Pt neutrão térmico *(m)*

thermal power En
De Heizkraft *(f)*
Es potencia térmica *(f)*
Fr puissance thermique *(f)*
It potenza termica *(f)*
Pt potência térmica *(f)*

thermal power station En
De Wärmekraftwerk *(n)*
Es central térmica *(f)*
Fr centrale thermique *(f)*
It centrale di energia termica *(f)*
Pt central de energia térmica *(f)*

thermal radiation En
De Wärmestrahlung *(f)*
Es radiación térmica *(f)*
Fr rayonnement thermique *(f)*
It radiazione termica *(f)*
Pt radiação térmica *(f)*

thermal reactor En
De thermischer Reaktor *(m)*
Es reactor térmico *(m)*
Fr réacteur thermique *(m)*
It reattore termico *(m)*
Pt reactor térmico *(m)*

thermal shield En
De Wärmeschutz *(m)*
Es pantalla térmica *(f)*
Fr bouclier thermique *(m)*
It schermo termico *(m)*
Pt protector térmico *(m)*

thermique *adj* Fr
De thermisch
En thermal
Es térmico
It termico
Pt térmico

thermisch *adj* De
En thermal
Es térmico
Fr thermique
It termico
Pt térmico

thermische Alterung *(f)* De
En thermal ageing
Es envejecimiento térmico *(m)*
Fr vieillissement thermique *(m)*
It invecchiamento termico *(m)*
Pt envelhecimento térmico *(m)*

thermischer Reaktor *(m)* De
En thermal reactor
Es reactor térmico *(m)*
Fr réacteur thermique *(m)*
It reattore termico *(m)*
Pt reactor térmico *(m)*

thermischer Wirkungsgrad *(m)* De
En thermal efficiency
Es rendimiento térmico *(m)*
Fr rendement thermique *(m)*
It rendimento termico *(m)*
Pt eficiência térmica *(f)*

thermisches Kracken *(n)* De
En thermal cracking
Es cráckíng térmico *(m)*
Fr craquage thermique *(m)*
It piroscissione termica *(f)*
Pt cracking térmico *(m)*

thermisches Neutron *(n)* De
En thermal neutron
Es neutrón térmico *(m)*
Fr neutron thermique *(m)*
It neutrone termico *(m)*
Pt neutrão térmico *(m)*

thermisch spaltbar De
En fissile
Es fissile
Fr fissile
It fissile
Pt físsil

thermistor *n* En; Fr *(m)*
De Thermistor *(m)*
Es termistor *(m)*
It termistore *(m)*
Pt termistor *(m)*

Thermistor *(m) n* De
En thermistor
Es termistor *(m)*
Fr thermistor *(m)*
It termistore *(m)*
Pt termistor *(m)*

thermocouple *n* En; Fr *(m)*
De Thermoelement *(n)*
Es termopar *(m)*
It termocoppia *(f)*
Pt termopar *(m)*

Thermodiffusion *(f) n* De
En thermal diffusion
Es difusión térmica *(f)*
Fr diffusion thermique *(f)*
It diffusione termica *(f)*
Pt difusão térmica *(f)*

thermodynamics *n* En
De Thermodynamik *(f)*
Es termodinámica *(f)*
Fr thermodynamique *(f)*
It termodinamica *(f)*
Pt termodinâmica *(f)*

thermodynamic temperature En
De thermodynamische Temperatur *(f)*
Es temperatura termodinámica *(f)*
Fr température thermodynamique *(f)*
It temperatura termodinamica *(f)*
Pt temperatura termodinâmica *(f)*

Thermodynamik *(f) n* De
En thermodynamics
Es termodinámica *(f)*
Fr thermodynamique *(f)*
It termodinamica *(f)*
Pt termodinâmica *(f)*

thermodynamique *(f) n* Fr
De Thermodynamik *(f)*
En thermodynamics
Es termodinámica *(f)*
It termodinamica *(f)*
Pt termodinâmica *(f)*

thermodynamische Temperatur *(f)* De
En thermodynamic temperature
Es temperatura termodinámica *(f)*
Fr température thermodynamique *(f)*
It temperatura termodinamica *(f)*
Pt temperatura termodinâmica *(f)*

thermoelectric *adj* En
De thermoelektrisch
Es termoeléctrico
Fr thermcélectrique
It termoelettrico
Pt termoeléctrico

thermoelectric effect En
De thermoelektrischer Effekt *(m)*
Es efecto termoeléctrico *(m)*
Fr effet thermoélectrique *(m)*
It effetto termoelettrico *(m)*
Pt efeito termo-eléctrico *(m)*

thermoélectricité *(f) n* Fr
De Thermoelektrizität *(f)*
En thermoelectricity
Es termoelectricidad *(f)*
It termoelettricità *(f)*
Pt termoelectricidade *(f)*

thermoelectricity *n* En
De Thermoelektrizität *(f)*
Es termoelectricidad *(f)*
Fr thermoélectricité *(f)*
It termoelettricità *(f)*
Pt termoelectricidade *(f)*

thermoélectrique *adj* Fr
De thermoelektrisch
En thermoelectric
Es termoeléctrico
It termoelettrico
Pt termoeléctrico

thermoelektrisch *adj* De
En thermoelectric
Es termoeléctrico
Fr thermoélectrique
It termoelettrico
Pt termoeléctrico

thermoelektrischer Effekt *(m)* De
En thermoelectric effect
Es efecto termoeléctrico *(m)*
Fr effet thermoélectrique *(m)*
It effetto termoelettrico *(m)*
Pt efeito termo-eléctrico *(m)*

Thermoelektrizität *(f) n* De
En thermoelectricity
Es termoelectricidad *(f)*
Fr thermoélectricité *(f)*
It termoelettricità *(f)*
Pt termoelectricidade *(f)*

Thermoelement *(n) n* De
En thermocouple
Es termopar *(m)*
Fr thermocouple *(m)*
It termocoppia *(f)*
Pt termopar *(m)*

thermometer *n* En
De Thermometer *(n)*
Es termómetro *(m)*
Fr thermomètre *(m)*
It termometro *(m)*
Pt termómetro *(m)*

Thermometer *(n)* *n* De
En thermometer
Es termómetro *(m)*
Fr thermomètre *(m)*
It termometro *(m)*
Pt termómetro *(m)*

thermomètre *(m)* *n* Fr
De Thermometer *(n)*
En thermometer
Es termómetro *(m)*
It termometro *(m)*
Pt termómetro *(m)*

thermomètre à gaz *(m)* Fr
De Gasthermometer *(n)*
En gas thermometer
Es termómetro de gas *(m)*
It termometro a gas *(m)*
Pt termómetro de gás *(m)*

thermonucléaire *adj* Fr
De thermonuklear
En thermonuclear
Es termonuclear
It termonucleare
Pt termonuclear

thermonuclear *adj* En
De thermonuklear
Es termonuclear
Fr thermonucléaire
It termonucleare
Pt termonuclear

thermonuklear *adj* De
En thermonuclear
Es termonuclear
Fr thermonucléaire
It termonucleare
Pt termonuclear

thermopile *n* En; Fr *(f)*
De Thermosäule *(f)*
Es termopila *(f)*
It termopila *(f)*
Pt pilha térmica *(f)*

thermoplongeur *(m)* *n* Fr
De Tauchsieder *(m)*
En immersion heater
Es calentador de inmersión *(m)*
It riscaldatore ad immersione *(m)*
Pt aquecedor por imersão *(m)*

thermopompe *(f)* *n* Fr
De Wärmepumpe *(f)*
En heat pump
Es bomba de calor *(f)*
It pompa di calore *(f)*
Pt bomba térmica *(f)*

Thermosäule *(f)* *n* De
En thermopile
Es termopila *(f)*
Fr thermopile *(f)*
It termopila *(f)*
Pt pilha térmica *(f)*

thermosiphon *n* En
De Thermosiphon *(n)*
Es termosifón *(m)*
Fr thermosyphon *(m)*
It termosifone *(m)*
Pt termosiffão *(m)*

Thermosiphon *(n)* *n* De
En thermosiphon
Es termosifón *(m)*
Fr thermosyphon *(m)*
It termosifone *(m)*
Pt termosiffão *(m)*

thermostat *n* En; Fr *(m)*
De Thermostat *(m)*
Es termostato *(m)*
It termostato *(m)*
Pt termostato *(m)*

Thermostat *(m)* *n* De
En thermostat
Es termostato *(m)*
Fr thermostat *(m)*
It termostato *(m)*
Pt termostato *(m)*

thermosyphon *(m)* *n* Fr
De Thermosiphon *(n)*
En thermosiphon
Es termosifón *(m)*
It termosifone *(m)*
Pt termosiffão *(m)*

thiol *n* En; Fr *(m)*
De Thiol *(n)*
Es tiol *(m)*
It tiolo *(m)*
Pt tiol *(m)*

Thiol *(n)* *n* De
En thiol
Es tiol *(m)*
Fr thiol *(m)*
It tiolo *(m)*
Pt tiol *(m)*

third rail *n* En
De Stromschiene *(f)*
Es riel conductor *(m)*
Fr rail de contact *(m)*
It terza rotaia *(f)*
Pt terceiro carril *(m)*

thorium *n* En; Fr *(m)*
De Thorium *(n)*
Es torio *(m)*
It torio *(m)*
Pt tório *(m)*

Thorium *(n)* *n* De
En thorium
Es torio *(m)*
Fr thorium *(m)*
It torio *(m)*
Pt tório *(m)*

three-phase *adj* En
De dreiphasig
Es trifásico
Fr triphasé
It trifase
Pt trifásico

threshold limit value (TLV) En
De Schwellenwert *(m)*
Es valor límite de umbral *(m)*
Fr valeur de seuil *(f)*
It valore del limite di soglia *(m)*
Pt valor de limite de limiar *(m)*

tidal barrage En
De Gezeitendamm *(m)*
Es presa de marea *(f)*
Fr barrage marémoteur *(m)*
It sbarramento di marea *(m)*
Pt barragem de marés *(f)*

tidal power En
De Gezeitenkraft *(f)*
Es potencia mareal *(f)*
Fr puissance marémotrice *(f)*
It potenza della marea *(f)*
Pt potência das marés *(f)*

Tiefbau *(m)* *n* De
En deep mining
Es explotación en profundidad *(f)*
Fr exploitation des niveaux inférieurs *(f)*
It scavo profondo *(m)*
Pt mineração profunda *(f)*

Tiefsiedepunkt-fraktionen *(pl)* De
En low-boiling fractions *(pl)*
Es fracciónes de bajo punto de ebullición *(f pl)*
Fr fractions à bas point d'ébullition *(f pl)*
It frazioni a basso punto di ebollizione *(f pl)* t fracções de baixo ponto de ebulição *(f pl)*

Tieftemperaturtechnik *(f)* *n* De
En cryogenics
Es criogénica *(f)*
Fr cryogénie *(f)*
It criogenia *(f)*
Pt criogenia *(f)*

Tieftemperaturteer *(m)* *n* De
En low-temperature tar
Es alquitrán de baja temperatura *(m)*
Fr goudron à basse température *(m)*
It catrame a bassa temperatura *(m)*
Pt alcatrão de baixa temperatura *(m)*

Tiegelofen *(m)* *n* De
En crucible furnace
Es horno de crisol *(m)*
Fr fourneau à creuset *(m)*
It forno a crogiolo *(m)*
Pt forno de cadinho *(m)*

tierra *(f)* *n* Es
Am ground
De Erde *(f)*
En earth
Fr terre *(f)*
It terra *(f)*
Pt massa *(f)*

tijolo refractário *(m)* Pt
De feuerfester Stein *(m)*
En firebrick

Es ladrillo refractario *(m)*
Fr brique réfractaire *(f)*
It mattone refrattario *(m)*

time switch En
De Zeitschalter *(m)*
Es temporizador *(m)*
Fr minuterie *(f)*
It interruttore a tempo *(m)*
Pt comutador cronométrico *(m)*

tiol *(m) n* Es, Pt
De Thiol *(n)*
En thiol
Fr thiol *(m)*
It tiolo *(m)*

tiolo *(m) n* It
De Thiol *(n)*
En thiol
Es tiol *(m)*
Fr thiol *(m)*
Pt tiol *(m)*

tira bimetálica *(f)* Es, Pt
De Bimetallstreifen *(m)*
En bimetallic strip
Fr bilame *(m)*
It nastro bimetallico *(m)*

tödliche Dosis *(f)* De
En lethal dose
Es dosis letal *(f)*
Fr dose mortelle *(f)*
It dose letale *(f)*
Pt dose letal *(f)*

toluene *n* En
De Toluol *(n)*
Es tolueno *(m)*
Fr toluène *(m)*
It toluolo *(m)*
Pt tolueno *(m)*

toluène *(m) n* Fr
De Toluol *(n)*
En toluene
Es tolueno *(m)*
It toluolo *(m)*
Pt tolueno *(m)*

tolueno *(m) n* Es, Pt
De Toluol *(n)*
En toluene
Fr toluène *(m)*
It toluolo *(m)*

Toluol *(n) n* De
En toluene
Es tolueno *(m)*
Fr toluène *(m)*
It toluolo *(m)*
Pt tolueno *(m)*

toluolo *(m) n* It
De Toluol *(n)*
En toluene
Es tolueno *(m)*
Fr toluène *(m)*
Pt tolueno *(m)*

toma de aire *(f)* Es
De Entlüftungsschlitz *(m)*
En air vent
Fr évent *(m)*
It sfiato aria *(m)*
Pt respirador de ar *(m)*

toma de tierra múltiple protectora *(f)* Es
De Mehrfach-Schutzerdung *(f)*
En protective multiple earthing
Fr mise à la terre multiple de protection *(f)*
It messa a terra multipla protettiva *(f)*
Pt ligação à terra múltipla protectora *(f)*

torba *(f) n* It
De Torf *(m)*
En peat
Es turba *(f)*
Fr tourbe *(f)*
Pt turfa *(f)*

torcia a plasma *(f)* It
De Plasmabrenner *(m)*
En plasma torch
Es soplete para plasma *(m)*
Fr chalumeau à plasma *(m)*
Pt maçarico de plasma *(m)*

Torf *(m) n* De
En peat
Es turba *(f)*
Fr tourbe *(f)*
It torba *(f)*
Pt turfa *(f)*

torio *(m) n* Es, It
De Thorium *(n)*
En thorium
Fr thorium *(m)*
Pt tório *(m)*

tório *(m) n* Pt
De Thorium *(n)*
En thorium
Es torio *(m)*
Fr thorium *(m)*
It torio *(m)*

torneira *(f) n* Pt
De Hahn *(m)*
En cock
Es espita *(f)*
Fr robinet *(m)*
It rubinetto *(m)*

torre de destilación *(f)* Es
De Destillationskolonne *(f)*
En distillation column
Fr colonne de distillation *(f)*
It colonna di distillazione *(f)*
Pt coluna de destilação *(f)*

total energy En
De Gesamtenergie *(f)*
Es energía total *(f)*
Fr énergie totale *(f)*
It energia totale *(f)*
Pt energia total *(f)*

tourbe *(f) n* Fr
De Torf *(m)*
En peat
Es turba *(f)*
It torba *(f)*
Pt turfa *(f)*

town gas En
De Stadtgas *(n)*
Es gas de ciudad *(m)*
Fr gaz de ville *(m)*
It gas per uso domestico *(m)*
Pt gás de cidade *(m)*

trabajo *(m) n* Es
De Arbeit *(f)*
En work
Fr travail *(m)*
It lavoro *(m)*
Pt trabalho *(m)*

trabajo externo *(m)* Es
De Außenarbeit *(f)*
En external work
Fr travail extérieur *(m)*
It lavoro esterno *(m)*
Pt trabalho externo *(m)*

trabajo interno *(m)* Es
De Innenarbeit *(f)*
En internal work
Fr travail interne *(m)*
It lavoro interno *(m)*
Pt trabalho interno *(m)*

trabajo virtual *(m)* Es
De virtuelle Arbeit *(f)*
En virtual work
Fr travail virtuel *(m)*
It lavoro virtuale *(m)*
Pt trabalho virtual *(m)*

trabalho *(m) n* Pt
De Arbeit *(f)*
En work
Es trabajo *(m)*
Fr travail *(m)*
It lavoro *(m)*

trabalho externo *(m)* Pt
De Außenarbeit *(f)*
En external work
Es trabajo externo *(m)*
Fr travail extérieur *(m)*
It lavoro esterno *(m)*

trabalho interno *(m)* Pt
De Innenarbeit *(f)*
En internal work
Es trabajo interno *(m)*
Fr travail interne *(m)*
It lavoro interno *(m)*

trabalho virtual *(m)* Pt
De virtuelle Arbeit *(f)*
En virtual work
Es trabajo virtual *(m)*
Fr travail virtuel *(m)*
It lavoro virtuale *(m)*

Trägheitsmoment *(n) m* De
En inertia
Es inercia *(f)*
Fr inertie *(f)*
It inerzia *(f)*
Pt inércia *(f)*

traitement à l'argile *(m)* Fr
De Lehmbehandlung *(f)*

En clay treatment
Es tratamiento con arcilla *(m)*
It trattamento all'argilla *(m)*
Pt tratamento de argila *(m)*

traitement des eaux *(m)* Fr
De Wasseraufbereitung *(f)*
En water treatment
Es tratamiento del agua *(m)*
It trattamento ad acqua *(m)*
Pt tratamento hidráulico *(m)*

traitement du combustible irradié *(m)* Fr
De Wiederaufbereitung *(f)*
En reprocessing
Es reelaboración *(f)*
It riutilazzione *(f)*
Pt reprocessamento *(m)*

traitement préalable *(m)* Fr
De Vorbehandlung *(f)*
En pretreatment
Es tratamiento previo *(m)*
It trattamento previo *(m)*
Pt tratamento prévio *(m)*

traitement thermique *(m)* Fr
De Warmbehandlung *(f)*
En heat treatment
Es tratamiento térmico *(m)*
It trattamento termico *(m)*
Pt tratamento térmico *(m)*

traliccio *(m) n* It
De Spaltstoffgitter *(n)*
En lattice
Es celosía *(f)*
Fr treillis *(m)*
Pt treliça *(f)*

transducer *n* En
De Meβwandler *(m)*
Es transductor *(m)*
Fr transducteur *(m)*
It trasduttore *(m)*
Pt transdutor *(m)*

transducteur *(m) n* Fr
De Meβwandler *(m)*
En transducer
Es transductor *(m)*
It trasduttore *(m)*
Pt transdutor *(m)*

transductor *(m) n* Es
De Meβwandler *(m)*
En transducer
Fr transducteur *(m)*
It trasduttore *(m)*
Pt transdutor *(m)*

transdutor *(m) n* Pt
De Meβwandler *(m)*
En transducer
Es transductor *(m)*
Fr transducteur *(m)*
It trasduttore *(m)*

transferencia térmica *(f)* Es
De Wärmeübertragung *(f)*
En heat transfer
Fr transfert de chaleur *(m)*
It scambio di calore *(m)*
Pt transferência térmica *(f)*

transferência térmica *(f)* Pt
De Wärmeübertragung *(f)*
En heat transfer
Es transferencia térmica *(f)*
Fr transfert de chaleur *(m)*
It scambio di calore *(m)*

transferencia térmica de dos flúidos *(f)* Es
De Wärmeübertragung mit zwei Flüssigkeiten *(f)*
En dual-fluid heat transfer
Fr transfert de chaleur à double fluide *(m)*
It trasferimento di calore a doppio fluido *(m)*
Pt termotransferência de dois fluidos *(f)*

transfert de chaleur *(m)* Fr
De Wärmeübertragung *(f)*
En heat transfer
Es transferencia térmica *(f)*
It scambio di calore *(m)*
Pt transferência térmica *(f)*

transfert de chaleur à double fluide *(m)* Fr
De Wärmeübertragung mit zwei Flüssigkeiten *(f)*
En dual-fluid heat transfer
Es transferencia térmica de dos flúidos *(f)*
It trasferimento di calore a doppio fluido *(m)*
Pt termotransferência de dois fluidos *(f)*

transformador *(m) n* Es, Pt
De Transformator *(m)*
En transformer
Fr transformateur *(m)*
It trasformatore *(m)*

transformateur *(m) n* Fr
De Transformator *(m)*
En transformer
Es transformador *(m)*
It trasformatore *(m)*
Pt transformador *(m)*

transformation d'énergie à bord *(f)* Fr
De Energieumwandlung an Bord *(f)*
En on-board energy conversion
Es conversión de energía a bordo *(f)*
It conversione di energia a bordo *(f)*
Pt conversão de energia a bordo *(m)*

Transformator *(m) n* De
En transformer
Es transformador *(m)*
Fr transformateur *(m)*
It trasformatore *(m)*
Pt transformador *(m)*

transformer *n* En
De Transformator *(m)*
Es transformador *(m)*
Fr transformateur *(m)*
It trasformatore *(m)*
Pt transformador *(m)*

transmisión con tensión muy alta *(f)* Es
De Höchstspannungs-transmission *(f)*
En extra-high voltage transmission
Fr transmission des ultra-hautes tensions *(f)*
It trasmissione a tensione extra elevata *(f)*
Pt transmissão de voltagem extra-alta *(f)*

transmissão de voltagem extra-alta *(f)* Pt
De Höchstspannungs-transmission *(f)*
En extra-high voltage transmission
Es transmisión con tensión muy alta *(f)*
Fr transmission des ultra-hautes tensions *(f)*
It trasmissione a tensione extra elevata *(f)*

transmission des ultra-hautes tensions *(f)* Fr
De Höchstspannungs-transmission *(f)*
En extra-high voltage transmission
Es transmisión con tensión muy alta *(f)*
It trasmissione a tensione extra elevata *(f)*
Pt transmissão de voltagem extra-alta *(f)*

transmission line En
De Transmissionsleitung *(f)*
Es línea de transmisión *(f)*
Fr ligne de transmission *(f)*
It linea di trasmissione *(f)*

Pt linha de transmissão (f)

transmission loss En
De Durchgangsdämpfung (f)
Es pérdida por transmisión (f)
Fr perte par transmission (f)
It perdita per trasmissione (f)
Pt perda por transmissão (f)

Transmissionsleitung (f) n De
En transmission line
Es línea de transmisión (f)
Fr ligne de transmission (f)
It linea di trasmissione (f)
Pt linha de transmissão (f)

transpiração (f) n Pt
De Transpiration (f)
En transpiration
Es transpiración (f)
Fr transpiration (f)
It traspirazione (f)

transpiración (f) n Es
De Transpiration (f)
En transpiration
Fr transpiration (f)
It traspirazione (f)
Pt transpiração (f)

transpiration n En; Fr (f)
De Transpiration (f)
Es transpiración (f)
It traspirazione (f)
Pt transpiração (f)

Transpiration (f) n De
En transpiration
Es transpiración (f)
Fr transpiration (f)
It traspirazione (f)
Pt transpiração (f)

trapanatura ad aria compressa (f) It
De Druckluftbohren (n)
En air drilling
Es perforación por aire comprimido (f)
Fr forage à l'air comprimé (m)
Pt perfuração pneumática (f)

trasduttore (m) n It
De Meßwandler (m)
En transducer
Es transductor (m)
Fr transducteur (m)
Pt transdutor (m)

trasferimento di calore a doppio fluido (m) It
De Wärmeübertragung mit zwei Flüssigkeiten (f)
En dual-fluid heat transfer
Es transferencia térmica de dos flúidos (f)
Fr transfert de chaleur à double fluide (m)
Pt termotransferência de dois fluidos (f)

trasformatore (m) n It
De Transformator (m)
En transformer
Es transformador (m)
Fr transformateur (m)
Pt transformador (m)

trasmissione a tensione extra elevata (f) It
De Höchstspannungs-transmission (f)
En extra-high voltage transmission
Es transmisión con tensión muy alta (f)
Fr transmission des ultra-hautes tensions (f)
Pt transmissão de voltagem extra-alta (f)

traspirazione (f) n It
De Transpiration (f)
En transpiration
Es transpiración (f)
Fr transpiration (f)
Pt transpiração (f)

tratamento de argila (m) Pt
De Lehmbehandlung (f)
En clay treatment
Es tratamiento con arcilla (m)
Fr traitement à l'argile (m)
It trattamento all'argilla (m)

tratamento hidráulico (m) Pt
De Wasseraufbereitung (f)
En water treatment
Es tratamiento del agua (m)
Fr traitement des eaux (m)
It trattamento ad acqua (m)

tratamento prévio (m) Pt
De Vorbehandlung (f)
En pretreatment
Es tratamiento previo (m)
Fr traitement préalable (m)
It trattamento previo (m)

tratamento térmico (m) Pt
De Warmbehandlung (f)
En heat treatment
Es tratamiento térmico (m)
Fr traitement thermique (m)
It trattamento termico (m)

tratamiento con arcilla (m) Es
De Lehmbehandlung (f)
En clay treatment
Fr traitement à l'argile (m)
It trattamento all'argilla (m)
Pt tratamento de argila (m)

tratamiento del agua (m) Es
De Wasseraufbereitung (f)
En water treatment
Fr traitement des eaux (m)
It trattamento ad acqua (m)
Pt tratamento hidráulico (m)

tratamiento previo (m) Es
De Vorbehandlung (f)
En pretreatment
Fr traitement préalable (m)
It trattamento previo (m)
Pt tratamento prévio (m)

tratamiento térmico (m) Es
De Warmbehandlung (f)
En heat treatment
Fr traitement thermique (m)
It trattamento termico (m)
Pt tratamento térmico (m)

trattamento ad acqua (m) It
De Wasseraufbereitung (f)
En water treatment
Es tratamiento del agua (m)
Fr traitement des eaux (m)
Pt tratamento hidráulico (m)

trattamento all'argilla (m) It
De Lehmbehandlung (f)
En clay treatment
Es tratamiento con arcilla (m)
Fr traitement à l'argile (m)
Pt tratamento de argila (m)

trattamento impermeabilizzante (m) It
De Feuchtigkeits-sperrschicht (f)
En damp-proof course
Es hilada hidrófuga (f)
Fr couche hydrofuge (f)
Pt curso a prova de imfietrações (m)

trattamento previo (m) It
De Vorbehandlung (f)
En pretreatment
Es tratamiento previo (m)

Fr traitement préalable *(m)*
Pt tratamento prévio *(m)*

trattamento termico *(m)* It
De Warmbehandlung *(f)*
En heat treatment
Es tratamiento térmico *(m)*
Fr traitement thermique *(m)*
Pt tratamento térmico *(m)*

travail *(m)* *n* Fr
De Arbeit *(f)*
En work
Es trabajo *(m)*
It lavoro *(m)*
Pt trabalho *(m)*

travail de sortie *(m)* Fr
De Arbeitsfunktion *(f)*
En work function
Es función de trabajo *(f)*
It funzione di lavoro *(f)*
Pt função de trabalho *(f)*

travail extérieur *(m)* Fr
De Außenarbeit *(f)*
En external work
Es trabajo externo *(m)*
It lavoro esterno *(m)*
Pt trabalho externo *(m)*

travail interne *(m)* Fr
De Innenarbeit *(f)*
En internal work
Es trabajo interno *(m)*
It lavoro interno *(m)*
Pt trabalho interno *(m)*

travail virtuel *(m)* Fr
De virtuelle Arbeit *(f)*
En virtual work
Es trabajo virtual *(m)*
It lavoro virtuale *(m)*
Pt trabalho virtual *(m)*

Treibflüssigkeit *(f)* *n* De
En working fluid
Es flúido motor *(m)*
Fr fluide moteur *(m)*
It fluido operante *(m)*
Pt fluido operativo *(m)*

Treiböl *(n)* *n* De
En fuel oil
Es fuel oil *(m)*
Fr mazout *(m)*
It nafta *(f)*
Pt fuel oil *(m)*

Treibstoff *(m)* *n* De
En propellant
Es propulsante *(m)*
Fr propergol *(m)*
It propellente *(m)*
Pt combustível propulsor *(m)*

treillis *(m)* *n* Fr
De Spaltstoffgitter *(n)*
En lattice
Es celosía *(f)*
It traliccio *(m)*
Pt treliça *(f)*

treliça *(f)* *n* Pt
De Spaltstoffgitter *(n)*
En lattice
Es celosía *(f)*
Fr treillis *(m)*
It traliccio *(m)*

tren de perforación de petróleo *(m)* Es
De Ölbohrturm *(m)*
En oilrig
Fr installation de forage *(f)*
It impianto di trivellazione per petrolio *(m)*
Pt plataforma petroleira *(f)*

trépan *(m)* *n* Fr
De Bohrmeißel *(m)*
En bit (drilling)
Es broca *(f)*
It punta *(f)*
Pt broca *(f)*

trifase *adj* It
De dreiphasig
En three-phase
Es trifásico
Fr triphasé
Pt trifásico

trifásico *adj* Es, Pt
De dreiphasig
En three-phase
Fr triphasé
It trifase

trinitrotoluene (TNT) *n* En
De Trinitrotoluol (TNT) *(n)*
Es trinitrotolueno (TNT) *(m)*
Fr trinitrotoluène (TNT) *(m)*
It trinitrotoluolo (TNT) *(m)*
Pt trinitrotolueno (TNT) *(m)*

trinitrotoluène (TNT) *(m)* *n* Fr
De Trinitrotoluol (TNT) *(n)*
En trinitrotoluene (TNT)
Es trinitrotolueno (TNT) *(m)*
It trinitrotoluolo (TNT) *(m)*
Pt trinitrotolueno (TNT) *(m)*

trinitrotolueno (TNT) *(m)* *n* Es, Pt
De Trinitrotoluol (TNT) *(n)*
En trinitrotoluene (TNT)
Fr trinitrotoluène (TNT) *(m)*
It trinitrotoluolo (TNT) *(m)*

Trinitrotoluol (TNT) *(n)* *n* De
En trinitrotoluene (TNT)
Es trinitrotolueno (TNT) *(m)*
Fr trinitrotoluène (TNT) *(m)*
It trinitrotoluolo (TNT) *(m)*
Pt trinitrotolueno (TNT) *(m)*

trinitrotoluolo (TNT) *(m)* *n* It
De Trinitrotoluol (TNT) *(n)*
En trinitrotoluene (TNT)
Es trinitrotolueno (TNT) *(m)*
Fr trinitrotoluène (TNT) *(m)*
Pt trinitrotolueno (TNT) *(m)*

triphasé *adj* Fr
De dreiphasig
En three-phase
Es trifásico
It trifase
Pt trifásico

tripla vetrata *(f)* It
De Dreifachverglasung *(f)*
En triple glazing n
Es encristalado triple *(m)*
Fr triple vitrage *(m)*
Pt vitrificação tripla *(f)*

triple glazing n En
De Dreifachverglasung *(f)*
Es encristalado triple *(m)*
Fr triple vitrage *(m)*
It tripla vetrata *(f)*
Pt vitrificação tripla *(f)*

triple vitrage *(m)* Fr
De Dreifachverglasung *(f)*
En triple glazing n
Es encristalado triple *(m)*
It tripla vetrata *(f)*
Pt vitrificação tripla *(f)*

tritio *(m)* *n* Es, It
De Tritium *(n)*
En tritrium
Fr tritium *(m)*
Pt trítio *(m)*

trítio *(m)* *n* Pt
De Tritium *(n)*
En tritrium
Es tritio *(m)*
Fr tritium *(m)*
It tritio *(m)*

tritium *n* En; Fr *(m)*
De Tritium *(n)*
Es tritio *(m)*
It tritio *(m)*
Pt trítio *(m)*

Tritium *(n)* *n* De
En tritrium
Es tritio *(m)*
Fr tritium *(m)*
It tritio *(m)*
Pt trítio *(m)*

trivellazione *(f)* *n* It
De Bohren *(n)*
En drilling

Es perforación *(f)*
Fr forage *(m)*
Pt perfuração *(f)*

trivellazione con utensile a fune *(f)* It
De Seilbohren *(n)*
En cable-tool drilling
Es perforación a cable *(f)*
Fr forage au câble *(m)*
Pt perfuração a cabo *(f)*

trivellazione direzionale *(f)* It
De gerichtetes Bohren *(n)*
En directional drilling
Es perforación direccional *(f)*
Fr forage dirigé *(m)*
Pt perfuração direccional *(f)*

Trockenasche-Ofen *(m)* *n* De
En dry-ash furnace
Es hogar de cenizas pulverulentas *(m)*
Fr four à cendres sèches *(m)*
It forno di cenere secca *(m)*
Pt forno de cinze seca *(m)*

Trockenmeβgerät *(n)* *n* De
En dry meter
Es medidor seco *(m)*
Fr dessicomètre *(m)*
It contatore a secco *(m)*
Pt medidor de seco *(m)*

Trockner *(m)* *n* De
En dryer
Es secador *(m)*
Fr sécheur *(m)*
It essiccatore *(m)*
Pt secador *(m)*

tubazione *(f)* *n* It
De Rohrleitung *(f)*
En pipeline
Es oleoducto *(m)*
Fr oléoduc *(m)*
Pt oleoduto *(m)*

tube à rayons cathodiques *(m)* Fr
De Kathodenstrahlröhre *(f)*
En cathode-ray tube
Es tubo de rayos catódicos *(m)*
It tubo a raggi catodici *(m)*
Pt válvula de raios catódicos *(f)*

tube de chaudière aileté *(m)* Fr
De Kesselrippenrohr *(n)*
En ribbed boiler tube
Es tubo de aletas de caldera *(m)*
It tubo nervato per caldaia *(m)*
Pt tubo de caldeira com nervuras *(m)*

tubo a raggi catodici *(m)* It
De Kathodenstrahlröhre *(f)*
En cathode-ray tube
Es tubo de rayos catódicos *(m)*
Fr tube à rayons cathodiques *(m)*
Pt válvula de raios catódicos *(f)*

tubo calefactor *(m)* Es
De Wärmeübertragungsrohr *(n)*
En heat pipe
Fr caloduc *(m)*
It tubo di calore *(m)*
Pt tubo de calor *(m)*

tubo de aletas de caldera *(m)* Es
De Kesselrippenrohr *(n)*
En ribbed boiler tube
Fr tube de chaudière aileté *(m)*
It tubo nervato per caldaia *(m)*
Pt tubo de caldeira com nervuras *(m)*

tubo de caldeira com nervuras *(m)* Pt
De Kesselrippenrohr *(n)*
En ribbed boiler tube
Es tubo de aletas de caldera *(m)*
Fr tube de chaudière aileté *(m)*
It tubo nervato per caldaia *(m)*

tubo de calor *(m)* Pt
De Wärmeübertragungsrohr *(n)*
En heat pipe
Es tubo calefactor *(m)*
Fr caloduc *(m)*
It tubo di calore *(m)*

tubo de rayos catódicos *(m)* Es
De Kathodenstrahlröhre *(f)*
En cathode-ray tube
Fr tube à rayons cathodiques *(m)*
It tubo a raggi catodici *(m)*
Pt válvula de raios catódicos *(f)*

tubo di calore *(m)* It
De Wärmeübertragungsrohr *(n)*
En heat pipe
Es tubo calefactor *(m)*
Fr caloduc *(m)*
Pt tubo de calor *(m)*

tubo nervato per caldaia *(m)* It
De Kesselrippenrohr *(n)*
En ribbed boiler tube
Es tubo de aletas de caldera *(m)*
Fr tube de chaudière aileté *(m)*
Pt tubo de caldeira com nervuras *(m)*

tubular recuperator En
De Röhrenrekuperator *(m)*
Es recuperador tubulare *(m)*
Fr récupérateur tubulaire *(m)*
It ricuperatore tubolare *(m)*
Pt recuperador tubulare *(m)*

turba *(f)* *n* Es
De Torf *(m)*
En peat
Fr tourbe *(f)*
It torba *(f)*
Pt turfa *(f)*

turbina *(f)* *n* Es, It, Pt
De Turbine *(f)*
En turbine
Fr turbine *(f)*

turbina a contropressione *(f)* It
De Gegendruckturbine *(f)*
En back-pressure turbine
Es turbina de contrapresión *(f)*
Fr turbine à contre-pression *(f)*
Pt turbina de contrapressão *(f)*

turbina ad acqua *(f)* It
De Wasserturbine *(f)*
En water turbine
Es turbina de agua *(f)*
Fr turbine hydraulique *(f)*
Pt turbina hidraúlica *(f)*

turbina a flusso assiale *(f)* It
De Axialturbine *(f)*
En axial flow turbine
Es turbina de flujo axial *(f)*
Fr turbine axiale *(f)*
Pt turbina de fluxo axial *(f)*

turbina a flusso assiale a due vie *(f)* It
De Zweiwege-Axialturbine *(f)*
En two-way axial flow turbine
Es turbina de flujo axial de doble dirección *(f)*
Fr turbine axiale à double sens *(f)*
Pt turbina de fluxo axial de duas vias *(f)*

turbina a gas *(f)* It
De Gasturbine *(f)*
En gas turbine
Es turbina de gas *(f)*
Fr turbine à gaz *(f)*
Pt turbina a gás *(f)*

turbina a gás *(f)* Pt
De Gasturbine *(f)*
En gas turbine
Es turbina de gas *(f)*
Fr turbine à gaz *(f)*
It turbina a gas *(f)*

turbina a vapor *(f)* Pt
De Dampfturbine *(f)*
En steam turbine

Es turbina de vapor *(f)*
Fr turbine à vapeur *(f)*
It turbina a vapore *(f)*

turbina a vapore *(f)* It
De Dampfturbine *(f)*
En steam turbine
Es turbina de vapor *(f)*
Fr turbine à vapeur *(f)*
Pt turbina a vapor *(f)*

turbina-bomba *(f) n* Es
De Pumpenturbine *(f)*
En pump-turbine
Fr pompe-turbine *(f)*
It turbopompa *(f)*
Pt turbobomba *(f)*

turbina de agua *(f)* Es
De Wasserturbine *(f)*
En water turbine
Fr turbine hydraulique *(f)*
It turbina ad acqua *(f)*
Pt turbina hidraúlica *(f)*

turbina de contrapresión *(f)* Es
De Gegendruckturbine *(f)*
En back-pressure turbine
Fr turbine à contre-pression *(f)*
It turbina a contropressione *(f)*
Pt turbina de contrapressão *(f)*

turbina de contrapressão *(f)* Pt
De Gegendruckturbine *(f)*
En back-pressure turbine
Es turbina de contrapresión *(f)*
Fr turbine à contre-pression *(f)*
It turbina a contropressione *(f)*

turbina de flujo axial *(f)* Es
De Axialturbine *(f)*
En axial flow turbine
Fr turbine axiale *(f)*
It turbina a flusso assiale *(f)*
Pt turbina de fluxo axial *(f)*

turbina de flujo axial de doble dirección *(f)* Es
De Zweiwege-Axialturbine *(f)*
En two-way axial flow turbine
Fr turbine axiale à double sens *(f)*
It turbina a flusso assiale a due vie *(f)*
Pt turbina de fluxo axial de duas vias *(f)*

turbina de fluxo axial *(f)* Pt
De Axialturbine *(f)*
En axial flow turbine
Es turbina de flujo axial *(f)*
Fr turbine axiale *(f)*
It turbina a flusso assiale *(f)*

turbina de fluxo axial de duas vias *(f)* Pt
De Zweiwege-Axialturbine *(f)*
En two-way axial flow turbine
Es turbina de flujo axial de doble dirección *(f)*
Fr turbine axiale à double sens *(f)*
It turbina a flusso assiale a due vie *(f)*

turbina de gas *(f)* Es
De Gasturbine *(f)*
En gas turbine
Fr turbine à gaz *(f)*
It turbina a gas *(f)*
Pt turbina a gás *(f)*

turbina de vapor *(f)* Es
De Dampfturbine *(f)*
En steam turbine
Fr turbine à vapeur *(f)*
It turbina a vapore *(f)*
Pt turbina a vapor *(f)*

turbina hidraúlica *(f)* Pt
De Wasserturbine *(f)*
En water turbine
Es turbina de agua *(f)*
Fr turbine hydraulique *(f)*
It turbina ad acqua *(f)*

turbine *n* En; Fr *(f)*
De Turbine *(f)*
Es turbina *(f)*
It turbina *(f)*
Pt turbina *(f)*

Turbine *(f) n* De
En turbine
Es turbina *(f)*
Fr turbine *(f)*
It turbina *(f)*
Pt turbina *(f)*

turbine à contre-pression *(f)* Fr
De Gegendruckturbine *(f)*
En back-pressure turbine
Es turbina de contrapresión *(f)*
It turbina a contropressione *(f)*
Pt turbina de contrapressão *(f)*

turbine à gaz *(f)* Fr
De Gasturbine *(f)*
En gas turbine
Es turbina de gas *(f)*
It turbina a gas *(f)*
Pt turbina a gás *(f)*

turbine à vapeur *(f)* Fr
De Dampfturbine *(f)*
En steam turbine
Es turbina de vapor *(f)*
It turbina a vapore *(f)*
Pt turbina a vapor *(f)*

turbine axiale *(f)* Fr
De Axialturbine *(f)*
En axial flow turbine
Es turbina de flujo axial *(f)*
It turbina a flusso assiale *(f)*
Pt turbina de fluxo axial *(f)*

turbine axiale à double sens *(f)* Fr
De Zweiwege-Axialturbine *(f)*
En two-way axial flow turbine
Es turbina de flujo axial de doble dirección *(f)*
It turbina a flusso assiale a due vie *(f)*
Pt turbina de fluxo axial de duas vias *(f)*

turbine hydraulique *(f)* Fr
De Wasserturbine *(f)*
En water turbine
Es turbina de agua *(f)*
It turbina ad acqua *(f)*
Pt turbina hidraúlica *(f)*

turboalternador *(m) n* Es
De Turbo-Wechselstromgenerator *(m)*
En turbo-alternator
Fr turbo-alternateur *(m)*
It turboalternatore *(m)*
Pt turbo-alternador *(m)*

turbo-alternador *(m) n* Pt
De Turbo-Wechselstromgenerator *(m)*
En turbo-alternator
Es turboalternador *(m)*
Fr turbo-alternateur *(m)*
It turboalternatore *(m)*

turbo-alternateur *(m) n* Fr
De Turbo-Wechselstromgenerator *(m)*
En turbo-alternator
Es turboalternador *(m)*
It turboalternatore *(m)*
Pt turbo-alternador *(m)*

turbo-alternator *n* En
De Turbo-Wechselstromgenerator *(m)*
Es turboalternador *(m)*
Fr turbo-alternateur *(m)*
It turboalternatore *(m)*
Pt turbo-alternador *(m)*

turboalternatore *(m) n* It
De Turbo-Wechselstromgenerator *(m)*
En turbo-alternator
Es turboalternador *(m)*
Fr turbo-alternateur *(m)*
Pt turbo-alternador *(m)*

turbobomba *(f) n* Pt
De Pumpenturbine *(f)*
En pump-turbine
Es turbina-bomba *(f)*
Fr pompe-turbine *(f)*
It turbopompa *(f)*

turboelica *(f) n* It
De PTL-Triebwerk *(n)*
En turboprop
Es turbohélice *(f)*

Fr turbopropulseur *(m)*
Pt turbo-hélice *(f)*

turbohélice *(f) n* Es
De PTL-Triebwerk *(n)*
En turboprop
Fr turbopropulseur *(m)*
It turboelica *(f)*
Pt turbo-hélice *(f)*

turbo-hélice *(f) n* Pt
De PTL-Triebwerk *(n)*
En turboprop
Es turbohélice *(f)*
Fr turbopropulseur *(m)*
It turboelica *(f)*

turbolenza *(f) n* It
De Turbulenz *(f)*
En turbulence
Es turbulencia *(f)*
Fr turbulence *(f)*
Pt turbulência *(f)*

turbopompa *(f) n* It
De Pumpenturbine *(f)*
En pump-turbine
Es turbina-bomba *(f)*
Fr pompe-turbine *(f)*
Pt turbobomba *(f)*

turboprop *n* En
De PTL-Triebwerk *(n)*
Es turbohélice *(f)*
Fr turbopropulseur *(m)*
It turboelica *(f)*
Pt turbo-hélice *(f)*

turbopropulseur *(m) n* Fr
De PTL-Triebwerk *(n)*
En turboprop
Es turbohélice *(f)*
It turboelica *(f)*
Pt turbo-hélice *(f)*

Turbo-Wechselstrom-generator *(m)* De
En turbo-alternator
Es turboalternador *(m)*
Fr turbo-alternateur *(m)*
It turboalternatore *(m)*
Pt turbo-alternador *(m)*

turbulence *n* En; Fr *(f)*
De Turbulenz *(f)*
Es turbulencia *(f)*
It turbolenza *(f)*
Pt turbulência *(f)*

turbulencia *(f) n* Es
De Turbulenz *(f)*
En turbulence
Fr turbulence *(f)*
It turbolenza *(f)*
Pt turbulência *(f)*

turbulência *(f) n* Pt
De Turbulenz *(f)*
En turbulence
Es turbulencia *(f)*
Fr turbulence *(f)*
It turbolenza *(f)*

Turbulenz *(f) n* De
En turbulence
Es turbulencia *(f)*
Fr turbulence *(f)*
It turbolenza *(f)*
Pt turbulência *(f)*

turfa *(f) n* Pt
De Torf *(m)*
En peat
Es turba *(f)*
Fr tourbe *(f)*
It torba *(f)*

turfização *(f) n* Pt
De Vertorfung *(f)*
En peatification
Es formación de turba *(f)*
Fr formation de la tourbe *(f)*
It produzione di torba *(f)*

two-stage combustion En
De Zweistufenver-brennung *(f)*
Es combustión en dos etapas *(f)*
Fr combustion à deux étages *(f)*
It combustione a due stadi *(f)*
Pt combustão de duas fases *(f)*

two-stroke engine En
De Zweitaktmotor *(m)*
Es motor de dos tiempos *(m)*
Fr moteur à deux temps *(m)*
It motore a due tempi *(m)*
Pt motor de dois tempos *(m)*

two-way axial flow turbine En
De Zweiwege-Axialturbine *(f)*
Es turbina de flujo axial de doble dirección *(f)*
Fr turbine axiale à double sens *(f)*
It turbina a flusso assiale a due vie *(f)*
Pt turbina de fluxo axial de duas vias *(f)*

U

Überhitzer *(m) n* De
En superheater
Es supercalentador *(m)*
Fr surchauffeur *(m)*
It surriscaldatore *(m)*
Pt superaquecedor *(m)*

überhitzter Dampf *(m)* De
En superheated steam
Es vapor supercalentado *(m)*
Fr vapeur surchauffée *(f)*
It vapore surriscaldato *(m)*
Pt vapor superaquecido *(m)*

überkritisch *adj* De
En supercritical
Es supercrítico
Fr supercritique
It supercritico
Pt supercrítico

Überlandleitungsnetz *(n) n* De
En grid (electrical power)
Es red nacional de energía eléctrica *(f)*
Fr réseau électrique national *(m)*
It rete nazionale *(f)*
Pt rede eléctrica *(f)*

Überlastung *(f) n* De
En overload
Es sobrecarga *(f)*
Fr sucharge *(f)*
It sovraccarico *(m)*
Pt sobrecarga *(f)*

Überspannung *(f) n* De
En overvoltage
Es sobretensión *(f)*
Fr surtension *(f)*
It sovratensione *(f)*
Pt sobrevoltagem *(f)*

ultimate recovery En
De Ausbeutefaktor *(m)*
Es producción final *(f)*
Fr récupération finale *(f)*
It recupero ultimo *(m)*
Pt recuperação final *(f)*

ultracentrifuga *(f) n* It
De Ultrazentrifuge *(f)*
En ultracentrifuge
Es ultracentrífuga *(f)*
Fr ultracentrifugeuse *(f)*
Pt ultracentrifugadora *(f)*

ultracentrífuga *(f) n* Es
De Ultrazentrifuge *(f)*
En ultracentrifuge
Fr ultracentrifugeuse *(f)*
It ultracentrifuga *(f)*
Pt ultracentrifugadora *(f)*

ultracentrifugadora *(f) n* Pt
De Ultrazentrifuge *(f)*
En ultracentrifuge
Es ultracentrífuga *(f)*
Fr ultracentrifugeuse *(f)*
It ultracentrifuga *(f)*

ultracentrifuge *n* En
De Ultrazentrifuge *(f)*
Es ultracentrífuga *(f)*
Fr ultracentrifugeuse *(f)*
It ultracentrifuga *(f)*
Pt ultracentrifugadora *(f)*

ultracentrifugeuse *(f) n* Fr
De Ultrazentrifuge *(f)*
En ultracentrifuge
Es ultracentrífuga *(f)*
It ultracentrifuga *(f)*
Pt ultracentrifugadora *(f)*

ultra-haute tension *(f)* Fr
De Ultrahochspannung *(f)*
En ultrahigh voltage (UHV)
Es tensión ultraelevada *(f)*
It tensione ultraelevata *(f)*
Pt voltagem ultra-elevada *(f)*

ultrahigh frequency (UHF) En
De Ultrahochfrequenz *(f)*
Es frecuencia ultraelevada *(f)*
Fr hyperfréquence *(f)*
It frequenza ultraelevata *(f)*
Pt frequência ultra-elevada *(f)*

ultrahigh voltage (UHV) En
De Ultrahochspannung *(f)*
Es tensión ultraelevada *(f)*
Fr ultra-haute tension *(f)*
It tensione ultraelevata *(f)*
Pt voltagem ultra-elevada *(f)*

Ultrahochfrequenz *(f) n* De
En ultrahigh frequency (UHF)
Es frecuencia ultraelevada *(f)*
Fr hyperfréquence *(f)*
It frequenza ultraelevata *(f)*
Pt frequência ultra-elevada *(f)*

Ultrahochspannung *(f) n* De
En ultrahigh voltage (UHV)
Es tensión ultraelevada *(f)*
Fr ultra-haute tension *(f)*
It tensione ultraelevata *(f)*
Pt voltagem ultra-elevada *(f)*

Ultraschall- De
En ultrasonic
Es ultrasónico
Fr ultrasonore
It ultrasonico
Pt ultra-sónico

ultrasonic *adj* En
De Ultraschall-
Es ultrasónico
Fr ultrasonore
It ultrasonico
Pt ultra-sónico

ultrasonico *adj* It
De Ultraschall-
En ultrasonic
Es ultrasónico
Fr ultrasonore
Pt ultra-sónico

ultrasónico *adj* Es
De Ultraschall-
En ultrasonic
Fr ultrasonore
It ultrasonico
Pt ultra-sónico

ultra-sónico *adj* Pt
De Ultraschall-
En ultrasonic
Es ultrasónico
Fr ultrasonore
It ultrasonico

ultrasonore *adj* Fr
De Ultraschall-
En ultrasonic
Es ultrasónico
It ultrasonico
Pt ultra-sónico

ultraviolet radiation (UV) En
De Ultraviolettstrahlung *(f)*
Es radiación ultravioleta *(f)*
Fr radiation ultraviolette *(f)*
It radiazione ultravioletta *(f)*
Pt radiação ultravioleta *(f)*

Ultraviolettstrahlung *(f) n* De
En ultraviolet radiation (UV)
Es radiación ultravioleta *(f)*
Fr radiation ultraviolette *(f)*
It radiazione ultravioletta *(f)*
Pt radiação ultravioleta *(f)*

Ultrazentrifuge *(f) n* De
En ultracentrifuge
Es ultracentrífuga *(f)*
Fr ultracentrifugeuse *(f)*
It ultracentrifuga *(f)*
Pt ultracentrifugadora *(f)*

umgekehrte Osmose *(f)* De
En reverse osmosis
Es ósmosis inversa *(f)*
Fr osmose inverse *(f)*
It osmosi inversa *(f)*
Pt osmose inversa *(f)*

umgekehrter Carnot-Prozeß *(m)* De
Am vapor compression cycle
En vapour compression cycle
Es ciclo de compresión del vapor *(m)*
Fr cycle de compression de la vapeur *(m)*
It ciclo di compressione del vapore *(m)*
Pt ciclo de compressão de vapor *(m)*

Umhüllung *(f) n* De
En cladding
Es encamisado *(m)*
Fr gainage *(m)*
It incamiciatura *(f)*
Pt revestimento *(m)*

umidità *(f) n* It
De Feuchtigkeit *(f)*
En humidity
Es humedad *(f)*
Fr humidité *(f)*
Pt humidade *(f)*

umidità assoluta *(f)* It
De absolute Feuchtigkeit *(f)*
En absolute humidity
Es humedad absoluta *(f)*
Fr humidité absolue *(f)*
Pt humidade absoluta *(f)*

umidità relativa *(f)* It
De relative Feuchtigkeit *(f)*
En relative humidity
Es humedad relativa *(f)*
Fr humidité relative *(f)*
Pt humidade relativa *(f)*

umkehrbarer Zyklus *(m)* De
En reversible cycle
Es ciclo reversible *(m)*
Fr cycle réversible *(m)*
It ciclo reversibile *(m)*
Pt ciclo reversível *(m)*

Ummantellung *(f) n* De
En lagging
Es revestimiento *(m)*
Fr revêtement calorifuge *(m)*
It rivestimento isolante *(m)*
Pt revestimento *(m)*

Umspannwerk *(n) n* De
En substation
Es subestación *(f)*
Fr sous-station *(f)*
It sottostazione *(f)*
Pt subestação *(f)*

Umstellung auf normalen Luftdruck *(f)* De
En depressurization
Es depresurización *(f)*
Fr dépressurisation *(f)*
It depressurizzazione *(f)*
Pt despressurização *(f)*

Umwelt- De
En environmental
Es ambiental
Fr de l'environnement
It ambientale
Pt ambiente

unavailable heat En
De nicht verfügbare Wärme *(f)*
Es calor no disponible *(m)*
Fr chaleur non disponible *(f)*
It calore non disponibile *(m)*
Pt calor não disponível *(m)*

unbelüftete Flamme *(f)* De
En nonaerated flame
Es llama no aireada *(f)*
Fr flamme non aérée *(f)*
It fiamma non aerata *(f)*
Pt chama não arejada *(f)*

underfrequency load-shedding En
De Unterfrequenz-Lastabschaltung *(f)*
Es restricción de la carga a baja frecuencia *(f)*
Fr délestage à sous-fréquence *(m)*
It spargimento di carica a sottofrequenza *(m)*
Pt restrição de carga a sobfrequência *(f)*

underground gasification En
De unterirdische Gasifizierung *(f)*
Es gasificación subterránea *(f)*
Fr gazéification souterraine *(f)*
It gassificazione sotterranea *(f)*
Pt gaseificação subterrânea *(f)*

underground mining En
De Grubenbetrieb *(m)*
Es minería subterránea *(f)*
Fr exploitation minière souterraine *(f)*
It scavo sotteraneo *(m)*
Pt mineração subterrânea *(f)*

Unfall durch Kühlmittelverlust *(m)* De
En loss of coolant accident (LOCA)
Es accidente por pérdida de refrigerante *(m)*
Fr accident par perte de fluide réfrigérant *(m)*
It incidente di perdita del refrigerante *(m)*
Pt acidente por perda de refrigerante *(m)*

universal gas constant En
De universelle Gaskonstante *(f)*
Es constante universal de gas *(f)*
Fr constante universelle des gaz *(f)*
It costante universale dei gas *(f)*
Pt constante universal de gás *(f)*

universal motor En
De Universalmotor *(m)*
Es motor universal *(m)*
Fr moteur universel *(m)*
It motore universale *(m)*
Pt motor universal *(m)*

Universalmotor *(m) n* De
En universal motor
Es motor universal *(m)*
Fr moteur universel *(m)*
It motore universale *(m)*
Pt motor universal *(m)*

universelle Gaskonstante *(f)* De
En universal gas constant
Es constante universal de gas *(f)*
Fr constante universelle des gaz *(f)*
It costante universale dei gas *(f)*
Pt constante universal de gás *(f)*

Unterboden *(m) n* De
En mantle (geology)
Es manto *(m)*
Fr manteau terrestre *(m)*
It manto *(m)*
Pt manto *(m)*

Unterfrequenz-Lastabschaltung *(f) n* De
En underfrequency load-shedding
Es restricción de la carga a baja frecuencia *(f)*
Fr délestage à sous-fréquence *(m)*
It spargimento di carico a sottofrequenza *(m)*
Pt restrição de carga a sobfrequência *(f)*

unterirdische Gasifizierung *(f)* De
En underground gasification
Es gasificación subterránea *(f)*
Fr gazéification souterraine *(f)*
It gassificazione sotterranea *(f)*
Pt gaseificção subterrãnea *(f)*

Unterwasser-Stromapparat *(m) n* De
En hydroseparator
Es hidroseparador *(m)*
Fr hydroséparateur *(m)*
It idroseparatore *(m)*
Pt hidroseparador *(m)*

upgrading *n* En
De Aufkonzentrierung *(f)*
Es mejoramiento *(m)*
Fr amélioration *(f)*
It promovimento a grado superiore *(m)*
Pt mejoramento *(m)*

Uraninit *(n) n* De
En uraninite
Es uraninita *(f)*
Fr uraninite *(f)*
It uraninite *(f)*
Pt uraninite *(f)*

uraninita *(f) n* Es
De Uraninit *(n)*
En uraninite
Fr uraninite *(f)*
It uraninite *(f)*
Pt uraninite *(f)*

uraninite *n* En; Fr, It, Pt *(f)*
De Uraninit *(n)*
Es uraninita *(f)*

uranio *(m) n* Es, It
De Uranium *(n)*
En uranium
Fr uranium *(m)*
Pt urânio *(m)*

urânio *(m) n* Pt
De Uranium *(n)*
En uranium
Es uranio *(m)*
Fr uranium *(m)*
It uranio *(m)*

uranium *n* En; Fr *(m)*
De Uranium *(n)*
Es uranio *(m)*
It uranio *(m)*
Pt urânio *(m)*

Uranium *(n) n* De
En uranium
Es uranio *(m)*
Fr uranium *(m)*
It uranio *(m)*
Pt urânio *(m)*

Uraniumdioxid *(n) n* De
En uranium dioxide
Es dióxido de uranio *(m)*
Fr bioxyde d'uranium *(m)*
It biossido di uranio *(m)*
Pt dióxido de urânio

uranium dioxide En
De Uraniumdioxid *(n)*
Es dióxido de uranio *(m)*
Fr bioxyde d'uranium *(m)*
It biossido di uranio *(m)*
Pt dióxido de urânio

urban solar flux En
De Stadtsonnenfluß *(m)*
Es flujo solar urbano *(m)*
Fr flux solaire urbain *(m)*
It flusso solare urbano *(m)*
Pt fluxo solar urbano *(m)*

urea-formaldehyde foam En
De Harnstoff-Formaldehyd-schaumstoff *(m)*
Es espuma de urea-formaldehido *(f)*
Fr mousse urée et formaldéhyde *(f)*
It espanso di urea-formaldeide *(m)*
Pt espuma de ureia-formaldeído *(f)*

useful heat En
De Wärmeleistung *(f)*
Es calor útil *(m)*

Fr chaleur utile *(f)*
It calore utile *(m)*
Pt calor útil *(m)*

U-value *n* En
De Wärmedurchgangs-zahl *(f)*
Es valor U *(m)*
Fr valeur U *(f)*
It valore U *(m)*
Pt valor de U *(m)*

V

vaciamiento *(m) n* Es
De Kippen *(n)*
En dumping
Fr déversement *(m)*
It pressatura *(f)*
Pt esvaziamento *(m)*

vacío *(m) n* Es
De Vakuum *(n)*
En vacuum
Fr vide *(m)*
It vuoto *(m)*
Pt vácuo *(m)*

vácuo *(m) n* Pt
De Vakuum *(n)*
En vacuum
Es vacío *(m)*
Fr vide *(m)*
It vuoto *(m)*

vacuum *n* En
De Vakuum *(n)*
Es vacío *(m)*
Fr vide *(m)*
It vuoto *(m)*
Pt vácuo *(m)*

vacuum arc furnace En
De Vakuum-Lichtbogenofen *(m)*
Es horno de arco eléctrico en vacío *(m)*
Fr four à arc dans le vide *(m)*
It forno ad arco a depressione *(m)*
Pt forno de arco voltáico em vácuo *(m)*

vacuum electric furnace En
De Vakuum-Elektroofen *(m)*
Es horno eléctrico en vacío *(m)*
Fr four électrique à vide *(m)*
It forno elettrico a depressione *(m)*
Pt forno eléctrico em vácuo *(m)*

vacuum induction furnace En
De Vakuuminduktions-ofen *(m)*
Es horno de inducción en vacío *(m)*
Fr four à induction à vide *(m)*
It forno ad induzione depressione *(m)*
Pt forno de indução em vácuo *(m)*

vacuum pump En
De Vakuumpumpe *(f)*
Es bomba de vacío *(f)*
Fr pompe à vide *(f)*
It depressore *(m)*
Pt bomba de vácuo *(f)*

Vakuum *(n) n* De
En vacuum
Es vacío *(m)*
Fr vide *(m)*
It vuoto *(m)*
Pt vácuo *(m)*

Vakuum-Elektroofen *(m) n* De
En vacuum electric furnace
Es horno eléctrico en vacío *(m)*
Fr four électrique à vide *(m)*
It forno elettrico a depressione *(m)*
Pt forno eléctrico em vácuo *(m)*

Vakuuminduktionsofen *(m) n* De
En vacuum induction furnace
Es horno de inducción en vacío *(m)*
Fr four à induction à vide *(m)*
It forno ad induzione depressione *(m)*
Pt forno de indução em vácuo *(m)*

Vakuum-Lichtbogenofen *(m) n* De
En vacuum arc furnace
Es horno de arco eléctrico en vacío *(m)*
Fr four à arc dans le vide *(m)*
It forno ad arco a depressione *(m)*
Pt forno de arco voltáico em vácuo *(m)*

Vakuumpumpe *(f) n* De
En vacuum pump
Es bomba de vacío *(f)*
Fr pompe à vide *(f)*
It depressore *(m)*
Pt bomba de vácuo *(f)*

valeur calorifique *(f)* Fr
De Heizwert *(m)*
En calorific value
Es potencia calorífica *(f)*
It potere calorifico *(m)*
Pt valor calorífico *(m)*

valeur de seuil *(f)* Fr
De Schwellenwert *(m)*
En threshold limit value (TLV)
Es valor límite de umbral *(m)*
It valore del limite di soglia *(m)*
Pt valor de limite de limiar *(m)*

valeur efficace *(f)* Fr
De Effektivwert *(m)*
En effective value
Es valor eficaz *(m)*
It valore efficace *(m)*
Pt valor eficaz *(m)*

valeur K *(f)* Fr
De K-Wert *(m)*
En K-value
Es valor K *(m)*
It valore K *(m)*
Pt valor de K *(m)*

valeur R *(f)* Fr
De R-Wert *(m)*
En R-value
Es valor R *(m)*
It valore R *(m)*
Pt valor de R *(m)*

valeur U *(f)* Fr
De Wärmedurchgangs-zahl *(f)*
En U-value
Es valor U *(m)*
It valore U *(m)*
Pt valor de U *(m)*

valor calorífico *(m)* Pt
De Heizwert *(m)*
En calorific value
Es potencia calorífica *(f)*
Fr valeur calorifique *(f)*
It potere calorifico *(m)*

valor de K *(m)* Pt
De K-Wert *(m)*
En K-value
Es valor K *(m)*
Fr valeur K *(f)*
It valore K *(m)*

valor de limite de limiar *(m)* Pt
De Schwellenwert *(m)*
En threshold limit value (TLV)
Es valor límite de umbral *(m)*
Fr valeur de seuil *(f)*
It valore del limite di soglia *(m)*

valor de R *(m)* Pt
De R-Wert *(m)*
En R-value
Es valor R *(m)*
Fr valeur R *(f)*
It valore R *(m)*

valor de U *(m)* Pt
De Wärmedurchgangs-zahl *(f)*
En U-value
Es valor U *(m)*
Fr valeur U *(f)*
It valore U *(m)*

valore del limite di soglia *(m)* It
De Schwellenwert *(m)*
En threshold limit value (TLV)

Es valor límite de umbral *(m)*
Fr valeur de seuil *(f)*
Pt valor de limite de limiar *(m)*

valore efficace *(m)* It
De Effektivwert *(m)*
En effective value
Es valor eficaz *(m)*
Fr valeur efficace *(f)*
Pt valor eficaz *(m)*

valor eficaz *(m)* Es, Pt
De Effektivwert *(m)*
En effective value
Fr valeur efficace *(f)*
It valore efficace *(m)*

valore K *(m)* It
De K-Wert *(m)*
En K-value
Es valor K *(m)*
Fr valeur K *(f)*
Pt valor de K *(m)*

valore R *(m)* It
De R-Wert *(m)*
En R-value
Es valor R *(m)*
Fr valeur R *(f)*
Pt valor de R *(m)*

valore U *(m)* It
De Wärmedurchgangs-zahl *(f)*
En U-value
Es valor U *(m)*
Fr valeur U *(f)*
Pt valor de U *(m)*

valor K *(m)* Es
De K-Wert *(m)*
En K-value
Fr valeur K *(f)*
It valore K *(m)*
Pt valor de K *(m)*

valor límite de umbral *(m)* Es
De Schwellenwert *(m)*
En threshold limit value (TLV)
Fr valeur de seuil *(f)*
It valore del limite di soglia *(m)*
Pt valor de limite de limiar *(m)*

valor R *(m)* Es
De R-Wert *(m)*
En R-value
Fr valeur R *(f)*
It valore R *(m)*
Pt valor de R *(m)*

valor U *(m)* Es
De Wärmedurchgangs-zahl *(f)*
En U-value
Fr valeur U *(f)*
It valore U *(m)*
Pt valor de U *(m)*

valve *n* En
De Ventil *(n)*
Es válvula *(f)*
Fr soupape *(f)*
It valvola *(f)*
Pt válvula *(f)*

valvola *(f) n* It
De Ventil *(n)*
En valve
Es válvula *(f)*
Fr soupape *(f)*
Pt válvula *(f)*

valvola di sicurezza *(f)* It
De Sicherheitsventil *(n)*
En safety valve
Es válvula de seguridad *(f)*
Fr soupape de sûreté *(f)*
Pt válvula de segurança *(f)*

válvula *(f) n* Es
De Ventil *(n)*
En valve
Fr soupape *(f)*
It valvola *(f)*
Pt válvula *(f)*

válvula *(f) n* Pt
De Ventil *(n)*
En valve
Es válvula *(f)*
Fr soupape *(f)*
It valvola *(f)*

válvula de raios catódicos *(f)* Pt
De Kathodenstrahlröhre *(f)*
En cathode-ray tube
Es tubo de rayos catódicos *(m)*
Fr tube à rayons cathodiques *(m)*
It tubo a raggi catodici *(m)*

válvula de segurança *(f)* Pt
De Sicherheitsventil *(n)*
En safety valve
Es válvula de seguridad *(f)*
Fr soupape de sûreté *(f)*
It valvola di sicurezza *(f)*

válvula de seguridad *(f)* Es
De Sicherheitsventil *(n)*
En safety valve
Fr soupape de sûreté *(f)*
It valvola di sicurezza *(f)*
Pt válvula de segurança *(f)*

vanadio *(m) n* Es, It
De Vanadium *(n)*
En vanadium
Fr vanadium *(m)*
Pt vanádio *(m)*

vanádio *(m) n* Pt
De Vanadium *(n)*
En vanadium
Es vanadio *(m)*
Fr vanadium *(m)*
It vanadio *(m)*

vanadium *n* En; Fr *(m)*
De Vanadium *(n)*
Es vanadio *(m)*
It vanadio *(m)*
Pt vanádio *(m)*

Vanadium *(n) n* De
En vanadium
Es vanadio *(m)*
Fr vanadium *(m)*
It vanadio *(m)*
Pt vanádio *(m)*

vane anemometer En
De Flügelrad-Windmesser *(m)*
Es anemómetro de catavientos *(m)*
Fr anémomètre à ailettes *(m)*
It anemometro a mulinello *(m)*
Pt anemómetro de aletas *(m)*

vapeur *(f) n* Fr
Am steam; vapor
De Dampf *(m)*
En steam; vapour
Es vapor *(m)*
It vapore *(m)*
Pt vapor *(m)*

vapeur magmatique *(f)* Fr
De Magmadampf *(f)*
En magmatic steam
Es vapor magmático *(m)*
It vapore magmatico *(m)*
Pt vapor magmático *(m)*

vapeur météorique *(f)* Fr
De meteoritischer Dampf *(m)*
En meteoritic steam
Es vapor meteorítico *(m)*
It vapore meteoritico *(m)*
Pt vapor meteórico *(m)*

vapeur saturée *(f)* Fr
De gesättigter Dampf *(m)*
En saturated steam
Es vapor saturado *(m)*
It vapore saturo *(m)*
Pt vapor saturado *(m)*

vapeur surchauffée *(f)* Fr
De überhitzter Dampf *(m)*
En superheated steam
Es vapor supercalentado *(m)*
It vapore surriscaldato *(m)*
Pt vapor superaquecido *(m)*

vapor *n* Am
De Dampf *(m)*
En vapour
Es vapor *(m)*
Fr vapeur *(f)*
It vapore *(m)*
Pt vapor *(m)*

vapor *(m) n* Es, Pt
Am steam; vapor
De Dampf *(m)*
En steam; vapour
Fr vapeur *(f)*
It vapore *(m)*

vapor barrier Am
De Dampfsperre *(f)*
En vapour barrier
Es barrera del vapor *(f)*
Fr écran d'étanchéité à la vapeur *(m)*
It schermo per il vapore *(m)*
Pt barreira de vapor *(f)*

vapor compression cycle Am
De umgekehrter Carnot-Prozeβ *(m)*
En vapour compression cycle
Es ciclo de compresión del vapor *(m)*
Fr cycle de compression de la vapeur *(m)*
It ciclo di compressione del vapore *(m)*
Pt ciclo de compressão de vapor *(m)*

vapor density Am
De Dampfdichte *(f)*
En vapour density
Es densidad del vapor *(f)*
Fr densité de vapeur *(f)*
It densità del vapore *(f)*
Pt densidade de vapor *(f)*

vapore *(m) n* It
Am steam; vapor
De Dampf *(m)*
En steam; vapour
Es vapor *(m)*
Fr vapeur *(f)*
Pt vapor *(m)*

vapore magmatico *(m)* It
De Magmadampf *(f)*
En magmatic steam
Es vapor magmático *(m)*
Fr vapeur magmatique *(f)*
Pt vapor magmático *(m)*

vapore meteoritico *(m)* It
De meteoritischer Dampf *(m)*
En meteoritic steam
Es vapor meteorítico *(m)*
Fr vapeur météorique *(f)*
Pt vapor meteórico *(m)*

vapore saturo *(m)* It
De gesättigter Dampf *(m)*
En saturated steam
Es vapor saturado *(m)*
Fr vapeur saturée *(f)*
Pt vapor saturado *(m)*

vapore surriscaldato *(m)* It
De überhitzter Dampf *(m)*
En superheated steam
Es vapor supercalentado *(m)*
Fr vapeur surchauffée *(f)*
Pt vapor superaquecido *(m)*

vapor magmático *(m)* Es, Pt
De Magmadampf *(f)*
En magmatic steam
Fr vapeur magmatique *(f)*
It vapore magmatico *(m)*

vapor meteórico *(m)* Pt
De meteoritischer Dampf *(m)*
En meteoritic steam
Es vapor meteorítico *(m)*
Fr vapeur météorique *(f)*
It vapore meteoritico *(m)*

vapor meteorítico *(m)* Es
De meteoritischer Dampf *(m)*
En meteoritic steam
Fr vapeur météorique *(f)*
It vapore meteoritico *(m)*
Pt vapor meteórico *(m)*

vapor pressure Am
De Dampfdruck *(m)*
En vapour pressure
Es presión del vapor *(f)*
Fr tension de vapeur *(f)*
It tensione di vapore *(f)*
Pt pressão de vapor *(f)*

vapor saturado *(m)* Es, Pt
De gesättigter Dampf *(m)*
En saturated steam
Fr vapeur saturée *(f)*
It vapore saturo *(m)*

vapor superaquecido *(m)* Pt
De überhitzter Dampf *(m)*
En superheated steam
Es vapor supercalentado *(m)*
Fr vapeur surchauffée *(f)*
It vapore surriscaldato *(m)*

vapor supercalentado *(m)* Es
De überhitzter Dampf *(m)*
En superheated steam
Fr vapeur surchauffée *(f)*
It vapore surriscaldato *(m)*
Pt vapor superaquecido *(m)*

vapour *n* En
Am vapor
De Dampf *(m)*
Es vapor *(m)*
Fr vapeur *(f)*
It vapore *(m)*
Pt vapor *(m)*

vapour barrier En
Am vapor barrier
De Dampfsperre *(f)*
Es barrera del vapor *(f)*
Fr écran d'étanchéité à la vapeur *(m)*
It schermo per il vapore *(m)*
Pt barreira de vapor *(f)*

vapour compression cycle En
Am vapor compression cycle
De umgekehrter Carnot-Prozeβ *(m)*
Es ciclo de compresión del vapor *(m)*
Fr cycle de compression de la vapeur *(m)*
It ciclo di compressione del vapore *(m)*
Pt ciclo de compressão de vapor *(m)*

vapour density En
Am vapor density
De Dampfdichte *(f)*
Es densidad del vapor *(f)*
Fr densité de vapeur *(f)*
It densità del vapore *(f)*
Pt densidade de vapor *(f)*

vapour pressure En
Am vapor pressure
De Dampfdruck *(m)*
Es presión del vapor *(f)*
Fr tension de vapeur *(f)*
It tensione di vapore *(f)*
Pt pressão de vapor *(f)*

vara de combustível *(f)* Pt
De Brennstoffstab *(m)*
En fuel rod
Es barra de combustible *(f)*
Fr barre de combustible *(f)*
It barra combustibile *(f)*

varilla de control *(f)* Es
De Kontrollstab *(m)*
En control rod
Fr barre de commande *(f)*
It barra di controllo *(f)*
Pt haste de controle *(f)*

vector quantity En
De Vektorquantität *(f)*
Es cantidad vectorial *(f)*
Fr quantité vectorielle *(f)*
It quantità di vettore *(f)*
Pt quantidade de vector *(f)*

vedação *(f) n* Pt
De Abdichtung *(f)*
En seal
Es cierre *(m)*
Fr joint *(m)*
It dispositivo di tenuta *(m)*

véhicule diesel routier *(m)* Fr
De Straβenfahrzeug mit Dieselmotor *(n)*
En diesel engine road vehicle (DERV)
Es vehículo para carretera con motor diesel *(m)*
It automezzo con motore a nafta *(m)*
Pt veículo rodoviário de motor diesel *(m)*

véhicule électrique *(m)* Fr
De Elektrofahrzeug *(n)*
En electric vehicle
Es vehículo eléctrico *(m)*
It veicolo elettrico *(m)*
Pt veículo eléctrico *(m)*

vehículo eléctrico *(m)* Es
De Elektrofahrzeug *(n)*
En electric vehicle
Fr véhicule électrique *(m)*
It veicolo elettrico *(m)*
Pt veículo eléctrico *(m)*

vehículo para carretera con motor diesel *(m)* Es
De Straβenfahrzeug mit Dieselmotor *(n)*
En diesel engine road vehicle (DERV)
Fr véhicule diesel routier *(m)*
It automezzo con motore a nafta *(m)*
Pt veículo rodoviário de motor diesel *(m)*

veicolo elettrico *(m)* It
De Elektrofahrzeug *(n)*
En electric vehicle
Es vehículo eléctrico *(m)*
Fr véhicule électrique *(m)*
Pt veículo eléctrico *(m)*

veículo eléctrico *(m)* Pt
De Elektrofahrzeug *(n)*
En electric vehicle
Es vehículo eléctrico *(m)*
Fr véhicule électrique *(m)*
It veicolo elettrico *(m)*

veículo rodoviário de motor diesel *(m)* Pt
De Straβenfahrzeug mit Dieselmotor *(n)*
En diesel engine road vehicle (DERV)
Es vehículo para carretera con motor diesel *(m)*
Fr véhicule diesel routier *(m)*
It automezzo con motore a nafta *(m)*

veilleuse *(f) n* Fr
De Zündflamme *(f)*
En pilot flame
Es llama piloto *(f)*
It semprevivo *(m)*
Pt chama piloto *(f)*

Vektorquantität *(f) n* De
En vector quantity
Es cantidad vectorial *(f)*
Fr quantité vectorielle *(f)*
It quantità di vettore *(f)*
Pt quantidade de vector *(f)*

velocidad *(f) n* Es
De Geschwindigkeit *(f)*
En speed; velocity
Fr vitesse *(f)*
It velocità *(f)*
Pt velocidade *(f)*

velocidad *(f) n* Es
De Geschwindigkeit *(f)*
En velocity
Fr vitesse *(f)*
It velocità *(f)*
Pt velocidade *(f)*

velocidad de la llama *(f)* Es
De Flammengeschwindigkeit *(f)*
En flame speed
Fr vitesse de flamme *(f)*
It velocità della fiamma *(f)*
Pt velocidade de chama *(f)*

velocidade *(f) n* Pt
De Geschwindigkeit *(f)*
En speed; velocity
Es velocidad *(f)*
Fr vitesse *(f)*
It velocità *(f)*

velocidade de chama *(f)* Pt
De Flammengeschwindigkeit *(f)*
En flame speed
Es velocidad de la llama *(f)*
Fr vitesse de flamme *(f)*
It velocità della fiamma *(f)*

velocità *(f) n* It
De Geschwindigkeit *(f)*
En speed; velocity
Es velocidad *(f)*
Fr vitesse *(f)*
Pt velocidade *(f)*

velocità della fiamma *(f)* It
De Flammengeschwindigkeit *(f)*
En flame speed
Es velocidad de la llama *(f)*
Fr vitesse de flamme *(f)*
Pt velocidade de chama *(f)*

velocity *n* En
De Geschwindigkeit *(f)*
Es velocidad *(f)*
Fr vitesse *(f)*
It velocità *(f)*
Pt velocidade *(f)*

vent *(m) n* Fr
De Wind *(m)*
En wind
Es viento *(m)*
It vento *(m)*
Pt vento *(m)*

ventanilla de absorción *(f)* Es
De Absorptionsfenster *(n)*
En absorption window
Fr fenêtre d'absorption *(f)*
It finestra di assorbimento *(f)*
Pt janela de absorção *(f)*

Ventil *(n) n* De
En valve
Es válvula *(f)*
Fr soupape *(f)*
It valvola *(f)*
Pt válvula *(f)*

ventilação *(f) n* Pt
De Belüftung *(f)*
En ventilation
Es ventilación *(f)*
Fr ventilation *(f)*
It ventilazione *(f)*

ventilación *(f) n* Es
De Belüftung *(f)*
En ventilation
Fr ventilation *(f)*
It ventilazione *(f)*
Pt ventilação *(f)*

ventilador *(m) n* Es
De Gebläse *(n)*
En fan
Fr ventilateur *(m)*
It ventilatore *(m)*
Pt ventoinha *(f)*

ventilador de extracción *(m)* Es
De Absauggebläse *(n)*
En extraction fan
Fr ventilateur d'extraction *(m)*
It ventilatore ad estrazione *(m)*
Pt ventoínha de extracção *(f)*

ventilateur *(m) n* Fr
De Gebläse *(n)*
En fan
Es ventilador *(m)*
It ventilatore *(m)*
Pt ventoinha *(f)*

ventilateur d'extraction *(m)* Fr
De Absauggebläse *(n)*
En extraction fan
Es ventilador de extracción *(m)*
It ventilatore ad estrazione *(m)*
Pt ventoínha de extracção *(f)*

ventilation *n* En; Fr *(f)*
De Belüftung *(f)*
Es ventilación *(f)*
It ventilazione *(f)*
Pt ventilação *(f)*

ventilatore *(m) n* It
De Gebläse *(n)*
En fan
Es ventilador *(m)*
Fr ventilateur *(m)*
Pt ventoinha *(f)*

ventilatore ad estrazione *(m)* It
De Absauggebläse *(n)*
En extraction fan
Es ventilador de extracción *(m)*
Fr ventilateur d'extraction *(m)*
Pt ventoínha de extracção *(f)*

ventilazione *(f)* *n* It
De Belüftung *(f)*
En ventilation
Es ventilación *(f)*
Fr ventilation *(f)*
Pt ventilação *(f)*

vento *(m)* *n* It, Pt
De Wind *(m)*
En wind
Es viento *(m)*
Fr vent *(m)*

ventoinha *(f)* *n* Pt
De Gebläse *(n)*
En fan
Es ventilador *(m)*
Fr ventilateur *(m)*
It ventilatore *(m)*

ventoínha de extracção *(f)* Pt
De Absauggebläse *(n)*
En extraction fan
Es ventilador de extracción *(m)*
Fr ventilateur d'extraction *(m)*
It ventilatore ad estrazione *(m)*

vento solar *(m)* Pt
De Sonnenwind *(m)*
En solar wind
Es viento solar *(m)*
Fr vent solaire *(m)*
It vento solare *(m)*

vento solare *(m)* It
De Sonnenwind *(m)*
En solar wind
Es viento solar *(m)*
Fr vent solaire *(m)*
Pt vento solar *(m)*

vent solaire *(m)* Fr
De Sonnenwind *(m)*
En solar wind
Es viento solar *(m)*
It vento solare *(m)*
Pt vento solar *(m)*

Venturimesser *(m)* *n* De
En Venturi meter
Es medidor Venturi *(m)*
Fr compteur Venturi *(m)*
It venturimetro *(m)*
Pt contador de Venturi *(m)*

Venturi meter En
De Venturimesser *(m)*
Es medidor Venturi *(m)*
Fr compteur Venturi *(m)*
It venturimetro *(m)*
Pt contador de Venturi *(m)*

venturimetro *(m)* *n* It
De Venturimesser *(m)*
En Venturi meter
Es medidor Venturi *(m)*
Fr compteur Venturi *(m)*
Pt contador de Venturi *(m)*

Verbrennung *(f)* *n* De
En combustion
Es combustión *(f)*
Fr combustion *(f)*
It combustione *(f)*
Pt combustão *(f)*

Verbrennungsmotor *(m)* *n* De
En internal-combustion engine
Es motor de combustión interna *(m)*
Fr moteur à combustion interne *(m)*
It motore a scoppio *(m)*
Pt motor de combustão interna *(m)*

Verbrennungswärme *(f)* *n* De
En heat of combustion
Es calor de combustión *(m)*
Fr chaleur de combustion *(f)*
It calore di combustione *(m)*
Pt calor de combustão *(m)*

Verdunstung *(f)* *n* De
En evaporation
Es evaporación *(f)*
Fr évaporation *(f)*
It evaporazione *(f)*
Pt evaporação *(f)*

verflüssigter Feststoff *(m)* De
En fluidized solid
Es sólido fluidizado *(m)*
Fr solide fluidifié *(m)*
It solido fluidizzato *(m)*
Pt sólido fluidificado *(m)*

verflüssigtes Erdgas *(n)* De
En liquefied natural gas (LNG)
Es gas natural licuado (GNL) *(m)*
Fr gaz naturel liquéfié (GNL) *(m)*
It gas naturale liquido *(m)*
Pt gás natural liquefeito *(m)*

verflüssigtes Petroleumgas *(n)* De
En liquefied petroleum gas (LPG)
Es gas licuado del petróleo (GLP) *(m)*
Fr gaz de pétrole liquéfié (GPL) *(m)*
It gas liquido di petrolio *(m)*
Pt gás de petróleo liquefeito *(m)*

Verflüssigung *(f)* *n* De
En liquefaction
Es licuación *(f)*
Fr liquéfaction *(f)*
It liquefazione *(f)*
Pt liquefacção *(f)*

Vergaser *(n)* *n* De
En carburettor
Es carburador *(m)*
Fr carburateur *(m)*
It carburatore *(m)*
Pt carburador *(m)*

Vergasung *(f)* *n* De
En gasification
Es gasificación *(f)*
Fr gazéification *(f)*
It gassificazione *(f)*
Pt gaseificação *(f)*

verglast *adj* De
En glassified
Es vitrificado
Fr vitrifié
It vetrificato
Pt vitrificado

Verläslichkeitstheorie *(f)* *n* De
En reliability theory
Es teoría de la fiabilidad *(f)*
Fr théorie de fiabilité *(f)*
It teoria dell'affidabilità *(f)*
Pt teoria da segurança *(f)*

Vermiculit *(m)* *n* De
En vermiculite
Es vermiculita *(f)*
Fr vermiculite *(f)*
It vermiculite *(f)*
Pt vermiculite *(f)*

vermiculita *(f)* *n* Es
De Vermiculit *(m)*
En vermiculite
Fr vermiculite *(f)*
It vermiculite *(f)*
Pt vermiculite *(f)*

vermiculite *n* En; Fr, It, Pt *(f)*
De Vermiculit *(m)*
Es vermiculita *(f)*

Verschmutzung *(f)* *n* De
En pollution
Es contaminación *(f)*
Fr pollution *(f)*
It inquinamento *(m)*
Pt poluição *(f)*

Verstopfung *(f)* *n* De
En fouling
Es incrustación *(f)*
Fr encrassement *(m)*
It incrostazione *(f)*
Pt incrustação *(m)*

Versuchsanlage *(f)* *n* De
En pilot plant
Es planta piloto *(f)*
Fr installation pilote *(f)*
It impianto pilota *(m)*
Pt fábrica piloto *(f)*

vertical-axis windmill En
De Windmühle mit Hochachse *(f)*
Es molino de viento de eje vertical *(m)*
Fr moulin à vent à axe vertical *(m)*
It mulino a vento a asse verticale *(m)*
Pt moínho de vento de eixo vertical *(m)*

Vertorfung *(f) n* De
En peatification
Es formación de turba *(f)*
Fr formation de la tourbe *(f)*
It produzione di torba *(f)*
Pt turfização *(f)*

Vertriebskosten *(pl) n* De
En distribution cost
Es coste de distribución *(m)*
Fr frais de distribution *(m)*
It costo di distribuzione *(m)*
Pt custo de distribuição *(m)*

verzögertes Neutron *(n)* De
En delayed neutron
Es neutrón diferido *(m)*
Fr neutron retardé *(m)*
It neutrone ritardato *(m)*
Pt neutrão retardado *(m)*

verzögerte Verkokung *(f)* De
En delayed coking
Es coquización retardada *(f)*
Fr cokéification différée *(f)*
It caricamento ritardato *(m)*
Pt coqueação retardada *(f)*

vetrificato *adj* It
De verglast
En glassified (nuclear waste)
Es vitrificado
Fr vitrifié
Pt vitrificado

vetrificazione *(f) n* It
De Vitrifizierung *(f)*
En vitrification
Es vitrificación *(f)*
Fr vitrification *(f)*
Pt vitrificação *(f)*

vida de neutrão *(f)* Pt
De Neutronenlebensdauer *(f)*
En neutron lifetime
Es vida de neutrón *(f)*
Fr durée de vie des neutrons *(f)*
It vita dei neutroni *(f)*

vida de neutrón *(f)* Es
De Neutronenlebensdauer *(f)*
En neutron lifetime
Fr durée de vie des neutrons *(f)*
It vita dei neutroni *(f)*
Pt vida de neutrão *(f)*

vide *(m) n* Fr
De Vakuum *(n)*
En vacuum
Es vacío *(m)*
It vuoto *(m)*
Pt vácuo *(m)*

vieillissement thermique *(m)* Fr
De thermische Alterung *(f)*
En thermal ageing
Es envejecimiento térmico *(m)*
It invecchiamento termico *(m)*
Pt envelhecimento térmico *(m)*

viento *(m) n* Es
De Wind *(m)*
En wind
Fr vent *(m)*
It vento *(m)*
Pt vento *(m)*

viento solar *(m)* Es
De Sonnenwind *(m)*
En solar wind
Fr vent solaire *(m)*
It vento solare *(m)*
Pt vento solar *(m)*

Viertaktmotor *(m) n* De
En four-stroke engine
Es motor de cuatro tiempos *(m)*
Fr moteur à quatre temps *(m)*
It motore a quattro tempi *(m)*
Pt motor a quatro tempos *(m)*

virtual work En
De virtuelle Arbeit *(f)*
Es trabajo virtual *(m)*
Fr travail virtuel *(m)*
It lavoro virtuale *(m)*
Pt trabalho virtual *(m)*

virtuelle Arbeit *(f)* De
En virtual work
Es trabajo virtual *(m)*
Fr travail virtuel *(m)*
It lavoro virtuale *(m)*
Pt trabalho virtual *(m)*

viscosidad *(f) n* Es
De Viskosität *(f)*
En viscosity
Fr viscosité *(f)*
It viscosità *(f)*
Pt viscosidade *(f)*

viscosidade *(f) n* Pt
De Viskosität *(f)*
En viscosity
Es viscosidad *(f)*
Fr viscosité *(f)*
It viscosità *(f)*

viscosità *(f) n* It
De Viskosität *(f)*
En viscosity
Es viscosidad *(f)*
Fr viscosité *(f)*
Pt viscosidade *(f)*

viscosité *(f) n* Fr
De Viskosität *(f)*
En viscosity
Es viscosidad *(f)*
It viscosità *(f)*
Pt viscosidade *(f)*

viscosity *n* En
De Viskosität *(f)*
Es viscosidad *(f)*
Fr viscosité *(f)*
It viscosità *(f)*
Pt viscosidade *(f)*

visible spectrum En
De sichtbares Spektrum *(n)*
Es espectro visible *(m)*
Fr spectre visible *(m)*
It spettro visibile *(m)*
Pt espectro visível *(m)*

Viskosität *(f) n* De
En viscosity
Es viscosidad *(f)*
Fr viscosité *(f)*
It viscosità *(f)*
Pt viscosidade *(f)*

vita dei neutroni *(f)* It
De Neutronenlebensdauer *(f)*
En neutron lifetime
Es vida de neutrón *(f)*
Fr durée de vie des neutrons *(f)*
Pt vida de neutrão *(f)*

vitesse *(f) n* Fr
De Geschwindigkeit *(f)*
En speed; velocity
Es velocidad *(f)*
It velocità *(f)*
Pt velocidade *(f)*

vitesse de flamme *(f)* Fr
De Flammengeschwindigkeit *(f)*
En flame speed
Es velocidad de la llama *(f)*
It velocità della fiamma *(f)*
Pt velocidade de chama *(f)*

vitrificação *(f) n* Pt
De Vitrifizierung *(f)*
En vitrification
Es vitrificación *(f)*
Fr vitrification *(f)*
It vetrificazione *(f)*

vitrificação dupla *(f)* Pt
De Doppelverglasung *(f)*
En double glazing
Es encristalado doble *(m)*
Fr double vitrage *(m)*
It doppia vetratura *(f)*

vitrificação dupla de janela secundária *(f)* Pt
De Doppelverglasung für sekundäre Fenster *(f)*
En secondary-window double glazing
Es doble encristalado de ventana secundaria *(m)*
Fr double vitrage secondaire *(f)*
It doppia vetrata di finestre secondarie *(f)*

vitrificação dupla de unidade vedada *(f)* Pt
De luftdichte

Doppelverglasungs-elemente *(pl)*
En sealed-unit double glazing
Es doble encristalado de unidad sellada *(m)*
Fr double vitrage à éléments scellés *(m)*
It doppia vetrata sigillata *(f)*

vitrificação tripla *(f)* Pt
De Dreifachverglasung *(f)*
En triple glazing n
Es encristalado triple *(m)*
Fr triple vitrage *(m)*
It tripla vetrata *(f)*

vitrificación *(f) n* Es
De Vitrifizierung *(f)*
En vitrification
Fr vitrification *(f)*
It vetrificazione *(f)*
Pt vitrificação *(f)*

vitrificado *adj* Es, Pt
De verglast
En glassified
Fr vitrifié
It vetrificato

vitrification *n* En; Fr *(f)*
De Vitrifizierung *(f)*
Es vitrificación *(f)*
It vetrificazione *(f)*
Pt vitrificação *(f)*

vitrifié *adj* Fr
De verglast
En glassified
Es vitrificado
It vetrificato
Pt vitrificado

Vitrifizierung *(f) n* De
En vitrification
Es vitrificación *(f)*
Fr vitrification *(f)*
It vetrificazione *(f)*
Pt vitrificação *(f)*

volano *(m) n* It
De Schwungrad *(n)*
En flywheel
Es volante *(m)*
Fr volant *(m)*
Pt volante de inércia *(m)*

volant *(m) n* Fr
De Schwungrad *(n)*
En flywheel
Es volante *(m)*
It volano *(m)*
Pt volante de inércia *(m)*

volante *(m) n* Es
De Schwungrad *(n)*
En flywheel
Fr volant *(m)*
It volano *(m)*
Pt volante de inércia *(m)*

volante de inércia *(m)* Pt
De Schwungrad *(n)*
En flywheel
Es volante *(m)*
Fr volant *(m)*
It volano *(m)*

volatile matter En
De flüchtiger Bestandteil *(m)*
Es materia volátil *(f)*
Fr matière volatile *(f)*
It sostanza volatile *(f)*
Pt matéria volátil *(f)*

voltage *n* En
De Spannung *(f)*
Es tensión *(f)*
Fr tension *(f)*
It tensione *(f)*
Pt voltagem *(f)*

voltage divider En
De Spannungsteiler *(m)*
Es divisor de potencial; divisor de tensión *(m)*
Fr diviseur de tension *(m)*
It partitore di tensione *(m)*
Pt divisor de potencial; divisor de tensão *(m)*

voltagem *(f) n* Pt
De Spannung *(f)*
En voltage
Es tensión *(f)*
Fr tension *(f)*
It tensione *(f)*

voltagem ultra-elevada *(f)* Pt
De Ultrahochspannung *(f)*
En ultrahigh voltage (UHV)
Es tensión ultraelevada *(f)*
Fr ultra-haute tension *(f)*
It tensione ultraelevata *(f)*

voltage regulator En
De Spannungsregler *(m)*
Es regulador de tensión *(m)*
Fr régulateur de tension *(m)*
It regolatore di tensione *(m)*
Pt regulador de tensão *(m)*

volt-electrónico *(m) n* Pt
De Elektronenvolt (eV) *(n)*
En electronvolt (eV)
Es electronvoltio (eV) *(m)*
Fr électron-volt (eV) *(m)*
It volt-elettrone *(m)*

volt-elettrone *(m) n* It
De Elektronenvolt (eV) *(n)*
En electronvolt (eV)
Es electronvoltio (eV) *(m)*
Fr électron-volt (eV) *(m)*
Pt volt-electrónico *(m)*

voltimetro *(m) n* It
De Voltmesser *(m)*
En voltmeter
Es voltímetro *(m)*
Fr voltmètre *(m)*
Pt voltímetro *(m)*

voltímetro *(m) n* Es, Pt
De Voltmesser *(m)*
En voltmeter
Fr voltmètre *(m)*
It voltimetro *(m)*

Voltmesser *(m) n* De
En voltmeter
Es voltímetro *(m)*
Fr voltmètre *(m)*
It voltimetro *(m)*
Pt voltímetro *(m)*

voltmeter *n* En
De Voltmesser *(m)*
Es voltímetro *(m)*
Fr voltmètre *(m)*
It voltimetro *(m)*
Pt voltímetro *(m)*

voltmètre *(m) n* Fr
De Voltmesser *(m)*
En voltmeter
Es voltímetro *(m)*
It voltimetro *(m)*
Pt voltímetro *(m)*

Vorbehandlung *(f) n* De
En pretreatment
Es tratamiento previo *(m)*
Fr traitement préalable *(m)*
It trattamento previo *(m)*
Pt tratamento prévio *(m)*

vorbelüfteter Brenner *(m)* De
En pre-aerated burner
Es quemador preaireado *(m)*
Fr brûleur à pré-aération *(m)*
It bruciatore pre-aerato *(m)*
Pt queimador pré-arejado *(m)*

Vorheizer *(m) n* De
En preheater
Es precalentador *(m)*
Fr pré-chauffeur *(m)*
It preriscaldatore *(m)*
Pt pré-aquecedor *(m)*

Vorrat *(m) n* De
En stockpile
Es existencia para emergencia *(f)*
Fr stock de réserves *(m)*
It riserva *(f)*
Pt pilha de armazenagem *(f)*

vuoto *(m) n* It
De Vakuum *(n)*
En vacuum
Es vacío *(m)*
Fr vide *(m)*
Pt vácuo *(m)*

W

Wabenstruktur *(f) n* De
- En honeycomb structure
- Es estructura alveolar *(f)*
- Fr structure en nids d'abeilles *(f)*
- It structura a nido d'ape *(f)*
- Pt estrutura alveolar *(f)*

wahrscheinliche Reserven *(pl)* De
- En probable reserves *(pl)*
- Es reservas probables *(f pl)*
- Fr réserves probables *(f pl)*
- It riserve probabili *(f pl)*
- Pt reservas prováveis *(f pl)*

Wahrscheinlichkeit *(f) n* De
- En probability
- Es probabilidad *(f)*
- Fr probabilité *(f)*
- It probabilità *(f)*
- Pt probabilidade *(f)*

wall insulation En
- De Wandisolierung *(f)*
- Es aislamiento de paredes *(m)*
- Fr isolement des murs *(m)*
- It isolamento delle pareti *(m)*
- Pt isolamento de paredes *(m)*

Wandisolierung *(f) n* De
- En wall insulation
- Es aislamiento de paredes *(m)*
- Fr isolement des murs *(m)*
- It isolamento delle pareti *(m)*
- Pt isolamento de paredes *(m)*

Wankel engine En
- De Wankel-Motor *(m)*
- Es motor Wankel *(m)*
- Fr moteur Wankel *(m)*
- It motore di Wankel *(m)*
- Pt motor Wankel *(m)*

Wankel-Motor *(m) n* De
- En Wankel engine
- Es motor Wankel *(m)*
- Fr moteur Wankel *(m)*
- It motore di Wankel *(m)*
- Pt motor Wankel *(m)*

warm-air heater En
- De Warmluftheizung *(f)*
- Es calentador con aire caliente *(m)*
- Fr chauffage à air chaud *(m)*
- It riscaldatore ad aria tiepida *(m)*
- Pt aquecedor de ar quente *(m)*

Warmbehandlung *(f) n* De
- En heat treatment
- Es tratamiento térmico *(m)*
- Fr traitement thermique *(m)*
- It trattamento termico *(m)*
- Pt tratamento térmico *(m)*

Wärme *(f) n* De
- En heat
- Es calor *(m)*
- Fr chaleur *(f)*
- It calore *(m)*
- Pt calor *(m)*

Wärmearbeitszyklus *(m) n* De
- En heat-work cycle
- Es ciclo de calor-trabajo *(m)*
- Fr cycle de chaleur-travail *(m)*
- It ciclo calore-lavoro *(m)*
- Pt ciclo de calor-trabalho *(m)*

Wärmeausgleich *(m) n* De
- En heat balance
- Es equilibrio térmico *(m)*
- Fr bilan calorifique *(m)*
- It bilancio termico *(m)*
- Pt equilíbrio térmico *(m)*

Wärmeaustauscher *(m) n* De
- En heat exchanger
- Es intercambiador de calor *(m)*
- Fr échangeur de chaleur *(m)*
- It scambiatore di calore *(m)*
- Pt termo-permutador *(m)*

Wärmeaustauscher bestehend aus Rohrschlangen im Bad *(m)* De
- En coil-in-bath heat exchanger
- Es intercambiador de calor con serpentín en baño *(m)*
- Fr échangeur de chaleur à serpentin immergé *(m)*
- It scambiatore di calore a serpentino in bagno *(m)*
- Pt termo-permutador de bobina em banho *(m)*

Wärmeaustauscher mit konzentrischen Röhren *(m)* De
- En concentric-tube heat exchanger
- Es intercambiador de calor de tubos concéntricos *(m)*
- Fr échangeur de chaleur à tubes concentriques (m)
- It scambiatore di calore a tubo concentrico *(m)*
- Pt termo-permutador de tubo concêntrico *(m)*

Wärmedurchgangszahl *(f) n* De
- En U-value
- Es valor U *(m)*
- Fr valeur U *(f)*
- It valore U *(m)*
- Pt valor de U *(m)*

Wärmeenergie *(f) n* De
- En heat energy; thermal energy
- Es energía térmica *(f)*
- Fr énergie calorifique; énergie thermique *(f)*
- It energia termica *(f)*
- Pt energia térmica *(f)*

Wärmeenergie-speicherung *(f) n* De
- En thermal-energy storage
- Es almacenamiento de energía térmica *(m)*
- Fr accumulation d'énergie thermique *(n)*
- It immagazzinaggio di energia termica *(m)*
- Pt armazenagem de energia térmica *(f)*

Wärmefluß *(m) n* De
- En thermal flux
- Es flujo térmico *(m)*
- Fr flux thermique *(m)*
- It flusso termico *(m)*
- Pt fluxo térmico *(m)*

Wärmekapazität *(f) n* De
- En heat capacity
- Es capacidad térmica *(f)*
- Fr capacité calorifique *(f)*
- It capacità termica *(f)*
- Pt capacidade térmica *(f)*

Wärmekraftmaschine *(f) n* De
- En heat engine
- Es motor térmico *(m)*
- Fr moteur thermique *(m)*
- It motore termico *(m)*
- Pt motor térmico *(m)*

Wärmekraftwerk *(n) n* De
- En thermal power station
- Es central térmica *(f)*
- Fr centrale thermique *(f)*
- It centrale di energia termica *(f)*
- Pt central de energia térmica *(f)*

Wärmeleistung *(f) n* De
- En useful heat
- Es calor útil *(m)*
- Fr chaleur utile *(f)*
- It calore utile *(m)*
- Pt calor útil *(m)*

Wärmeleitfähigkeit *(f) n* De
- En thermal conductivity
- Es conductividad térmica *(f)*
- Fr conductivité thermique *(f)*

It conduttività termica (f)
Pt condutividade térmica (f)

Wärmemengenmessung (f) n De
En calorimetry
Es calorimetría (f)
Fr calorimétrie (f)
It calorimetria (f)
Pt calorimetria (f)

Wärmepumpe (f) n De
En heat pump
Es bomba de calor (f)
Fr thermopompe (f)
It pompa di calore (f)
Pt bomba térmica (f)

Wärmerückgewinnung (f) n De
En heat recovery
Es recuperación del calor (f)
Fr récupération de chaleur (f)
It recupero di calore (m)
Pt recuperação térmica (f)

Wärmeschutz (m) n De
En heat shield; thermal shield
Es pantalla térmica (f)
Fr bouclier thermique (m)
It schermo termico (m)
Pt protector térmico (m)

Wärmespeicher (m) n De
En heat reservoir
Es depósito de calor (m)
Fr réservoir de chaleur (m)
It serbatoio di calore (m)
Pt reservatório de calor (m)

Wärmespeicherung (f) n De
En heat storage
Es almacenamiento de calor (m)
Fr accumulation de chaleur (f)
It immagazzinaggio di calore (m)
Pt armazenagem de calor (f)

Wärmestrahlung (f) n De
En thermal radiation
Es radiación térmica (f)
Fr rayonnement thermique (f)
It radiazione termica (f)
Pt radiação térmica (f)

Wärmeträgheit (f) n De
En thermal inertia
Es inercia térmica (f)
Fr inertie thermique (f)
It inerzia termica (f)
Pt inércia térmica (f)

Wärmeübertragung (f) n De
En heat transfer
Es transferencia térmica (f)
Fr transfert de chaleur (m)
It scambio di calore (m)
Pt transferência térmica (f)

Wärmeübertragung mit zwei Flüssigkeiten (f) De
En dual-fluid heat transfer
Es transferencia térmica de dos flúidos (f)
Fr transfert de chaleur à double fluide (m)
It trasferimento di calore a doppio fluido (m)
Pt termotransferência de dois fluidos (f)

Wärmeübertragungsflüssigkeit (f) n De
En heat-transfer fluid
Es flúido de transferencia térmica (m)
Fr fluide de transfert de chaleur (m)
It fluido di scambio del calore (m)
Pt fluido de transferência térmica (m)

Wärmeübertragungsrohr (n) n De
En heat pipe
Es tubo calefactor (m)
Fr caloduc (m)
It tubo di calore (m)
Pt tubo de calor (m)

Wärmeübertragungszahl (f) n De
En heat-transfer coefficient
Es coeficiente de transferencia térmica (m)
Fr coefficient de transfert de chaleur (m)
It coefficiente di scambio di calore (m)
Pt coeficiente de transferência térmica (m)

Wärmeverlust (m) n De
En heat loss
Es pérdida de calor (f)
Fr perte de chaleur (f)
It perdita di calore (f)
Pt perda de calor (f)

Warmluftheizung (f) n De
En warm-air heater
Es calentador con aire caliente (m)
Fr chauffage à air chaud (m)
It riscaldatore ad aria tiepida (m)
Pt aquecedor de ar quente (m)

Warmwasser (n) n De
En hot water
Es agua caliente (f)
Fr eau chaude (f)
It acqua calda (f)
Pt água quente (f)

Warmwasserspeicherung (f) n De
En hot-water storage
Es almacenamiento de agua caliente (m)
Fr accumulation d'eau chaude (f)
It immagazzinaggio di acqua calda (m)
Pt armazenagem de água quente (f)

Waschmittelzusätze (pl) n De
En detergent additives (pl)
Es aditivos detergentes (m pl)
Fr additifs détergents (m pl)
It additivi detergenti (m pl)
Pt aditivos detergentes (m pl)

Wasseraufbereitung (f) n De
En water treatment
Es tratamiento del agua (m)
Fr traitement des eaux (m)
It trattamento ad acqua (m)
Pt tratamento hidráulico (m)

Wassererhitzer (m) n De
En water heater
Es calentador de agua (m)
Fr chauffe-eau (m)
It riscaldatore di acqua (m)
Pt aquecedor hidráulico (m)

Wassergas (n) n De
En water gas
Es gas de agua (m)
Fr gaz à l'eau (m)
It gas d'acqua (m)
Pt gás de água (m)

wassergekühlter Reaktor (m) De
En water-cooled reactor
Es reactor refrigerado con agua (m)
Fr réacteur à refroidissement par eau (m)
It reattore raffreddato ad acqua (m)
Pt reactor arrefecido por água (m)

Wassermühle (f) n De
En watermill
Es molino hidráulico (m)
Fr moulin à eau (m)
It mulino ad acqua (m)
Pt moínho hidráulico (m)

Wasserrad *(n)* *n* De
En waterwheel
Es rueda hidráulica *(f)*
Fr roue à eau *(f)*
It ruota idraulica *(f)*
Pt roda hidráulica *(f)*

Wasserrohrkessel *(m)* *n* De
En water-tube boiler
Es caldera de tubos de agua *(f)*
Fr chaudière tubulaire *(f)*
It caldaia a tubo d'acqua *(f)*
Pt caldeira de tubos de água *(f)*

Wasser-Schnellerhitzer *(m)* *n* De
En instantaneous water heater
Es calentador de agua instantáneo *(m)*
Fr chauffe-eau instantané *(m)*
It riscaldatore istantaneo di acqua *(m)*
Pt aquecedor de água instantâneo *(m)*

Wasserschwerkraft-Energie *(f)* *n* De
En hydrogravitational energy
Es energía hidrogravitacional *(f)*
Fr énergie d'hydrogravitation *(f)*
It energia idrogravitazionale *(f)*
Pt energia hidrogravitacional *(f)*

Wasserspeicherung *(f)* *n* De
En hydrostorage
Es hidroalmacenamiento *(m)*
Fr hydrostockage *(m)*
It idroimmagazzinaggio *(m)*
Pt hidroarmazenamento *(m)*

Wasserstoff *(m)* *n* De
En hydrogen
Es hidrógeno *(m)*
Fr hydrogène *(m)*
It idrogeno *(m)*
Pt hidrogénio *(m)*

Wasserstoffbombe *(f)* *n* De
En hydrogen bomb
Es bomba de hidrógeno *(f)*
Fr bombe à hydrogène *(f)*
It bomba all'idrogeno *(f)*
Pt bomba de hidrogénio *(f)*

wasserstoffgekühlt *adj* De
En hydrogen-cooled
Es enfriada con hidrógeno
Fr refroidi par l'hydrogène
It raffreddata ad idrogeno
Pt arrefecido a hidrogénio

Wasserstoff-Luftbrennstoffzelle *(f)* *n* De
En hydrogen-air fuel cell
Es célula de combustible de hidrógeno-aire *(f)*
Fr pile à combustible hydrogène-air *(f)*
It cellula combustibile idrogeno-aria *(f)*
Pt célula de combustível hidrogénio-ar *(f)*

Wasserstoff-Ölverhältnis *(n)* *n* De
En hydrogen-oil ratio
Es relación hidrógeno-aceite *(f)*
Fr rapport hydrogène-huile *(m)*
It rapporto idrogeno-olio *(m)*
Pt razão hidrogénio-óleo *(f)*

Wasserstoffwirtschaft *(f)* *n* De
En hydrogen economy
Es economía de hidrógeno *(f)*
Fr économie d'hydrogène *(f)*
It economia dell'idrogeno *(f)*
Pt economia de hidrogénio *(f)*

Wasserturbine *(f)* *n* De
En water turbine
Es turbina de agua *(f)*
Fr turbine hydraulique *(f)*
It turbina ad acqua *(f)*
Pt turbina hidraúlica *(f)*

waste fuels *(pl)* En
De Abfalltreibstoffe *(pl)*
Es combustibles de desecho *(m pl)*
Fr combustibles de récupération *(m pl)*
It combustibili di scarico *(m pl)*
Pt combustíveis perdidos *(m pl)*

waste gases *(pl)* En
De Abgase *(pl)*
Es gases de desecho *(m pl)*
Fr gaz perdus *(m pl)*
It gas di scarico *(m pl)*
Pt gases peridos *(m pl)*

waste heat En
De Abwärme *(f)*
Es calor residual *(m)*
Fr chaleur perdue *(f)*
It calore perduto *(m)*
Pt calor residual *(m)*

waste-heat boiler En
De Abwärmekessel *(m)*
Es caldera de calor residual *(f)*
Fr chaudière de récupération des chaleurs perdues *(f)*
It caldaia a recupero di calore perduto *(f)*
Pt caldeira aquecida com calor residual *(f)*

waste-heat recovery En
De Abwärmerückgewinnung *(f)*
Es recuperación de calor residual *(f)*
Fr récupération des chaleurs perdues *(f)*
It recupero di calore perduto *(m)*
Pt recuperação de calor residual *(f)*

water-cooled reactor En
De wassergekühlter Reaktor *(m)*
Es reactor refrigerado con agua *(m)*
Fr réacteur à refroidissement par eau *(m)*
It reattore raffreddato ad acqua *(m)*
Pt reactor arrefecido por água *(m)*

water gas En
De Wassergas *(n)*
Es gas de agua *(m)*
Fr gaz à l'eau *(m)*
It gas d'acqua *(m)*
Pt gás de água *(m)*

water heater En
De Wassererhitzer *(m)*
Es calentador de agua *(m)*
Fr chauffe-eau *(m)*
It riscaldatore di acqua *(m)*
Pt aquecedor hidráulico *(m)*

watermill *n* En
De Wassermühle *(f)*
Es molino hidráulico *(m)*
Fr moulin à eau *(m)*
It mulino ad acqua *(m)*
Pt moínho hidráulico *(m)*

water treatment En
De Wasseraufbereitung *(f)*
Es tratamiento del agua *(m)*
Fr traitement des eaux *(m)*
It trattamento ad acqua *(m)*
Pt tratamento hidráulico *(m)*

water-tube boiler En
De Wasserrohrkessel *(m)*
Es caldera de tubos de agua *(f)*
Fr chaudière tubulaire *(f)*
It caldaia a tubo d'acqua *(f)*
Pt caldeira de tubos de água *(f)*

water turbine En
De Wasserturbine *(f)*
Es turbina de agua *(f)*
Fr turbine hydraulique *(f)*
It turbina ad acqua *(f)*
Pt turbina hidraúlica *(f)*

waterwheel *n* En
De Wasserrad *(n)*
Es rueda hidráulica *(f)*
Fr roue à eau *(f)*
It ruota idraulica *(f)*
Pt roda hidráulica *(f)*

watímetro *(m) n* Es
De Wattmesser *(m)*
En wattmeter
Fr wattmètre *(m)*
It wattometro *(m)*
Pt medidor de watt *(m)*

Wattmesser *(m) n* De
En wattmeter
Es watímetro *(m)*
Fr wattmètre *(m)*
It wattometro *(m)*
Pt medidor de watt *(m)*

wattmeter *n* En
De Wattmesser *(m)*
Es watímetro *(m)*
Fr wattmètre *(m)*
It wattometro *(m)*
Pt medidor de watt *(m)*

wattmètre *(m) n* Fr
De Wattmesser *(m)*
En wattmeter
Es watímetro *(m)*
It wattometro *(m)*
Pt medidor de watt *(m)*

wattometro *(m) n* It
De Wattmesser *(m)*
En wattmeter
Es watímetro *(m)*
Fr wattmètre *(m)*
Pt medidor de watt *(m)*

wave *n* En
De Welle *(f)*
Es onda *(f)*
Fr onde *(f)*
It onda *(f)*
Pt onda *(f)*

wave energy En
De Wellenenergie *(f)*
Es energía de la onda *(f)*
Fr énergie ondulatoire *(f)*
It energia d'onda *(f)*
Pt energia ondulatória *(f)*

waveform *n* En
De Wellenform *(f)*
Es forma de onda *(f)*
Fr forme d'onde *(f)*
It forma d'onda *(f)*
Pt forma de onda *(f)*

wavefront *n* En
De Wellenfront *(f)*
Es frente de onda *(m)*
Fr front d'onde *(m)*
It fronte d'onda *(f)*
Pt frente ondulatória *(f)*

wave generator En
De Wellengenerator *(m)*
Es generador de ondas *(m)*
Fr générateur d'ondes *(f)*
It generatore d'onda *(m)*
Pt gerador de ondas *(m)*

wavelength *n* En
De Wellenlänge *(f)*
Es longitud de onda *(f)*
Fr longueur d'onde *(f)*
It lunghezza d'onda *(f)*
Pt comprimento de onda *(m)*

wavepower *n* En
De Wellenkraft *(f)*
Es potencia de la onda *(f)*
Fr puissance ondulatoire *(f)*
It potenza d'onda *(f)*
Pt potência ondulatória *(f)*

weatherproofing *n* En
De Wetterschutz *(m)*
Es impermeabilización *(f)*
Fr protection contre les intempéries *(f)*
It resistenza alle intemperie *(f)*
Pt protecção contra a intempérie *(f)*

Wechselrichter *(m) n* De
En inverter
Es inversor *(m)*
Fr inverseur *(m)*
It invertitore *(m)*
Pt inversor *(m)*

Wechselstrom *(m) n* De
En alternating current (a.c.)
Es corriente alterna (c.a.) *(f)*
Fr courant alternatif (c.a.) *(m)*
It corrente alternata (c.a.) *(f)*
Pt corrente alterna (c.a.) *(f)*

Wechselstrom-generator *(m) n* De
En alternator
Es alternador *(m)*
Fr alternateur *(m)*
It alternatore *(m)*
Pt alternador *(m)*

Weichstrahlung *(f) n* De
En soft radiation
Es radiación blanda *(f)*
Fr radiation molle *(f)*
It radiazione soffice *(f)*
Pt radiação suave *(f)*

weight *n* En
De Gewicht *(n)*
Es peso *(m)*
Fr poids *(m)*
It peso *(m)*
Pt peso *(m)*

weighted mean En
De gewogener Durchschnitt *(m)*
Es media compensada *(f)*
Fr moyenne pondérée *(f)*
It media ponderata *(f)*
Pt média ponderada *(f)*

weißglühend *adj* De
En incandescent
Es incandescente
Fr incandescent
It incandescente
Pt incandescente

well *n* En
De Bohrung *(f)*
Es pozo *(m)*
Fr puits *(m)*
It pozzo *(m)*
Pt poço *(m)*

Welle *(f) n* De
En wave
Es onda *(f)*
Fr onde *(f)*
It onda *(f)*
Pt onda *(f)*

Wellenenergie *(f) n* De
En wave energy
Es energía de la onda *(f)*
Fr énergie ondulatoire *(f)*
It energia d'onda *(f)*
Pt energia ondulatória *(f)*

Wellenform *(f) n* De
En waveform
Es forma de onda *(f)*
Fr forme d'onde *(f)*
It forma d'onda *(f)*
Pt forma de onda *(f)*

Wellenfront *(f) n* De
En wavefront
Es frente de onda *(m)*
Fr front d'onde *(m)*
It fronte d'onda *(f)*
Pt frente ondulatória *(f)*

Wellengenerator *(m) n* De
En wave generator
Es generador de ondas *(m)*
Fr générateur d'ondes *(f)*
It generatore d'onda *(m)*
Pt gerador de ondas *(m)*

Wellenkraft *(f) n* De
En wavepower
Es potencia de la onda *(f)*
Fr puissance ondulatoire *(f)*
It potenza d'onda *(f)*
Pt potência ondulatória *(f)*

Wellenlänge *(f) n* De
En wavelength
Es longitud de onda *(f)*
Fr longueur d'onde *(f)*
It lunghezza d'onda *(f)*

Pt comprimento de onda *(m)*

wet meter En
De Naβmesser *(m)*
Es medidor húmedo *(m)*
Fr compteur humide *(m)*
It contatore a liquido *(m)*
Pt medidor de humidade *(m)*

wet scrubbing En
De Naβreinigen *(n)*
Es lavado húmedo *(m)*
Fr épuration humide *(f)*
It lavaggio *(m)*
Pt limpeza com lavagem *(f)*

Wetterschutz *(m) n* De
En weatherproofing
Es impermeabilización *(f)*
Fr protection contre les intempéries *(f)*
It resistenza alle intemperie *(f)*
Pt protecção contra a intempérie *(f)*

Wicklung *(f) n* De
En winding
Es devanado *(m)*
Fr enroulement *(m)*
It avvolgimento *(m)*
Pt enrolamento *(m)*

Widerstand *(m) n* De
En resistance
Es resistencia *(f)*
Fr résistance *(f)*
It resistenza *(f)*
Pt resistência *(f)*

Widerstandsbremsung *(f) n* De
En rheostatic braking
Es frenado reostático *(m)*
Fr freinage par résistance *(m)*
It frenatura reostatica *(f)*
Pt quebra reostática *(f)*

Wiederaufbereitung *(f) n* De
En reprocessing
Es reelaboración *(f)*
Fr traitement du combustible irradié *(m)*
It riutilazzione *(f)*
Pt reprocessamento *(m)*

wind *n* En
De Wind *(m)*
Es viento *(m)*
Fr vent *(m)*
It vento *(m)*
Pt vento *(m)*

Wind *(m) n* De
En wind
Es viento *(m)*
Fr vent *(m)*
It vento *(m)*
Pt vento *(m)*

windage loss En
De Luftreibungsverlust *(m)*
Es pérdida por resistencia aerodinámica *(f)*
Fr perte due au jeu *(f)*
It perdita per resistenza aerodinamica *(f)*
Pt perda por fricção do vento *(f)*

Windbrenner *(m) n* De
En air-blast burner
Es quemador de chorro de aire *(m)*
Fr brûleur à air soufflé *(m)*
It bruciatore a ventilazione forzata *(m)*
Pt queimador de corrente de ar *(m)*

wind-driven generator En
De windgetriebener Generator *(m)*
Es generador eólico *(m)*
Fr générateur éolien *(m)*
It generatore a vento *(m)*
Pt gerador accionado pelo vento *(m)*

windgetriebener Generator *(m)* De
En wind-driven generator
Es generador eólico *(m)*
Fr générateur éolien *(m)*
It generatore a vento *(m)*
Pt gerador accionado pelo vento *(m)*

winding *n* En
De Wicklung *(f)*
Es devanado *(m)*
Fr enroulement *(m)*
It avvolgimento *(m)*
Pt enrolamento *(m)*

Windkraft *(f) n* De
En windpower
Es potencia del viento *(f)*
Fr puissance éolienne *(f)*
It potenza del vento *(f)*
Pt potência eólica *(f)*

Windmesser *(m) n* De
En anemometer
Es anemómetro *(m)*
Fr anémomètre *(m)*
It anemometro *(m)*
Pt anemómetro *(m)*

windmill *n* En
De Windmühle *(f)*
Es molino de viento *(m)*
Fr moulin à vent *(m)*
It mulino a vento *(m)*
Pt moínho de vento *(m)*

Windmühle *(f) n* De
En windmill
Es molino de viento *(m)*
Fr moulin à vent *(m)*
It mulino a vento *(m)*
Pt moínho de vento *(m)*

Windmühle mit Hochachse *(f)* De
En vertical-axis windmill
Es molino de viento de eje vertical *(m)*
Fr moulin à vent à axe vertical *(m)*
It mulino a vento a asse verticale *(m)*
Pt moínho de vento de eixo vertical *(m)*

window insulation En
De Fensterisolierung *(f)*
Es aislamiento de ventanas *(m)*
Fr isolement des fenêtres *(m)*
It isolamento delle finestre *(m)*
Pt isolamento de janela *(m)*

windpower *n* En
De Windkraft *(f)*
Es potencia del viento *(f)*
Fr puissance éolienne *(f)*
It potenza del vento *(f)*
Pt potência eólica *(f)*

Wirbelofen *(m) n* De
En cyclone furnace
Es hogar de turbulencia *(m)*
Fr four à cyclone *(m)*
It forno a ciclone *(m)*
Pt forno de ciclone *(m)*

Wirbelstrom *(m) n* De
En eddy current
Es corriente de Foucault *(f)*
Fr courant de Foucault *(m)*
It corrente parassita *(f)*
Pt corrente parasítico *(m)*

Wobbe number En
De Wobbe-Zahl *(f)*
Es número Wobbe *(m)*
Fr nombre de Wobbe *(m)*
It numero di Wobbe *(m)*
Pt número de Wobbe *(m)*

Wobbe-Zahl *(f) n* De
En Wobbe number
Es número Wobbe *(m)*
Fr nombre de Wobbe *(m)*
It numero di Wobbe *(m)*
Pt número de Wobbe *(m)*

work *n* En
De Arbeit *(f)*
Es trabajo *(m)*
Fr travail *(m)*
It lavoro *(m)*
Pt trabalho *(m)*

work function En
De Arbeitsfunktion *(f)*
Es función de trabajo *(f)*
Fr travail de sortie *(m)*
It funzione di lavoro *(f)*
Pt função de trabalho *(f)*

working fluid En
De Treibflüssigkeit *(f)*
Es flúido motor *(m)*
Fr fluide moteur *(m)*
It fluido operante *(m)*
Pt fluido operativo *(m)*

X

xileno *(m) n* Es, Pt
De Xylol *(n)*
En xylene
Fr xylène *(m)*
It xilolo *(m)*

xilolo *(m) n* It
De Xylol *(n)*
En xylene
Es xileno *(m)*
Fr xylène *(m)*
Pt xileno *(m)*

xistos betuminosos *(m pl)* Pt
De Ölschiefer *(pl)*
En oil shales *(pl)*
Es pizarras bituminosas *(f pl)*
Fr schistes bitumineux *(m pl)*
It schisti bituminosi *(m pl)*

X-rays *(pl) n* En
De Röntgen-Strahlen *(pl)*
Es rayos X *(m pl)*
Fr rayons X *(m pl)*
It raggi X *(m pl)*
Pt raios X *(m pl)*

xylene *n* En
De Xylol *(n)*
Es xileno *(m)*
Fr xylène *(m)*
It xilolo *(m)*
Pt xileno *(m)*

xylène *(m) n* Fr
De Xylol *(n)*
En xylene
Es xileno *(m)*
It xilolo *(m)*
Pt xileno *(m)*

Xylol *(n) n* De
En xylene
Es xileno *(m)*
Fr xylène *(m)*
It xilolo *(m)*
Pt xileno *(m)*

Y

yacimiento de petróleo *(m)* Es
De Ölfeld *(n)*
En oilfield
Fr gisement pétrolifère *(m)*
It giacimento petrolifero *(m)*
Pt jazigo de petróleo *(m)*

yield *n* En
De Ertrag *(m)*
Es rendimiento *(m)*
Fr rendement *(m)*
It rendimento *(m)*
Pt rendimento *(m)*

Z

zavorra pulita *(f)* It
De sauberer Ballast *(m)*
En clean ballast
Es lastre limpio *(m)*
Fr ballast propre *(m)*
Pt balastro puro *(m)*

Zeitschalter *(m) n* De
En time switch
Es temporizador *(m)*
Fr minuterie *(f)*
It interruttore a tempo *(m)*
Pt comutador cronométrico *(m)*

Zelle (Batterie) *(f) n* De
En cell (battery)
Es elemento (batería) *(m)*
Fr élément (batterie) *(m)*
It elemento (batteria) *(m)*
Pt célula (bateria) *(f)*

Zellulose *(f) n* De
En cellulose
Es celulosa *(f)*
Fr cellulose *(f)*
It cellulosa *(f)*
Pt celulose *(f)*

zenit *(m) n* It
De Zenit *(n)*
En zenith
Es cénit *(m)*
Fr zénith *(m)*
Pt zénite *(m)*

Zenit *(n) n* De
En zenith
Es cénit *(m)*
Fr zénith *(m)*
It zenit *(m)*
Pt zénite *(m)*

zénite *(m) n* Pt
De Zenit *(n)*
En zenith
Es cénit *(m)*
Fr zénith *(m)*
It zenit *(m)*

zenith *n* En
De Zenit *(n)*
Es cénit *(m)*
Fr zénith *(m)*
It zenit *(m)*
Pt zénite *(m)*

zénith *(m) n* Fr
De Zenit *(n)*
En zenith
Es cénit *(m)*
It zenit *(m)*
Pt zénite *(m)*

zenith angle En
De Zenitwinkel *(m)*
Es ángulo cenital *(m)*
Fr angle zénithal *(m)*
It angolo dello zenit *(m)*
Pt ângulo zenital *(m)*

zenith solar flux En
De Zenitsonnenfluβ *(m)*
Es flujo solar cenital *(m)*
Fr flux solaire zénithal *(m)*
It flusso solare allo zenit *(m)*
Pt fluxo solar zenital *(m)*

Zenitsonnenfluβ *(m) n* De
En zenith solar flux
Es flujo solar cenital *(m)*
Fr flux solaire zénithal *(m)*
It flusso solare allo zenit *(m)*
Pt fluxo solar zenital *(m)*

Zenitwinkel *(m) n* De
En zenith angle
Es ángulo cenital *(m)*
Fr angle zénithal *(m)*
It angolo dello zenit *(m)*
Pt ângulo zenital *(m)*

Zentralheizung *(f) n* De
En central heating
Es calefacción central *(f)*
Fr chauffage central *(m)*
It riscaldamento centrale *(m)*
Pt aquecimento central *(m)*

Zentralheizung mit Ölfeuerung *(f)* De
En oil-fired central heating
Es calefacción central con petróleo *(f)*
Fr chauffage central au mazout *(m)*
It riscaldamento centrale a nafta *(m)*
Pt aquecimento central a óleo *(m)*

Zentrifugalkraft *(f) n* De
En centrifugal force
Es fuerza contrífuga *(f)*
Fr force centrifuge *(f)*
It forza centrifuga *(f)*
Pt força centrífuga *(f)*

Zentrifuge *(f) n* De
En centrifuge
Es centrífuga *(f)*
Fr centrifugeuse *(f)*
It centrifuga *(f)*
Pt centrifugadora *(f)*

Zentripetalkraft *(f) n* De
En centripetal force
Es fuerza centrípeta *(f)*
Fr force centripète *(f)*
It forza centripeta *(f)*
Pt força centrípeda *(f)*

zéro absolu *(m)* Fr
De absoluter Nullpunkt *(m)*
En absolute zero
Es cero absoluto *(m)*
It zero assoluto *(m)*
Pt zero absoluto *(m)*

zero absoluto *(m)* Pt
De absoluter Nullpunkt *(m)*
En absolute zero
Es cero absoluto *(m)*
Fr zéro absolu *(m)*
It zero assoluto *(m)*

zero assoluto *(m)* It
De absoluter Nullpunkt *(m)*
En absolute zero
Es cero absoluto *(m)*
Fr zéro absolu *(m)*
Pt zero absoluto *(m)*

Zersetzungskonstante *(f) n* De
En decay constant
Es constante de desintegración *(f)*
Fr constante de désintégration *(f)*
It constante di disintegrazione *(f)*
Pt constante de descomposição *(f)*

Zersetzungswärme *(f) n* De
En decay heat
Es calor de desintegración *(m)*
Fr chaleur de désintégration *(f)*
It calore di disintegrazione *(m)*
Pt calor de descomposição *(m)*

Zerstäuber *(m) n* De
En atomizer
Es atomizador *(m)*
Fr atomiseur *(m)*
It nebulizzatore *(m)*
Pt atomizador *(m)*

zinc-chloride battery En
De Zinkchloridbatterie *(f)*
Es batería de cloruro de zinc *(f)*
Fr batterie au chlorure de zinc *(f)*
It batteria al cloruro di zinco *(f)*
Pt bateria de cloreto de zinco *(f)*

Zinkchloridbatterie *(f) n* De
En zinc-chloride battery
Es batería de cloruro de zinc *(f)*
Fr batterie au chlorure de zinc *(f)*
It batteria al cloruro di zinco *(f)*
Pt bateria de cloreto de zinco *(f)*

zircalloy *n* En
De Zirkonlegierung *(f)*
Es aleación de zirconio *(f)*
Fr alliage de zirconium *(m)*
It lega di zirconio *(f)*
Pt liga de zircónio *(f)*

zircone *(f) n* Fr
De Zirkonerde *(f)*
En zirconia
Es bióxido de zirconio *(m)*
It zirconia *(f)*
Pt zircónia *(f)*

zirconia *n* En; It *(f)*
De Zirkonerde *(f)*
Es bióxido de zirconio *(m)*
Fr zircone *(f)*
Pt zircónia *(f)*

zircónia *(f) n* Pt
De Zirkonerde *(f)*
En zirconia
Es bióxido de zirconio *(m)*
Fr zircone *(f)*
It zirconia *(f)*

Zirkonerde *(f) n* De
En zirconia
Es bióxido de zirconio *(m)*
Fr zircone *(f)*
It zirconia *(f)*
Pt zircónia *(f)*

Zirkonlegierung *(f) n* De
En zircalloy
Es aleación de zirconio *(f)*
Fr alliage de zirconium *(m)*
It lega di zirconio *(f)*
Pt liga de zircónio *(f)*

zolfo *(m) n* It
Am sulfur
De Schwefel *(m)*
En sulphur
Es azufre *(m)*
Fr soufre *(m)*
Pt enxofre *(m)*

zona aria pulita *(f)* It
De Sauberluftzone *(f)*
En clean-air zone
Es zona de aire limpio *(f)*
Fr zone d'air pur *(f)*
Pt zona de ar puro *(f)*

zona de aire limpio *(f)* Es
De Sauberluftzone *(f)*
En clean-air zone
Fr zone d'air pur *(f)*
It zona aria pulita *(f)*
Pt zona de ar puro *(f)*

zona de ar puro *(f)* Pt
De Sauberluftzone *(f)*
En clean-air zone
Es zona de aire limpio *(f)*
Fr zone d'air pur *(f)*
It zona aria pulita *(f)*

zona fértil *(f)* Es
De Brutzone *(f)*
En blanket
Fr couche fertile *(f)*
It copertura *(f)*
Pt cortina *(f)*

zone d'air pur *(f)* Fr
De Sauberluftzone *(f)*
En clean-air zone
Es zona de aire limpio *(f)*
It zona aria pulita *(f)*
Pt zona de ar puro *(f)*

zufällige Wärmezunahme *(f)* De
En incidental heat gain
Es ganancia de calor incidental *(f)*
Fr gain de chaleur incident *(m)*
It guadagno di calore incidentale *(m)*
Pt ganho de calor acessório *(m)*

Zündflamme *(f) n* De
En pilot flame
Es llama piloto *(f)*
Fr veilleuse *(f)*
It semprevivo *(m)*
Pt chama piloto *(f)*

Zündtemperatur *(f) n* De
En ignition temperature
Es temperatura de encendido *(f)*
Fr température d'allumage *(f)*
It temperatura di accensione *(f)*
Pt temperatura de ignição *(f)*

Zündung *(f) n* De
En ignition
Es encendido *(m)*
Fr allumage *(m)*
It accensione *(f)*
Pt ignição *(f)*

Zusatzstoff *(m) n* De
En additive
Es aditivo *(m)*
Fr additif *(m)*
It additivo *(m)*
Pt aditivo *(m)*

Zwangsdurchlaufkessel *(m) n* De
En once-through boiler
Es caldera de proceso directo *(f)*
Fr chaudière sans recyclage *(f)*
It caldaia a processo diretto *(f)*
Pt caldeira de uma só passagem *(f)*

Zweifachtreibstoff *(m) n* De
En bipropellant
Es bipropulsante *(m)*
Fr biergol *(m)*
It bipropellente *(m)*
Pt combustível bipropulsor *(m)*

Zweikreisverbrennung *(f) n* De

En dual-cycle combustion
Es combustión de ciclo doble *(f)*
Fr combustion à deux temps *(f)*
It combustione a ciclo doppio *(f)*
Pt combustão de ciclo duplo *(f)*

Zweistufen-verbrennung *(f)* *n* De
En two-stage combustion
Es combustión en dos etapas *(f)*
Fr combustion à deux étages *(f)*
It combustione a due stadi *(f)*
Pt combustão de duas fases *(f)*

Zweitaktmotor *(m)* *n* De
En two-stroke engine
Es motor de dos tiempos *(m)*
Fr moteur à deux temps *(m)*
It motore a due tempi *(m)*
Pt motor de dois tempos *(m)*

ZweiwegeAxialturbine *(f)* *n* De
En two-way axial flow turbine
Es turbina de flujo axial de doble dirección *(f)*
Fr turbine axiale à double sens *(f)*
It turbina a flusso assiale a due vie *(f)*
Pt turbina de fluxo axial de duas vias *(f)*

Zwischenkühler *(m)* *n* De
En intercooler
Es enfriador intermedio *(m)*
Fr refroidisseur intermédiaire *(m)*
It refrigeratore intermedio *(m)*
Pt inter-arrefecedor *(m)*

Zyklohexan *(n)* *n* De
En cyclohexane
Es ciclohexano *(m)*
Fr cyclohexane *(m)*
It cicloesano *(m)*
Pt ciclohexano *(m)*

Zyklon *(m)* *n* De
En cyclone
Es ciclón *(m)*
Fr cyclone *(m)*
It ciclone *(m)*
Pt ciclone *(m)*